Ballistic Performance of Rifle Bullets

Second Edition

By Bryan Litz

Ballistic Performance of Rifle Bullets

Second Edition

By Bryan Litz

ISBN 978-0-9909206-2-5

2nd Edition

Published by:
Applied Ballistics, LLC
PO Box 195
Cedar Springs, MI 49319

Table of Contents

Acknowledgments

This book was created with a lot of assistance from numerous individuals. I would like to thank the following people for their efforts in helping to make this work a reality.

Bill Litz. Bill hand loads the ammo for numerous projects and conducts a great deal of the live fire testing in the Applied Ballistics Laboratory.

Jennifer Fiorenzo. Jennifer worked tirelessly to apply common formatting to the book, including data entry, error checking, and proof reading.

Mitch Fitzpatrick. Mitch created and updated a number of the 2-D bullet drawings which characterize the bullet dimensions in a common format.

I would like to say a sincere thank you to the above individuals who all performed vital roles in the creation of this book.

Introduction

This is a reference book, filled with hundreds of pages of data. Before we get into that, this brief introduction will discuss the nature of this book, as in, why it was created and why it's useful.

The second part of the introduction will explain how to interpret and use the information to assist the ultimate objective, which is hitting targets.

Why this book?

Since the advent of modern rifles, shooters have relied on bullet manufacturers for performance data. The primary measure of ballistic flight performance for rifle bullets is the Ballistic Coefficient, or BC. BC is a number which is neither well understood, nor easy to measure for average shooters, but a number which is vitally important to understanding your downrange performance.

Each bullet manufacturer takes different approaches to measuring or calculating the BC's of their bullets. Some test at 100 yards which only covers the high velocity portion of bullet flight. Others test at longer ranges and provide averaged BC's or segmented BC's. Some of the smaller companies don't have the resources to test and rely on theoretical predictions of BC.

Combined with the obvious marketing pressures to deliberately inflate BC's, ***it's difficult for a shooter to know how accurate the bullet companies' advertised BC's are***. Downrange trajectory predictions of bullet drop and drift can't be any more accurate than the BC's used to calculate them. Furthermore; apples-to-apples comparisons between various bullets aren't possible when the BC's are determined by different methods.

Up thru most of the 20th century, this problem of inaccurate and inconsistent BC data wasn't a huge problem because the rifles and optics that were common in that time usually weren't accurate enough to discern trajectory errors related to inaccurate BC's. However, in modern times, factory rifles and optics have become extremely capable to the point where ***small errors in BC can be the cause of missing targets due to trajectory prediction error***. In the shooting sports, competition is so tight that a couple percent advantage in BC could be the difference between silver and gold medals. For this reason, accurate BC data has become more important than ever before. Shooters can no longer rely on the various inconsistent methods of the manufacturers for their BC data[1].

This brings us to the purpose of this book. ***This book contains an extensive library of BC data for bullets of most common brands; all tested using a common method.*** This work expands on the bullet library that was first published in *Applied Ballistics for Long Range Shooting.* In addition to explaining the elements of external ballistics, that book includes an appendix of rifle bullet performance data for 225 bullets. That performance data is all based on live fire testing, using a common method which is repeatable within +/-1%. That library of performance data is widely regarded as the most reliably accurate and consistent set of BC data available for modern rifle bullets. Often referred to as the 'Litz' BC library, it is used by numerous software products available from Applied Ballistics, as well as other ballistics programs including; online, smart phone apps, as well as integrated into numerous other hardware products such as the Applied Ballistics Kestrel.

The BC library published in the Appendix of *Applied Ballistics for Long Range Shooting* covers 225 of the more popular bullets used in long range shooting, but clearly it's just a drop in the bucket compared to what's available, and what will be available. It became clear that ***the need for accurate bullet performance data has graduated from an appendix, to a book of its own, and that's what you're now holding in your hands***.

This book is the second edition in an ongoing series. New editions will be released as new bullets become available, and existing bullet performance assessments are updated based on the most recent production runs. In this second edition, ballistic performance data is published for 533 bullets from .224 thru .510 caliber.

---

[1] This is not to say that all manufacturers advertised BC's are inaccurate, but it's difficult for the average shooter to know which brands are reliable and which are not.

How to use the information in this book

There are two basic uses for the information in this book. One is to compare the various bullets available for your rifle and *select* the best option for your rifle and application. Then, once you've selected a bullet, you can use the data to support highly accurate long range trajectory modeling.

Of course there are other applications such as; using the dimensioned bullet drawings to help with custom chamber reamer designs, comparing stability requirements of different bullets, and gaining insight about the velocity dependence of G1 vs. G7 referenced BC's, along with many other tid-bits.

The following sections will explain where the information comes from on the bullet data pages, how accurate it is, and why it's important.

The book is broken into sections based on caliber, and each caliber is blocked off by brand. There is one full page of data per bullet, starting with a dimensioned drawing and geometric data.

Berger .243 caliber 95 grain VLD

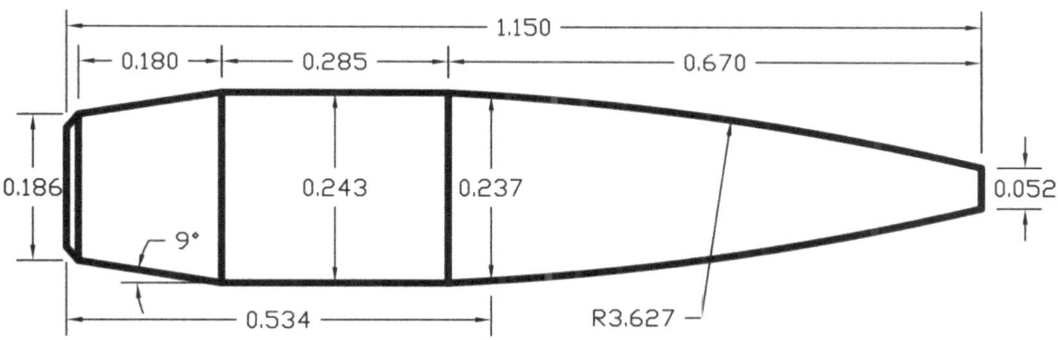

Lot#:	**556**	**Ogive Radius:**	**14.9 calibers**
Sectional Density:	**0.230 lb/in^2**	**Rt/R:**	**0.66**

Figure 1. Each data page includes a detailed dimensioned drawing of the bullet.

The dimensioned drawings are useful for many purposes including:

- If you're considering a bullet for hand loading and are interested to see how it needs to be seated to feed from your rifles magazine. In other words, will the bullet nose go down into the case neck when the bullet is seated to magazine length?
- You can use the drawings to assist in custom chamber design for a particular bullet.

- The drawings show how aggressive the ogive (nose) radius is, which is an indication for how sensitive a bullets' precision might be to seating depth.
- Anyone interested in doing aerodynamic or mass properties modeling of bullets will find the dimensions useful.

The bullets linear dimensions are measured with standard calipers. The overall length is a very accurate measurement, but bullets can typically vary by +/-0.010" in a given lot, more or less depending on the nature of the tip. Open tip, or hollow point bullets tend to have greater variation in length due to the ragged tip compared to Full Metal Jackets or plastic tipped bullets. Of course lot variation can result in length variation as well.

The meplat, or tip diameter is difficult to measure since you're measuring a ragged tip or radius, and as with length, the dimension can vary from bullet to bullet. For both length and meplat diameter, several bullets are measured, and a 'typical' bullet is chosen for full dimensioning. Note that the meplat diameter is the dimension most likely to vary between lots of bullets.

The Boat Tail angle and ogive radius have been measured in various ways, from mounting the bullet in a mini-lathe and collecting points with a dial indicator, to using a Forester datum dial comparator to measure the length to various diameters. Boat tail angle is easily calculated from two points, whereas the nose radius requires at least 3 points to be measured along the curve. Measurements of boat tail angle are good within +/- 0.5 degrees. Ogive radius dimensions are typically repeatable within +/-0.15 calibers, but I've not checked these measurements against many other reliable sources to really know how accurate they are. When dimensioned drawings are available from other sources, my measurements of ogive radius tends to be close.

You'll note on each bullet drawing that the groove diameter is dimensioned across the bearing surface (in Figure 1 its 0.243"). You can also see the point on the nose that's equal to the common *bore* diameter for that caliber is indicated as well (in Figure 1 it's 0.237").

This dimension is a useful reference for those designing a custom chamber reamer, or considering how a bullet might seat in a known chamber throat. Note that these dimensions are subject to measurement uncertainty, and are based on the particular lot of bullets measured. Lot variation and measurement uncertainty can result in the actual dimensions being somewhat different, so use these dimensions as a guide, with the knowledge they may not be exact for the latest lots of bullets. Also note that bullet comparators vary, and they may not be indicating the actual bore diameter where they encounter the bullet nose.

The information in the blocks below the bullet drawing are just a few metrics which further characterize the bullets geometry. The lot number is given for bullets where it's known, otherwise it's simply listed as *unknown*.

The bullets sectional density is shown, which characterizes the bullets ability to penetrate targets, as well as air (the relationship between Sectional Density and Ballistic coefficient is explained in *Applied Ballistics for Long Range Shooting*).

The bullets ogive radius in calibers is also given. This is simply the ogive radius in inches, divided by the caliber of the bullet. So for example, the bullet in Figure 1 has an ogive radius of 3.627", and a caliber of 0.243". So the ogive radius is 3.627" divided by 0.243" equals 14.9 calibers.

Bullet noses are typically sections of circular arcs. Circles are defined by the length of their radius, and that's all we're talking about here. Bullets having a long radius for their caliber will typically have lower drag.

Finally, the Rt/R ratio is shown. This is a measure of how tangent or secant a nose is. Rt/R ratios close to 1.0 will have a very smooth connection between the nose curvature and the bearing surface. This type of nose geometry is known as tangent. Bullets that have a very abrupt connection between the nose and bearing surface are known as secant (or VLD) and will have Rt/R ratios closer to 0.5.

Tangent bullets having Rt/R ratios close to 1.0 are known for easy precision, meaning they tend to be less sensitive to how they're loaded for different rifles. The downside of the tangent ogive design is that they're higher drag than a secant nose.

A secant nose, or a bullet with an Rt/R ratio closer to 0.5 (think VLD nose) has the advantage of lower drag, but are known for being more sensitive to seating depth.

This is all explained in greater detail in *Applied Ballistics for Long Range Shooting,* but for the purposes of using the information in this book, just think: Rt/R close to 1.0 means easy accuracy, but not necessarily high BC, whereas Rt/R ratios closer to 0.5 means finicky accuracy but high BC.

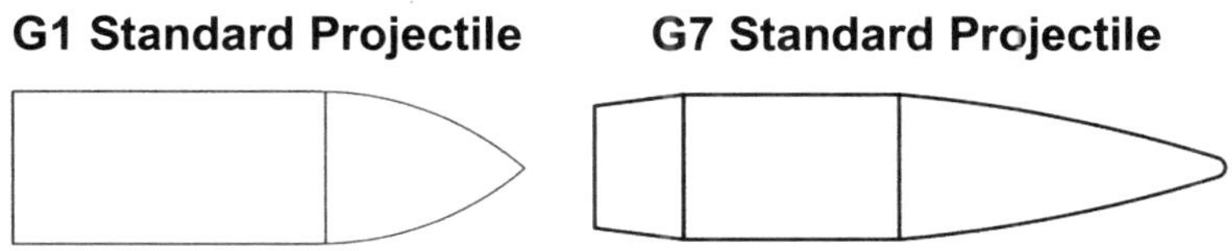

Figure 2. G1 and G7 standard projectiles

The next section of the bullet data pages presents the bullets external ballistic performance in terms of G1 and G7 BC's. For detailed reading about ballistic coefficients, refer to *Applied Ballistics for Long Range Shooting.* Here, I'll just say that the G1 standard was established in the industry many years ago when bullets looked more like that (flat based blunt nosed). With modern bullets having evolved with longer noses and boat tails, the G7 standard is now more appropriate because the G7 BC better represents the drag of modern bullets over greater ranges than the G1 model.

Ballistic Performance

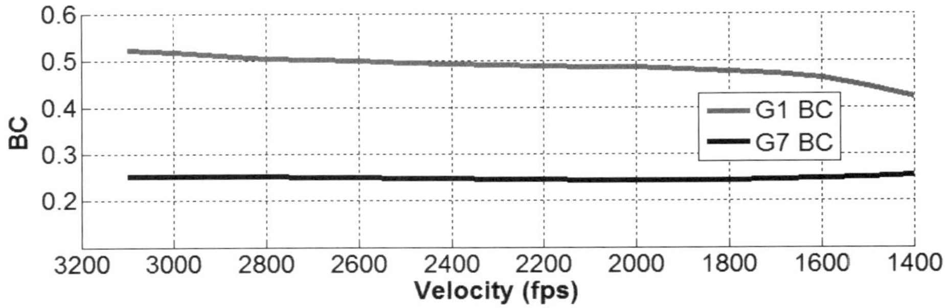

	Above 3000	3000-2500	2500-2000	2000-1500	Below 1500
G1 BC	0.533	0.506	0.490	0.465	0.445
G7 BC	0.252	0.251	0.246	0.247	0.251

Averages from 3000 fps to 1500 fps:
G1 BC: 0.485, variation of G1 BC: 0.071
G7 BC: 0.249, variation of G7 BC: 0.009
G7 form factor: 0.923

Figure 3. Ballistic Performance data.

The ballistic performance data shows a plot of G1 and G7 BC over a typical range of velocities from muzzle to long range. For modern long range bullets, you'll typically see more variation in the G1 BC than the G7 BC. ***This speed dependent variation of the G1 BC leads to inaccurate trajectory predictions and prevents valid apples-to-apples comparisons of bullet performance.***

For most bullets which have more variation in the G1 BC, the G1 data is shown in gray to indicate that it's a poor fit for that bullet. The G7 data is shown in black to indicate that it's a better fit. In rare cases when the G1 drag model is a better fit, that data is shown in black and the G7 will be gray. This is only true for several bullets which tend to have wide tips and shallow or no boat tail.

Below the BC graph is a table which shows what the G1 and G7 BC's are in velocity bands. This is just a numeric representation of the data shown in the plot, which can be entered into a ballistics program that takes segmented BC's. Again, gray numbers are a bad fit, and shouldn't be used.

Below the velocity segmented table, the performance is boiled down into single number averages. For those looking for that 'one number' they can use to model or compare bullet performance, this is it. All BC's are averaged over a common speed range of 3000 fps to 1500 fps. **The most important number on the page, the best fitting BC average, is bold and highlighted**. In the example shown in Figure 3, the best fitting model is G7, and the G7 BC is 0.249.

The G7 form factor is a measure of drag for the bullet compared to the G7 standard. A bullet with a form factor of 1.00 means that bullet has the same drag as the G7 standard projectile (shown in Figure 2 above). If a bullets form factor is less than 1.00, then it has less drag than the G7 standard, vice versa for a form factor greater than 1.00.

Typically, form factors vary from 0.80 for the most streamlined shapes, to 1.20+ for short nose, blunt, flat based bullets. For good long range performance, you want to look for form factors in the low 0.90's or lower.

BC is actually just sectional density divided by form factor, so for example, the bullet used in this example is a 95 grain .243 caliber bullet, so its SD is 0.230. The G7 form factor is 0.923, so it's G7 BC is 0.230/0.923 = 0.249.

The final block of data relates to the bullet stability requirements. Stability can be tricky because it's highly dependent on environmental conditions and velocity. A bullet fired from a rifle might be perfectly stable at 5000 feet altitude (where the air is relatively thin), but if you fire it down at sea level where the air is thicker, it will be less stable and may tumble.

Rifling Twist Rate and Stability

Barrel Twist	Best case	Stability Factor Nominal	Worst case
1:7"	2.47	2.16	1.87
1:8"	1.89	1.66	1.43
1:9"	1.50	1.31	1.13
1:10"	1.21	1.06	0.92

Minimum twist required for stability is 1:8.8"
Recommended twist for optimal performance is 1:7.9" or faster

The stability table is sort of mapping out the range of stability scenarios for you so you can make well informed decisions about barrel twist for a given bullet. The conditions which define Best, Nominal and Worst case stability scenarios are as follows:

- Best case: 1000 feet altitude, 100% humidity, 100 degrees F, and 3000 fps muzzle velocity.
- Nominal case: sea level, 50% humidity, 59 degrees F, and 2800 fps muzzle velocity.

- Worse case: sea level, 0% humidity, 0 degrees F, and 2600 fps muzzle velocity.

The numbers shown in the table as 'Stability Factor' is referring to the gyroscopic stability factor which defines how stable a bullet is. Mathematically, anything above 1.0 is defined as 'stable'. However in practice, due to error in the calculations, variations in velocity and atmospherics, you want your bullets to have a stability factor somewhat greater than 1.0. In general, a stability factor of 1.2 will get the bullets to fly straight and accurate, but may be flying with some wobble, which can reduce BC. Plus, if your stability is marginal in one condition, it may be inadequate for a slightly different condition.

A stability factor of 1.5 or higher is recommended to fully stabilize a bullet and get its full BC potential[2].

Finally, below the table of stability values is a summary which suggests a minimum and recommended twist rate to achieve certain stability conditions.

The 'Minimum twist required for stability' is based on achieving a stability factor of 1.2 in standard sea level conditions[3] and 2800 fps muzzle velocity. The 'Recommended twist for optimal performance' is based on achieving a stability factor of 1.5 in the same conditions.

Remember that if you're considering a bullet for use in a rifle which has a twist rate slower than recommended, it may work just fine if you're shooting in a favorable (low air density) environment. This is why the table is shown with the best, nominal, and Worst case scenarios; so you can determine stability in various conditions. The 'Minimum' and 'Recommended' barrel twists are just boiled down numbers which are conservative guidelines.

It's been stated numerous times in this introduction that; for more information you should consult *Applied Ballistics for Long Range Shooting.* That book goes into great detail on the science of accuracy, and spells out the elements of external ballistics. Understanding the fundamentals is important for long range shooting. Combined with the highly accurate ballistic performance data in this book, you're well equipped to maximize your hit percentage at long range.

---

[2] There are several chapters in *Modern Advancements in Long Range Shooting* which discusses the relationship between Stability and BC.

[3] 29.92 inHg, 59 degrees F, and 0% Humidity.

22 Rimfire

95 types of 22 Rimfire ammo were tested from 5 different barrels of various length and twist rate. Each ammo type was tested for: muzzle velocity, remaining velocity at 92 yards, and G1 BC as well as the standard deviation of each metric. Below is an explanation of all the metrics that appear in the following tables of live fire data:

- Barrel: This is the brand of barrel and what type of action it fits.
    - Vol-77/22 is a Volquartsen barrel for the Ruger 77/22 action.
    - Rug-10/22 is a Ruger brand barrel for the Ruger 10/22 action.
    - Rug-77/22 is a Ruger brand barrel for the Ruger 77/22 action.
- Length: Length of the barrel.
- Twist: Twist rate of the riflings in the barrel.
- Ave MV: Average muzzle velocity.
- SD of MV: Standard deviation of the muzzle velocity.
- Ave DRV: Average down range velocity (remaining velocity at 92 yards as measured by a chronograph).
- SD of DRV: Standard deviation of the down range velocity
- Ave G1 BC: The average BC of the bullet referenced to the G1 standard which is valid for the average flight speed of the bullet tested. BC's are corrected to standard ICAO sea level standard atmosphere so they're suitable for use in modern ballistics programs.
- SD of G1 BC: Standard deviation of the G1 BC. A high SD in the G1 BC can indicate marginal stability.
- # of shots: This is the number of shots for which good data was recorded. A total of 10 shots of each ammo type were tested but not all shots were read by both chronographs.

This book only gives the *raw test results* for all 95 ammo types. This information is very useful for those wishing to predict drop at extend range for their 22 rimfire rifles. Simply find your ammo type and note the average muzzle velocity and G1 BC for the barrel that's the closest match to the barrel on your rifle. Use this muzzle velocity and BC to calculate drop in any standard ballistics program. Trajectories calculated using this data have proven to be very accurate out to 300 yards. Typically the constraint in shooting past 200 yards with a 22 rimfire is having enough vertical adjustment in the scope to compensate for the large amount of drop.

More extensive analysis of this rimfire test data is published in *Modern Advancements in Long Range Shooting – Volume 2.* That book explores the relationships between barrel length and muzzle velocity, twist rate and BC, and other interesting relationships discovered in this testing.

Agulia Golden Eagle Match Rifle 40 grain

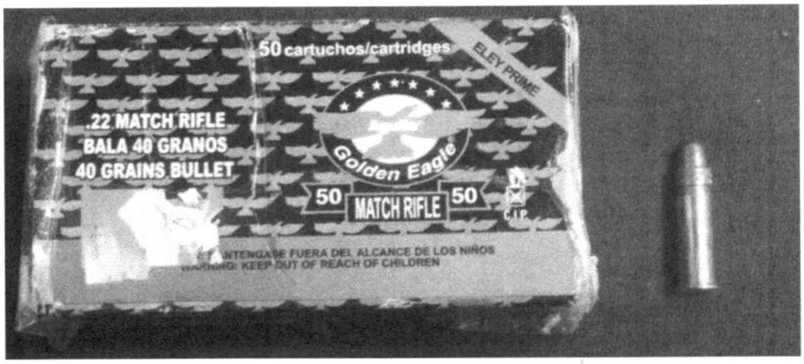

Barrel	Vol-77/22	Vol-77/22	Vol-77/22	Rug-10/22	Rug-77/22
Length	16.5"	18.5"	18.5"	18"	24"
Twist	1:11.25"	1:16"	1:9"	1:14"	1:16"
Ave MV	1055	1063	1052	1081	1065
SD of MV	9.5	15.7	16.4	5.4	4.0
Ave DRV	905	913	908	932	914
SD of DRV	8.5	11.3	13.9	6.0	8.0
Ave G1 BC	0.113	0.117	0.120	0.117	0.112
SD of G1 BC	0.004	0.005	0.006	0.005	0.006
# of shots	7	9	8	8	6

Agulia SSS 60 grain

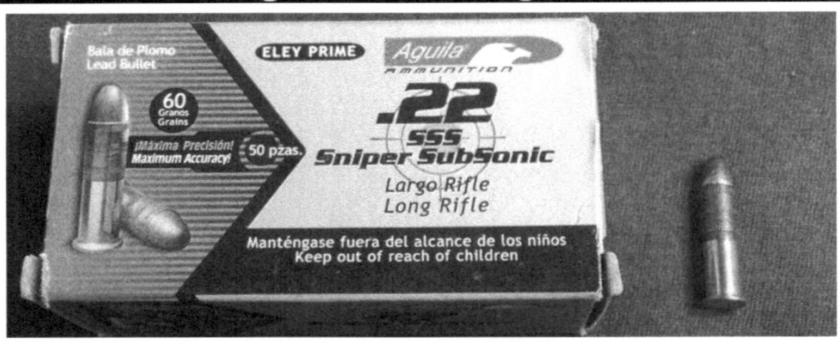

Barrel	Vol-77/22	Vol-77/22	Vol-77/22	Rug-10/22	Rug-77/22
Length	16.5"	18.5"	18.5"	18"	24"
Twist	1:11.25"	1:16"	1:9"	1:14"	1:16"
Ave MV	894	879	876	908	897
SD of MV	8.5		14.9	13.0	19.6
Ave DRV	816	755	797	817	794
SD of DRV	8.8		15.5	6.5	18.3
Ave G1 BC	0.144	0.082	0.138	0.116	0.104
SD of G1 BC	0.005		0.009	0.009	0.010
# of shots	6	1	7	9	9

Agulia SuperExtra 40 grain Cu Plated

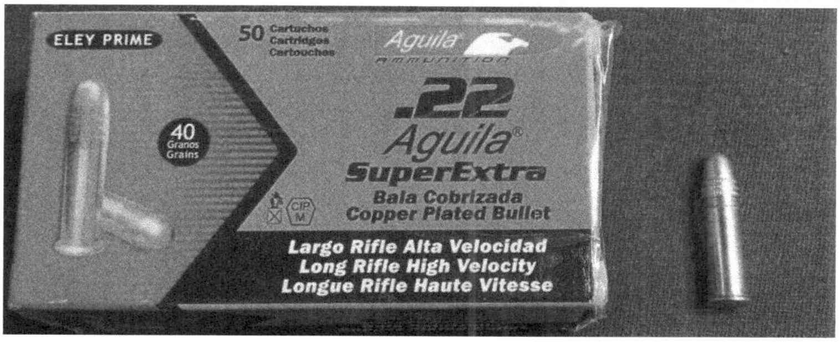

Barrel	Vol-77/22	Vol-77/22	Vol-77/22	Rug-10/22	Rug-77/22
Length	16.5"	18.5"	18.5"	18"	24"
Twist	1:11.25"	1:16"	1:9"	1:14"	1:16"
Ave MV	1185	1212	1203	1245	1206
SD of MV	15.8	17.2	16.9	18.0	11.4
Ave DRV	973	993	988	1019	992
SD of DRV	6.1	11.3	4.9	8.7	6.1
Ave G1 BC	0.114	0.122	0.124	0.121	0.118
SD of G1 BC	0.005	0.005	0.004	0.002	0.003
# of shots	8	9	6	8	8

Agulia SuperExtra 40 grain Lead Lot 100914006

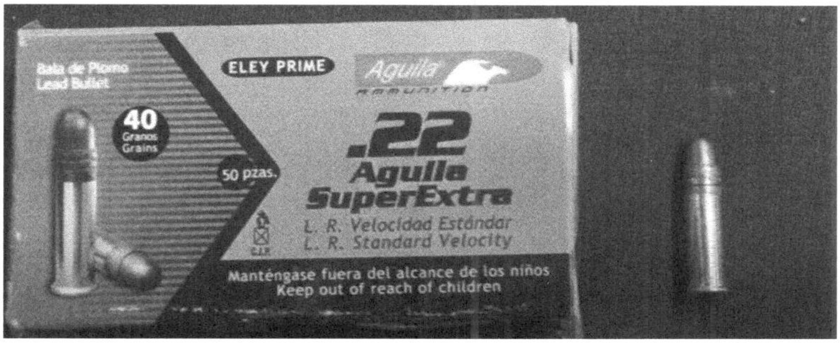

Barrel	Vol-77/22	Vol-77/22	Vol-77/22	Rug-10/22	Rug-77/22
Length	16.5"	18.5"	18.5"	18"	24"
Twist	1:11.25"	1:16"	1:9"	1:14"	1:16"
Ave MV	997	993	993	1016	998
SD of MV	20.4	8.4	6.6	9.4	5.5
Ave DRV	861	861	858	878	860
SD of DRV	15.0	8.3	6.1	11.0	10.1
Ave G1 BC	0.103	0.106	0.105	0.100	0.099
SD of G1 BC	0.004	0.004	0.005	0.008	0.007
# of shots	8	8	7	7	5

Agulia SuperExtra 40 grain Lead Lot 180914001

Barrel	Vol-77/22	Vol-77/22	Vol-77/22	Rug-10/22	Rug-77/22
Length	16.5"	18.5"	18.5"	18"	24"
Twist	1:11.25"	1:16"	1:9"	1:14"	1:16"
Ave MV	1075	1066	1064	1089	1079
SD of MV	15.8	14.2	10.4	9.0	6.7
Ave DRV	923	920	920	943	926
SD of DRV	11.4	11.3	7.2	9.6	7.1
Ave G1 BC	0.118	0.122	0.125	0.122	0.115
SD of G1 BC	0.004	0.004	0.005	0.006	0.004
# of shots	8	7	9	8	9

Agulia SuperExtra Subsonic 38 grain

Barrel	Vol-77/22	Vol-77/22	Vol-77/22	Rug-10/22	Rug-77/22
Length	16.5"	18.5"	18.5"	18"	24"
Twist	1:11.25"	1:16"	1:9"	1:14"	1:16"
Ave MV	985	970	969	994	973
SD of MV	11.4	12.3	10.5	12.5	12.9
Ave DRV	866	851	853	878	853
SD of DRV	11.1	8.1	10.5	11.2	11.0
Ave G1 BC	0.116	0.112	0.117	0.115	0.107
SD of G1 BC	0.005	0.005	0.008	0.009	0.004
# of shots	8	7	9	9	8

Agulia Target 40 grain

Barrel	Vol-77/22	Vol-77/22	Vol-77/22	Rug-10/22	Rug-77/22
Length	16.5"	18.5"	18.5"	18"	24"
Twist	1:11.25"	1:16"	1:9"	1:14"	1:16"
Ave MV	1096	1105	1085	1120	1106
SD of MV	12.8	11.8	10.0	15.2	10.8
Ave DRV	929	940	927	950	940
SD of DRV	9.5	8.3	4.9	8.3	12.3
Ave G1 BC	0.113	0.121	0.121	0.113	0.115
SD of G1 BC	0.006	0.005	0.002	0.005	0.008
# of shots	7	9	7	6	8

American Eagle 38 grain CU Plated

Barrel	Vol-77/22	Vol-77/22	Vol-77/22	Rug-10/22	Rug-77/22
Length	16.5"	18.5"	18.5"	18"	24"
Twist	1:11.25"	1:16"	1:9"	1:14"	1:16"
Ave MV	1162				
SD of MV	29.6				
Ave DRV	975				
SD of DRV	18.4				
Ave G1 BC	0.126				
SD of G1 BC	0.013				
# of shots	8				

American Eagle Suppressor Sub 45 grain

Barrel	Vol-77/22	Vol-77/22	Vol-77/22	Rug-10/22	Rug-77/22
Length	16.5"	18.5"	18.5"	18"	24"
Twist	1:11.25"	1:16"	1:9"	1:14"	1:16"
Ave MV	990	967	952	988	966
SD of MV	5.5	11.0	6.6	11.1	15.1
Ave DRV	877	860	849	883	862
SD of DRV	6.0	8.8	6.5	10.4	12.9
Ave G1 BC	0.125	0.125	0.127	0.127	0.123
SD of G1 BC	0.004	0.003	0.004	0.004	0.005
# of shots	8	9	8	8	9

Blazer 40 grain

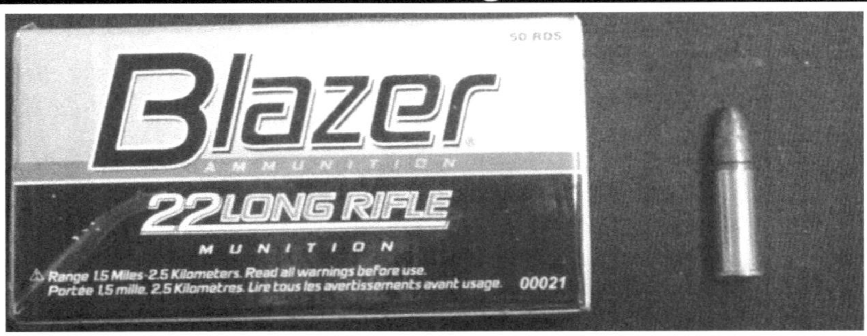

Barrel	Vol-77/22	Vol-77/22	Vol-77/22	Rug-10/22	Rug-77/22
Length	16.5"	18.5"	18.5"	18"	24"
Twist	1:11.25"	1:16"	1:9"	1:14"	1:16"
Ave MV	1201	1198	1217	1209	1205
SD of MV	13.4	9.2	11.8	14.3	19.3
Ave DRV	993	995	994	1012	1003
SD of DRV	6.6	2.9	3.8	12.6	7.7
Ave G1 BC	0.124	0.129	0.123	0.129	0.126
SD of G1 BC	0.002	0.004	0.004	0.006	0.003
# of shots	8	6	3	8	7

CCI AR Tactical 40 grain

Barrel	Vol-77/22	Vol-77/22	Vol-77/22	Rug-10/22	Rug-77/22
Length	16.5"	18.5"	18.5"	18"	24"
Twist	1:11.25"	1:16"	1:9"	1:14"	1:16"
Ave MV	1226	1220	1209		1235
SD of MV	16.0	13.2	15.1		24.6
Ave DRV	984	979	986		995
SD of DRV	13.9	20.4	16.7		17.1
Ave G1 BC	0.110	0.112	0.115		0.108
SD of G1 BC	0.007	0.012	0.010		0.008
# of shots	8	6	8		9

CCI 22 Long 29 grain

Barrel	Vol-77/22	Vol-77/22	Vol-77/22	Rug-10/22	Rug-77/22
Length	16.5"	18.5"	18.5"	18"	24"
Twist	1:11.25"	1:16"	1:9"	1:14"	1:16"
Ave MV	1204	1223	1188	1241	1210
SD of MV	15.3	32.5	15.1	7.2	24.7
Ave DRV	929	948	926	965	938
SD of DRV	5.7	20.4	6.9	6.6	6.6
Ave G1 BC	0.082	0.090	0.087	0.088	0.082
SD of G1 BC	0.001	0.009	0.002	0.003	0.002
# of shots	5	8	7	7	7

CCI Green Tag 40 grain

Barrel	Vol-77/22	Vol-77/22	Vol-77/22	Rug-10/22	Rug-77/22
Length	16.5"	18.5"	18.5"	18"	24"
Twist	1:11.25"	1:16"	1:9"	1:14"	1:16"
Ave MV	1057	1054	1053	1064	1058
SD of MV	10.0	12.8	11.6	10.0	13.1
Ave DRV	923	919	922	927	925
SD of DRV	8.1	10.2	11.9	12.6	11.7
Ave G1 BC	0.129	0.129	0.136	0.121	0.127
SD of G1 BC	0.004	0.004	0.008	0.009	0.005
# of shots	8	8	10	9	9

CCI Mini-Mag 36 grain

Barrel	Vol-77/22	Vol-77/22	Vol-77/22	Rug-10/22	Rug-77/22
Length	16.5"	18.5"	18.5"	18"	24"
Twist	1:11.25"	1:16"	1:9"	1:14"	1:16"
Ave MV	1224	1259	1218	1258	1231
SD of MV	40.7	18.4	15.5	15.3	27.2
Ave DRV	985	1002	986	1014	981
SD of DRV	13.1	8.3	5.0	7.2	29.3
Ave G1 BC	0.110	0.113	0.117	0.112	0.104
SD of G1 BC	0.005	0.004	0.005	0.004	0.019
# of shots	8	9	8	8	6

CCI Mini-Mag 40 grain

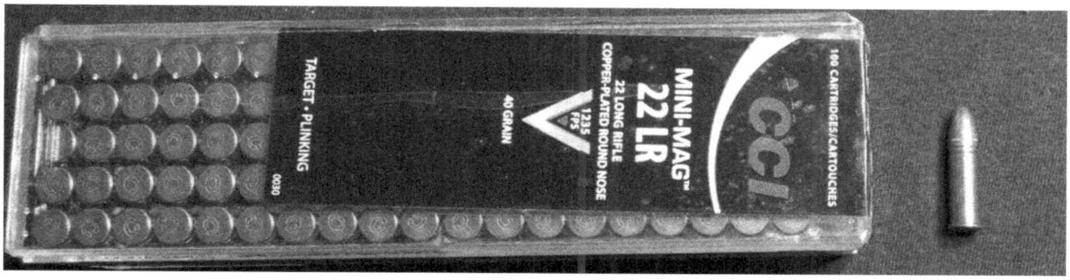

Barrel	Vol-77/22	Vol-77/22	Vol-77/22	Rug-10/22	Rug-77/22
Length	16.5"	18.5"	18.5"	18"	24"
Twist	1:11.25"	1:16"	1:9"	1:14"	1:16"
Ave MV	1214	1213	1199	1220	1211
SD of MV	12.5	8.7	16.9	15.4	12.2
Ave DRV	998	1001	984	1019	1004
SD of DRV	6.5	7.5	18.6	7.8	5.6
Ave G1 BC	0.123	0.129	0.123	0.131	0.124
SD of G1 BC	0.003	0.004	0.012	0.005	0.004
# of shots	8	10	8	6	9

CCI Pistol Match 40 grain

Barrel	Vol-77/22	Vol-77/22	Vol-77/22	Rug-10/22	Rug-77/22
Length	16.5"	18.5"	18.5"	18"	24"
Twist	1:11.25"	1:16"	1:9"	1:14"	1:16"
Ave MV	1076	1062	1061	1090	1073
SD of MV	16.5	9.1	10.6	9.4	14.2
Ave DRV	931	920	926	946	931
SD of DRV	16.3	8.9	7.4	9.9	11.9
Ave G1 BC	0.125	0.125	0.134	0.125	0.122
SD of G1 BC	0.009	0.008	0.006	0.006	0.006
# of shots	7	7	8	8	8

CCI Quiet-22 40 grain

Barrel	Vol-77/22	Vol-77/22	Vol-77/22	Rug-10/22	Rug-77/22
Length	16.5"	18.5"	18.5"	18"	24"
Twist	1:11.25"	1:16"	1:9"	1:14"	1:16"
Ave MV	746			739	
SD of MV	15.6			12.0	
Ave DRV	663			659	
SD of DRV	16.4			10.1	
Ave G1 BC	0.097			0.097	
SD of G1 BC	0.005			0.004	
# of shots	4			10	

CCI Select 40 grain

Barrel	Vol-77/22	Vol-77/22	Vol-77/22	Rug-10/22	Rug-77/22
Length	16.5"	18.5"	18.5"	18"	24"
Twist	1:11.25"	1:16"	1:9"	1:14"	1:16"
Ave MV	1189	1207	1187	1212	1217
SD of MV	25.6	11.1	24.8	27.3	28.3
Ave DRV	988	1002	992	1018	1001
SD of DRV	16.2	8.9	15.4	17.2	10.7
Ave G1 BC	0.123	0.133	0.134	0.133	0.120
SD of G1 BC	0.005	0.004	0.006	0.005	0.011
# of shots	8	10	9	9	9

CCI SGB 40 grain

Barrel	Vol-77/22	Vol-77/22	Vol-77/22	Rug-10/22	Rug-77/22
Length	16.5"	18.5"	18.5"	18"	24"
Twist	1:11.25"	1:16"	1:9"	1:14"	1:16"
Ave MV	1185	1184	1182		1173
SD of MV	11.2	18.1	10.7		18.8
Ave DRV	987	988	987		989
SD of DRV	5.5	9.7	6.1		6.5
Ave G1 BC	0.124	0.129	0.131		0.127
SD of G1 BC	0.001	0.002	0.003		0.005
# of shots	8	8	9		5

CCI Standard Velocity 40 grain

Barrel	Vol-77/22	Vol-77/22	Vol-77/22	Rug-10/22	Rug-77/22
Length	16.5"	18.5"	18.5"	18"	24"
Twist	1:11.25"	1:16"	1:9"	1:14"	1:16"
Ave MV	1043	1050	1050	1075	1069
SD of MV	15.5	11.6	10.4	9.9	20.1
Ave DRV	909	913	919	934	926
SD of DRV	14.3	8.2	9.1	10.5	13.9
Ave G1 BC	0.122	0.124	0.133	0.121	0.119
SD of G1 BC	0.008	0.007	0.004	0.007	0.005
# of shots	8	7	8	3	6

CCI Suppressor 45 grain

Barrel	Vol-77/22	Vol-77/22	Vol-77/22	Rug-10/22	Rug-77/22
Length	16.5"	18.5"	18.5"	18"	24"
Twist	1:11.25"	1:16"	1:9"	1:14"	1:16"
Ave MV	991	983	982	996	982
SD of MV	15.3	14.3	6.9	9.1	17.6
Ave DRV	879	871	872	885	871
SD of DRV	11.5	14.0	6.4	9.5	14.6
Ave G1 BC	0.124	0.125	0.128	0.123	0.118
SD of G1 BC	0.004	0.007	0.004	0.005	0.003
# of shots	9	10	9	9	6

CCI Velocitor 40 grain

Barrel	Vol-77/22	Vol-77/22	Vol-77/22	Rug-10/22	Rug-77/22
Length	16.5"	18.5"	18.5"	18"	24"
Twist	1:11.25"	1:16"	1:9"	1:14"	1:16"
Ave MV	1360	1396	1386	1423	1409
SD of MV	18.6	28.9	15.7	12.4	24.8
Ave DRV	1053	1070	1065	1096	1078
SD of DRV	9.0	15.7	7.3	6.4	13.1
Ave G1 BC	0.113	0.116	0.117	0.115	0.111
SD of G1 BC	0.002	0.003	0.002	0.002	0.003
# of shots	7	7	8	6	9

Eley Club 40 grain

Barrel	Vol-77/22	Vol-77/22	Vol-77/22	Rug-10/22	Rug-77/22
Length	16.5"	18.5"	18.5"	18"	24"
Twist	1:11.25"	1:16"	1:9"	1:14"	1:16"
Ave MV	1096	1097	1077	1116	1102
SD of MV	9.2	6.6	8.5	7.0	8.9
Ave DRV	967	966	956	987	975
SD of DRV	7.0	6.5	6.7	5.1	7.9
Ave G1 BC	0.155	0.156	0.164	0.159	0.157
SD of G1 BC	0.004	0.004	0.004	0.004	0.005
# of shots	8	6	8	9	9

Eley Edge 40 grain Lot 30154

Barrel	Vol-77/22	Vol-77/22	Vol-77/22	Rug-10/22	Rug-77/22
Length	16.5"	18.5"	18.5"	18"	24"
Twist	1:11.25"	1:16"	1:9"	1:14"	1:16"
Ave MV	1097	1086	1082	1113	1100
SD of MV	9.8	10.7	6.4	6.1	8.5
Ave DRV	966	955	955	983	969
SD of DRV	7.7	9.7	8.4	5.9	5.9
Ave G1 BC	0.153	0.152	0.156	0.156	0.149
SD of G1 BC	0.006	0.006	0.008	0.006	0.005
# of shots	9	8	9	10	9

Eley Edge 40 grain Lot 30323

Barrel	Vol-77/22	Vol-77/22	Vol-77/22	Rug-10/22	Rug-77/22
Length	16.5"	18.5"	18.5"	18"	24"
Twist	1:11.25"	1:16"	1:9"	1:14"	1:16"
Ave MV	1111	1092	1086	1119	1113
SD of MV	3.5	9.5	8.5	7.1	7.5
Ave DRV	979	962	961	989	976
SD of DRV	3.7	9.8	6.2	5.6	3.8
Ave G1 BC	0.159	0.158	0.162	0.159	0.149
SD of G1 BC	0.005	0.007	0.003	0.008	0.006
# of shots	9	9	10	10	7

Eley Match EPS 40 grain

Barrel	Vol-77/22	Vol-77/22	Vol-77/22	Rug-10/22	Rug-77/22
Length	16.5"	18.5"	18.5"	18"	24"
Twist	1:11.25"	1:16"	1:9"	1:14"	1:16"
Ave MV	1174	1164	1158	1180	1164
SD of MV	8.0	11.7	7.1	7.0	19.2
Ave DRV	995	988	984	1011	996
SD of DRV	1.9	5.3	3.2	4.5	7.5
Ave G1 BC	0.136	0.140	0.140	0.143	0.140
SD of G1 BC	0.004	0.005	0.003	0.005	0.008
# of shots	9	8	9	8	8

Eley Match OSP 40 grain

Barrel	Vol-77/22	Vol-77/22	Vol-77/22	Rug-10/22	Rug-77/22
Length	16.5"	18.5"	18.5"	18"	24"
Twist	1:11.25"	1:16"	1:9"	1:14"	1:16"
Ave MV	1052	1044	1039	1067	1059
SD of MV	7.8	3.7	6.4	4.3	5.3
Ave DRV	930	925	921	946	937
SD of DRV	6.8	7.7	7.6	8.5	4.8
Ave G1 BC	0.139	0.145	0.145	0.142	0.139
SD of G1 BC	0.004	0.008	0.006	0.010	0.003
# of shots	9	8	9	10	9

Eley Match Pistol 40 grain

Barrel	Vol-77/22	Vol-77/22	Vol-77/22	Rug-10/22	Rug-77/22
Length	16.5"	18.5"	18.5"	18"	24"
Twist	1:11.25"	1:16"	1:9"	1:14"	1:16"
Ave MV	1042	1036	1035	1055	1053
SD of MV	4.1	9.5	3.7	9.8	5.1
Ave DRV	919	913	914	935	928
SD of DRV	3.3	12.2	3.1	9.9	4.1
Ave G1 BC	0.131	0.135	0.139	0.138	0.133
SD of G1 BC	0.002	0.007	0.003	0.006	0.004
# of shots	8	8	9	10	9

Eley Subsonic HP 40 grain					

Barrel	Vol-77/22	Vol-77/22	Vol-77/22	Rug-10/22	Rug-77/22
Length	16.5"	18.5"	18.5"	18"	24"
Twist	1:11.25"	1:16"	1:9"	1:14"	1:16"
Ave MV	1035	1031	1026	1069	1046
SD of MV	7.7	9.8	9.4	7.4	7.1
Ave DRV	904	905	900	940	918
SD of DRV	4.8	7.8	10.8	9.0	4.9
Ave G1 BC	0.118	0.128	0.127	0.133	0.125
SD of G1 BC	0.007	0.003	0.008	0.008	0.004
# of shots	8	7	9	9	9

Eley Target 40 grain					

Barrel	Vol-77/22	Vol-77/22	Vol-77/22	Rug-10/22	Rug-77/22
Length	16.5"	18.5"	18.5"	18"	24"
Twist	1:11.25"	1:16"	1:9"	1:14"	1:16"
Ave MV	1094	1084	1073	1109	1105
SD of MV	13.1	15.8	6.0	6.4	11.3
Ave DRV	964	961	951	984	971
SD of DRV	8.4	14.0	4.8	5.4	8.0
Ave G1 BC	0.151	0.162	0.158	0.161	0.150
SD of G1 BC	0.005	0.009	0.003	0.007	0.013
# of shots	9	8	9	10	9

Eley Tenex 40 grain Lot 05123					

Barrel	Vol-77/22	Vol-77/22	Vol-77/22	Rug-10/22	Rug-77/22
Length	16.5"	18.5"	18.5"	18"	24"
Twist	1:11.25"	1:16"	1:9"	1:14"	1:16"
Ave MV	1090	1088	1082	1100	1098
SD of MV	7.6	9.4	7.3	8.1	6.3
Ave DRV	963	961	956	976	972
SD of DRV	5.8	10.1	4.9	9.3	4.5
Ave G1 BC	0.152	0.159	0.158	0.157	0.156
SD of G1 BC	0.003	0.007	0.005	0.008	0.004
# of shots	9	8	9	10	9

Eley Tenex 40 grain Lot 01081					

Barrel	Vol-77/22	Vol-77/22	Vol-77/22	Rug-10/22	Rug-77/22
Length	16.5"	18.5"	18.5"	13"	24"
Twist	1:11.25"	1:16"	1:9"	1:14"	1:16"
Ave MV	1105	1104	1091	1116	1116
SD of MV	7.3	3.7	3.9	8.8	4.2
Ave DRV	973	975	967	988	983
SD of DRV	7.9	4.3	4.6	6.8	4.7
Ave G1 BC	0.153	0.162	0.161	0.160	0.156
SD of G1 BC	0.006	0.005	0.005	0.005	0.006
# of shots	9	8	9	10	10

Eley Tenex Pistol 40 grain

Barrel	Vol-77/22	Vol-77/22	Vol-77/22	Rug-10/22	Rug-77/22
Length	16.5"	18.5"	18.5"	18"	24"
Twist	1:11.25"	1:16"	1:9"	1:14"	1:16"
Ave MV	1068	1052	1040	1074	1072
SD of MV	11.2	5.8	15.3	7.0	10.0
Ave DRV	946	932	924	956	950
SD of DRV	12.1	5.3	14.7	10.4	10.4
Ave G1 BC	0.147	0.146	0.144	0.151	0.147
SD of G1 BC	0.009	0.006	0.008	0.010	0.009
# of shots	9	9	9	9	9

Eley Tenex Ultimate EPS 40 grain Lot UFW4382 1068

Barrel	Vol-77/22	Vol-77/22	Vol-77/22	Rug-10/22	Rug-77/22
Length	16.5"	18.5"	18.5"	18"	24"
Twist	1:11.25"	1:16"	1:9"	1:14"	1:16"
Ave MV	1079	1072	1067	1083	1082
SD of MV	8.6	9.2	6.8	7.7	6.3
Ave DRV	948	940	942	953	954
SD of DRV	5.8	8.9	5.4	5.4	3.1
Ave G1 BC	0.140	0.140	0.145	0.138	0.144
SD of G1 BC	0.003	0.006	0.003	0.004	0.003
# of shots	8	10	9	6	9

Eley Tenex Ultimate EPS 40 grain Lot UGY2047 1054

Barrel	Vol-77/22	Vol-77/22	Vol-77/22	Rug-10/22	Rug-77/22
Length	16.5"	18.5"	18.5"	18"	24"
Twist	1:11.25"	1:16"	1:9"	1:14"	1:16"
Ave MV	1062	1077	1073	1082	1087
SD of MV	28.7	11.7	7.6	8.9	8.8
Ave DRV	940	948	949	956	961
SD of DRV	23.7	7.8	3.7	6.5	4.0
Ave G1 BC	0.144	0.146	0.150	0.144	0.150
SD of G1 BC	0.009	0.004	0.003	0.004	0.004
# of shots	8	10	7	8	10

Federal Auto Match 40 grain

Barrel	Vol-77/22	Vol-77/22	Vol-77/22	Rug-10/22	Rug-77/22
Length	16.5"	18.5"	18.5"	18"	24"
Twist	1:11.25"	1:16"	1:9"	1:14"	1:16"
Ave MV	1205	1218	1215		1222
SD of MV	16.2	6.5	7.2		11.0
Ave DRV	999	1006	1014		1019
SD of DRV	8.0	7.2	10.0		8.9
Ave G1 BC	0.130	0.133	0.136		0.132
SD of G1 BC	0.006	0.005	0.007		0.008
# of shots	8	9	9		9

Federal Champion 40 grain

Barrel	Vol-77/22	Vol-77/22	Vol-77/22	Rug-10/22	Rug-77/22
Length	16.5"	18.5"	18.5"	18"	24"
Twist	1:11.25"	1:16"	1:9"	1:14"	1:16"
Ave MV	1234	1227	1232	1237	1223
SD of MV	7.9	11.3	7.4	0.6	17.9
Ave DRV	1027	1021	1027	1029	1019
SD of DRV	6.7	6.3	3.6	7.6	7.9
Ave G1 BC	0.136	0.138	0.139	0.132	0.132
SD of G1 BC	0.005	0.007	0.002	0.007	0.004
# of shots	7	4	8	3	6

Federal Game Shok 38 grain Cu Plated HP

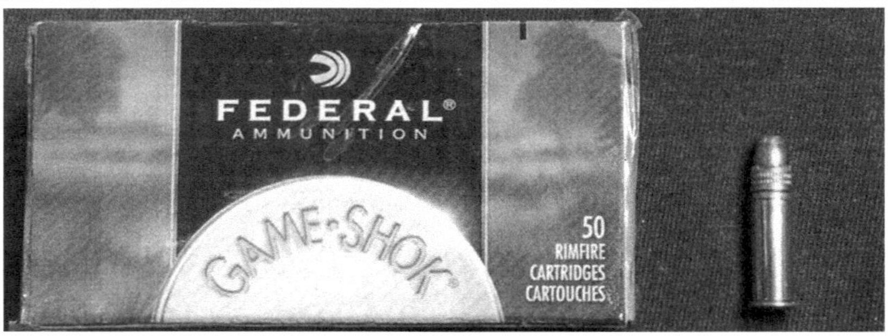

Barrel	Vol-77/22	Vol-77/22	Vol-77/22	Rug-10/22	Rug-77/22
Length	16.5"	18.5"	18.5"	18"	24"
Twist	1:11.25"	1:16"	1:9"	1:14"	1:16"
Ave MV	1217	1218	1208	1228	1219
SD of MV	11.0	15.7	9.0	9.0	8.8
Ave DRV	997	999	988	1007	998
SD of DRV	13.2	15.7	13.5	14.2	12.8
Ave G1 BC	0.117	0.122	0.117	0.116	0.116
SD of G1 BC	0.009	0.008	0.009	0.009	0.009
# of shots	9	8	6	3	8

Federal Game Shok 40 grain Cu Plated Solid

Barrel	Vol-77/22	Vol-77/22	Vol-77/22	Rug-10/22	Rug-77/22
Length	16.5"	18.5"	18.5"	18"	24"
Twist	1:11.25"	1:16"	1:9"	1:14"	1:16"
Ave MV	1272	1276	1262	1273	1273
SD of MV	14.4	18.6	19.7	12.5	21.8
Ave DRV	1004	1029	1012	1011	1034
SD of DRV	10.2	17.8	20.7	20.8	12.3
Ave G1 BC	0.105	0.124	0.116	0.106	0.125
SD of G1 BC	0.006	0.007	0.010	0.014	0.007
# of shots	8	6	6	4	8

Federal Gold Medal Match 40 grain

Barrel	Vol-77/22	Vol-77/22	Vol-77/22	Rug-10/22	Rug-77/22
Length	16.5"	18.5"	18.5"	18"	24"
Twist	1:11.25"	1:16"	1:9"	1:14"	1:16"
Ave MV	1169	1173	1169	1184	1177
SD of MV	2.8	4.4	3.9	9.9	2.5
Ave DRV	999	996	1007	1021	1006
SD of DRV	5.1	5.4	3.0	8.7	5.8
Ave G1 BC	0.141	0.140	0.154	0.153	0.143
SD of G1 BC	0.006	0.005	0.004	0.006	0.006
# of shots	8	8	8	6	9

Federal Gold Medal Target 40 grain

Barrel	Vol-77/22	Vol-77/22	Vol-77/22	Rug-10/22	Rug-77/22
Length	16.5"	18.5"	18.5"	18"	24"
Twist	1:11.25"	1:16"	1:9"	1:14"	1:16"
Ave MV	1089	1098	1096	1109	1097
SD of MV	11.3	11.5	9.0	7.5	20.3
Ave DRV	949	955	963	965	956
SD of DRV	10.6	10.3	8.0	3.4	15.4
Ave G1 BC	0.133	0.138	0.151	0.135	0.135
SD of G1 BC	0.006	0.007	0.007	0.003	0.008
# of shots	9	9	9	7	9

Federal Gold Medal UltraMatch 40 grain

Barrel	Vol-77/22	Vol-77/22	Vol-77/22	Rug-10/22	Rug-77/22
Length	16.5"	18.5"	18.5"	18"	24"
Twist	1:11.25"	1:16"	1:9"	1:14"	1:16"
Ave MV	1150	1147	1146		
SD of MV	10.2	7.2	9.2		
Ave DRV	993	984	999		
SD of DRV	5.3	7.0	11.3		
Ave G1 BC	0.146	0.142	0.161		
SD of G1 BC	0.006	0.006	0.010		
# of shots	9	9	5		

Federal Lightning 40 grain

Barrel	Vol-77/22	Vol-77/22	Vol-77/22	Rug-10/22	Rug-77/22
Length	16.5"	18.5"	18.5"	18"	24"
Twist	1:11.25"	1:16"	1:9"	1:14"	1:16"
Ave MV	1224	1234	1229		
SD of MV	15.9	12.8	11.4		
Ave DRV	1021	1009	1029		
SD of DRV	8.8	7.1	3.4		
Ave G1 BC	0.134	0.133	0.144		
SD of G1 BC	0.003	0.004	0.004		
# of shots	7	9	5		

Fiocchi Official 300 40 grain

Barrel	Vol-77/22	Vol-77/22	Vol-77/22	Rug-10/22	Rug-77/22
Length	16.5"	18.5"	18.5"	18"	24"
Twist	1:11.25"	1:16"	1:9"	1:14"	1:16"
Ave MV	1028	1015	1022	1057	1028
SD of MV	4.7	8.8	11.2	16.9	9.6
Ave DRV	919	904	916	941	919
SD of DRV	4.4	6.8	10.0	9.4	8.9
Ave G1 BC	0.145	0.148	0.149	0.145	0.144
SD of G1 BC	0.004	0.006	0.006	0.009	0.006
# of shots	8	9	9	9	9

Geco Semi-Auto 40 grain

Barrel	Vol-77/22	Vol-77/22	Vol-77/22	Rug-10/22	Rug-77/22
Length	16.5"	18.5"	18.5"	18"	24"
Twist	1:11.25"	1:16"	1:9"	1:14"	1:16"
Ave MV	1158	1153	1170	1166	1178
SD of MV	12.4	13.8	4.8	9.9	5.5
Ave DRV	999	987	1008	1015	1006
SD of DRV	8.1	6.6	3.4	5.2	3.2
Ave G1 BC	0.148	0.152	0.155	0.158	0.142
SD of G1 BC	0.005	0.003	0.004	0.004	0.003
# of shots	8	9	9	8	8

Gemtech Silencer Subsonic 42 grain

Barrel	Vol-77/22	Vol-77/22	Vol-77/22	Rug-10/22	Rug-77/22
Length	16.5"	18.5"	18.5"	18"	24"
Twist	1:11.25"	1:16"	1:9"	1:14"	1:16"
Ave MV	1049	1037	1040	1061	1044
SD of MV	15.2	12.6	14.3	14.7	11.7
Ave DRV	923	911	924	938	923
SD of DRV	12.8	14.3	13.3	15.2	9.7
Ave G1 BC	0.130	0.139	0.144	0.138	0.134
SD of G1 BC	0.006	0.010	0.008	0.010	0.005
# of shots	8	9	8	9	10

Golden Eagle Target 40 grain

Barrel	Vol-77/22	Vol-77/22	Vol-77/22	Rug-10/22	Rug-77/22
Length	16.5"	18.5"	18.5"	18"	24"
Twist	1:11.25"	1:16"	1:9"	1:14"	1:16"
Ave MV	1038	1047	1040	1065	1051
SD of MV	12.3	7.3	11.5	6.3	7.9
Ave DRV	904	909	915	930	920
SD of DRV	8.2	6.0	9.9	5.5	6.8
Ave G1 BC	0.121	0.125	0.132	0.124	0.123
SD of G1 BC	0.005	0.004	0.005	0.005	0.004
# of shots	10	8	9	8	9

Lapua Center-X 40 grain

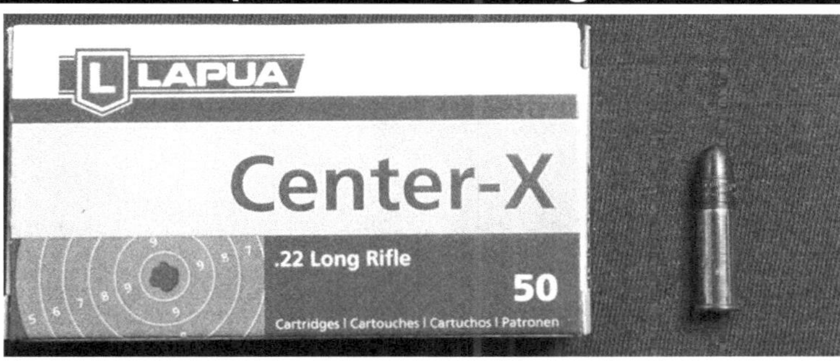

Barrel	Vol-77/22	Vol-77/22	Vol-77/22	Rug-10/22	Rug-77/22
Length	16.5"	18.5"	18.5"	18"	24"
Twist	1:11.25"	1:16"	1:9"	1:14"	1:16"
Ave MV	1066	1050	1060	1071	1075
SD of MV	9.4	5.9	7.0	9.8	4.6
Ave DRV	939	933	948	954	960
SD of DRV	6.9	6.6	6.9	8.9	3.8
Ave G1 BC	0.139	0.160	0.162	0.152	0.161
SD of G1 BC	0.011	0.006	0.005	0.010	0.005
# of shots	8	9	8	8	10

Lapua Midas+ 40 grain Lot 5/707/5 2012

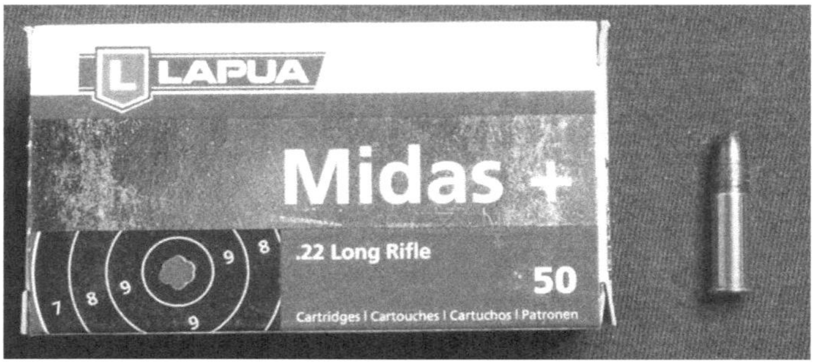

Barrel	Vol-77/22	Vol-77/22	Vol-77/22	Rug-10/22	Rug-77/22
Length	16.5"	18.5"	18.5"	18"	24"
Twist	1:11.25"	1:16"	1:9"	1:14"	1:16"
Ave MV	1073	1057	1064	1076	1076
SD of MV	5.3	7.0	8.1	6.0	5.5
Ave DRV	953	937	952	959	955
SD of DRV	4.5	7.0	6.4	7.1	8.0
Ave G1 BC	0.151	0.159	0.163	0.155	0.151
SD of G1 BC	0.003	0.006	0.003	0.007	0.010
# of shots	7	9	9	10	10

Lapua Midas+ 40 grain Lot 353 44 5 972 1

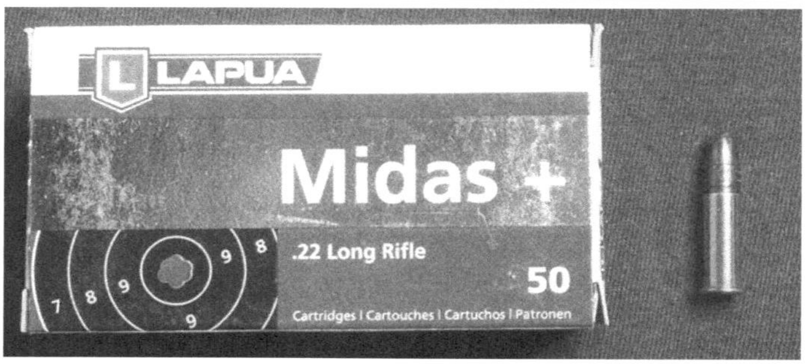

Barrel	Vol-77/22	Vol-77/22	Vol-77/22	Rug-10/22	Rug-77/22
Length	16.5"	18.5"	18.5"	18"	24"
Twist	1:11.25"	1:16"	1:9"	1:14"	1:16"
Ave MV	1073	1057	1064	1076	1076
SD of MV	5.3	7.0	8.1	6.0	6
Ave DRV	953	937	952	959	955
SD of DRV	4.5	7.0	6.4	7.1	8
Ave G1 BC	0.151	0.159	0.163	0.155	0.151
SD of G1 BC	0.003	0.006	0.003	0.007	0.010
# of shots	7	9	9	10	10

Lapua Midas+ (Neal Waltz Die, Eley EPS nose)

Barrel	Vol-77/22	Vol-77/22	Vol-77/22	Rug-10/22	Rug-77/22
Length	16.5"	18.5"	18.5"	18"	24"
Twist	1:11.25"	1:16"	1:9"	1:14"	1:16"
Ave MV	1082	1077	1084	1094	1090
SD of MV	12.6	9.7	10.6	11.6	10.8
Ave DRV	947	938	952	953	947
SD of DRV	13.0	9.2	6.2	11.8	8.6
Ave G1 BC	0.136	0.142	0.144	0.131	0.128
SD of G1 BC	0.010	0.007	0.005	0.010	0.006
# of shots	8	9	7	7	9

Lapua Midas+ (Neal Waltz Die LHP)

Barrel	Vol-77/22	Vol-77/22	Vol-77/22	Rug-10/22	Rug-77/22
Length	16.5"	18.5"	18.5"	18"	24"
Twist	1:11.25"	1:16"	1:9"	1:14"	1:16"
Ave MV	1084	1079	1084	1095	1087
SD of MV	8.8	7.6	11.1	9.2	9.4
Ave DRV	948	921	945	956	943
SD of DRV	6.8	18.1	8.6	15.4	10.3
Ave G1 BC	0.135	0.122	0.135	0.133	0.127
SD of G1 BC	0.006	0.014	0.009	0.012	0.010
# of shots	8	10	9	8	9

Lapua Pistol King 40 grain

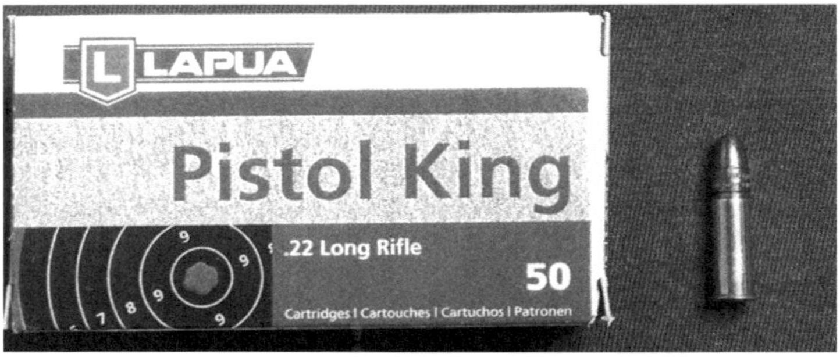

Barrel	Vol-77/22	Vol-77/22	Vol-77/22	Rug-10/22	Rug-77/22
Length	16.5"	18.5"	18.5"	18"	24"
Twist	1:11.25"	1:16"	1:9"	1:14"	1:16"
Ave MV	1068	1058	1063	1072	1077
SD of MV	6.2	6.2	7.0	11.6	7.5
Ave DRV	949	940	948	952	959
SD of DRV	5.9	5.9	7.0	5.8	8.8
Ave G1 BC	0.150	0.162	0.158	0.147	0.155
SD of G1 BC	0.004	0.005	0.005	0.005	0.009
# of shots	9	9	9	7	9

Lapua Pistol OSP 40 grain

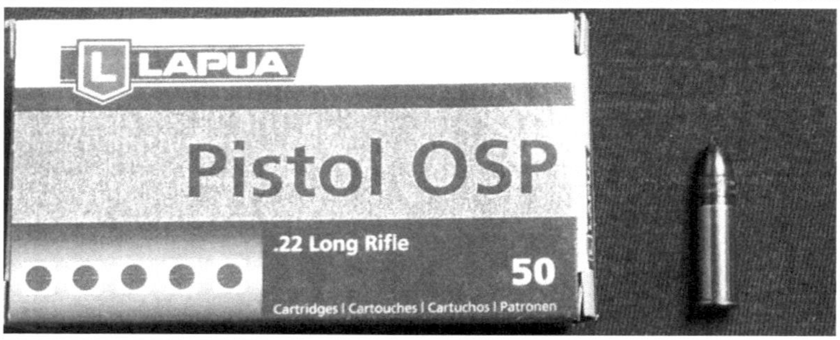

Barrel	Vol-77/22	Vol-77/22	Vol-77/22	Rug-10/22	Rug-77/22
Length	16.5"	18.5"	18.5"	18"	24"
Twist	1:11.25"	1:16"	1:9"	1:14"	1:16"
Ave MV	1043	1046	1042	1053	1058
SD of MV	2.9	8.6	9.4	9.7	9.1
Ave DRV	925	929	928	940	944
SD of DRV	4.1	6.9	6.6	8.8	8.3
Ave G1 BC	0.138	0.156	0.148	0.148	0.151
SD of G1 BC	0.004	0.003	0.010	0.005	0.004
# of shots	8	9	8	10	10

Lapua Polar Biathalon 40 grain

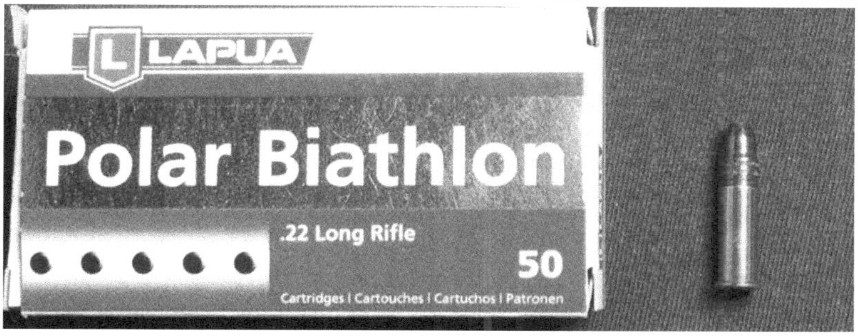

Barrel	Vol-77/22	Vol-77/22	Vol-77/22	Rug-10/22	Rug-77/22
Length	16.5"	18.5"	18.5"	18"	24"
Twist	1:11.25"	1:16"	1:9"	1:14"	1:16"
Ave MV	1099	1093	1091	1106	1103
SD of MV	8.5	4.4	3.6	8.2	8.2
Ave DRV	978	957	970	981	980
SD of DRV	5.3	5.6	2.6	9.3	3.5
Ave G1 BC	0.166	0.155	0.166	0.161	0.163
SD of G1 BC	0.003	0.006	0.002	0.010	0.005
# of shots	7	8	9	10	10

Lapua X-Act 40 grain

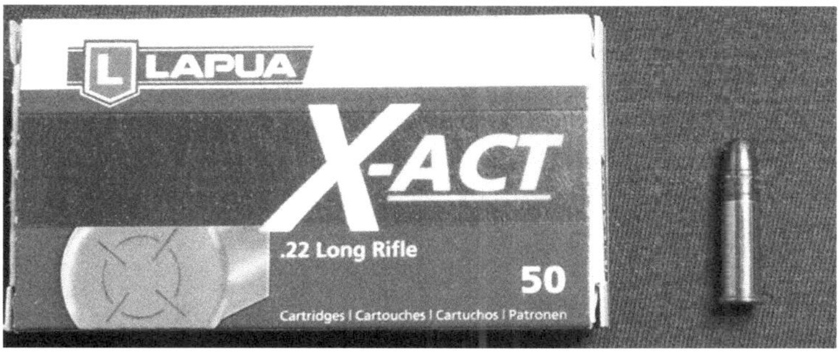

Barrel	Vol-77/22	Vol-77/22	Vol-77/22	Rug-10/22	Rug-77/22
Length	16.5"	18.5"	18.5"	18"	24"
Twist	1:11.25"	1:16"	1:9"	1:14"	1:16"
Ave MV	1062	1058	1061	1068	1062
SD of MV	7.3	5.7	8.6	8.6	9.4
Ave DRV	945	940	949	953	950
SD of DRV	6.2	4.9	7.5	11.1	9.6
Ave G1 BC	0.150	0.161	0.163	0.152	0.157
SD of G1 BC	0.003	0.003	0.004	0.011	0.007
# of shots	8	9	9	10	10

Magtech 40 grain

Barrel	Vol-77/22	Vol-77/22	Vol-77/22	Rug-10/22	Rug-77/22
Length	16.5"	18.5"	18.5"	18"	24"
Twist	1:11.25"	1:16"	1:9"	1:14"	1:16"
Ave MV	1145	1128	1156	1147	1154
SD of MV	20.9	12.5	31.1	12.2	39.0
Ave DRV	976	955	982	985	982
SD of DRV	9.4	12.1	21.4	12.9	22.1
Ave G1 BC	0.129	0.130	0.134	0.134	0.129
SD of G1 BC	0.012	0.009	0.008	0.011	0.009
# of shots	7	9	6	8	8

Norma USA Match-22 40 grain

Barrel	Vol-77/22	Vol-77/22	Vol-77/22	Rug-10/22	Rug-77/22
Length	16.5"	18.5"	18.5"	18"	24"
Twist	1:11.25"	1:16"	1:9"	1:14"	1:16"
Ave MV	1072	1086	1098	1092	1102
SD of MV	9.9	10.6	9.6	11.5	6.4
Ave DRV	945	953	972	971	968
SD of DRV	7.4	5.6	8.0	10.7	6.3
Ave G1 BC	0.140	0.156	0.161	0.154	0.145
SD of G1 BC	0.003	0.005	0.005	0.008	0.006
# of shots	8	9	9	9	9

Norma USA TAC-22 40 grain

Barrel	Vol-77/22	Vol-77/22	Vol-77/22	Rug-10/22	Rug-77/22
Length	16.5"	18.5"	18.5"	18"	24"
Twist	1:11.25"	1:16"	1:9"	1:14"	1:16"
Ave MV	1044	1080	1079	1078	1093
SD of MV	17.9	14.2	7.1	15.0	9.5
Ave DRV	922	948	959	957	963
SD of DRV	14.0	10.9	5.8	15.1	7.9
Ave G1 BC	0.133	0.153	0.161	0.149	0.146
SD of G1 BC	0.004	0.007	0.004	0.011	0.005
# of shots	8	9	9	9	10

PMC Moderator Subsonic HP 38 grain

Barrel	Vol-77/22	Vol-77/22	Vol-77/22	Rug-10/22	Rug-77/22
Length	16.5"	18.5"	18.5"	18"	24"
Twist	1:11.25"	1:16"	1:9"	1:14"	1:16"
Ave MV	1004	1000	997	1029	1002
SD of MV	9.5	12.1	16.0	23.3	9.5
Ave DRV	884	876	881	899	879
SD of DRV	8.9	12.1	14.0	9.8	7.9
Ave G1 BC	0.117	0.121	0.122	0.115	0.112
SD of G1 BC	0.006	0.007	0.005	0.014	0.005
# of shots	7	9	9	6	8

Remington 22 Subsonic 38 grain

Barrel	Vol-77/22	Vol-77/22	Vol-77/22	Rug-10/22	Rug-77/22
Length	16.5"	18.5"	18.5"	18"	24"
Twist	1:11.25"	1:16"	1:9"	1:14"	1:16"
Ave MV	1034	1061	1058	1050	1060
SD of MV	26.6	13.6	33.5	25.0	23.0
Ave DRV	915	932	938	922	934
SD of DRV	18.6	19.9	25.7	48.8	20.7
Ave G1 BC	0.133	0.147	0.148	0.130	0.135
SD of G1 BC	0.005	0.018	0.011	0.032	0.012
# of shots	8	8	7	8	9

Remington 22 Viper 36 grain

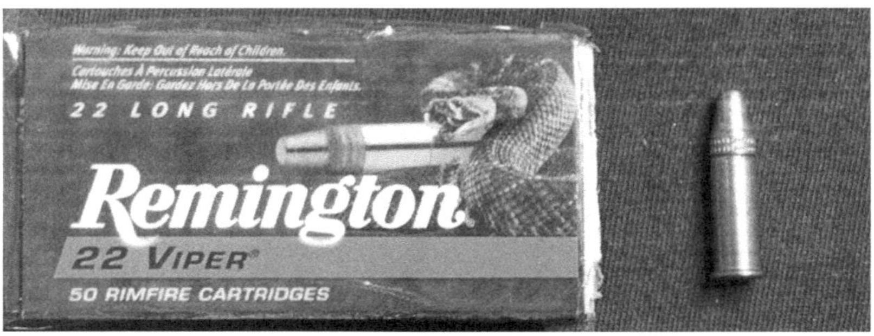

Barrel	Vol-77/22	Vol-77/22	Vol-77/22	Rug-10/22	Rug-77/22
Length	16.5"	18.5"	18.5"	18"	24"
Twist	1:11.25"	1:16"	1:9"	1:14"	1:16"
Ave MV	1297	1297	1285	1326	1304
SD of MV	29.4	25.6	14.8	35.1	35.3
Ave DRV	982	969	970	1008	981
SD of DRV	15.0	11.0	11.0	20.3	10.4
Ave G1 BC	0.088	0.090	0.086	0.092	0.085
SD of G1 BC	0.003	0.004	0.003	0.005	0.003
# of shots	7	8	5	7	9

Remington 22 Yellow Jacket 33 grain

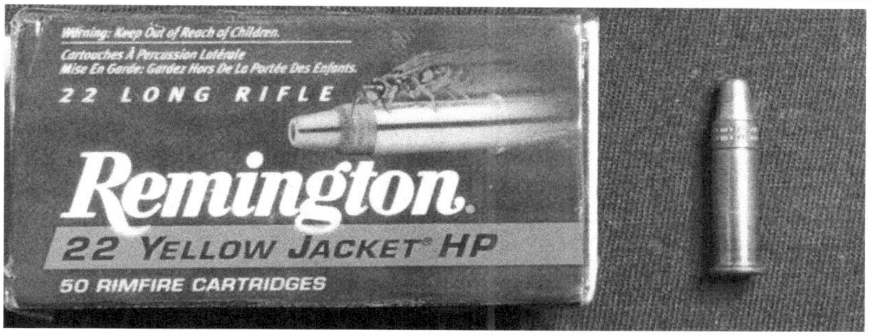

Barrel	Vol-77/22	Vol-77/22	Vol-77/22	Rug-10/22	Rug-77/22
Length	16.5"	18.5"	18.5"	18"	24"
Twist	1:11.25"	1:16"	1:9"	1:14"	1:16"
Ave MV	1391	1376	1404	1405	1405
SD of MV	26.8	12.0	28.7	20.8	29.1
Ave DRV	1016	993	1018	1038	1014
SD of DRV	13.6	10.6	16.5	18.7	47.5
Ave G1 BC	0.087	0.087	0.088	0.091	0.084
SD of G1 BC	0.003	0.003	0.005	0.005	0.013
# of shots	8	8	8	6	7

Remington Club Extra 40 grain

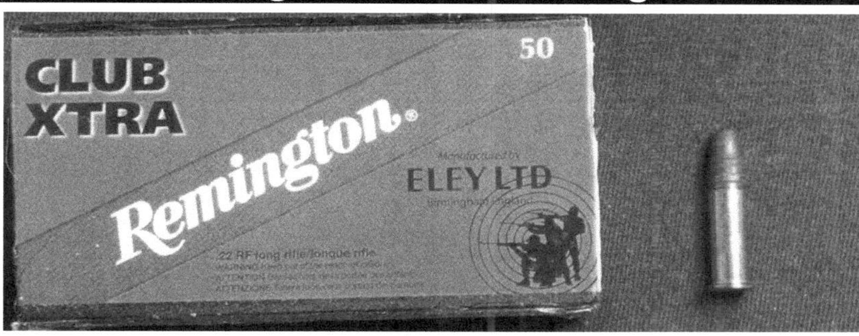

Barrel	Vol-77/22	Vol-77/22	Vol-77/22	Rug-10/22	Rug-77/22
Length	16.5"	18.5"	18.5"	18"	24"
Twist	1:11.25"	1:16"	1:9"	1:14"	1:16"
Ave MV	1096	1075	1087	1112	1097
SD of MV	10.0	10.2	4.1	27.0	12.1
Ave DRV	971	949	966	981	974
SD of DRV	5.8	8.2	2.4	5.8	11.9
Ave G1 BC	0.157	0.159	0.163	0.157	0.159
SD of G1 BC	0.005	0.005	0.003	0.018	0.007
# of shots	7	9	9	8	9

Remington Cyclone 36 grain

Barrel	Vol-77/22	Vol-77/22	Vol-77/22	Rug-10/22	Rug-77/22
Length	16.5"	18.5"	18.5"	18"	24"
Twist	1:11.25"	1:16"	1:9"	1:14"	1:16"
Ave MV	1240	1253	1247	1274	1269
SD of MV	40.4	23.3	16.1	38.0	21.2
Ave DRV	1020	1006	1015	1044	1023
SD of DRV	9.3	5.6	7.2	12.7	5.2
Ave G1 BC	0.127	0.122	0.123	0.129	0.119
SD of G1 BC	0.009	0.006	0.005	0.007	0.006
# of shots	8	9	9	6	7

Remington Golden Bullet (40 grain)

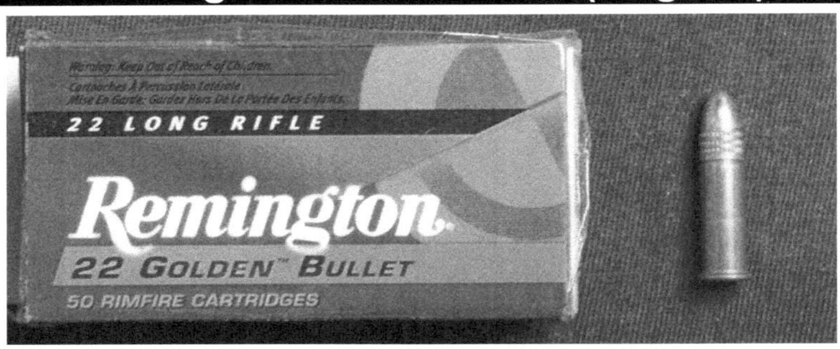

Barrel	Vol-77/22	Vol-77/22	Vol-77/22	Rug-10/22	Rug-77/22
Length	16.5"	18.5"	18.5"	18"	24"
Twist	1:11.25"	1:16"	1:9"	1:14"	1:16"
Ave MV	1179	1178	1176	1198	1160
SD of MV	47.4	25.3	11.3	13.2	38.9
Ave DRV	1008	991	1000	1016	994
SD of DRV	14.3	10.6	6.0	8.5	25.6
Ave G1 BC	0.146	0.141	0.142	0.139	0.139
SD of G1 BC	0.014	0.007	0.007	0.008	0.015
# of shots	6	9	9	7	8

Remington High Velocity 40 grain

Barrel	Vol-77/22	Vol-77/22	Vol-77/22	Rug-10/22	Rug-77/22
Length	16.5"	18.5"	18.5"	18"	24"
Twist	1:11.25"	1:16"	1:9"	1:14"	1:16"
Ave MV	1200	1215	1202		
SD of MV	25.3	18.4	20.8		
Ave DRV	1025	1011	1013		
SD of DRV	10.6	5.3	13.5		
Ave G1 BC	0.151	0.141	0.142		
SD of G1 BC	0.004	0.007	0.015		
# of shots	8	9	4		

Remington Match EPS 40 grain

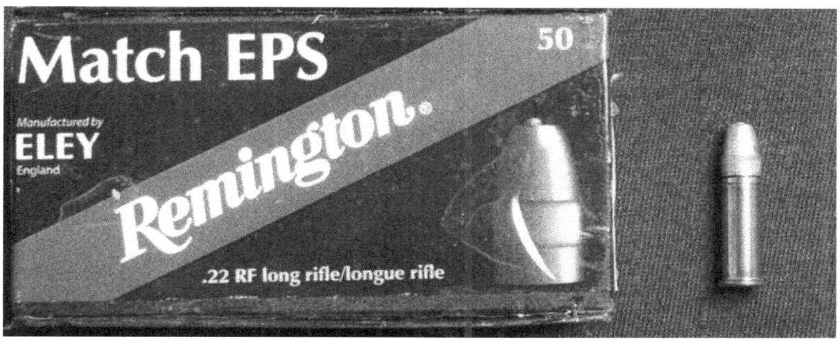

Barrel	Vol-77/22	Vol-77/22	Vol-77/22	Rug-10/22	Rug-77/22
Length	16.5"	18.5"	18.5"	18"	24"
Twist	1:11.25"	1:16"	1:9"	1:14"	1:16"
Ave MV	1104	1089	1094	1109	1099
SD of MV	5.1	3.8	6.5	4.5	6.6
Ave DRV	972	955	966	978	969
SD of DRV	4.8	4.0	6.5	6.2	6.1
Ave G1 BC	0.152	0.154	0.157	0.151	0.149
SD of G1 BC	0.003	0.003	0.006	0.006	0.004
# of shots	7	9	9	9	9

Remington Target 40 grain

Barrel	Vol-77/22	Vol-77/22	Vol-77/22	Rug-10/22	Rug-77/22
Length	16.5"	18.5"	18.5"	18"	24"
Twist	1:11.25"	1:16"	1:9"	1:14"	1:16"
Ave MV	1097	1073	1096	1090	1085
SD of MV	18.1	26.9	10.4	24.8	21.6
Ave DRV	971	947	972	971	960
SD of DRV	14.0	17.9	2.1	17.5	16.5
Ave G1 BC	0.161	0.157	0.164	0.159	0.149
SD of G1 BC	0.024	0.011	0.013	0.007	0.008
# of shots	8	9	4	9	9

Remington Target Rifle 40 grain

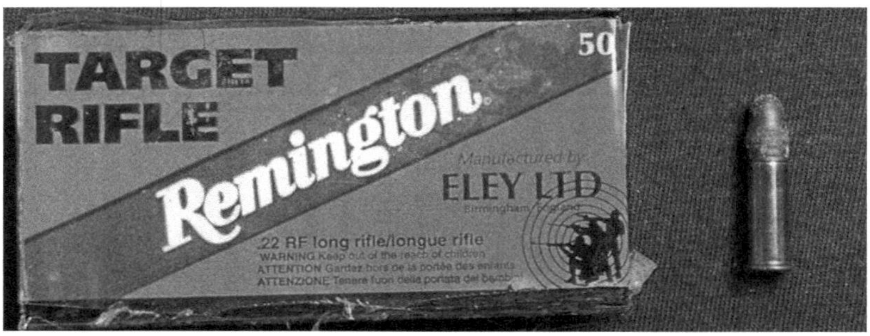

Barrel	Vol-77/22	Vol-77/22	Vol-77/22	Rug-10/22	Rug-77/22
Length	16.5"	18.5"	18.5"	18"	24"
Twist	1:11.25"	1:16"	1:9"	1:14"	1:16"
Ave MV	1066	1059	1058	1082	1078
SD of MV	3.9	6.5	10.1	9.1	6.6
Ave DRV	946	931	941	962	951
SD of DRV	6.2	7.5	11.2	10.0	8.9
Ave G1 BC	0.146	0.146	0.153	0.151	0.142
SD of G1 BC	0.006	0.007	0.010	0.008	0.008
# of shots	8	8	7	8	9

Remington Thunderbolt 40 grain

Barrel	Vol-77/22	Vol-77/22	Vol-77/22	Rug-10/22	Rug-77/22
Length	16.5"	18.5"	18.5"	18"	24"
Twist	1:11.25"	1:16"	1:9"	1:14"	1:16"
Ave MV	1215	1200	1247	1219	1256
SD of MV	21.0	58.6	22.6	32.1	22.5
Ave DRV	1022	1012	1032	1033	1036
SD of DRV	10.1	11.6	9.9	12.7	6.2
Ave G1 BC	0.139	0.160	0.138	0.144	0.133
SD of G1 BC	0.004	0.045	0.006	0.006	0.006
# of shots	7	8	7	7	9

RWS R100 40 grain

Barrel	Vol-77/22	Vol-77/22	Vol-77/22	Rug-10/22	Rug-77/22
Length	16.5"	18.5"	18.5"	18'	24"
Twist	1:11.25"	1:16"	1:9"	1:14"	1:16"
Ave MV	1112	1115	1118	1110	1126
SD of MV	9.0	10.8	7.5	7.1	8.0
Ave DRV	978	971	985	980	985
SD of DRV	5.7	6.7	5.3	7.6	7.2
Ave G1 BC	0.153	0.157	0.163	0.154	0.177
SD of G1 BC	0.003	0.002	0.004	0.007	0.078
# of shots	7	8	9	9	9

Sellier & Bellot HV HP 38 grain

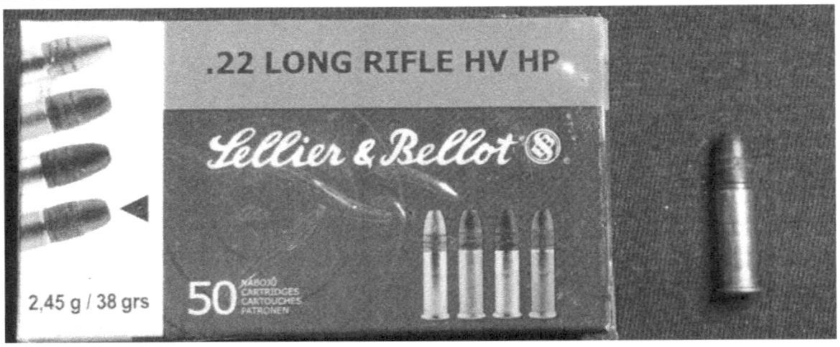

Barrel	Vol-77/22	Vol-77/22	Vol-77/22	Rug-10/22	Rug-77/22
Length	16.5"	18.5"	18.5"	18"	24"
Twist	1:11.25"	1:16"	1:9"	1:14"	1:16"
Ave MV	1194	1229	1207	1229	1221
SD of MV	19.6	31.9	25.1	26.8	24.1
Ave DRV	980	984	987	1013	998
SD of DRV	6.1	17.5	10.5	12.6	6.8
Ave G1 BC	0.119	0.113	0.115	0.121	0.115
SD of G1 BC	0.008	0.004	0.003	0.005	0.005
# of shots	6	8	5	6	9

SK High Velocity HP 40 grain

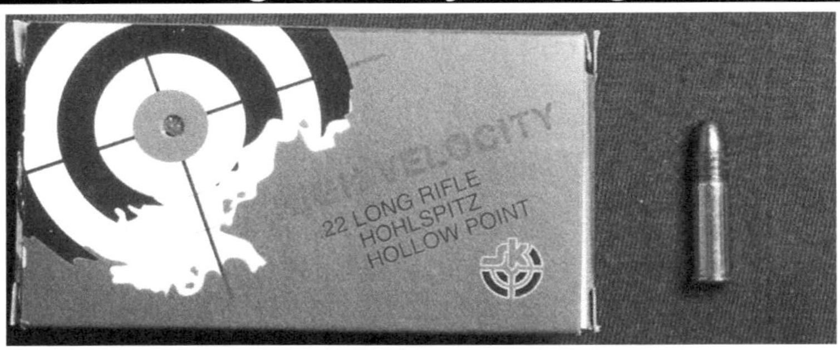

Barrel	Vol-77/22	Vol-77/22	Vol-77/22	Rug-10/22	Rug-77/22
Length	16.5"	18.5"	18.5"	18"	24"
Twist	1:11.25"	1:16"	1:9"	1:14"	1:16"
Ave MV	1192	1242	1257	1287	1271
SD of MV	21.2	17.3	7.4	22.5	22.2
Ave DRV	996	1017	1038	1062	1050
SD of DRV	8.9	9.0	5.7	8.1	7.9
Ave G1 BC	0.134	0.134	0.138	0.139	0.139
SD of G1 BC	0.007	0.008	0.006	0.005	0.005
# of shots	7	8	8	9	9

SK High Velocity Solid 40 grain

Barrel	Vol-77/22	Vol-77/22	Vol-77/22	Rug-10/22	Rug-77/22
Length	16.5"	18.5"	18.5"	18"	24"
Twist	1:11.25"	1:16"	1:9"	1:14"	1:16"
Ave MV	1187	1227	1229	1264	1260
SD of MV	18.8	13.3	21.9	14.0	20.4
Ave DRV	989	1007	1053	1049	1049
SD of DRV	7.4	10.2	71.8	8.4	7.6
Ave G1 BC	0.129	0.131	0.141	0.138	0.142
SD of G1 BC	0.005	0.007	0.003	0.005	0.005
# of shots	6	9	8	10	10

SK Pistol Match 40 grain

Barrel	Vol-77/22	Vol-77/22	Vol-77/22	Rug-10/22	Rug-77/22
Length	16.5"	18.5"	18.5"	18"	24"
Twist	1:11.25"	1:16"	1:9"	1:14"	1:16"
Ave MV	1071	1072	1074	1092	1092
SD of MV	9.2	8.0	10.3	7.8	6.0
Ave DRV	948	953	958	975	970
SD of DRV	8.8	5.9	8.9	8.7	6.8
Ave G1 BC	0.156	0.167	0.163	0.164	0.158
SD of G1 BC	0.006	0.006	0.006	0.008	0.006
# of shots	8	9	6	10	10

SK Rifle Match 40 grain

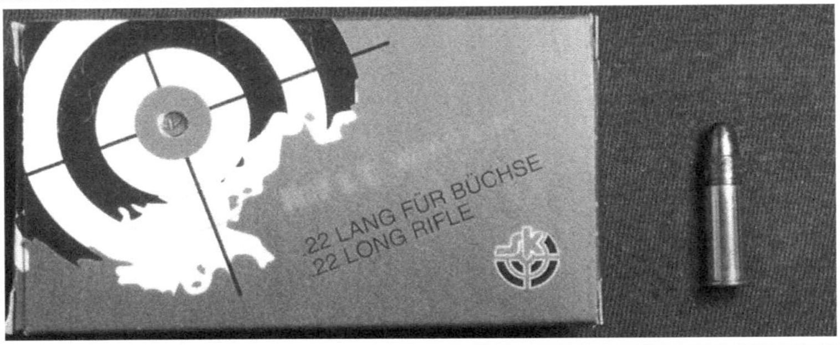

Barrel	Vol-77/22	Vol-77/22	Vol-77/22	Rug-10/22	Rug-77/22
Length	16.5"	18.5"	18.5"	18"	24"
Twist	1:11.25"	1:16"	1:9"	1:14"	1:16"
Ave MV	1057	1058	1055	1073	1072
SD of MV	8.2	9.7	6.9	8.5	8.5
Ave DRV	934	940	942	960	956
SD of DRV	8.1	10.8	8.7	7.3	9.9
Ave G1 BC	0.147	0.158	0.155	0.158	0.155
SD of G1 BC	0.005	0.008	0.008	0.005	0.008
# of shots	8	9	9	9	10

SK Rifle Match (Neal Waltz Die, Eley EPS Nose)

Barrel	Vol-77/22	Vol-77/22	Vol-77/22	Rug-10/22	Rug-77/22
Length	16.5"	18.5"	18.5"	18"	24"
Twist	1:11.25"	1:16"	1:9"	1:14"	1:16"
Ave MV	1068	1068	1071	1083	1080
SD of MV	7.5	3.2	14.1	8.1	9.0
Ave DRV	933	928	931	949	943
SD of DRV	15.6	5.5	13.0	14.7	15.1
Ave G1 BC	0.138	0.133	0.127	0.134	0.131
SD of G1 BC	0.017	0.005	0.009	0.017	0.015
# of shots	10	9	8	8	10

SK Rifle Match (Neal Waltz Die, LHP)

Barrel	Vol-77/22	Vol-77/22	Vol-77/22	Rug-10/22	Rug-77/22
Length	16.5"	18.5"	18.5"	18"	24"
Twist	1:11.25"	1:16"	1:9"	1:14"	1:16"
Ave MV	1074	1075	1064	1087	1082
SD of MV	9.9	7.9	10.7	10.1	8.9
Ave DRV	943	930	939	964	952
SD of DRV	10.9	21.1	11.2	10.7	6.1
Ave G1 BC	0.146	0.133	0.147	0.150	0.139
SD of G1 BC	0.011	0.020	0.009	0.009	0.008
# of shots	10	8	9	8	10

SK Standard Plus 40 grain

Barrel	Vol-77/22	Vol-77/22	Vol-77/22	Rug-10/22	Rug-77/22
Length	16.5"	18.5"	18.5"	18"	24"
Twist	1:11.25"	1:16"	1:9"	1:14"	1:16"
Ave MV	1044	1070	1041	1064	1073
SD of MV	19.3	17.5	18.1	16.5	12.7
Ave DRV	921	944	928	949	955
SD of DRV	15.9	13.7	15.2	13.2	12.2
Ave G1 BC	0.140	0.153	0.152	0.149	0.152
SD of G1 BC	0.007	0.006	0.007	0.004	0.008
# of shots	7	9	9	9	10

SK Subsonic 40 grain

Barrel	Vol-77/22	Vol-77/22	Vol-77/22	Rug-10/22	Rug-77/22
Length	16.5"	18.5"	18.5"	18"	24"
Twist	1:11.25"	1:16"	1:9"	1:14"	1:16"
Ave MV	1046	1039	1039	1056	1054
SD of MV	12.4	7.4	17.5	11.1	6.4
Ave DRV	926	923	928	945	941
SD of DRV	9.8	7.0	15.0	10.4	6.2
Ave G1 BC	0.145	0.151	0.154	0.152	0.150
SD of G1 BC	0.004	0.005	0.007	0.006	0.006
# of shots	8	9	9	9	10

Winchester Subsonic Max 42 grain

Barrel	Vol-77/22	Vol-77/22	Vol-77/22	Rug-10/22	Rug-77/22
Length	16.5"	18.5"	18.5"	18"	24"
Twist	1:11.25"	1:16"	1:9"	1:14"	1:16"
Ave MV	1057	1061	1049	1067	1066
SD of MV	16.5	9.6	11.0	12.0	11.8
Ave DRV	920	916	921	934	927
SD of DRV	14.2	9.9	11.8	12.8	11.3
Ave G1 BC	0.129	0.123	0.134	0.127	0.120
SD of G1 BC	0.008	0.006	0.009	0.008	0.008
# of shots	9	8	6	9	9

Winchester Super-X HP Cu Plated 36 grain

Barrel	Vol-77/22	Vol-77/22	Vol-77/22	Rug-10/22	Rug-77/22
Length	16.5"	18.5"	18.5"	18"	24"
Twist	1:11.25"	1:16"	1:9"	1:14"	1:16"
Ave MV	1254	1263	1265	1280	1269
SD of MV	18.8	10.6	12.0	12.6	14.3
Ave DRV	990	996	1008	1018	1012
SD of DRV	8.8	9.6	5.7	6.5	7.0
Ave G1 BC	0.107	0.111	0.114	0.109	0.109
SD of G1 BC	0.002	0.004	0.002	0.002	0.002
# of shots	8	9	9	7	9

Winchester Super-X HV 40 grain

Barrel	Vol-77/22	Vol-77/22	Vol-77/22	Rug-10/22	Rug-77/22
Length	16.5"	18.5"	18.5"	18"	24"
Twist	1:11.25"	1:16"	1:9"	1:14"	1:16"
Ave MV	1177	1209	1208	1205	1213
SD of MV	11.4	4.4	13.4	9.9	21.5
Ave DRV	973	992	1006	1002	1009
SD of DRV	3.4	4.9	8.1	9.0	13.9
Ave G1 BC	0.119	0.125	0.134	0.123	0.124
SD of G1 BC	0.003	0.005	0.003	0.005	0.006
# of shots	8	8	9	8	8

Winchester Super-X HVHP 37 grain

Barrel	Vol-77/22	Vol-77/22	Vol-77/22	Rug-10/22	Rug-77/22
Length	16.5"	18.5"	18.5"	18"	24"
Twist	1:11.25"	1:16"	1:9"	1:14"	1:16"
Ave MV	1240	1261	1264	1266	1275
SD of MV	6.9	8.0	8.7	11.6	5.9
Ave DRV	1014	1026	1036	1043	1044
SD of DRV	4.1	2.7	5.2	6.7	8.1
Ave G1 BC	0.129	0.133	0.135	0.132	0.130
SD of G1 BC	0.005	0.003	0.003	0.002	0.006
# of shots	9	9	8	6	10

Winchester Super-X Subsonic 40 grain

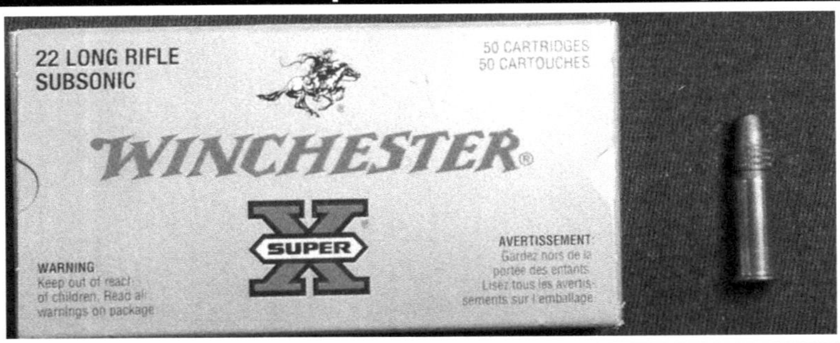

Barrel	Vol-77/22	Vol-77/22	Vol-77/22	Rug-10/22	Rug-77/22
Length	16.5"	18.5"	18.5"	18"	24"
Twist	1:11.25"	1:16"	1:9"	1:14"	1:16"
Ave MV	1037	1037	1045	1052	1048
SD of MV	11.5	10.5	28.4	4.9	6.9
Ave DRV	903	895	906	912	913
SD of DRV	13.0	12.6	24.5	11.9	6.4
Ave G1 BC	0.122	0.116	0.119	0.113	0.118
SD of G1 BC	0.009	0.009	0.010	0.010	0.005
# of shots	8	9	9	7	9

Winchester Super-X Super Speed 40 grain

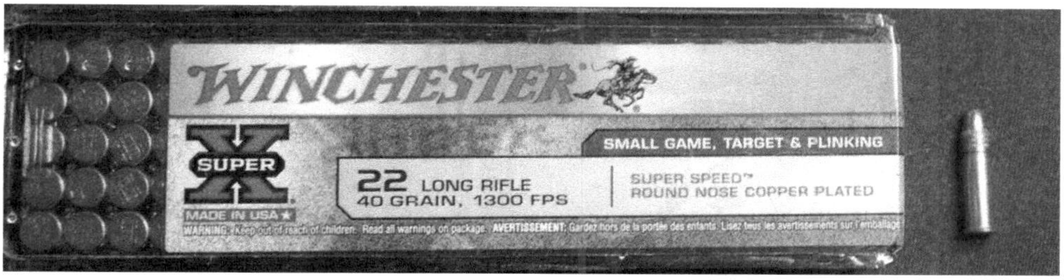

Barrel	Vol-77/22	Vol-77/22	Vol-77/22	Rug-10/22	Rug-77/22
Length	16.5"	18.5"	18.5"	18"	24"
Twist	1:11.25"	1:16"	1:9"	1:14"	1:16"
Ave MV	1280	1319	1323	1323	1332
SD of MV	16.6	10.5	9.6	10.5	16.7
Ave DRV	1014	1039	1057	1055	1052
SD of DRV	19.2	10.4	5.6	7.5	14.4
Ave G1 BC	0.114	0.122	0.128	0.120	0.115
SD of G1 BC	0.008	0.006	0.004	0.006	0.009
# of shots	10	7	7	7	10

Wolf Match Gold 40 grain

Barrel	Vol-77/22	Vol-77/22	Vol-77/22	Rug-10/22	Rug-77/22
Length	16.5"	18.5"	18.5"	18"	24"
Twist	1:11.25"	1:16"	1:9"	1:14"	1:16"
Ave MV	1026	1029	1029	1035	1039
SD of MV	10.9	10.9	20.4	11.4	12.9
Ave DRV	911	914	922	925	927
SD of DRV	10.3	6.4	19.0	9.5	11.3
Ave G1 BC	0.140	0.144	0.153	0.144	0.142
SD of G1 BC	0.006	0.004	0.009	0.004	0.006
# of shots	8	8	9	9	9

Wolf Match Target

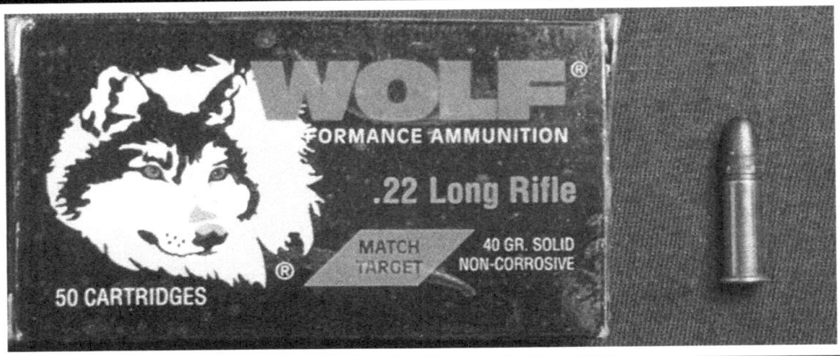

Barrel	Vol-77/22	Vol-77/22	Vol-77/22	Rug-10/22	Rug-77/22
Length	16.5"	18.5"	18.5"	18"	24"
Twist	1:11.25"	1:16"	1:9"	1:14"	1:16"
Ave MV	1024	1032	971	1034	1026
SD of MV	21.0	26.9	12.4	18.6	30.1
Ave DRV	913	924	871	928	920
SD of DRV	16.5	22.5	12.2	17.2	27.9
Ave G1 BC	0.146	0.162	0.135	0.149	0.145
SD of G1 BC	0.006	0.029	0.006	0.009	0.013
# of shots	9	8	8	5	8

.224 Caliber (5.56mm)

.224 caliber bullets typically range from 50 to 90 grains. The image below relates bullet weight to sectional density for .224 caliber bullets.

Range of Weight and Sectional Density for 0.224 Caliber

Bullet Weight (grains)

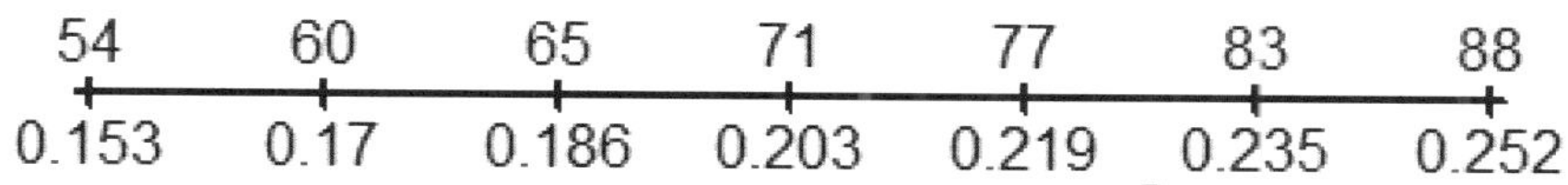

Sectional Density (lb/in^2)

Sectional Density (SD) is a measure of how much mass is behind the bullets frontal area. Higher SD means a greater ability for the bullet to penetrate targets on impact, and also retain velocity in flight.

Spectrum of G7 form factors

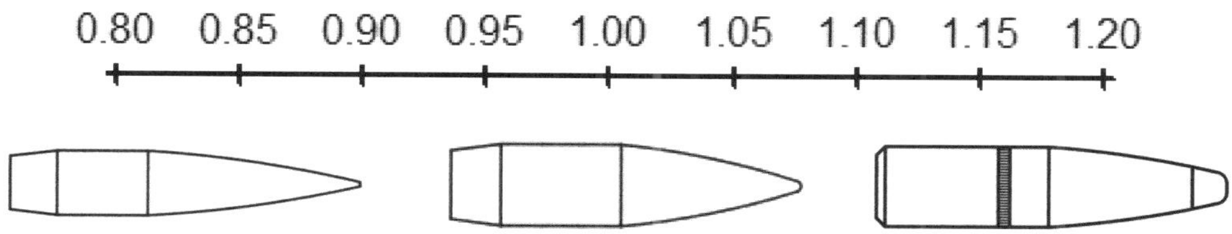

G7 Form factors range from 0.80 for the lowest drag shapes, up to 1.20+ for blunter shapes.

G7 Ballistic Coefficients are simply Sectional Density (SD) divided by G7 form factor.

You can determine a bullets' suitability for long range performance by seeing where it lies on the SD and form factor lines. Ideally, you want a bullet with a high SD and a low G7 form factor.

As an example, a 77 grain .224 bullet has an SD of 0.203. If that bullet had a G7 form factor of 1.10, it's G7 BC would be:

G7 BC = SD/FF = 0.203/1.10 = 0.185.

Barnes .224 caliber 69 grain Match Burner

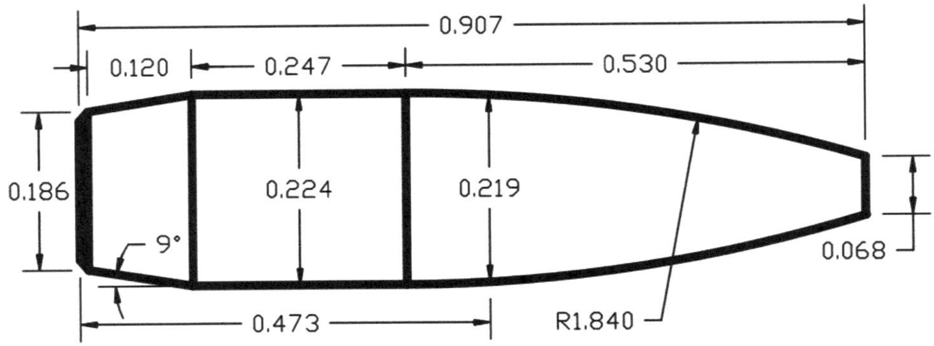

Lot #:	5247A-1 353-M CDW-M	Ogive Radius:	8.21 calibers
Sectional Density:	0.197 lb/in^2	Rt/R:	0.95

Ballistic Performance

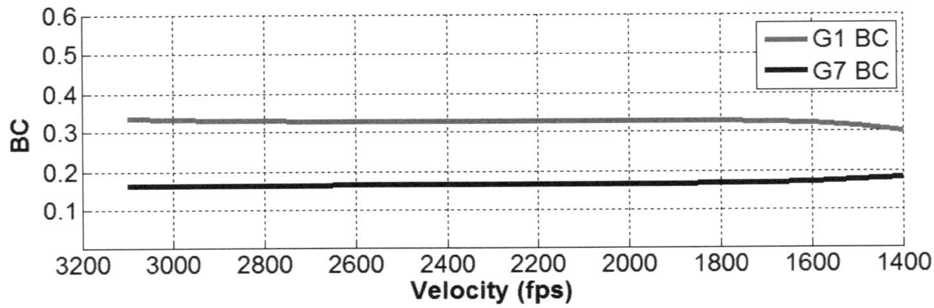

	Above 3000	3000-2500	2500-2000	2000-1500	Below 1500
G1 BC	0.341	0.331	0.329	0.321	0.314
G7 BC	0.162	0.164	0.165	0.171	0.177

Averages from 3000 fps to 1500 fps:
G1 BC: 0.326, variation of G1 BC: 0.021
G7 BC: 0.168, variation of G7 BC: 0.013
G7 form factor: 1.172

Rifling Twist Rate and Stability

Barrel Twist	Stability Factor Best case	Nominal	Worst case
1:6"	4.17	3.64	3.15
1:7"	3.06	2.68	2.32
1:8"	2.35	2.05	1.77
1:9"	1.85	1.62	1.40

Minimum twist required for stability is 1:10.5"
Recommended twist for optimal performance is 1:9.4" or faster

Barnes .224 caliber 85 grain Match Burner

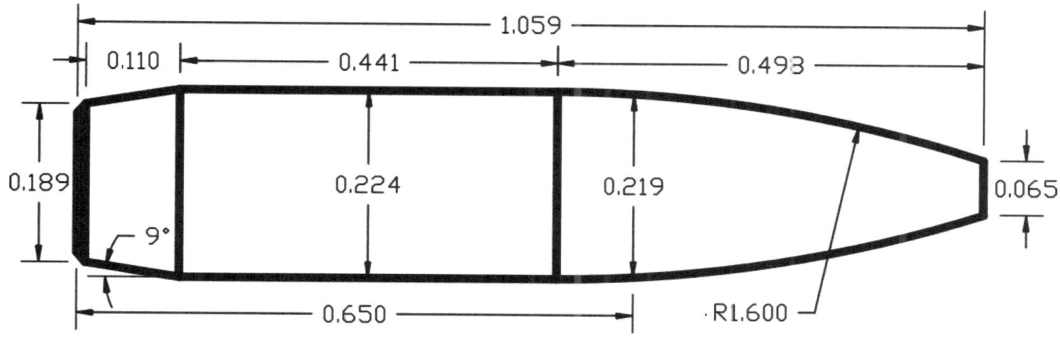

Lot #:	5529A-1 343-M CDW-M	**Ogive Radius:**	**7.14 calibers**
Sectional Density:	**0.242 lb/in²**	**Rt/R:**	**0.95**

Ballistic Performance

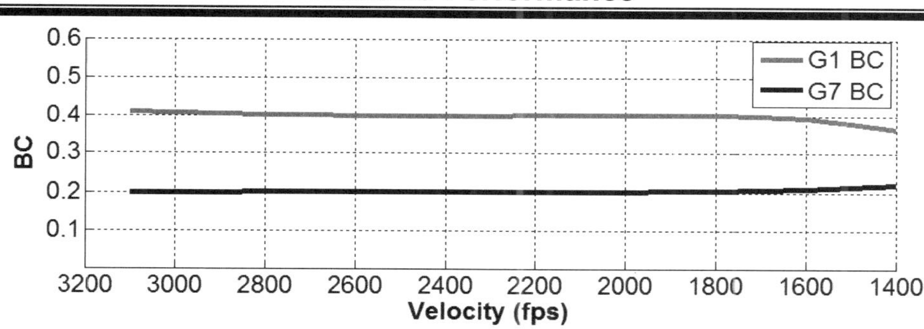

	Above 3000	3000-2500	2500-2000	2000-1500	Below 1500
G1 BC	0.415	0.403	0.400	0.391	0.381
G7 BC	0.197	0.200	0.201	0.208	0.215

Averages from 3000 fps to 1500 fps:
G1 BC: 0.397, variation of G1 BC: 0.026
G7 BC: 0.204, variation of G7 BC: 0.015
G7 form factor: 1.188

Rifling Twist Rate and Stability

	Stability Factor		
Barrel Twist	**Best case**	**Nominal**	**Worst case**
1:6"	3.28	2.86	2.48
1:7"	2.41	2.10	1.82
1:8"	1.84	1.61	1.39
1:9"	1.46	1.27	1.10

Minimum twist required for stability is 1:9.3"
Recommended twist for optimal performance is 1:8.3" or faster

Barnes .224 caliber 70 grain TacX MIL/LE

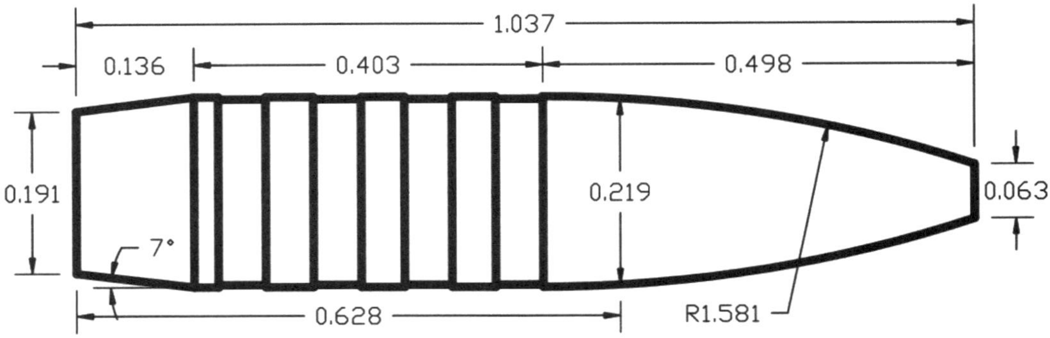

Lot #:	5384-3 260M PKG-M	Ogive Radius:	7.06 calibers
Sectional Density:	0.199 lb/in^2	Rt/R:	0.80

Ballistic Performance

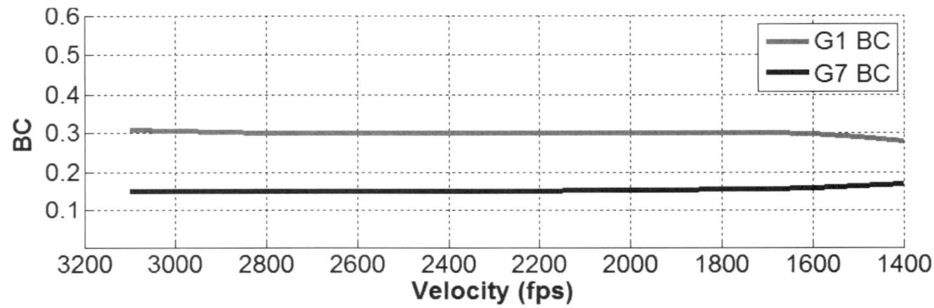

	Above 3000	3000-2500	2500-2000	2000-1500	Below 1500
G1 BC	0.313	0.303	0.301	0.296	0.289
G7 BC	0.148	0.151	0.152	0.158	0.163

Averages from 3000 fps to 1500 fps:
G1 BC: 0.300, variation of G1 BC: 0.017
G7 BC: 0.154, variation of G7 BC: 0.013
G7 form factor: 1.295

Rifling Twist Rate and Stability

Barrel Twist	Stability Factor Best case	Nominal	Worst case
1:6"	2.87	2.51	2.17
1:7"	2.11	1.84	1.59
1:8"	1.61	1.41	1.22
1:9"	1.28	1.11	0.96

Minimum twist required for stability is 1:8.7"
Recommended twist for optimal performance is 1:7.8" or faster

Barnes .224 caliber 70 grain TSXBT

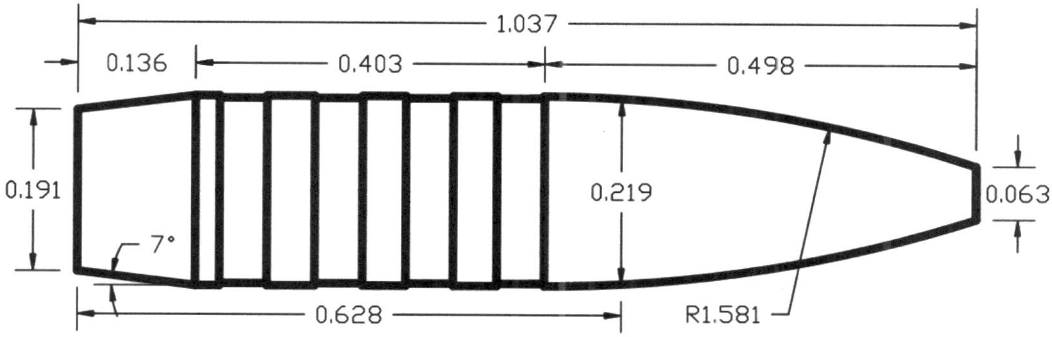

Lot #:	5686-2 149N KDJ	**Ogive Radius:**	**7.06 calibers**
Sectional Density:	**0.199 lb/in²**	**Rt/R:**	**0.80**

Ballistic Performance

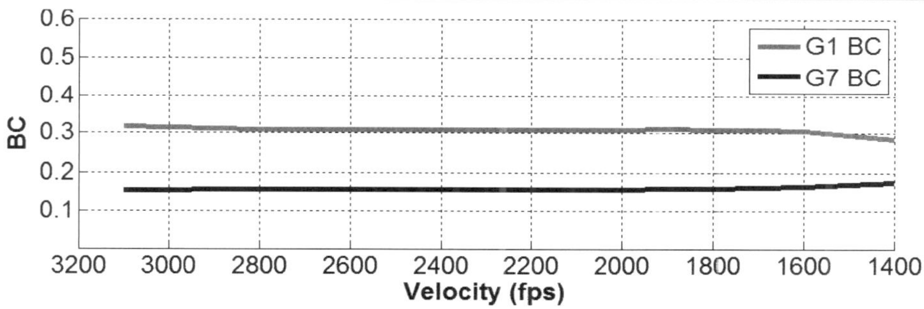

	Above 3000	3000-2500	2500-2000	2000-1500	Below 1500
G1 BC	0.322	0.312	0.310	0.304	0.297
G7 BC	0.152	0.155	0.156	0.162	0.168

Averages from 3000 fps to 1500 fps:
G1 BC: 0.308, variation of G1 BC: 0.018
G7 BC: 0.158, variation of G7 BC: 0.014
G7 form factor: 1.260

Rifling Twist Rate and Stability

Barrel Twist	Stability Factor Best case	Nominal	Worst case
1:6"	2.87	2.51	2.17
1:7"	2.11	1.84	1.59
1:8"	1.61	1.41	1.22
1:9"	1.28	1.11	0.96

Minimum twist required for stability is 1:8.7"
Recommended twist for optimal performance is 1:7.8" or faster

Barnes .224 caliber 62 grain TTSXBT

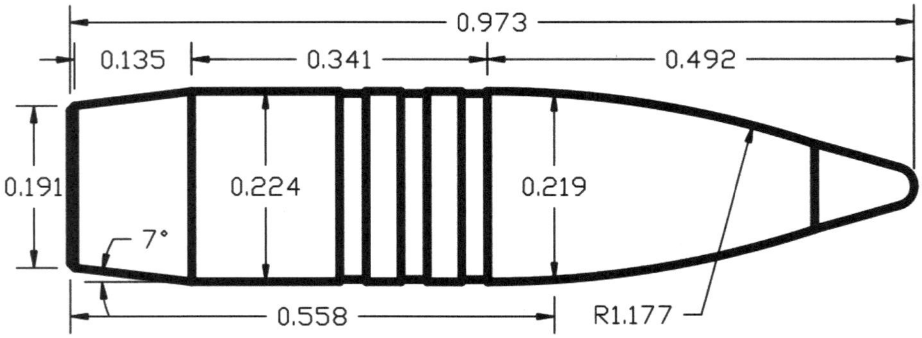

Lot #:	5558-1 125N AH-M	Ogive Radius:	5.26 calibers
Sectional Density:	0.177 lb/in^2	Rt/R:	0.85

Ballistic Performance

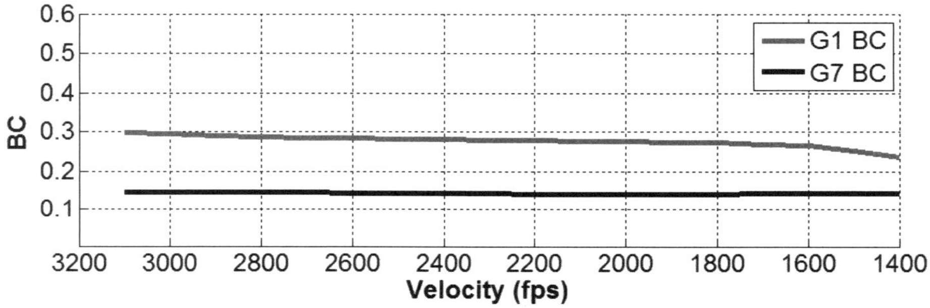

	Above 3000	3000-2500	2500-2000	2000-1500	Below 1500
G1 BC	0.305	0.289	0.280	0.265	0.253
G7 BC	0.144	0.143	0.141	0.141	0.142

Averages from 3000 fps to 1500 fps:
G1 BC: 0.277, variation of G1 BC: 0.042
G7 BC: 0.142, variation of G7 BC: 0.005
G7 form factor: 1.243

Rifling Twist Rate and Stability

Barrel Twist	Stability Factor Best case	Nominal	Worst case
1:6"	3.06	2.67	2.31
1:7"	2.25	1.96	1.70
1:8"	1.72	1.50	1.30
1:9"	1.36	1.19	1.03

Minimum twist required for stability is 1:9.0"
Recommended twist for optimal performance is 1:8.0" or faster

Berger .224 caliber 77 grain Tactical OTM

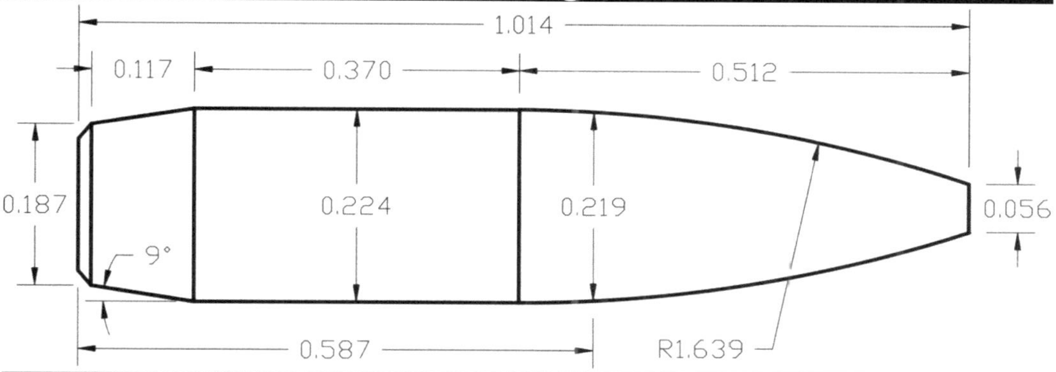

Lot #:	5348	Ogive Radius:	7.32 calibers
Sectional Density:	0.219 lb/in^2	Rt/R:	1.00

Ballistic Performance

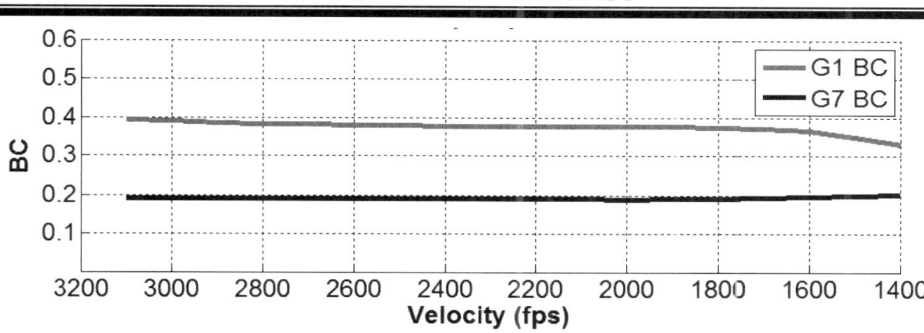

	Above 3000	3000-2500	2500-2000	2000-1500	Below 1500
G1 BC	0.399	0.384	0.378	0.364	0.351
G7 BC	0.189	0.191	0.190	0.193	0.198

Averages from 3000 fps to 1500 fps:
G1 BC: 0.374, variation of G1 BC: 0.038
G7 BC: 0.192, variation of G7 BC: 0.009
G7 form factor: 1.142

Rifling Twist Rate and Stability

Barrel Twist	Stability Factor Best case	Stability Factor Nominal	Stability Factor Worst case
1:7"	2.47	2.16	1.87
1:8"	1.89	1.65	1.43
1:9"	1.49	1.30	1.13
1:10"	1.21	1.06	0.91

Minimum twist required for stability is 1:9.4"
Recommended twist for optimal performance is 1:8.4" or faster

Berger .224 caliber 52 grain Target FB

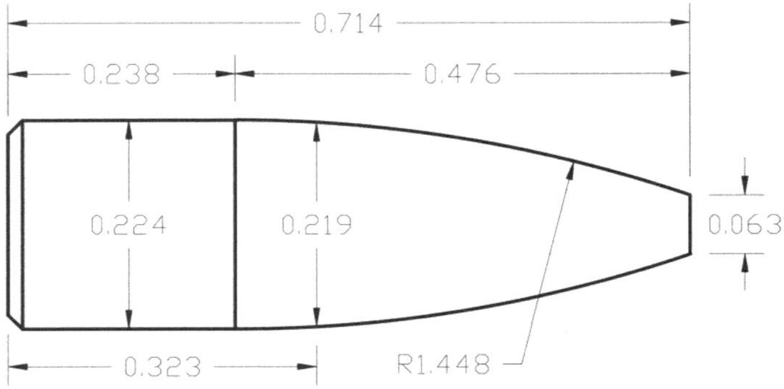

Lot #:	**5504**	**Ogive Radius:**	**6.46 calibers**
Sectional Density:	**0.148 lb/in^2**	**Rt/R:**	**1.00**

Ballistic Performance

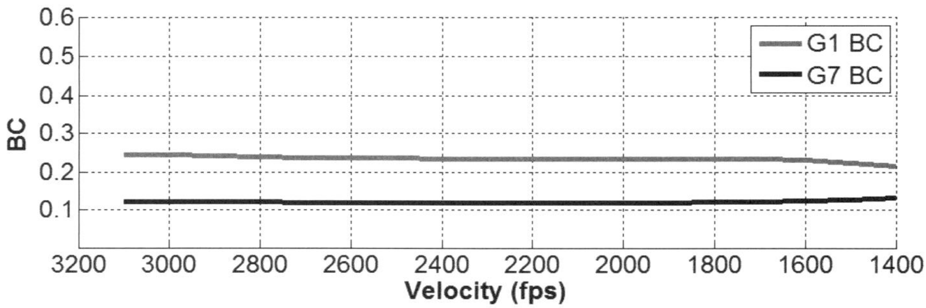

	Above 3000	3000-2500	2500-2000	2000-1500	Below 1500
G1 BC	0.250	0.239	0.235	0.229	0.224
G7 BC	0.119	0.119	0.118	0.122	0.126

Averages from 3000 fps to 1500 fps:
G1 BC: 0.234, variation of G1 BC: 0.019
G7 BC: 0.120, variation of G7 BC: 0.009
G7 form factor: 1.231

Barrel Twist Rate and Stability

Barrel Twist	Stability Factor Best case	Nominal	Worst case
1:10"	2.24	1.96	1.69
1:11"	1.85	1.62	1.40
1:12"	1.56	1.36	1.18
1:13"	1.33	1.16	1.00

Minimum twist required for stability is 1:12.8"
Recommended twist for optimal performance is 1:11.4" or faster

Berger .224 caliber 55 grain Target FB

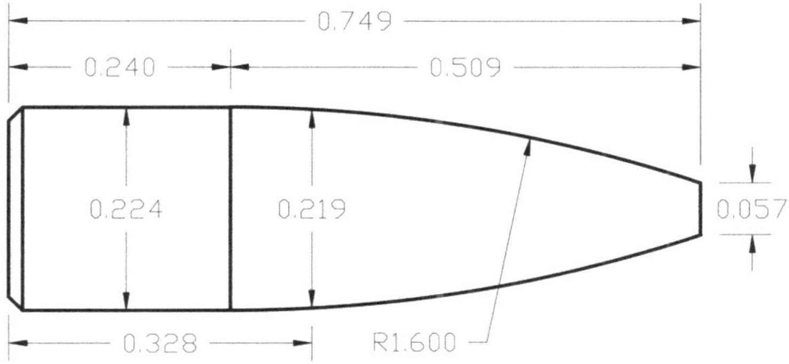

Lot #:	5684	Ogive Radius:	7.14 calibers
Sectional Density:	0.157 lb/in^2	Rt/R:	1.00

Ballistic Performance

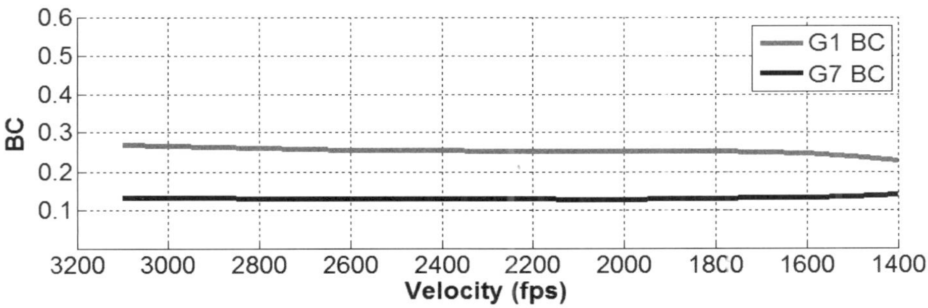

	Above 3000	3000-2500	2500-2000	2000-1500	Below 1500
G1 BC	0.274	0.260	0.254	0.246	0.239
G7 BC	0.130	0.129	0.128	0.131	0.135

Averages from 3000 fps to 1500 fps:
G1 BC: 0.253, variation of G1 BC: 0.026
G7 BC: 0.130, variation of G7 BC: 0.008
G7 form factor: 1.205

Barrel Twist Rate and Stability

Barrel Twist	Stability Factor Best case	Nominal	Worst case
1:10"	2.07	1.81	1.56
1:11"	1.71	1.49	1.29
1:12"	1.44	1.26	1.09
1:13"	1.23	1.07	0.93

Minimum twist required for stability is 1:12.3"
Recommended twist for optimal performance is 1:11.0" or faster

Berger .224 caliber 70 grain Target VLD

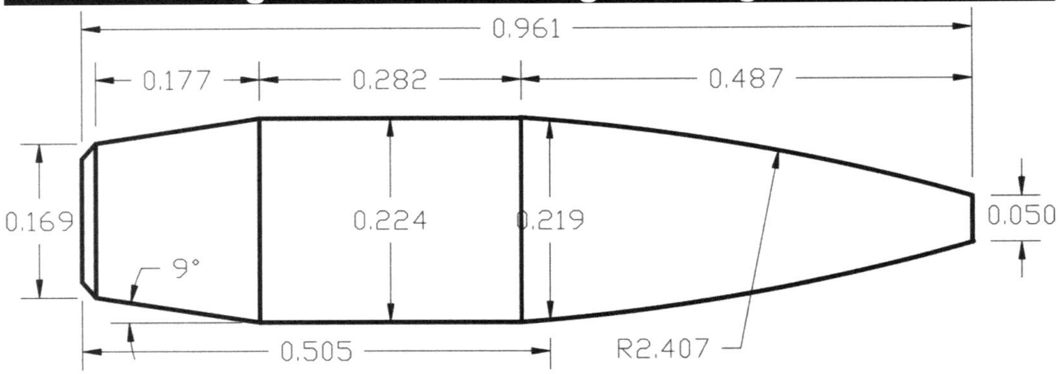

Lot #:	5364	Ogive Radius:	10.75 calibers
Sectional Density:	0.199 lb/in²	Rt/R:	0.58

Ballistic Performance

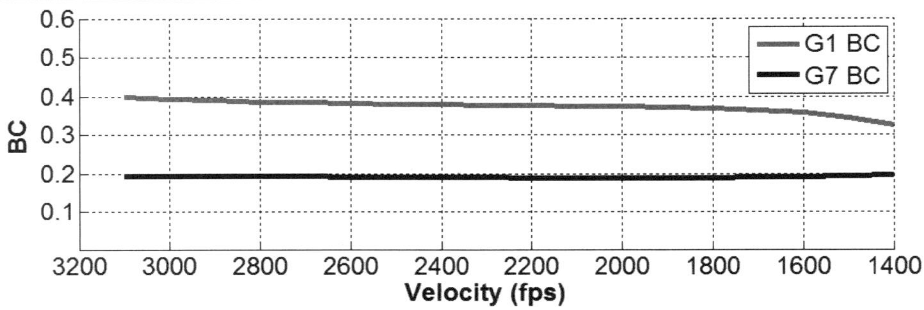

	Above 3000	3000-2500	2500-2000	2000-1500	Below 1500
G1 BC	0.405	0.387	0.377	0.359	0.344
G7 BC	0.192	0.192	0.189	0.191	0.194

Averages from 3000 fps to 1500 fps:
G1 BC: 0.373, variation of G1 BC: 0.050
G7 BC: 0.191, variation of G7 BC: 0.006
G7 form factor: 1.042

Barrel Twist Rate and Stability

Barrel Twist	Stability Factor Best case	Nominal	Worst case
1:7"	2.63	2.30	1.99
1:8"	2.01	1.76	1.52
1:9"	1.59	1.39	1.20
1:10"	1.29	1.13	0.97

Minimum twist required for stability is 1:9.7"
Recommended twist for optimal performance is 1:8.7" or faster

Berger .224 caliber 73 grain Target Boat Tail

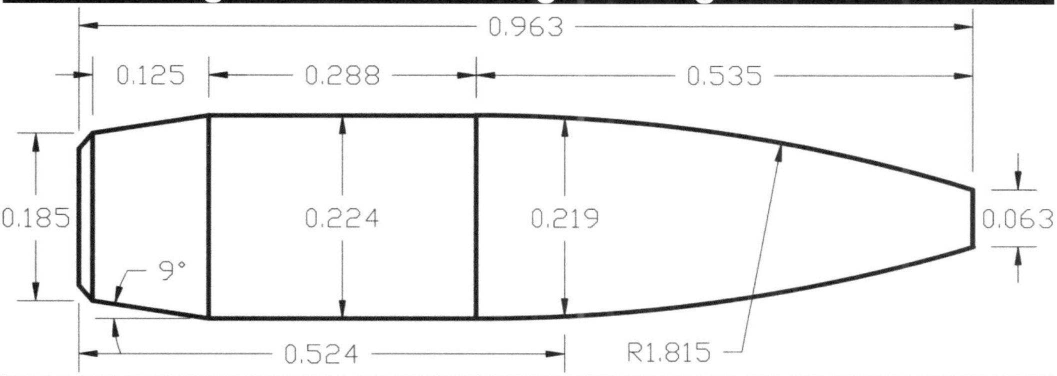

Lot #:	5861	Ogive Radius:	8.10 calibers
Sectional Density:	0.208 lb/in^2	Rt/R:	1.00

Ballistic Performance

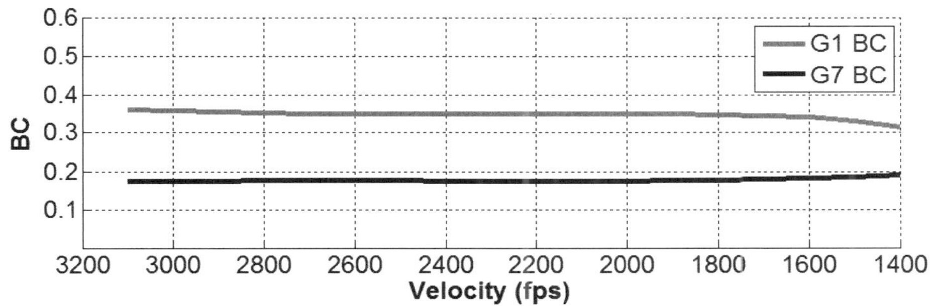

	Above 3000	3000-2500	2500-2000	2000-1500	Below 1500
G1 BC	0.366	0.353	0.349	0.340	0.330
G7 BC	0.173	0.175	0.176	0.181	0.186

Averages from 3000 fps to 1500 fps:
G1 BC: 0.347, variation of G1 BC: 0.027
G7 BC: 0.178, variation of G7 BC: 0.011
G7 form factor: 1.168

Barrel Twist Rate and Stability

	Stability Factor		
Barrel Twist	Best case	Nominal	Worst case
1:7"	2.73	2.38	2.06
1:8"	2.09	1.82	1.58
1:9"	1.65	1.44	1.25
1:10"	1.34	1.17	1.01

Minimum twist required for stability is 1:9.9"
Recommended twist for optimal performance is 1:8.8" or faster

Berger .224 caliber 75 grain Target VLD

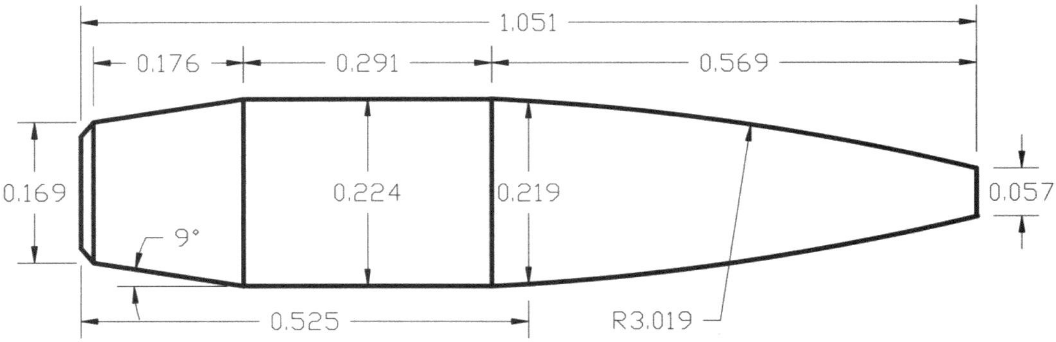

Lot #:	3838	Ogive Radius:	13.48 calibers
Sectional Density:	0.214 lb/in^2	Rt/R:	0.66

Ballistic Performance

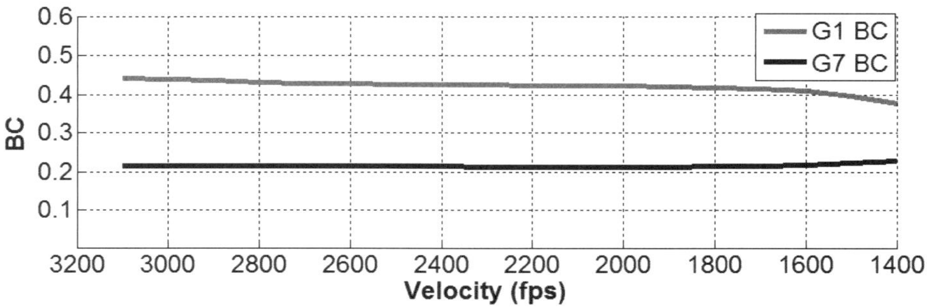

	Above 3000	3000-2500	2500-2000	2000-1500	Below 1500
G1 BC	0.450	0.433	0.424	0.409	0.395
G7 BC	0.213	0.215	0.213	0.217	0.223

Averages from 3000 fps to 1500 fps:
G1 BC: 0.421, variation of G1 BC: 0.044
G7 BC: 0.216, variation of G7 BC: 0.010
G7 form factor: 0.989

Rifling Twist Rate and Stability

Barrel Twist	Stability Factor Best case	Nominal	Worst case
1:6"	2.96	2.58	2.23
1:7"	2.17	1.90	1.64
1:8"	1.66	1.45	1.26
1:9"	1.31	1.15	0.99

Minimum twist required for stability is 1:8.8"
Recommended twist for optimal performance is 1:7.9" or faster

Berger .224 caliber 80 grain Target VLD

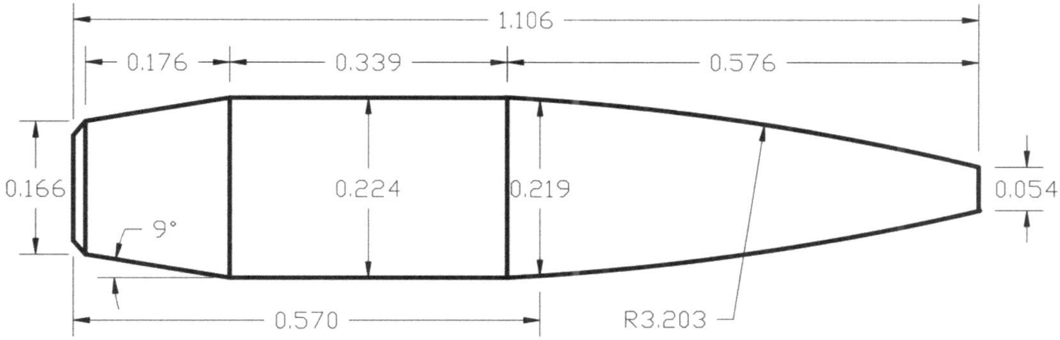

Lot #:	7646	Ogive Radius:	14.30 calibers
Sectional Density:	0.228 lb/in^2	Rt/R:	0.62

Ballistic Performance

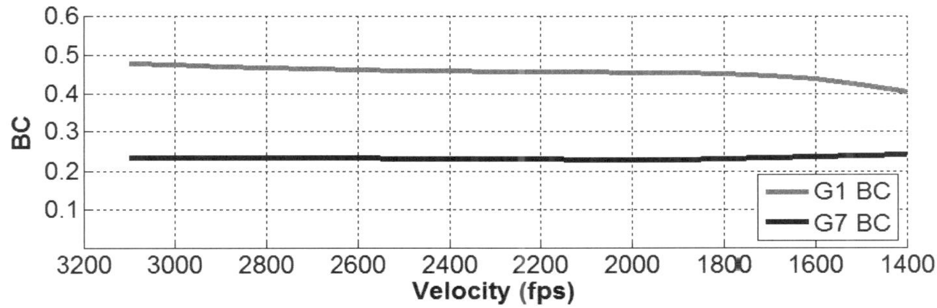

	Above 3000	3000-2500	2500-2000	2000-1500	Below 1500
G1 BC	0.489	0.469	0.459	0.440	0.424
G7 BC	0.232	0.233	0.231	0.234	0.239

Averages from 3000 fps to 1500 fps:
G1 BC: 0.455, variation of G1 BC: 0.053
G7 BC: 0.233, variation of G7 BC: 0.010
G7 form factor: 0.977

Rifling Twist Rate and Stability

Barrel Twist	Stability Factor Best case	Nominal	Worst case
1:6"	2.72	2.38	2.05
1:7"	2.00	1.75	1.51
1:8"	1.53	1.34	1.16
1:9"	1.21	1.06	0.91

Minimum twist required for stability is 1:8.4"
Recommended twist for optimal performance is 1:7.5" or faster

Berger .224 caliber 80.5 grain Target FULLBORE

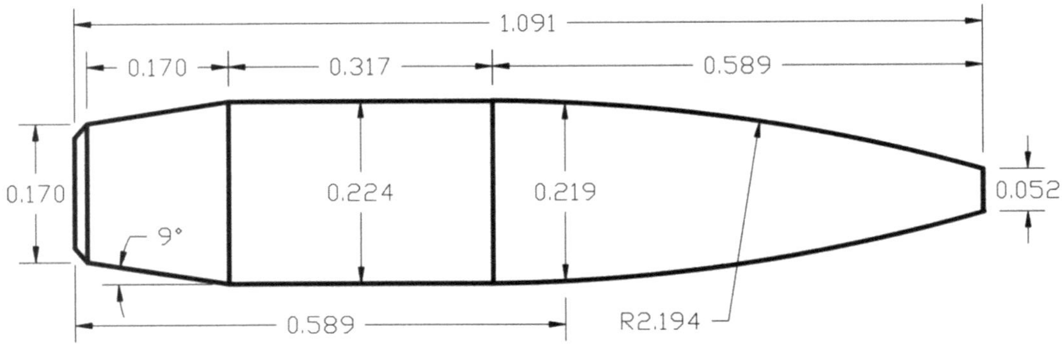

Lot #:	5715	Ogive Radius:	9.79 calibers
Sectional Density:	0.229 lb/in²	Rt/R:	0.94

Ballistic Performance

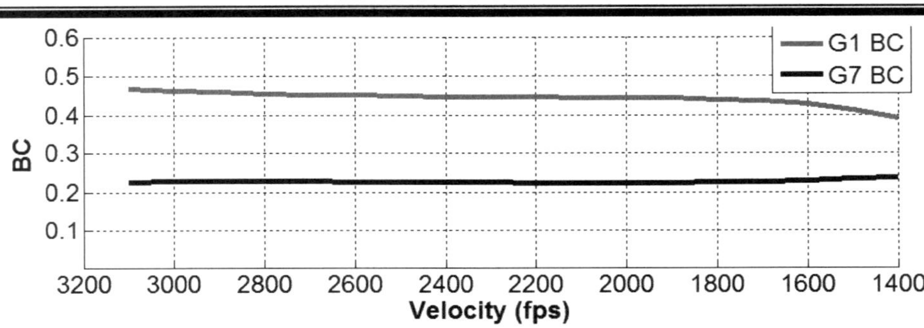

	Above 3000	3000-2500	2500-2000	2000-1500	Below 1500
G1 BC	0.477	0.457	0.447	0.429	0.413
G7 BC	0.226	0.227	0.225	0.228	0.233

Averages from 3000 fps to 1500 fps:
G1 BC: 0.443, variation of G1 BC: 0.051
G7 BC: 0.227, variation of G7 BC: 0.009
G7 form factor: 1.015

Rifling Twist Rate and Stability

Barrel Twist	Stability Factor Best case	Nominal	Worst case
1:6"	2.86	2.50	2.16
1:7"	2.10	1.84	1.59
1:8"	1.61	1.41	1.22
1:9"	1.27	1.11	0.96

Minimum twist required for stability is 1:8.6"
Recommended twist for optimal performance is 1:7.7" or faster

Berger .224 caliber 82 grain Target Boat Tail

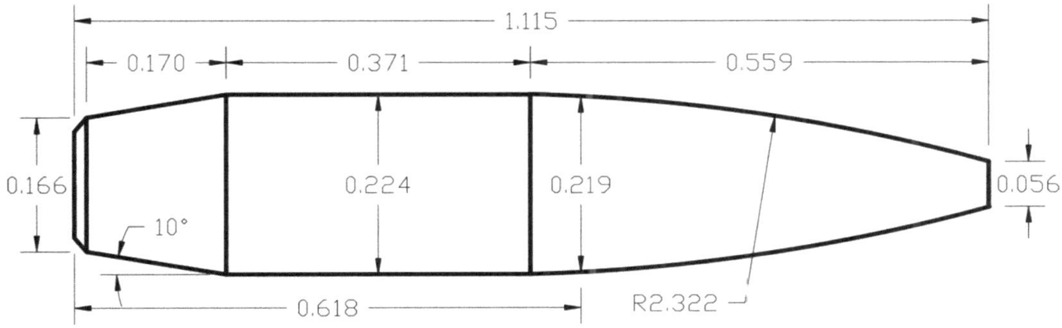

Lot #:	7611	Ogive Radius:	10.37 calibers
Sectional Density:	0.233 lb/in²	Rt/R:	0.82

Ballistic Performance

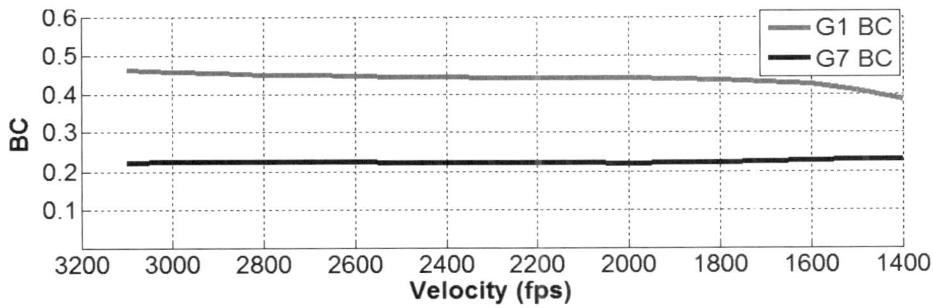

	Above 3000	3000-2500	2500-2000	2000-1500	Below 1500
G1 BC	0.471	0.452	0.443	0.425	0.409
G7 BC	0.223	0.224	0.223	0.226	0.230

Averages from 3000 fps to 1500 fps:
G1 BC: 0.438, variation of G1 BC: 0.051
G7 BC: 0.225, variation of G7 BC: 0.009
G7 form factor: 1.038

Rifling Twist Rate and Stability

Barrel Twist	Best case	Stability Factor Nominal	Worst case
1:6"	2.72	2.38	2.06
1:7"	2.00	1.75	1.51
1:8"	1.53	1.34	1.16
1:9"	1.21	1.06	0.91

Minimum twist required for stability is 1:8.4"
Recommended twist for optimal performance is 1:7.6" or faster

Berger .224 caliber 90 grain Target VLD

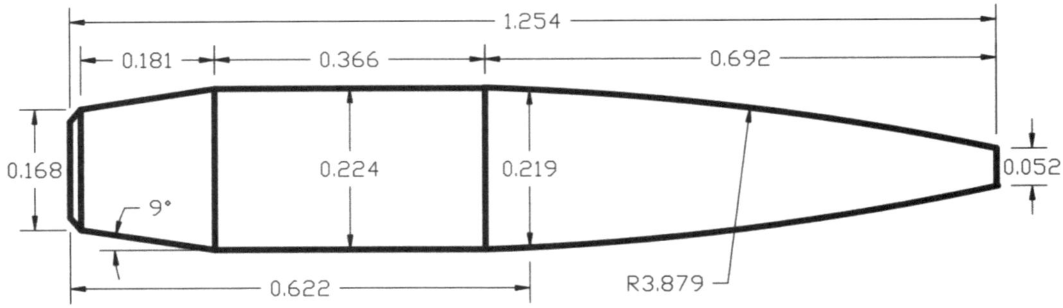

Lot #:	5397	Ogive Radius:	17.31 calibers
Sectional Density:	0.256 lb/in^2	Rt/R:	0.73

Ballistic Performance

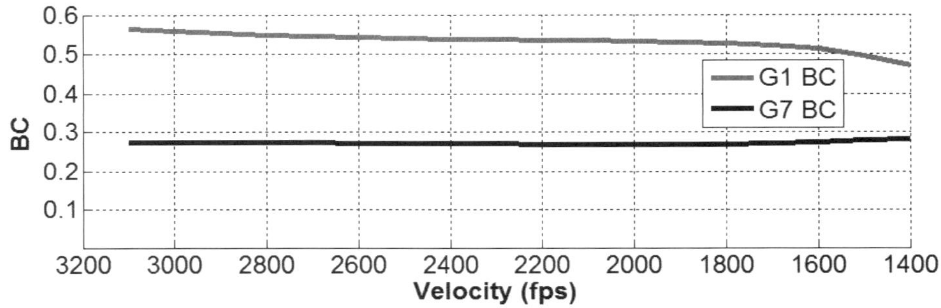

	Above 3000	3000-2500	2500-2000	2000-1500	Below 1500
G1 BC	0.576	0.551	0.538	0.515	0.496
G7 BC	0.273	0.273	0.270	0.274	0.279

Averages from 3000 fps to 1500 fps:
G1 BC: 0.533, variation of G1 BC: 0.064
G7 BC: 0.273, variation of G7 BC: 0.011
G7 form factor: 0.937

Rifling Twist Rate and Stability

Barrel Twist	Stability Factor Best case	Nominal	Worst case
1:5"	3.05	2.66	2.30
1:6"	2.12	1.85	1.60
1:7"	1.56	1.36	1.17
1:8"	1.19	1.04	0.90

Minimum twist required for stability is 1:7.4"
Recommended twist for optimal performance is 1:6.7" or faster

Berger .224 caliber 52 grain Varmint FB

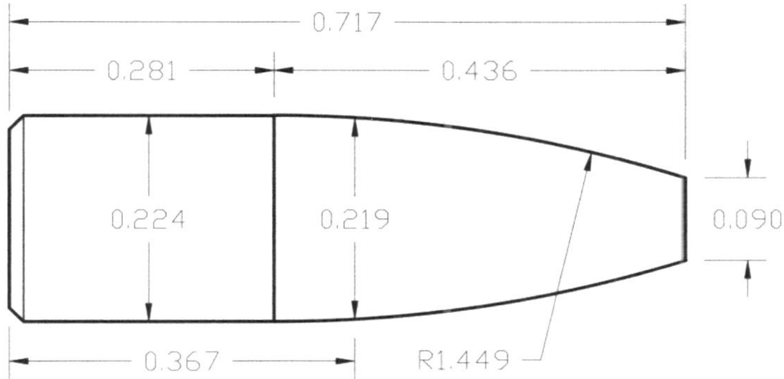

Lot #:	5514	Ogive Radius:	6.47 calibers
Sectional Density:	0.148 lb/in^2	Rt/R:	1.00

Ballistic Performance

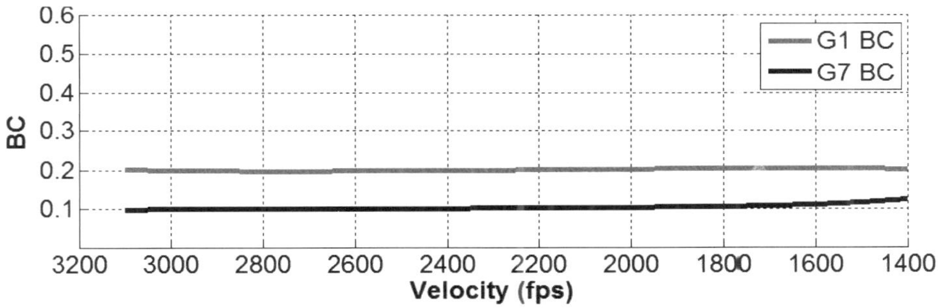

	Above 3000	3000-2500	2500-2000	2000-1500	Below 1500
G1 BC	0.202	0.198	0.200	0.203	0.204
G7 BC	0.096	0.098	0.101	0.109	0.115

Averages from 3000 fps to 1500 fps:
G1 BC: 0.201, variation of G1 BC: 0.006
G7 BC: 0.103, variation of G7 BC: 0.018
G7 form factor: 1.437

Barrel Twist Rate and Stability

Barrel Twist	Stability Factor Best case	Nominal	Worst case
1:10"	2.21	1.93	1.67
1:11"	1.83	1.60	1.38
1:12"	1.54	1.34	1.16
1:13"	1.31	1.14	0.99

Minimum twist required for stability is 1:12.7"
Recommended twist for optimal performance is 1:11.4" or faster

Berger .224 caliber 55 grain Varmint FB

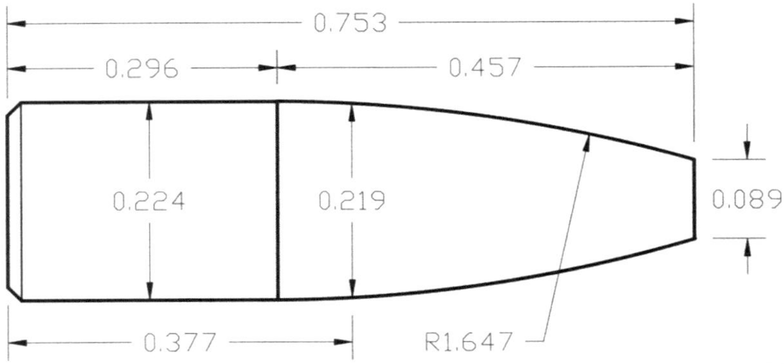

Lot #:	5514	Ogive Radius:	7.35 calibers
Sectional Density:	0.157 lb/in^2	Rt/R:	0.96

Ballistic Performance

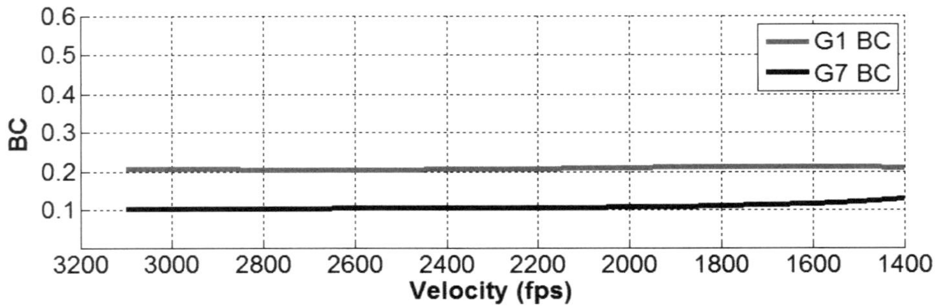

	Above 3000	3000-2500	2500-2000	2000-1500	Below 1500
G1 BC	0.210	0.206	0.208	0.211	0.212
G7 BC	0.100	0.102	0.104	0.113	0.120

Averages from 3000 fps to 1500 fps:
G1 BC: 0.209, variation of G1 BC: 0.006
G7 BC: 0.107, variation of G7 BC: 0.019
G7 form factor: 1.463

Barrel Twist Rate and Stability

Barrel Twist	Best case	Stability Factor Nominal	Worst case
1:9"	2.52	2.20	1.90
1:10"	2.04	1.78	1.54
1:11"	1.69	1.47	1.27
1:12"	1.42	1.24	1.07

Minimum twist required for stability is 1:12.2"
Recommended twist for optimal performance is 1:10.9" or faster

Berger .224 caliber 60 grain Varmint FB

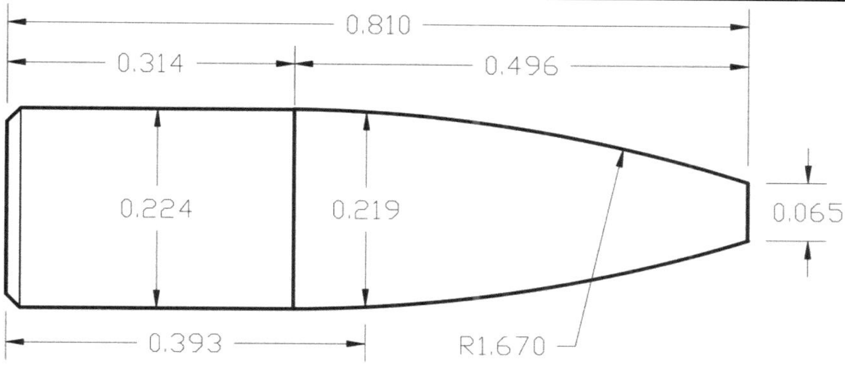

Lot #:	**6040**	**Ogive Radius:**	**7.46 calibers**
Sectional Density:	**0.171 lb/in²**	**Rt/R:**	**0.95**

Ballistic Performance

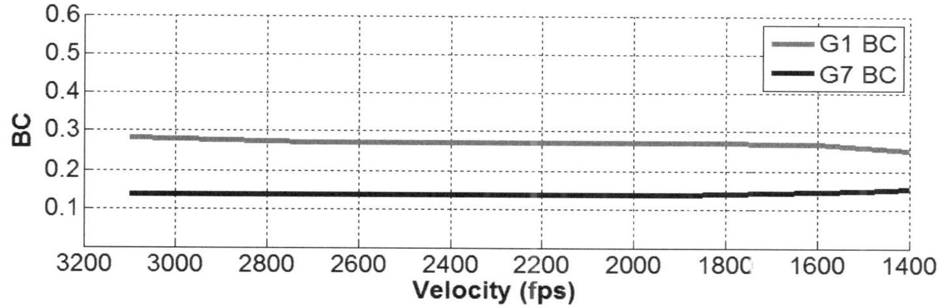

	Above 3000	3000-2500	2500-2000	2000-1500	Below 1500
G1 BC	0.287	0.275	0.271	0.266	0.261
G7 BC	0.136	0.136	0.136	0.142	0.147

Averages from 3000 fps to 1500 fps:
G1 BC: 0.271, variation of G1 BC: 0.018
G7 BC: 0.139, variation of G7 BC: 0.011
G7 form factor: 1.229

Barrel Twist Rate and Stability

Barrel Twist	Stability Factor Best case	Nominal	Worst case
1:9"	2.23	1.95	1.69
1:10"	1.81	1.58	1.37
1:11"	1.49	1.30	1.13
1:12"	1.25	1.10	0.95

Minimum twist required for stability is 1:11.5"
Recommended twist for optimal performance is 1:10.3" or faster

DRT .224 caliber 79 grain Frangible

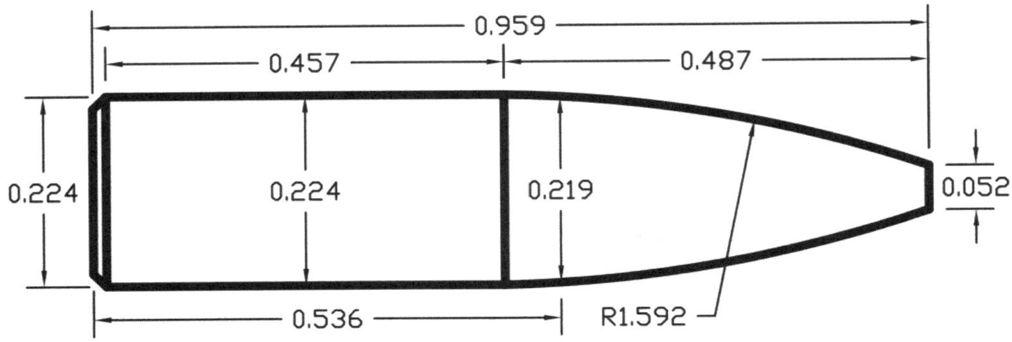

Lot #:	Unknown	Ogive Radius:	7.11 calibers
Sectional Density:	0.225 lb/in^2	Rt/R:	0.90

Ballistic Performance

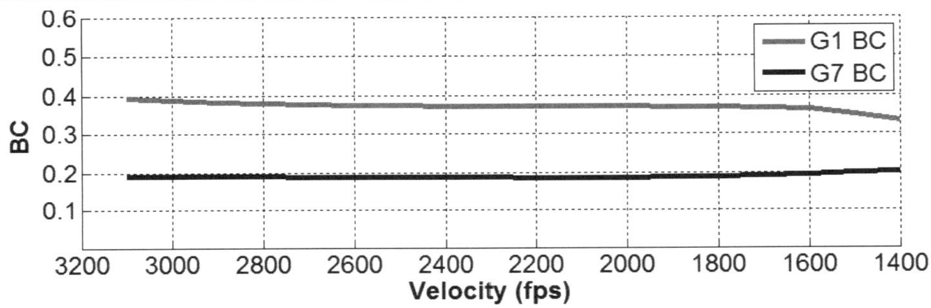

	Above 3000	3000-2500	2500-2000	2000-1500	Below 1500
G1 BC	0.400	0.381	0.372	0.361	0.351
G7 BC	0.190	0.189	0.187	0.192	0.198

Averages from 3000 fps to 1500 fps:
G1 BC: 0.371, variation of G1 BC: 0.037
G7 BC: 0.190, variation of G7 BC: 0.012
G7 form factor: 1.182

Rifling Twist Rate and Stability

Barrel Twist	Stability Factor Best case	Nominal	Worst case
1:6"	2.94	2.56	2.22
1:7"	2.25	1.96	1.70
1:8"	1.78	1.55	1.34
1:9"	1.44	1.26	1.09

Minimum twist required for stability is 1:10.3"
Recommended twist for optimal performance is 1:9.2" or faster

GS Custom .224 caliber 57 grain SP

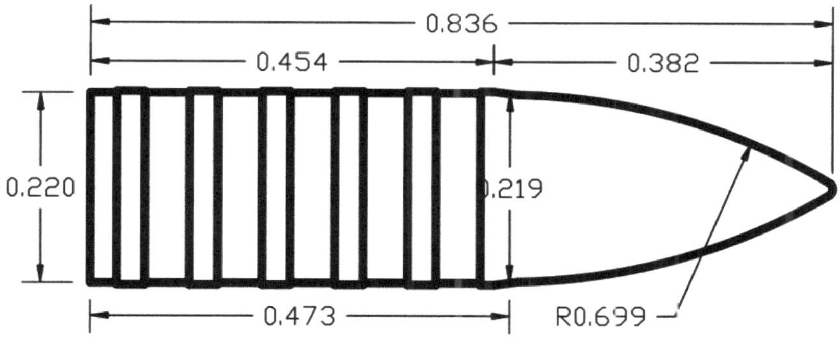

Lot #:	**Unknown**	**Ogive Radius:**	**3.12 calibers**
Sectional Density:	**0.162 lb/in^2**	**Rt/R:**	**1.00**

Ballistic Performance

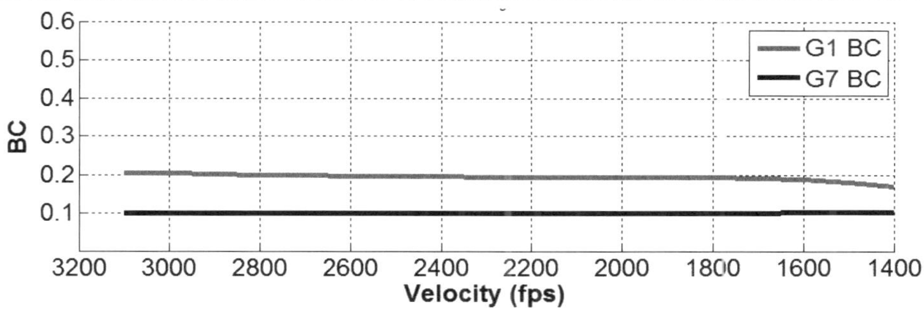

	Above 3000	3000-2500	2500-2000	2000-1500	Below 1500
G1 BC	0.209	0.199	0.195	0.187	0.180
G7 BC	0.099	0.099	0.098	0.099	0.101

Averages from 3000 fps to 1500 fps:
G1 BC: 0.193, variation of G1 BC: 0.023
G7 BC: 0.099, variation of G7 BC: 0.004
G7 form factor: 1.639

Rifling Twist Rate and Stability

	Stability Factor		
Barrel Twist	**Best case**	**Nominal**	**Worst case**
1:6"	4.36	3.81	3.29
1:7"	3.20	2.80	2.42
1:8"	2.45	2.14	1.85
1:9"	1.94	1.69	1.46

Minimum twist required for stability is 1:10.7"
Recommended twist for optimal performance is 1:9.6" or faster

GS Custom .224 caliber 69 grain SP

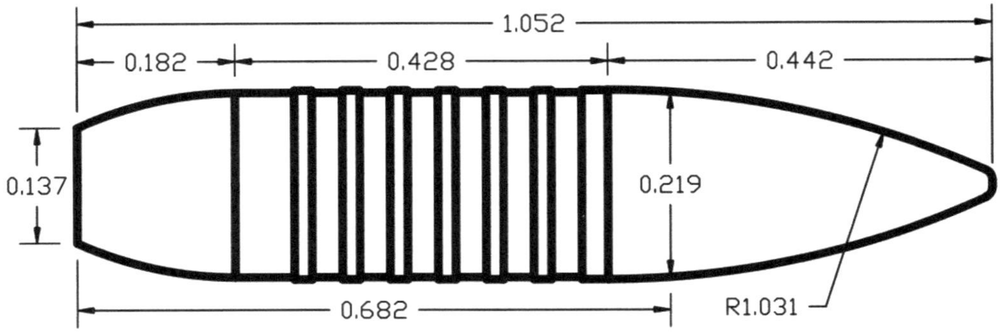

Lot #:	Unknown	Ogive Radius:	4.60 calibers
Sectional Density:	0.197 lb/in^2	Rt/R:	1.00

Ballistic Performance

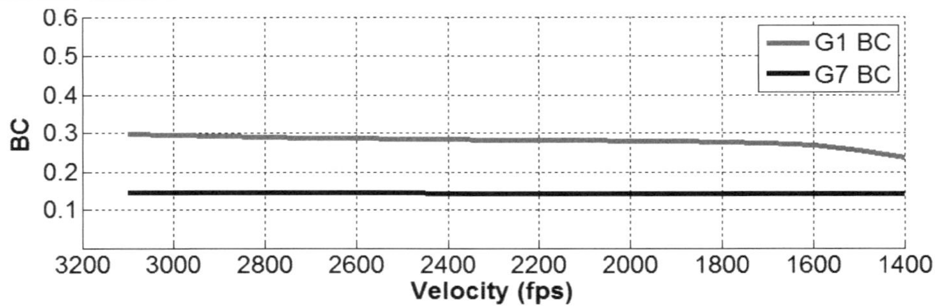

	Above 3000	3000-2500	2500-2000	2000-1500	Below 1500
G1 BC	0.304	0.291	0.283	0.268	0.255
G7 BC	0.144	0.144	0.142	0.142	0.143

Averages from 3000 fps to 1500 fps:
G1 BC: 0.279, variation of G1 BC: 0.041
G7 BC: 0.143, variation of G7 BC: 0.004
G7 form factor: 1.373

Rifling Twist Rate and Stability

Barrel Twist	Stability Factor Best case	Nominal	Worst case
1:6"	2.71	2.37	2.05
1:7"	1.99	1.74	1.51
1:8"	1.53	1.33	1.15
1:9"	1.21	1.05	0.91

Minimum twist required for stability is 1:8.4"
Recommended twist for optimal performance is 1:7.5" or faster

Hornady .224 caliber 52 grain Amax

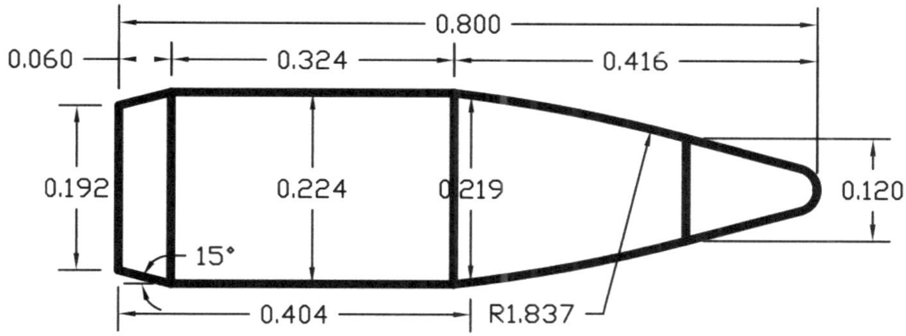

Lot #:	2070674	Ogive Radius:	8.20 calibers
Sectional Density:	0.148 lb/in^2	Rt/R:	0.55

Ballistic Performance

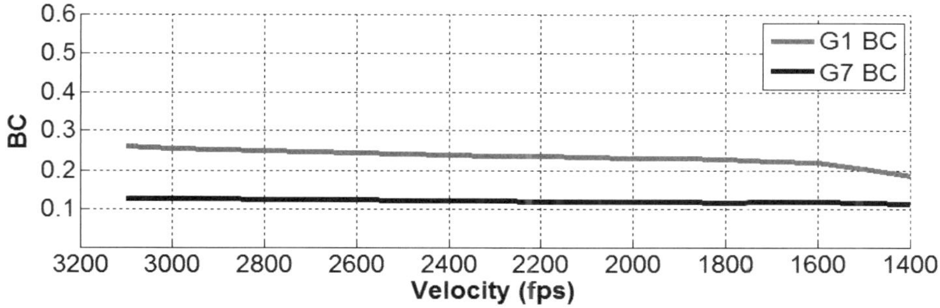

	Above 3000	3000-2500	2500-2000	2000-1500	Below 1500
G1 BC	0.266	0.249	0.237	0.219	0.205
G7 BC	0.126	0.123	0.119	0.116	0.115

Averages from 3000 fps to 1500 fps:

G1 BC: 0.234, variation of G1 BC: 0.051

G7 BC: 0.120, variation of G7 BC: 0.010

G7 form factor: 1.240

Rifling Twist Rate and Stability

Barrel Twist	Stability Factor Best case	Nominal	Worst case
1:6"	4.51	3.94	3.41
1:7"	3.31	2.89	2.50
1:8"	2.54	2.22	1.92
1:9"	2.00	1.75	1.51

Minimum twist required for stability is 1:10.9"

Recommended twist for optimal performance is 1:9.7" or faster

Hornady .224 caliber 75 grain Amax

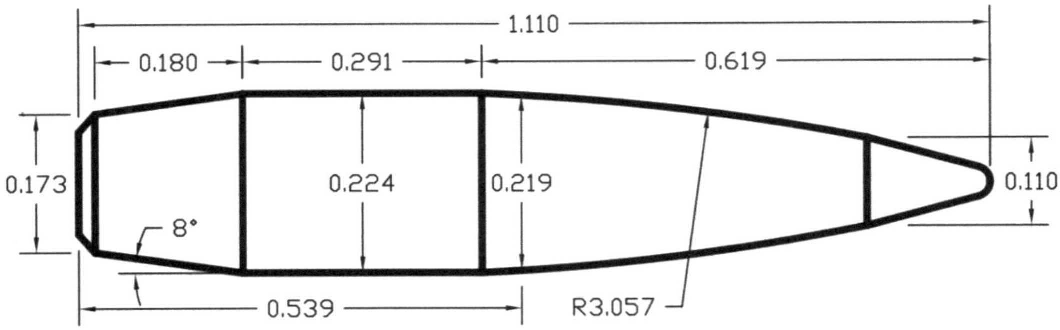

Lot #:	22792	Ogive Radius:	13.65 calibers
Sectional Density:	0.214 lb/in^2	Rt/R:	0.68

Ballistic Performance

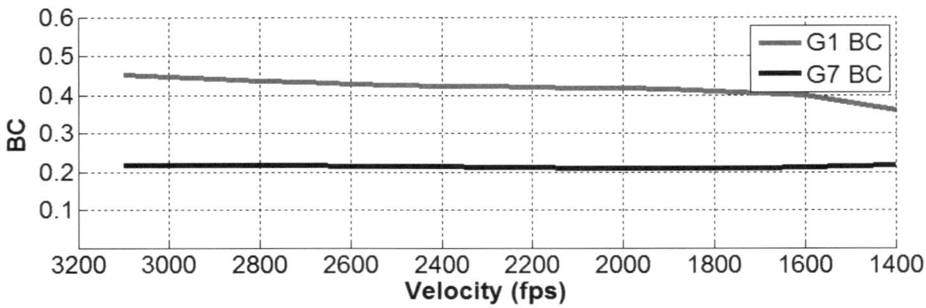

	Above 3000	3000-2500	2500-2000	2000-1500	Below 1500
G1 BC	0.462	0.437	0.422	0.399	0.381
G7 BC	0.219	0.217	0.212	0.212	0.214

Averages from 3000 fps to 1500 fps:
G1 BC: 0.418, variation of G1 BC: 0.066
G7 BC: 0.214, variation of G7 BC: 0.009
G7 form factor: 0.997

Rifling Twist Rate and Stability

Barrel Twist	Stability Factor Best case	Nominal	Worst case
1:6"	2.52	2.20	1.91
1:7"	1.85	1.62	1.40
1:8"	1.42	1.24	1.07
1:9"	1.12	0.98	0.85

Minimum twist required for stability is 1:8.1"
Recommended twist for optimal performance is 1:7.3" or faster

Hornady .224 caliber 80 grain Amax

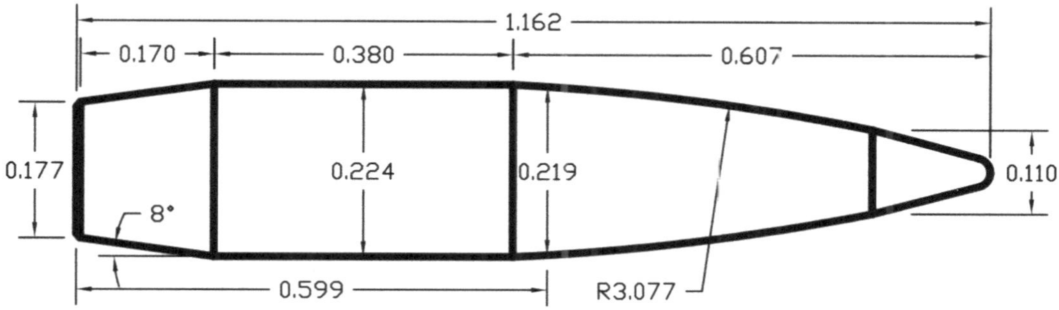

Lot #:	22832	Ogive Radius:	13.74 calibers
Sectional Density:	0.228 lb/in^2	Rt/R:	0.67

Ballistic Performance

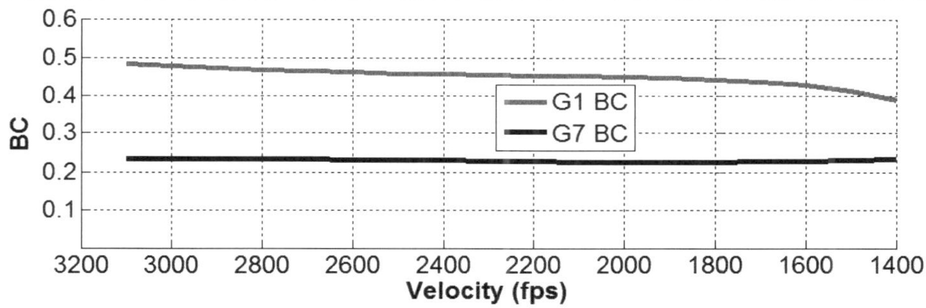

	Above 3000	3000-2500	2500-2000	2000-1500	Below 1500
G1 BC	0.496	0.470	0.454	0.430	0.411
G7 BC	0.235	0.233	0.228	0.229	0.231

Averages from 3000 fps to 1500 fps:
G1 BC: 0.450, variation of G1 BC: 0.069
G7 BC: 0.231, variation of G7 BC: 0.009
G7 form factor: 0.987

Rifling Twist Rate and Stability

Barrel Twist	Stability Factor Best case	Nominal	Worst case
1:6"	2.35	2.06	1.78
1:7"	1.73	1.51	1.31
1:8"	1.32	1.16	1.00
1:9"	1.05	0.91	0.79

Minimum twist required for stability is 1:7.9"
Recommended twist for optimal performance is 1:7.0" or faster

Hornady .224 caliber 68 grain BTHP

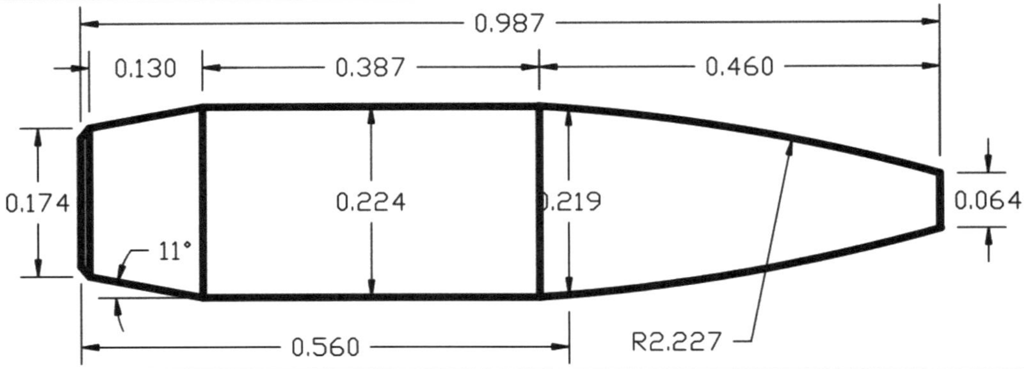

Lot #:	**2070697**	**Ogive Radius:**	**9.94 calibers**
Sectional Density:	**0.194 lb/in²**	**Rt/R:**	**0.66**

Ballistic Performance

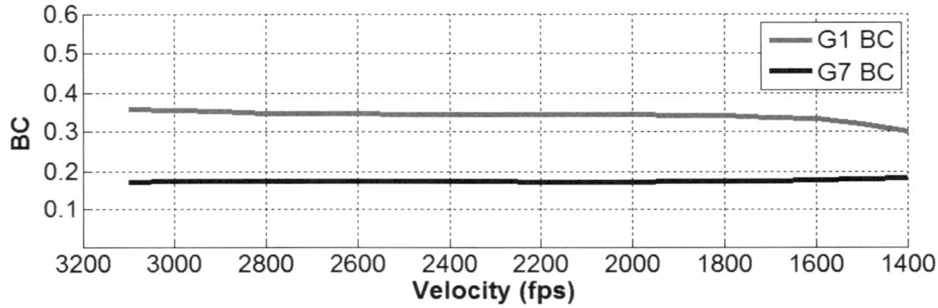

	Above 3000	3000-2500	2500-2000	2000-1500	Below 1500
G1 BC	0.362	0.350	0.344	0.332	0.320
G7 BC	0.172	0.173	0.173	0.176	0.180

Averages from 3000 fps to 1500 fps:
G1 BC: 0.341, variation of G1 BC: 0.034
G7 BC: 0.175, variation of G7 BC: 0.008
G7 form factor: 1.107

Rifling Twist Rate and Stability

Barrel Twist	Stability Factor Best case	Nominal	Worst case
1:6"	3.22	2.81	2.43
1:7"	2.37	2.07	1.79
1:8"	1.81	1.58	1.37
1:9"	1.43	1.25	1.08

Minimum twist required for stability is 1:9.2"
Recommended twist for optimal performance is 1:8.2" or faster

Hornady .224 caliber 75 grain BTHP

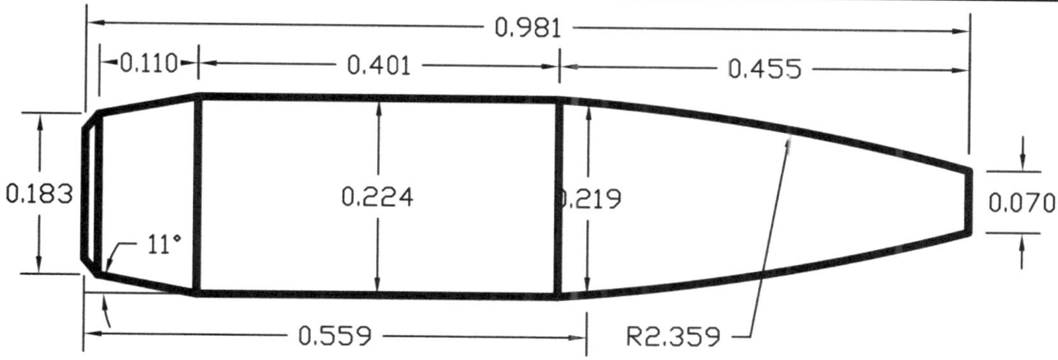

Lot #:	**2100145**	**Ogive Radius:**	**10.53 calibers**
Sectional Density:	**0.214 lb/in^2**	**Rt/R:**	**0.64**

Ballistic Performance

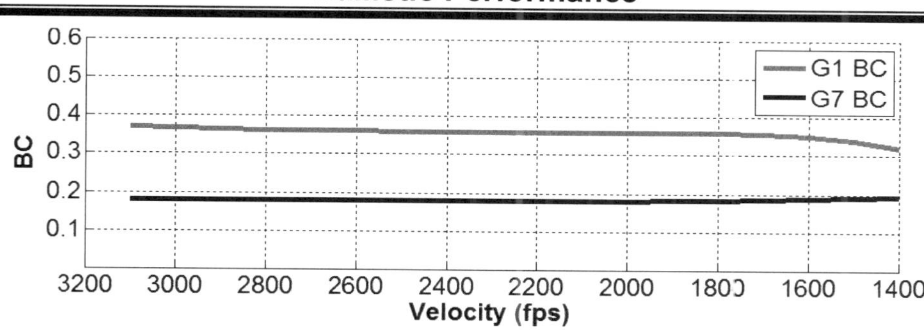

	Above 3000	3000-2500	2500-2000	2000-1500	Below 1500
G1 BC	0.374	0.363	0.359	0.349	0.338
G7 BC	0.177	0.180	0.180	0.186	0.191

Averages from 3000 fps to 1500 fps:
G1 BC: 0.356, variation of G1 BC: 0.028
G7 BC: 0.183, variation of G7 BC: 0.011
G7 form factor: 1.170

Rifling Twist Rate and Stability

Barrel Twist	Stability Factor Best case	Nominal	Worst case
1:6"	3.61	3.16	2.73
1:7"	2.66	2.32	2.01
1:8"	2.03	1.78	1.54
1:9"	1.61	1.40	1.21

Minimum twist required for stability is 1:9.7"
Recommended twist for optimal performance is 1:8.7" or faster

Hornady .224 caliber 55 grain FMJBT

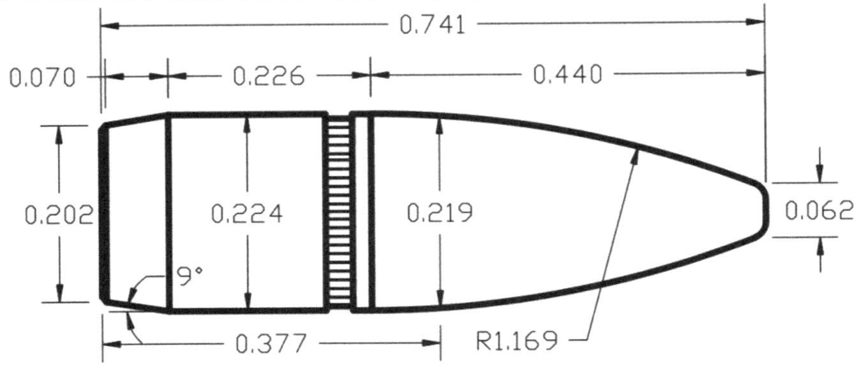

Lot #:	2140687	**Ogive Radius:**	**5.22 calibers**
Sectional Density:	**0.157 lb/in^2**	**Rt/R:**	**0.95**

Ballistic Performance

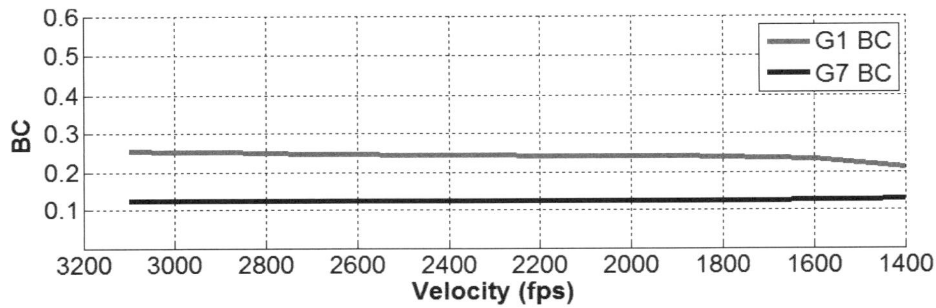

	Above 3000	3000-2500	2500-2000	2000-1500	Below 1500
G1 BC	0.261	0.250	0.244	0.233	0.224
G7 BC	0.124	0.124	0.123	0.124	0.126

Averages from 3000 fps to 1500 fps:
G1 BC: 0.241, variation of G1 BC: 0.030
G7 BC: 0.124, variation of G7 BC: 0.005
G7 form factor: 1.264

Rifling Twist Rate and Stability

Barrel Twist	Stability Factor Best case	Nominal	Worst case
1:6"	5.93	5.18	4.48
1:7"	4.36	3.81	3.29
1:8"	3.33	2.91	2.52
1:9"	2.63	2.30	1.99

Minimum twist required for stability is 1:12.5"
Recommended twist for optimal performance is 1:11.1" or faster

Hornady .224 caliber 40 grain Vmax

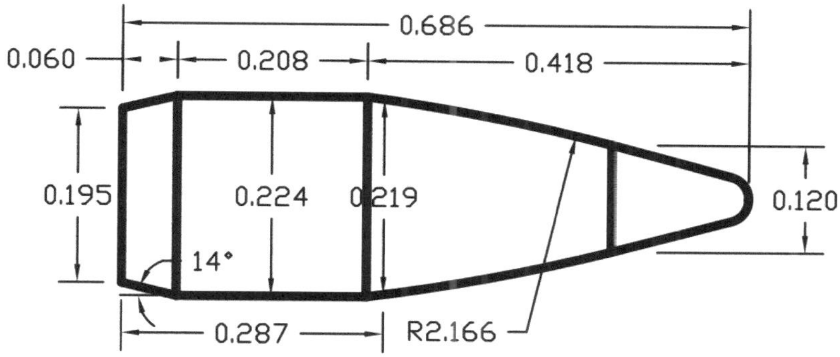

Lot #:	20800111	Ogive Radius:	9.67 calibers
Sectional Density:	0.114 lb/in^2	Rt/R:	0.50

Ballistic Performance

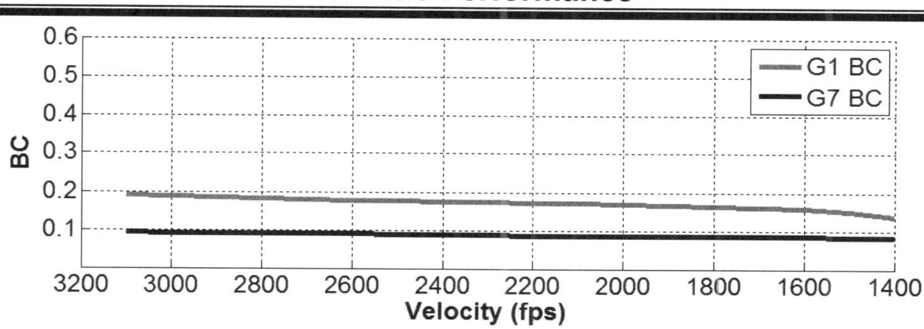

	Above 3000	3000-2500	2500-2000	2000-1500	Below 1500
G1 BC	0.195	0.182	0.173	0.159	0.149
G7 BC	0.092	0.090	0.087	0.084	0.083

Averages from 3000 fps to 1500 fps:
G1 BC: 0.170, variation of G1 BC: 0.039
G7 BC: 0.087, variation of G7 BC: 0.008
G7 form factor: 1.308

Rifling Twist Rate and Stability

Barrel Twist	Stability Factor Best case	Nominal	Worst case
1:6"	5.36	4.68	4.05
1:7"	3.94	3.44	2.97
1:8"	3.01	2.63	2.28
1:9"	2.38	2.08	1.80

Minimum twist required for stability is 1:11.9"
Recommended twist for optimal performance is 1:10.6" or faster

Hornady .224 caliber 50 grain Vmax

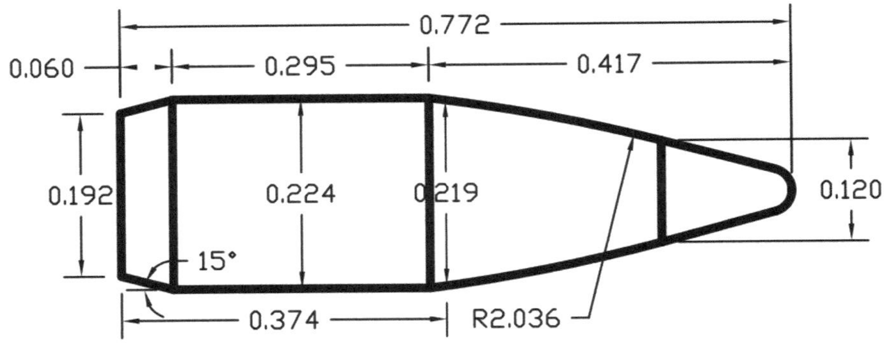

Lot #:	2080112	Ogive Radius:	9.09 calibers
Sectional Density:	0.142 lb/in^2	Rt/R:	0.51

Ballistic Performance

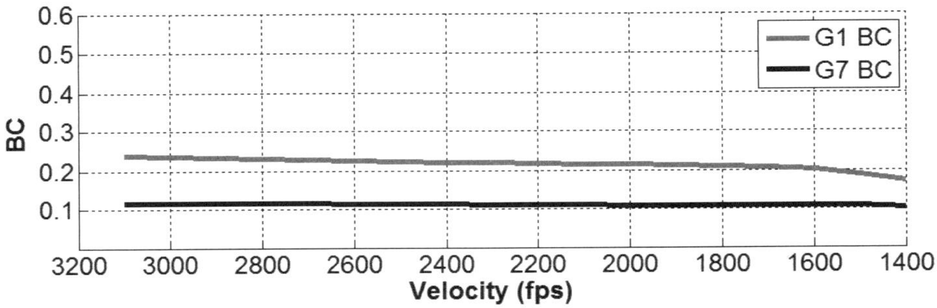

	Above 3000	3000-2500	2500-2000	2000-1500	Below 1500
G1 BC	0.246	0.230	0.219	0.201	0.189
G7 BC	0.116	0.114	0.110	0.107	0.106

Averages from 3000 fps to 1500 fps:
G1 BC: 0.216, variation of G1 BC: 0.048
G7 BC: 0.110, variation of G7 BC: 0.010
G7 form factor: 1.293

Rifling Twist Rate and Stability

Barrel Twist	Stability Factor Best case	Nominal	Worst case
1:6"	4.80	4.19	3.62
1:7"	3.52	3.08	2.66
1:8"	2.70	2.36	2.04
1:9"	2.13	1.86	1.61

Minimum twist required for stability is 1:11.2"
Recommended twist for optimal performance is 1:10.0" or faster

Hornady .224 caliber 53 grain Vmax

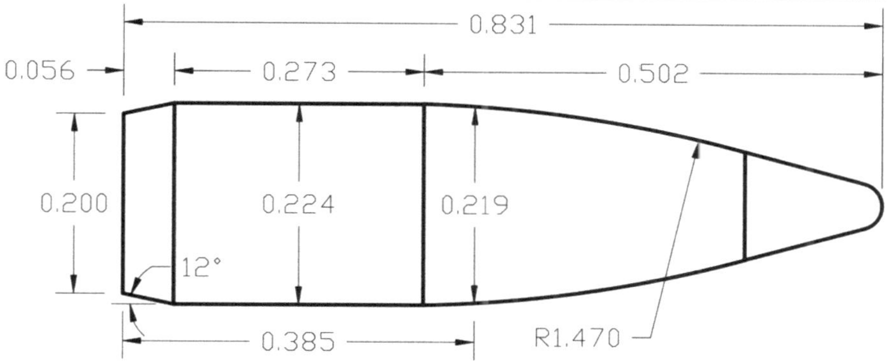

Lot #:	**2110689**	**Ogive Radius:**	**6.56 calibers**
Sectional Density:	**0.151 lb/in^2**	**Rt/R:**	**0.60**

Ballistic Performance

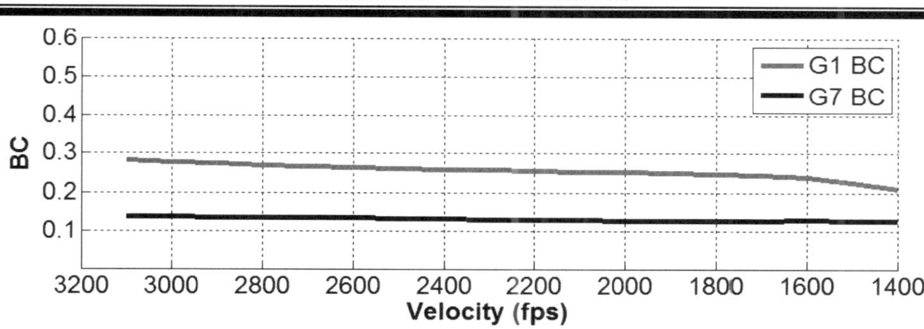

	Above 3000	3000-2500	2500-2000	2000-1500	Below 1500
G1 BC	0.288	0.269	0.257	0.239	0.226
G7 BC	0.136	0.134	0.129	0.127	0.127

Averages from 3000 fps to 1500 fps:

G1 BC: 0.254, variation of G1 BC: 0.052

G7 BC: 0.130, variation of G7 BC: 0.009

G7 form factor: 1.161

Rifling Twist Rate and Stability

Barrel Twist	Stability Factor Best case	Nominal	Worst case
1:8"	2.32	2.03	1.75
1:9"	1.83	1.60	1.38
1:10"	1.48	1.30	1.12
1:11"	1.23	1.07	0.93

Minimum twist required for stability is 1:10.4"

Recommended twist for optimal performance is 1:9.3" or faster

Hornady .224 caliber 55 grain Vmax

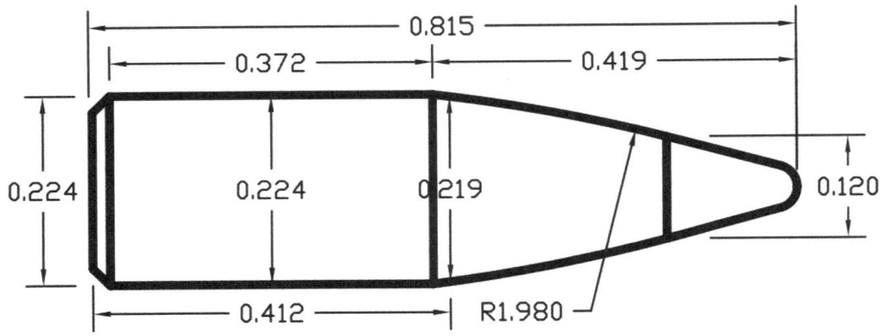

Lot #:	2080019	Ogive Radius:	8.84 calibers
Sectional Density:	0.157 lb/in^2	Rt/R:	0.52

Ballistic Performance

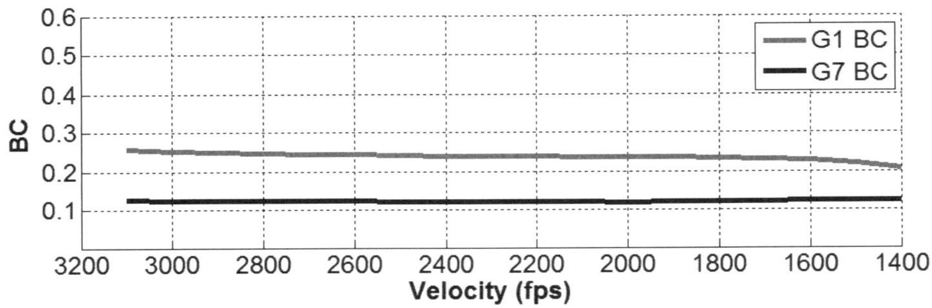

	Above 3000	3000-2500	2500-2000	2000-1500	Below 1500
G1 BC	0.263	0.248	0.239	0.228	0.219
G7 BC	0.125	0.123	0.120	0.121	0.123

Averages from 3000 fps to 1500 fps:
G1 BC: 0.238, variation of G1 BC: 0.035
G7 BC: 0.122, variation of G7 BC: 0.005
G7 form factor: 1.284

Rifling Twist Rate and Stability

Barrel Twist	Stability Factor Best case	Nominal	Worst case
1:6"	4.59	4.01	3.46
1:7"	3.37	2.94	2.54
1:8"	2.58	2.25	1.95
1:9"	2.04	1.78	1.54

Minimum twist required for stability is 1:11.0"
Recommended twist for optimal performance is 1:9.8" or faster

JLK .224 caliber 80 grain VLD

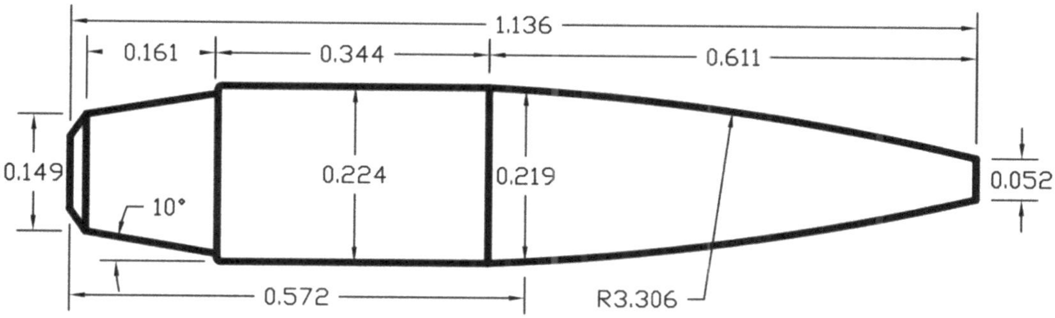

Lot #:	Unknown	Ogive Radius:	14.76 calibers
Sectional Density:	0.228 lb/in²	Rt/R:	0.70

Ballistic Performance

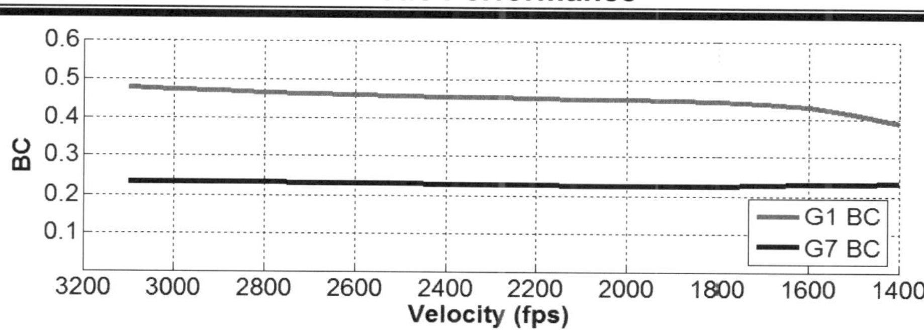

	Above 3000	3000-2500	2500-2000	2000-1500	Below 1500
G1 BC	0.489	0.467	0.454	0.431	0.413
G7 BC	0.232	0.231	0.228	0.229	0.232

Averages from 3000 fps to 1500 fps:
G1 BC: 0.449, variation of G1 BC: 0.062
G7 BC: 0.230, variation of G7 BC: 0.007
G7 form factor: 0.989

Rifling Twist Rate and Stability

Barrel Twist	Stability Factor Best case	Nominal	Worst case
1:6"	2.51	2.20	1.90
1:7"	1.85	1.61	1.40
1:8"	1.41	1.24	1.07
1:9"	1.12	0.98	0.84

Minimum twist required for stability is 1:8 1"
Recommended twist for optimal performance is 1:7.3" or faster

Lake City .224 caliber 62 grain FMJBT

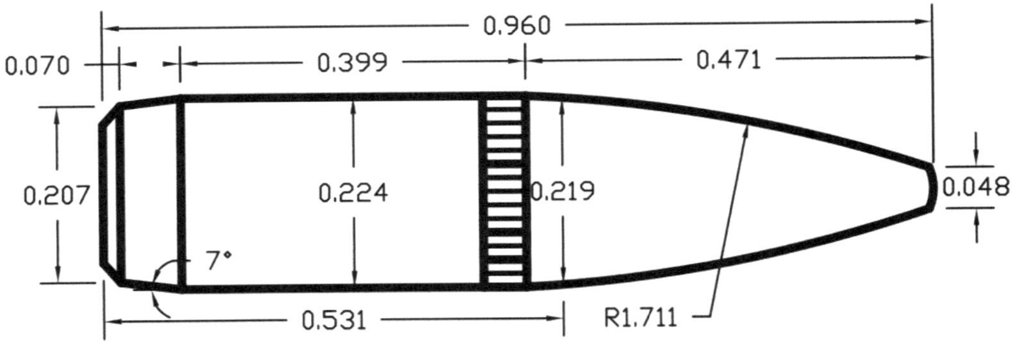

Lot #:	**Unknown**	**Ogive Radius:**	**7.64 calibers**
Sectional Density:	**0.177 lb/in²**	**Rt/R:**	**0.90**

Ballistic Performance

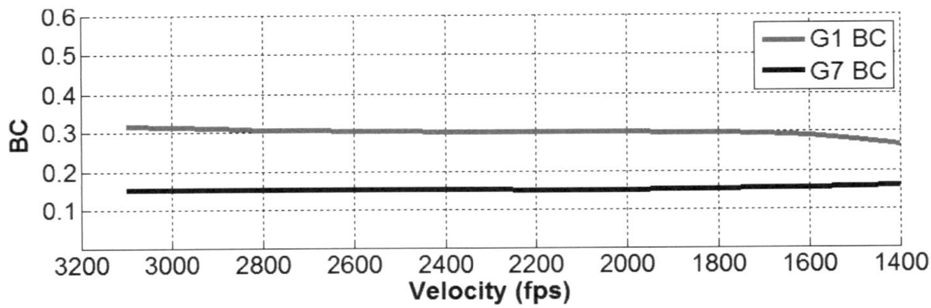

	Above 3000	3000-2500	2500-2000	2000-1500	Below 1500
G1 BC	0.323	0.309	0.302	0.291	0.280
G7 BC	0.153	0.153	0.152	0.154	0.158

Averages from 3000 fps to 1500 fps:
G1 BC: 0.300, variation of G1 BC: 0.034
G7 BC: 0.154, variation of G7 BC: 0.007
G7 form factor: 1.148

Rifling Twist Rate and Stability

	Stability Factor		
Barrel Twist	**Best case**	**Nominal**	**Worst case**
1:6"	3.18	2.78	2.40
1:7"	2.34	2.04	1.77
1:8"	1.79	1.56	1.35
1:9"	1.41	1.24	1.07

Minimum twist required for stability is 1:9.1"
Recommended twist for optimal performance is 1:8.2" or faster

Lapua .224 caliber 69 grain Scenar

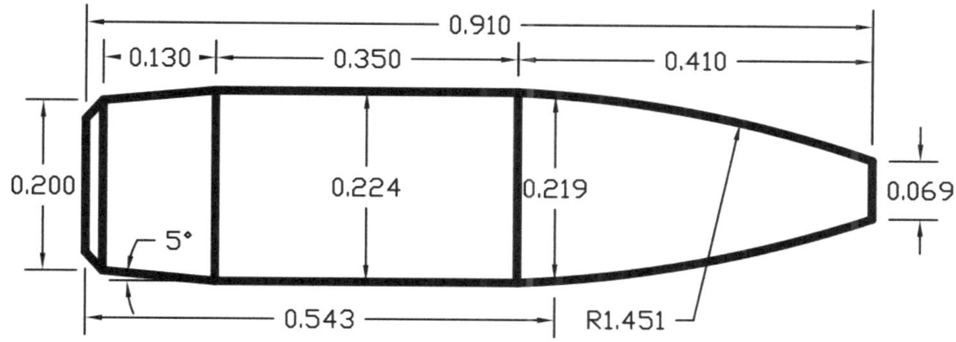

Lot #:	4PL5010	Ogive Radius:	6.48 calibers
Sectional Density:	0.197 lb/in^2	Rt/R:	0.81

Ballistic Performance

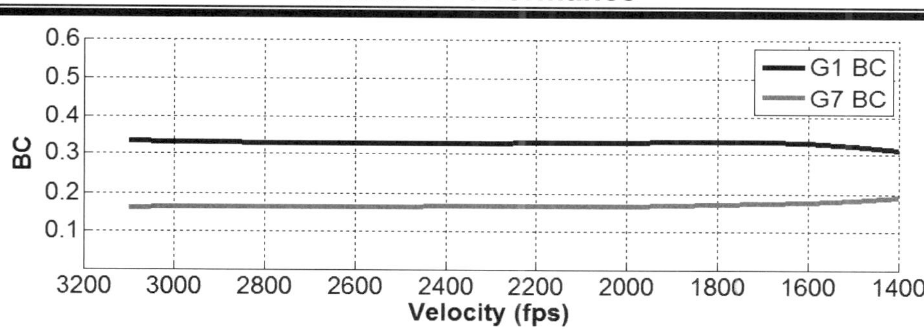

	Above 3000	3000-2500	2500-2000	2000-1500	Below 1500
G1 BC	0.338	0.330	0.330	0.328	0.323
G7 BC	0.160	0.164	0.166	0.175	0.182

Averages from 3000 fps to 1500 fps:
G1 BC: 0.329, variation of G1 BC: 0.009
G7 BC: 0.169, variation of G7 BC: 0.020
G7 form factor: 1.164

Rifling Twist Rate and Stability

Barrel Twist	Stability Factor Best case	Nominal	Worst case
1:6"	4.13	3.61	3.12
1:7"	3.04	2.65	2.29
1:8"	2.32	2.03	1.76
1:9"	1.84	1.60	1.39

Minimum twist required for stability is 1:9.8"
Recommended twist for optimal performance is 1:8.8" or faster

Lapua .224 caliber 77 grain Scenar

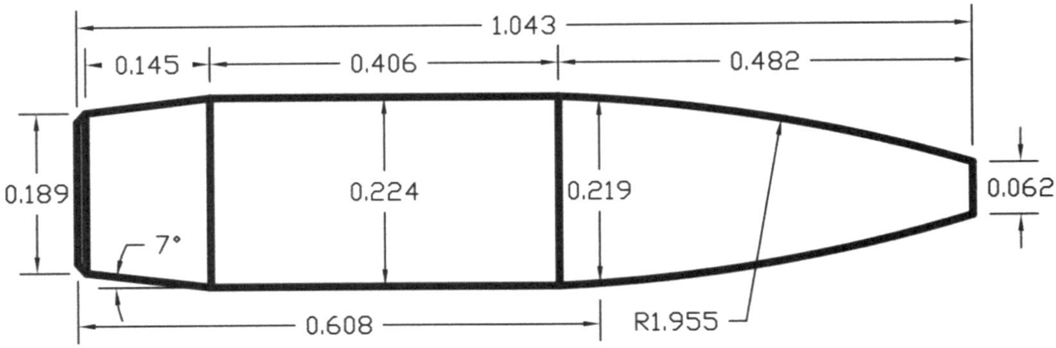

Lot #:	P00454601	Ogive Radius:	8.73 calibers
Sectional Density:	0.219 lb/in^2	Rt/R:	0.78

Ballistic Performance

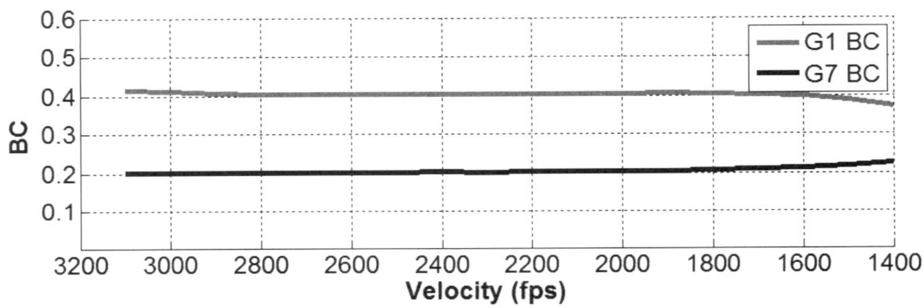

	Above 3000	3000-2500	2500-2000	2000-1500	Below 1500
G1 BC	0.419	0.407	0.404	0.396	0.387
G7 BC	0.199	0.202	0.203	0.211	0.218

Averages from 3000 fps to 1500 fps:
G1 BC: 0.401, variation of G1 BC: 0.024
G7 BC: 0.206, variation of G7 BC: 0.017
G7 form factor: 1.063

Rifling Twist Rate and Stability

Barrel Twist	Stability Factor Best case	Nominal	Worst case
1:6"	3.10	2.71	2.35
1:7"	2.28	1.99	1.72
1:8"	1.75	1.53	1.32
1:9"	1.38	1.21	1.04

Minimum twist required for stability is 1:9.0"
Recommended twist for optimal performance is 1:8.1" or faster

Lapua .224 caliber 69 grain Scenar L

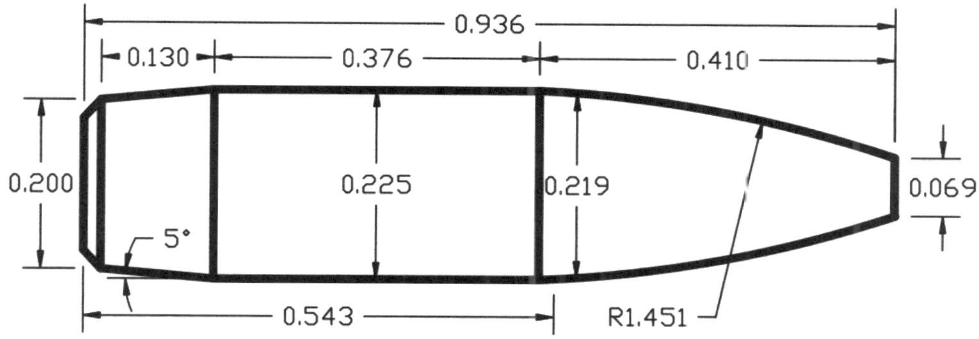

Lot #:	**Unknown**	**Ogive Radius:**	**6.48 calibers**
Sectional Density:	**0.197 lb/in²**	**Rt/R:**	**0.81**

Ballistic Performance

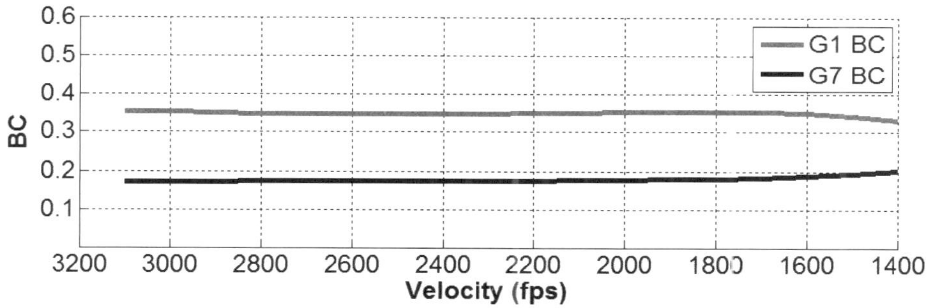

	Above 3000	3000-2500	2500-2000	2000-1500	Below 1500
G1 BC	0.358	0.349	0.350	0.347	0.342
G7 BC	0.170	0.173	0.176	0.185	0.193

Averages from 3000 fps to 1500 fps:

G1 BC: 0.348, variation of G1 BC: 0.010

G7 BC: 0.179, variation of G7 BC: 0.021

G7 form factor: 1.098

Rifling Twist Rate and Stability

Barrel Twist	**Stability Factor** Best case	Nominal	Worst case
1:6"	3.81	3.33	2.88
1:7"	2.80	2.44	2.11
1:8"	2.14	1.87	1.62
1:9"	1.69	1.48	1.28

Minimum twist required for stability is 1:10.4"

Recommended twist for optimal performance is 1:9.3" or faster

Lapua .224 caliber 77 grain Scenar L

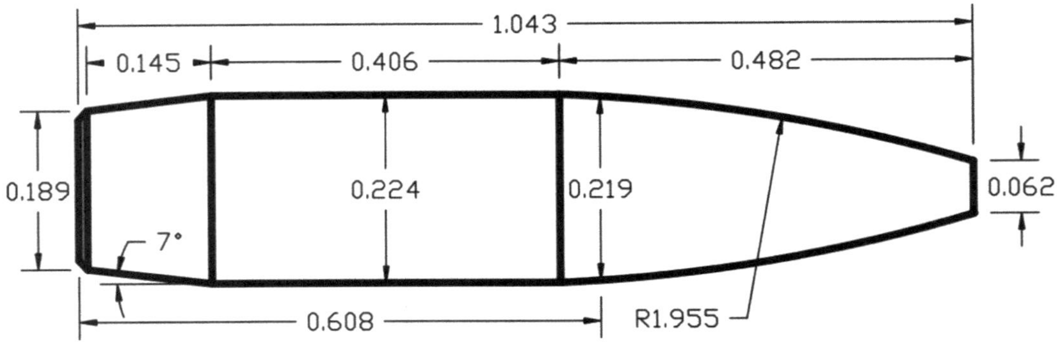

Lot #:	**Unknown**	**Ogive Radius:**	**8.73 calibers**
Sectional Density:	**0.219 lb/in^2**	**Rt/R:**	**0.78**

Ballistic Performance

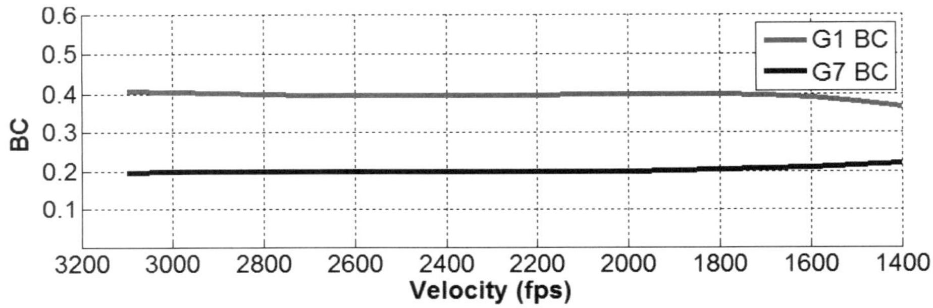

	Above 3000	3000-2500	2500-2000	2000-1500	Below 1500
G1 BC	0.412	0.400	0.397	0.389	0.380
G7 BC	0.195	0.198	0.200	0.207	0.215

Averages from 3000 fps to 1500 fps:
G1 BC: 0.394, variation of G1 BC: 0.023
G7 BC: 0.203, variation of G7 BC: 0.017
G7 form factor: 1.082

Rifling Twist Rate and Stability

Barrel Twist	Stability Factor Best case	Nominal	Worst case
1:6"	3.10	2.71	2.35
1:7"	2.28	1.99	1.72
1:8"	1.75	1.53	1.32
1:9"	1.38	1.21	1.04

Minimum twist required for stability is 1:9.0"
Recommended twist for optimal performance is 1:8.1" or faster

Nosler .224 caliber 55 grain Ballistic Tip

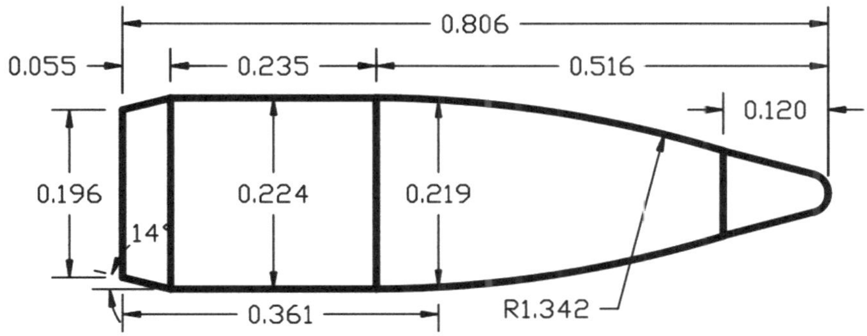

Lot #:	**95311881165**	**Ogive Radius:**	**5.99 calibers**
Sectional Density:	**0.157 lb/in^2**	**Rt/R:**	**1.00**

Ballistic Performance

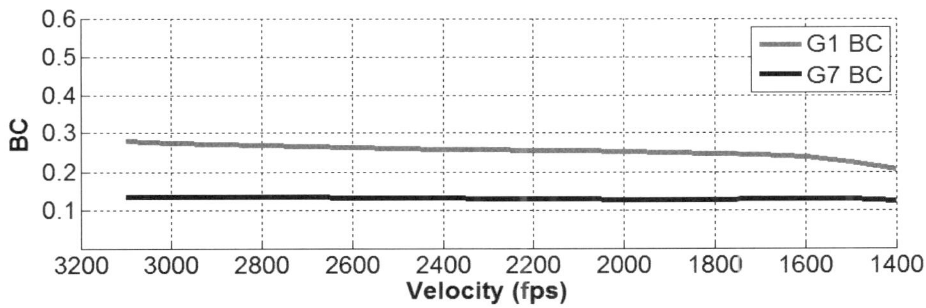

	Above 3000	3000-2500	2500-2000	2000-1500	Below 1500
G1 BC	0.285	0.268	0.257	0.240	0.227
G7 BC	0.135	0.133	0.129	0.127	0.127

Averages from 3000 fps to 1500 fps:

G1 BC: 0.254, variation of G1 BC: 0.049

G7 BC: 0.130, variation of G7 BC: 0.008

G7 form factor: 1.205

Rifling Twist Rate and Stability

Barrel Twist	Stability Factor Best case	Nominal	Worst case
1:6"	4.67	4.08	3.53
1:7"	3.43	3.00	2.59
1:8"	2.63	2.29	1.98
1:9"	2.07	1.81	1.57

Minimum twist required for stability is 1:11.1"

Recommended twist for optimal performance is 1:9.9" or faster

Nosler .224 caliber 60 grain Ballistic Tip

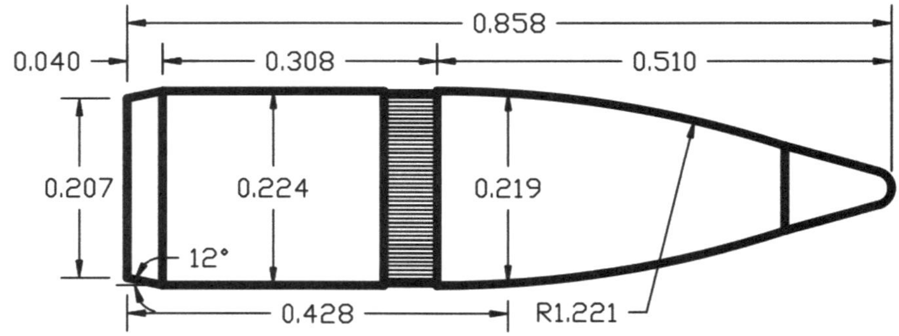

Lot #:	AL24F5	**Ogive Radius:**	**5.45 calibers**
Sectional Density:	**0.171 lb/in^2**	**Rt/R:**	**1.00**

Ballistic Performance

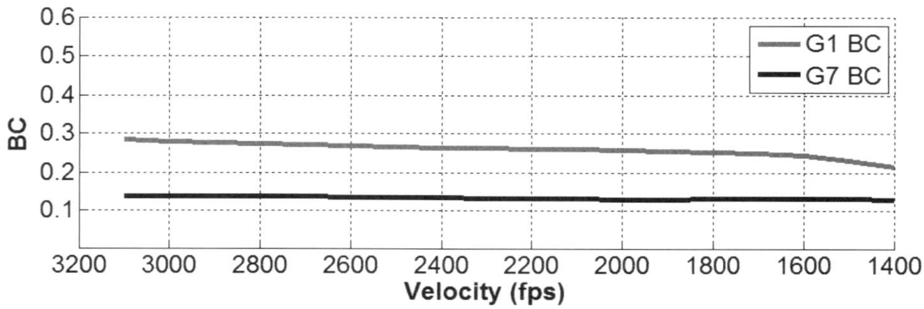

	Above 3000	3000-2500	2500-2000	2000-1500	Below 1500
G1 BC	0.291	0.273	0.262	0.245	0.232
G7 BC	0.138	0.136	0.132	0.130	0.130

Averages from 3000 fps to 1500 fps:
G1 BC: 0.259, variation of G1 BC: 0.049
G7 BC: 0.133, variation of G7 BC: 0.008
G7 form factor: 1.288

Rifling Twist Rate and Stability

Barrel Twist	Stability Factor Best case	Stability Factor Nominal	Stability Factor Worst case
1:6"	4.26	3.72	3.22
1:7"	3.13	2.73	2.36
1:8"	2.39	2.09	1.81
1:9"	1.89	1.65	1.43

Minimum twist required for stability is 1:10.6"
Recommended twist for optimal performance is 1:9.4" or faster

Nosler .224 caliber 52 grain Custom Competition

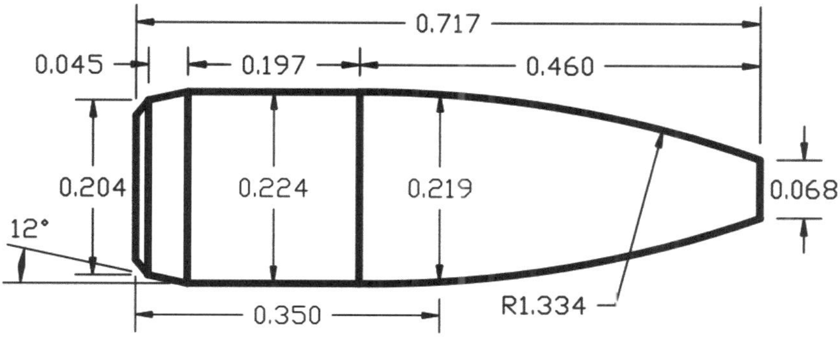

Lot #:	HEO9F12-1	Ogive Radius:	5.95 calibers
Sectional Density:	0.148 lb/in^2	Rt/R:	1.00

Ballistic Performance

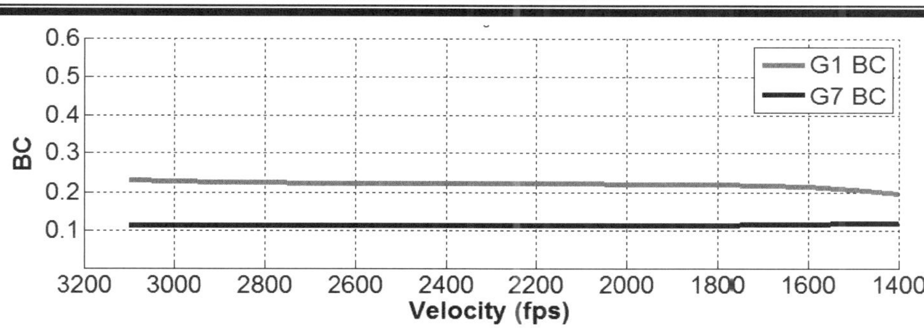

	Above 3000	3000-2500	2500-2000	2000-1500	Below 1500
G1 BC	0.234	0.226	0.222	0.215	0.208
G7 BC	0.111	0.112	0.112	0.114	0.117

Averages from 3000 fps to 1500 fps:
G1 BC: 0.220, variation of G1 BC: 0.021
G7 BC: 0.113, variation of G7 BC: 0.006
G7 form factor: 1.311

Rifling Twist Rate and Stability

Barrel Twist	Stability Factor Best case	 Nominal	 Worst case
1:6"	6.15	5.37	4.65
1:7"	4.52	3.95	3.41
1:8"	3.46	3.02	2.61
1:9"	2.73	2.39	2.07

Minimum twist required for stability is 1:12.7"
Recommended twist for optimal performance is 1:11.4" or faster

Nosler .224 caliber 69 grain Custom Competition

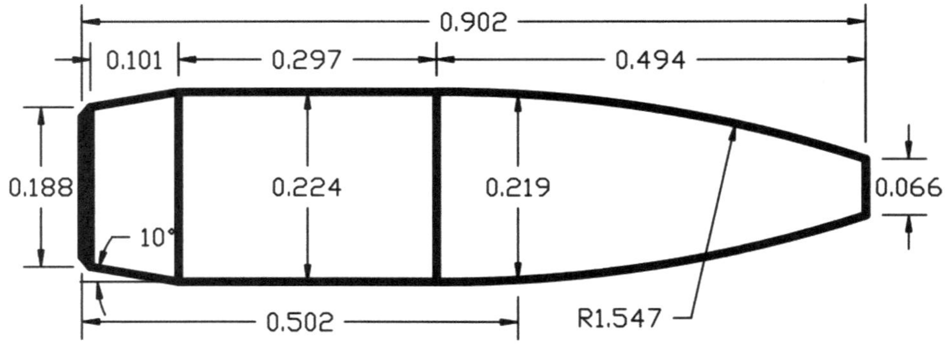

Lot #:	BL10F12	Ogive Radius:	6.91 calibers
Sectional Density:	0.197 lb/in^2	Rt/R:	0.95

Ballistic Performance

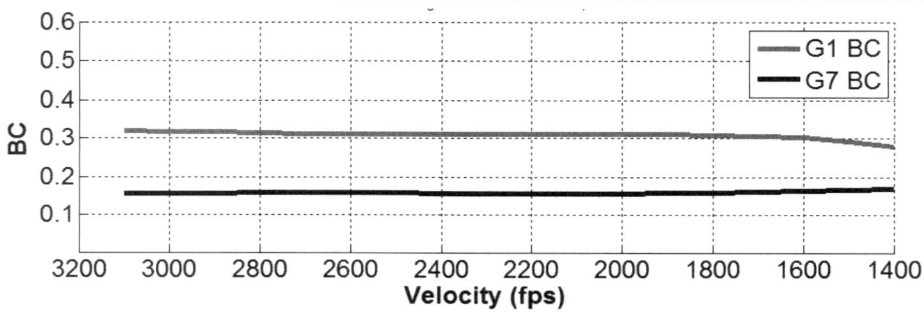

	Above 3000	3000-2500	2500-2000	2000-1500	Below 1500
G1 BC	0.326	0.315	0.312	0.303	0.294
G7 BC	0.154	0.156	0.157	0.161	0.166

Averages from 3000 fps to 1500 fps:
G1 BC: 0.309, variation of G1 BC: 0.025
G7 BC: 0.159, variation of G7 BC: 0.010
G7 form factor: 1.239

Rifling Twist Rate and Stability

Barrel Twist	Stability Factor Best case	Nominal	Worst case
1:6"	4.24	3.70	3.20
1:7"	3.11	2.72	2.35
1:8"	2.38	2.08	1.80
1:9"	1.88	1.65	1.42

Minimum twist required for stability is 1:10.5"
Recommended twist for optimal performance is 1:9.4" or faster

Nosler .224 caliber 77 grain Custom Competition

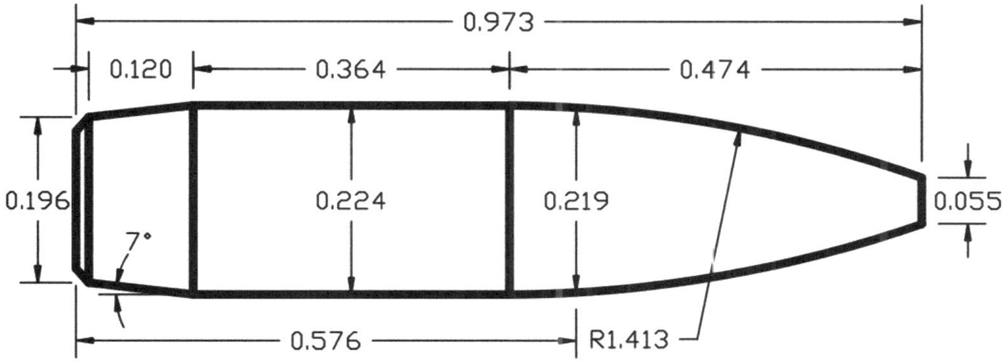

Lot #:	FE08F12-1	Ogive Radius:	6.3 calibers
Sectional Density:	0.219 lb/in^2	Rt/R:	0.98

Ballistic Performance

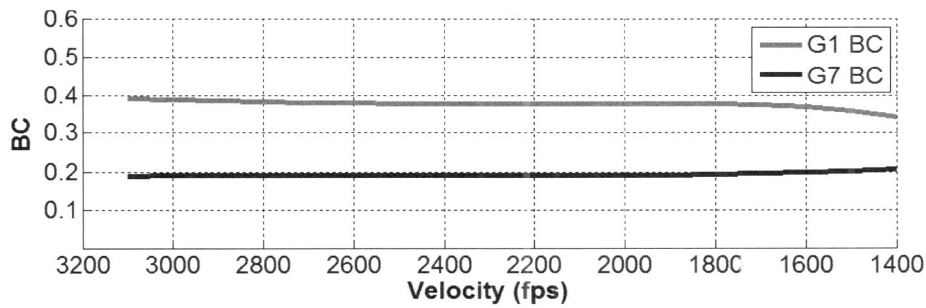

	Above 3000	3000-2500	2500-2000	2000-1500	Below 1500
G1 BC	0.397	0.383	0.378	0.368	0.357
G7 BC	0.188	0.190	0.190	0.196	0.201

Averages from 3000 fps to 1500 fps:
G1 BC: 0.375, variation of G1 BC: 0.030
G7 BC: 0.193, variation of G7 BC: 0.012
G7 form factor: 1.138

Rifling Twist Rate and Stability

Barrel Twist	Stability Factor Best case	Nominal	Worst case
1:6"	3.80	3.32	2.87
1:7"	2.79	2.44	2.11
1:8"	2.14	1.87	1.61
1:9"	1.69	1.48	1.28

Minimum twist required for stability is 1:10.0"
Recommended twist for optimal performance is 1:8.9" or faster

PMC .224 caliber 62 grain FMJBT Green Tip Ammo

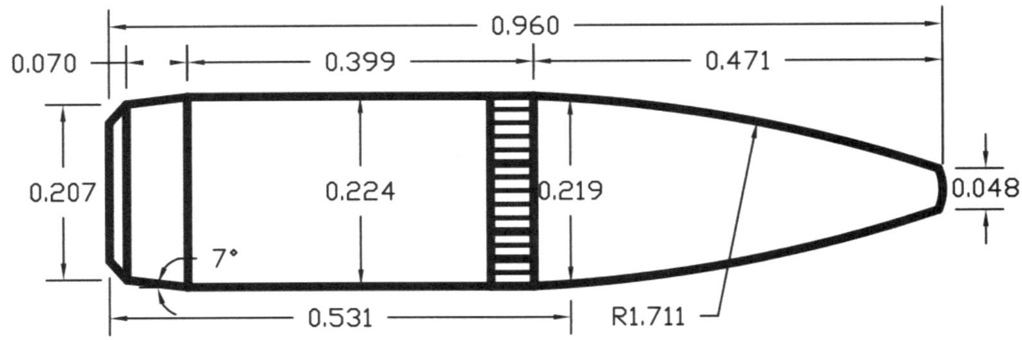

Lot #:	unknown	Ogive Radius:	7.64 calibers
Sectional Density:	0.177 lb/in^2	Rt/R:	0.90

Ballistic Performance

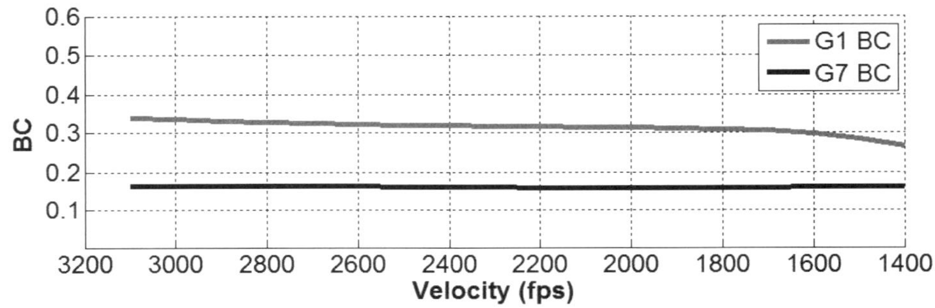

	Above 3000	3000-2500	2500-2000	2000-1500	Below 1500
G1 BC	0.347	0.329	0.318	0.300	0.286
G7 BC	0.164	0.163	0.160	0.159	0.161

Averages from 3000 fps to 1500 fps:
G1 BC: 0.314, variation of G1 BC: 0.050
G7 BC: 0.161, variation of G7 BC: 0.006
G7 form factor: 1.096

Rifling Twist Rate and Stability

Barrel Twist	Stability Factor Best case	Nominal	Worst case
1:7"	2.69	2.35	2.03
1:8"	2.06	1.80	1.55
1:9"	1.62	1.42	1.23
1:10"	1.32	1.15	0.99

Minimum twist required for stability is 1:9.8"
Recommended twist for optimal performance is 1:8.8" or faster

RUAG .224 caliber 69 grain Swiss P Target

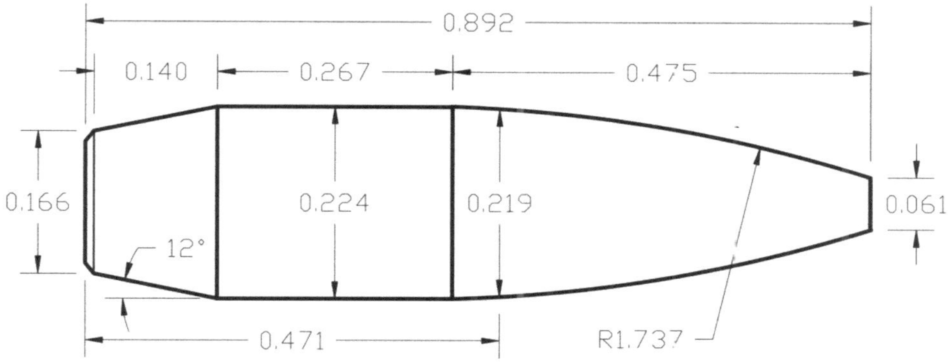

Lot #:	unknown	Ogive Radius:	5.08 calibers
Sectional Density:	0.197 lb/in^2	Rt/R:	0.82

Ballistic Performance

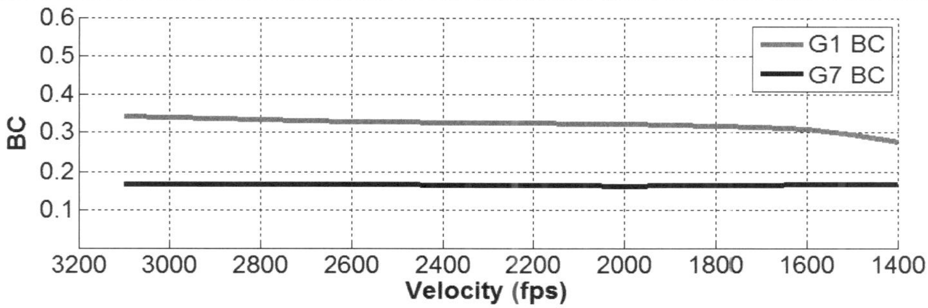

	Above 3000	3000-2500	2500-2000	2000-1500	Below 1500
G1 BC	0.348	0.334	0.325	0.310	0.296
G7 BC	0.165	0.165	0.164	0.164	0.167

Averages from 3000 fps to 1500 fps:
G1 BC: 0.322, variation of G1 BC: 0.043
G7 BC: 0.165, variation of G7 BC: 0.005
G7 form factor: 1.191

Rifling Twist Rate and Stability

Barrel Twist	Stability Factor Best case	Nominal	Worst case
1:8"	2.46	2.15	1.86
1:9"	1.95	1.70	1.47
1:10"	1.58	1.38	1.19
1:11"	1.30	1.14	0.98

Minimum twist required for stability is 1:10.7"
Recommended twist for optimal performance is 1:9.6" or faster

Sierra .224 caliber 55 grain BlitzKing

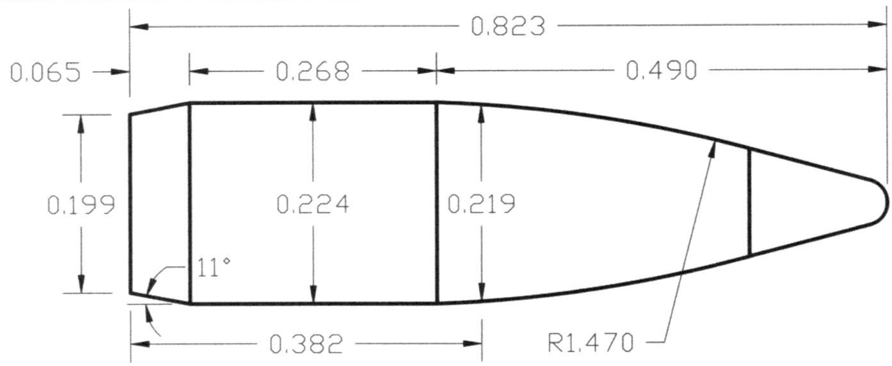

Lot #:	5211785	Ogive Radius:	6.56 calibers
Sectional Density:	0.157 lb/in²	Rt/R:	1.00

Ballistic Performance

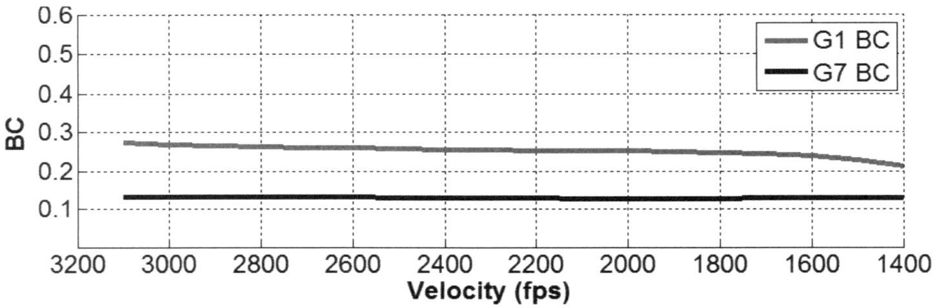

	Above 3000	3000-2500	2500-2000	2000-1500	Below 1500
G1 BC	0.279	0.264	0.255	0.240	0.228
G7 BC	0.132	0.131	0.128	0.127	0.128

Averages from 3000 fps to 1500 fps:
G1 BC: 0.252, variation of G1 BC: 0.042
G7 BC: 0.129, variation of G7 BC: 0.006
G7 form factor: 1.214

Rifling Twist Rate and Stability

Barrel Twist	Stability Factor Best case	Nominal	Worst case
1:8"	2.47	2.16	1.87
1:9"	1.95	1.71	1.48
1:10"	1.58	1.38	1.20
1:11"	1.31	1.14	0.99

Minimum twist required for stability is 1:10.7"
Recommended twist for optimal performance is 1:9.6" or faster

Sierra .224 caliber 55 grain FMJ

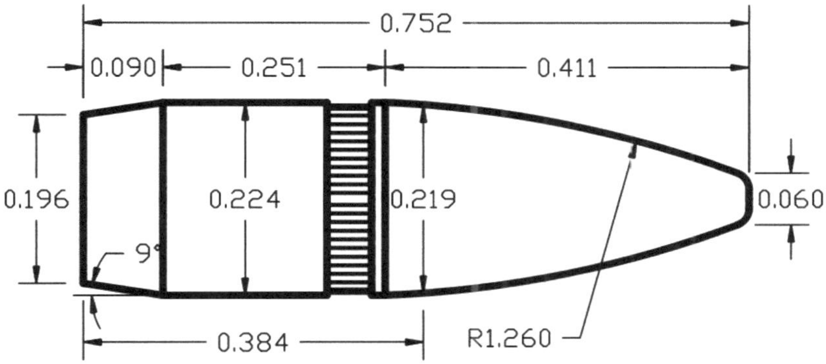

Lot #:	Fed loaded ammo	Ogive Radius:	5.62 calibers
Sectional Density:	0.157 lb/in^2	Rt/R:	0.80

Ballistic Performance

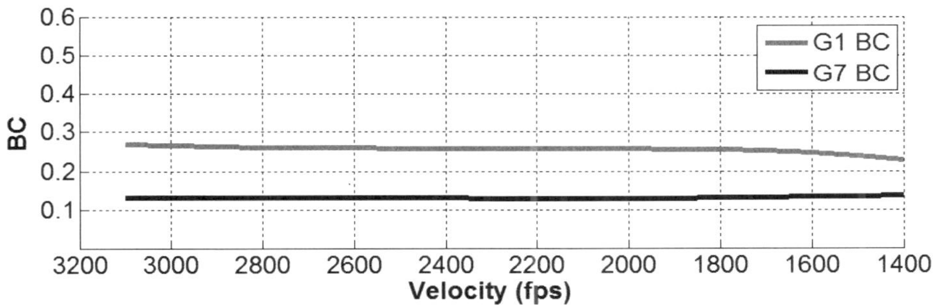

	Above 3000	3000-2500	2500-2000	2000-1500	Below 1500
G1 BC	0.273	0.263	0.258	0.248	0.240
G7 BC	0.130	0.130	0.130	0.132	0.135

Averages from 3000 fps to 1500 fps:
G1 BC: 0.255, variation of G1 BC: 0.027
G7 BC: 0.131, variation of G7 BC: 0.006
G7 form factor: 1.194

Rifling Twist Rate and Stability

Barrel Twist	Stability Factor Best case	Nominal	Worst case
1:6"	5.69	4.97	4.30
1:7"	4.18	3.65	3.16
1:8"	3.20	2.79	2.42
1:9"	2.53	2.21	1.91

Minimum twist required for stability is 1:12.2"
Recommended twist for optimal performance is 1:10.9" or faster

Sierra .224 caliber 65 grain GameKing

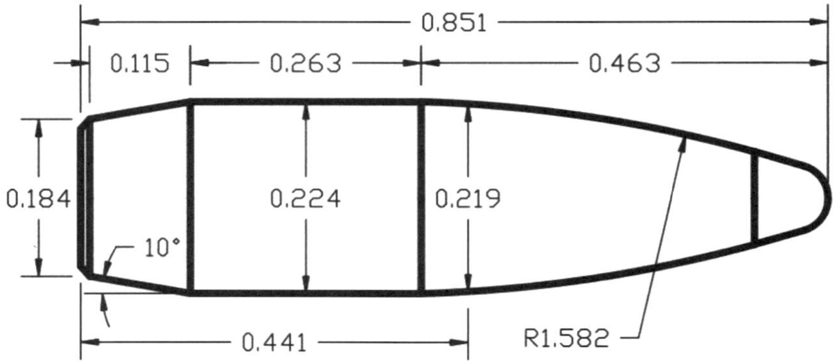

Lot #:	0089421334545	Ogive Radius:	7.06 calibers
Sectional Density:	0.185 lb/in^2	Rt/R:	0.90

Ballistic Performance

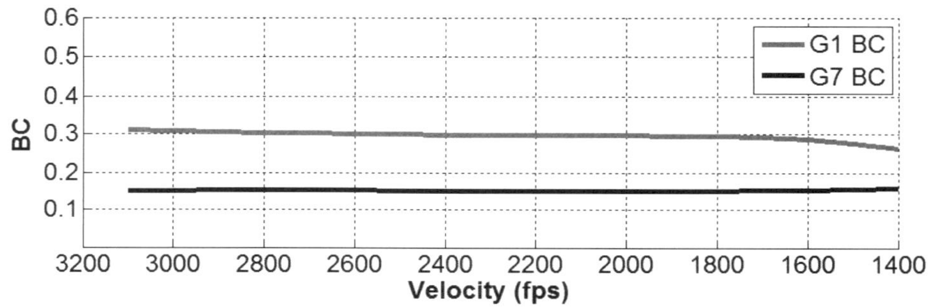

	Above 3000	3000-2500	2500-2000	2000-1500	Below 1500
G1 BC	0.317	0.305	0.299	0.288	0.277
G7 BC	0.150	0.151	0.150	0.153	0.156

Averages from 3000 fps to 1500 fps:
G1 BC: 0.296, variation of G1 BC: 0.032
G7 BC: 0.152, variation of G7 BC: 0.006
G7 form factor: 1.217

Rifling Twist Rate and Stability

Barrel Twist	Stability Factor Best case	Nominal	Worst case
1:6"	4.72	4.12	3.57
1:7"	3.47	3.03	2.62
1:8"	2.66	2.32	2.01
1:9"	2.10	1.83	1.59

Minimum twist required for stability is 1:11.1"
Recommended twist for optimal performance is 1:9.9" or faster

Sierra .224 caliber 52 grain MatchKing

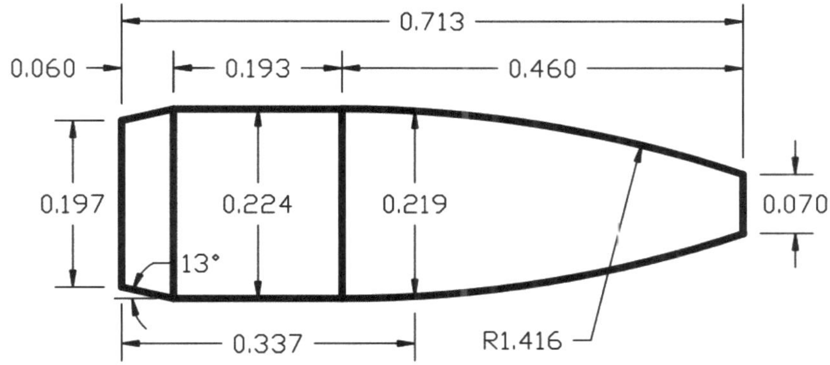

Lot #:	0033972241648	Ogive Radius:	6.32 calibers
Sectional Density:	0.148 lb/in²	Rt/R:	1.00

Ballistic Performance

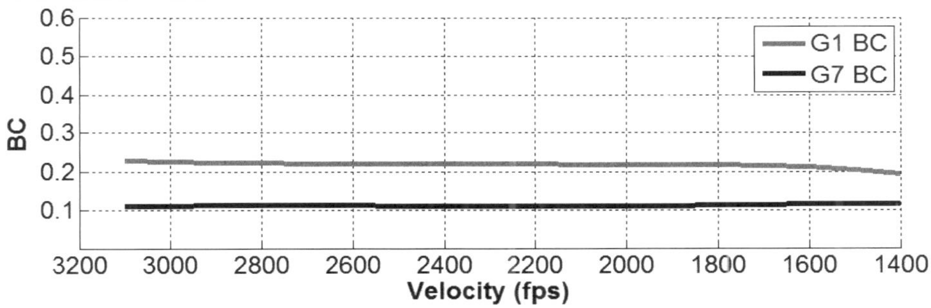

	Above 3000	3000-2500	2500-2000	2000-1500	Below 1500
G1 BC	0.232	0.223	0.219	0.212	0.205
G7 BC	0.110	0.111	0.110	0.112	0.115

Averages from 3000 fps to 1500 fps:
G1 BC: 0.217, variation of G1 BC: 0.022
G7 BC: 0.112, variation of G7 BC: 0.005
G7 form factor: 1.328

Rifling Twist Rate and Stability

Barrel Twist	Stability Factor Best case	Nominal	Worst case
1:6"	6.25	5.46	4.72
1:7"	4.59	4.01	3.47
1:8"	3.52	3.07	2.66
1:9"	2.78	2.43	2.10

Minimum twist required for stability is 1:12.8"
Recommended twist for optimal performance is 1:11.4" or faster

Sierra .224 caliber 69 grain MatchKing

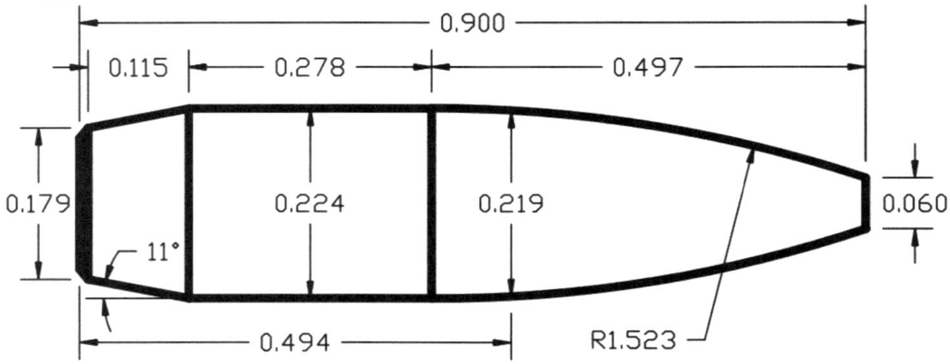

Lot #:	003807376	Ogive Radius:	6.8 calibers
Sectional Density:	0.197 lb/in^2	Rt/R:	1.00

Ballistic Performance

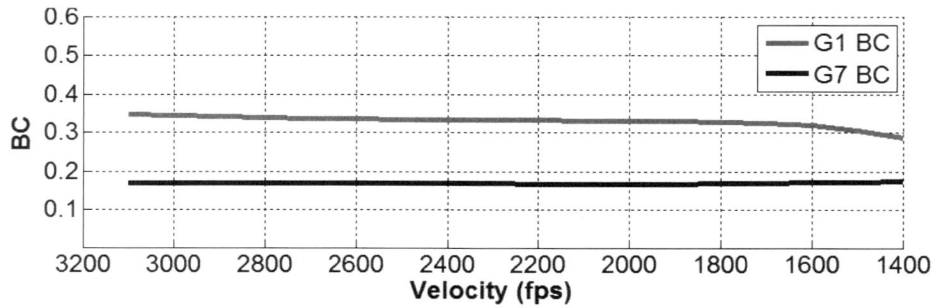

	Above 3000	3000-2500	2500-2000	2000-1500	Below 1500
G1 BC	0.354	0.340	0.333	0.319	0.307
G7 BC	0.168	0.169	0.168	0.170	0.173

Averages from 3000 fps to 1500 fps:
G1 BC: 0.330, variation of G1 BC: 0.038
G7 BC: 0.169, variation of G7 BC: 0.006
G7 form factor: 1.161

Rifling Twist Rate and Stability

Barrel Twist	Stability Factor Best case	Nominal	Worst case
1:6"	4.27	3.73	3.22
1:7"	3.13	2.74	2.37
1:8"	2.40	2.10	1.81
1:9"	1.90	1.66	1.43

Minimum twist required for stability is 1:10.6"
Recommended twist for optimal performance is 1:9.5" or faster

Sierra .224 caliber 77 grain MatchKing

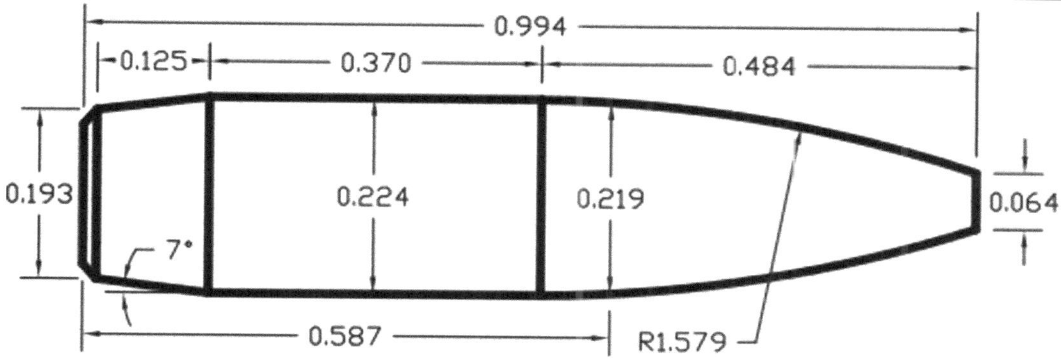

Lot #:	4123666	Ogive Radius:	7.05 calibers
Sectional Density:	0.219 lb/in^2	Rt/R:	0.96

Ballistic Performance

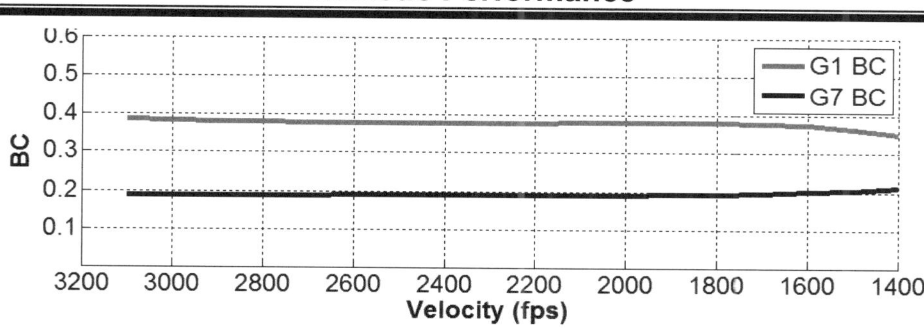

	Above 3000	3000-2500	2500-2000	2000-1500	Below 1500
G1 BC	0.391	0.380	0.378	0.372	0.364
G7 BC	0.186	0.189	0.190	0.198	0.205

Averages from 3000 fps to 1500 fps:
G1 BC: 0.376, variation of G1 BC: 0.020
G7 BC: 0.193, variation of G7 BC: 0.018
G7 form factor: 1.135

Rifling Twist Rate and Stability

Barrel Twist	Stability Factor Best case	Nominal	Worst case
1:6"	3.57	3.12	2.70
1:7"	2.62	2.29	1.98
1:8"	2.01	1.75	1.52
1:9"	1.59	1.39	1.20

Minimum twist required for stability is 1:9.7"
Recommended twist for optimal performance is 1:8.7" or faster

Sierra .224 caliber 80 grain MatchKing

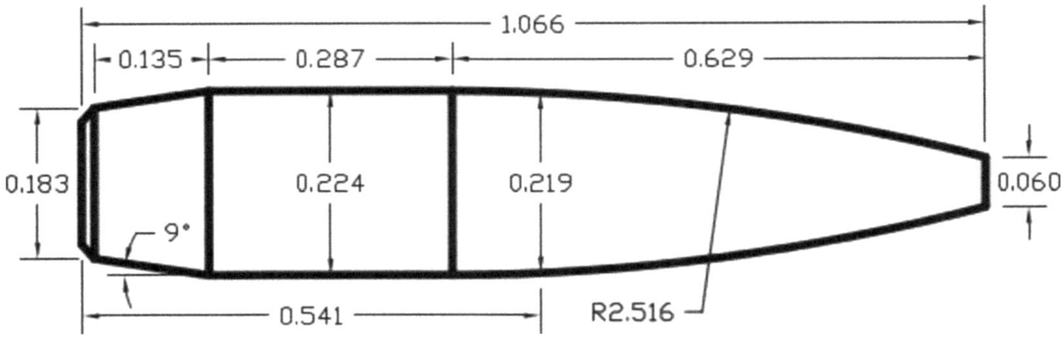

Lot #:	0028603914848	Ogive Radius:	11.23 calibers
Sectional Density:	0.228 lb/in²	Rt/R:	0.98

Ballistic Performance

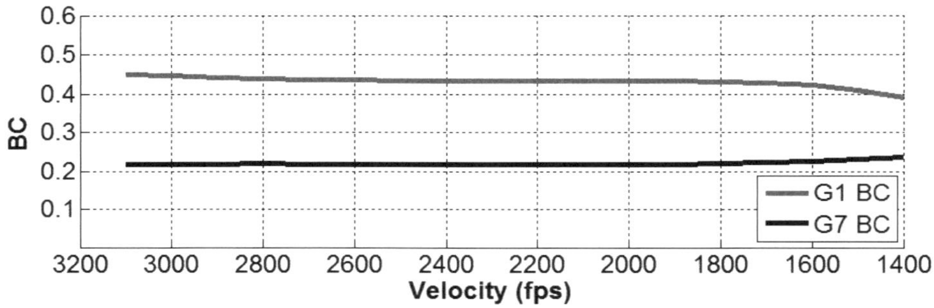

	Above 3000	3000-2500	2500-2000	2000-1500	Below 1500
G1 BC	0.457	0.440	0.434	0.421	0.409
G7 BC	0.217	0.218	0.218	0.224	0.230

Averages from 3000 fps to 1500 fps:
G1 BC: 0.431, variation of G1 BC: 0.037
G7 BC: 0.221, variation of G7 BC: 0.013
G7 form factor: 1.030

Rifling Twist Rate and Stability

Barrel Twist	Stability Factor Best case	Nominal	Worst case
1:6"	3.03	2.64	2.29
1:7"	2.22	1.94	1.68
1:8"	1.70	1.49	1.29
1:9"	1.35	1.18	1.02

Minimum twist required for stability is 1:8.9"
Recommended twist for optimal performance is 1:8.0" or faster

Sierra .224 caliber 90 grain MatchKing

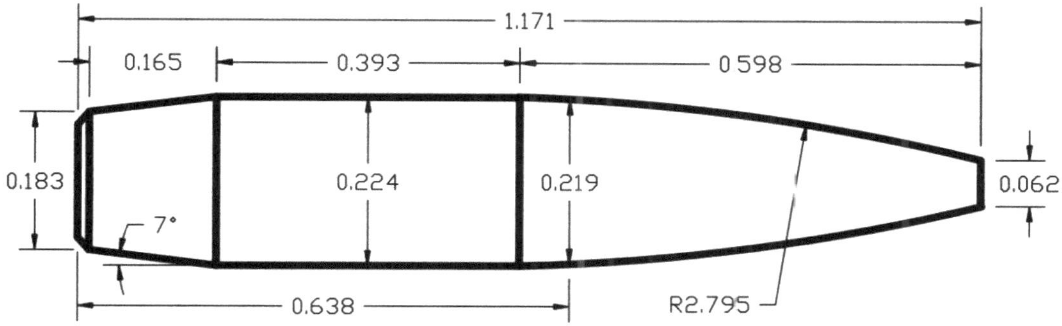

Lot #:	**Unknown**	**Ogive Radius:**	**12.48 calibers**
Sectional Density:	**0.256 lb/in^2**	**Rt/R:**	**0.80**

Ballistic Performance

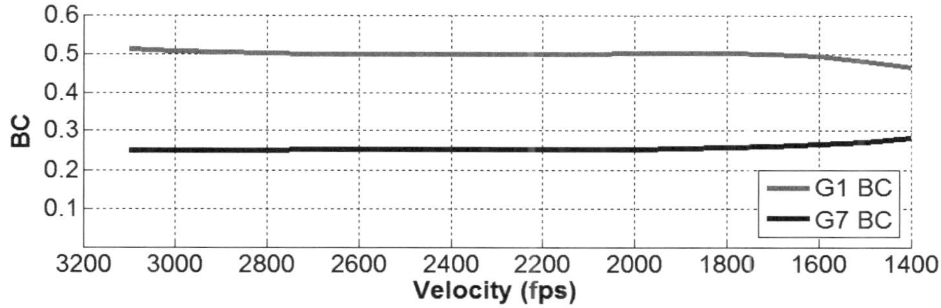

	Above 3000	3000-2500	2500-2000	2000-1500	Below 1500
G1 BC	0.521	0.505	0.502	0.493	0.483
G7 BC	0.247	0.251	0.252	0.263	0.273

Averages from 3000 fps to 1500 fps:
G1 BC: 0.499, variation of G1 BC: 0.027
G7 BC: 0.257, variation of G7 BC: 0.023
G7 form factor: 0.999

Rifling Twist Rate and Stability

Barrel Twist	Stability Factor Best Case	Nominal	Worst case
1:6"	2.59	2.26	1.95
1:7"	1.90	1.66	1.44
1:8"	1.46	1.27	1.10
1:9"	1.15	1.00	0.87

Minimum twist required for stability is 1:8.2"
Recommended twist for optimal performance is 1:7.4" or faster

Sierra .224 caliber 69 grain Tipped MatchKing

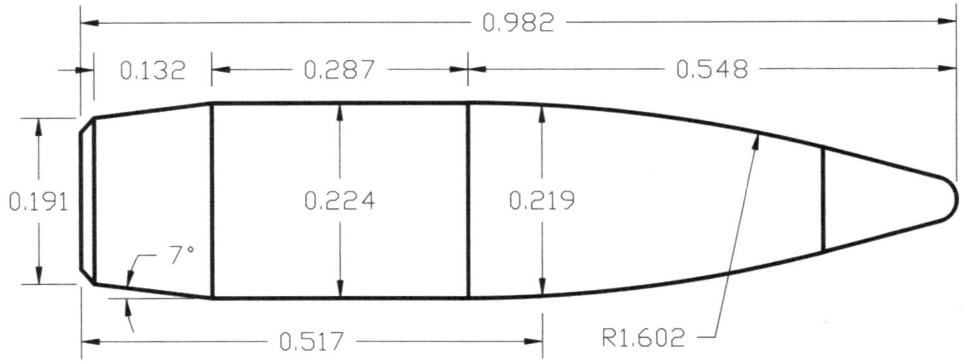

Lot #:	9886820	Ogive Radius:	7.15 calibers
Sectional Density:	0.197 lb/in²	Rt/R:	1.00

Ballistic Performance

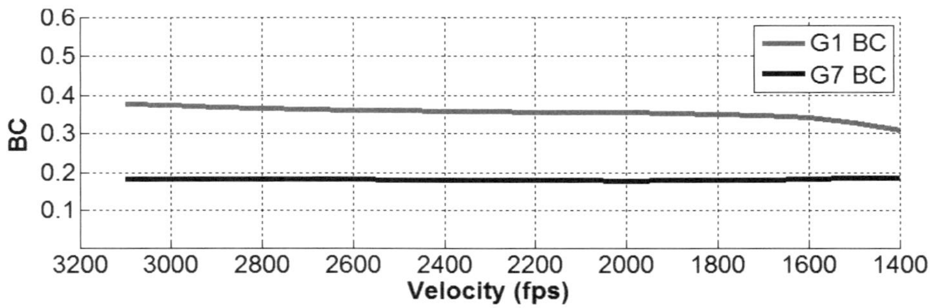

	Above 3000	3000-2500	2500-2000	2000-1500	Below 1500
G1 BC	0.385	0.367	0.357	0.341	0.327
G7 BC	0.183	0.182	0.180	0.181	0.184

Averages from 3000 fps to 1500 fps:
G1 BC: 0.354, variation of G1 BC: 0.047
G7 BC: 0.182, variation of G7 BC: 0.006
G7 form factor: 1.082

Rifling Twist Rate and Stability

Barrel Twist	Best case	Stability Factor Nominal	Worst case
1:7"	2.44	2.13	1.84
1:8"	1.86	1.63	1.41
1:9"	1.47	1.29	1.11
1:10"	1.19	1.04	0.90

Minimum twist required for stability is 1:9.3"
Recommended twist for optimal performance is 1:8.3" or faster

Sierra .224 caliber 77 grain Tipped MatchKing

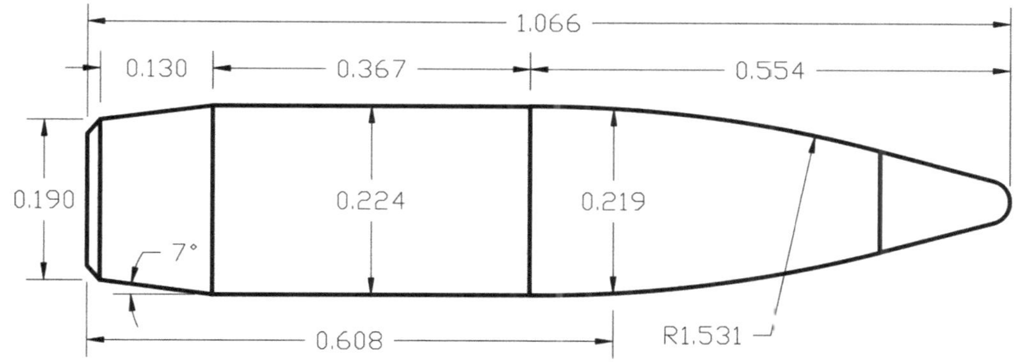

Lot #:	7258618761	Ogive Radius:	6.84 calibers
Sectional Density:	0.219 lb/in^2	Rt/R:	0.96

Ballistic Performance

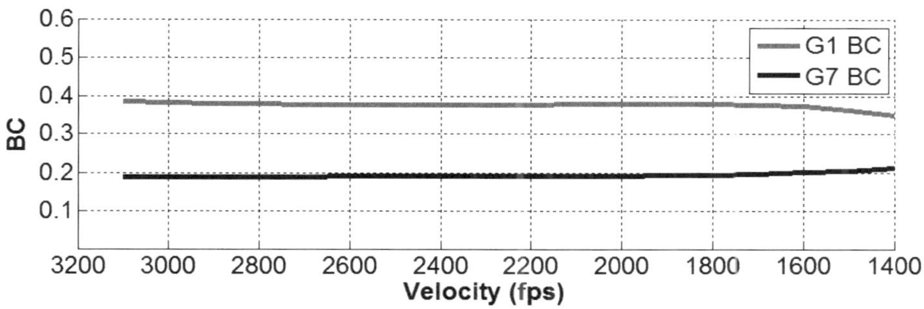

	Above 3000	3000-2500	2500-2000	2000-1500	Below 1500
G1 BC	0.426	0.407	0.398	0.381	0.366
G7 BC	0.202	0.202	0.200	0.202	0.206

Averages from 3000 fps to 1500 fps:
G1 BC: 0.394, variation of G1 BC: 0.048
G7 BC: 0.202, variation of G7 BC: 0.007
G7 form factor: 1.085

Rifling Twist Rate and Stability

Barrel Twist	Stability Factor Best case	Nominal	Worst case
1:7"	2.62	2.29	1.98
1:8"	2.01	1.75	1.52
1:9"	1.59	1.39	1.20
1:10"	1.29	1.12	0.97

Minimum twist required for stability is 1:9.7"
Recommended twist for optimal performance is 1:8.7" or faster

.243 Caliber (6mm)

.243 caliber bullets typically range from 70 to 115 grains. The image below relates bullet weight to sectional density for .243 caliber bullets.

Range of Weight and Sectional Density for 0.243 Caliber

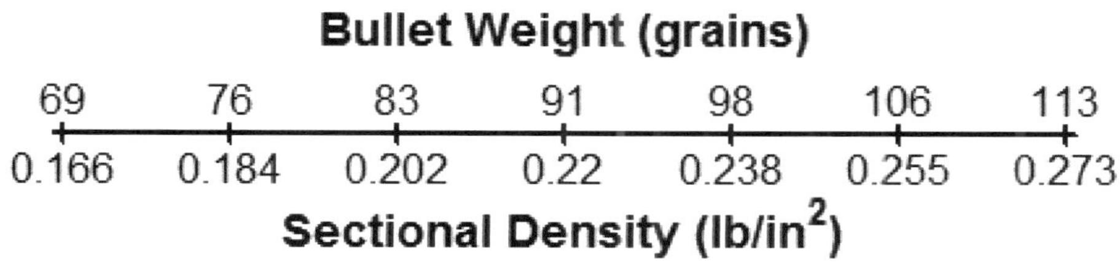

Sectional Density (SD) is a measure of how much mass is behind the bullets frontal area. Higher SD means a greater ability for the bullet to penetrate targets on impact, and also retain velocity in flight.

Spectrum of G7 form factors

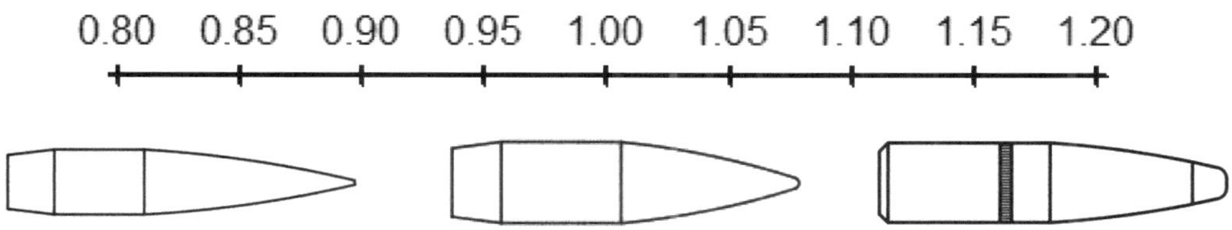

G7 Form factors range from 0.80 for the lowest drag shapes, up to 1.20+ for blunter shapes.

G7 Ballistic Coefficients are simply Sectional Density (SD) divided by G7 form factor.

You can determine a bullets' suitability for long range performance by seeing where it lies on the SD and form factor lines. Ideally, you want a bullet with a high SD and a low G7 form factor.

As an example, a 91 grain .243 bullet has an SD of 0.220. If that bullet had a G7 form factor of 0.95, it's G7 BC would be:

G7 BC = SD/FF = 0.220/0.95 = 0.232.

Barnes .243 caliber 80 grain TTSXBT

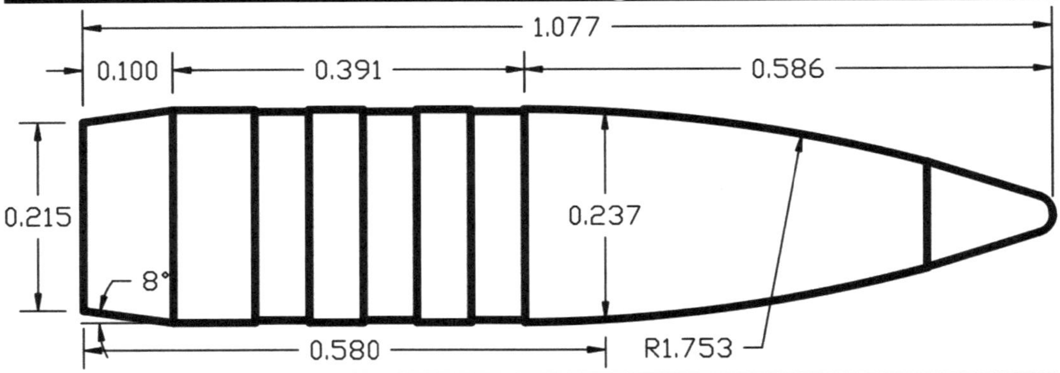

Lot #:	5523-1 063-N BS-M	**Ogive Radius:**	**7.21 calibers**
Sectional Density:	**0.194 lb/ in²**	**Rt/R:**	**1.00**

Ballistic Performance

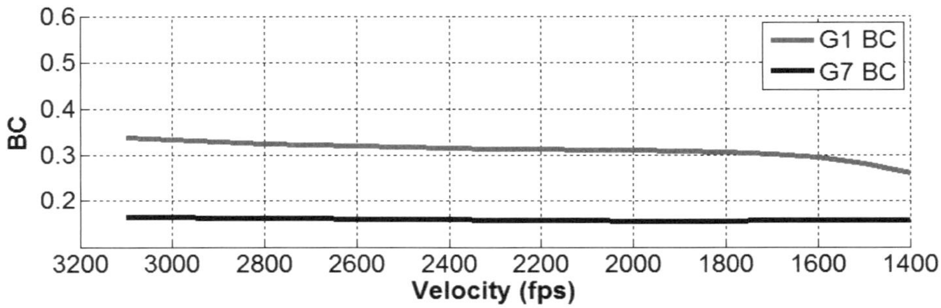

	Above 3000	3000-2500	2500-2000	2000-1500	Below 1500
G1 BC	0.346	0.326	0.314	0.296	0.282
G7 BC	0.164	0.162	0.158	0.157	0.158

Averages from 3000 fps to 1500 fps:
G1 BC: 0.311, variation of G1 BC: 0.053
G7 BC: 0.159, variation of G7 BC: 0.007
G7 form factor: 1.214

Rifling Twist Rate and Stability

Barrel Twist	Stability Factor Best case	Nominal	Worst case
1:6"	2.52	2.20	1.90
1:7"	1.93	1.69	1.46
1:8"	1.53	1.33	1.15
1:9"	1.24	1.08	0.93

Minimum twist required for stability is 1:9.5"
Recommended twist for optimal performance is 1:8.5" or faster

Barts .243 caliber 105 grain RBT

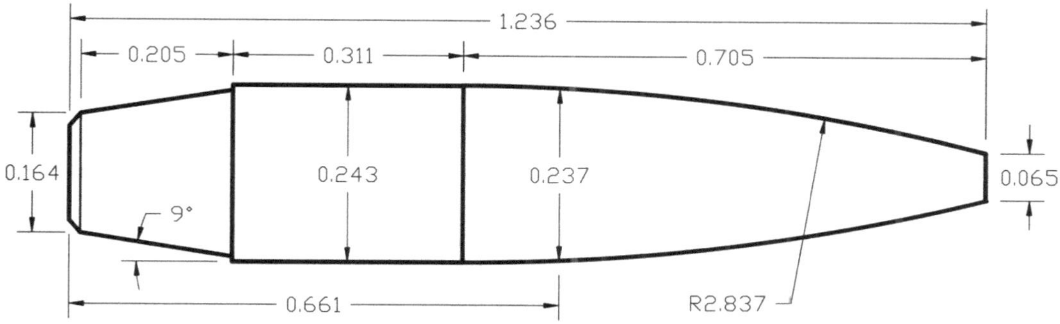

Lot #:	unknown	Ogive Radius:	11.68 calibers
Sectional Density:	0.254 lb/in^2	Rt/R:	1.0

Ballistic Performance

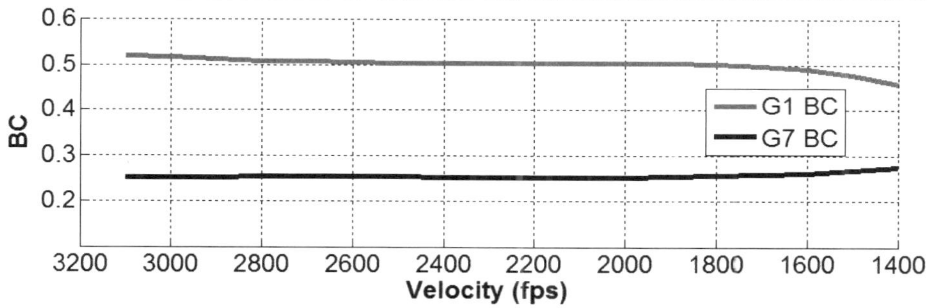

	Above 3000	3000-2500	2500-2000	2000-1500	Below 1500
G1 BC	0.528	0.511	0.504	0.490	0.477
G7 BC	0.250	0.253	0.254	0.261	0.269

Averages from 3000 fps to 1500 fps:
G1 BC: 0.501, variation of G1 BC: 0 040
G7 BC: 0.257, variation of G7 BC: 0.016
G7 form factor: 0.988

Rifling Twist Rate and Stability

Barrel Twist	Stability Factor Best case	Nominal	Worst case
1:6"	3.02	2.63	2.28
1:7"	2.22	1.94	1.67
1:8"	1.70	1.48	1.28
1:9"	1.34	1.17	1.01

Minimum twist required for stability is 1:8.9'
Recommended twist for optimal performance is 1:8.0" or faster

Berger .243 caliber 87 grain Hunting VLD

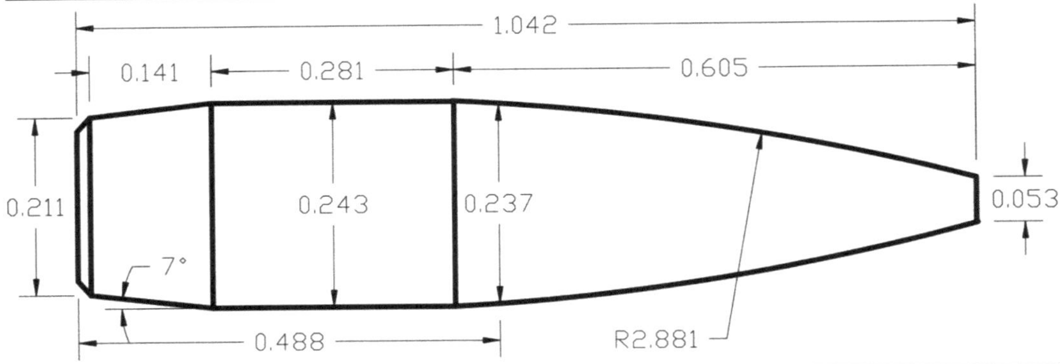

Lot #:	7600	Ogive Radius:	11.86 calibers
Sectional Density:	0.211 lb/in^2	Rt/R:	0.69

Ballistic Performance

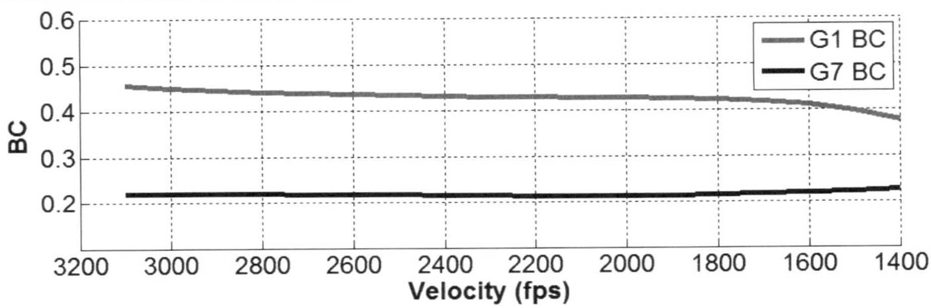

	Above 3000	3000-2500	2500-2000	2000-1500	Below 1500
G1 BC	0.464	0.442	0.430	0.412	0.397
G7 BC	0.220	0.219	0.216	0.219	0.223

Averages from 3000 fps to 1500 fps:
G1 BC: 0.427, variation of G1 BC: 0.054
G7 BC: 0.219, variation of G7 BC: 0.009
G7 form factor: 0.961

Rifling Twist Rate and Stability

Barrel Twist	Stability Factor Best case	Stability Factor Nominal	Stability Factor Worst case
1:8"	2.31	2.02	1.75
1:9"	1.83	1.60	1.38
1:10"	1.48	1.29	1.12
1:11"	1.22	1.07	0.92

Minimum twist required for stability is 1:10.4"
Recommended twist for optimal performance is 1:9.3" or faster

Berger .243 caliber 95 grain Classic Hunter

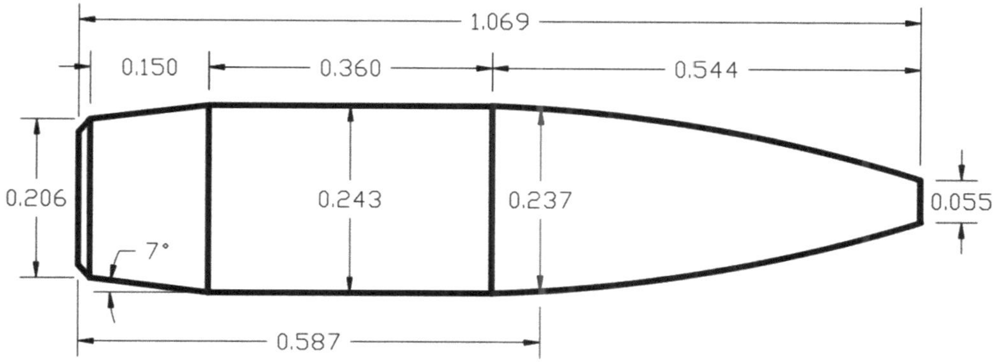

Lot #:	6805	Ogive Radius:	Hybrid
Sectional Density:	0.230 lb/in^2	Rt/R:	Hybrid

Ballistic Performance

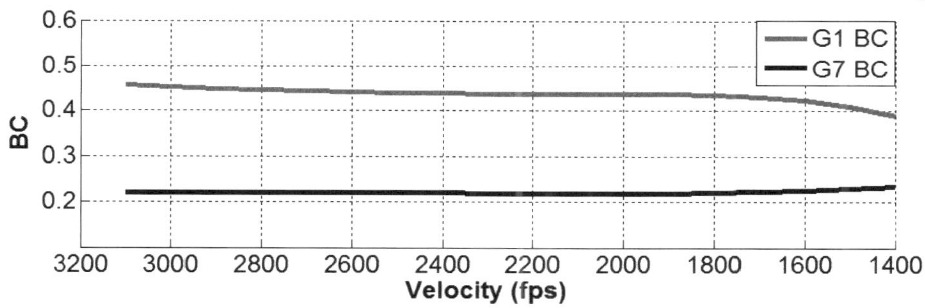

	Above 3000	3000-2500	2500-2000	2000-1500	Below 1500
G1 BC	0.464	0.446	0.437	0.422	0.408
G7 BC	0.220	0.221	0.220	0.224	0.230

Averages from 3000 fps to 1500 fps:
G1 BC: 0.434, variation of G1 BC: 0.045
G7 BC: 0.223, variation of G7 BC: 0.011
G7 form factor: 1.033

Rifling Twist Rate and Stability

	Stability Factor		
Barrel Twist	**Best case**	**Nominal**	**Worst case**
1:8"	2.34	2.05	1.77
1:9"	1.85	1.62	1.40
1:10"	1.50	1.31	1.13
1:11"	1.24	1.08	0.94

Minimum twist required for stability is 1:10.4"
Recommended twist for optimal performance is 1:9.3" or faster

Berger .243 caliber 95 grain Hunting VLD

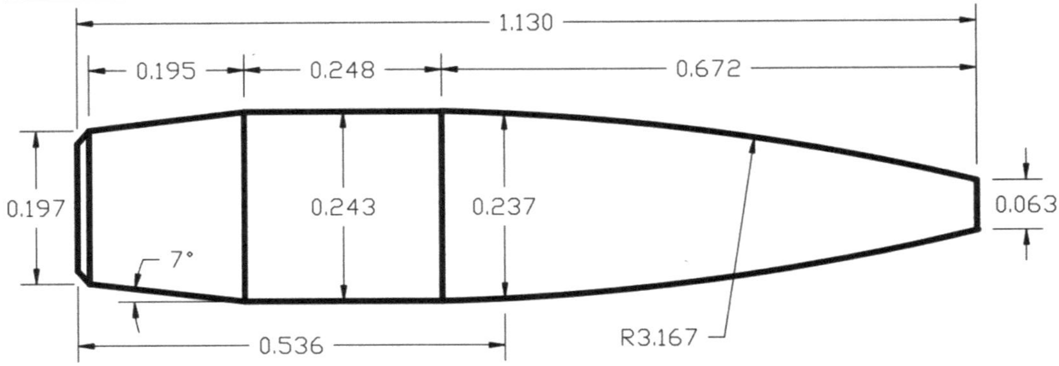

Lot #:	5564	Ogive Radius:	13.07 calibers
Sectional Density:	0.230 lb/in²	Rt/R:	0.80

Ballistic Performance

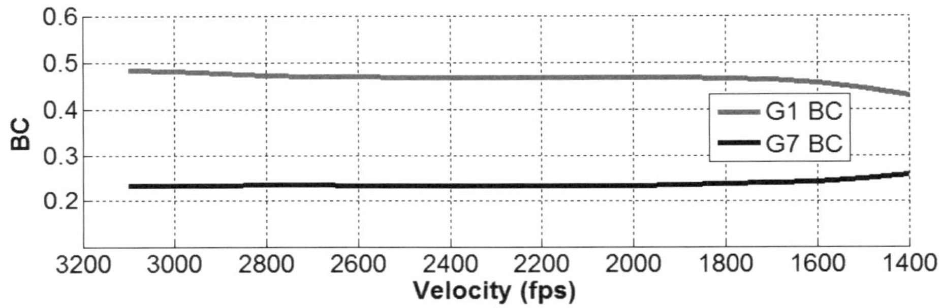

	Above 3000	3000-2500	2500-2000	2000-1500	Below 1500
G1 BC	0.492	0.474	0.468	0.456	0.445
G7 BC	0.233	0.235	0.235	0.243	0.251

Averages from 3000 fps to 1500 fps:
G1 BC: 0.465, variation of G1 BC: 0.035
G7 BC: 0.239, variation of G7 BC: 0.016
G7 form factor: 0.962

Rifling Twist Rate and Stability

Barrel Twist	Stability Factor Best case	Nominal	Worst case
1:7"	2.60	2.28	1.97
1:8"	1.99	1.74	1.51
1:9"	1.58	1.38	1.19
1:10"	1.28	1.11	0.96

Minimum twist required for stability is 1:9.6"
Recommended twist for optimal performance is 1:8.6" or faster

Berger .243 caliber 105 grain Hunting VLD

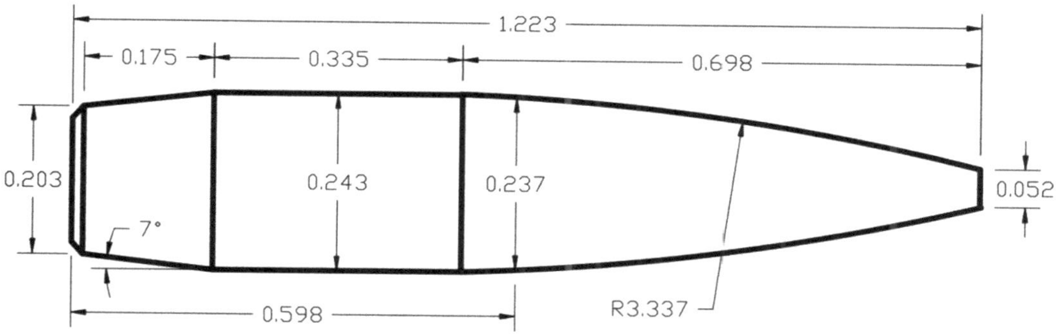

Lot #:	5744	Ogive Radius:	13.73 calibers
Sectional Density:	0.254 lb/in²	Rt/R:	0.78

Ballistic Performance

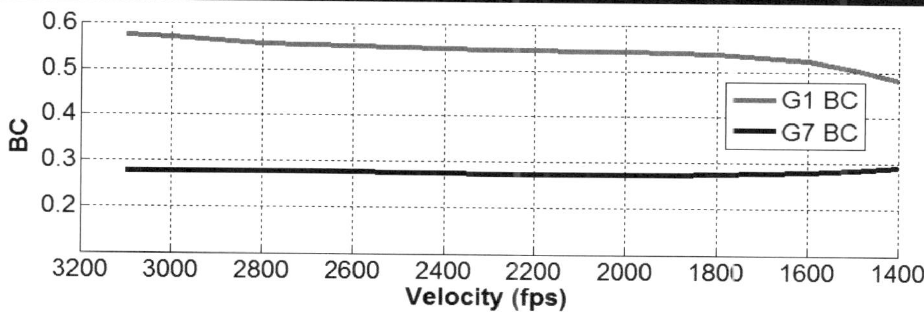

	Above 3000	3000-2500	2500-2000	2000-1500	Below 1500
G1 BC	0.586	0.558	0.545	0.523	0.505
G7 BC	0.278	0.277	0.274	0.278	0.285

Averages from 3000 fps to 1500 fps:

G1 BC: 0.541, variation of G1 BC: 0.064

G7 BC: 0.278, variation of G7 BC: 0.013

G7 form factor: 0.915

Rifling Twist Rate and Stability

Barrel Twist	Stability Factor Best case	Nominal	Worst case
1:7"	2.29	2.00	1.73
1:8"	1.75	1.53	1.32
1:9"	1.38	1.21	1.04
1:10"	1.12	0.98	0.85

Minimum twist required for stability is 1:9.0"

Recommended twist for optimal performance is 1:8.1" or faster

Berger .243 caliber 62 grain Target EUWIN

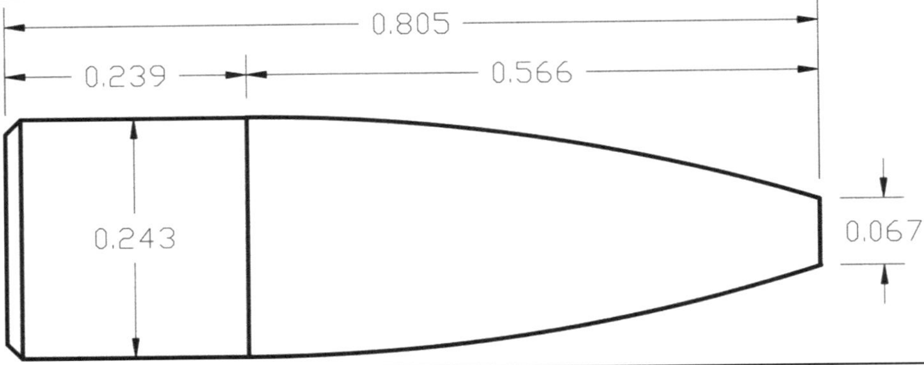

Lot #:	5807	Ogive Radius:	Hybrid
Sectional Density:	0.150 lb/in^2	Rt/R:	Hybrid

Ballistic Performance

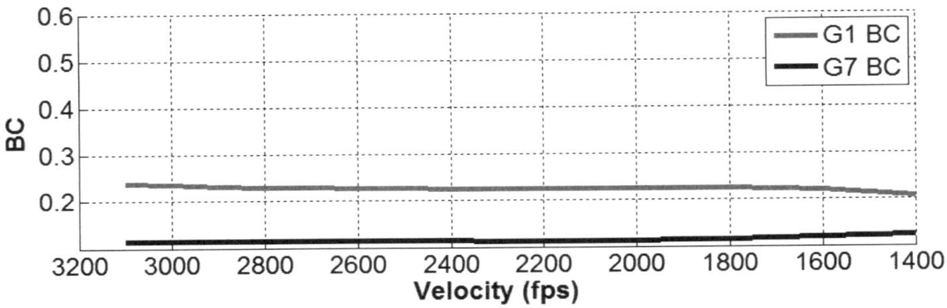

	Above 3000	3000-2500	2500-2000	2000-1500	Below 1500
G1 BC	0.243	0.231	0.226	0.220	0.215
G7 BC	0.115	0.115	0.114	0.117	0.121

Averages from 3000 fps to 1500 fps:
G1 BC: 0.226, variation of G1 BC: 0.021
G7 BC: 0.116, variation of G7 BC: 0.008
G7 form factor: 1.293

Rifling Twist Rate and Stability

Barrel Twist	Stability Factor Best case	Nominal	Worst case
1:10"	2.21	1.93	1.67
1:11"	1.83	1.59	1.38
1:12"	1.53	1.34	1.16
1:13"	1.31	1.14	0.99

Minimum twist required for stability is 1:12.7"
Recommended twist for optimal performance is 1:11.3" or faster

Berger .243 caliber 64.25 grain Target Column

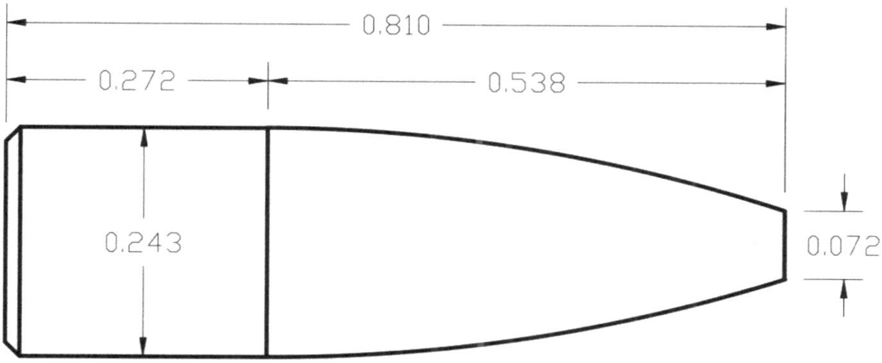

Lot #:	6650	Ogive Radius:	Hybrid
Sectional Density:	0.155 lb/in^2	Rt/R:	Hybrid

Ballistic Performance

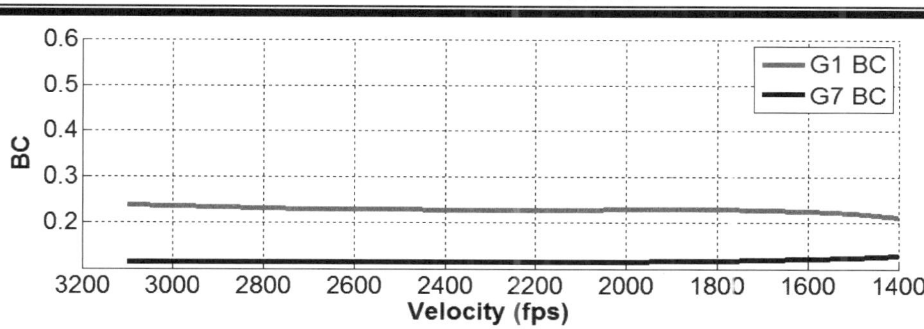

	Above 3000	3000-2500	2500-2000	2000-1500	Below 1500
G1 BC	0.243	0.233	0.229	0.225	0.221
G7 BC	0.115	0.115	0.115	0.120	0.125

Averages from 3000 fps to 1500 fps:
G1 BC: 0.229, variation of G1 BC: 0.016
G7 BC: 0.118, variation of G7 BC: 0.009
G7 form factor: 1.318

Rifling Twist Rate and Stability

Barrel Twist	Stability Factor Best case	Nominal	Worst case
1:10"	2.24	1.96	1.69
1:11"	1.85	1.62	1.40
1:12"	1.56	1.36	1.18
1:13"	1.33	1.16	1.00

Minimum twist required for stability is 1:12.8"
Recommended twist for optimal performance is 1:11.4" or faster

Berger .243 caliber 65 grain Target BT

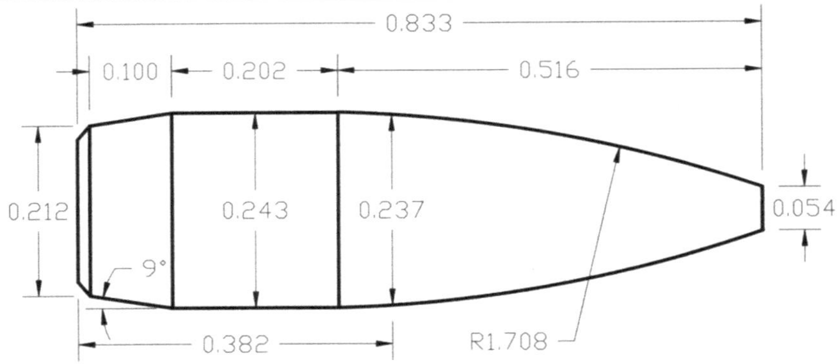

Lot #:	**5762**		**Ogive Radius:**	**7.03 calibers**
Sectional Density:	**0.246 lb/in^2**		**Rt/R:**	**0.85**

Ballistic Performance

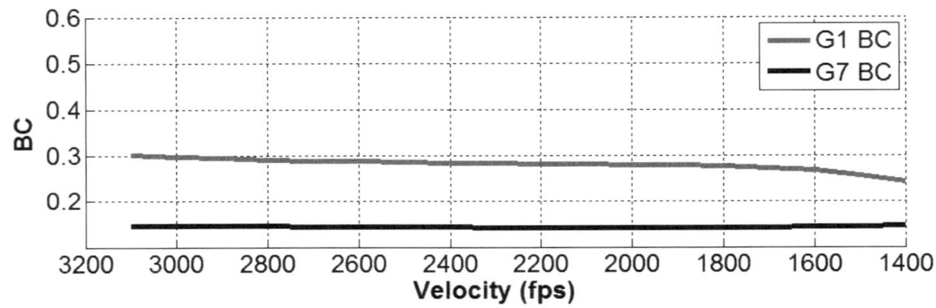

	Above 3000	3000-2500	2500-2000	2000-1500	Below 1500
G1 BC	0.263	0.251	0.245	0.240	0.235
G7 BC	0.124	0.124	0.123	0.128	0.133

Averages from 3000 fps to 1500 fps:
G1 BC: 0.245, variation of G1 BC: 0.020
G7 BC: 0.126, variation of G7 BC: 0.010
G7 form factor: 1.248

Rifling Twist Rate and Stability

	Stability Factor		
Barrel Twist	**Best case**	**Nominal**	**Worst case**
1:10"	2.10	1.84	1.59
1:11"	1.74	1.52	1.31
1:12"	1.46	1.27	1.10
1:13"	1.24	1.09	0.94

Minimum twist required for stability is 1:12.4"
Recommended twist for optimal performance is 1:11.1" or faster

Berger .243 caliber 65 grain Target WEB BR

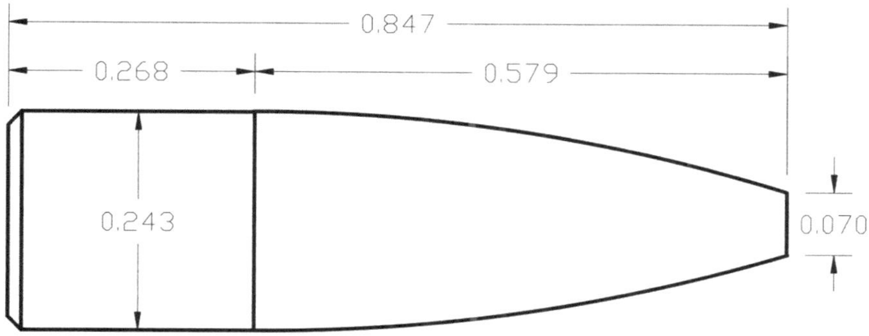

Lot #:	2525	Ogive Radius:	Hybrid
Sectional Density:	0.246 lb/in^2	Rt/R:	Hybrid

Ballistic Performance

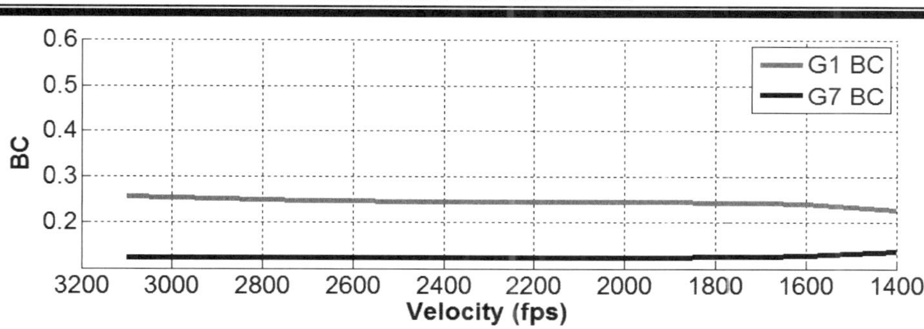

	Above 3000	3000-2500	2500-2000	2000-1500	Below 1500
G1 BC	0.263	0.251	0.245	0.240	0.235
G7 BC	0.124	0.124	0.123	0.128	0.133

Averages from 3000 fps to 1500 fps:
G1 BC: 0.245, variation of G1 BC: 0.020
G7 BC: 0.126, variation of G7 BC: 0.010
G7 form factor: 1.248

Rifling Twist Rate and Stability

Barrel Twist	Stability Factor Best case	Nominal	Worst case
1:9"	2.47	2.16	1.87
1:10"	2.00	1.75	1.51
1:11"	1.66	1.45	1.25
1:12"	1.39	1.22	1.05

Minimum twist required for stability is 1:12.1"
Recommended twist for optimal performance is 1:10.8" or faster

Berger .243 caliber 68 grain Target FB

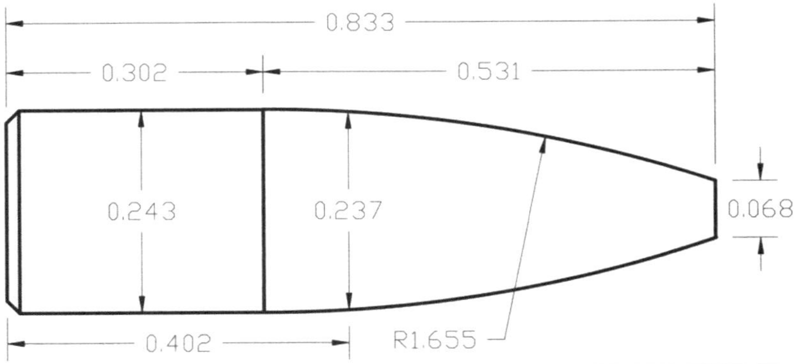

Lot #:	**7630**	**Ogive Radius:**	**6.81**
Sectional Density:	**0.165 lb/in^2**	**Rt/R:**	**1.00**

Ballistic Performance

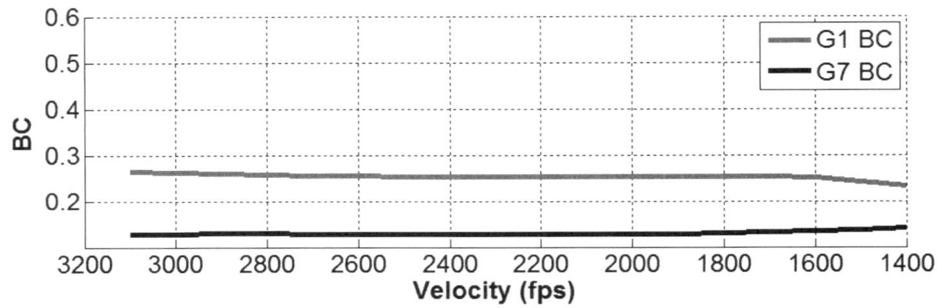

	Above 3000	3000-2500	2500-2000	2000-1500	Below 1500
G1 BC	0.272	0.260	0.256	0.250	0.245
G7 BC	0.129	0.129	0.128	0.133	0.138

Averages from 3000 fps to 1500 fps:
G1 BC: 0.255, variation of G1 BC: 0.020
G7 BC: 0.131, variation of G7 BC: 0.010
G7 form factor: 1.256

Rifling Twist Rate and Stability

Barrel Twist	Stability Factor Best case	Nominal	Worst case
1:10"	2.20	1.92	1.66
1:11"	1.82	1.59	1.37
1:12"	1.53	1.33	1.15
1:13"	1.30	1.14	0.98

Minimum twist required for stability is 1:12.7"
Recommended twist for optimal performance is 1:11.3" or faster

Berger .243 caliber 90 grain Target BT

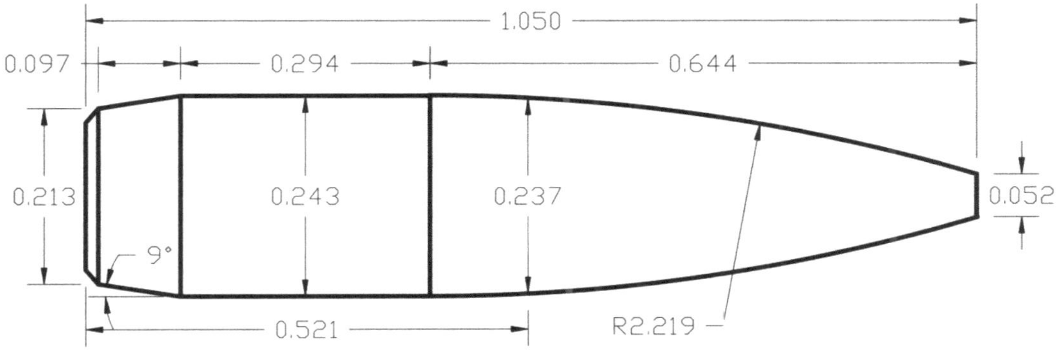

	Lot #:	3778	Ogive Radius:	9.13 calibers
	Sectional Density:	0.218 lb/in^2	Rt/R:	1.00

Ballistic Performance

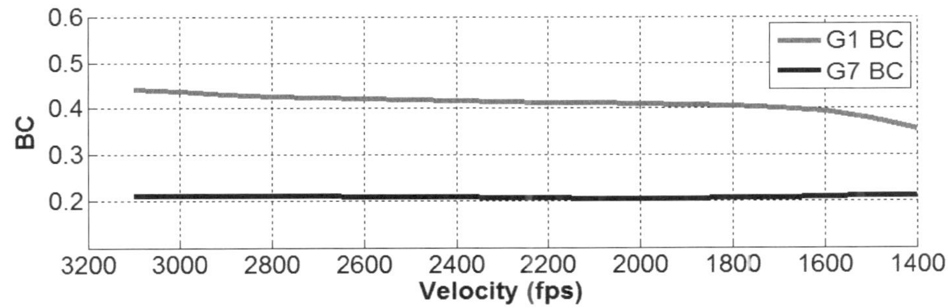

	Above 3000	3000-2500	2500-2000	2000-1500	Below 1500
G1 BC	0.449	0.426	0.413	0.393	0.377
G7 BC	0.213	0.211	0.208	0.209	0.212

Averages from 3000 fps to 1500 fps:
G1 BC: 0.409, variation of G1 BC: 0.059
G7 BC: 0.210, variation of G7 BC: 0.007
G7 form factor: 1.037

Rifling Twist Rate and Stability

Barrel Twist	Stability Factor Best case	Nominal	Worst case
1:8"	2.34	2.04	1.77
1:9"	1.85	1.61	1.40
1:10"	1.50	1.31	1.13
1:11"	1.24	1.08	0.93

Minimum twist required for stability is 1:10.4"
Recommended twist for optimal performance is 1:9.3" or faster

Berger .243 caliber 95 grain Target VLD

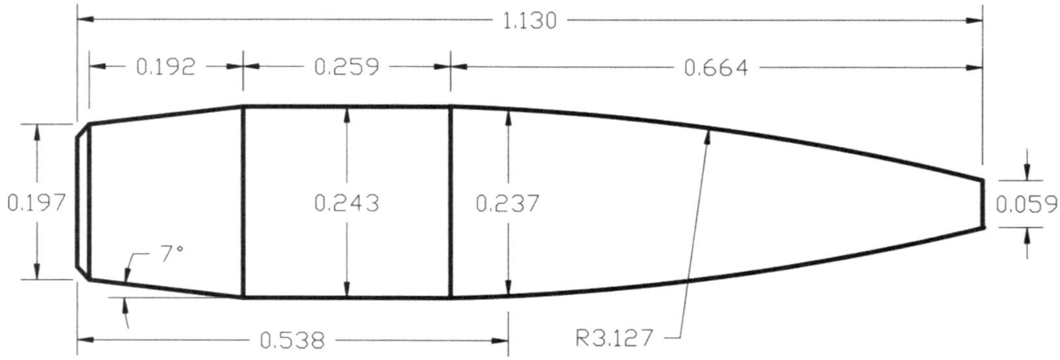

Lot #:	**3793**	**Ogive Radius:**	**12.87 calibers**
Sectional Density:	**0.230 lb/in²**	**Rt/R:**	**0.78**

Ballistic Performance

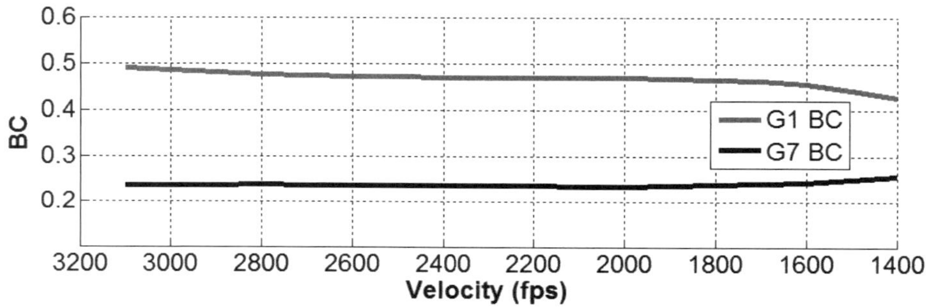

	Above 3000	3000-2500	2500-2000	2000-1500	Below 1500
G1 BC	0.499	0.479	0.470	0.456	0.443
G7 BC	0.236	0.237	0.236	0.243	0.250

Averages from 3000 fps to 1500 fps:
G1 BC: 0.467, variation of G1 BC: 0.043
G7 BC: 0.240, variation of G7 BC: 0.014
G7 form factor: 0.958

Rifling Twist Rate and Stability

Barrel Twist	Stability Factor: Best case	Stability Factor: Nominal	Stability Factor: Worst case
1:7"	2.60	2.28	1.97
1:8"	1.99	1.74	1.51
1:9"	1.58	1.38	1.19
1:10"	1.28	1.11	0.96

Minimum twist required for stability is 1:9.6"
Recommended twist for optimal performance is 1:8.6" or faster

Berger .243 caliber 105 grain Target Hybrid

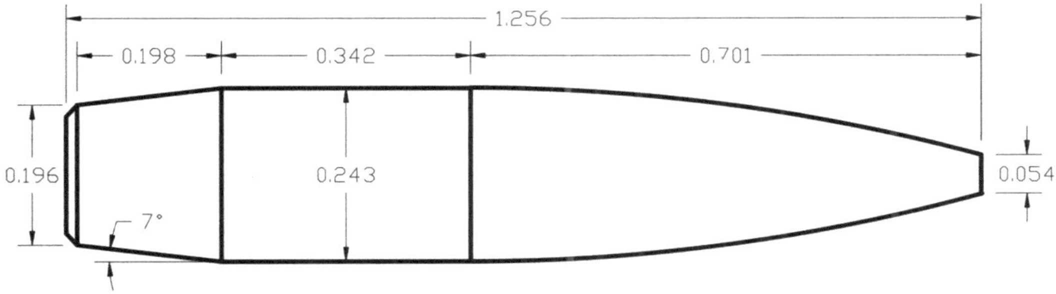

Lot #:	6836	Ogive Radius:	Hybrid
Sectional Density:	0.254 lb/in²	Rt/R:	Hybrid

Ballistic Performance

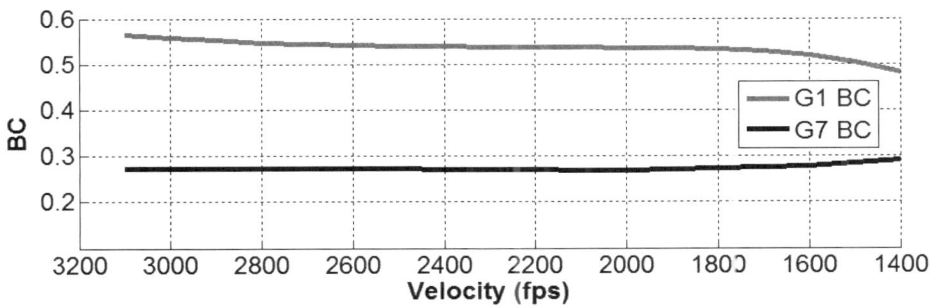

	Above 3000	3000-2500	2500-2000	2000-1500	Below 1500
G1 BC	0.574	0.549	0.539	0.520	0.504
G7 BC	0.272	0.273	0.271	0.277	0.284

Averages from 3000 fps to 1500 fps:
G1 BC: 0.535, variation of G1 BC: 0.055
G7 BC: 0.275, variation of G7 BC: 0.014
G7 form factor: 0.925

Rifling Twist Rate and Stability

Barrel Twist	Stability Factor Best case	Nominal	Worst case
1:6"	2.88	2.51	2.17
1:7"	2.11	1.85	1.60
1:8"	1.62	1.41	1.22
1:9"	1.28	1.12	0.97

Minimum twist required for stability is 1:8.7"
Recommended twist for optimal performance is 1:7.8" or faster

Berger .243 caliber 105 grain Target Boat Tail

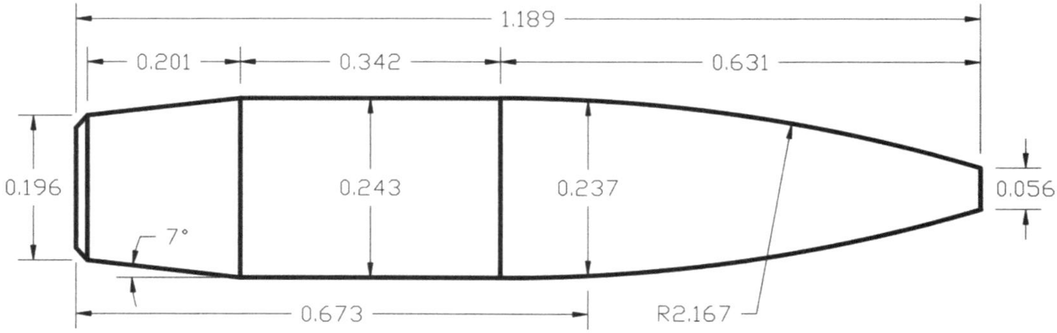

Lot #:	7649	Ogive Radius:	8.92 calibers
Sectional Density:	0.254 lb/in^2	Rt/R:	1.00

Ballistic Performance

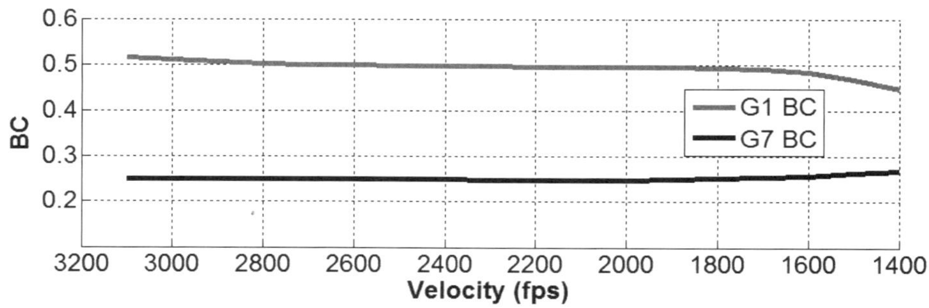

	Above 3000	3000-2500	2500-2000	2000-1500	Below 1500
G1 BC	0.523	0.504	0.496	0.481	0.467
G7 BC	0.248	0.250	0.250	0.256	0.263

Averages from 3000 fps to 1500 fps:
G1 BC: 0.493, variation of G1 BC: 0.044
G7 BC: 0.253, variation of G7 BC: 0.014
G7 form factor: 1.004

Rifling Twist Rate and Stability

Barrel Twist	Stability Factor Best case	Nominal	Worst case
1:7"	2.48	2.17	1.87
1:8"	1.90	1.66	1.44
1:9"	1.50	1.31	1.13
1:10"	1.22	1.06	0.92

Minimum twist required for stability is 1:9.4"
Recommended twist for optimal performance is 1:8.4" or faster

Berger .243 caliber 105 grain Target VLD

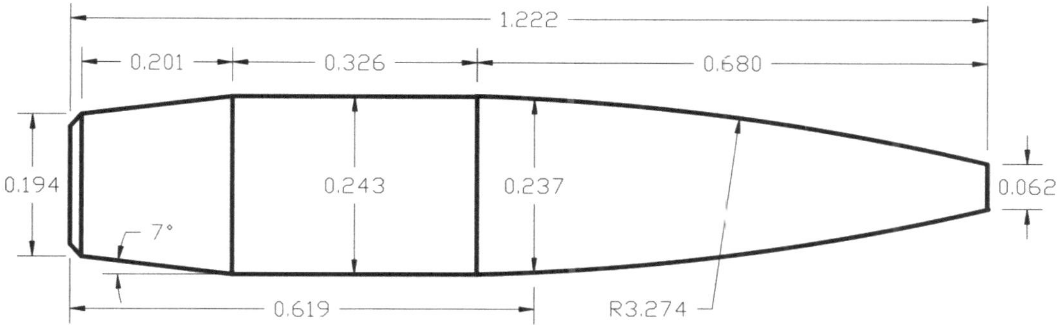

Lot #:	5625	Ogive Radius:	13.47 calibers
Sectional Density:	0.254 lb/in^2	Rt/R:	0.79

Ballistic Performance

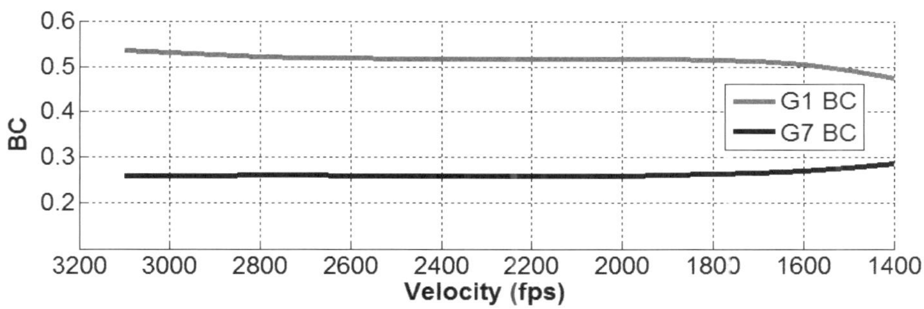

	Above 3000	3000-2500	2500-2000	2000-1500	Below 1500
G1 BC	0.544	0.524	0.518	0.506	0.493
G7 BC	0.258	0.260	0.260	0.269	0.278

Averages from 3000 fps to 1500 fps:
G1 BC: 0.515, variation of G1 BC: 0.038
G7 BC: 0.265, variation of G7 BC: 0.019
G7 form factor: 0.961

Rifling Twist Rate and Stability

Barrel Twist	Stability Factor Best case	Nominal	Worst case
1:7"	2.29	2.00	1.73
1:8"	1.75	1.53	1.32
1:9"	1.39	1.21	1.05
1:10"	1.12	0.98	0.85

Minimum twist required for stability is 1:9.0"
Recommended twist for optimal performance is 1:8.1" or faster

Berger .243 caliber 108 grain Target Boat Tail

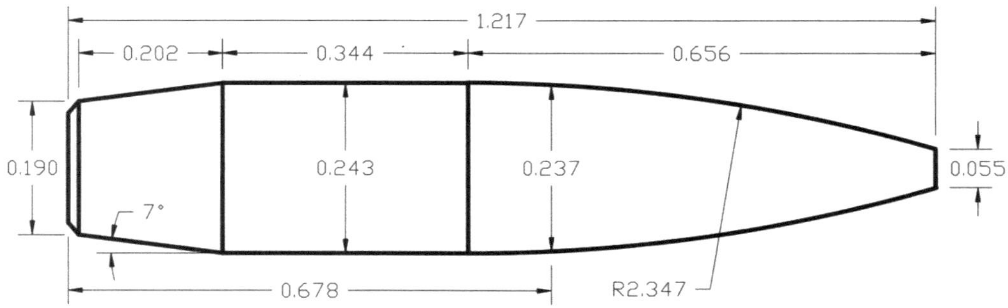

Lot #:	5638	Ogive Radius:	9.66 calibers
Sectional Density:	0.261 lb/in²	Rt/R:	0.99

Ballistic Performance

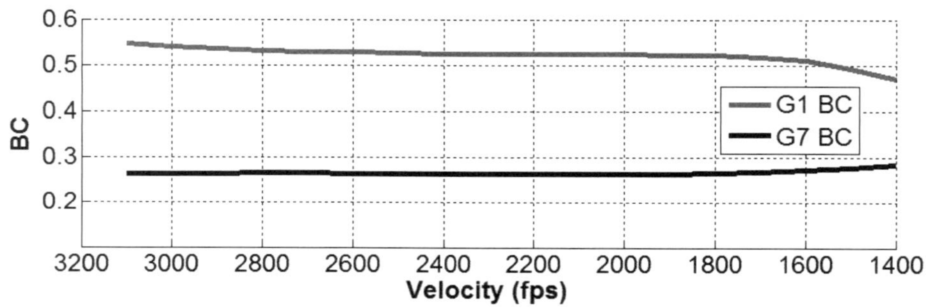

	Above 3000	3000-2500	2500-2000	2000-1500	Below 1500
G1 BC	0.555	0.533	0.525	0.508	0.493
G7 BC	0.263	0.265	0.264	0.270	0.278

Averages from 3000 fps to 1500 fps:
G1 BC: 0.521, variation of G1 BC: 0.048
G7 BC: 0.268, variation of G7 BC: 0.015
G7 form factor: 0.977

Rifling Twist Rate and Stability

Barrel Twist	Stability Factor Best case	Nominal	Worst case
1:7"	2.38	2.08	1.80
1:8"	1.83	1.59	1.38
1:9"	1.44	1.26	1.09
1:10"	1.17	1.02	0.88

Minimum twist required for stability is 1:9.2"
Recommended twist for optimal performance is 1:8.2" or faster

Berger .243 caliber 115 grain Target VLD

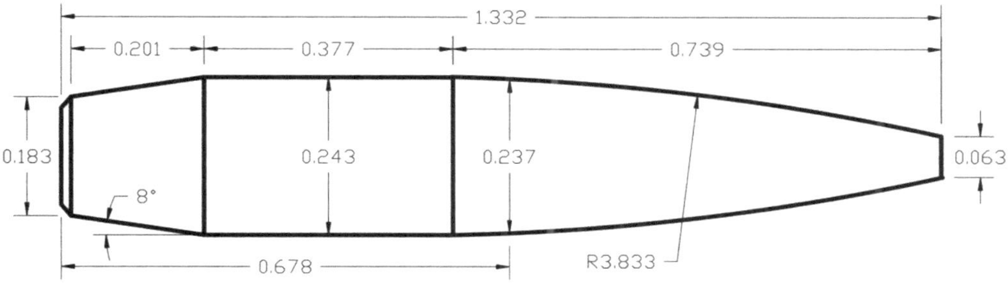

Lot #:	**2877**	**Ogive Radius:**	**15.77 calibers**
Sectional Density:	**0.278 lb/in^2**	**Rt/R:**	**0.80**

Ballistic Performance

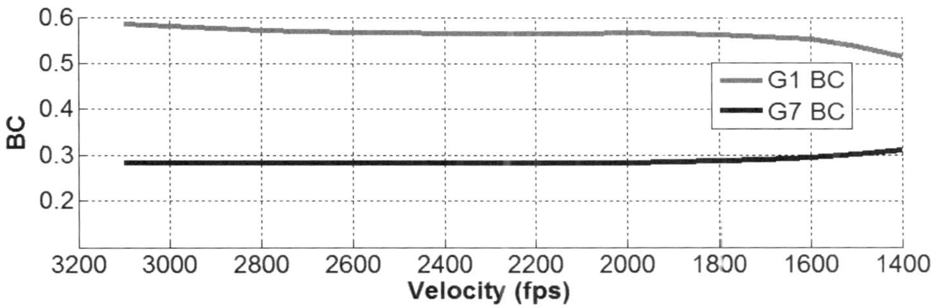

	Above 3000	3000-2500	2500-2000	2000-1500	Below 1500
G1 BC	0.595	0.574	0.566	0.552	0.537
G7 BC	0.282	0.285	0.285	0.294	0.303

Averages from 3000 fps to 1500 fps:
G1 BC: 0.563, variation of G1 BC: 0.044
G7 BC: 0.289, variation of G7 BC: 0.019
G7 form factor: 0.963

Rifling Twist Rate and Stability

Barrel Twist	Stability Factor Best case	Nominal	Worst case
1:6"	2.65	2.32	2.00
1:7"	1.95	1.70	1.47
1:8"	1.49	1.30	1.13
1:9"	1.18	1.03	0.89

Minimum twist required for stability is 1:8.3"
Recommended twist for optimal performance is 1:7.5" or faster

Berger .243 caliber 69 grain Varmint High BC FB

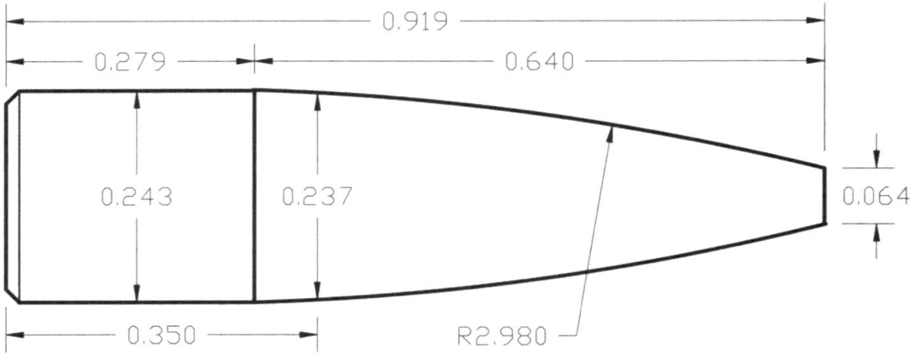

Lot #:	7424	Ogive Radius:	12.26 calibers
Sectional Density:	0.167 lb/in^2	Rt/R:	0.78

Ballistic Performance

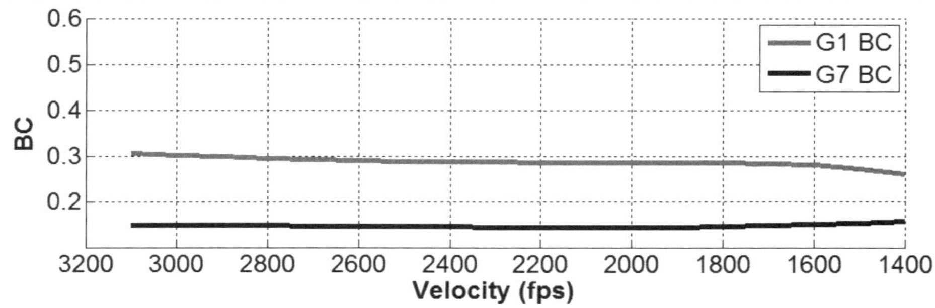

	Above 3000	3000-2500	2500-2000	2000-1500	Below 1500
G1 BC	0.313	0.297	0.288	0.280	0.273
G7 BC	0.148	0.147	0.145	0.149	0.154

Averages from 3000 fps to 1500 fps:
G1 BC: 0.288, variation of G1 BC: 0.030
G7 BC: 0.148, variation of G7 BC: 0.010
G7 form factor: 1.128

Rifling Twist Rate and Stability

Barrel Twist	Stability Factor Best case	Nominal	Worst case
1:8"	2.63	2.30	1.99
1:9"	2.08	1.82	1.57
1:10"	1.68	1.47	1.27
1:11"	1.39	1.22	1.05

Minimum twist required for stability is 1:11.1"
Recommended twist for optimal performance is 1:9.9" or faster

Berger .243 caliber 80 grain Varmint FB

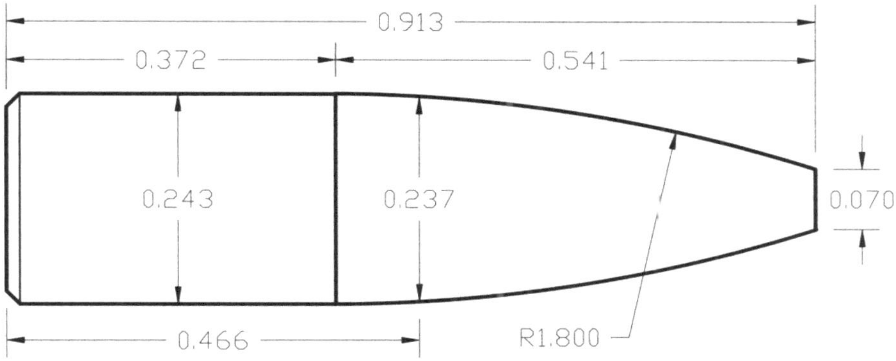

Lot #:	7417	Ogive Radius:	7.41 calibers
Sectional Density:	0.194 lb/in^2	Rt/R:	0.96

Ballistic Performance

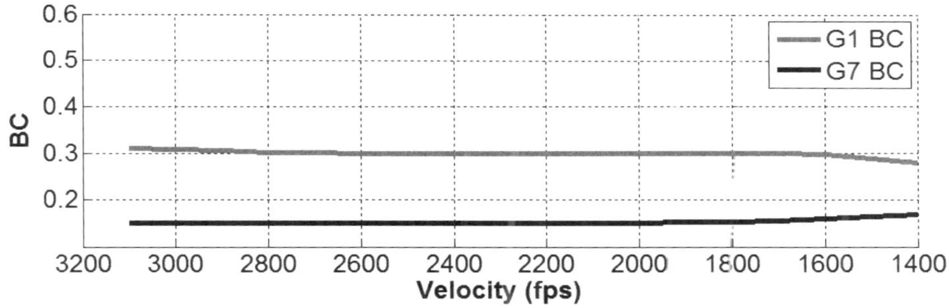

	Above 3000	3000-2500	2500-2000	2000-1500	Below 1500
G1 BC	0.318	0.305	0.300	0.295	0.290
G7 BC	0.151	0.151	0.151	0.157	0.163

Averages from 3000 fps to 1500 fps:
G1 BC: 0.300, variation of G1 BC: 0.020
G7 BC: 0.154, variation of G7 BC: 0.013
G7 form factor: 1.257

Rifling Twist Rate and Stability

Barrel Twist	Best case	Stability Factor Nominal	Worst case
1:9"	2.46	2.15	1.86
1:10"	1.99	1.74	1.50
1:11"	1.65	1.44	1.24
1:12"	1.38	1.21	1.04

Minimum twist required for stability is 1:12.0"
Recommended twist for optimal performance is 1:10.8" or faster

Berger .243 caliber 88 grain Varmint HBC FB

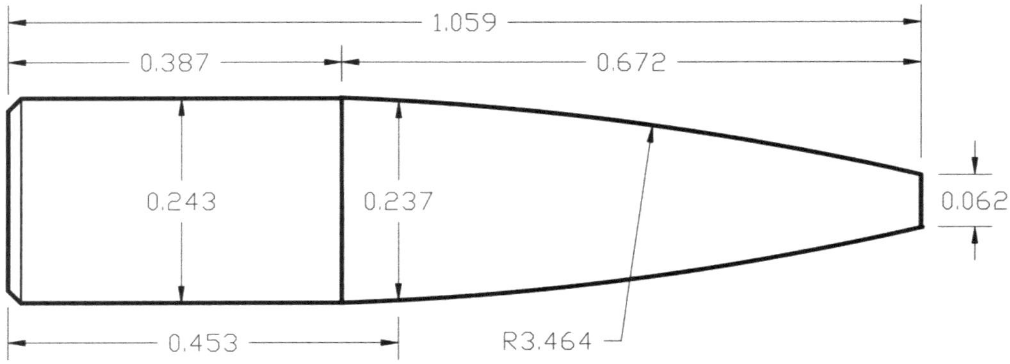

Lot #:	5565	Ogive Radius:	14.26 calibers
Sectional Density:	0.213 lb/in^2	Rt/R:	0.73

Ballistic Performance

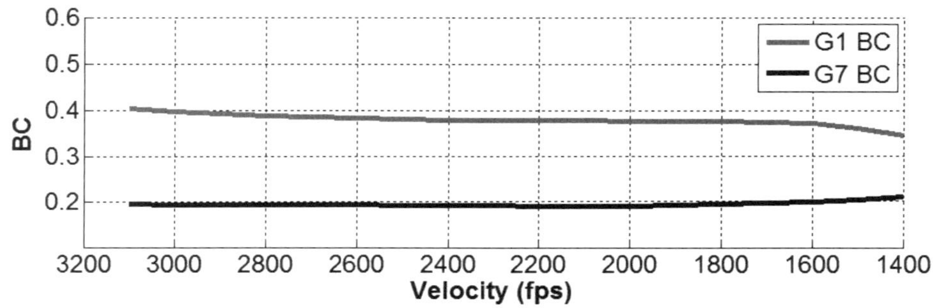

	Above 3000	3000-2500	2500-2000	2000-1500	Below 1500
G1 BC	0.412	0.390	0.380	0.370	0.361
G7 BC	0.195	0.193	0.191	0.197	0.204

Averages from 3000 fps to 1500 fps:
G1 BC: 0.380, variation of G1 BC: 0.038
G7 BC: 0.195, variation of G7 BC: 0.014
G7 form factor: 1.092

Rifling Twist Rate and Stability

Barrel Twist	Stability Factor Best case	Nominal	Worst case
1:8"	2.23	1.95	1.68
1:9"	1.76	1.54	1.33
1:10"	1.43	1.25	1.08
1:11"	1.18	1.03	0.89

Minimum twist required for stability is 1:10.2"
Recommended twist for optimal performance is 1:9.1" or faster

DTAC .243 caliber 117 grain BT

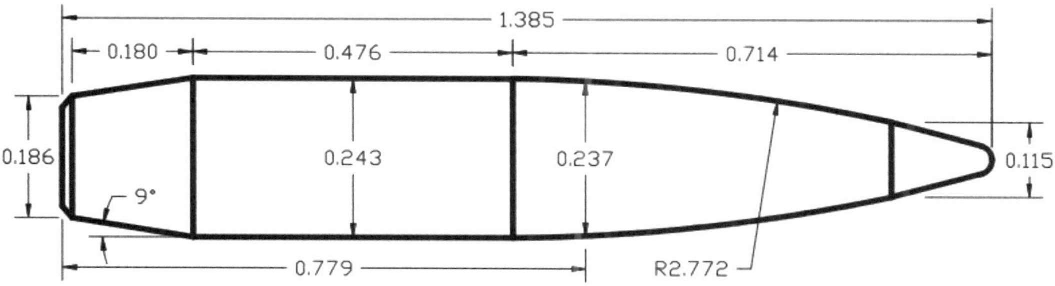

Lot #:	Unknown	Ogive Radius:	11.56 calibers
Sectional Density:	0.283 lb/in^2	Rt/R:	0.90

Ballistic Performance

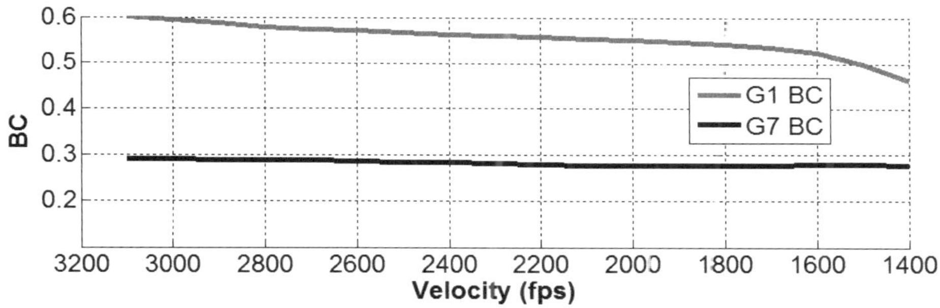

	Above 3000	3000-2500	2500-2000	2000-1500	Below 1500
G1 BC	0.614	0.580	0.559	0.525	0.499
G7 BC	0.291	0.288	0.281	0.279	0.280

Averages from 3000 fps to 1500 fps:
G1 BC: 0.553, variation of G1 BC: 0.095
G7 BC: 0.283, variation of G7 BC: 0.013
G7 form factor: 1.000

Rifling Twist Rate and Stability

Barrel Twist	Best case	Stability Factor Nominal	Worst case
1:7"	1.77	1.54	1.34
1:8"	1.35	1.18	1.02
1:9"	1.07	0.93	0.81
1:10"	0.87	0.76	0.65

Minimum twist required for stability is 1:7.9"
Recommended twist for optimal performance is 1:7.1" or faster

DTAC .243 caliber 115 grain BTHP

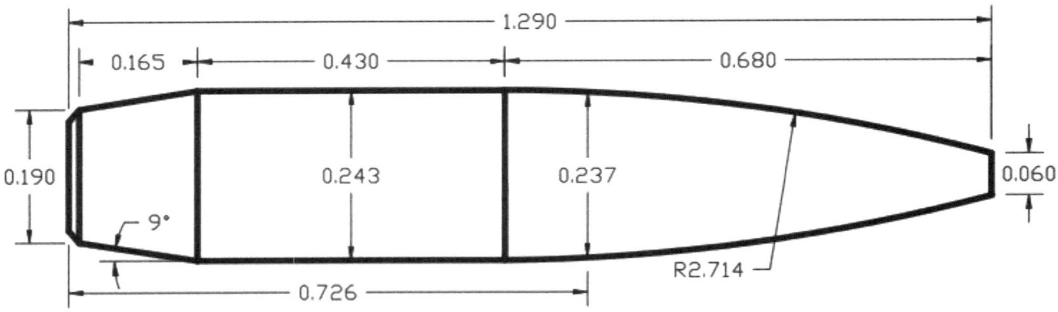

Lot #:	Unknown	Ogive Radius:	11.30 calibers
Sectional Density:	0.278 lb/in^2	Rt/R:	0.91

Ballistic Performance

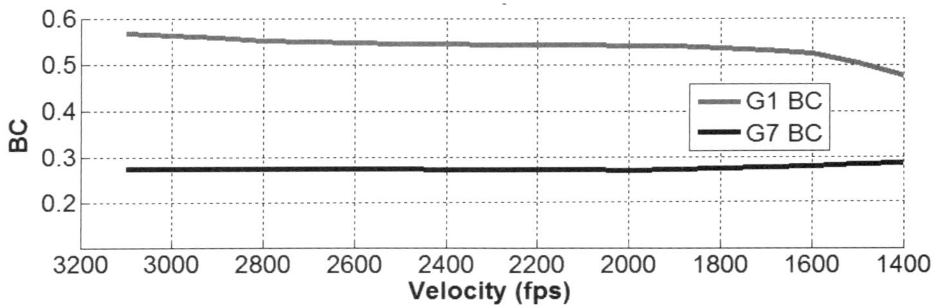

	Above 3000	3000-2500	2500-2000	2000-1500	Below 1500
G1 BC	0.577	0.554	0.543	0.523	0.504
G7 BC	0.273	0.275	0.273	0.278	0.284

Averages from 3000 fps to 1500 fps:
G1 BC: 0.538, variation of G1 BC: 0.058
G7 BC: 0.276, variation of G7 BC: 0.012
G7 form factor: 1.007

Rifling Twist Rate and Stability

Barrel Twist	Stability Factor Best case	Stability Factor Nominal	Stability Factor Worst case
1:7"	2.14	1.87	1.62
1:8"	1.64	1.43	1.24
1:9"	1.30	1.13	0.98
1:10"	1.05	0.92	0.79

Minimum twist required for stability is 1:8.7"
Recommended twist for optimal performance is 1:7.8" or faster

GS Custom .243 caliber 81 grain SP

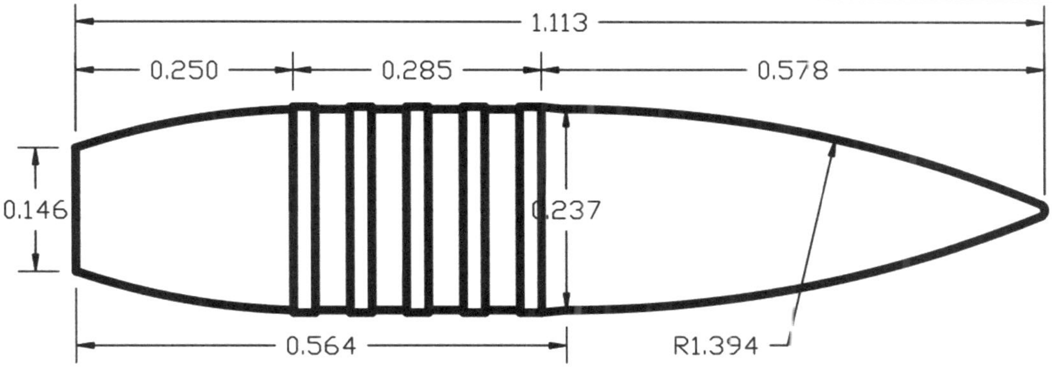

Lot #:	Unknown	Ogive Radius:	5.74 calibers
Sectional Density:	0.196 lb/in^2	Rt/R:	1.00

Ballistic Performance

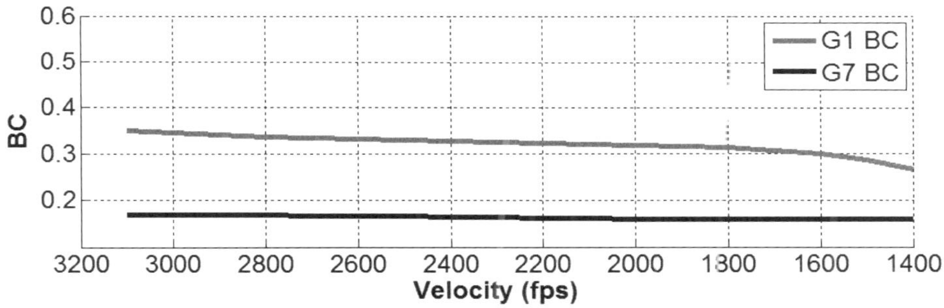

	Above 3000	3000-2500	2500-2000	2000-1500	Below 1500
G1 BC	0.357	0.338	0.325	0.303	0.286
G7 BC	0.169	0.167	0.163	0.161	0.161

Averages from 3000 fps to 1500 fps:
G1 BC: 0.320, variation of G1 BC: 0.059
G7 BC: 0.164, variation of G7 BC: 0.009
G7 form factor: 1.195

Rifling Twist Rate and Stability

Barrel Twist	Stability Factor Best case	Stability Factor Nominal	Stability Factor Worst case
1:7"	2.32	2.03	1.75
1:8"	1.78	1.55	1.34
1:9"	1.40	1.23	1.06
1:10"	1.14	0.99	0.86

Minimum twist required for stability is 1:9.1"
Recommended twist for optimal performance is 1:8.1" or faster

GS Custom .243 caliber 109 grain SP

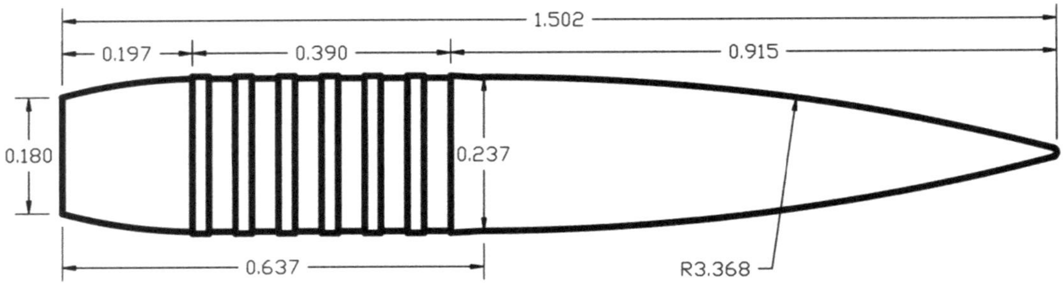

Lot #:	Unknown	Ogive Radius:	13.86 calibers
Sectional Density:	0.264 lb/in^2	Rt/R:	1.00

Ballistic Performance

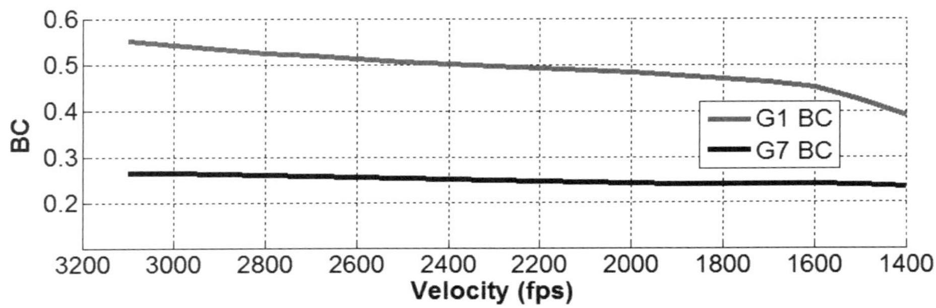

	Above 3000	3000-2500	2500-2000	2000-1500	Below 1500
G1 BC	0.567	0.525	0.495	0.454	0.426
G7 BC	0.269	0.260	0.249	0.241	0.239

Averages from 3000 fps to 1500 fps:
G1 BC: 0.490, variation of G1 BC: 0.117
G7 BC: 0.251, variation of G7 BC: 0.027
G7 form factor: 1.054

Rifling Twist Rate and Stability

Barrel Twist	Stability Factor Best case	Nominal	Worst case
1:6"	1.77	1.54	1.33
1:7"	1.30	1.13	0.98
1:8"	0.99	0.87	0.75
1:9"	0.78	0.69	0.59

Minimum twist required for stability is 1:6.8"
Recommended twist for optimal performance is 1:6.1" or faster

Hornady .243 caliber 105 grain Amax

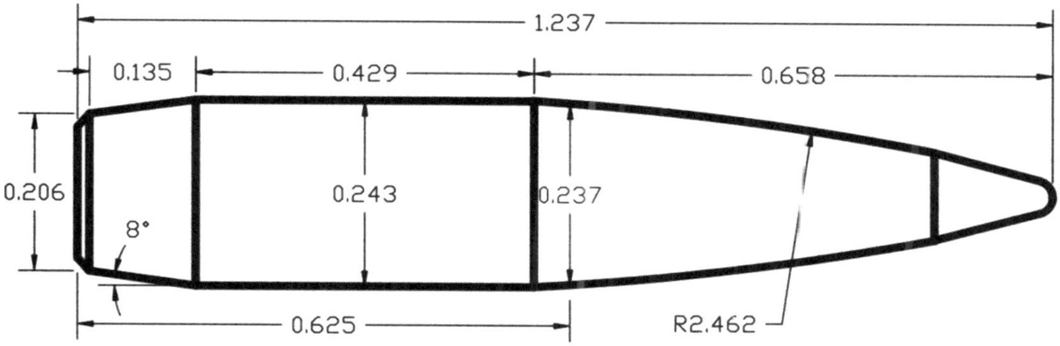

Lot #:	208199	Ogive Radius:	10.09 calibers
Sectional Density:	0.254 lb/in^2	Rt/R:	0.88

Ballistic Performance

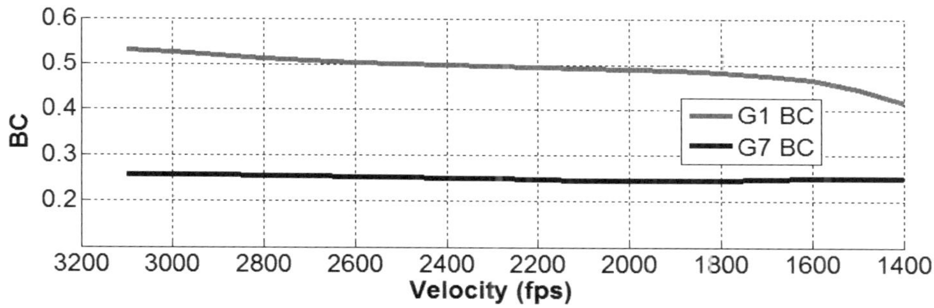

	Above 3000	3000-2500	2500-2000	2000-1500	Below 1500
G1 BC	0.544	0.513	0.496	0.468	0.447
G7 BC	0.257	0.255	0.249	0.249	0.251

Averages from 3000 fps to 1500 fps:
G1 BC: 0.491, variation of G1 BC: 0.079
G7 BC: 0.252, variation of G7 BC: 0.011
G7 form factor: 1.010

Rifling Twist Rate and Stability

Barrel Twist	Stability Factor Best case	Nominal	Worst case
1:7"	2.21	1.93	1.67
1:8"	1.69	1.48	1.28
1:9"	1.34	1.17	1.01
1:10"	1.08	0.95	0.82

Minimum twist required for stability is 1:8.9"
Recommended twist for optimal performance is 1:7.9" or faster

Hornady .243 caliber 87 grain BTHP

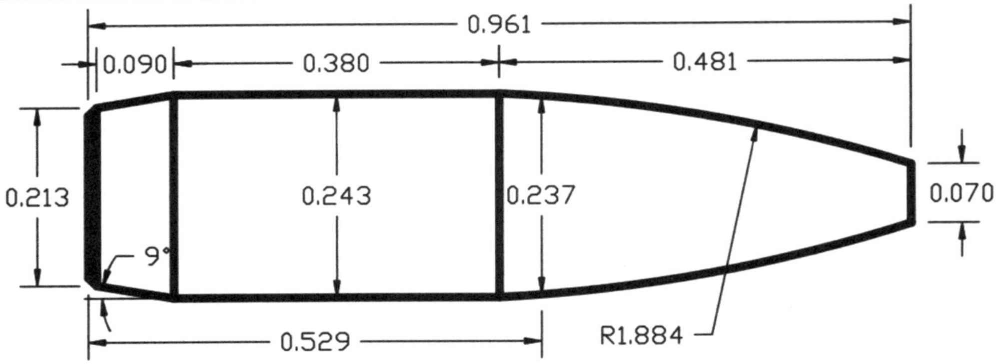

Lot #:	2070749	Ogive Radius:	7.72 calibers
Sectional Density:	0.211 lb/in^2	Rt/R:	0.77

Ballistic Performance

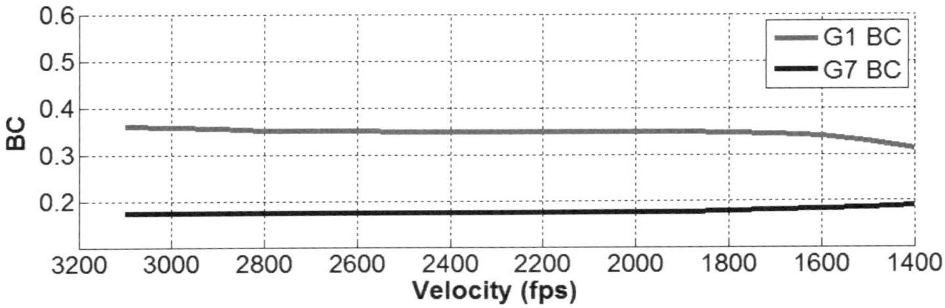

	Above 3000	3000-2500	2500-2000	2000-1500	Below 1500
G1 BC	0.367	0.355	0.350	0.340	0.330
G7 BC	0.174	0.176	0.176	0.181	0.186

Averages from 3000 fps to 1500 fps:
G1 BC: 0.347, variation of G1 BC: 0.029
G7 BC: 0.178, variation of G7 BC: 0.010
G7 form factor: 1.181

Rifling Twist Rate and Stability

Barrel Twist	Stability Factor Best case	Nominal	Worst case
1:7"	3.81	3.33	2.88
1:8"	2.92	2.55	2.21
1:9"	2.31	2.02	1.74
1:10"	1.87	1.63	1.41

Minimum twist required for stability is 1:11.7"
Recommended twist for optimal performance is 1:10.4" or faster

Hornady .243 caliber 105 grain BTHP Match

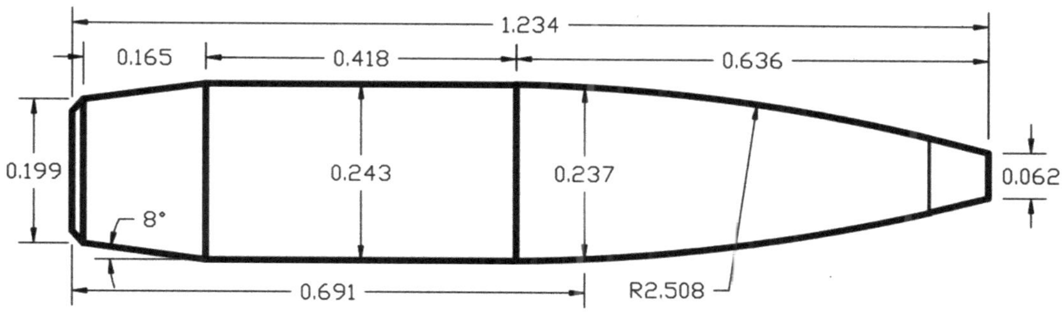

Lot #:	2140591	**Ogive Radius:**	**10.32 calibers**
Sectional Density:	**0.254 lb/ in²**	**Rt/R:**	**1.00**

Ballistic Performance

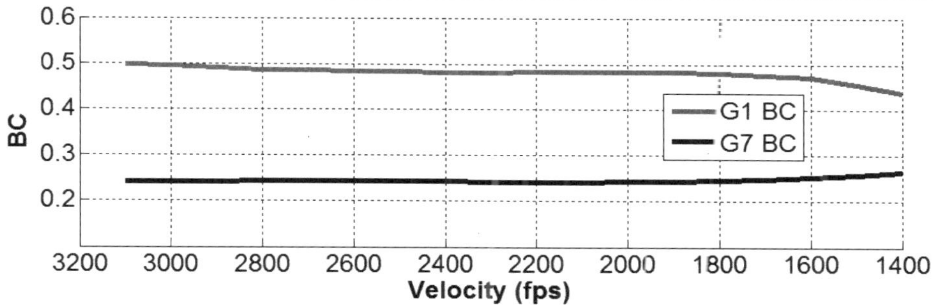

	Above 3000	3000-2500	2500-2000	2000-1500	Below 1500
G1 BC	0.506	0.488	0.483	0.470	0.457
G7 BC	0.240	0.242	0.243	0.250	0.258

Averages from 3000 fps to 1500 fps:
G1 BC: 0.479, variation of G1 BC: 0.037
G7 BC: 0.246, variation of G7 BC: 0.016
G7 form factor: 1.032

Rifling Twist Rate and Stability

	Stability Factor		
Barrel Twist	**Best case**	**Nominal**	**Worst case**
1:7"	2.23	1.94	1.68
1:8"	1.70	1.49	1.29
1:9"	1.35	1.18	1.02
1:10"	1.09	0.95	0.82

Minimum twist required for stability is 1:8.9"
Recommended twist for optimal performance is 1:8.0" or faster

Hornady .243 caliber 100 grain BTSP

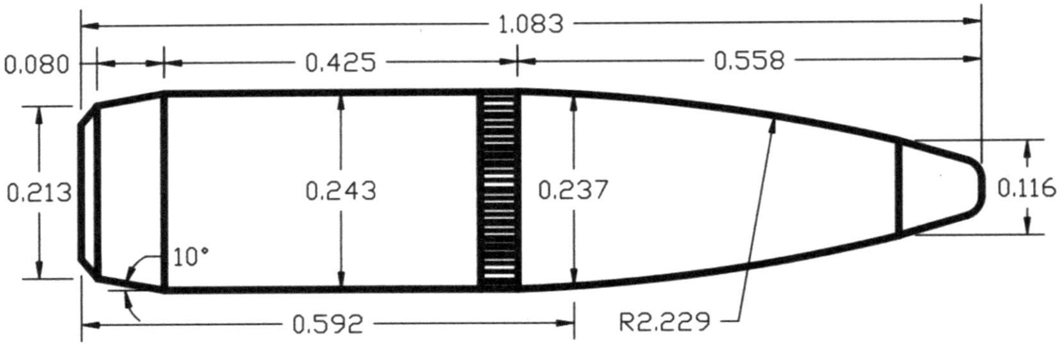

Lot #:	**2070146**	**Ogive Radius:**	**9.33 calibers**
Sectional Density:	**0.242 lb/in^2**	**Rt/R:**	**0.77**

Ballistic Performance

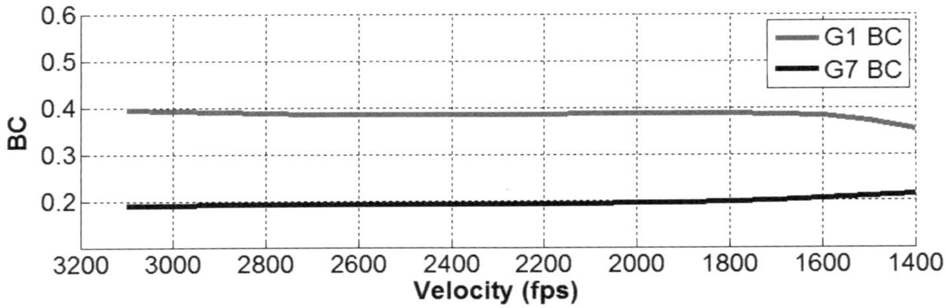

	Above 3000	3000-2500	2500-2000	2000-1500	Below 1500
G1 BC	0.401	0.389	0.387	0.380	0.372
G7 BC	0.190	0.193	0.195	0.202	0.210

Averages from 3000 fps to 1500 fps:
G1 BC: 0.385, variation of G1 BC: 0.021
G7 BC: 0.198, variation of G7 BC: 0.017
G7 form factor: 1.225

Rifling Twist Rate and Stability

	Stability Factor		
Barrel Twist	**Best case**	**Nominal**	**Worst case**
1:7"	3.10	2.71	2.34
1:8"	2.37	2.07	1.79
1:9"	1.88	1.64	1.42
1:10"	1.52	1.33	1.15

Minimum twist required for stability is 1:10.5"
Recommended twist for optimal performance is 1:9.4" or faster

Hornady .243 caliber 80 grain GMX

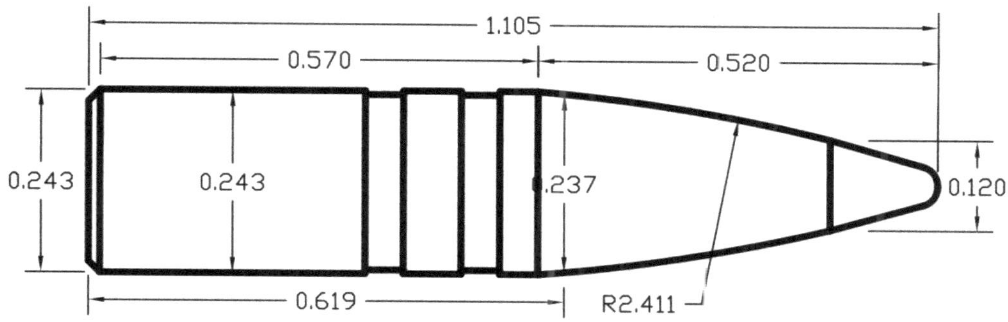

Lot #:	**Unknown**	**Ogive Radius:**	**9.92 calibers**
Sectional Density:	**0.194 lb/in²**	**Rt/R:**	**0.60**

Ballistic Performance

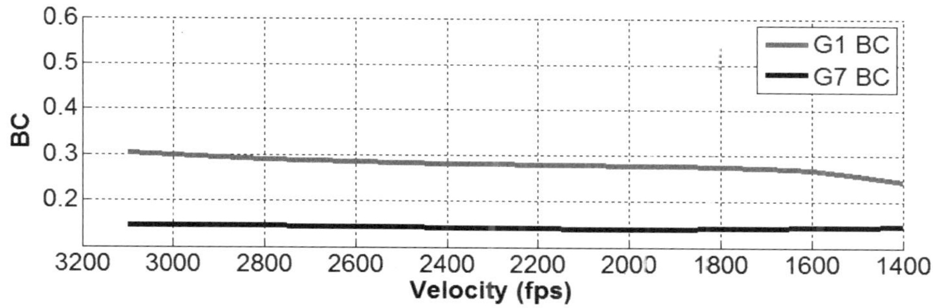

	Above 3000	3000-2500	2500-2000	2000-1500	Below 1500
G1 BC	0.311	0.292	0.282	0.268	0.258
G7 BC	0.147	0.145	0.142	0.143	0.145

Averages from 3000 fps to 1500 fps:
G1 BC: 0.280, variation of G1 BC: 0.042
G7 BC: 0.144, variation of G7 BC: 0.007
G7 form factor: 1.347

Rifling Twist Rate and Stability

Barrel Twist	**Stability Factor** Best case	Nominal	Worst case
1:7"	2.34	2.04	1.77
1:8"	1.79	1.57	1.35
1:9"	1.42	1.24	1.07
1:10"	1.15	1.00	0.87

Minimum twist required for stability is 1:9.1"
Recommended twist for optimal performance is 1:8.2" or faster

Hornady .243 caliber 85 grain Interbond

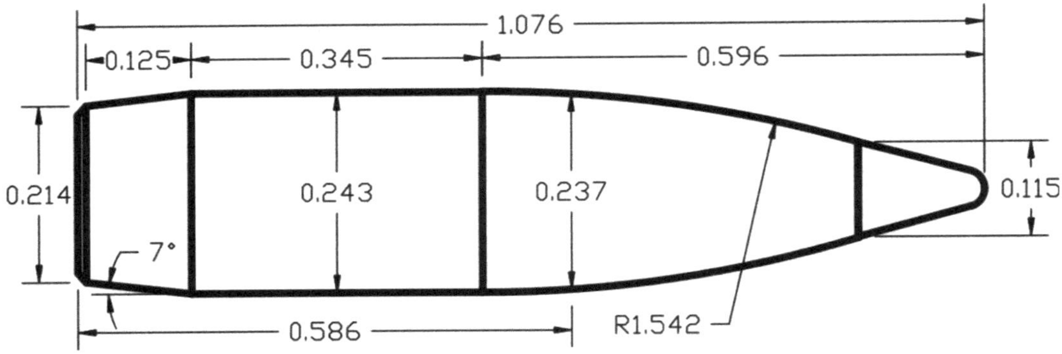

Lot #:	2060676	Ogive Radius:	6.42 calibers
Sectional Density:	0.206 lb/in^2	Rt/R:	0.95

Ballistic Performance

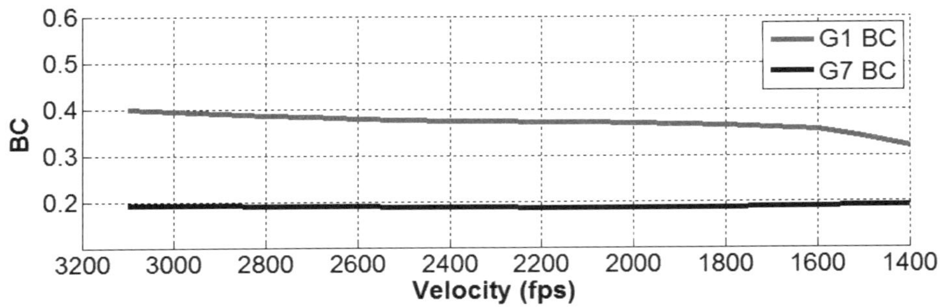

	Above 3000	3000-2500	2500-2000	2000-1500	Below 1500
G1 BC	0.408	0.387	0.374	0.355	0.340
G7 BC	0.193	0.192	0.188	0.189	0.191

Averages from 3000 fps to 1500 fps:

G1 BC: 0.371, variation of G1 BC: 0.055

G7 BC: 0.190, variation of G7 BC: 0.007

G7 form factor: 1.081

Rifling Twist Rate and Stability

Barrel Twist	Stability Factor Best case	Nominal	Worst case
1:7"	2.69	2.35	2.03
1:8"	2.06	1.80	1.55
1:9"	1.63	1.42	1.23
1:10"	1.32	1.15	0.99

Minimum twist required for stability is 1:9.8"
Recommended twist for optimal performance is 1:8.8" or faster

Hornady .243 caliber 58 grain Vmax

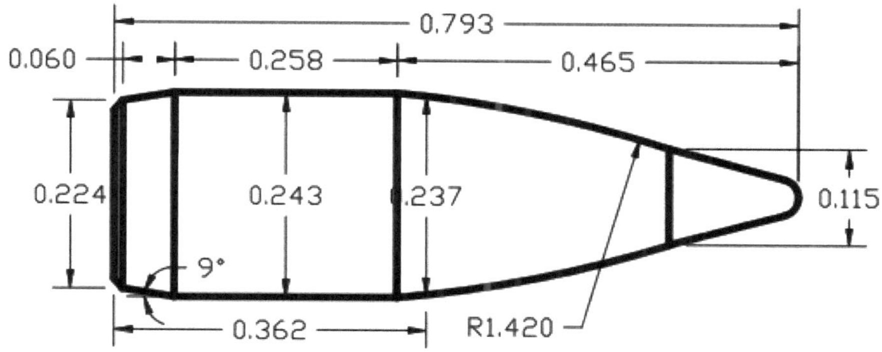

Lot #:	2080522	Ogive Radius:	5.98 calibers
Sectional Density:	0.140 lb/in^2	Rt/R:	0.50

Ballistic Performance

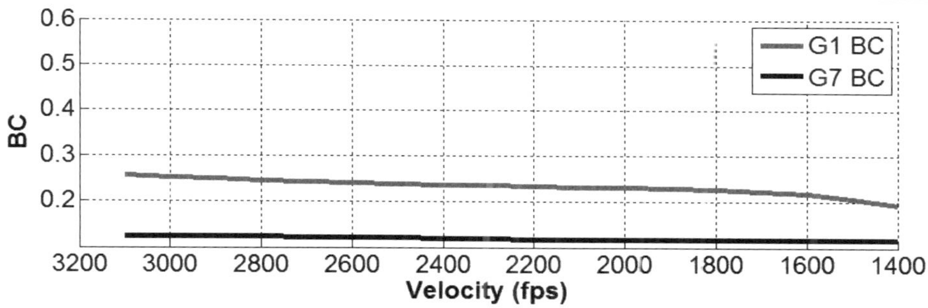

	Above 3000	3000-2500	2500-2000	2000-1500	Below 1500
G1 BC	0.262	0.246	0.235	0.220	0.208
G7 BC	0.124	0.122	0.118	0.117	0.117

Averages from 3000 fps to 1500 fps:
G1 BC: 0.233, variation of G1 BC: 0.045
G7 BC: 0.119, variation of G7 BC: 0.007
G7 form factor: 1.177

Rifling Twist Rate and Stability

Barrel Twist	Stability Factor Best case	Nominal	Worst case
1:7"	4.40	3.84	3.32
1:8"	3.37	2.94	2.54
1:9"	2.66	2.33	2.01
1:10"	2.16	1.88	1.63

Minimum twist required for stability is 1:12.5"
Recommended twist for optimal performance is 1:11.2" or faster

Hornady .243 caliber 65 grain Vmax

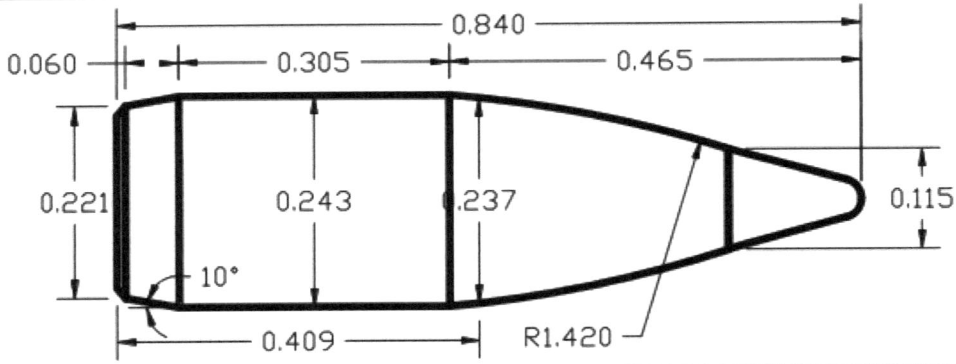

Lot #:	2070855	Ogive Radius:	5.98 calibers
Sectional Density:	0.157 lb/in²	Rt/R:	0.50

Ballistic Performance

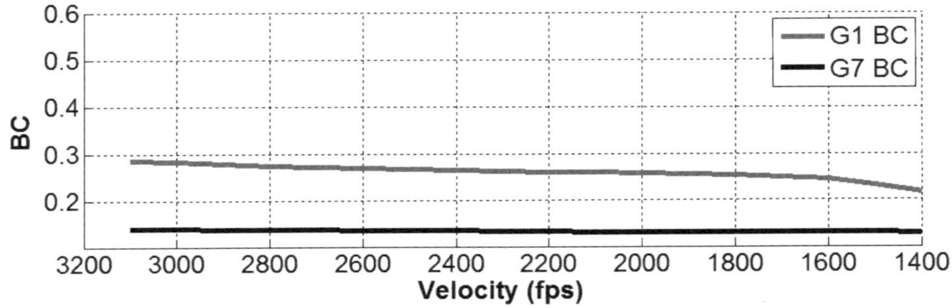

	Above 3000	3000-2500	2500-2000	2000-1500	Below 1500
G1 BC	0.294	0.276	0.264	0.246	0.233
G7 BC	0.139	0.137	0.133	0.130	0.131

Averages from 3000 fps to 1500 fps:
G1 BC: 0.261, variation of G1 BC: 0.051
G7 BC: 0.134, variation of G7 BC: 0.008
G7 form factor: 1.178

Rifling Twist Rate and Stability

Barrel Twist	Stability Factor Best case	Nominal	Worst case
1:7"	4.19	3.66	3.16
1:8"	3.21	2.80	2.42
1:9"	2.53	2.21	1.91
1:10"	2.05	1.79	1.55

Minimum twist required for stability is 1:12.2"
Recommended twist for optimal performance is 1:10.9" or faster

Hornady .243 caliber 75 grain Vmax

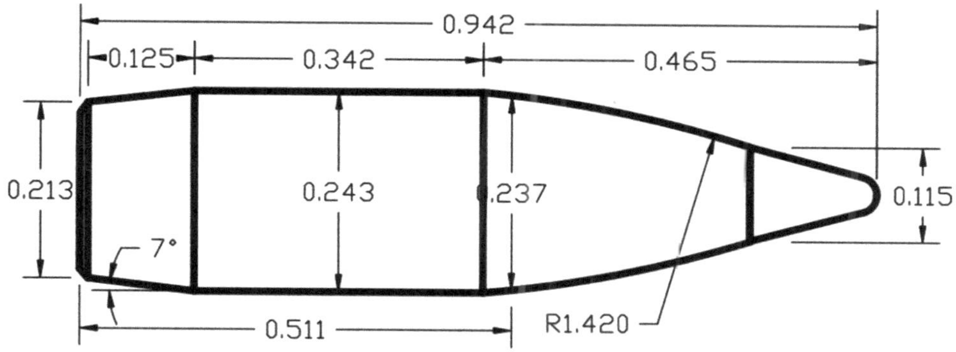

Lot #:	**2070619**	**Ogive Radius:**	**5.98 calibers**
Sectional Density:	**0.182 lb/in^2**	**Rt/R:**	**0.50**

Ballistic Performance

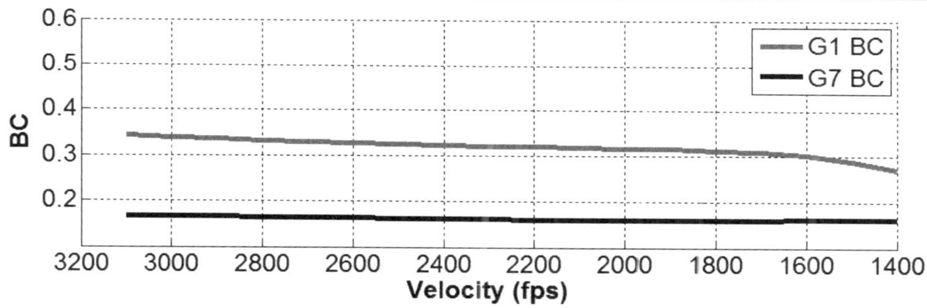

	Above 3000	3000-2500	2500-2000	2000-1500	Below 1500
G1 BC	0.351	0.332	0.321	0.303	0.289
G7 BC	0.166	0.165	0.161	0.161	0.162

Averages from 3000 fps to 1500 fps:
G1 BC: 0.317, variation of G1 BC: 0.050
G7 BC: 0.163, variation of G7 BC: 0.007
G7 form factor: 1.115

Rifling Twist Rate and Stability

Barrel Twist	Stability Factor Best case	Stability Factor Nominal	Stability Factor Worst case
1:7"	3.48	3.04	2.63
1:8"	2.67	2.33	2.01
1:9"	2.11	1.84	1.59
1:10"	1.71	1.49	1.29

Minimum twist required for stability is 1:11.1"
Recommended twist for optimal performance is 1:10.0" or faster

Hornady .243 caliber 87 grain Vmax

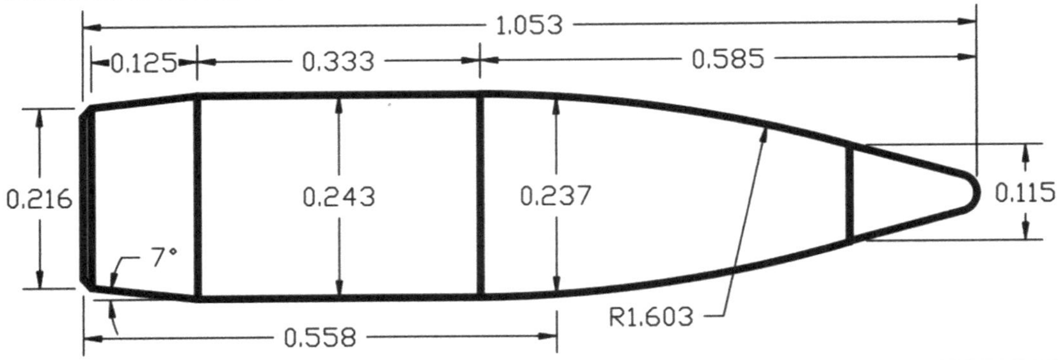

Lot #:	2070711	Ogive Radius:	6.66 calibers
Sectional Density:	0.211 lb/in^2	Rt/R:	0.87

Ballistic Performance

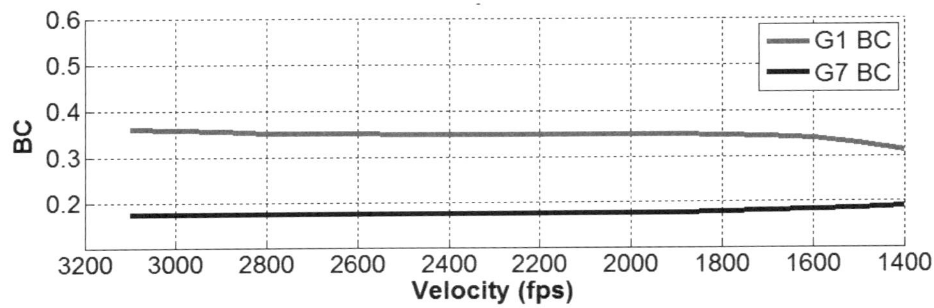

	Above 3000	3000-2500	2500-2000	2000-1500	Below 1500
G1 BC	0.421	0.398	0.385	0.365	0.350
G7 BC	0.199	0.197	0.193	0.194	0.197

Averages from 3000 fps to 1500 fps:
G1 BC: 0.381, variation of G1 BC: 0.057
G7 BC: 0.196, variation of G7 BC: 0.008
G7 form factor: 1.076

Rifling Twist Rate and Stability

Barrel Twist	Stability Factor: Best case	Stability Factor: Nominal	Stability Factor: Worst case
1:7"	2.95	2.58	2.23
1:8"	2.26	1.97	1.71
1:9"	1.78	1.56	1.35
1:10"	1.45	1.26	1.09

Minimum twist required for stability is 1:10.2"
Recommended twist for optimal performance is 1:9.1" or faster

JLK .243 caliber 105 grain VLD

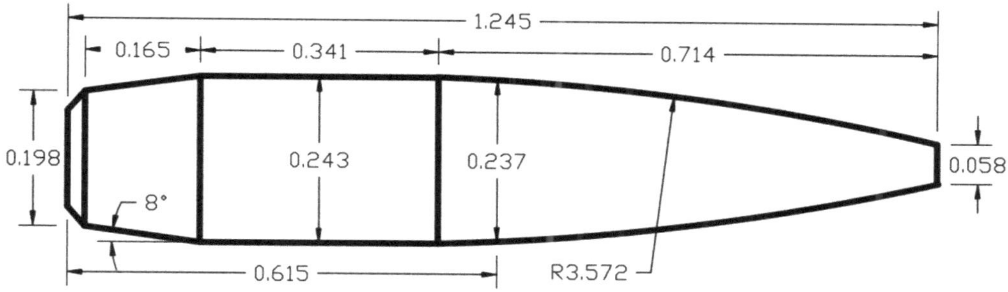

Lot #:	Unknown	Ogive Radius:	14.84 calibers
Sectional Density:	0.254 lb/in^2	Rt/R:	0.80

Ballistic Performance

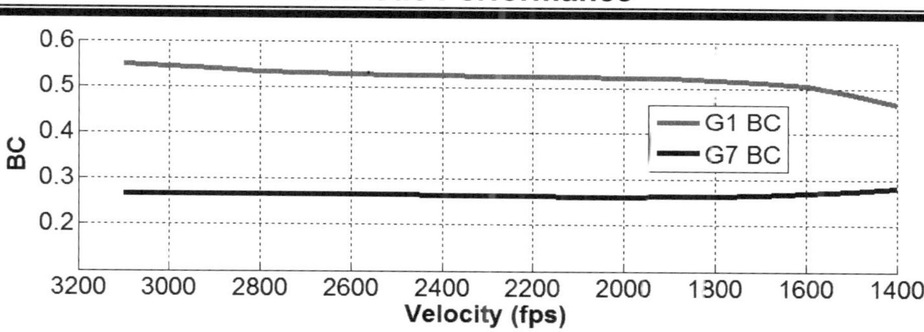

	Above 3000	3000-2500	2500-2000	2000-1500	Below 1500
G1 BC	0.560	0.536	0.525	0.506	0.489
G7 BC	0.265	0.266	0.264	0.269	0.276

Averages from 3000 fps to 1500 fps:
G1 BC: 0.521, variation of G1 BC: 0.056
G7 BC: 0.267, variation of G7 BC: 0.013
G7 form factor: 0.950

Rifling Twist Rate and Stability

Barrel Twist	Stability Factor Best case	Nominal	Worst case
1:7"	2.17	1.89	1.64
1:8"	1.66	1.45	1.25
1:9"	1.31	1.15	0.99
1:10"	1.06	0.93	0.80

Minimum twist required for stability is 1:8.8"
Recommended twist for optimal performance is 1:7.9" or faster

Lapua .243 caliber 90 grain FMJBT

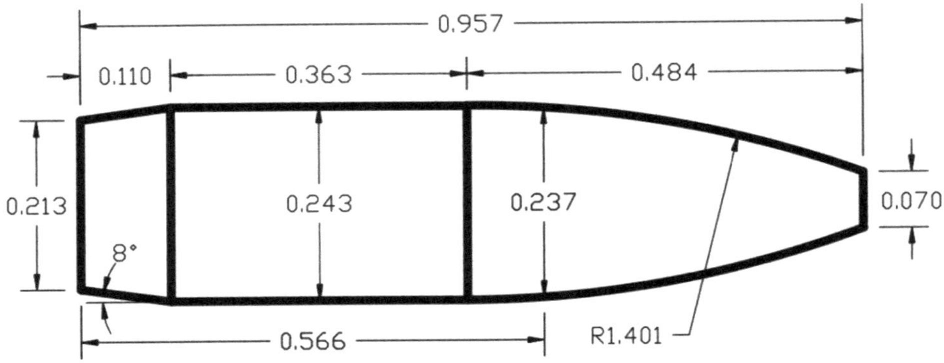

Lot #:	4PL6006	Ogive Radius:	5.74 calibers
Sectional Density:	0.218 lb/in²	Rt/R:	0.99

Ballistic Performance

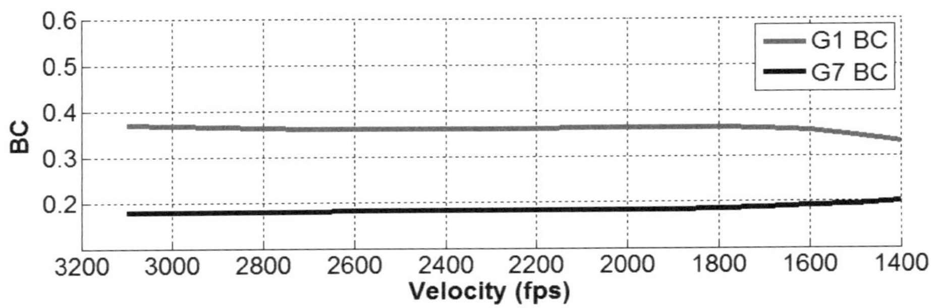

	Above 3000	3000-2500	2500-2000	2000-1500	Below 1500
G1 BC	0.376	0.365	0.362	0.354	0.346
G7 BC	0.178	0.181	0.182	0.189	0.195

Averages from 3000 fps to 1500 fps:
G1 BC: 0.360, variation of G1 BC: 0.023
G7 BC: 0.185, variation of G7 BC: 0.015
G7 form factor: 1.180

Rifling Twist Rate and Stability

Barrel Twist	Stability Factor Best case	Nominal	Worst case
1:7"	3.99	3.49	3.02
1:8"	3.06	2.67	2.31
1:9"	2.41	2.11	1.82
1:10"	1.96	1.71	1.48

Minimum twist required for stability is 1:11.9"
Recommended twist for optimal performance is 1:10.7" or faster

Lapua .243 caliber 90 grain Scenar

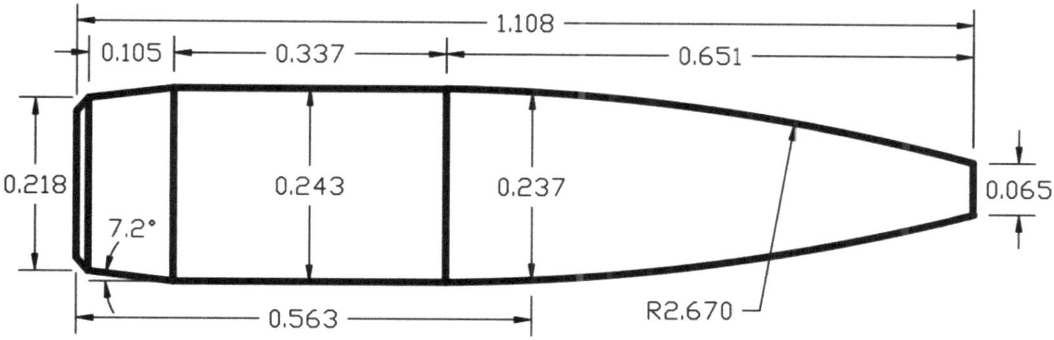

Lot #:	P00365701	Ogive Radius:	10.94 calibers
Sectional Density:	0.218 lb/in^2	Rt/R:	0.79

Ballistic Performance

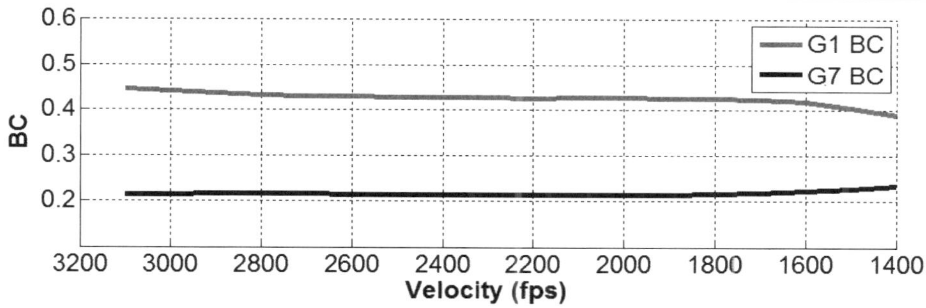

	Above 3000	3000-2500	2500-2000	2000-1500	Below 1500
G1 BC	0.452	0.434	0.426	0.415	0.404
G7 BC	0.214	0.215	0.214	0.221	0.228

Averages from 3000 fps to 1500 fps:
G1 BC: 0.424, variation of G1 BC: 0.036
G7 BC: 0.218, variation of G7 BC: 0.014
G7 form factor: 0.999

Rifling Twist Rate and Stability

Barrel Twist	Stability Factor Best case	Nominal	Worst case
1:7"	2.61	2.28	1.97
1:8"	2.00	1.75	1.51
1:9"	1.58	1.38	1.19
1:10"	1.28	1.12	0.97

Minimum twist required for stability is 1:9.7"
Recommended twist for optimal performance is 1:8.6" or faster

Lapua .243 caliber 105 grain Scenar

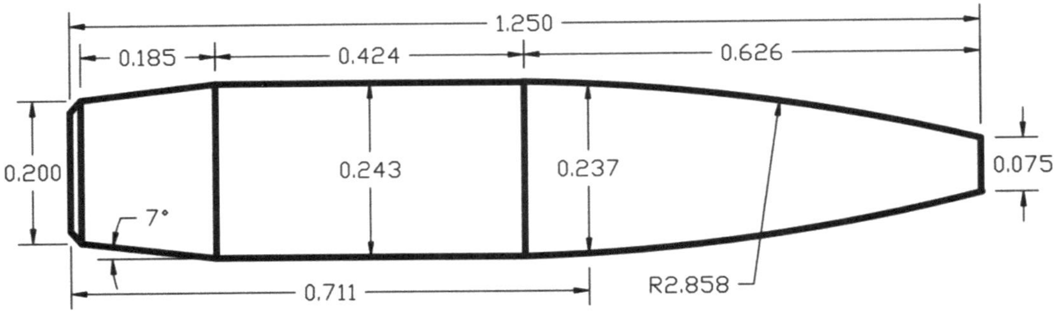

Lot #:	4PL6045	Ogive Radius:	11.88 calibers
Sectional Density:	0.254 lb/in^2	Rt/R:	0.86

Ballistic Performance

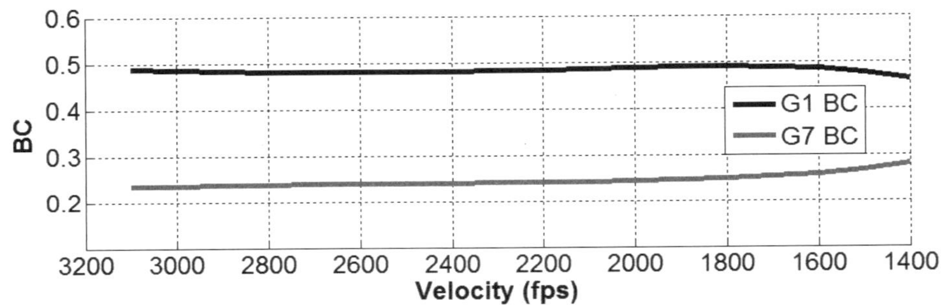

	Above 3000	3000-2500	2500-2000	2000-1500	Below 1500
G1 BC	0.494	0.483	0.484	0.482	0.476
G7 BC	0.234	0.240	0.243	0.257	0.269

Averages from 3000 fps to 1500 fps:
G1 BC: 0.483, variation of G1 BC: 0.012
G7 BC: 0.248, variation of G7 BC: 0.031
G7 form factor: 1.025

Rifling Twist Rate and Stability

Barrel Twist	Stability Factor Best case	Stability Factor Nominal	Stability Factor Worst case
1:7"	2.14	1.87	1.62
1:8"	1.64	1.43	1.24
1:9"	1.30	1.13	0.98
1:10"	1.05	0.92	0.79

Minimum twist required for stability is 1:8.7"
Recommended twist for optimal performance is 1:7.8" or faster

Lapua .243 caliber 90 grain Scenar L

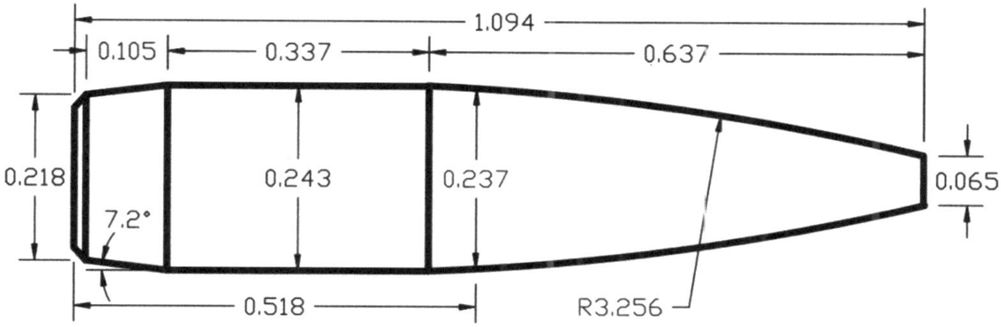

Lot #:	Unknown	Ogive Radius:	13.40 calibers
Sectional Density:	0.218 lb/in²	Rt/R:	0.79

Ballistic Performance

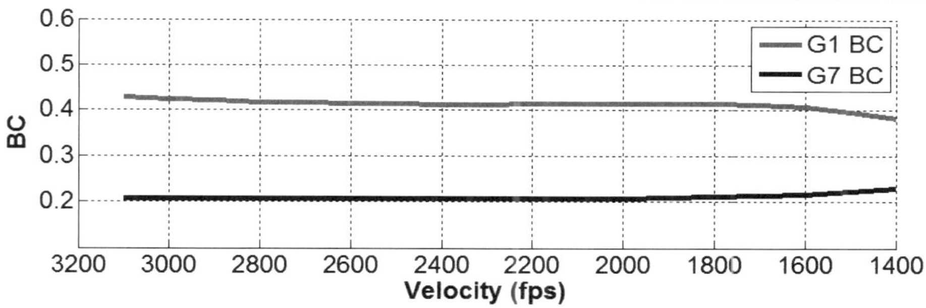

	Above 3000	3000-2500	2500-2000	2000-1500	Below 1500
G1 BC	0.434	0.418	0.413	0.404	0.395
G7 BC	0.205	0.207	0.208	0.215	0.223

Averages from 3000 fps to 1500 fps:
G1 BC: 0.411, variation of G1 BC: 0.028
G7 BC: 0.211, variation of G7 BC: 0.016
G7 form factor: 1.031

Rifling Twist Rate and Stability

Barrel Twist	Stability Factor Best case	Nominal	Worst case
1:7"	2.71	2.37	2.05
1:8"	2.08	1.81	1.57
1:9"	1.64	1.43	1.24
1:10"	1.33	1.16	1.00

Minimum twist required for stability is 1:9.8"
Recommended twist for optimal performance is 1:8.8" or faster

Matrix .243 caliber 107 grain VLD

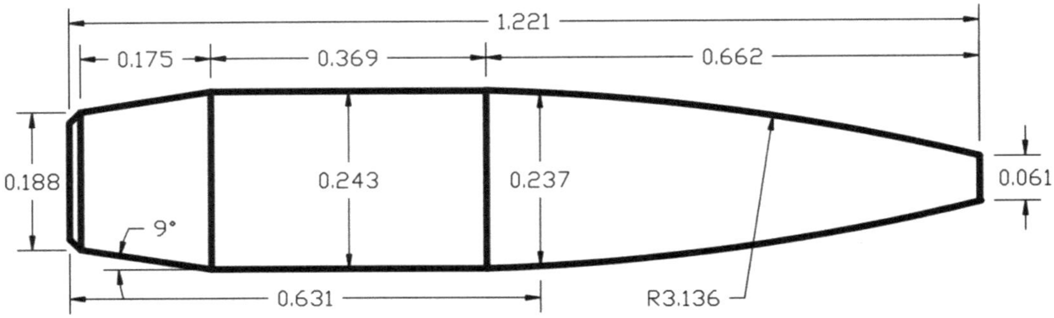

Lot #:	Unknown	Ogive Radius:	12.85 calibers
Sectional Density:	0.259 lb/in^2	Rt/R:	0.77

Ballistic Performance

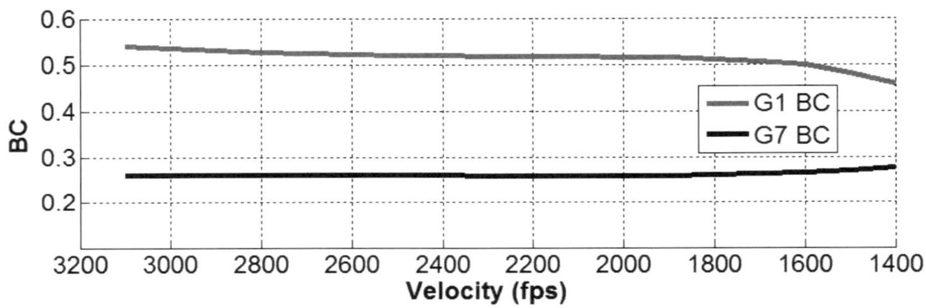

	Above 3000	3000-2500	2500-2000	2000-1500	Below 1500
G1 BC	0.550	0.528	0.518	0.499	0.482
G7 BC	0.261	0.262	0.260	0.265	0.272

Averages from 3000 fps to 1500 fps:
G1 BC: 0.513, variation of G1 BC: 0.054
G7 BC: 0.264, variation of G7 BC: 0.012
G7 form factor: 0.982

Rifling Twist Rate and Stability

Barrel Twist	Stability Factor Best case	Nominal	Worst case
1:7"	2.36	2.06	1.78
1:8"	1.81	1.58	1.36
1:9"	1.43	1.25	1.08
1:10"	1.16	1.01	0.87

Minimum twist required for stability is 1:9.1"
Recommended twist for optimal performance is 1:8.2" or faster

Nosler .243 caliber 90 grain Accubond

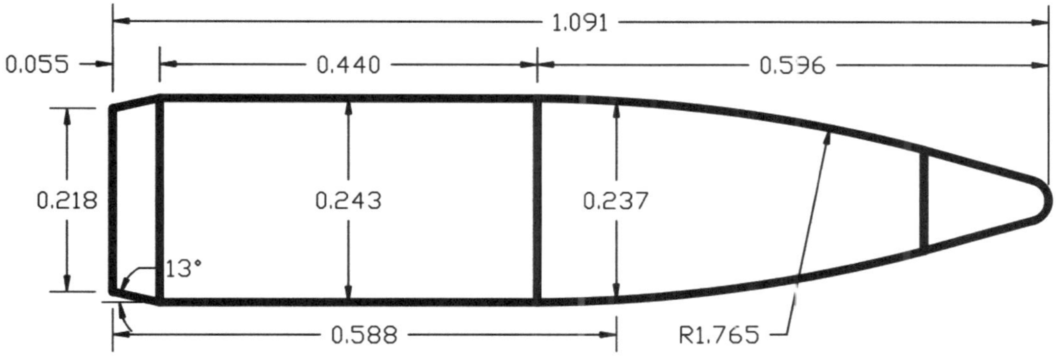

Lot #:	EL27F6	**Ogive Radius:**	**7.26 calibers**
Sectional Density:	**0.218 lb/ in²**	**Rt/R:**	**1.00**

Ballistic Performance

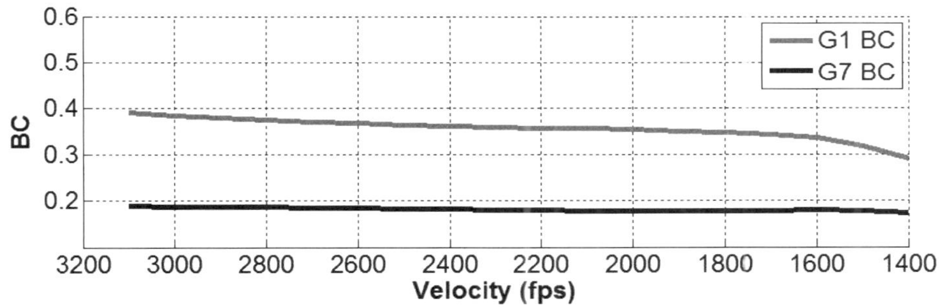

	Above 3000	3000-2500	2500-2000	2000-1500	Below 1500
G1 BC	0.399	0.375	0.359	0.335	0.317
G7 BC	0.189	0.186	0.181	0.178	0.178

Averages from 3000 fps to 1500 fps:

G1 BC: 0.355, variation of G1 BC: 0.068

G7 BC: 0.182, variation of G7 BC: 0.011

G7 form factor: 1.198

Rifling Twist Rate and Stability

Barrel Twist	**Stability Factor** Best case	Nominal	Worst case
1:7"	2.73	2.39	2.06
1:8"	2.09	1.83	1.58
1:9"	1.65	1.44	1.25
1:10"	1.34	1.17	1.01

Minimum twist required for stability is 1:9.9"

Recommended twist for optimal performance is 1:8.8" or faster

Nosler .243 caliber 70 grain Ballistic Tip

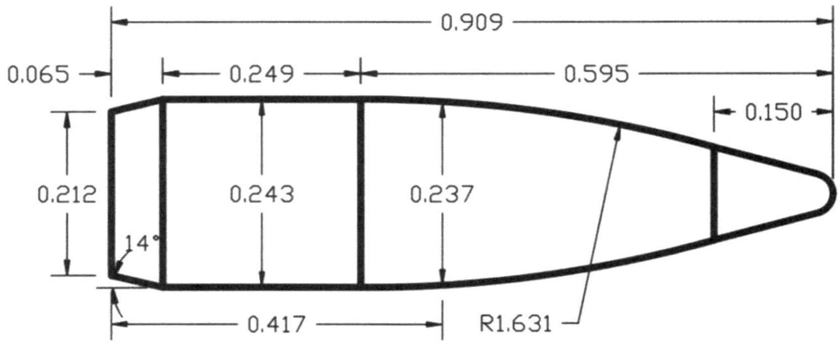

Lot #:	CE20F7-1	Ogive Radius:	6.69 calibers
Sectional Density:	0.169 lb/in^2	Rt/R:	1.00

Ballistic Performance

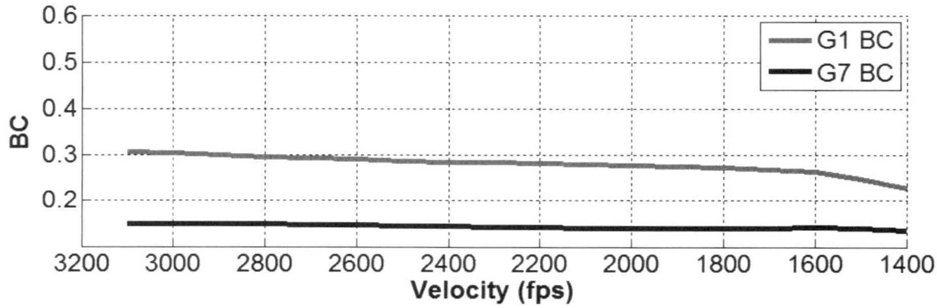

	Above 3000	3000-2500	2500-2000	2000-1500	Below 1500
G1 BC	0.315	0.296	0.283	0.263	0.249
G7 BC	0.149	0.147	0.142	0.140	0.140

Averages from 3000 fps to 1500 fps:
G1 BC: 0.280, variation of G1 BC: 0.055
G7 BC: 0.143, variation of G7 BC: 0.009
G7 form factor: 1.184

Rifling Twist Rate and Stability

Barrel Twist	Stability Factor Best case	Nominal	Worst case
1:7"	3.60	3.14	2.72
1:8"	2.76	2.41	2.08
1:9"	2.18	1.90	1.64
1:10"	1.76	1.54	1.33

Minimum twist required for stability is 1:11.3"
Recommended twist for optimal performance is 1:10.1" or faster

Nosler .243 caliber 80 grain Ballistic Tip

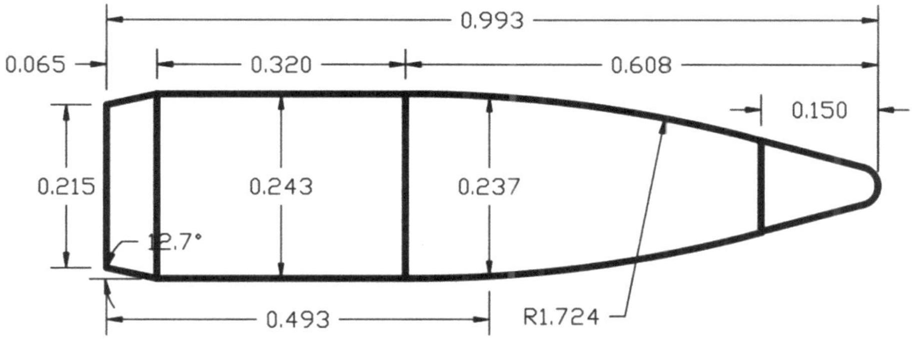

Lot #:	FE21F7-1	Ogive Radius:	7.07 calibers
Sectional Density:	0.194 lb/in²	Rt/R:	1.00

Ballistic Performance

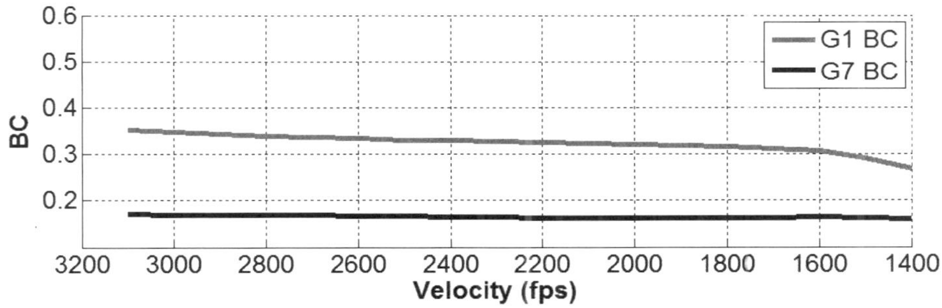

	Above 3000	3000-2500	2500-2000	2000-1500	Below 1500
G1 BC	0.360	0.339	0.326	0.306	0.291
G7 BC	0.170	0.168	0.164	0.162	0.163

Averages from 3000 fps to 1500 fps:
G1 BC: 0.323, variation of G1 BC: 0.057
G7 BC: 0.165, variation of G7 BC: 0.008
G7 form factor: 1.172

Rifling Twist Rate and Stability

Barrel Twist	Stability Factor Best case	Nominal	Worst case
1:7"	3.19	2.79	2.41
1:8"	2.44	2.13	1.85
1:9"	1.93	1.69	1.46
1:10"	1.56	1.37	1.18

Minimum twist required for stability is 1:10.7"
Recommended twist for optimal performance is 1:9.5" or faster

Nosler .243 caliber 90 grain E-Tip

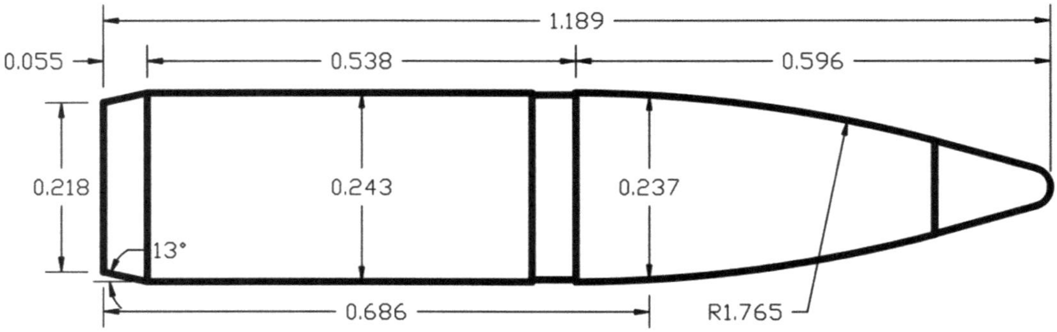

Lot #:	CL04F14	**Ogive Radius:**	**7.26 calibers**
Sectional Density:	**0.218 lb/ in²**	**Rt/R:**	**1.00**

Ballistic Performance

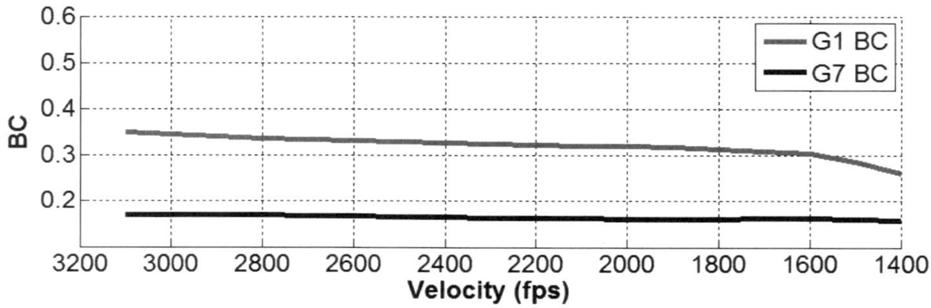

	Above 3000	3000-2500	2500-2000	2000-1500	Below 1500
G1 BC	0.358	0.337	0.324	0.303	0.287
G7 BC	0.170	0.167	0.163	0.161	0.161

Averages from 3000 fps to 1500 fps:
G1 BC: 0.320, variation of G1 BC: 0.059
G7 BC: 0.164, variation of G7 BC: 0.009
G7 form factor: 1.328

Rifling Twist Rate and Stability

Barrel Twist	Stability Factor Best case	Stability Factor Nominal	Stability Factor Worst case
1:7"	2.22	1.94	1.68
1:8"	1.70	1.48	1.28
1:9"	1.34	1.17	1.01
1:10"	1.09	0.95	0.82

Minimum twist required for stability is 1:8.9"
Recommended twist for optimal performance is 1:8.0" or faster

Nosler .243 caliber 95 grain Partition

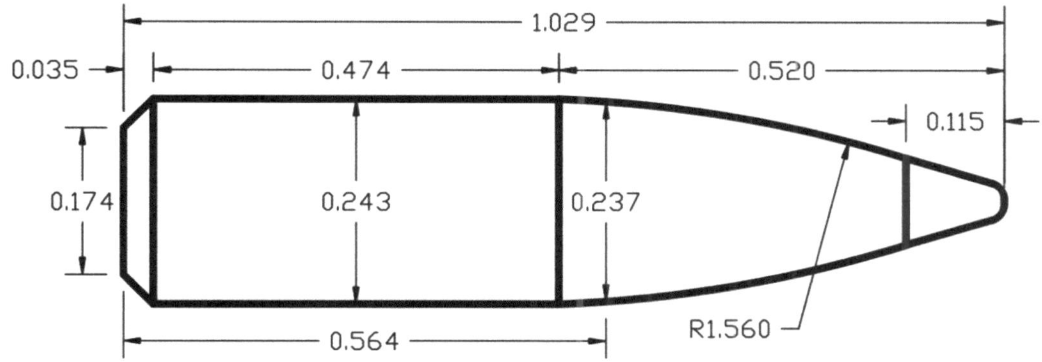

Lot #:	LC20F 41	Ogive Radius:	6.39 calibers
Sectional Density:	0.230 lb/in^2	Rt/R:	0.97

Ballistic Performance

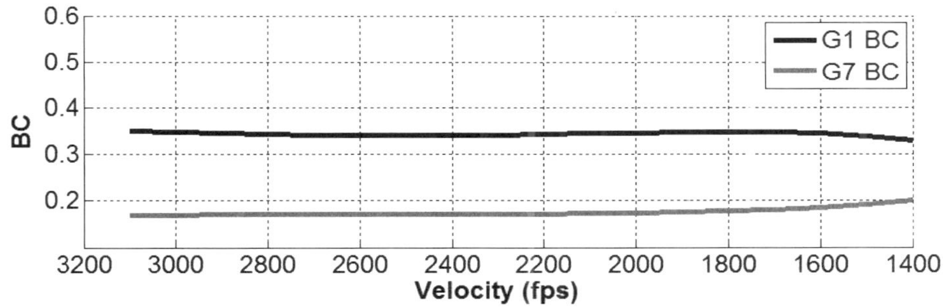

	Above 3000	3000-2500	2500-2000	2000-1500	Below 1500
G1 BC	0.355	0.344	0.343	0.343	0.340
G7 BC	0.168	0.171	0.173	0.183	0.192

Averages from 3000 fps to 1500 fps:
G1 BC: 0.343, variation of G1 BC: 0.008
G7 BC: 0.177, variation of G7 BC: 0.022
G7 form factor: 1.303

Rifling Twist Rate and Stability

Barrel Twist	Stability Factor Best case	Nominal	Worst case
1:7"	3.42	2.99	2.58
1:8"	2.62	2.29	1.98
1:9"	2.07	1.81	1.56
1:10"	1.67	1.46	1.27

Minimum twist required for stability is 1:11.0"
Recommended twist for optimal performance is 1:9.9" or faster

Precision Ballistics .243 caliber 103 grain VLD

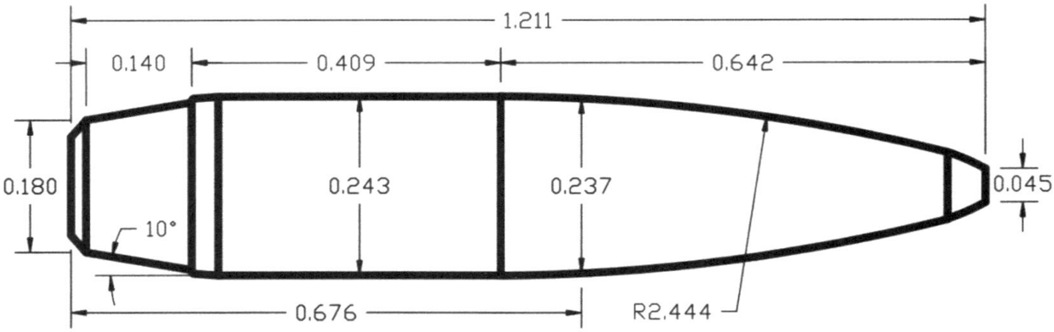

Lot #:	2267	Ogive Radius:	10.16 calibers
Sectional Density:	0.247 lb/ in²	Rt/R:	0.95

Ballistic Performance

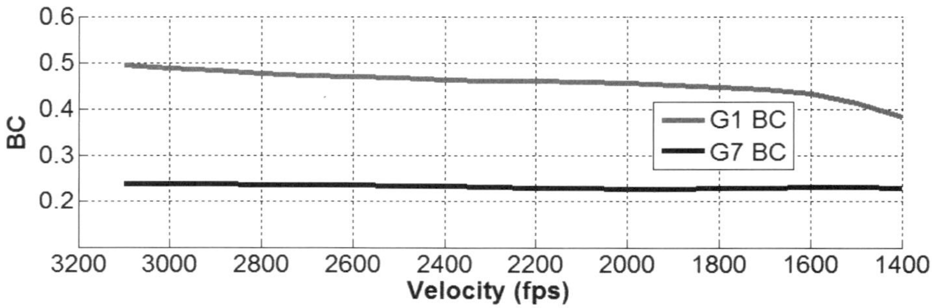

	Above 3000	3000-2500	2500-2000	2000-1500	Below 1500
G1 BC	0.504	0.477	0.461	0.434	0.413
G7 BC	0.239	0.237	0.232	0.230	0.232

Averages from 3000 fps to 1500 fps:
G1 BC: 0.456, variation of G1 BC: 0.075
G7 BC: 0.234, variation of G7 BC: 0.010
G7 form factor: 1.067

Rifling Twist Rate and Stability

Barrel Twist	Stability Factor Best case	Stability Factor Nominal	Stability Factor Worst case
1:6"	3.37	2.95	2.55
1:7"	2.48	2.16	1.87
1:8"	1.90	1.66	1.43
1:9"	1.50	1.31	1.13

Minimum twist required for stability is 1:9.4"
Recommended twist for optimal performance is 1:8.4" or faster

Precision Ballistics .243 caliber 105 grain VLD

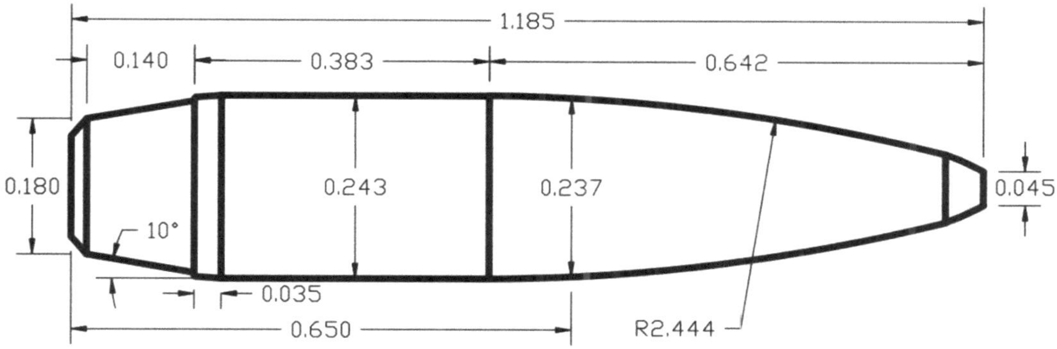

Lot #:	2268	Ogive Radius:	10.16 calibers
Sectional Density:	0.254 lb/ in^2	Rt/R:	0.95

Ballistic Performance

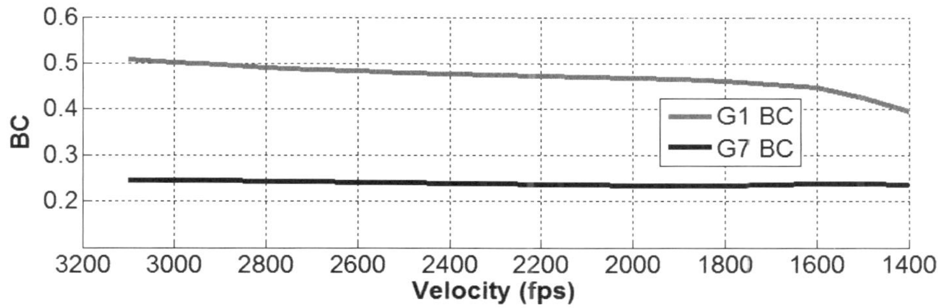

	Above 3000	3000-2500	2500-2000	2000-1500	Below 1500
G1 BC	0.518	0.491	0.474	0.446	0.425
G7 BC	0.245	0.243	0.238	0.237	0.239

Averages from 3000 fps to 1500 fps:
G1 BC: 0.469, variation of G1 BC: 0.077
G7 BC: 0.240, variation of G7 BC: 0.010
G7 form factor: 1.058

Rifling Twist Rate and Stability

Barrel Twist	Stability Factor: Best case	Stability Factor: Nominal	Stability Factor: Worst case
1:6"	3.44	3.00	2.60
1:7"	2.53	2.21	1.91
1:8"	1.93	1.69	1.46
1:9"	1.53	1.33	1.15

Minimum twist required for stability is 1:9.5"
Recommended twist for optimal performance is 1:8.5" or faster

Precision Ballistics .243 caliber 108 grain VLD

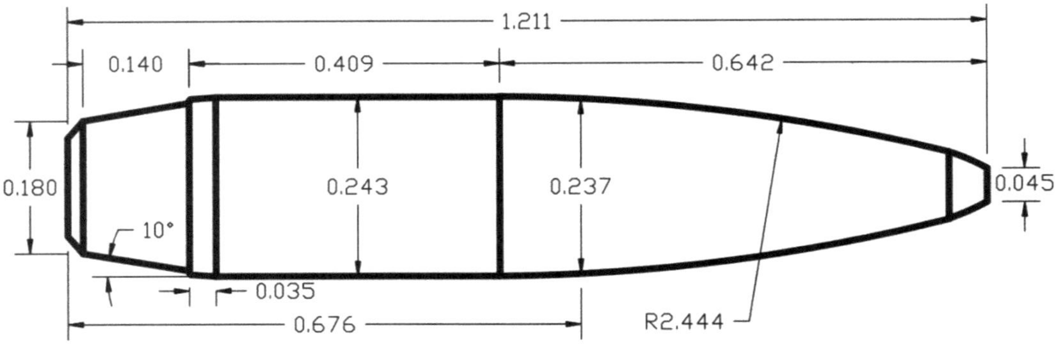

Lot #:	2231	Ogive Radius:	10.16 calibers
Sectional Density:	0.261 lb/ in^2	Rt/R:	0.95

Ballistic Performance

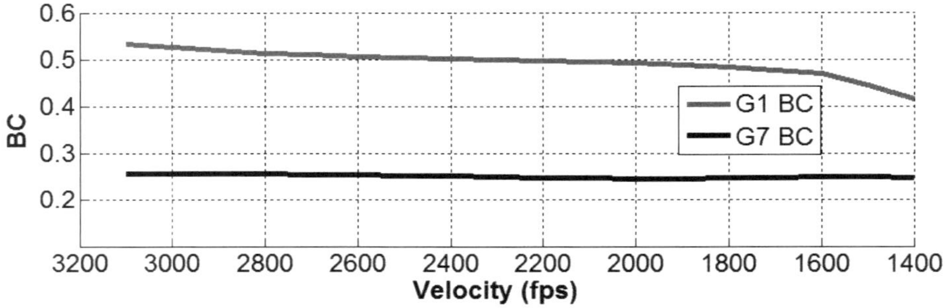

	Above 3000	3000-2500	2500-2000	2000-1500	Below 1500
G1 BC	0.543	0.515	0.497	0.469	0.446
G7 BC	0.257	0.255	0.250	0.249	0.251

Averages from 3000 fps to 1500 fps:
G1 BC: 0.492, variation of G1 BC: 0.080
G7 BC: 0.252, variation of G7 BC: 0.010
G7 form factor: 1.037

Rifling Twist Rate and Stability

Barrel Twist	Stability Factor Best case	Nominal	Worst case
1:6"	3.32	2.90	2.51
1:7"	2.44	2.13	1.84
1:8"	1.87	1.63	1.41
1:9"	1.48	1.29	1.11

Minimum twist required for stability is 1:9.3"
Recommended twist for optimal performance is 1:8.3" or faster

Precision Ballistics .243 caliber 113 grain VLD

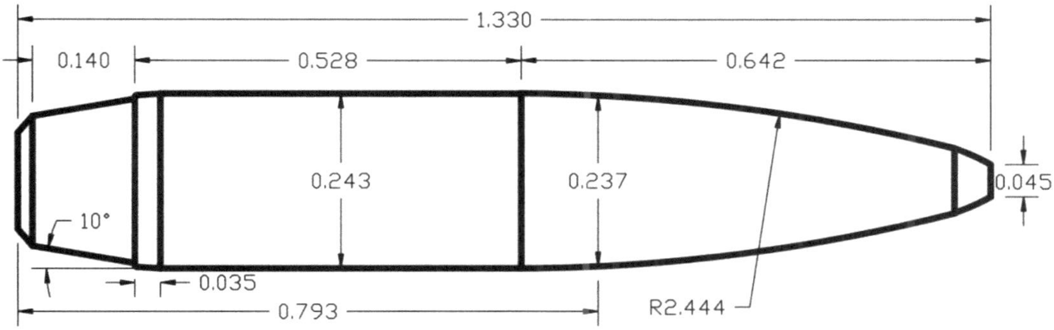

Lot #:	1198	Ogive Radius:	10.16 calibers
Sectional Density:	0.274 lb/ in²	Rt/R:	0.95

Ballistic Performance

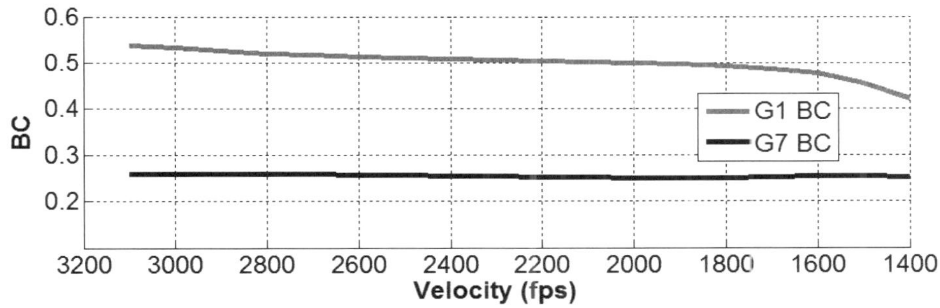

	Above 3000	3000-2500	2500-2000	2000-1500	Below 1500
G1 BC	0.549	0.521	0.505	0.476	0.453
G7 BC	0.260	0.258	0.254	0.253	0.255

Averages from 3000 fps to 1500 fps:
G1 BC: 0.499, variation of G1 BC: 0.079
G7 BC: 0.256, variation of G7 BC: 0.010
G7 form factor: 1.070

Rifling Twist Rate and Stability

Barrel Twist	Stability Factor Best case	Nominal	Worst case
1:6"	2.64	2.31	1.99
1:7"	1.94	1.69	1.46
1:8"	1.48	1.30	1.12
1:9"	1.17	1.02	0.89

Minimum twist required for stability is 1:8.3"
Recommended twist for optimal performance is 1:7.4" or faster

Precision Ballistics .243 caliber 115 grain VLD

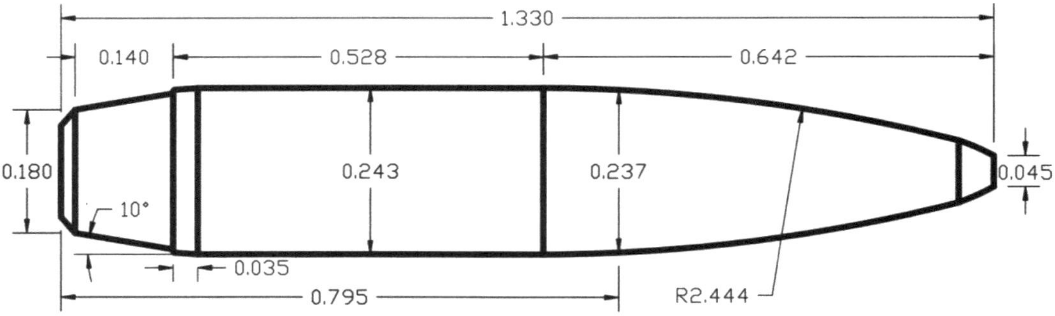

Lot #:	2306	Ogive Radius:	10.16 calibers
Sectional Density:	0.278 lb/ in^2	Rt/R:	0.95

Ballistic Performance

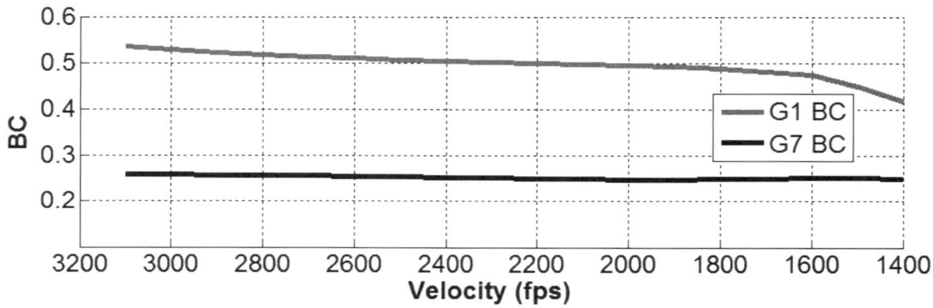

	Above 3000	3000-2500	2500-2000	2000-1500	Below 1500
G1 BC	0.546	0.518	0.501	0.473	0.450
G7 BC	0.259	0.257	0.252	0.251	0.253

Averages from 3000 fps to 1500 fps:
G1 BC: 0.495, variation of G1 BC: 0.079
G7 BC: 0.254, variation of G7 BC: 0.010
G7 form factor: 1.096

Rifling Twist Rate and Stability

Barrel Twist	Stability Factor Best case	Nominal	Worst case
1:6"	2.69	2.35	2.03
1:7"	1.97	1.72	1.49
1:8"	1.51	1.32	1.14
1:9"	1.19	1.04	0.90

Minimum twist required for stability is 1:8.4"
Recommended twist for optimal performance is 1:7.5" or faster

Sierra .243 caliber 80 grain Blitz

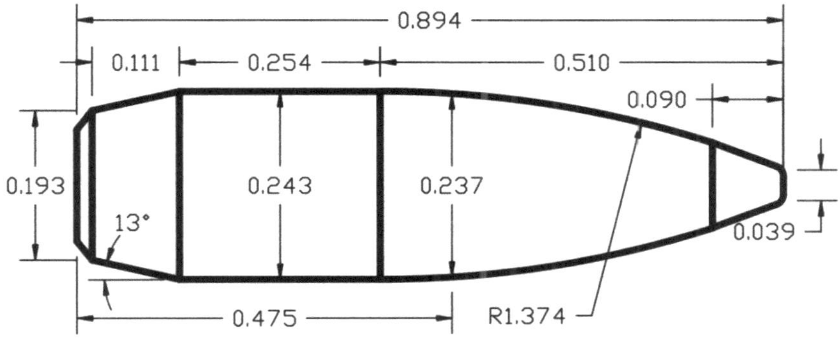

Lot #:	0013827141818	Ogive Radius:	5.63 calibers
Sectional Density:	0.194 lb/in^2	Rt/R:	0.99

Ballistic Performance

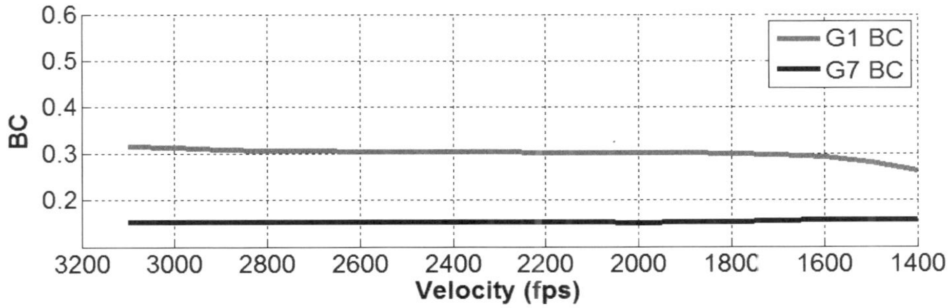

	Above 3000	3000-2500	2500-2000	2000-1500	Below 1500
G1 BC	0.320	0.309	0.304	0.292	0.281
G7 BC	0.152	0.153	0.153	0.155	0.158

Averages from 3000 fps to 1500 fps:
G1 BC: 0.300, variation of G1 BC: 0.032
G7 BC: 0.154, variation of G7 BC: 0.006
G7 form factor: 1.256

Rifling Twist Rate and Stability

Barrel Twist	Stability Factor Best case	Nominal	Worst case
1:7"	4.31	3.77	3.26
1:8"	3.30	2.89	2.50
1:9"	2.61	2.28	1.97
1:10"	2.11	1.85	1.60

Minimum twist required for stability is 1:12.4"
Recommended twist for optimal performance is 1:11.1" or faster

Sierra .243 caliber 70 grain MatchKing

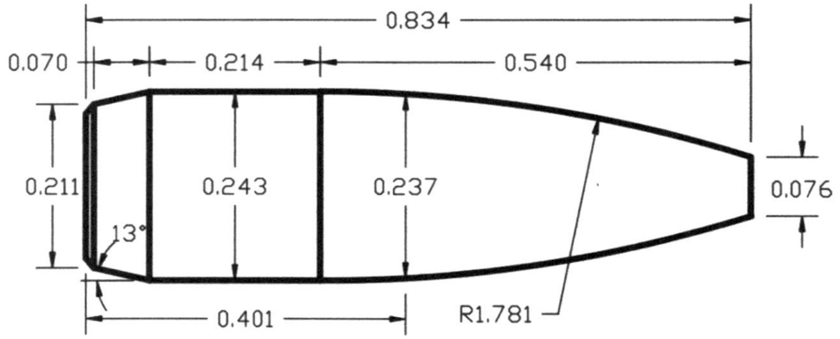

Lot #:	Unknown	Ogive Radius:	7.3 calibers
Sectional Density:	0.169 lb/in²	Rt/R:	1.00

Ballistic Performance

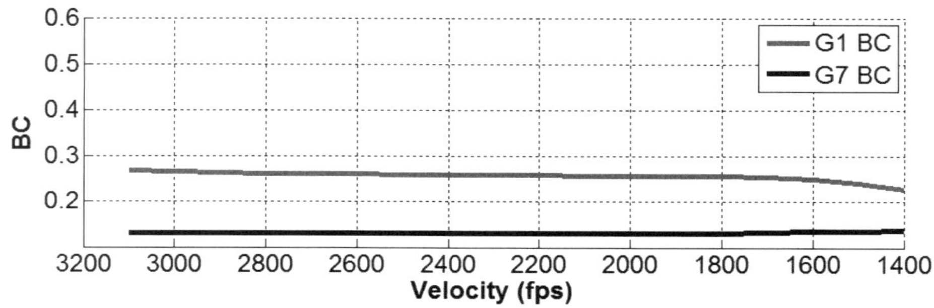

	Above 3000	3000-2500	2500-2000	2000-1500	Below 1500
G1 BC	0.274	0.263	0.259	0.250	0.241
G7 BC	0.130	0.131	0.130	0.133	0.136

Averages from 3000 fps to 1500 fps:
G1 BC: 0.257, variation of G1 BC: 0.026
G7 BC: 0.132, variation of G7 BC: 0.006
G7 form factor: 1.286

Rifling Twist Rate and Stability

Barrel Twist	Stability Factor Best case	Nominal	Worst case
1:7"	4.60	4.02	3.48
1:8"	3.52	3.08	2.66
1:9"	2.78	2.43	2.10
1:10"	2.26	1.97	1.70

Minimum twist required for stability is 1:12.8"
Recommended twist for optimal performance is 1:11.5" or faster

Sierra .243 caliber 95 grain MatchKing

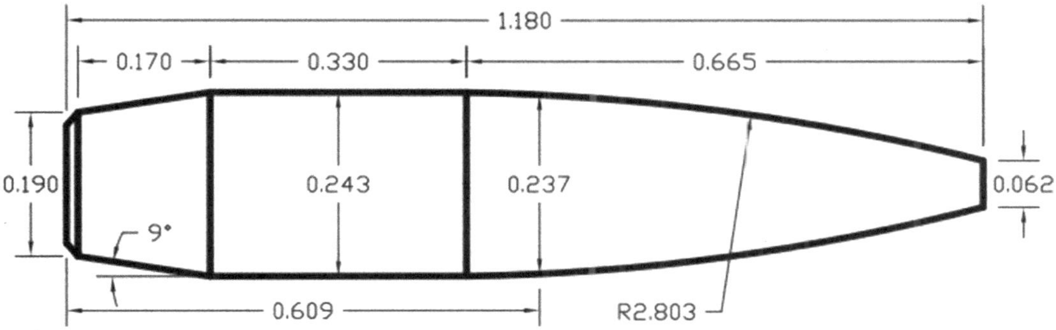

Lot #:	003811983	Ogive Radius:	11.5 calibers
Sectional Density:	0.230 lb/in^2	Rt/R:	0.90

Ballistic Performance

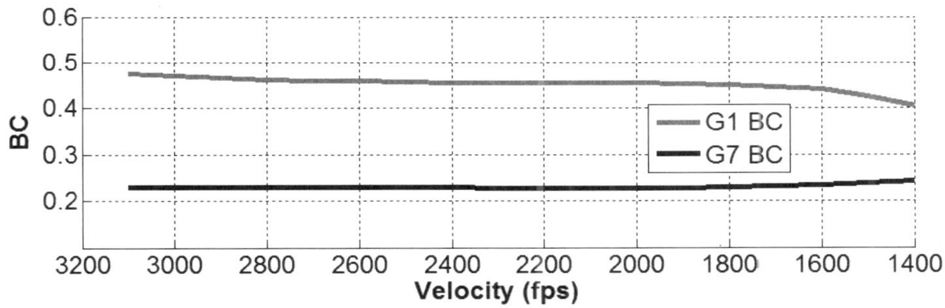

	Above 3000	3000-2500	2500-2000	2000-1500	Below 1500
G1 BC	0.481	0.463	0.455	0.439	0.424
G7 BC	0.228	0.230	0.229	0.233	0.239

Averages from 3000 fps to 1500 fps:
G1 BC: 0.451, variation of G1 BC: 0.045
G7 BC: 0.231, variation of G7 BC: 0.012
G7 form factor: 0.993

Rifling Twist Rate and Stability

	Stability Factor		
Barrel Twist	Best case	Nominal	Worst case
1:7"	2.30	2.01	1.73
1:8"	1.76	1.54	1.33
1:9"	1.39	1.21	1.05
1:10"	1.12	0.98	0.85

Minimum twist required for stability is 1:9.0"
Recommended twist for optimal performance is 1:8.1" or faster

Sierra .243 caliber 107 grain MatchKing

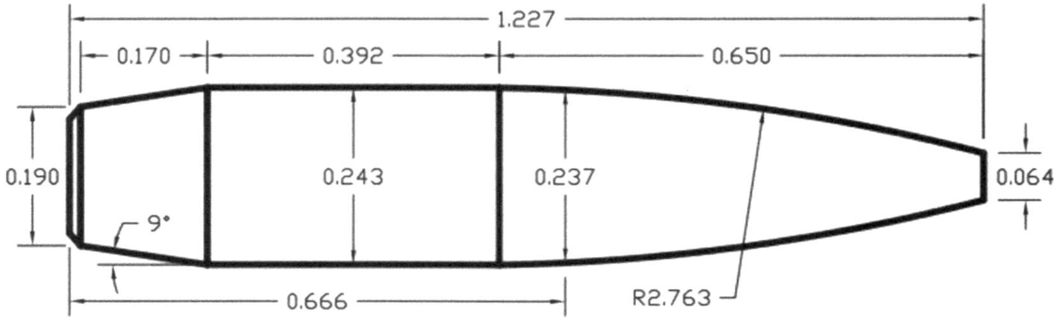

Lot #:	003842744	Ogive Radius:	11.32 calibers
Sectional Density:	0.259 lb/in^2	Rt/R:	0.89

Ballistic Performance

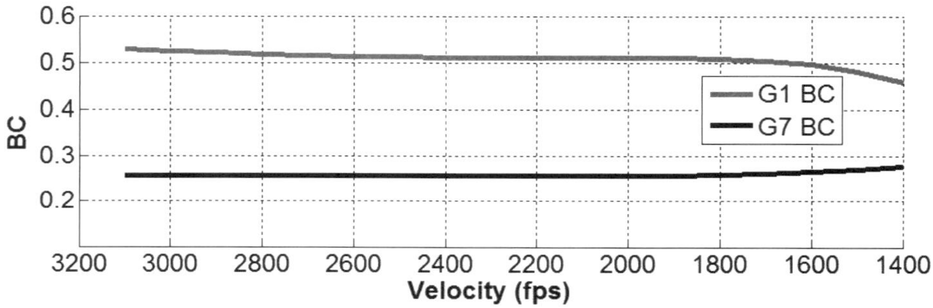

	Above 3000	3000-2500	2500-2000	2000-1500	Below 1500
G1 BC	0.538	0.519	0.511	0.496	0.481
G7 BC	0.255	0.257	0.257	0.264	0.271

Averages from 3000 fps to 1500 fps:
G1 BC: 0.507, variation of G1 BC: 0.045
G7 BC: 0.260, variation of G7 BC: 0.014
G7 form factor: 0.994

Rifling Twist Rate and Stability

Barrel Twist	Stability Factor Best case	Nominal	Worst case
1:7"	2.31	2.02	1.74
1:8"	1.77	1.54	1.33
1:9"	1.40	1.22	1.05
1:10"	1.13	0.99	0.85

Minimum twist required for stability is 1:9.1"
Recommended twist for optimal performance is 1:8.1" or faster

Sierra .243 caliber 100 grain ProHunter

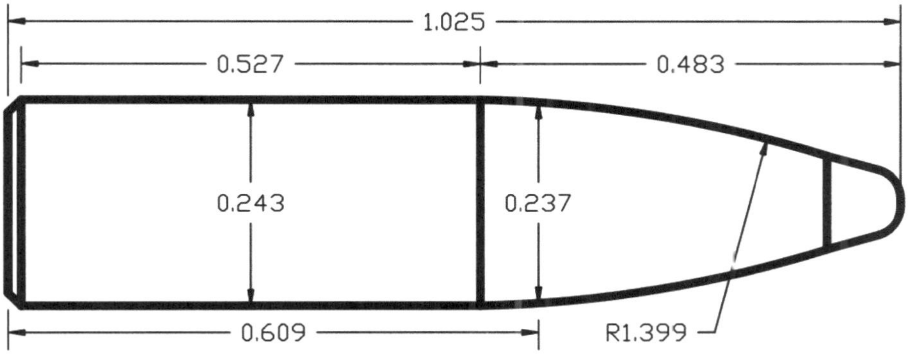

Lot #:	91717465633	Ogive Radius:	5.76 calibers
Sectional Density:	0.242 lb/ in^2	Rt/R:	0.78

Ballistic Performance

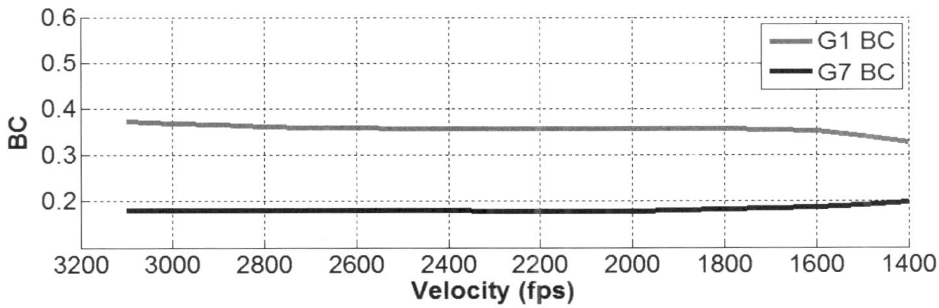

	Above 3000	3000-2500	2500-2000	2000-1500	Below 1500
G1 BC	0.380	0.363	0.357	0.349	0.341
G7 BC	0.180	0.180	0.179	0.186	0.192

Averages from 3000 fps to 1500 fps:
G1 BC: 0.356, variation of G1 BC: 0.028
G7 BC: 0.183, variation of G7 BC: 0.013
G7 form factor: 1.323

Rifling Twist Rate and Stability

Barrel Twist	Stability Factor Best case	Stability Factor Nominal	Stability Factor Worst case
1:7"	3.64	3.18	2.75
1:8"	2.79	2.43	2.10
1:9"	2.20	1.92	1.66
1:10"	1.78	1.56	1.35

Minimum twist required for stability is 1:11.4"
Recommended twist for optimal performance is 1:10.2" or faster

Vapor Trail .243 caliber 103 grain BTHP

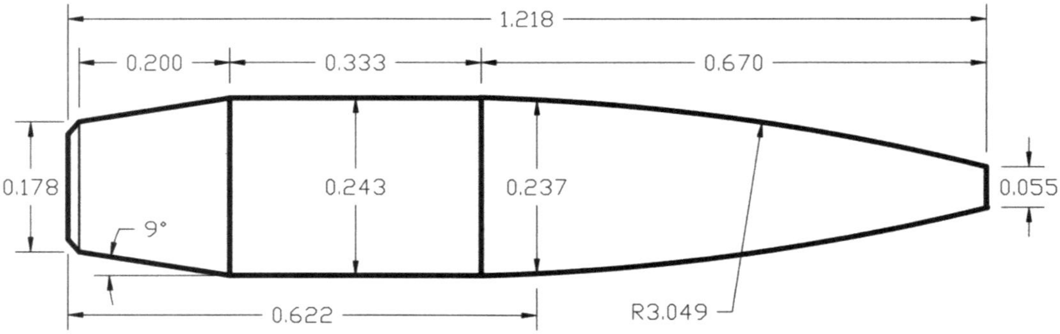

Lot #:	unknown	Ogive Radius:	12.55 calibers
Sectional Density:	0.249 lb/in^2	Rt/R:	0.80

Ballistic Performance

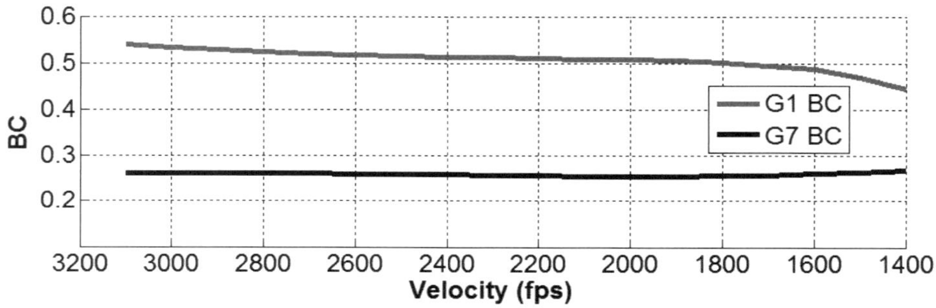

	Above 3000	3000-2500	2500-2000	2000-1500	Below 1500
G1 BC	0.549	0.525	0.512	0.489	0.470
G7 BC	0.260	0.260	0.257	0.260	0.265

Averages from 3000 fps to 1500 fps:
G1 BC: 0.507, variation of G1 BC: 0.064
G7 BC: 0.260, variation of G7 BC: 0.010
G7 form factor: 0.958

Rifling Twist Rate and Stability

Barrel Twist	Stability Factor Best case	Stability Factor Nominal	Stability Factor Worst case
1:7"	2.27	1.98	1.71
1:8"	1.74	1.52	1.31
1:9"	1.37	1.20	1.04
1:10"	1.11	0.97	0.84

Minimum twist required for stability is 1:9.0"
Recommended twist for optimal performance is 1:8.0" or faster

25 Caliber (.257" diameter)

.257 caliber bullets typically range from 80 to 115 grains (the traditional 1:10" twist rate doesn't stabilize heavy 25 caliber bullets). The image below relates bullet weight to sectional density for .257 caliber bullets.

Range of Weight and Sectional Density for 0.257 Caliber

Bullet Weight (grains)	81	90	99	107	116	125	134
Sectional Density (lb/in^2)	0.176	0.195	0.214	0.232	0.251	0.27	0.289

Sectional Density (SD) is a measure of how much mass is behind the bullets frontal area. Higher SD means a greater ability for the bullet to penetrate targets on impact, and also retain velocity in flight.

Spectrum of G7 form factors

0.80 0.85 0.90 0.95 1.00 1.05 1.10 1.15 1.20

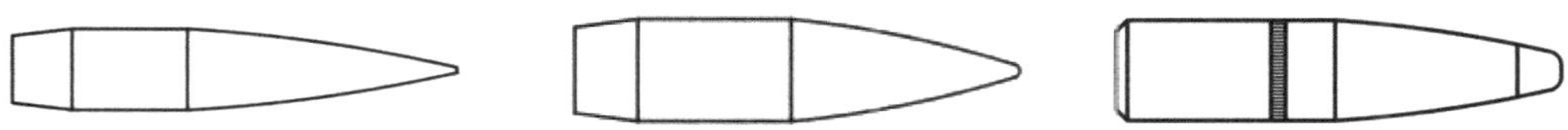

G7 Form factors range from 0.80 for the lowest drag shapes, up to 1.20+ for blunter shapes.

G7 Ballistic Coefficients are simply Sectional Density (SD) divided by G7 form factor.

You can determine a bullets' suitability for long range performance by seeing where it lies on the SD and form factor lines. Ideally, you want a bullet with a high SD and a low G7 form factor.

As an example, a 116 grain .257 caliber bullet has an SD of 0.251. If that bullet had a G7 form factor of 0.95, it's G7 BC would be:

G7 BC = SD/FF = 0.251/0.95 = 0.264.

Barnes 25 caliber 80 grain TTSXBT

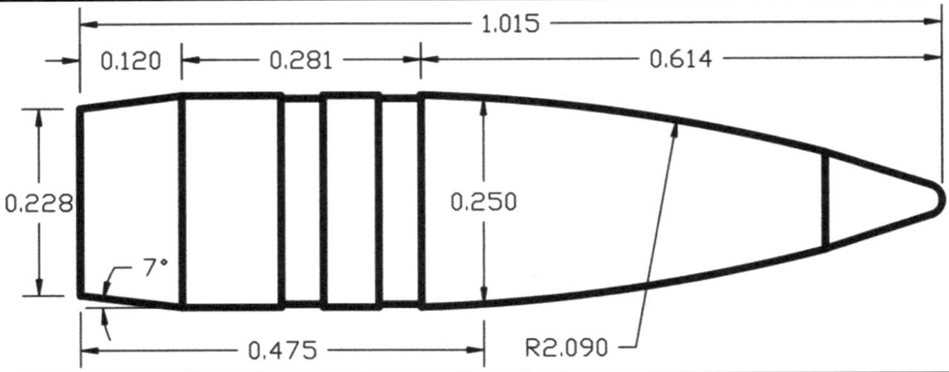

Lot #:	5281-1 091N BB-M	**Ogive Radius:**	**2.09 calibers**
Sectional Density:	**0.173 lb/ in²**	**Rt/R:**	**0.54**

Ballistic Performance

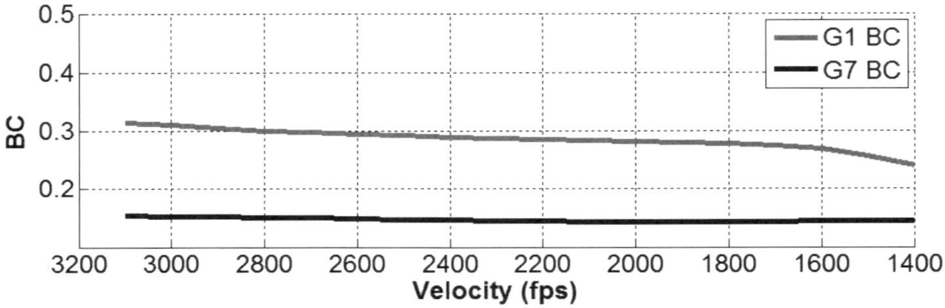

	Above 3000	3000-2500	2500-2000	2000-1500	Below 1500
G1 BC	0.324	0.301	0.287	0.270	0.257
G7 BC	0.153	0.149	0.144	0.143	0.144

Averages from 3000 fps to 1500 fps:
G1 BC: 0.285, variation of G1 BC: 0.054
G7 BC: 0.146, variation of G7 BC: 0.010
G7 form factor: 1.184

Rifling Twist Rate and Stability

Barrel Twist	Best case	Stability Factor Nominal	Worst case
1:9"	2.01	1.76	1.52
1:10"	1.63	1.42	1.23
1:11"	1.35	1.18	1.02
1:12"	1.13	0.99	0.86

Minimum twist required for stability is 1:10.9"
Recommended twist for optimal performance is 1:9.8" or faster

Barnes 25 caliber 115 grain TSX FB

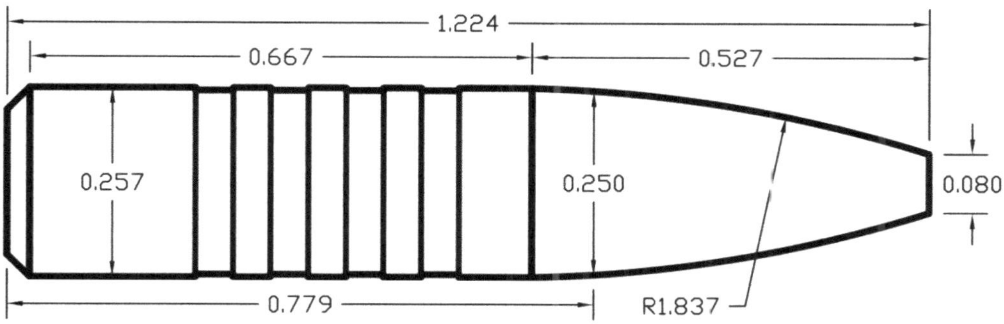

Lot #:	Unknown	Ogive Radius:	7.15 calibers
Sectional Density:	0.249 lb/in^2	Rt/R:	0.89

Ballistic Performance

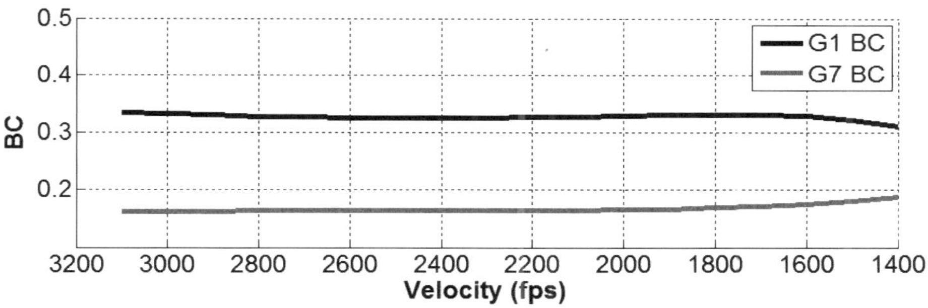

	Above 3000	3000-2500	2500-2000	2000-1500	Below 1500
G1 BC	0.340	0.329	0.327	0.325	0.321
G7 BC	0.161	0.163	0.164	0.173	0.181

Averages from 3000 fps to 1500 fps:
G1 BC: 0.327, variation of G1 BC: 0.011
G7 BC: 0.168, variation of G7 BC: 0.019
G7 form factor: 1.481

Rifling Twist Rate and Stability

Barrel Twist	Stability Factor Best case	Nominal	Worst case
1:9"	1.68	1.47	1.27
1:10"	1.36	1.19	1.03
1:11"	1.13	0.98	0.85
1:12"	0.95	0.83	0.71

Minimum twist required for stability is 1:10.0"
Recommended twist for optimal performance is 1:8.9" or faster

Berger 25 caliber 115 grain VLD

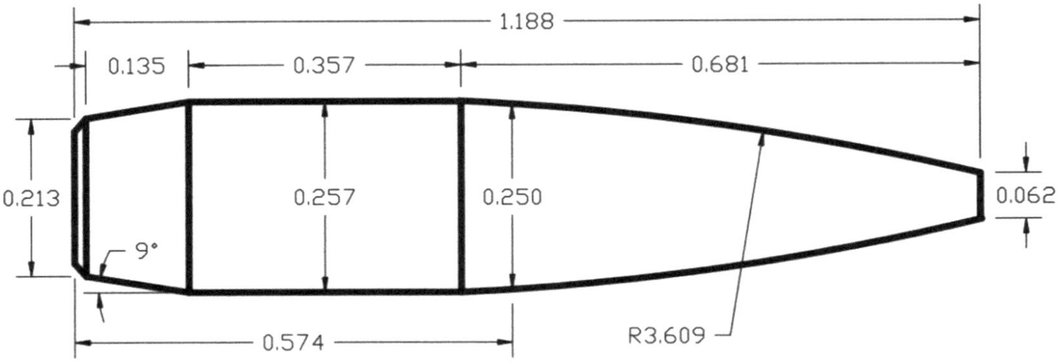

Lot #:	**5960**	**Ogive Radius:**	**14.04 calibers**
Sectional Density:	**0.249 lb/in²**	**Rt/R:**	**0.67**

Ballistic Performance

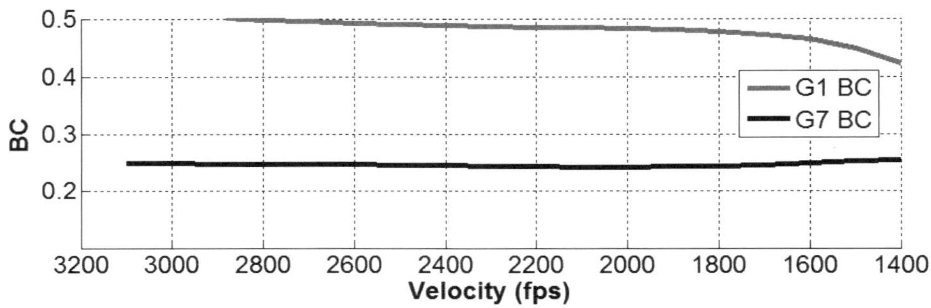

	Above 3000	3000-2500	2500-2000	2000-1500	Below 1500
G1 BC	0.523	0.499	0.486	0.465	0.448
G7 BC	0.248	0.247	0.244	0.247	0.252

Averages from 3000 fps to 1500 fps:
G1 BC: 0.482, variation of G1 BC: 0.060
G7 BC: 0.247, variation of G7 BC: 0.010
G7 form factor: 1.006

Rifling Twist Rate and Stability

Barrel Twist	**Stability Factor** Best case	Nominal	Worst case
1:8"	2.32	2.03	1.75
1:9"	1.83	1.60	1.39
1:10"	1.49	1.30	1.12
1:11"	1.23	1.07	0.93

Minimum twist required for stability is 1:10.4"
Recommended twist for optimal performance is 1:9.3" or faster

GS Custom 25 caliber 90 grain SP

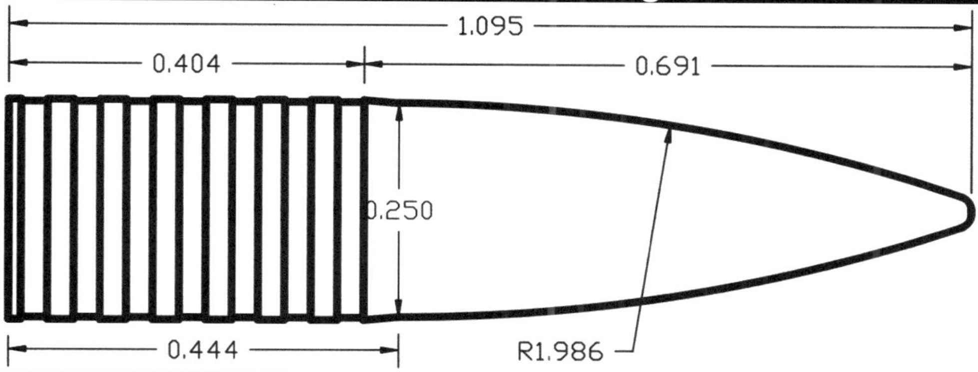

Lot #:	**Unknown**	**Ogive Radius:**	**7.23 calibers**
Sectional Density:	**0.195 lb/ in²**	**Rt/R:**	**0.83**

Ballistic Performance

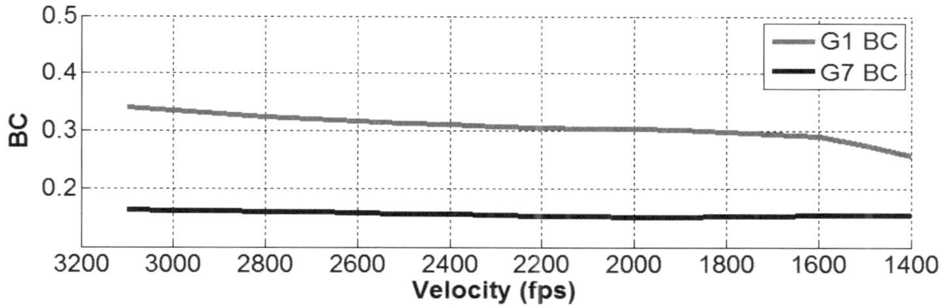

	Above 3000	3000-2500	2500-2000	2000-1500	Below 1500
G1 BC	0.349	0.323	0.308	0.290	0.277
G7 BC	0.165	0.160	0.155	0.154	0.156

Averages from 3000 fps to 1500 fps:
G1 BC: 0.307, variation of G1 BC: 0.057
G7 BC: 0.157, variation of G7 BC: 0.011
G7 form factor: 1.240

Rifling Twist Rate and Stability

	Stability Factor		
Barrel Twist	**Best case**	**Nominal**	**Worst case**
1:9"	1.82	1.59	1.37
1:10"	1.47	1.29	1.11
1:11"	1.22	1.06	0.92
1:12"	1.02	0.89	0.77

Minimum twist required for stability is 1:10.4"
Recommended twist for optimal performance is 1:9.3" or faster

Hornady 25 caliber 110 grain Interbond

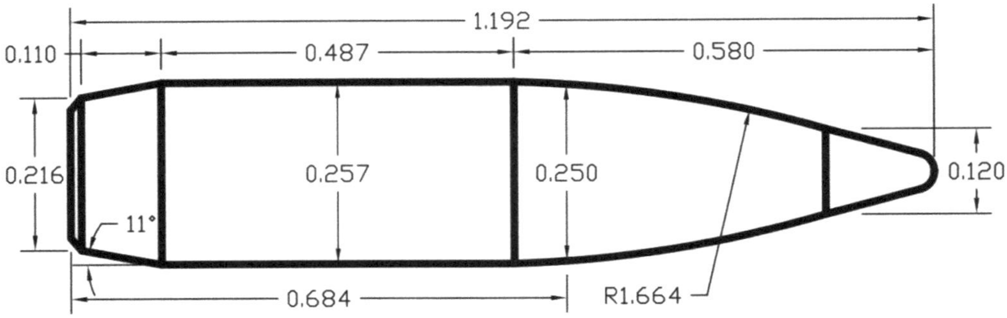

Lot #:	2070200	Ogive Radius:	6.48 calibers
Sectional Density:	0.238 lb/in²	Rt/R:	0.87

Ballistic Performance

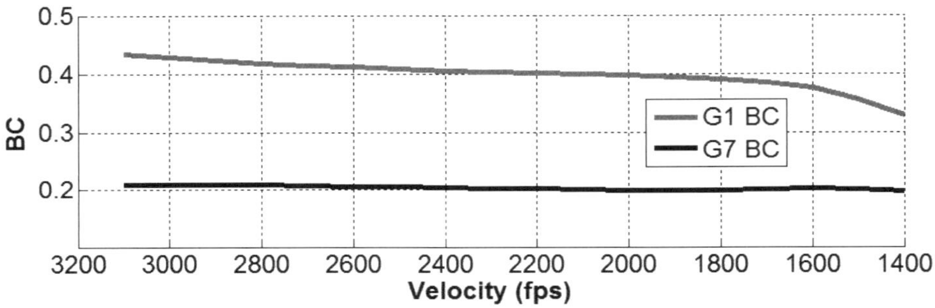

	Above 3000	3000-2500	2500-2000	2000-1500	Below 1500
G1 BC	0.444	0.419	0.403	0.377	0.356
G7 BC	0.210	0.208	0.202	0.200	0.200

Averages from 3000 fps to 1500 fps:
G1 BC: 0.398, variation of G1 BC: 0.072
G7 BC: 0.204, variation of G7 BC: 0.010
G7 form factor: 1.169

Rifling Twist Rate and Stability

Barrel Twist	Stability Factor Best case	Nominal	Worst case
1:9"	1.74	1.52	1.31
1:10"	1.41	1.23	1.06
1:11"	1.16	1.02	0.88
1:12"	0.98	0.85	0.74

Minimum twist required for stability is 1:10.1"
Recommended twist for optimal performance is 1:9.1" or faster

Hornady 25 caliber 87 grain Spitzer

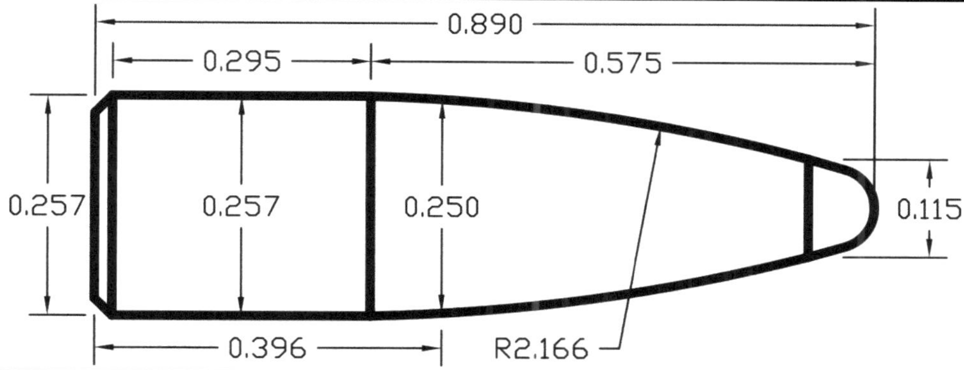

Lot #:	2060236		Ogive Radius:	8.43 calibers
Sectional Density:	0.188 lb/in^2		Rt/R:	0.86

Ballistic Performance

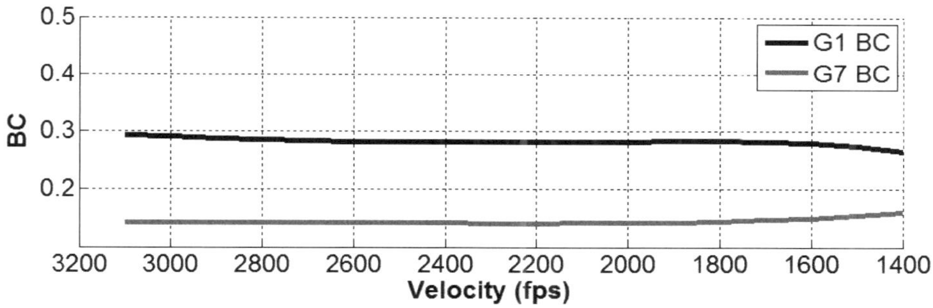

	Above 3000	3000-2500	2500-2000	2000-1500	Below 1500
G1 BC	0.298	0.286	0.282	0.279	0.274
G7 BC	0.141	0.142	0.142	0.148	0.155

Averages from 3000 fps to 1500 fps:
G1 BC: 0.282, variation of G1 BC: 0.016
G7 BC: 0.145, variation of G7 BC: 0.013
G7 form factor: 1.298

Rifling Twist Rate and Stability

	Stability Factor		
Barrel Twist	**Best case**	**Nominal**	**Worst case**
1:9"	3.19	2.79	2.41
1:10"	2.58	2.26	1.95
1:11"	2.14	1.87	1.61
1:12"	1.79	1.57	1.36

Minimum twist required for stability is 1:13.7"
Recommended twist for optimal performance is 1:12.3" or faster

Hornady 25 caliber 117 grain SST

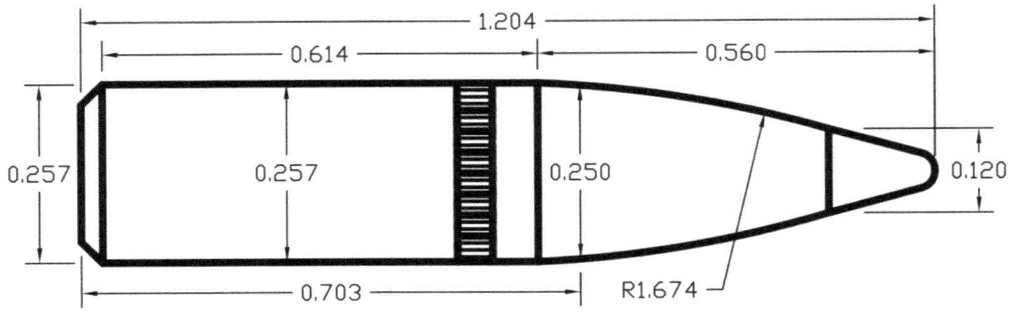

Lot #:	**2070719**	**Ogive Radius:**	**6.51 calibers**
Sectional Density:	**0.253 lb/in^2**	**Rt/R:**	**0.70**

Ballistic Performance

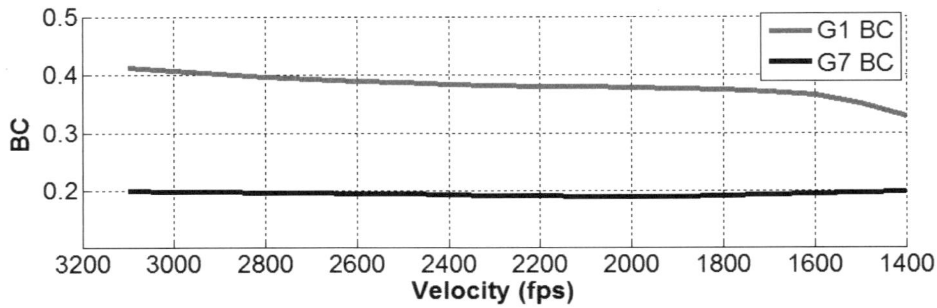

	Above 3000	3000-2500	2500-2000	2000-1500	Below 1500
G1 BC	0.423	0.396	0.382	0.364	0.350
G7 BC	0.200	0.197	0.192	0.193	0.197

Averages from 3000 fps to 1500 fps:
G1 BC: 0.380, variation of G1 BC: 0.057
G7 BC: 0.195, variation of G7 BC: 0.009
G7 form factor: 1.298

Rifling Twist Rate and Stability

Barrel Twist	Stability Factor Best case	Nominal	Worst case
1:9"	1.80	1.57	1.36
1:10"	1.45	1.27	1.10
1:11"	1.20	1.05	0.91
1:12"	1.01	0.88	0.76

Minimum twist required for stability is 1:10.3"
Recommended twist for optimal performance is 1:9.2" or faster

Nosler 25 caliber 110 grain Accubond

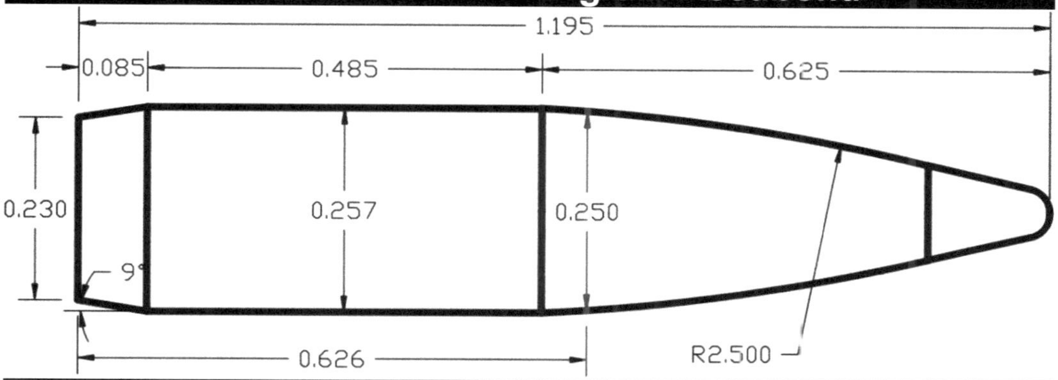

Lot #:	Unknown	Ogive Radius:	9.73 calibers
Sectional Density:	0.238 lb/ in²	Rt/R:	0.69

Ballistic Performance

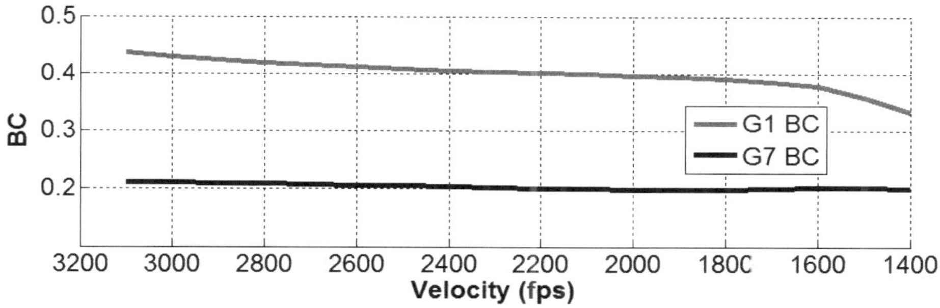

	Above 3000	3000-2500	2500-2000	2000-1500	Below 1500
G1 BC	0.447	0.419	0.402	0.378	0.359
G7 BC	0.212	0.208	0.202	0.200	0.201

Averages from 3000 fps to 1500 fps:
G1 BC: 0.399, variation of G1 BC: 0.072
G7 BC: 0.204, variation of G7 BC: 0.011
G7 form factor: 1.166

Rifling Twist Rate and Stability

Barrel Twist	Best case	Stability Factor Nominal	Worst case
1:9"	1.73	1.51	1.30
1:10"	1.40	1.22	1.06
1:11"	1.15	1.01	0.87
1:12"	0.97	0.85	0.73

Minimum twist required for stability is 1:10.1"
Recommended twist for optimal performance is 1:9.0" or faster

Nosler 25 caliber 100 grain Ballistic Tip

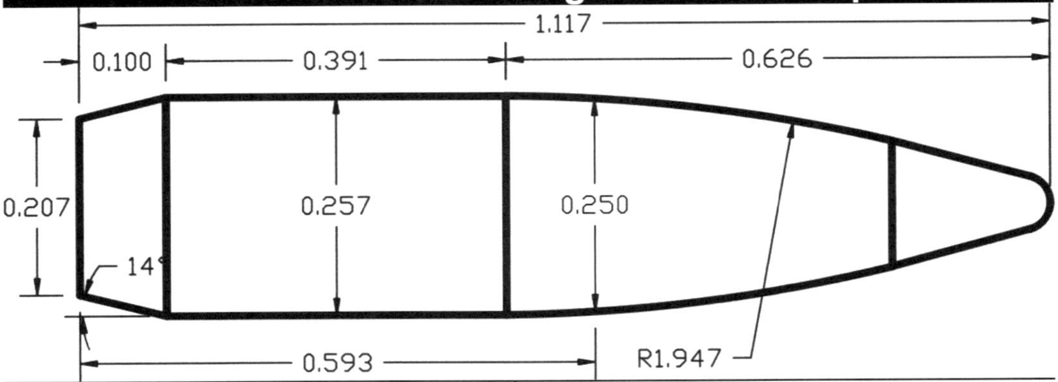

Lot #:	DL15F7	Ogive Radius:	7.58 calibers
Sectional Density:	0.216 lb/ in^2	Rt/R:	1.00

Ballistic Performance

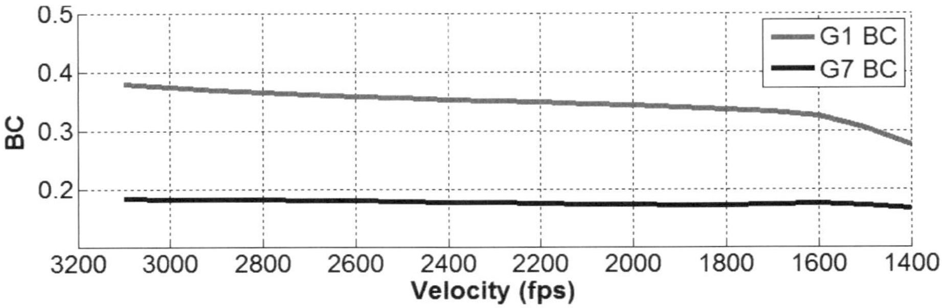

	Above 3000	3000-2500	2500-2000	2000-1500	Below 1500
G1 BC	0.387	0.365	0.350	0.325	0.306
G7 BC	0.183	0.181	0.176	0.172	0.171

Averages from 3000 fps to 1500 fps:
G1 BC: 0.345, variation of G1 BC: 0.068
G7 BC: 0.176, variation of G7 BC: 0.012
G7 form factor: 1.227

Rifling Twist Rate and Stability

Barrel Twist	Stability Factor Best case	Nominal	Worst case
1:9"	1.91	1.67	1.44
1:10"	1.55	1.35	1.17
1:11"	1.28	1.12	0.96
1:12"	1.07	0.94	0.81

Minimum twist required for stability is 1:10.6"
Recommended twist for optimal performance is 1:9.5" or faster

Nosler 25 caliber 115 grain Ballistic Tip

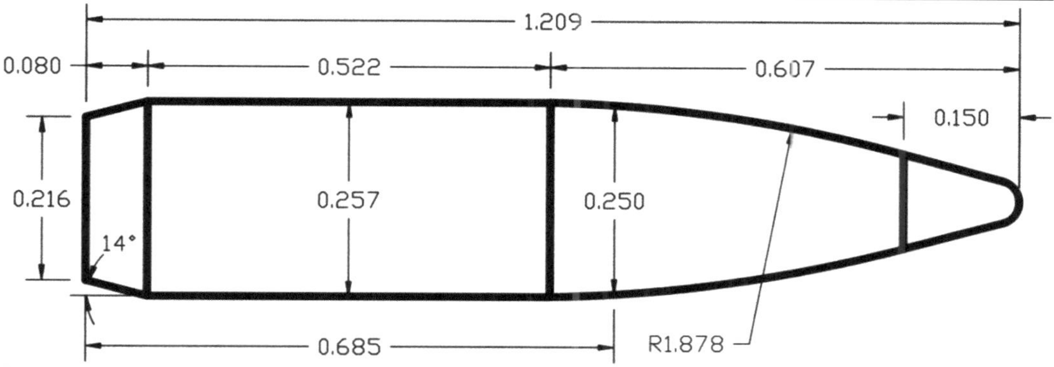

Lot #:	CEO1F7-1	Ogive Radius:	7.31 calibers
Sectional Density:	0.249 lb/in^2	Rt/R:	0.90

Ballistic Performance

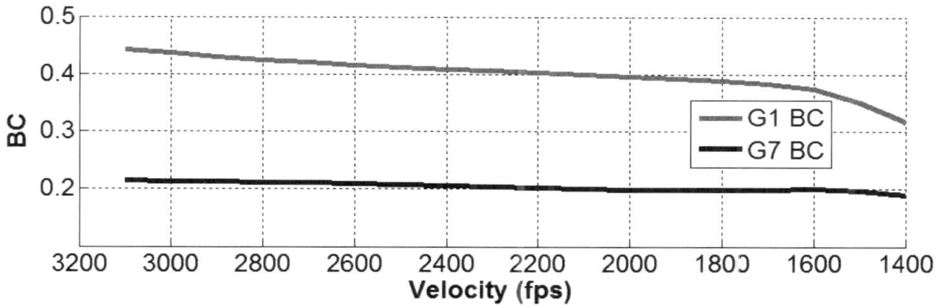

	Above 3000	3000-2500	2500-2000	2000-1500	Below 1500
G1 BC	0.454	0.425	0.405	0.374	0.351
G7 BC	0.215	0.211	0.203	0.198	0.196

Averages from 3000 fps to 1500 fps:
G1 BC: 0.399, variation of G1 BC: 0.086
G7 BC: 0.204, variation of G7 BC: 0.017
G7 form factor: 1.219

Rifling Twist Rate and Stability

Barrel Twist	Stability Factor Best case	Nominal	Worst case
1:9"	1.74	1.52	1.32
1:10"	1.41	1.23	1.07
1:11"	1.17	1.02	0.88
1:12"	0.98	0.86	0.74

Minimum twist required for stability is 1:10.1"
Recommended twist for optimal performance is 1:9.1" or faster

Nosler 25 caliber 115 grain Partition

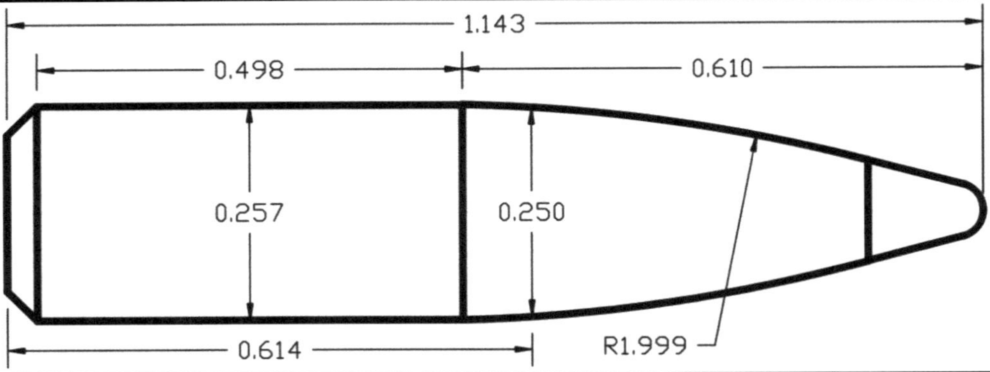

Lot #:	EL21F1	Ogive Radius:	7.78 calibers
Sectional Density:	0.249 lb/ in²	Rt/R:	1.00

Ballistic Performance

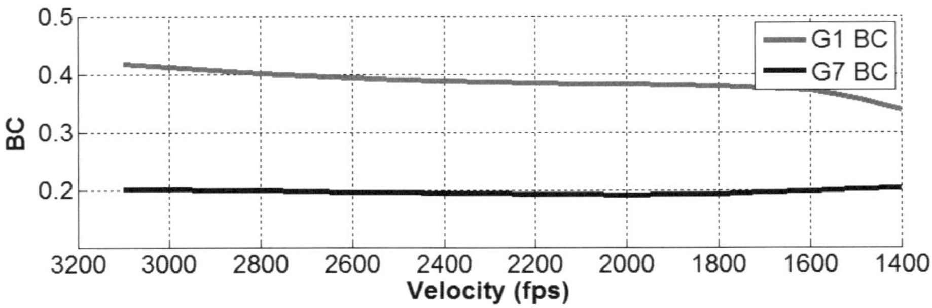

	Above 3000	3000-2500	2500-2000	2000-1500	Below 1500
G1 BC	0.428	0.401	0.387	0.370	0.358
G7 BC	0.203	0.199	0.194	0.197	0.201

Averages from 3000 fps to 1500 fps:
G1 BC: 0.386, variation of G1 BC: 0.054
G7 BC: 0.198, variation of G7 BC: 0.009
G7 form factor: 1.257

Rifling Twist Rate and Stability

Barrel Twist	Stability Factor Best case	Nominal	Worst case
1:9"	2.05	1.79	1.55
1:10"	1.66	1.45	1.26
1:11"	1.37	1.20	1.04
1:12"	1.15	1.01	0.87

Minimum twist required for stability is 1:11.0"
Recommended twist for optimal performance is 1:9.8" or faster

Sierra 25 caliber 70 grain BlitzKing

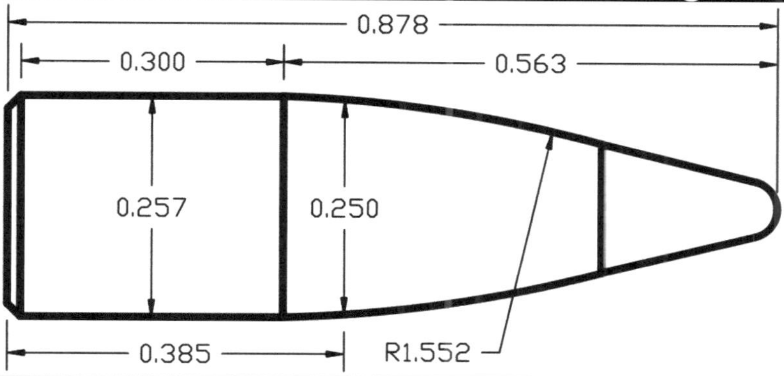

Lot #:	0061258379845GI	**Ogive Radius:**	**6.04 calibers**
Sectional Density:	**0.151 lb/ in^2**	**Rt/R:**	**1.00**

Ballistic Performance

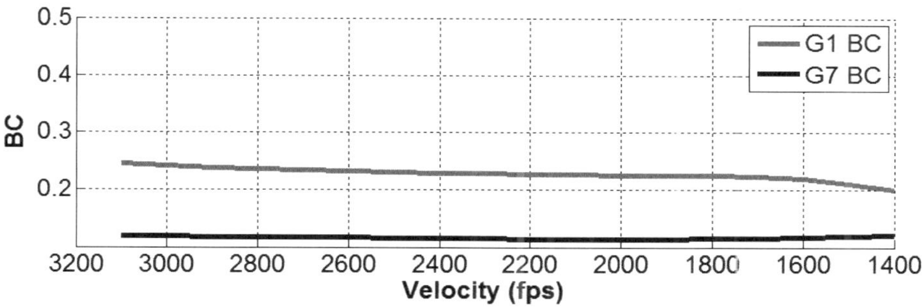

	Above 3000	3000-2500	2500-2000	2000-1500	Below 1500
G1 BC	0.250	0.236	0.229	0.219	0.212
G7 BC	0.119	0.117	0.115	0.117	0.119

Averages from 3000 fps to 1500 fps:
G1 BC: 0.228, variation of G1 BC: 0.030
G7 BC: 0.117, variation of G7 BC: 0.005
G7 form factor: 1.296

Rifling Twist Rate and Stability

	Stability Factor		
Barrel Twist	**Best case**	**Nominal**	**Worst case**
1:9"	2.67	2.33	2.02
1:10"	2.16	1.89	1.63
1:11"	1.79	1.56	1.35
1:12"	1.50	1.31	1.13

Minimum twist required for stability is 1:12.5"
Recommended twist for optimal performance is 1:11.2" or faster

Sierra 25 caliber 100 grain MatchKing

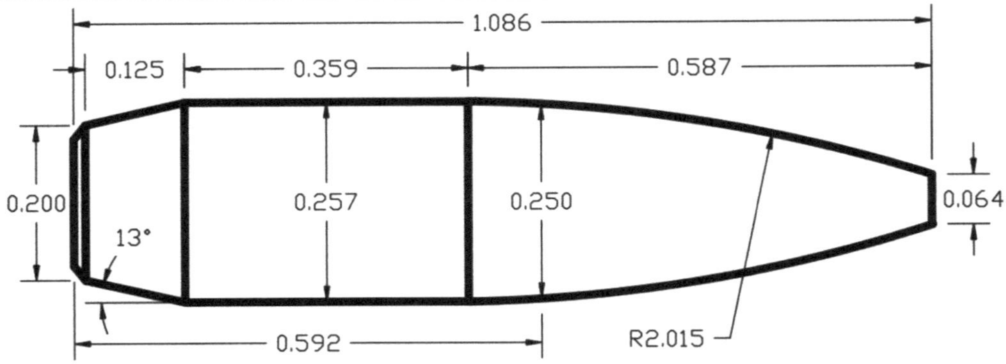

Lot #:	**003984405**	**Ogive Radius:**	**7.84 calibers**
Sectional Density:	**0.216 lb/in²**	**Rt/R:**	**0.95**

Ballistic Performance

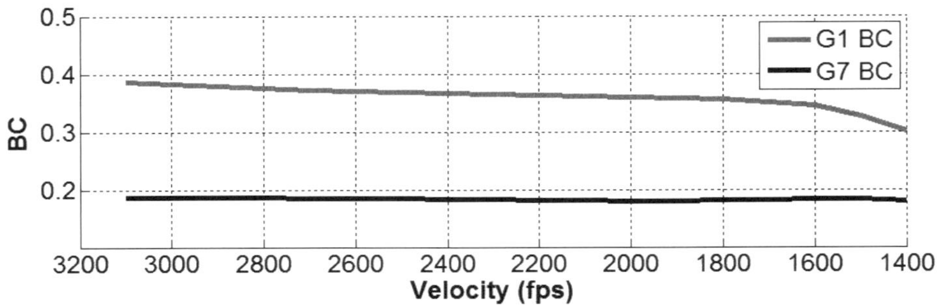

	Above 3000	3000-2500	2500-2000	2000-1500	Below 1500
G1 BC	0.395	0.376	0.365	0.344	0.327
G7 BC	0.187	0.186	0.183	0.182	0.184

Averages from 3000 fps to 1500 fps:
G1 BC: 0.360, variation of G1 BC: 0.056
G7 BC: 0.184, variation of G7 BC: 0.006
G7 form factor: 1.173

Rifling Twist Rate and Stability

Barrel Twist	Stability Factor Best case	Stability Factor Nominal	Stability Factor Worst case
1:9"	2.07	1.81	1.56
1:10"	1.68	1.46	1.27
1:11"	1.39	1.21	1.05
1:12"	1.16	1.02	0.88

Minimum twist required for stability is 1:11.0"
Recommended twist for optimal performance is 1:9.9" or faster

Sierra 25 caliber 117 grain ProHunter

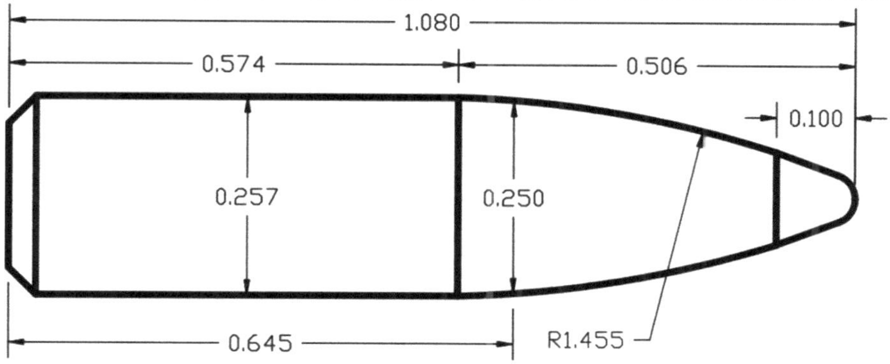

Lot #:	003962598	Ogive Radius:	5.66 calibers
Sectional Density:	0.253 lb/in^2	Rt/R:	0.96

Ballistic Performance

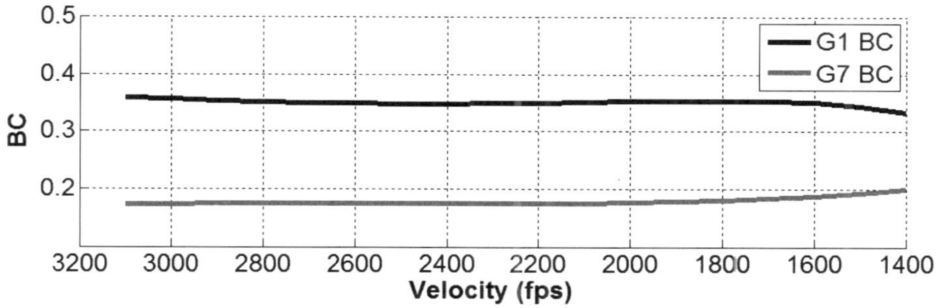

	Above 3000	3000-2500	2500-2000	2000-1500	Below 1500
G1 BC	0.363	0.352	0.350	0.347	0.343
G7 BC	0.172	0.174	0.176	0.185	0.194

Averages from 3000 fps to 1500 fps:
G1 BC: 0.349, variation of G1 BC: 0.013
G7 BC: 0.180, variation of G7 BC: 0.020
G7 form factor: 1.410

Rifling Twist Rate and Stability

Barrel Twist	Stability Factor Best case	Nominal	Worst case
1:9"	2.46	2.15	1.86
1:10"	1.99	1.74	1.51
1:11"	1.65	1.44	1.24
1:12"	1.38	1.21	1.05

Minimum twist required for stability is 1:12.0"
Recommended twist for optimal performance is 1:10.8" or faster

Swift 25 caliber 100 grain Scirocco II

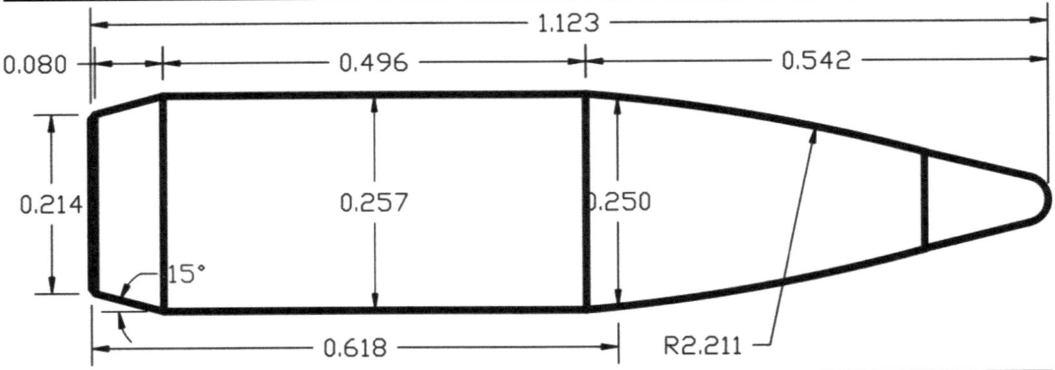

Lot #:	884W	Ogive Radius:	8.60 calibers
Sectional Density:	0.216 lb/ in²	Rt/R:	0.50

Ballistic Performance

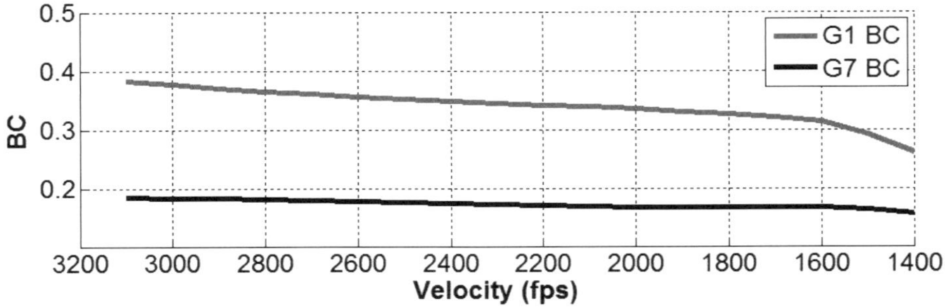

	Above 3000	3000-2500	2500-2000	2000-1500	Below 1500
G1 BC	0.393	0.365	0.344	0.314	0.293
G7 BC	0.186	0.181	0.173	0.166	0.164

Averages from 3000 fps to 1500 fps:
G1 BC: 0.340, variation of G1 BC: 0.084
G7 BC: 0.174, variation of G7 BC: 0.020
G7 form factor: 1.249

Rifling Twist Rate and Stability

Barrel Twist	Stability Factor Best case	Nominal	Worst case
1:9"	1.88	1.64	1.42
1:10"	1.52	1.33	1.15
1:11"	1.26	1.10	0.95
1:12"	1.06	0.92	0.80

Minimum twist required for stability is 1:10.5"
Recommended twist for optimal performance is 1:9.4" or faster

.264 Caliber (6.5mm)

.264 caliber bullets typically range from 90 to 145 grains. The image below relates bullet weight to sectional density for .264 caliber bullets.

Range of Weight and Sectional Density for 0.264 Caliber

Bullet Weight (grains)

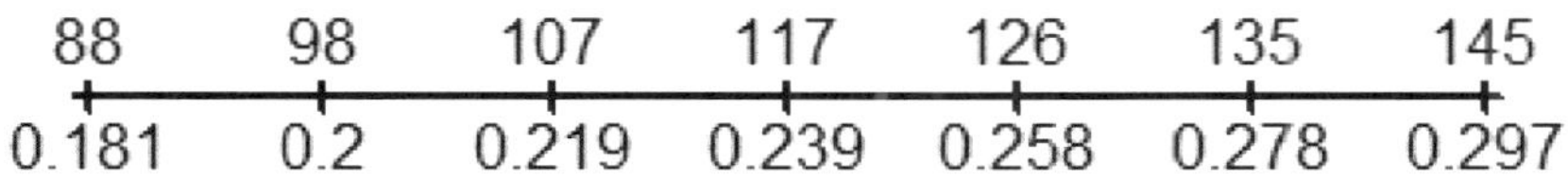

Sectional Density (lb/in^2)

Sectional Density (SD) is a measure of how much mass is behind the bullets frontal area. Higher SD means a greater ability for the bullet to penetrate targets on impact, and also retain velocity in flight.

Spectrum of G7 form factors

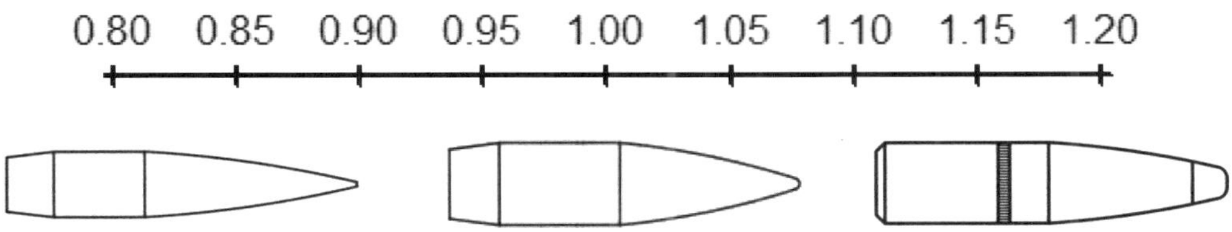

G7 Form factors range from 0.80 for the lowest drag shapes, up to 1.20+ for blunter shapes.

G7 Ballistic Coefficients are simply Sectional Density (SD) divided by G7 form factor.

You can determine a bullets' suitability for long range performance by seeing where it lies on the SD and form factor lines. Ideally, you want a bullet with a high SD and a low G7 form factor.

As an example, a 140 grain .264 caliber bullet has an SD of 0.287. If that bullet had a G7 form factor of 0.95, it's G7 BC would be:

G7 BC = SD/FF = 0.287/0.95 = 0.302.

Barnes .264 caliber 127 grain LRXBT

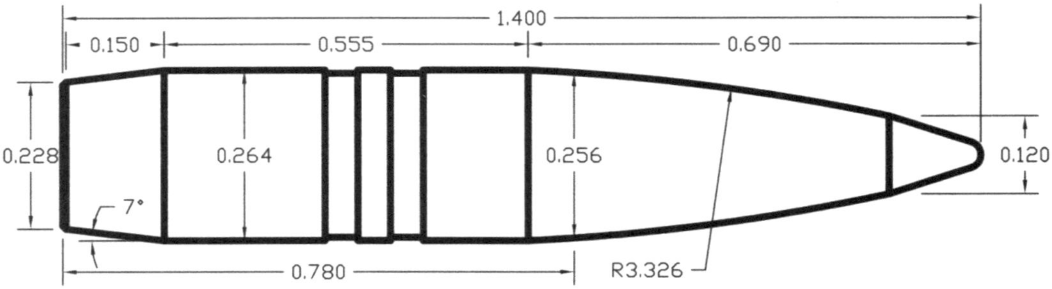

Lot #:	Unknown	Ogive Radius:	12.60 calibers
Sectional Density:	0.260 lb/in^2	Rt/R:	0.66

Ballistic Performance

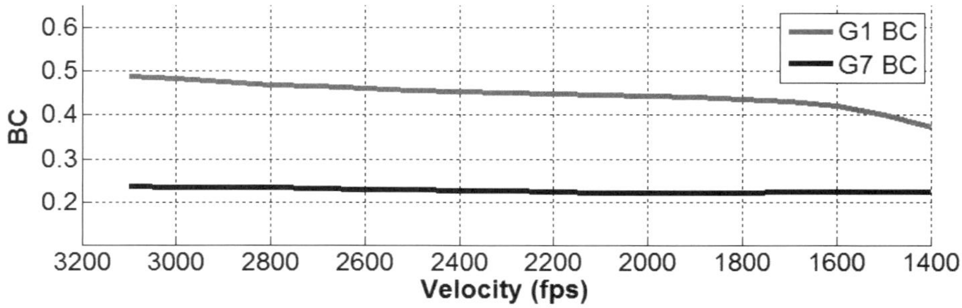

	Above 3000	3000-2500	2500-2000	2000-1500	Below 1500
G1 BC	0.501	0.469	0.450	0.422	0.401
G7 BC	0.237	0.233	0.226	0.224	0.225

Averages from 3000 fps to 1500 fps:
G1 BC: 0.446, variation of G1 BC: 0.081
G7 BC: 0.228, variation of G7 BC: 0.013
G7 form factor: 1.141

Rifling Twist Rate and Stability

Barrel Twist	Stability Factor Best case	Nominal	Worst case
1:8"	1.68	1.47	1.27
1:9"	1.33	1.16	1.00
1:10"	1.08	0.94	0.81
1:11"	0.89	0.78	0.67

Minimum twist required for stability is 1:8.8"
Recommended twist for optimal performance is 1:7.9" or faster

Barnes .264 caliber 100 grain TTSXBT

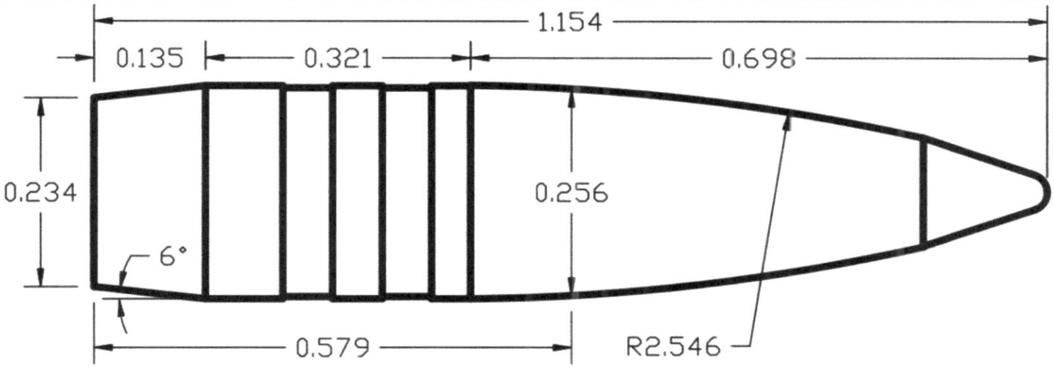

Lot #:	6387 091 113N AH-M	**Ogive Radius:**	**9.64 calibers**
Sectional Density:	**0.205 lb/in²**	**Rt/R:**	**1.00**

Ballistic Performance

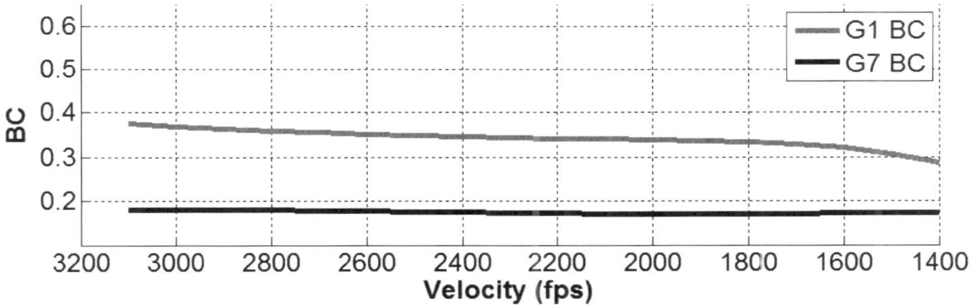

	Above 3000	3000-2500	2500-2000	2000-1500	Below 1500
G1 BC	0.384	0.360	0.344	0.323	0.308
G7 BC	0.182	0.178	0.173	0.172	0.173

Averages from 3000 fps to 1500 fps:
G1 BC: 0.341, variation of G1 BC: 0.062
G7 BC: 0.175, variation of G7 BC: 0.01
G7 form factor: 1.172

Rifling Twist Rate and Stability

Barrel Twist	Stability Factor Best case	Stability Factor Nominal	Stability Factor Worst case
1:8"	2.31	2.02	1.75
1:9"	1.83	1.60	1.38
1:10"	1.48	1.29	1.12
1:11"	1.22	1.07	0.92

Minimum twist required for stability is 1:10.4"
Recommended twist for optimal performance is 1:9.3" or faster

Barnes .264 caliber 120 grain TTSXBT

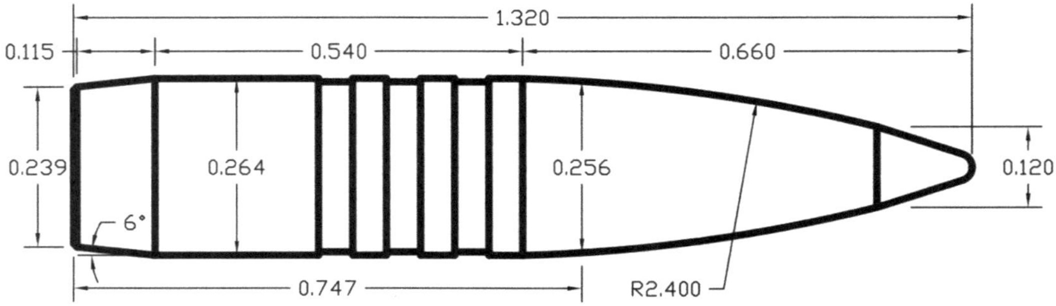

Lot #:	Unknown	Ogive Radius:	9.09 calibers
Sectional Density:	0.246 lb/in^2	Rt/R:	0.80

Ballistic Performance

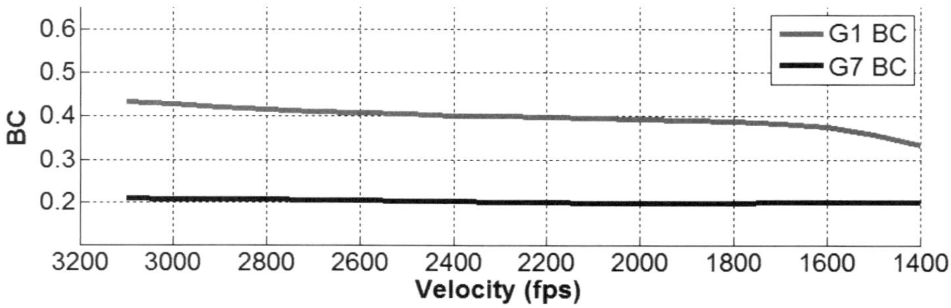

	Above 3000	3000-2500	2500-2000	2000-1500	Below 1500
G1 BC	0.443	0.416	0.400	0.376	0.359
G7 BC	0.210	0.206	0.201	0.200	0.202

Averages from 3000 fps to 1500 fps:
G1 BC: 0.396, variation of G1 BC: 0.069
G7 BC: 0.203, variation of G7 BC: 0.011
G7 form factor: 1.212

Rifling Twist Rate and Stability

Barrel Twist	Stability Factor Best case	Nominal	Worst case
1:8"	1.88	1.64	1.42
1:9"	1.48	1.30	1.12
1:10"	1.20	1.05	0.91
1:11"	0.99	0.87	0.75

Minimum twist required for stability is 1:9.3"
Recommended twist for optimal performance is 1:8.4" or faster

Berger .264 caliber 130 grain Hunting VLD

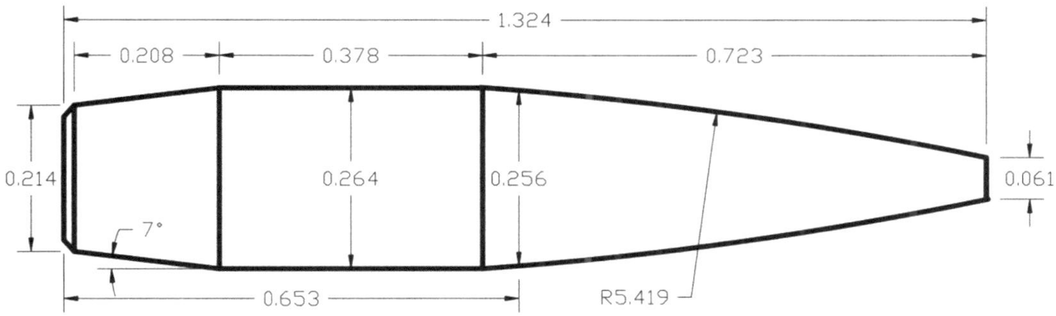

Lot #:	7618	Ogive Radius:	20.53 calibers
Sectional Density:	0.266 lb/in^2	Rt/R:	0.49

Ballistic Performance

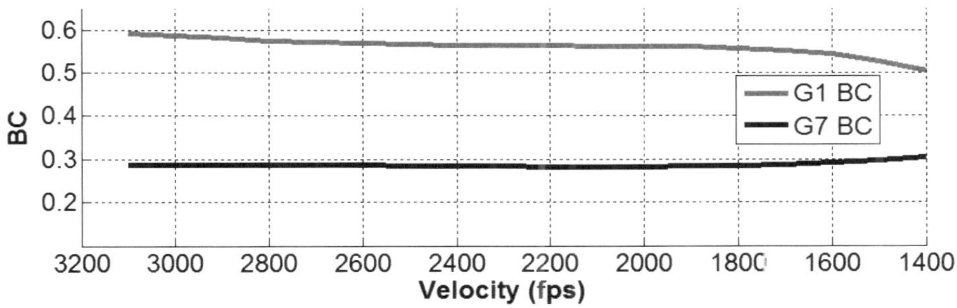

	Above 3000	3000-2500	2500-2000	2000-1500	Below 1500
G1 BC	0.603	0.577	0.565	0.545	0.528
G7 BC	0.286	0.286	0.284	0.290	0.297

Averages from 3000 fps to 1500 fps:
G1 BC: 0.561, variation of G1 BC: 0.059
G7 BC: 0.288, variation of G7 BC: 0.015
G7 form factor: 0.926

Rifling Twist Rate and Stability

Barrel Twist	Stability Factor Best case	Nominal	Worst case
1:7"	2.63	2.30	1.99
1:8"	2.01	1.76	1.52
1:9"	1.59	1.39	1.20
1:10"	1.29	1.13	0.97

Minimum twist required for stability is 1:9.7"
Recommended twist for optimal performance is 1:8.7" or faster

Berger .264 caliber 140 grain Hunting VLD

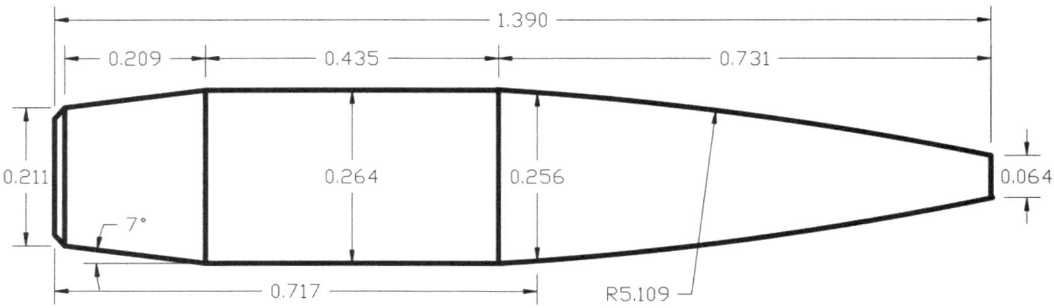

Lot #:	7636	Ogive Radius:	19.35 calibers
Sectional Density:	0.287 lb/in^2	Rt/R:	0.53

Ballistic Performance

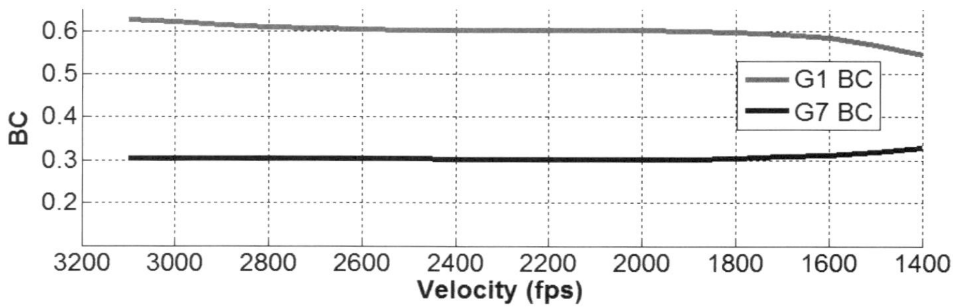

	Above 3000	3000-2500	2500-2000	2000-1500	Below 1500
G1 BC	0.637	0.612	0.602	0.584	0.567
G7 BC	0.302	0.303	0.302	0.311	0.320

Averages from 3000 fps to 1500 fps:
G1 BC: 0.598, variation of G1 BC: 0.054
G7 BC: 0.307, variation of G7 BC: 0.018
G7 form factor: 0.935

Rifling Twist Rate and Stability

Barrel Twist	Best case	Stability Factor Nominal	Worst case
1:7"	2.46	2.15	1.86
1:8"	1.88	1.64	1.42
1:9"	1.49	1.30	1.12
1:10"	1.20	1.05	0.91

Minimum twist required for stability is 1:9.4"
Recommended twist for optimal performance is 1:8.4" or faster

Berger .264 caliber 130 grain Tactical AR Hybrid

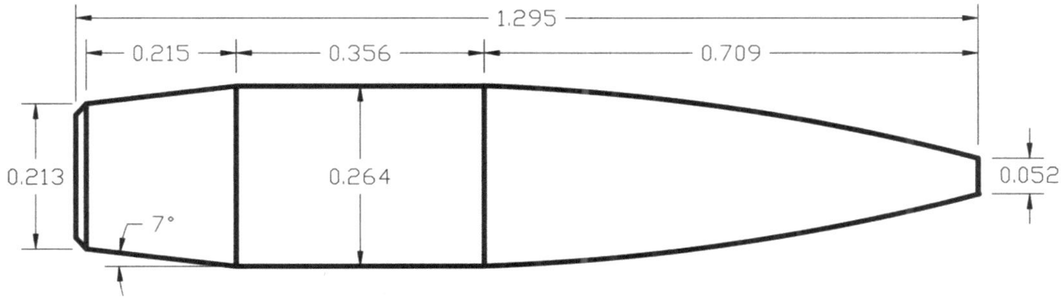

Lot #:	7543	Ogive Radius:	Hybrid
Sectional Density:	0.267 lb/in^2	Rt/R:	Hybrid

Ballistic Performance

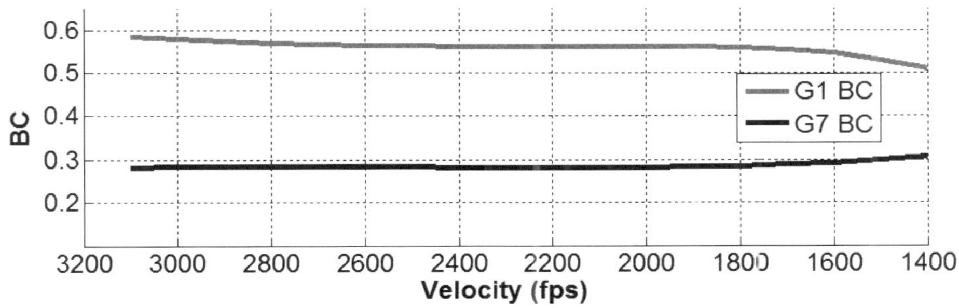

	Above 3000	3000-2500	2500-2000	2000-1500	Below 1500
G1 BC	0.594	0.571	0.563	0.546	0.531
G7 BC	0.281	0.283	0.283	0.291	0.299

Averages from 3000 fps to 1500 fps:
G1 BC: 0.559, variation of G1 BC: 0.049
G7 BC: 0.287, variation of G7 BC: 0.017
G7 form factor: 0.929

Rifling Twist Rate and Stability

Barrel Twist	Stability Factor Best case	Nominal	Worst case
1:7"	2.27	1.99	1.72
1:8"	1.74	1.52	1.31
1:9"	1.37	1.20	1.04
1:10"	1.11	0.97	0.84

Minimum twist required for stability is 1:9.0"
Recommended twist for optimal performance is 1:8.1" or faster

Berger .264 caliber 120 grain Target BT

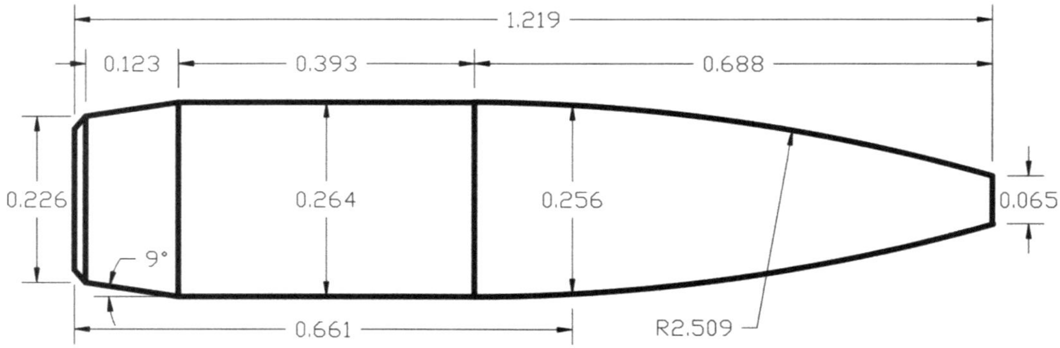

Lot #:	7582	Ogive Radius:	9.50 calibers
Sectional Density:	0.246 lb/in^2	Rt/R:	0.97

Ballistic Performance

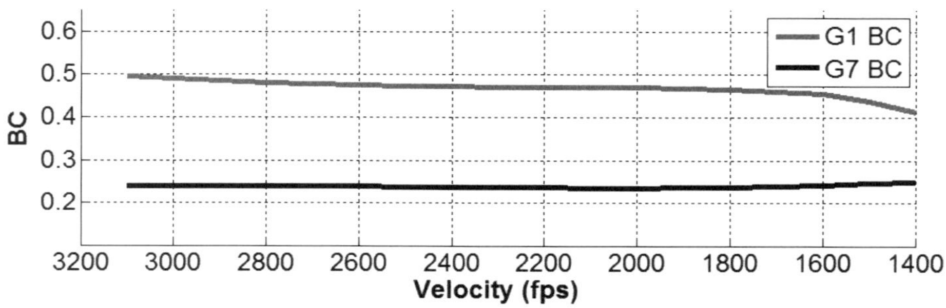

	Above 3000	3000-2500	2500-2000	2000-1500	Below 1500
G1 BC	0.503	0.482	0.472	0.454	0.438
G7 BC	0.238	0.239	0.237	0.241	0.246

Averages from 3000 fps to 1500 fps:
G1 BC: 0.468, variation of G1 BC: 0.052
G7 BC: 0.240, variation of G7 BC: 0.010
G7 form factor: 1.025

Rifling Twist Rate and Stability

Barrel Twist	Stability Factor Best case	Nominal	Worst case
1:8"	2.37	2.07	1.79
1:9"	1.87	1.63	1.41
1:10"	1.51	1.32	1.14
1:11"	1.25	1.09	0.95

Minimum twist required for stability is 1:10.5"
Recommended twist for optimal performance is 1:9.4" or faster

Berger .264 caliber 130 grain Target VLD

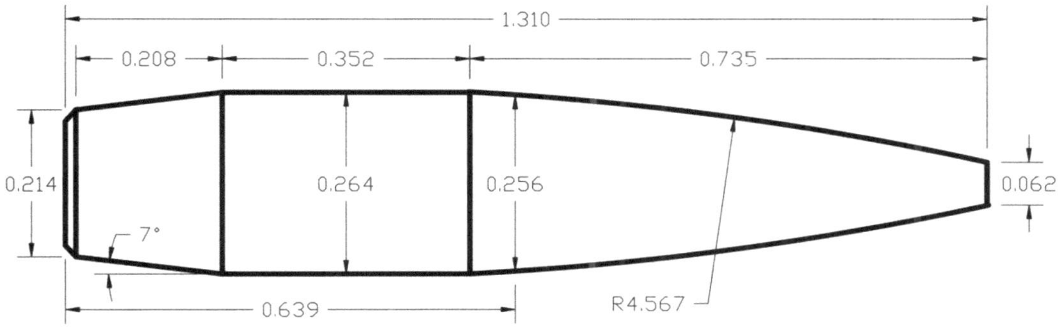

Lot #:	**7573**	**Ogive Radius:**	**17.30 calibers**
Sectional Density:	**0.267 lb/in^2**	**Rt/R:**	**0.60**

Ballistic Performance

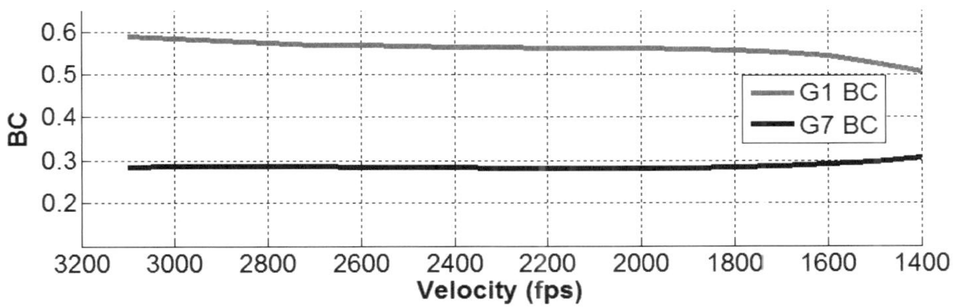

	Above 3000	3000-2500	2500-2000	2000-1500	Below 1500
G1 BC	0.600	0.575	0.564	0.545	0.529
G7 BC	0.285	0.285	0.283	0.290	0.298

Averages from 3000 fps to 1500 fps:
G1 BC: 0.560, variation of G1 BC: 0 056
G7 BC: 0.288, variation of G7 BC: 0.016
G7 form factor: 0.927

Rifling Twist Rate and Stability

Barrel Twist	Stability Factor Best case	Stability Factor Nominal	Stability Factor Worst case
1:7"	2.71	2.37	2.05
1:8"	2.08	1.82	1.57
1:9"	1.64	1.43	1.24
1:10"	1.33	1.16	1.00

Minimum twist required for stability is 1:9.8"
Recommended twist for optimal performance is 1:8.8" or faster

Berger .264 caliber 140 grain Target Hybrid

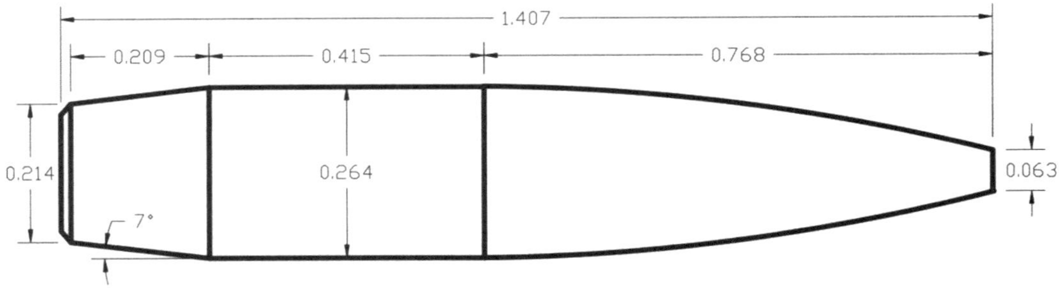

Lot #:	7506	Ogive Radius:	Hybrid
Sectional Density:	0.287 lb/in^2	Rt/R:	Hybrid

Ballistic Performance

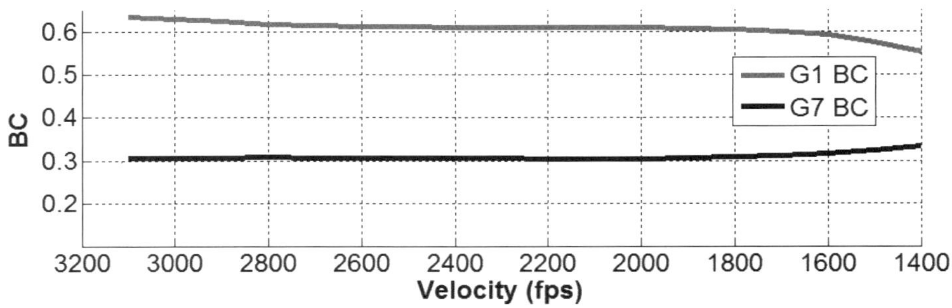

	Above 3000	3000-2500	2500-2000	2000-1500	Below 1500
G1 BC	0.645	0.620	0.609	0.592	0.575
G7 BC	0.306	0.307	0.306	0.315	0.324

Averages from 3000 fps to 1500 fps:
G1 BC: 0.606, variation of G1 BC: 0.054
G7 BC: 0.311, variation of G7 BC: 0.019
G7 form factor: 0.923

Rifling Twist Rate and Stability

Barrel Twist	Stability Factor Best case	Nominal	Worst case
1:7"	2.37	2.07	1.79
1:8"	1.82	1.59	1.37
1:9"	1.43	1.25	1.08
1:10"	1.16	1.02	0.88

Minimum twist required for stability is 1:9.2"
Recommended twist for optimal performance is 1:8.2" or faster

Berger .264 caliber 140 grain Target Long Range BT

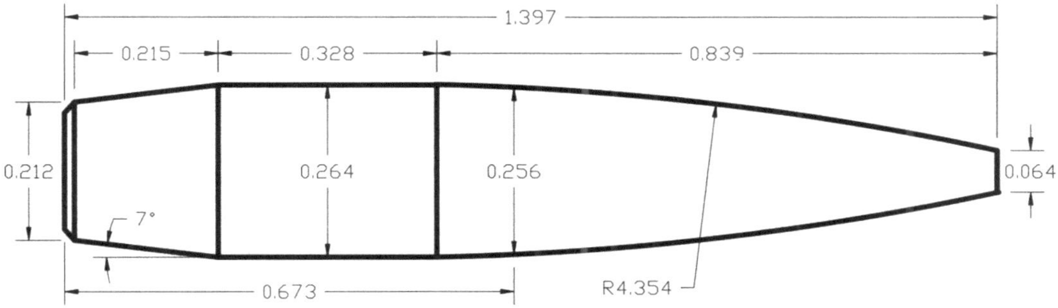

Lot #:	6015	Ogive Radius:	16.49 calibers
Sectional Density:	0.287 lb/in^2	Rt/R:	0.82

Ballistic Performance

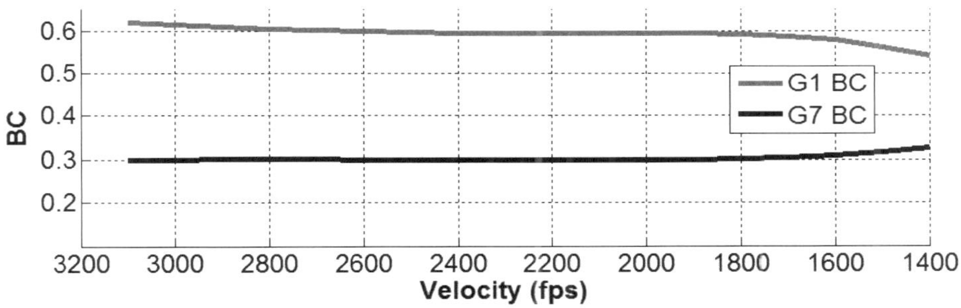

	Above 3000	3000-2500	2500-2000	2000-1500	Below 1500
G1 BC	0.630	0.605	0.595	0.578	0.563
G7 BC	0.299	0.300	0.299	0.308	0.318

Averages from 3000 fps to 1500 fps:
G1 BC: 0.592, variation of G1 BC: 0.052
G7 BC: 0.304, variation of G7 BC: 0.019
G7 form factor: 0.944

Rifling Twist Rate and Stability

Barrel Twist	Stability Factor Best case	Nominal	Worst case
1:7"	2.42	2.12	1.83
1:8"	1.85	1.62	1.40
1:9"	1.46	1.28	1.11
1:10"	1.19	1.04	0.90

Minimum twist required for stability is 1:9.3"
Recommended twist for optimal performance is 1:8.3" or faster

Berger .264 caliber 140 grain Target VLD

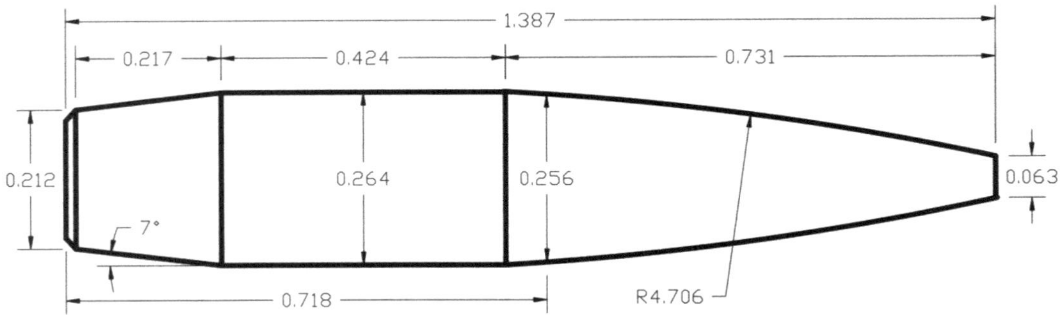

Lot #:	6072	Ogive Radius:	17.83 calibers
Sectional Density:	0.287 lb/in^2	Rt/R:	0.58

Ballistic Performance

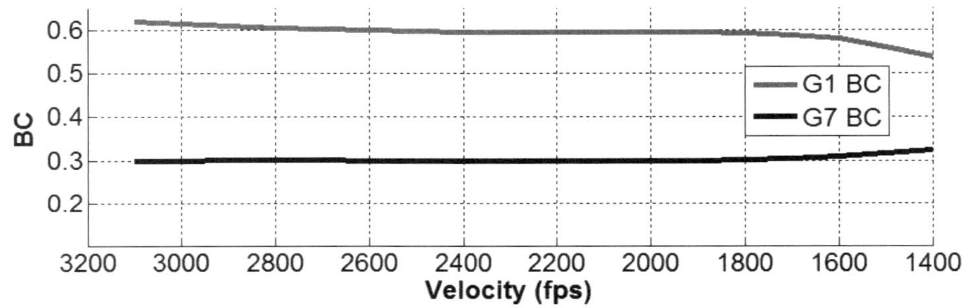

	Above 3000	3000-2500	2500-2000	2000-1500	Below 1500
G1 BC	0.630	0.606	0.595	0.577	0.561
G7 BC	0.299	0.300	0.299	0.307	0.316

Averages from 3000 fps to 1500 fps:
G1 BC: 0.591, variation of G1 BC: 0.054
G7 BC: 0.304, variation of G7 BC: 0.018
G7 form factor: 0.945

Rifling Twist Rate and Stability

Barrel Twist	Stability Factor Best case	Nominal	Worst case
1:7"	2.47	2.16	1.87
1:8"	1.89	1.65	1.43
1:9"	1.50	1.31	1.13
1:10"	1.21	1.06	0.92

Minimum twist required for stability is 1:9.4"
Recommended twist for optimal performance is 1:8.4" or faster

Cauterucio .264 caliber 119 grain 10-ogive

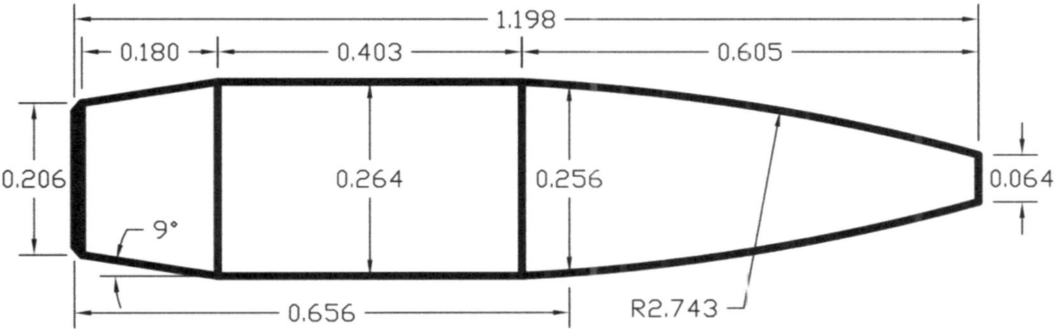

Lot #:	**Unknown**	**Ogive Radius:**	**10.39 calibers**
Sectional Density:	**0.244 lb/in^2**	**Rt/R:**	**0.73**

Ballistic Performance

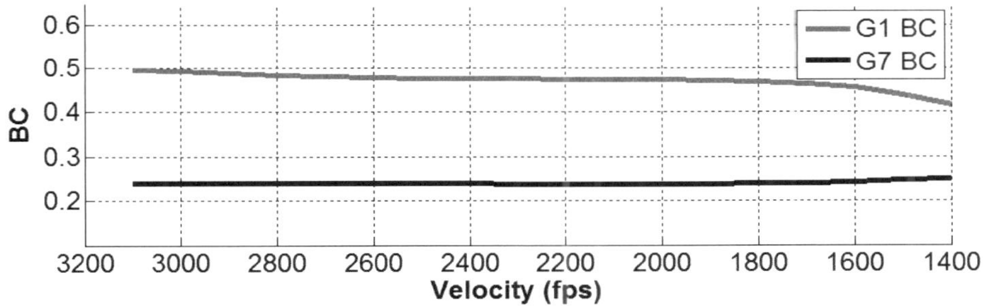

	Above 3000	3000-2500	2500-2000	2000-1500	Below 1500
G1 BC	0.504	0.484	0.474	0.455	0.438
G7 BC	0.239	0.240	0.238	0.242	0.247

Averages from 3000 fps to 1500 fps:
G1 BC: 0.470, variation of G1 BC: 0.053
G7 BC: 0.241, variation of G7 BC: 0.010
G7 form factor: 1.012

Rifling Twist Rate and Stability

Barrel Twist	Stability Factor Best case	Nominal	Worst case
1:8"	2.47	2.16	1.86
1:9"	1.95	1.70	1.47
1:10"	1.58	1.38	1.19
1:11"	1.31	1.14	0.99

Minimum twist required for stability is 1:10.7"
Recommended twist for optimal performance is 1:9.6" or faster

Cauterucio .264 caliber 119 grain 15-ogive

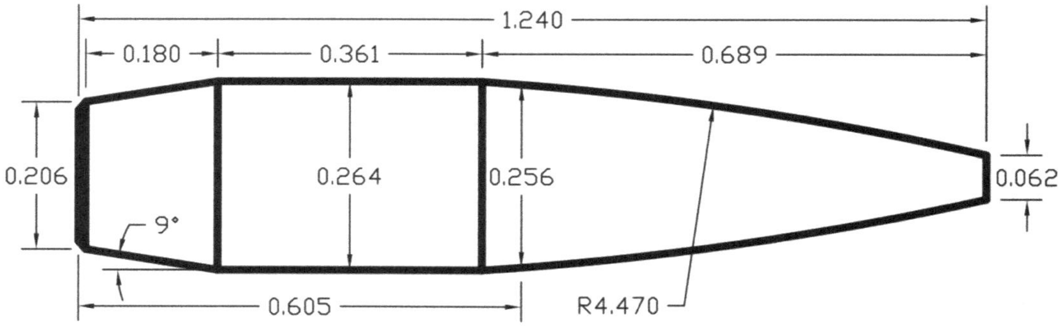

Lot #:	Unknown	Ogive Radius:	16.93 calibers
Sectional Density:	0.244 lb/in^2	Rt/R:	0.59

Ballistic Performance

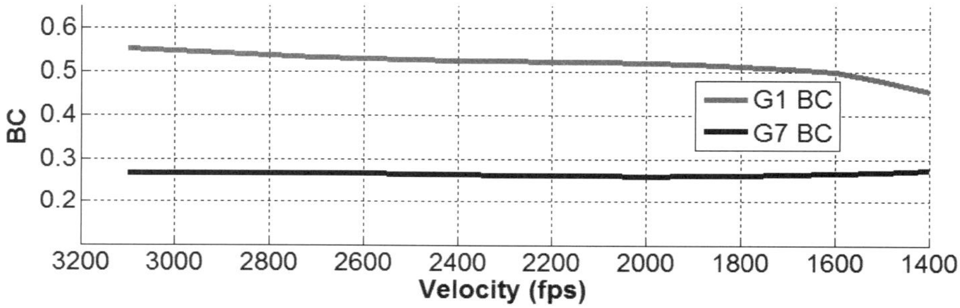

	Above 3000	3000-2500	2500-2000	2000-1500	Below 1500
G1 BC	0.563	0.538	0.524	0.500	0.481
G7 BC	0.267	0.267	0.263	0.266	0.271

Averages from 3000 fps to 1500 fps:
G1 BC: 0.519, variation of G1 BC: 0.067
G7 BC: 0.266, variation of G7 BC: 0.010
G7 form factor: 0.916

Rifling Twist Rate and Stability

Barrel Twist	Stability Factor Best case	Nominal	Worst case
1:8"	2.23	1.95	1.69
1:9"	1.76	1.54	1.33
1:10"	1.43	1.25	1.08
1:11"	1.18	1.03	0.89

Minimum twist required for stability is 1:10.2"
Recommended twist for optimal performance is 1:9.1" or faster

Cauterucio .264 caliber 131 grain VLD

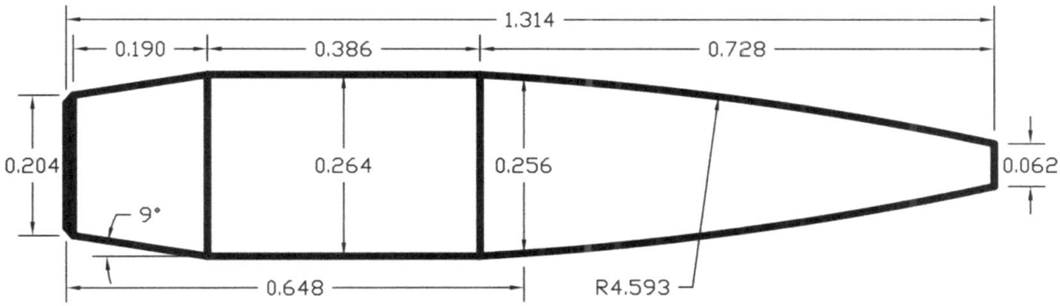

Lot #:	**Unknown**	**Ogive Radius:**	**17.40 calibers**
Sectional Density:	**0.269 lb/in²**	**Rt/R:**	**0.62**

Ballistic Performance

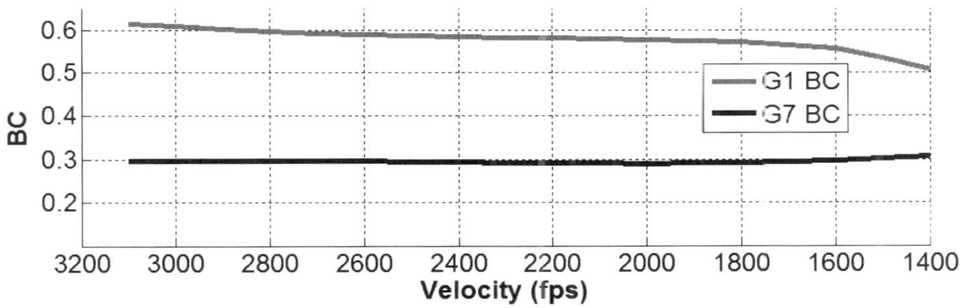

	Above 3000	3000-2500	2500-2000	2000-1500	Below 1500
G1 BC	0.625	0.597	0.582	0.557	0.535
G7 BC	0.296	0.296	0.293	0.296	0.302

Averages from 3000 fps to 1500 fps:
G1 BC: 0.577, variation of G1 BC: 0.072
G7 BC: 0.296, variation of G7 BC: 0.011
G7 form factor: 0.907

Rifling Twist Rate and Stability

	Stability Factor		
Barrel Twist	**Best case**	**Nominal**	**Worst case**
1:8"	2.08	1.81	1.57
1:9"	1.64	1.43	1.24
1:10"	1.33	1.16	1.00
1:11"	1.10	0.96	0.83

Minimum twist required for stability is 1:9.8"
Recommended twist for optimal performance is 1:8.8" or faster

Cauterucio .264 caliber 155 grain VLD

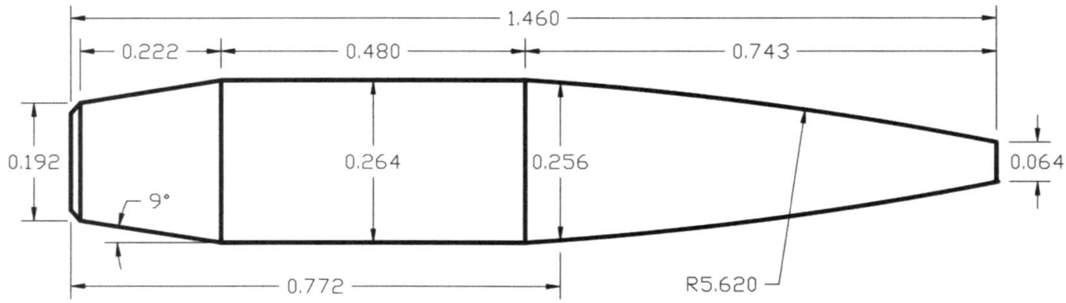

Lot #:	Unknown	Ogive Radius:	21.29 calibers
Sectional Density:	0.318 lb/in^2	Rt/R:	0.44

Ballistic Performance

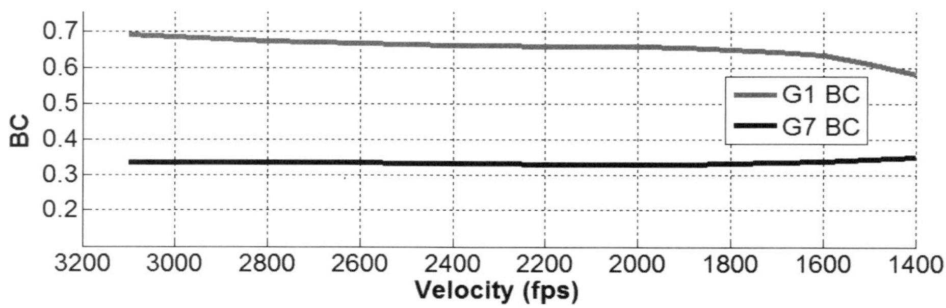

	Above 3000	3000-2500	2500-2000	2000-1500	Below 1500
G1 BC	0.703	0.674	0.660	0.634	0.612
G7 BC	0.333	0.334	0.332	0.337	0.345

Averages from 3000 fps to 1500 fps:
G1 BC: 0.654, variation of G1 BC: 0.073
G7 BC: 0.336, variation of G7 BC: 0.015
G7 form factor: 0.946

Rifling Twist Rate and Stability

Barrel Twist	Stability Factor Best case	Nominal	Worst case
1:7"	2.36	2.06	1.78
1:8"	1.80	1.58	1.36
1:9"	1.43	1.24	1.08
1:10"	1.15	1.01	0.87

Minimum twist required for stability is 1:9.2"
Recommended twist for optimal performance is 1:8.2" or faster

GS Custom .264 caliber 112 grain SP

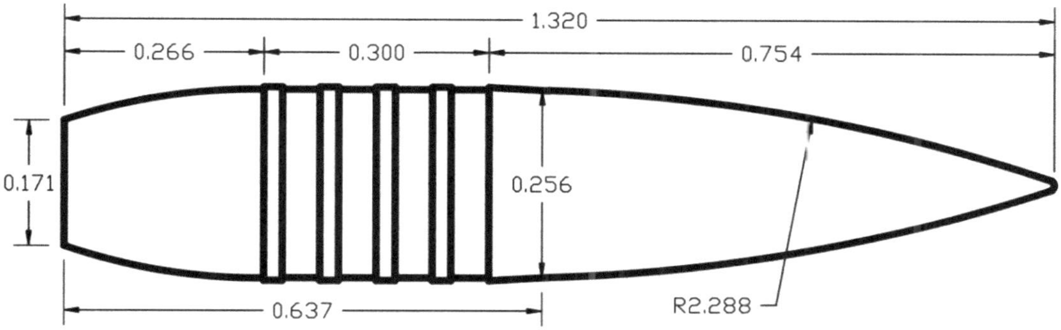

Lot #:	**Unknown**		**Ogive Radius:**	**8.67 calibers**
Sectional Density:	**0.230 lb/in^2**		**Rt/R:**	**1.00**

Ballistic Performance

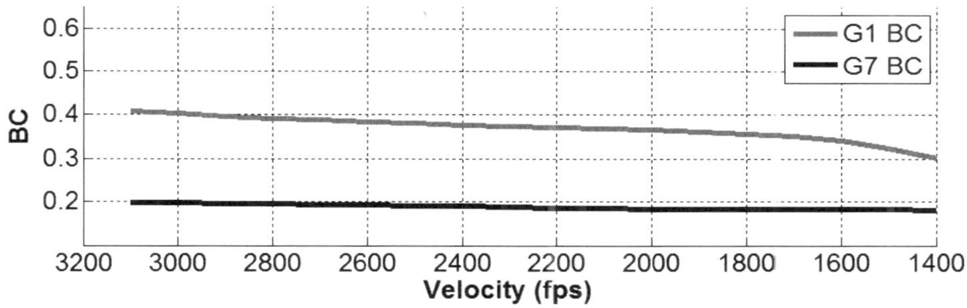

	Above 3000	3000-2500	2500-2000	2000-1500	Below 1500
G1 BC	0.418	0.391	0.373	0.345	0.325
G7 BC	0.198	0.194	0.187	0.183	0.182

Averages from 3000 fps to 1500 fps:
G1 BC: 0.368, variation of G1 BC: 0.078
G7 BC: 0.189, variation of G7 BC: 0.014
G7 form factor: 1.219

Rifling Twist Rate and Stability

	Stability Factor		
Barrel Twist	**Best case**	**Nominal**	**Worst case**
1:8"	1.75	1.53	1.32
1:9"	1.38	1.21	1.05
1:10"	1.12	0.98	0.85
1:11"	0.93	0.81	0.70

Minimum twist required for stability is 1:9.0"
Recommended twist for optimal performance is 1:8.1" or faster

GS Custom .264 caliber 119 grain SP

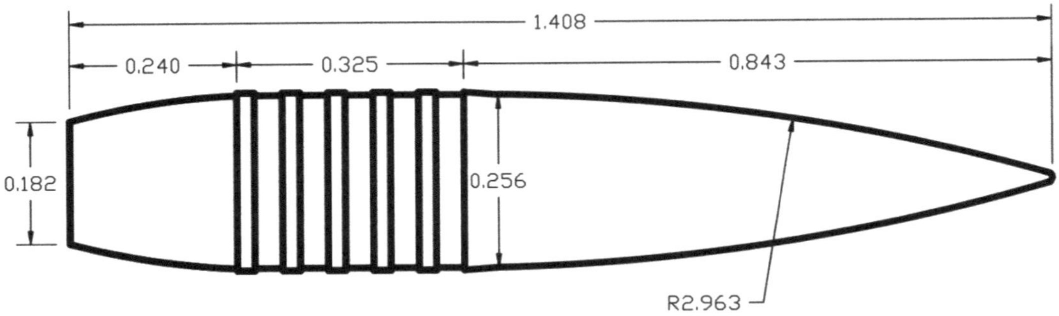

Lot #:	**Unknown**	**Ogive Radius:**	**11.22 calibers**
Sectional Density:	**0.244 lb/in^2**	**Rt/R:**	**0.97**

Ballistic Performance

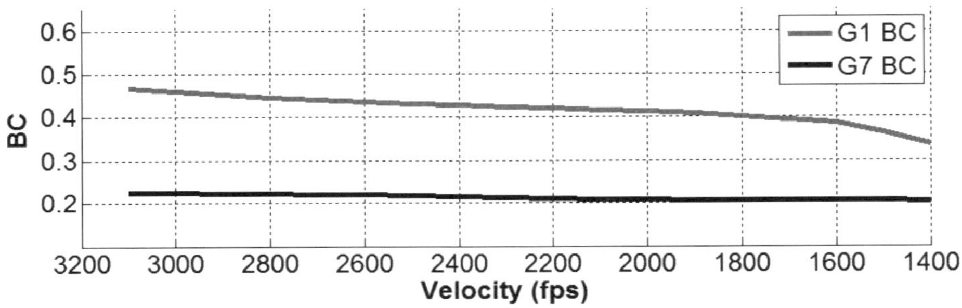

	Above 3000	3000-2500	2500-2000	2000-1500	Below 1500
G1 BC	0.479	0.446	0.422	0.388	0.364
G7 BC	0.227	0.221	0.212	0.206	0.205

Averages from 3000 fps to 1500 fps:
G1 BC: 0.417, variation of G1 BC: 0.096
G7 BC: 0.213, variation of G7 BC: 0.020
G7 form factor: 1.145

Rifling Twist Rate and Stability

Barrel Twist	**Stability Factor** Best case	Nominal	Worst case
1:8"	1.54	1.35	1.16
1:9"	1.22	1.06	0.92
1:10"	0.99	0.86	0.74
1:11"	0.81	0.71	0.62

Minimum twist required for stability is 1:8.5"
Recommended twist for optimal performance is 1:7.6" or faster

GS Custom .264 caliber 128 grain SP

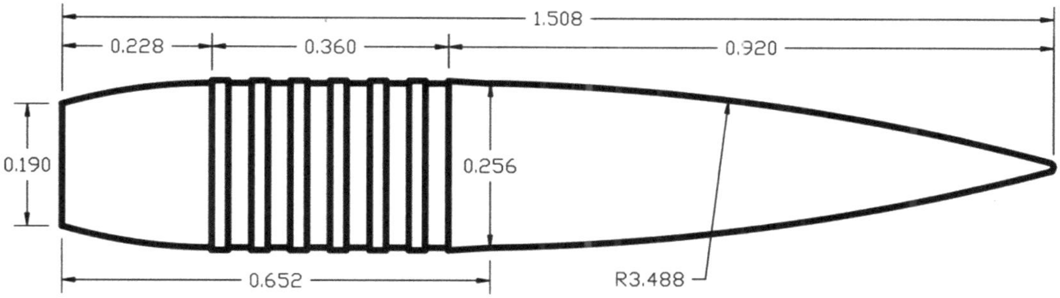

Lot #:	Unknown	Ogive Radius:	13.21 calibers
Sectional Density:	0.262 lb/in^2	Rt/R:	0.97

Ballistic Performance

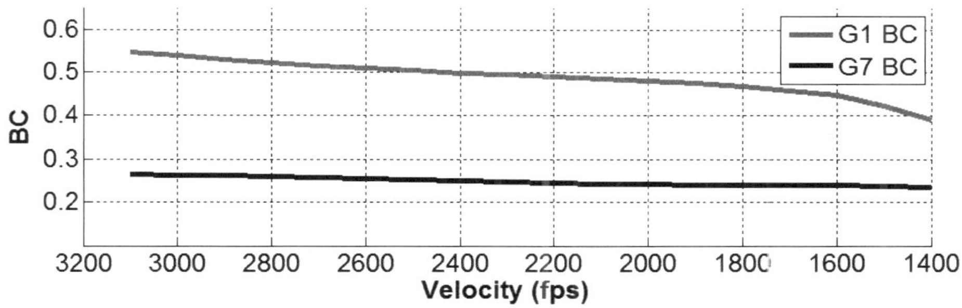

	Above 3000	3000-2500	2500-2000	2000-1500	Below 1500
G1 BC	0.563	0.521	0.492	0.452	0.423
G7 BC	0.266	0.259	0.247	0.239	0.238

Averages from 3000 fps to 1500 fps:
G1 BC: 0.487, variation of G1 BC: 0.115
G7 BC: 0.249, variation of G7 BC: 0.026
G7 form factor: 1.056

Rifling Twist Rate and Stability

Barrel Twist	Stability Factor Best case	Nominal	Worst case
1:7"	1.77	1.55	1.34
1:8"	1.35	1.18	1.02
1:9"	1.07	0.93	0.81
1:10"	0.87	0.76	0.65

Minimum twist required for stability is 1:7.9"
Recommended twist for optimal performance is 1:7.1" or faster

Hoover .264 caliber 136 grain

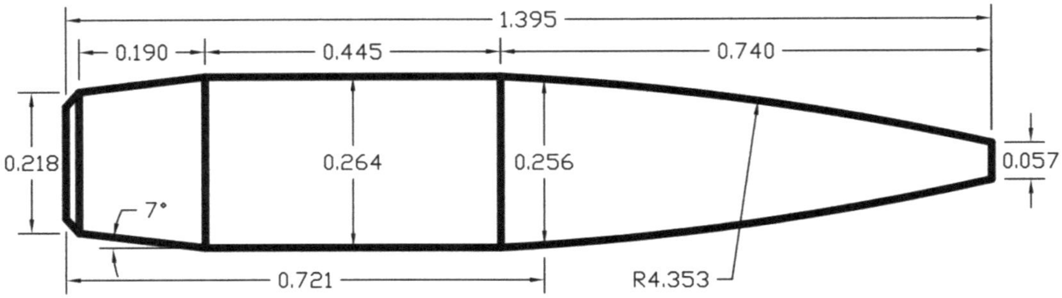

Lot #:	Unknown	Ogive Radius:	16.49 calibers
Sectional Density:	0.279 lb/in^2	Rt/R:	0.50

Ballistic Performance

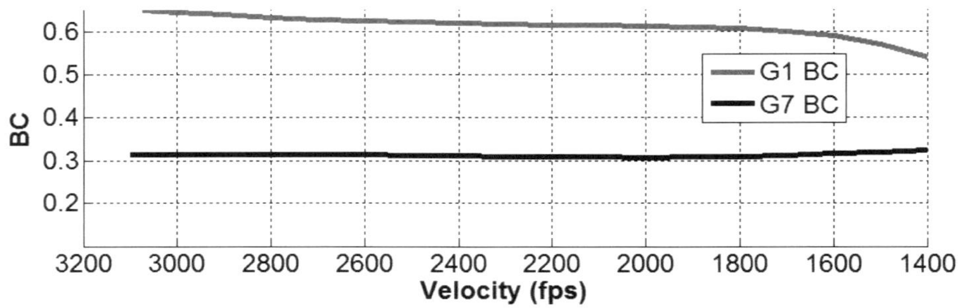

	Above 3000	3000-2500	2500-2000	2000-1500	Below 1500
G1 BC	0.664	0.633	0.617	0.590	0.569
G7 BC	0.315	0.314	0.310	0.314	0.320

Averages from 3000 fps to 1500 fps:
G1 BC: 0.612, variation of G1 BC: 0.076
G7 BC: 0.314, variation of G7 BC: 0.013
G7 form factor: 0.888

Rifling Twist Rate and Stability

Barrel Twist	Stability Factor Best case	Nominal	Worst case
1:8"	1.81	1.58	1.37
1:9"	1.43	1.25	1.08
1:10"	1.16	1.01	0.87
1:11"	0.96	0.84	0.72

Minimum twist required for stability is 1:9.2"
Recommended twist for optimal performance is 1:8.2" or faster

Hornady .264 caliber 123 grain Amax

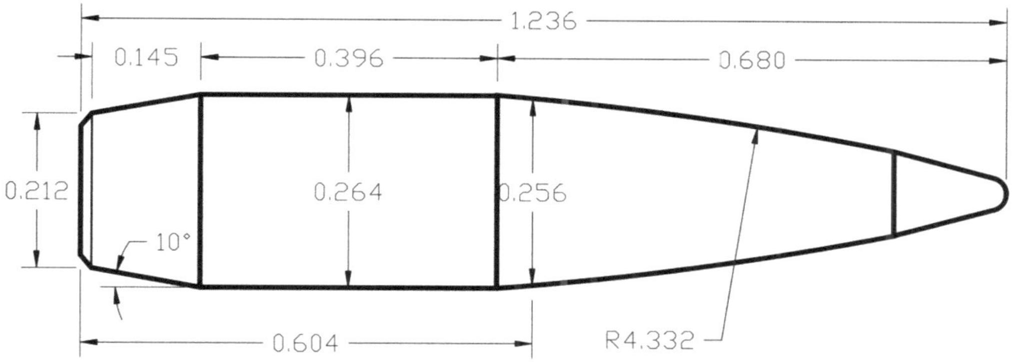

Lot #:	**2150068**	**Ogive Radius:**	**16.4 calibers**
Sectional Density:	**0.252 lb/in^2**	**Rt/R:**	**0.43**

Ballistic Performance

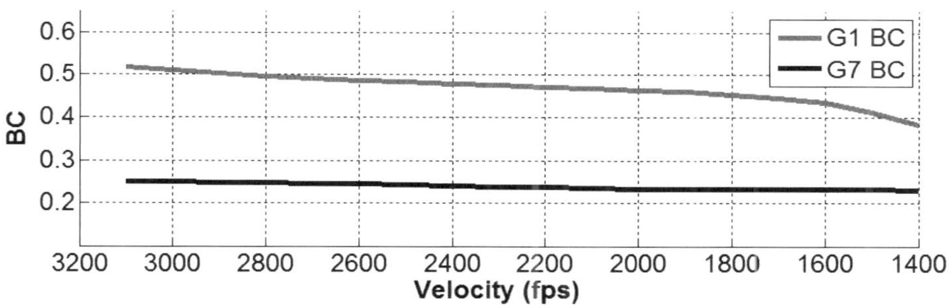

	Above 3000	3000-2500	2500-2000	2000-1500	Below 1500
G1 BC	0.529	0.496	0.473	0.439	0.414
G7 BC	0.250	0.246	0.238	0.233	0.232

Averages from 3000 fps to 1500 fps:
G1 BC: 0.467, variation of G1 BC: 0.096
G7 BC: 0.239, variation of G7 BC: 0.017
G7 form factor: 1.055

Rifling Twist Rate and Stability

Barrel Twist	Stability Factor: Best case	Stability Factor: Nominal	Stability Factor: Worst case
1:8"	2.33	2.04	1.76
1:9"	1.84	1.61	1.39
1:10"	1.49	1.30	1.13
1:11"	1.23	1.08	0.93

Minimum twist required for stability is 1:10.4"
Recommended twist for optimal performance is 1:9.3" or faster

Hornady .264 caliber 140 grain Amax

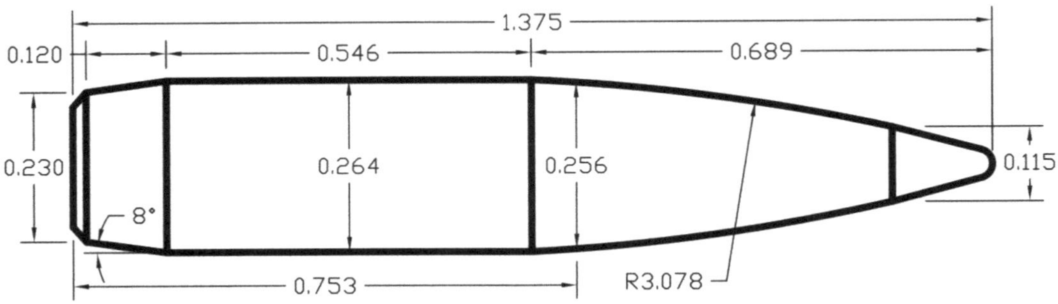

Lot #:	2080069	Ogive Radius:	11.66 calibers
Sectional Density:	0.287 lb/in²	Rt/R:	0.72

Ballistic Performance

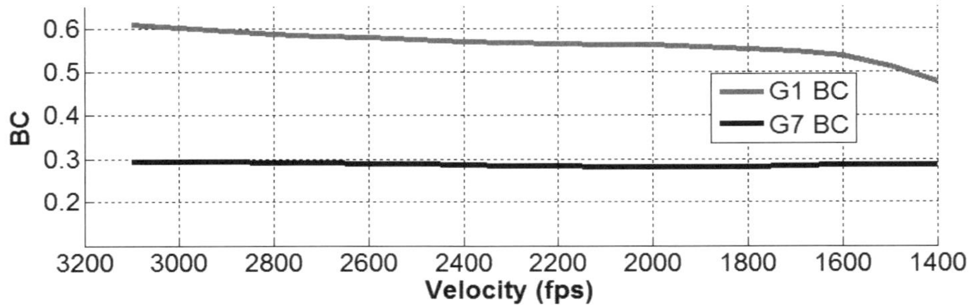

	Above 3000	3000-2500	2500-2000	2000-1500	Below 1500
G1 BC	0.624	0.588	0.567	0.536	0.512
G7 BC	0.295	0.292	0.285	0.285	0.288

Averages from 3000 fps to 1500 fps:
G1 BC: 0.562, variation of G1 BC: 0.091
G7 BC: 0.288, variation of G7 BC: 0.013
G7 form factor: 0.996

Rifling Twist Rate and Stability

Barrel Twist	Stability Factor Best case	Nominal	Worst case
1:8"	1.94	1.70	1.47
1:9"	1.53	1.34	1.16
1:10"	1.24	1.09	0.94
1:11"	1.03	0.90	0.78

Minimum twist required for stability is 1:9.5"
Recommended twist for optimal performance is 1:8.5" or faster

Hornady .264 caliber 140 grain BTHP Match

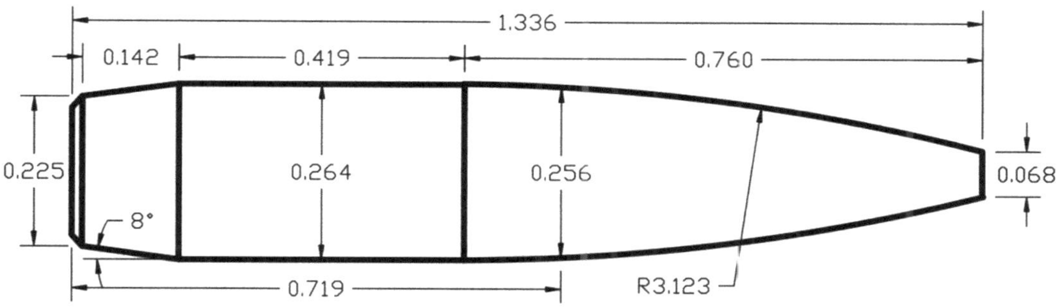

Lot #:	2140799	**Ogive Radius:**	**11.83 calibers**
Sectional Density:	**0.287 lb/in²**	**Rt/R:**	**0.96**

Ballistic Performance

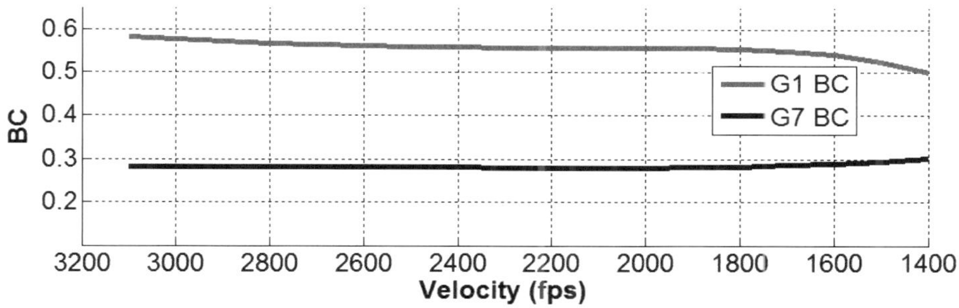

	Above 3000	3000-2500	2500-2000	2000-1500	Below 1500
G1 BC	0.592	0.568	0.558	0.541	0.524
G7 BC	0.280	0.282	0.281	0.288	0.295

Averages from 3000 fps to 1500 fps:
G1 BC: 0.554, variation of G1 BC: 0.053
G7 BC: 0.285, variation of G7 BC: 0.016
G7 form factor: 1.008

Rifling Twist Rate and Stability

Barrel Twist	Stability Factor Best case	Nominal	Worst case
1:8"	2.11	1.85	1.60
1:9"	1.67	1.46	1.26
1:10"	1.35	1.18	1.02
1:11"	1.12	0.98	0.84

Minimum twist required for stability is 1:9.9"
Recommended twist for optimal performance is 1:8.9" or faster

Hornady .264 caliber 120 grain GMX

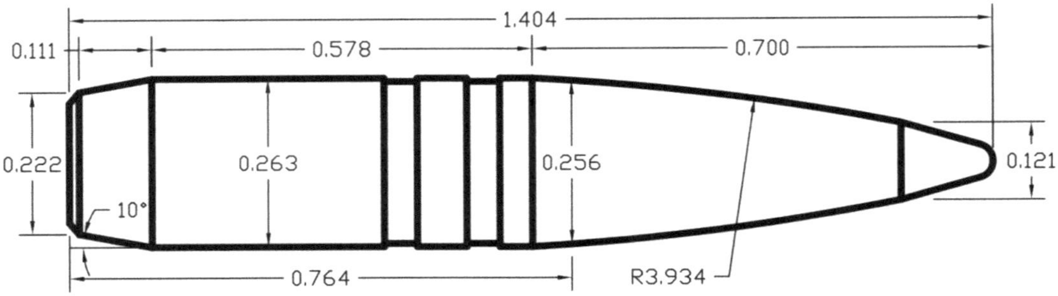

Lot #:	**Unknown**	**Ogive Radius:**	**14.90 calibers**
Sectional Density:	**0.246 lb/in²**	**Rt/R:**	**0.53**

Ballistic Performance

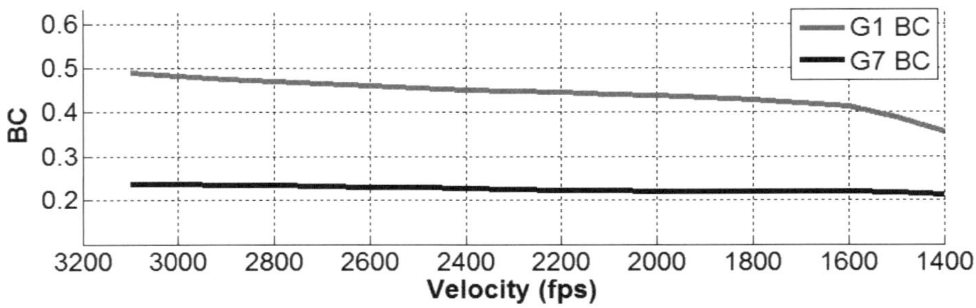

	Above 3000	3000-2500	2500-2000	2000-1500	Below 1500
G1 BC	0.503	0.470	0.447	0.414	0.389
G7 BC	0.238	0.233	0.225	0.219	0.218

Averages from 3000 fps to 1500 fps:
G1 BC: 0.442, variation of G1 BC: 0.094
G7 BC: 0.226, variation of G7 BC: 0.018
G7 form factor: 1.089

Rifling Twist Rate and Stability

Barrel Twist	**Stability Factor**		
	Best case	**Nominal**	**Worst case**
1:8"	1.57	1.37	1.18
1:9"	1.24	1.08	0.93
1:10"	1.00	0.88	0.76
1:11"	0.83	0.72	0.63

Minimum twist required for stability is 1:8.5"
Recommended twist for optimal performance is 1:7.6" or faster

Hornady .264 caliber 129 grain Interbond

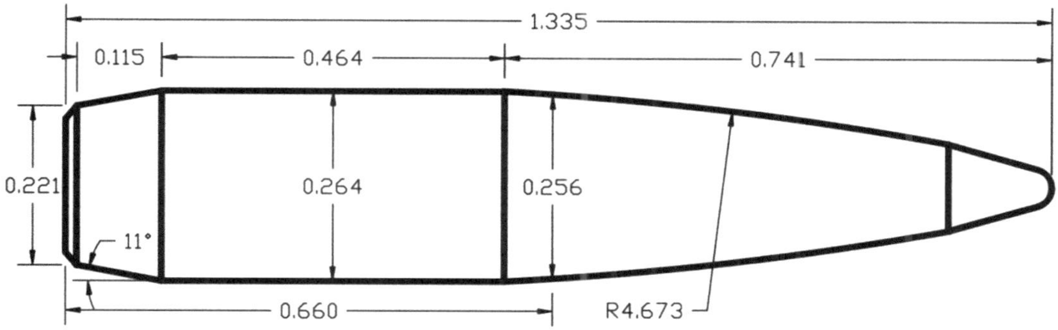

Lot #:	**Unknown**	**Ogive Radius:**	**17.70 calibers**
Sectional Density:	**0.264 lb/in²**	**Rt/R:**	**0.53**

Ballistic Performance

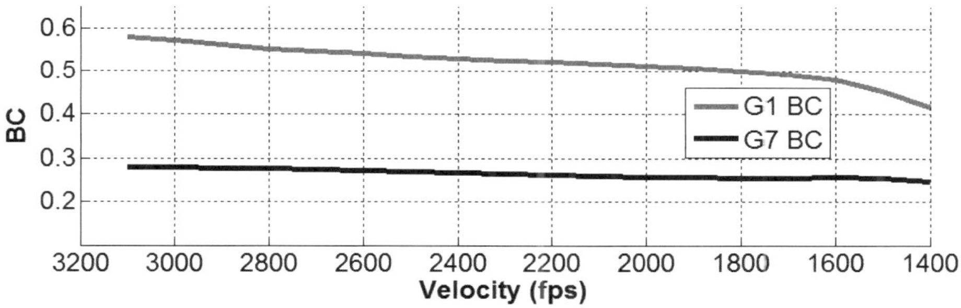

	Above 3000	3000-2500	2500-2000	2000-1500	Below 1500
G1 BC	0.595	0.553	0.524	0.483	0.453
G7 BC	0.282	0.274	0.263	0.256	0.254

Averages from 3000 fps to 1500 fps:
G1 BC: 0.518, variation of G1 BC: 0.118
G7 BC: 0.265, variation of G7 BC: 0.025
G7 form factor: 0.999

Rifling Twist Rate and Stability

Barrel Twist	Stability Factor Best case	Nominal	Worst case
1:8"	1.95	1.70	1.47
1:9"	1.54	1.35	1.16
1:10"	1.25	1.09	0.94
1:11"	1.03	0.90	0.78

Minimum twist required for stability is 1:9.5"
Recommended twist for optimal performance is 1:8.5" or faster

Hornady .264 caliber 123 grain SST BT

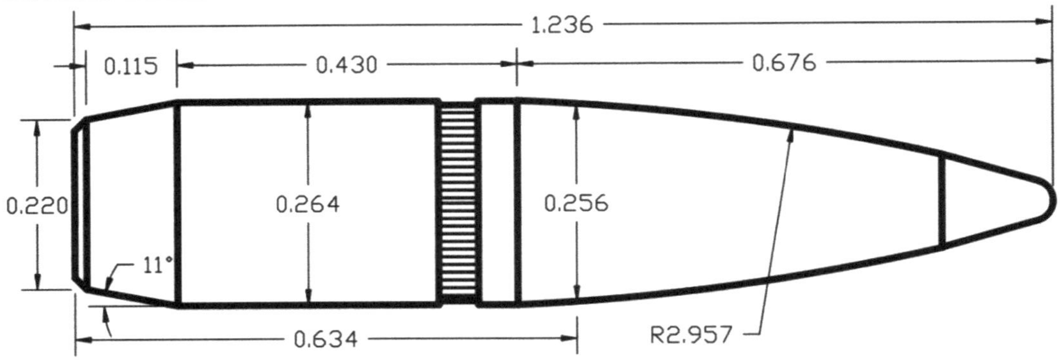

Lot #:	Unknown	Ogive Radius:	11.20 calibers
Sectional Density:	0.252 lb/in^2	Rt/R:	0.67

Ballistic Performance

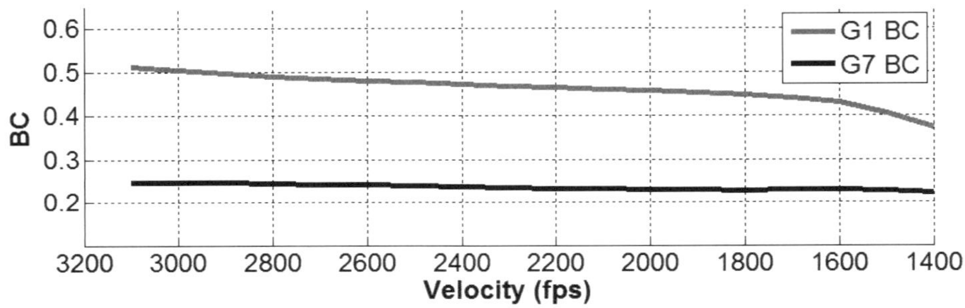

	Above 3000	3000-2500	2500-2000	2000-1500	Below 1500
G1 BC	0.526	0.491	0.467	0.432	0.406
G7 BC	0.249	0.244	0.235	0.229	0.228

Averages from 3000 fps to 1500 fps:
G1 BC: 0.462, variation of G1 BC: 0.100
G7 BC: 0.236, variation of G7 BC: 0.020
G7 form factor: 1.069

Rifling Twist Rate and Stability

Barrel Twist	Stability Factor Best case	Stability Factor Nominal	Stability Factor Worst case
1:8"	2.27	1.99	1.72
1:9"	1.80	1.57	1.36
1:10"	1.45	1.27	1.10
1:11"	1.20	1.05	0.91

Minimum twist required for stability is 1:10.4"
Recommended twist for optimal performance is 1:9.3" or faster

Hornady .264 caliber 129 grain SST BT

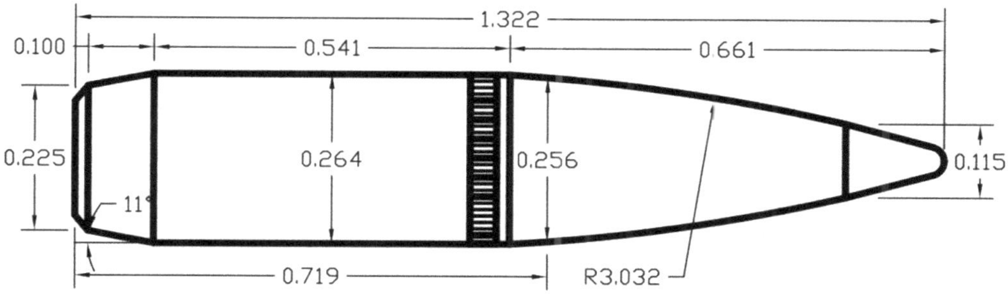

Lot #:	2070526	Ogive Radius:	11.49 calibers
Sectional Density:	0.264 lb/in²	Rt/R:	0.67

Ballistic Performance

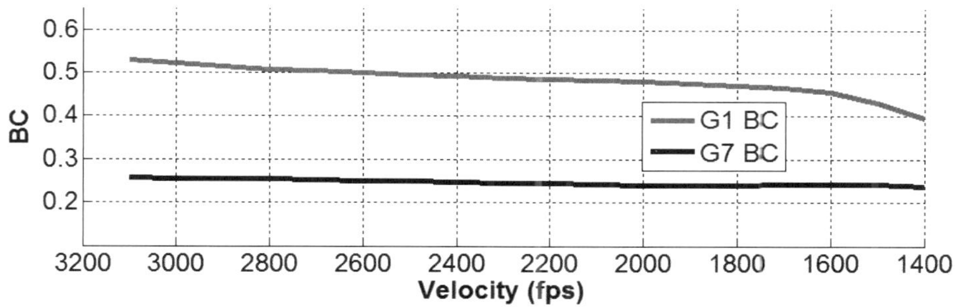

	Above 3000	3000-2500	2500-2000	2000-1500	Below 1500
G1 BC	0.542	0.509	0.488	0.455	0.430
G7 BC	0.257	0.252	0.245	0.241	0.242

Averages from 3000 fps to 1500 fps:
G1 BC: 0.482, variation of G1 BC: 0.092
G7 BC: 0.247, variation of G7 BC: 0.015
G7 form factor: 1.072

Rifling Twist Rate and Stability

Barrel Twist	Stability Factor Best case	Nominal	Worst case
1:8"	2.01	1.75	1.52
1:9"	1.59	1.39	1.20
1:10"	1.29	1.12	0.97
1:11"	1.06	0.93	0.80

Minimum twist required for stability is 1:9.7"
Recommended twist for optimal performance is 1:8.7" or faster

Hornady .264 caliber 140 grain SST BT

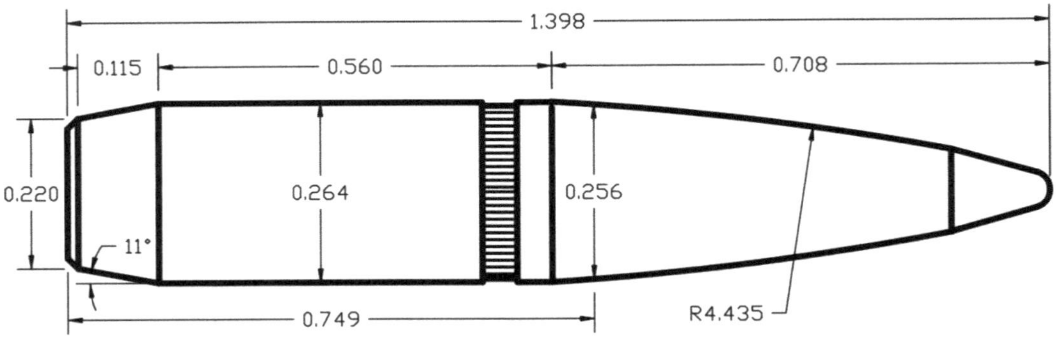

Lot #:	Unknown	Ogive Radius:	16.80 calibers
Sectional Density:	0.287 lb/in^2	Rt/R:	0.49

Ballistic Performance

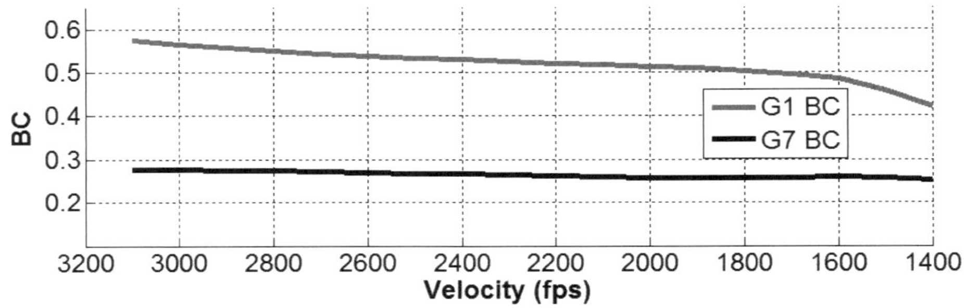

	Above 3000	3000-2500	2500-2000	2000-1500	Below 1500
G1 BC	0.588	0.549	0.523	0.486	0.458
G7 BC	0.278	0.272	0.263	0.257	0.257

Averages from 3000 fps to 1500 fps:
G1 BC: 0.517, variation of G1 BC: 0.108
G7 BC: 0.265, variation of G7 BC: 0.020
G7 form factor: 1.085

Rifling Twist Rate and Stability

Barrel Twist	Stability Factor Best case	Nominal	Worst case
1:8"	1.85	1.62	1.40
1:9"	1.46	1.28	1.10
1:10"	1.18	1.03	0.89
1:11"	0.98	0.85	0.74

Minimum twist required for stability is 1:9.3"
Recommended twist for optimal performance is 1:8.3" or faster

Hornady .264 caliber 95 grain Vmax

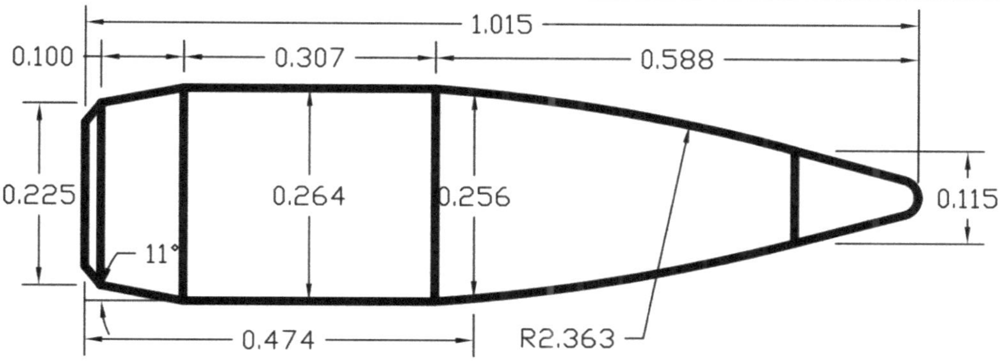

Lot #:	2070928	Ogive Radius:	8.95 calibers
Sectional Density:	0.195 lb/in²	Rt/R:	0.64

Ballistic Performance

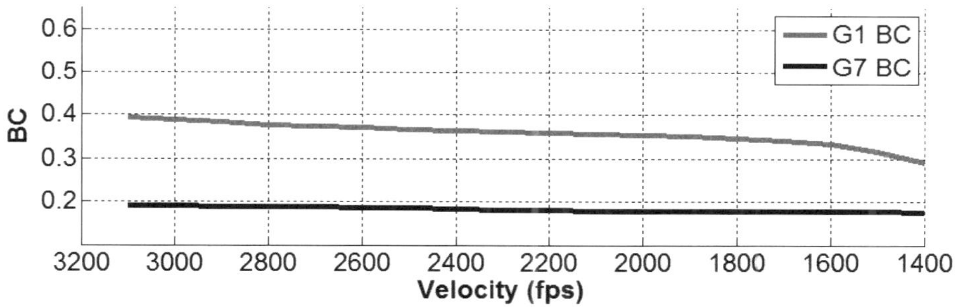

	Above 3000	3000-2500	2500-2000	2000-1500	Below 1500
G1 BC	0.403	0.377	0.360	0.335	0.316
G7 BC	0.191	0.187	0.181	0.178	0.177

Averages from 3000 fps to 1500 fps:
G1 BC: 0.356, variation of G1 BC: 0.072
G7 BC: 0.182, variation of G7 BC: 0.012
G7 form factor: 1.069

Rifling Twist Rate and Stability

Barrel Twist	Stability Factor Best case	Nominal	Worst case
1:8"	3.18	2.78	2.40
1:9"	2.51	2.20	1.90
1:10"	2.04	1.78	1.54
1:11"	1.68	1.47	1.27

Minimum twist required for stability is 1:12.2"
Recommended twist for optimal performance is 1:10.9" or faster

JLK .264 caliber 140 grain VLD

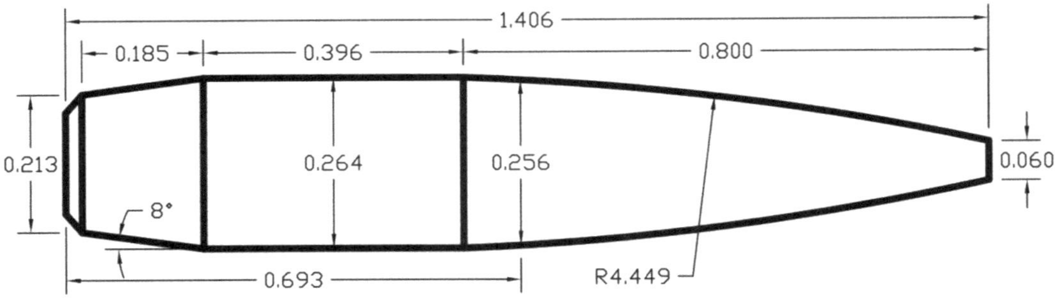

Lot #:	53-1	Ogive Radius:	16.85 calibers
Sectional Density:	0.287 lb/in^2	Rt/R:	0.75

Ballistic Performance

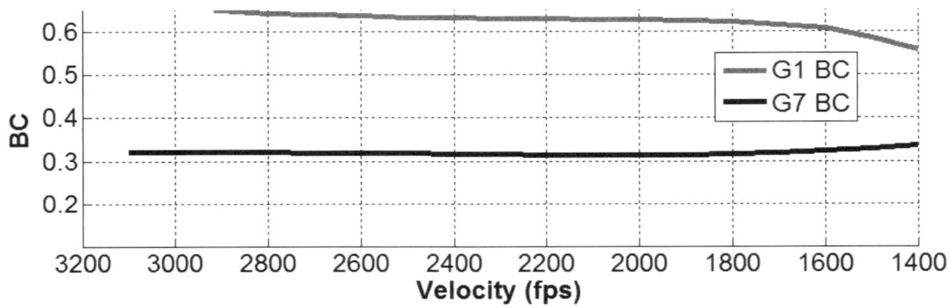

	Above 3000	3000-2500	2500-2000	2000-1500	Below 1500
G1 BC	0.675	0.644	0.629	0.605	0.584
G7 BC	0.320	0.319	0.316	0.322	0.329

Averages from 3000 fps to 1500 fps:
G1 BC: 0.625, variation of G1 BC: 0.071
G7 BC: 0.321, variation of G7 BC: 0.015
G7 form factor: 0.895

Rifling Twist Rate and Stability

Barrel Twist	Stability Factor Best case	Stability Factor Nominal	Stability Factor Worst case
1:8"	1.82	1.59	1.37
1:9"	1.44	1.26	1.09
1:10"	1.16	1.02	0.88
1:11"	0.96	0.84	0.73

Minimum twist required for stability is 1:9.2"
Recommended twist for optimal performance is 1:8.2" or faster

Lapua .264 caliber 144 grain FMJBT

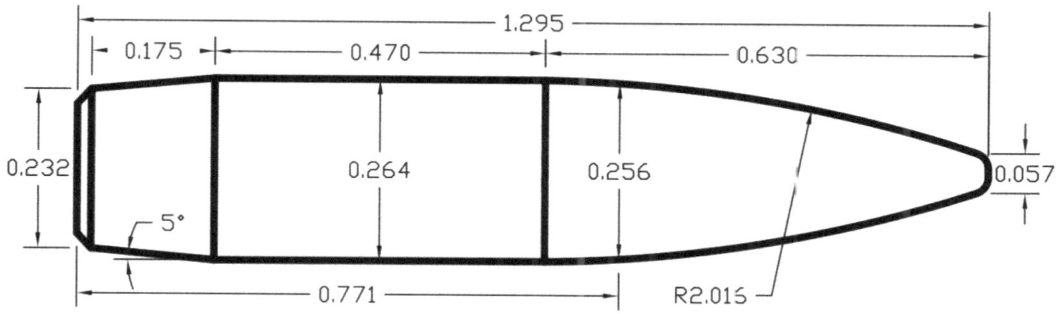

Lot #:	4PL6000	Ogive Radius:	7.63 calibers
Sectional Density:	0.295 lb/in^2	Rt/R:	1.00

Ballistic Performance

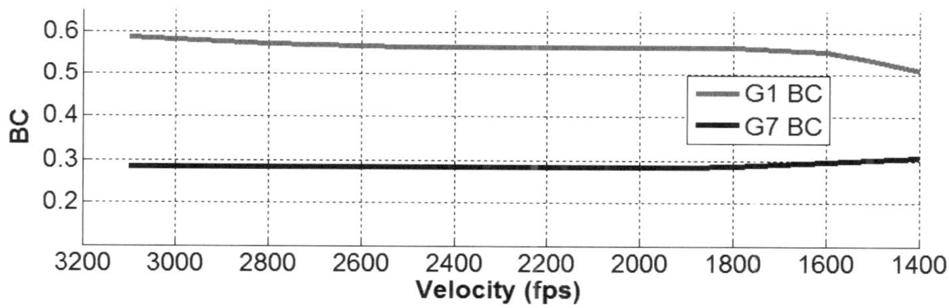

	Above 3000	3000-2500	2500-2000	2000-1500	Below 1500
G1 BC	0.596	0.573	0.565	0.550	0.535
G7 BC	0.283	0.284	0.284	0.293	0.301

Averages from 3000 fps to 1500 fps:
G1 BC: 0.562, variation of G1 BC: 0.046
G7 BC: 0.289, variation of G7 BC: 0.017
G7 form factor: 1.023

Rifling Twist Rate and Stability

Barrel Twist	Stability Factor Best case	Nominal	Worst case
1:8"	2.38	2.08	1.80
1:9"	1.88	1.64	1.42
1:10"	1.52	1.33	1.15
1:11"	1.26	1.10	0.95

Minimum twist required for stability is 1:10.5"
Recommended twist for optimal performance is 1:9.4" or faster

Lapua .264 caliber 108 grain Scenar

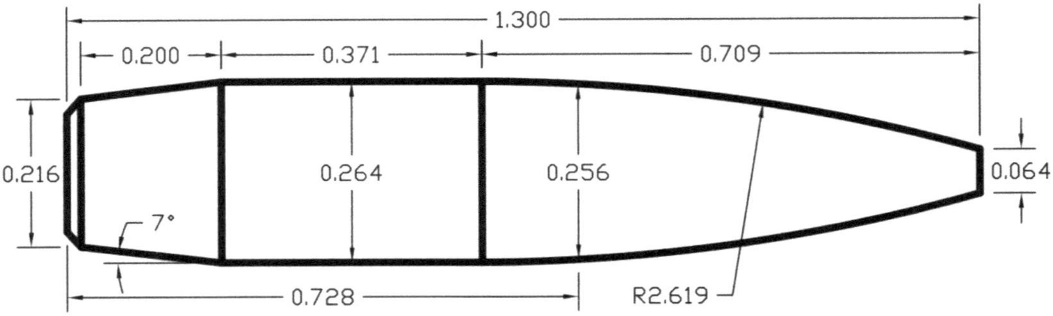

Lot #:	4PL6020	Ogive Radius:	9.92 calibers
Sectional Density:	0.221 lb/in^2	Rt/R:	1.00

Ballistic Performance

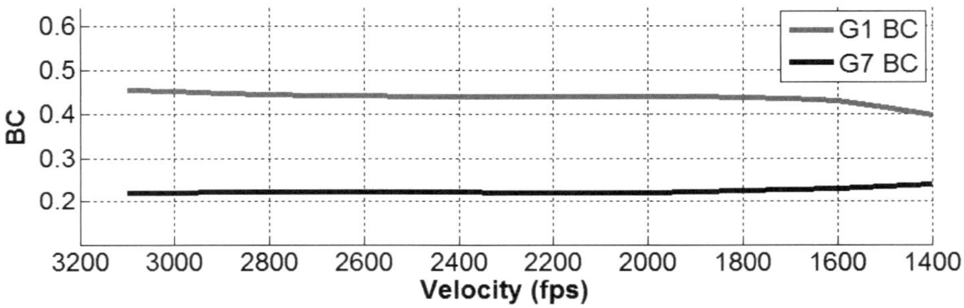

	Above 3000	3000-2500	2500-2000	2000-1500	Below 1500
G1 BC	0.463	0.446	0.441	0.429	0.417
G7 BC	0.220	0.221	0.221	0.228	0.235

Averages from 3000 fps to 1500 fps:

G1 BC: 0.438, variation of G1 BC: 0.035

G7 BC: 0.225, variation of G7 BC: 0.014

G7 form factor: 0.985

Rifling Twist Rate and Stability

Barrel Twist	Stability Factor Best case	Nominal	Worst case
1:8"	1.77	1.54	1.33
1:9"	1.39	1.22	1.05
1:10"	1.13	0.99	0.85
1:11"	0.93	0.82	0.71

Minimum twist required for stability is 1:9.1"

Recommended twist for optimal performance is 1:8.1" or faster

Lapua .264 caliber 123 grain Scenar

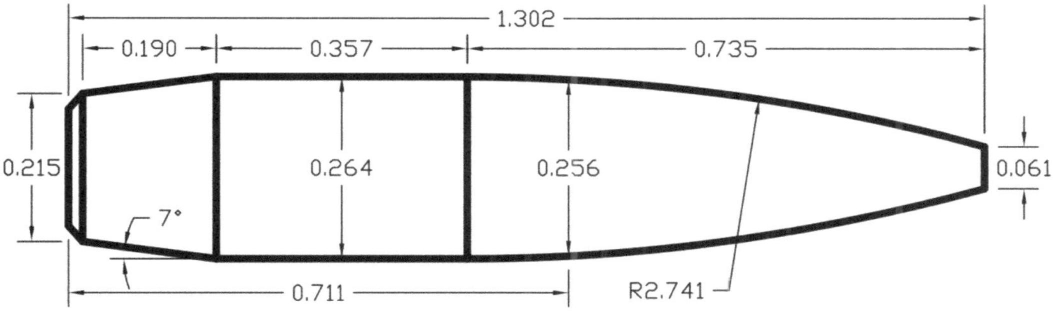

Lot #:	Unknown	Ogive Radius:	10.38 calibers
Sectional Density:	0.252 lb/in^2	Rt/R:	0.94

Ballistic Performance

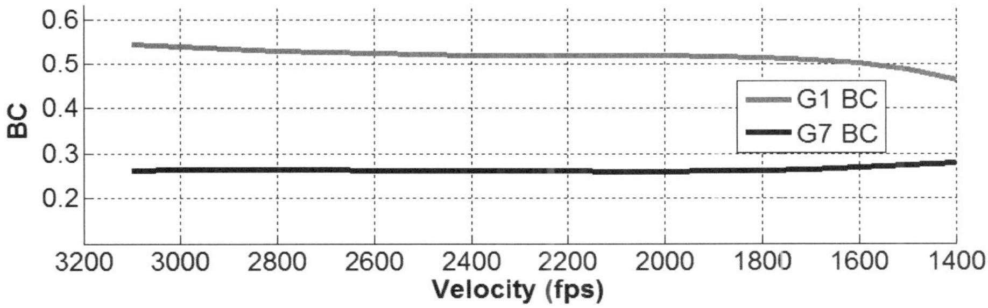

	Above 3000	3000-2500	2500-2000	2000-1500	Below 1500
G1 BC	0.554	0.531	0.521	0.503	0.487
G7 BC	0.262	0.263	0.262	0.267	0.274

Averages from 3000 fps to 1500 fps:
G1 BC: 0.517, variation of G1 BC: 0.053
G7 BC: 0.265, variation of G7 BC: 0.014
G7 form factor: 0.950

Rifling Twist Rate and Stability

Barrel Twist	Stability Factor Best case	Nominal	Worst case
1:8"	2.00	1.75	1.51
1:9"	1.58	1.38	1.19
1:10"	1.28	1.12	0.97
1:11"	1.06	0.92	0.80

Minimum twist required for stability is 1:9.7"
Recommended twist for optimal performance is 1:8.6" or faster

Lapua .264 caliber 139 grain Scenar

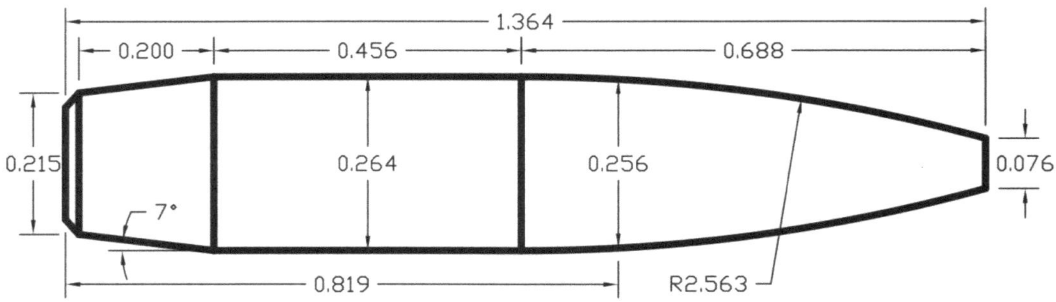

Lot #:	4PL6018	Ogive Radius:	9.71 calibers
Sectional Density:	0.285 lb/in^2	Rt/R:	0.95

Ballistic Performance

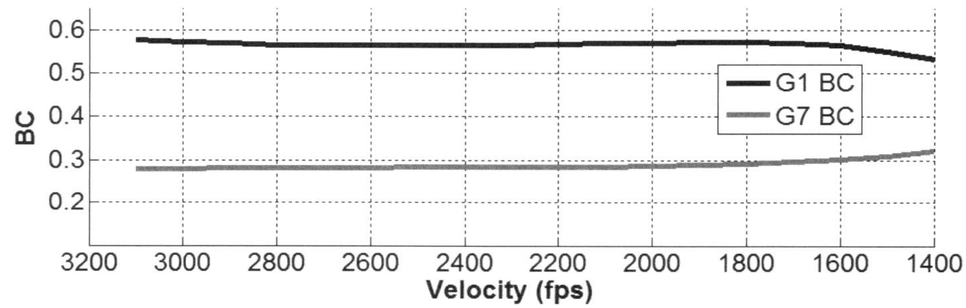

	Above 3000	3000-2500	2500-2000	2000-1500	Below 1500
G1 BC	0.584	0.568	0.567	0.560	0.550
G7 BC	0.277	0.282	0.285	0.298	0.311

Averages from 3000 fps to 1500 fps:
G1 BC: 0.564, variation of G1 BC: 0.023
G7 BC: 0.290, variation of G7 BC: 0.030
G7 form factor: 0.983

Rifling Twist Rate and Stability

Barrel Twist	Stability Factor Best case	Stability Factor Nominal	Stability Factor Worst case
1:8"	1.97	1.72	1.49
1:9"	1.56	1.36	1.18
1:10"	1.26	1.10	0.95
1:11"	1.04	0.91	0.79

Minimum twist required for stability is 1:9.6"
Recommended twist for optimal performance is 1:8.6" or faster

Lapua .264 caliber 120 grain Scenar L

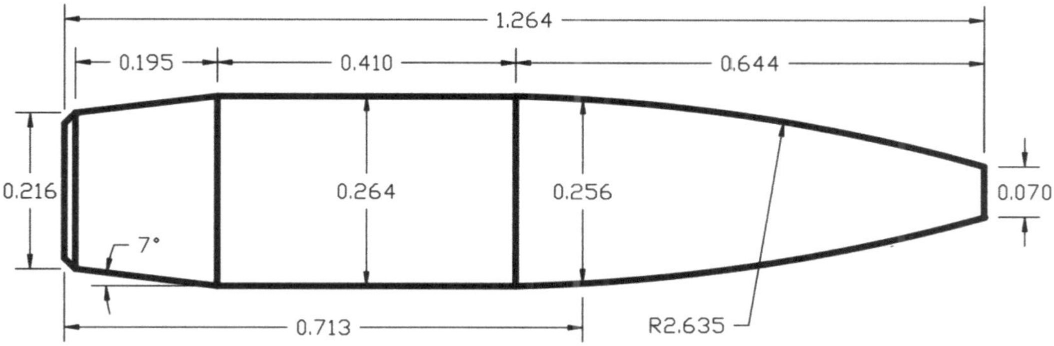

Lot #:	**Unknown**	**Ogive Radius:**	**9.98 calibers**
Sectional Density:	**0.246 lb/in^2**	**Rt/R:**	**0.83**

Ballistic Performance

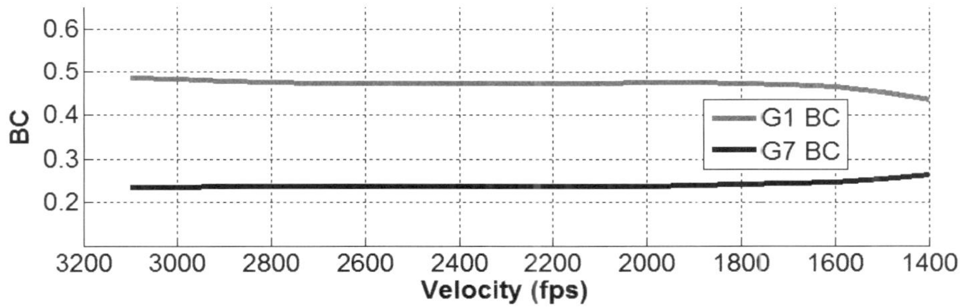

	Above 3000	3000-2500	2500-2000	2000-1500	Below 1500
G1 BC	0.493	0.477	0.473	0.463	0.453
G7 BC	0.234	0.237	0.238	0.247	0.255

Averages from 3000 fps to 1500 fps:
G1 BC: 0.470, variation of G1 BC: 0.030
G7 BC: 0.242, variation of G7 BC: 0.019
G7 form factor: 1.018

Rifling Twist Rate and Stability

	Stability Factor		
Barrel Twist	**Best case**	**Nominal**	**Worst case**
1:8"	2.13	1.86	1.61
1:9"	1.68	1.47	1.27
1:10"	1.36	1.19	1.03
1:11"	1.13	0.98	0.85

Minimum twist required for stability is 1:10.0"
Recommended twist for optimal performance is 1:8.9" or faster

Lapua .264 caliber 136 grain Scenar L

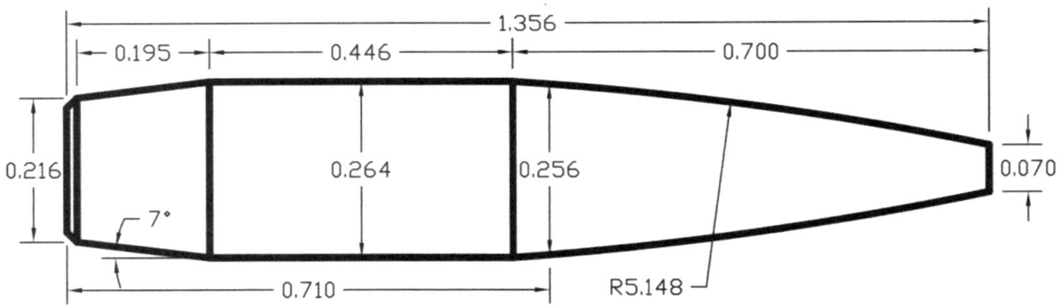

Lot #:	Unknown	Ogive Radius:	19.50 calibers
Sectional Density:	0.279 lb/in^2	Rt/R:	0.47

Ballistic Performance

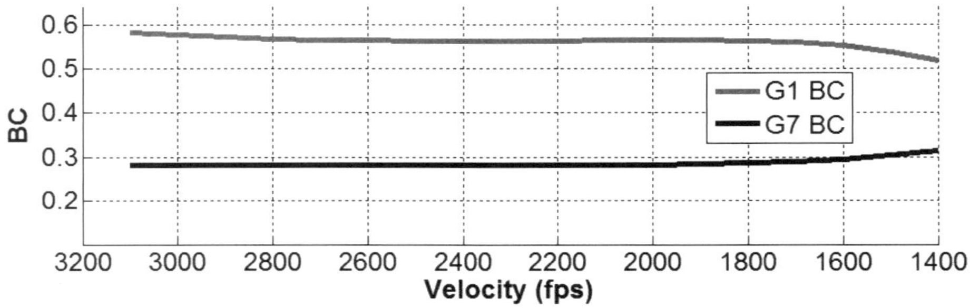

	Above 3000	3000-2500	2500-2000	2000-1500	Below 1500
G1 BC	0.590	0.569	0.563	0.551	0.537
G7 BC	0.279	0.282	0.283	0.293	0.303

Averages from 3000 fps to 1500 fps:
G1 BC: 0.560, variation of G1 BC: 0.039
G7 BC: 0.288, variation of G7 BC: 0.021
G7 form factor: 0.969

Rifling Twist Rate and Stability

Barrel Twist	Stability Factor Best case	Nominal	Worst case
1:6"	1.97	1.72	1.48
1:7"	1.55	1.36	1.17
1:8"	1.26	1.10	0.95
1:9"	1.04	0.91	0.79

Minimum twist required for stability is 1:9.6"
Recommended twist for optimal performance is 1:8.6" or faster

Matrix .264 caliber 150 grain BTHP

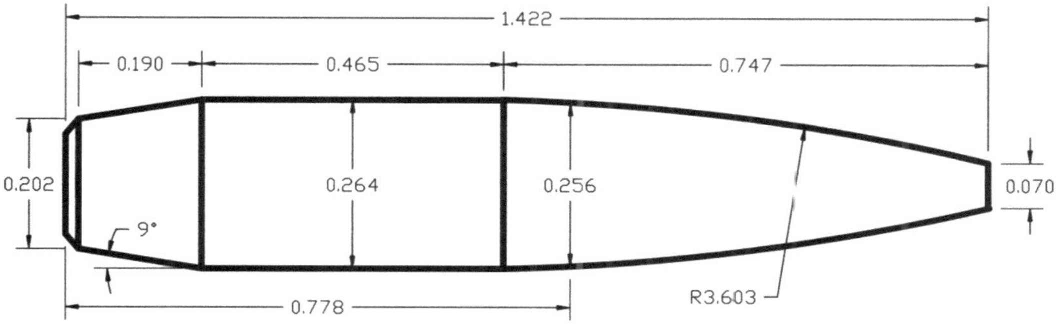

Lot #:	Unknown	Ogive Radius:	13.65 calibers
Sectional Density:	0.308 lb/in²	Rt/R:	0.81

Ballistic Performance

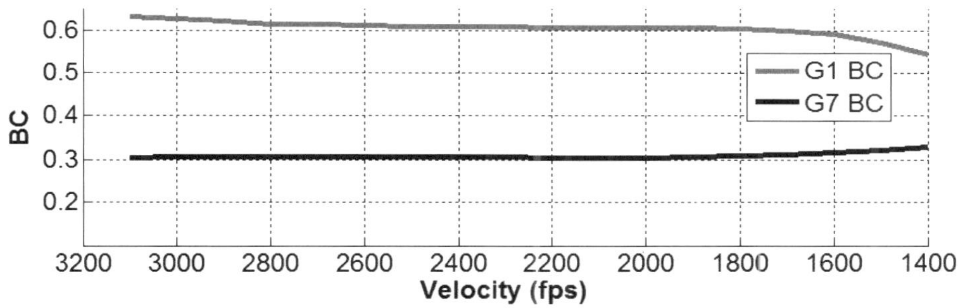

	Above 3000	3000-2500	2500-2000	2000-1500	Below 1500
G1 BC	0.640	0.617	0.608	0.590	0.572
G7 BC	0.303	0.306	0.306	0.314	0.322

Averages from 3000 fps to 1500 fps:
G1 BC: 0.604, variation of G1 BC: 0.054
G7 BC: 0.310, variation of G7 BC: 0.017
G7 form factor: 0.992

Rifling Twist Rate and Stability

Barrel Twist	Stability Factor Best case	Nominal	Worst case
1:8"	1.89	1.65	1.42
1:9"	1.49	1.30	1.13
1:10"	1.21	1.05	0.91
1:11"	1.00	0.87	0.75

Minimum twist required for stability is 1:9.4"
Recommended twist for optimal performance is 1:8.4" or faster

Matrix .264 caliber 160 grain BTHP

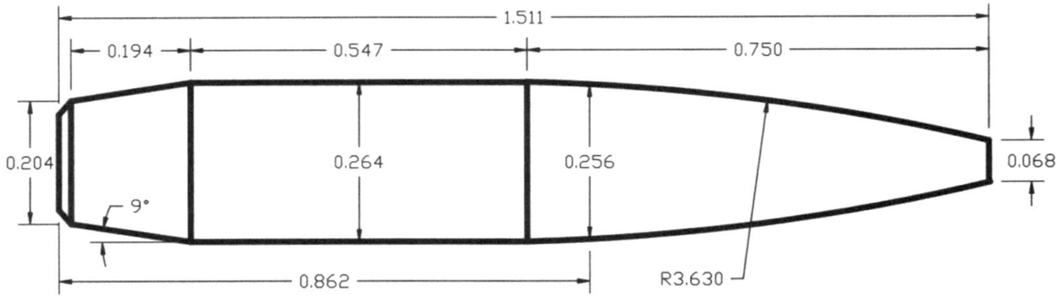

Lot #:	Unknown	Ogive Radius:	13.75 calibers
Sectional Density:	0.328 lb/in^2	Rt/R:	0.80

Ballistic Performance

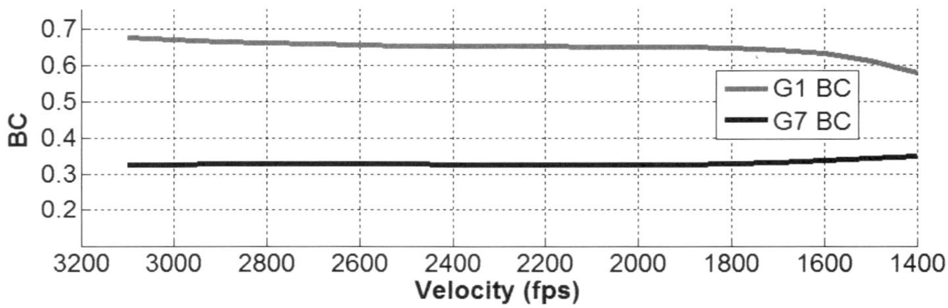

	Above 3000	3000-2500	2500-2000	2000-1500	Below 1500
G1 BC	0.687	0.662	0.651	0.630	0.610
G7 BC	0.325	0.328	0.327	0.335	0.344

Averages from 3000 fps to 1500 fps:
G1 BC: 0.646, variation of G1 BC: 0.060
G7 BC: 0.332, variation of G7 BC: 0.017
G7 form factor: 0.989

Rifling Twist Rate and Stability

Barrel Twist	Stability Factor Best case	Nominal	Worst case
1:8"	1.68	1.47	1.27
1:9"	1.33	1.16	1.00
1:10"	1.08	0.94	0.81
1:11"	0.89	0.78	0.67

Minimum twist required for stability is 1:8.9"
Recommended twist for optimal performance is 1:7.9" or faster

Norma .264 caliber 130 grain Diamond

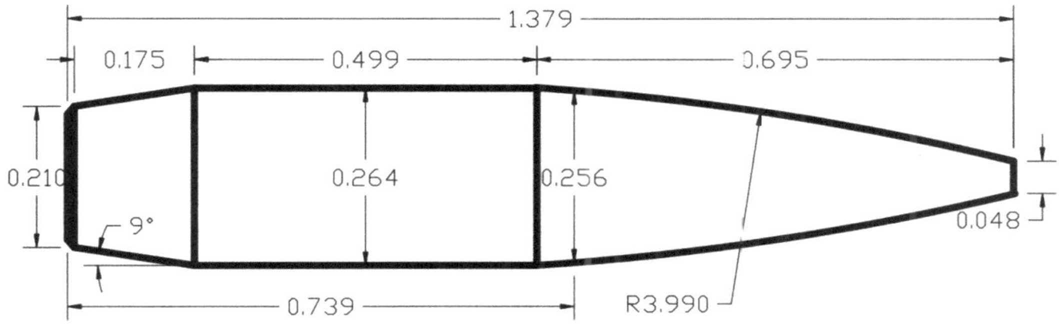

Lot #:	Unknown	Ogive Radius:	15.11 calibers
Sectional Density:	0.267 lb/in^2	Rt/R:	0.60

Ballistic Performance

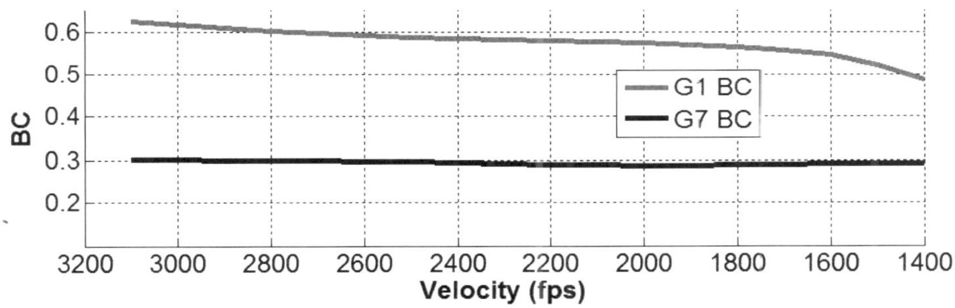

	Above 3000	3000-2500	2500-2000	2000-1500	Below 1500
G1 BC	0.636	0.602	0.581	0.547	0.521
G7 BC	0.301	0.298	0.292	0.291	0.293

Averages from 3000 fps to 1500 fps:
G1 BC: 0.575, variation of G1 BC: 0.094
G7 BC: 0.294, variation of G7 BC: 0.013
G7 form factor: 0.905

Rifling Twist Rate and Stability

Barrel Twist	Stability Factor: Best case	Stability Factor: Nominal	Stability Factor: Worst case
1:8"	1.79	1.56	1.35
1:9"	1.41	1.23	1.07
1:10"	1.14	1.00	0.86
1:11"	0.95	0.83	0.71

Minimum twist required for stability is 1:9.1"
Recommended twist for optimal performance is 1:8.2" or faster

Nosler .264 caliber 130 grain Accubond

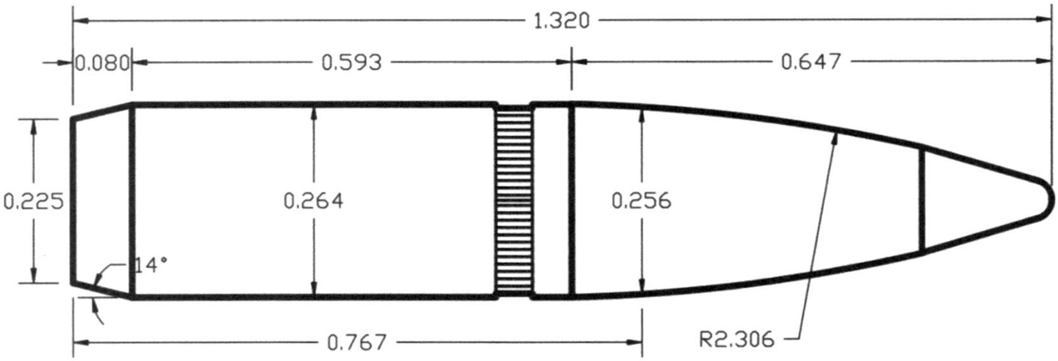

Lot #:	**Unknown**	**Ogive Radius:**	**8.74 calibers**
Sectional Density:	**0.267 lb/in²**	**Rt/R:**	**0.83**

Ballistic Performance

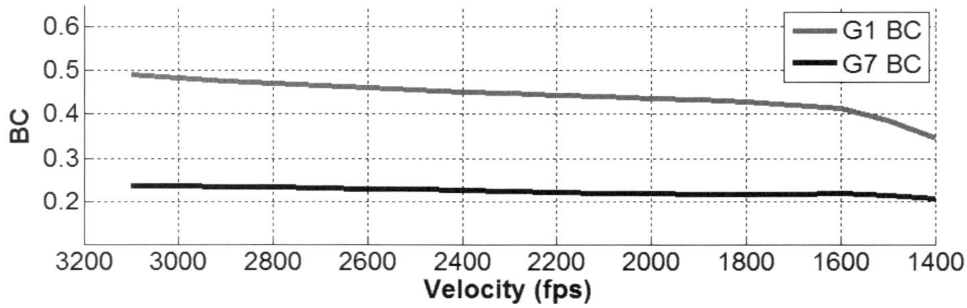

	Above 3000	3000-2500	2500-2000	2000-1500	Below 1500
G1 BC	0.503	0.469	0.446	0.411	0.385
G7 BC	0.238	0.233	0.224	0.218	0.216

Averages from 3000 fps to 1500 fps:
G1 BC: 0.440, variation of G1 BC: 0.098
G7 BC: 0.225, variation of G7 BC: 0.021
G7 form factor: 1.185

Rifling Twist Rate and Stability

Barrel Twist	Stability Factor Best case	Stability Factor Nominal	Stability Factor Worst case
1:8"	2.03	1.78	1.54
1:9"	1.61	1.40	1.21
1:10"	1.30	1.14	0.98
1:11"	1.08	0.94	0.81

Minimum twist required for stability is 1:9.7"
Recommended twist for optimal performance is 1:8.7" or faster

Nosler .264 caliber 140 grain Accubond

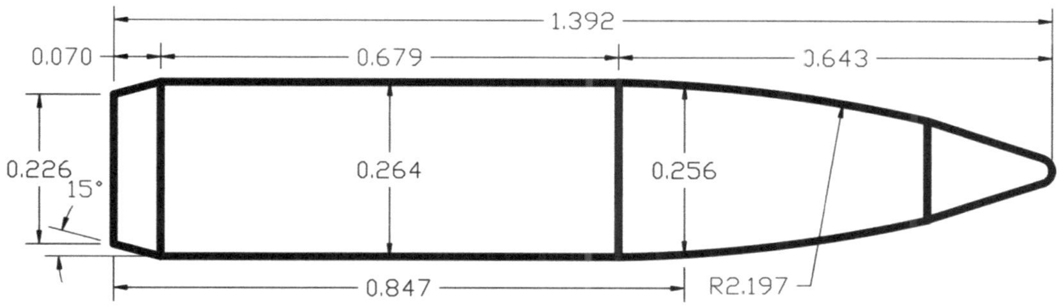

Lot #:	BH23F 6 1	Ogive Radius:	8.32 calibers
Sectional Density:	0.287 lb/in^2	Rt/R:	0.90

Ballistic Performance

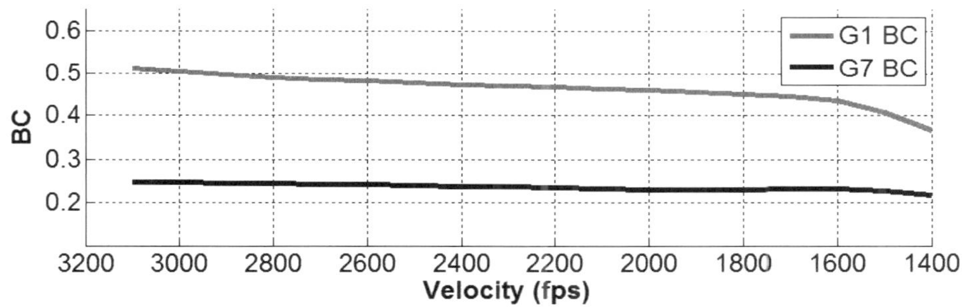

	Above 3000	3000-2500	2500-2000	2000-1500	Below 1500
G1 BC	0.525	0.491	0.469	0.434	0.408
G7 BC	0.248	0.244	0.236	0.230	0.228

Averages from 3000 fps to 1500 fps:
G1 BC: 0.463, variation of G1 BC: 0.097
G7 BC: 0.237, variation of G7 BC: 0.019
G7 form factor: 1.214

Rifling Twist Rate and Stability

Barrel Twist	Stability Factor Best case	Nominal	Worst case
1:8"	1.87	1.64	1.42
1:9"	1.48	1.29	1.12
1:10"	1.20	1.05	0.91
1:11"	0.99	0.87	0.75

Minimum twist required for stability is 1:9.3"
Recommended twist for optimal performance is 1:8.4" or faster

Nosler .264 caliber 129 grain Long Range Accubond

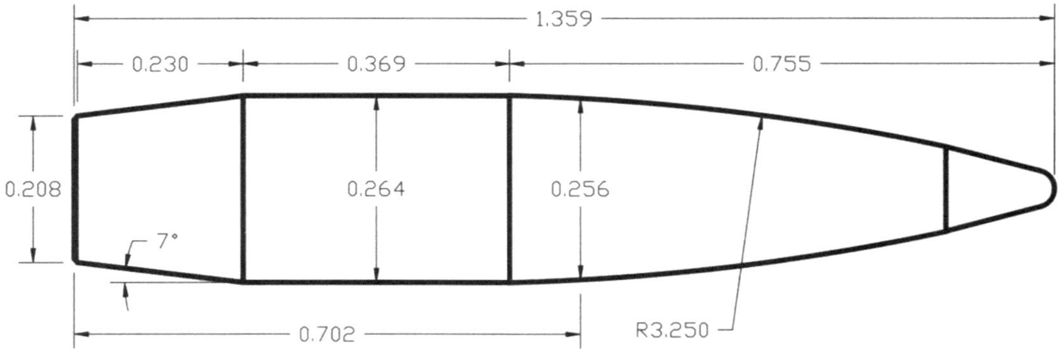

Lot #:	unknown	Ogive Radius:	12.31 calibers
Sectional Density:	0.264 lb/in^2	Rt/R:	0.82

Ballistic Performance

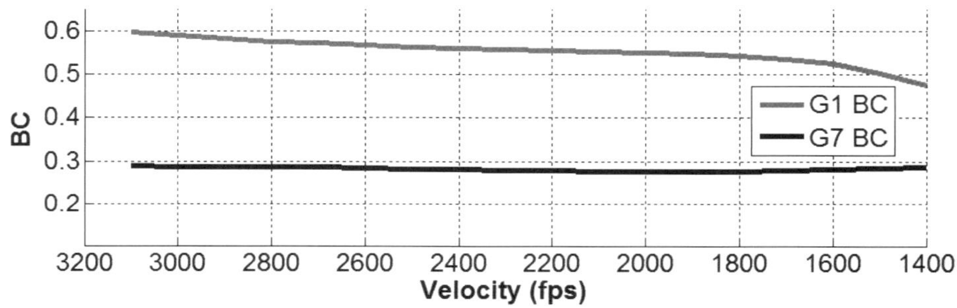

	Above 3000	3000-2500	2500-2000	2000-1500	Below 1500
G1 BC	0.610	0.576	0.556	0.527	0.504
G7 BC	0.289	0.286	0.280	0.280	0.283

Averages from 3000 fps to 1500 fps:
G1 BC: 0.552, variation of G1 BC: 0.086
G7 BC: 0.283, variation of G7 BC: 0.012
G7 form factor: 0.935

Rifling Twist Rate and Stability

Barrel Twist	Stability Factor Best case	Nominal	Worst case
1:7"	2.42	2.11	1.83
1:8"	1.85	1.62	1.40
1:9"	1.46	1.28	1.11
1:10"	1.19	1.04	0.90

Minimum twist required for stability is 1:9.3"
Recommended twist for optimal performance is 1:8.3" or faster

Nosler .264 caliber 142 grain Long Range Accubond

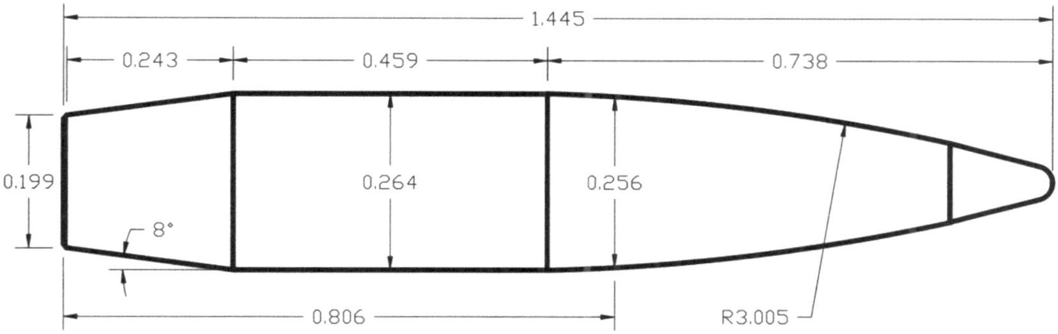

Lot #:	EM18F21	Ogive Radius:	11.38 calibers
Sectional Density:	0.291 lb/in^2	Rt/R:	0.83

Ballistic Performance

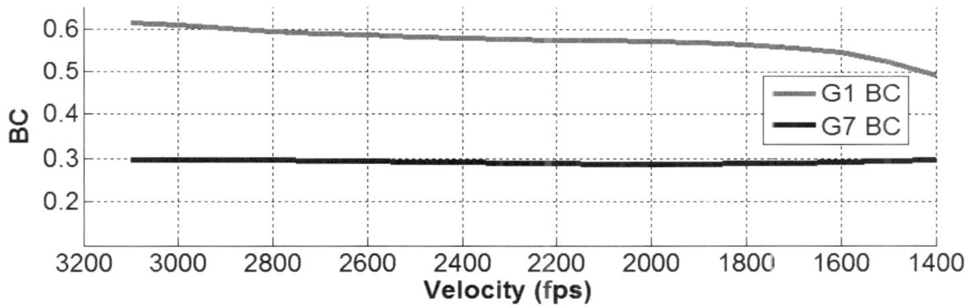

	Above 3000	3000-2500	2500-2000	2000-1500	Below 1500
G1 BC	0.628	0.595	0.577	0.548	0.524
G7 BC	0.297	0.295	0.290	0.291	0.295

Averages from 3000 fps to 1500 fps:
G1 BC: 0.572, variation of G1 BC: 0.084
G7 BC: 0.293, variation of G7 BC: 0.010
G7 form factor: 0.993

Rifling Twist Rate and Stability

Barrel Twist	Stability Factor Best case	Nominal	Worst case
1:6"	3.03	2.65	2.29
1:7"	2.22	1.94	1.68
1:8"	1.70	1.49	1.29
1:9"	1.35	1.18	1.02

Minimum twist required for stability is 1:8.9"
Recommended twist for optimal performance is 1:8.0" or faster

Nosler .264 caliber 100 grain Ballistic Tip

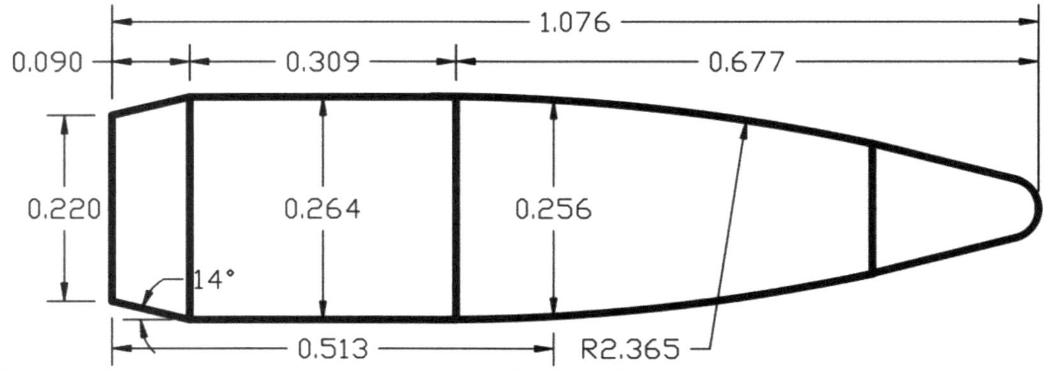

Lot #:	DL25F7	Ogive Radius:	8.96 calibers
Sectional Density:	.205 lb/in^2	Rt/R:	0.82

Ballistic Performance

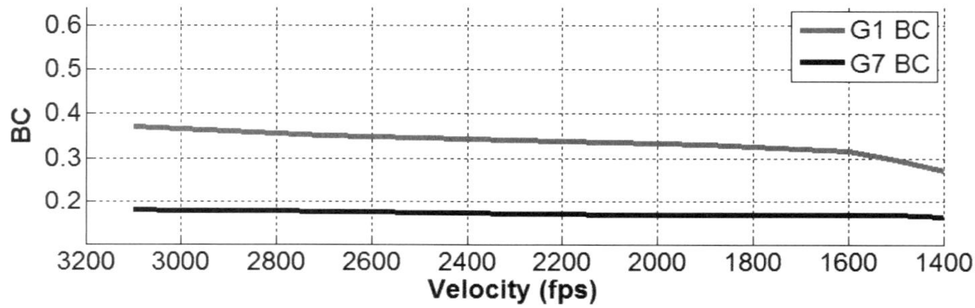

	Above 3000	3000-2500	2500-2000	2000-1500	Below 1500
G1 BC	0.379	0.356	0.340	0.316	0.298
G7 BC	0.180	0.177	0.171	0.167	0.167

Averages from 3000 fps to 1500 fps:

G1 BC: 0.336, variation of G1 BC: 0.068

G7 BC: 0.172, variation of G7 BC: 0.012

G7 form factor: 1.193

Rifling Twist Rate and Stability

Barrel Twist	Stability Factor Best case	Stability Factor Nominal	Stability Factor Worst case
1:8"	2.83	2.47	2.14
1:9"	2.24	1.95	1.69
1:10"	1.81	1.58	1.37
1:11"	1.50	1.31	1.13

Minimum twist required for stability is 1:11.5"

Recommended twist for optimal performance is 1:10.3" or faster

Nosler .264 caliber 120 grain Ballistic Tip

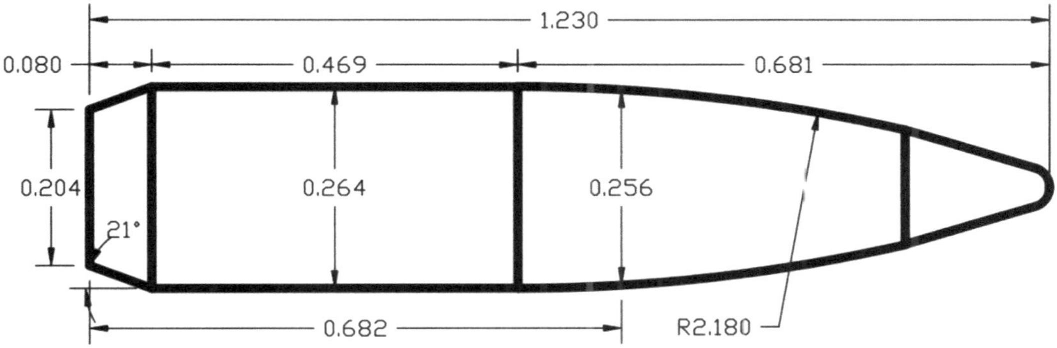

Lot #:	BE13F7-2	Ogive Radius:	8.26 calibers
Sectional Density:	0.246 lb/in^2	Rt/R:	1.00

Ballistic Performance

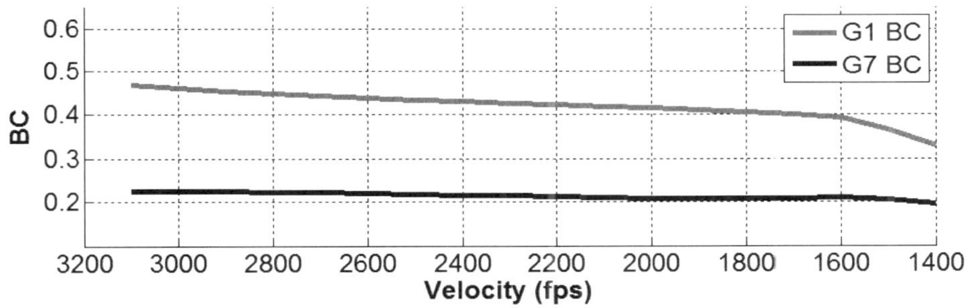

	Above 3000	3000-2500	2500-2000	2000-1500	Below 1500
G1 BC	0.479	0.447	0.425	0.391	0.367
G7 BC	0.227	0.222	0.214	0.207	0.205

Averages from 3000 fps to 1500 fps:
G1 BC: 0.419, variation of G1 BC: 0.094
G7 BC: 0.214, variation of G7 BC: 0.020
G7 form factor: 1.149

Rifling Twist Rate and Stability

Barrel Twist	Stability Factor Best case	Nominal	Worst case
1:8"	2.31	2.01	1.74
1:9"	1.82	1.59	1.38
1:10"	1.48	1.29	1.11
1:11"	1.22	1.07	0.92

Minimum twist required for stability is 1:10.4"
Recommended twist for optimal performance is 1:9.3" or faster

Nosler .264 caliber 123 grain Custom Competition

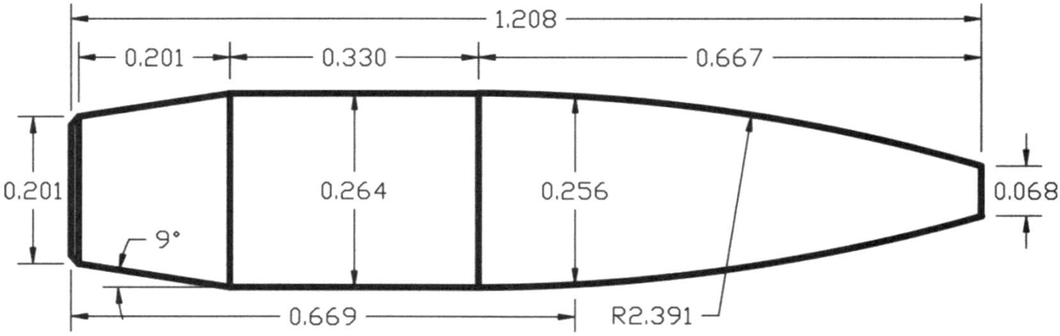

Lot #:	EL06F15	Ogive Radius:	9.06 calibers
Sectional Density:	0. 252 lb/in^2	Rt/R:	0.88

Ballistic Performance

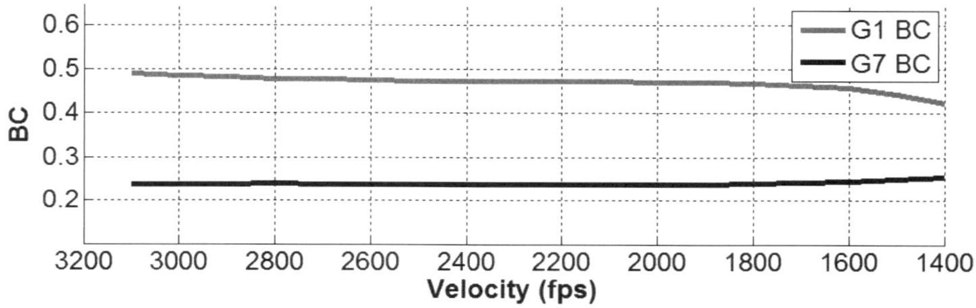

	Above 3000	3000-2500	2500-2000	2000-1500	Below 1500
G1 BC	0.497	0.480	0.473	0.457	0.442
G7 BC	0.236	0.238	0.238	0.243	0.249

Averages from 3000 fps to 1500 fps:
G1 BC: 0.468, variation of G1 BC: 0.044
G7 BC: 0.241, variation of G7 BC: 0.012
G7 form factor: 1.048

Rifling Twist Rate and Stability

Barrel Twist	Stability Factor Best case	Nominal	Worst case
1:8"	2.49	2.18	1.88
1:9"	1.97	1.72	1.49
1:10"	1.59	1.39	1.20
1:11"	1.32	1.15	1.00

Minimum twist required for stability is 1:10.8"
Recommended twist for optimal performance is 1:9.6" or faster

Nosler .264 caliber 140 grain Custom Competition

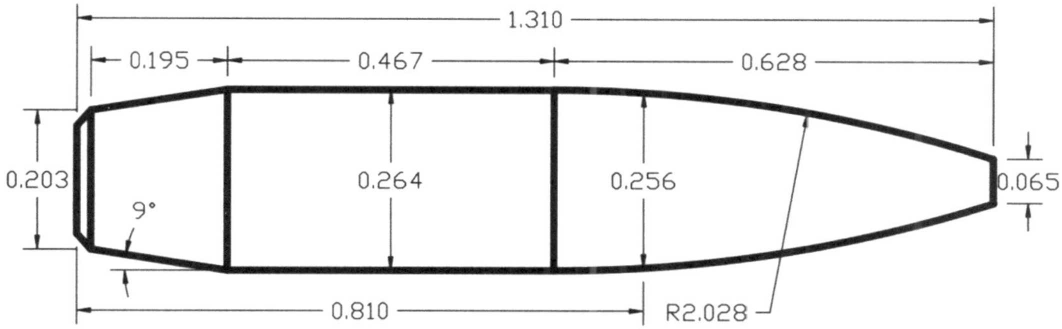

Lot #:	KE08F12-1	Ogive Radius:	7.68 calibers
Sectional Density:	0.287 lb/in^2	Rt/R:	1.00

Ballistic Performance

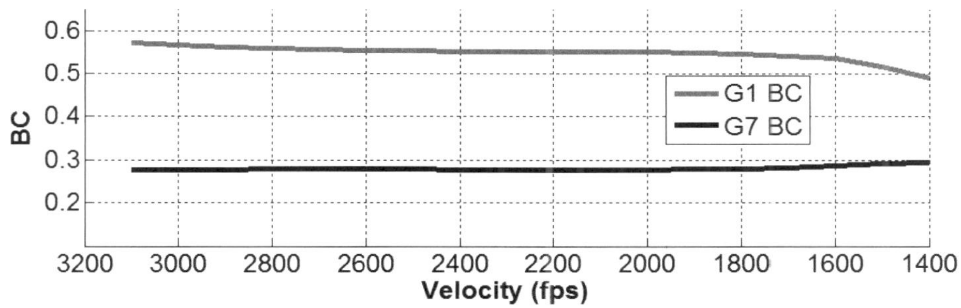

	Above 3000	3000-2500	2500-2000	2000-1500	Below 1500
G1 BC	0.581	0.561	0.553	0.534	0.516
G7 BC	0.275	0.278	0.278	0.284	0.291

Averages from 3000 fps to 1500 fps:
G1 BC: 0.547, variation of G1 BC: 0.051
G7 BC: 0.281, variation of G7 BC: 0.014
G7 form factor: 1.021

Rifling Twist Rate and Stability

Barrel Twist	Stability Factor Best case	Nominal	Worst case
1:8"	2.24	1.96	1.69
1:9"	1.77	1.54	1.34
1:10"	1.43	1.25	1.08
1:11"	1.18	1.03	0.89

Minimum twist required for stability is 1:10.2"
Recommended twist for optimal performance is 1:9.1" or faster

Sierra .264 caliber 107 grain MatchKing

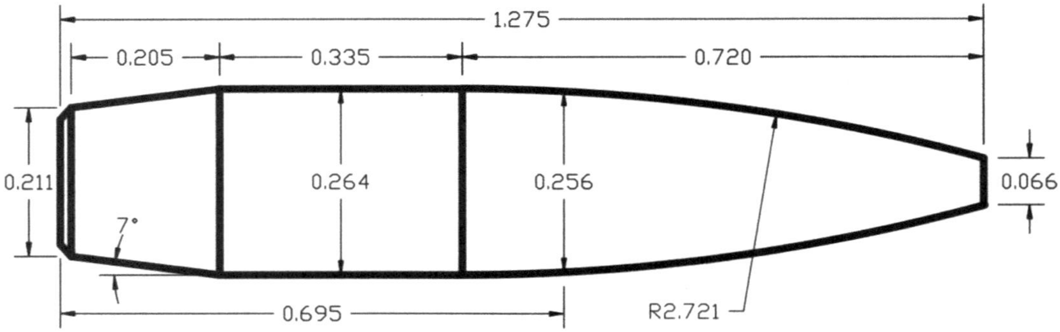

Lot #:	0041262744816	Ogive Radius:	10.3 calibers
Sectional Density:	0.219 lb/in^2	Rt/R:	0.96

Ballistic Performance

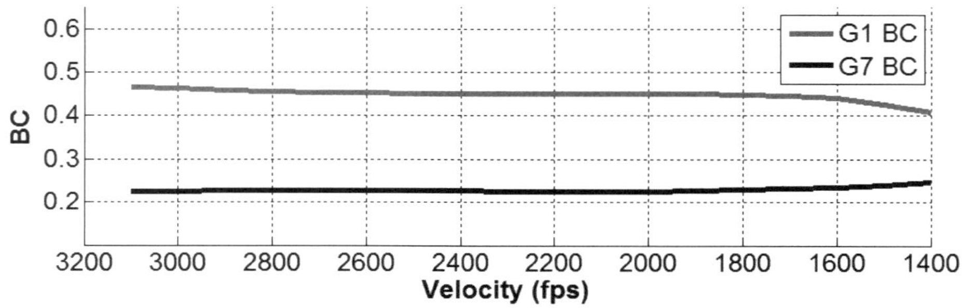

	Above 3000	3000-2500	2500-2000	2000-1500	Below 1500
G1 BC	0.474	0.457	0.451	0.438	0.426
G7 BC	0.225	0.227	0.227	0.233	0.240

Averages from 3000 fps to 1500 fps:
G1 BC: 0.448, variation of G1 BC: 0.036
G7 BC: 0.230, variation of G7 BC: 0.014
G7 form factor: 0.954

Rifling Twist Rate and Stability

Barrel Twist	Stability Factor Best case	Nominal	Worst case
1:8"	1.85	1.62	1.40
1:9"	1.46	1.28	1.10
1:10"	1.18	1.04	0.89
1:11"	0.98	0.86	0.74

Minimum twist required for stability is 1:9.3"
Recommended twist for optimal performance is 1:8.3" or faster

Sierra .264 caliber 123 grain MatchKing

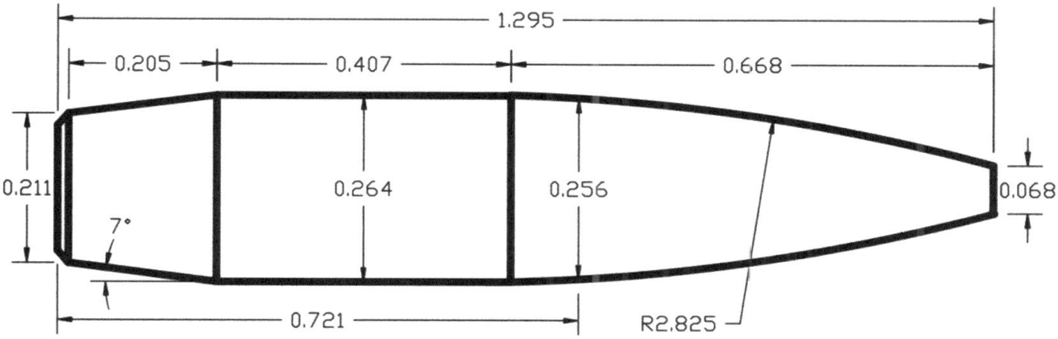

Lot #:	0038865256296	Ogive Radius:	10.7 calibers
Sectional Density:	0.252 lb/in^2	Rt/R:	0.85

Ballistic Performance

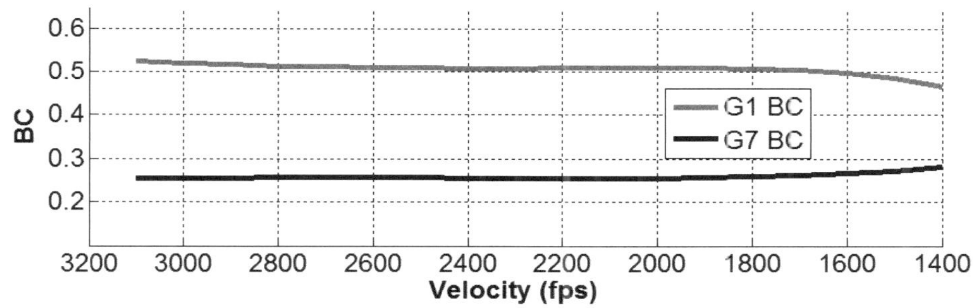

	Above 3000	3000-2500	2500-2000	2000-1500	Below 1500
G1 BC	0.533	0.515	0.509	0.497	0.484
G7 BC	0.253	0.256	0.256	0.265	0.273

Averages from 3000 fps to 1500 fps:
G1 BC: 0.506, variation of G1 BC: 0.037
G7 BC: 0.260, variation of G7 BC: 0.018
G7 form factor: 0.970

Rifling Twist Rate and Stability

Barrel Twist	Best case	Stability Factor Nominal	Worst case
1:8"	2.03	1.78	1.54
1:9"	1.61	1.40	1.21
1:10"	1.30	1.14	0.98
1:11"	1.08	0.94	0.81

Minimum twist required for stability is 1:9.7"
Recommended twist for optimal performance is 1:8.7" or faster

Sierra .264 caliber 142 grain MatchKing

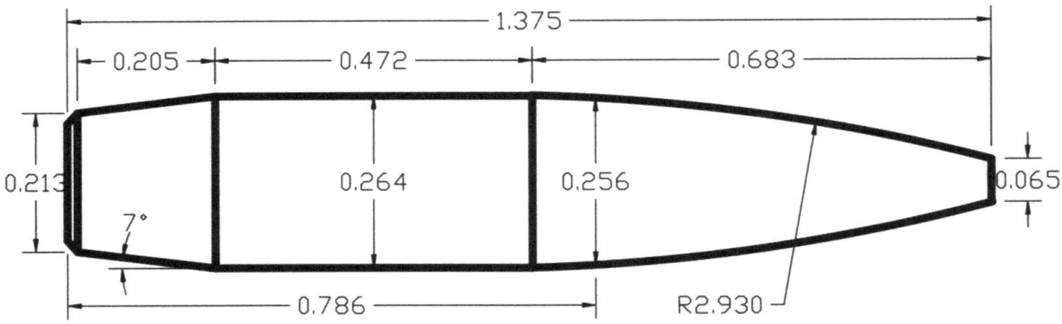

Lot #:	0039855964848	Ogive Radius:	11.1 calibers
Sectional Density:	0.291 lb/in^2	Rt/R:	0.84

Ballistic Performance

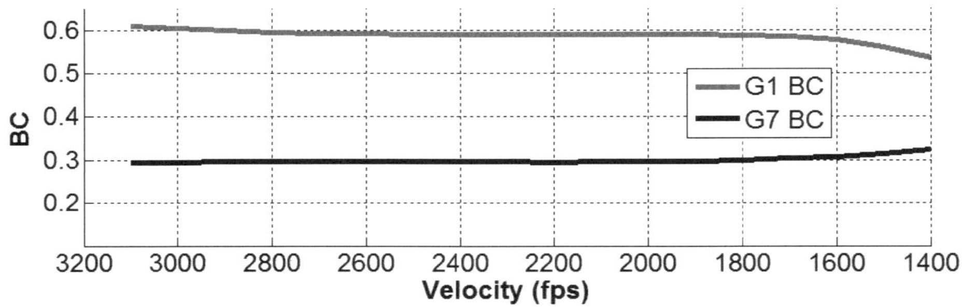

	Above 3000	3000-2500	2500-2000	2000-1500	Below 1500
G1 BC	0.619	0.597	0.589	0.574	0.558
G7 BC	0.293	0.296	0.296	0.305	0.315

Averages from 3000 fps to 1500 fps:
G1 BC: 0.585, variation of G1 BC: 0.046
G7 BC: 0.301, variation of G7 BC: 0.019
G7 form factor: 0.968

Rifling Twist Rate and Stability

Barrel Twist	Stability Factor Best case	Nominal	Worst case
1:8"	1.97	1.72	1.49
1:9"	1.56	1.36	1.18
1:10"	1.26	1.10	0.95
1:11"	1.04	0.91	0.79

Minimum twist required for stability is 1:9.6"
Recommended twist for optimal performance is 1:8.6" or faster

Sierra .264 caliber 120 grain ProHunter

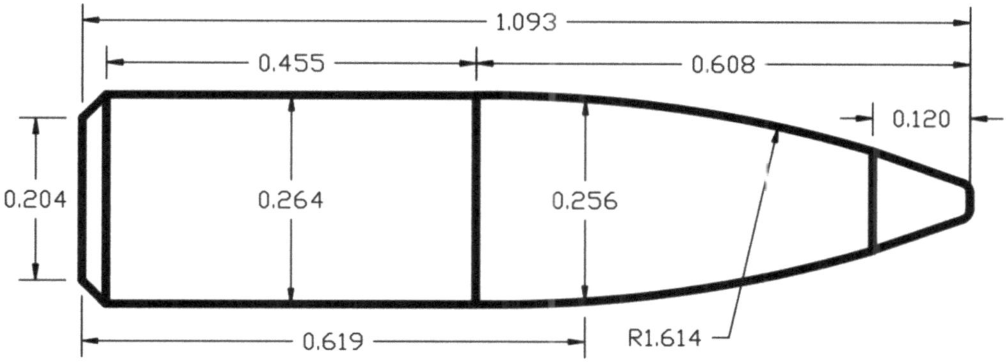

Lot #:	0039536739595	Ogive Radius:	6.11 calibers
Sectional Density:	0.246 lb/in^2	Rt/R:	1.00

Ballistic Performance

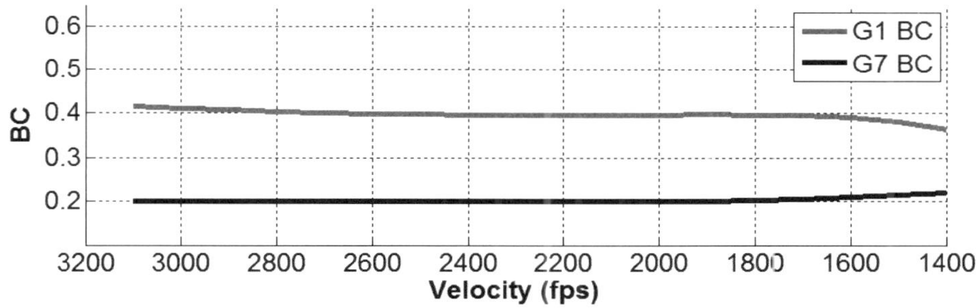

	Above 3000	3000-2500	2500-2000	2000-1500	Below 1500
G1 BC	0.423	0.405	0.398	0.389	0.381
G7 BC	0.201	0.201	0.200	0.207	0.215

Averages from 3000 fps to 1500 fps:
G1 BC: 0.397, variation of G1 BC: 0.031
G7 BC: 0.204, variation of G7 BC: 0.015
G7 form factor: 1.207

Rifling Twist Rate and Stability

Barrel Twist	Stability Factor Best case	Nominal	Worst case
1:8"	3.25	2.84	2.45
1:9"	2.57	2.24	1.94
1:10"	2.08	1.82	1.57
1:11"	1.72	1.50	1.30

Minimum twist required for stability is 1:12.3"
Recommended twist for optimal performance is 1:11.0" or faster

Swift .264 caliber 130 grain Scirocco

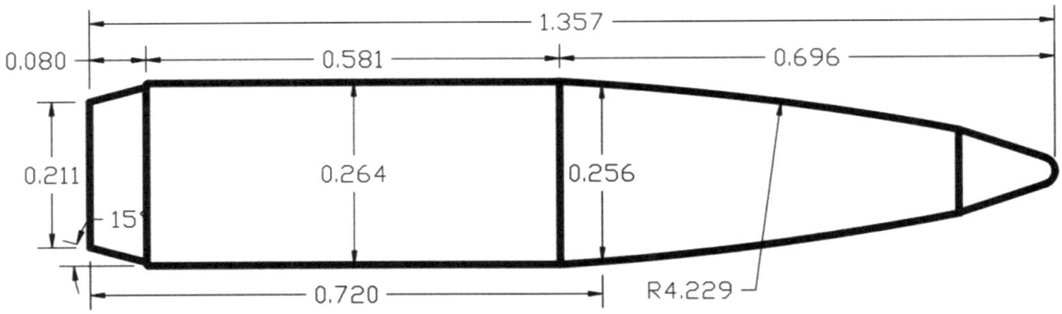

Lot #:	**Unknown**	**Ogive Radius:**	**16.02 calibers**
Sectional Density:	**0.267 lb/in²**	**Rt/R:**	**0.62**

Ballistic Performance

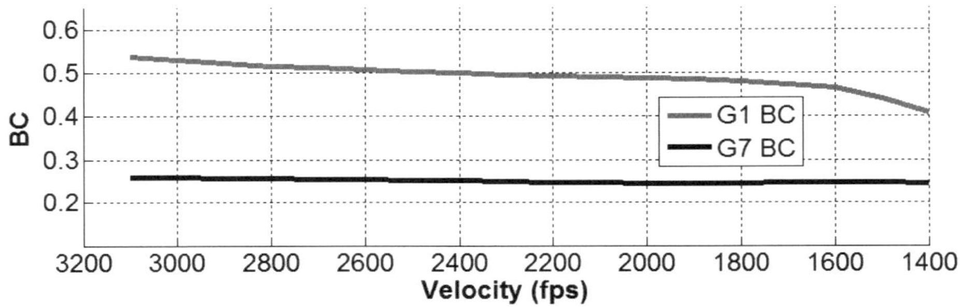

	Above 3000	3000-2500	2500-2000	2000-1500	Below 1500
G1 BC	0.551	0.517	0.495	0.465	0.441
G7 BC	0.261	0.256	0.249	0.246	0.248

Averages from 3000 fps to 1500 fps:
G1 BC: 0.491, variation of G1 BC: 0.090
G7 BC: 0.251, variation of G7 BC: 0.014
G7 form factor: 1.061

Rifling Twist Rate and Stability

Barrel Twist	**Stability Factor** Best case	Nominal	Worst case
1:8"	1.87	1.64	1.42
1:9"	1.48	1.29	1.12
1:10"	1.20	1.05	0.91
1:11"	0.99	0.87	0.75

Minimum twist required for stability is 1:9.3"
Recommended twist for optimal performance is 1:8.4" or faster

.270 Caliber (.277" Diameter)

.277 caliber bullets typically range from 100 to 150 grains (the traditional 1:10" twist rate doesn't stabilize heavier 25 caliber bullets). The image below relates bullet weight to sectional density for .277 caliber bullets.

Range of Weight and Sectional Density for 0.277 Caliber

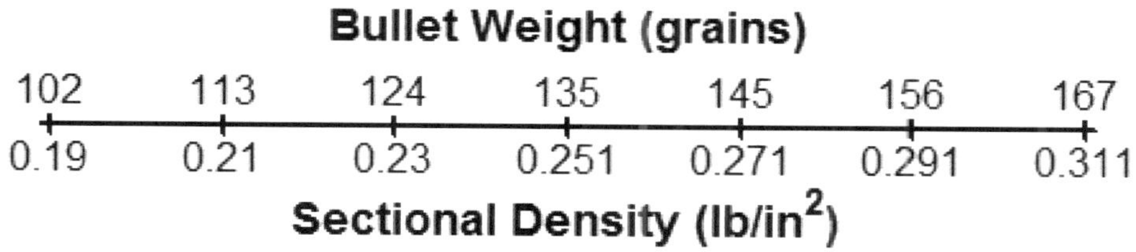

Sectional Density (SD) is a measure of how much mass is behind the bullets frontal area. Higher SD means a greater ability for the bullet to penetrate targets on impact, and also retain velocity in flight.

Spectrum of G7 form factors

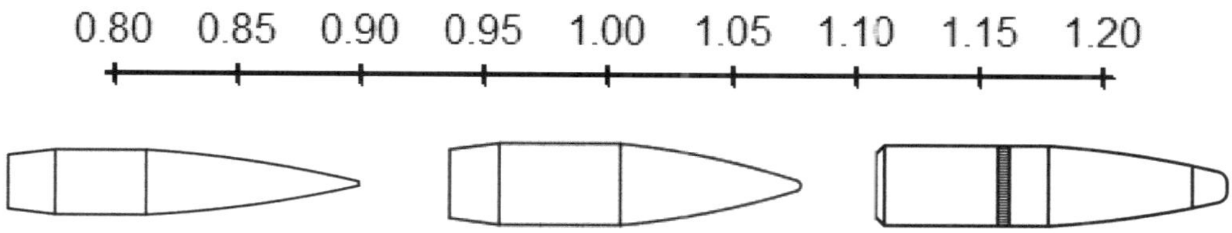

G7 Form factors range from 0.80 for the lowest drag shapes, up to 1.20+ for blunter shapes.

G7 Ballistic Coefficients are simply Sectional Density (SD) divided by G7 form factor.

You can determine a bullets' suitability for long range performance by seeing where it lies on the SD and form factor lines. Ideally, you want a bullet with a high SD and a low G7 form factor.

As an example, a 150 grain .277 caliber bullet has an SD of 0.279. If that bullet had a G7 form factor of 0.95, it's G7 BC would be:

G7 BC = SD/FF = 0.279/0.95 = 0.294.

Barnes 270 caliber 130 grain TSXBT

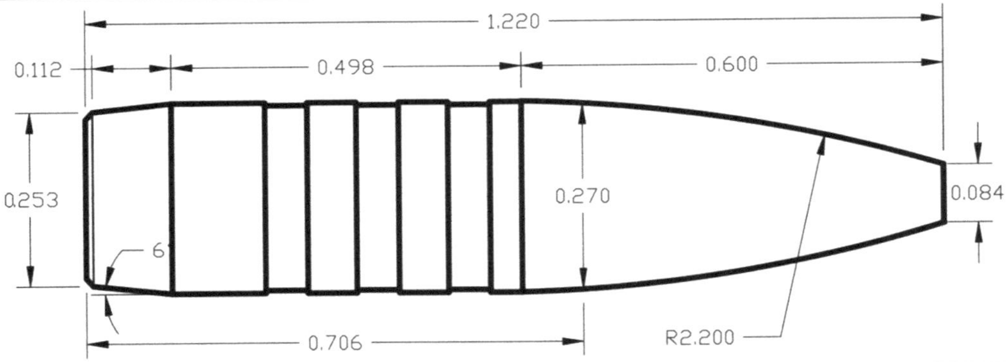

Lot #:	6480584-1	Ogive Radius:	7.94 calibers
Sectional Density:	0.242 lb/in^2	Rt/R:	0.83

Ballistic Performance

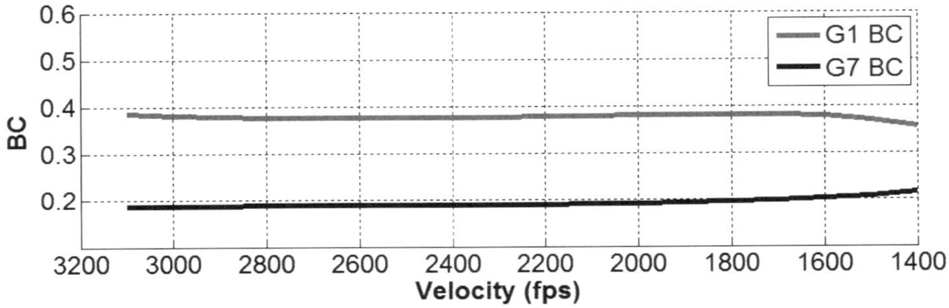

	Above 3000	3000-2500	2500-2000	2000-1500	Below 1500
G1 BC	0.390	0.380	0.379	0.376	0.369
G7 BC	0.185	0.188	0.191	0.200	0.209

Averages from 3000 fps to 1500 fps:
G1 BC: 0.378, variation of G1 BC: 0.013
G7 BC: 0.194, variation of G7 BC: 0.021
G7 form factor: 1.248

Rifling Twist Rate and Stability

Barrel Twist	Stability Factor Best case	Stability Factor Nominal	Stability Factor Worst case
1:9"	2.21	1.93	1.67
1:10"	1.79	1.57	1.36
1:11"	1.48	1.30	1.12
1:12"	1.25	1.09	0.94

Minimum twist required for stability is 1:11.4"
Recommended twist for optimal performance is 1:10.2" or faster

Barnes 270 caliber 140 grain TSXBT

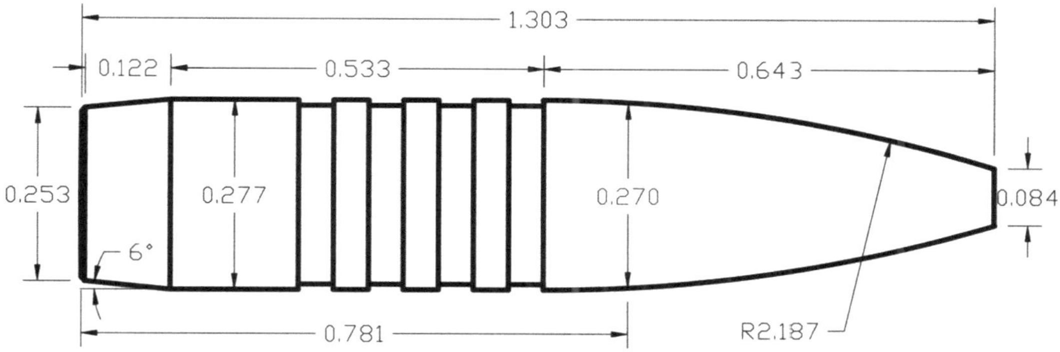

Lot #:	6480686-1	Ogive Radius:	7.90 calibers
Sectional Density:	0.261 lb/in^2	Rt/R:	1.00

Ballistic Performance

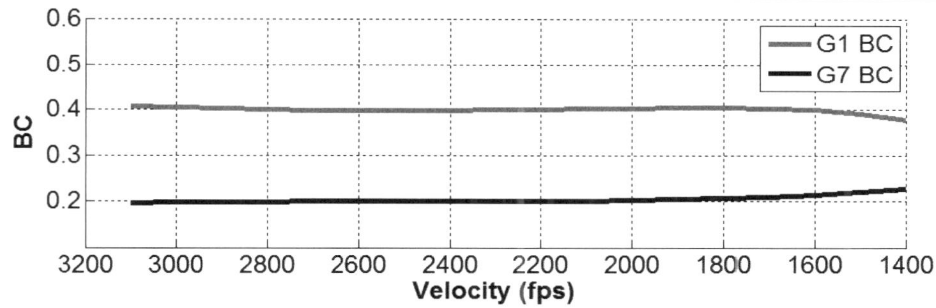

	Above 3000	3000-2500	2500-2000	2000-1500	Below 1500
G1 BC	0.413	0.401	0.400	0.397	0.391
G7 BC	0.196	0.199	0.201	0.211	0.221

Averages from 3000 fps to 1500 fps:
G1 BC: 0.399, variation of G1 BC: 0.014
G7 BC: 0.205, variation of G7 BC: 0.023
G7 form factor: 1.272

Rifling Twist Rate and Stability

	Stability Factor		
Barrel Twist	**Best case**	**Nominal**	**Worst case**
1:8"	2.49	2.18	1.88
1:9"	1.97	1.72	1.49
1:10"	1.60	1.39	1.21
1:11"	1.32	1.15	1.00

Minimum twist required for stability is 1:10.8"
Recommended twist for optimal performance is 1:9.6" or faster

Barnes 270 caliber 110 grain TTSXBT

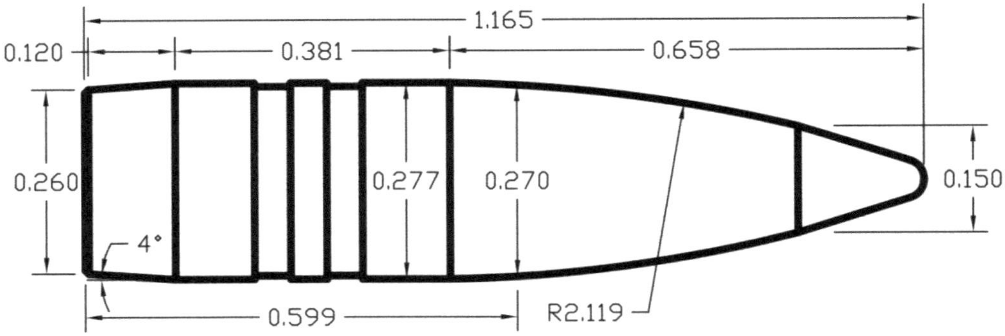

Lot #:	**Unknown**	**Ogive Radius:**	**7.65 calibers**
Sectional Density:	**0.205 lb/in²**	**Rt/R:**	**0.87**

Ballistic Performance

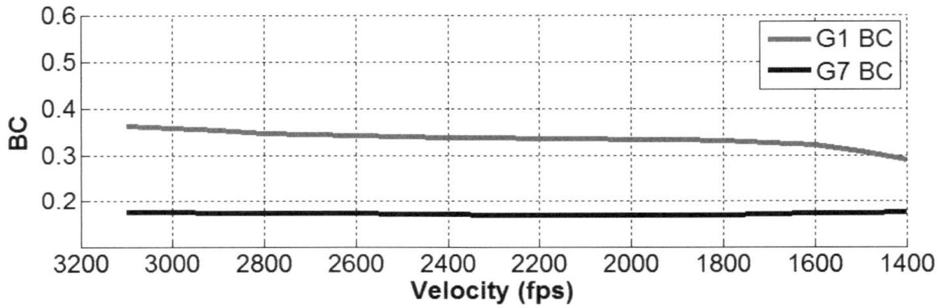

	Above 3000	3000-2500	2500-2000	2000-1500	Below 1500
G1 BC	0.371	0.350	0.338	0.322	0.310
G7 BC	0.176	0.173	0.170	0.171	0.174

Averages from 3000 fps to 1500 fps:
G1 BC: 0.336, variation of G1 BC: 0.048
G7 BC: 0.172, variation of G7 BC: 0.007
G7 form factor: 1.189

Rifling Twist Rate and Stability

Barrel Twist	Stability Factor Best case	Nominal	Worst case
1:9"	2.15	1.88	1.62
1:10"	1.74	1.52	1.31
1:11"	1.44	1.26	1.09
1:12"	1.21	1.06	0.91

Minimum twist required for stability is 1:11.3"
Recommended twist for optimal performance is 1:10.1" or faster

Barnes 270 caliber 130 grain TTSXBT

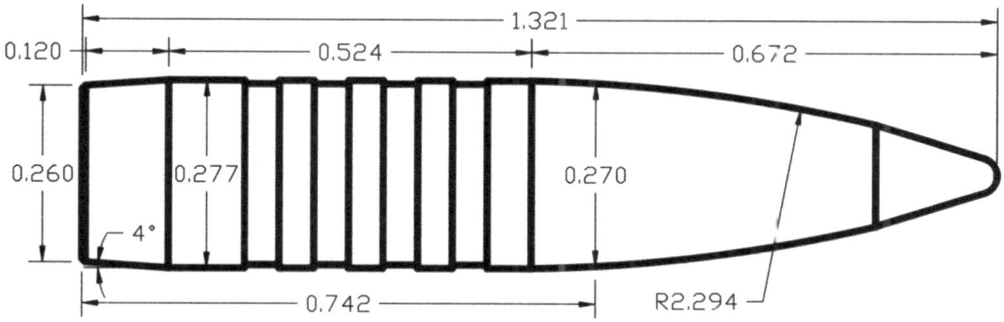

Lot #:	Unknown	Ogive Radius:	8.28 calibers
Sectional Density:	0.242 lb/in^2	Rt/R:	0.83

Ballistic Performance

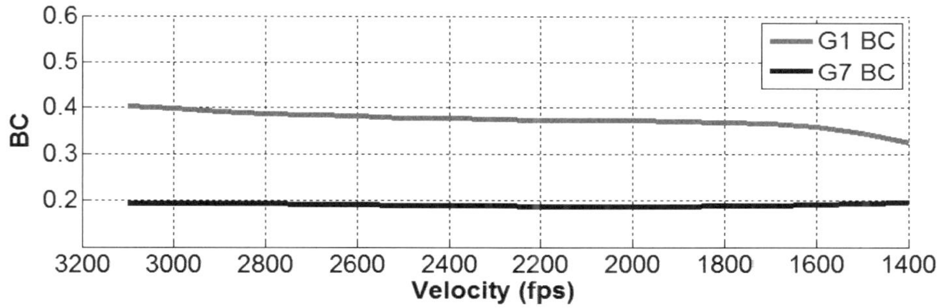

	Above 3000	3000-2500	2500-2000	2000-1500	Below 1500
G1 BC	0.411	0.387	0.375	0.358	0.345
G7 BC	0.194	0.192	0.188	0.190	0.194

Averages from 3000 fps to 1500 fps:
G1 BC: 0.373, variation of G1 BC: 0.052
G7 BC: 0.191, variation of G7 BC: 0.007
G7 form factor: 1.266

Rifling Twist Rate and Stability

Barrel Twist	Stability Factor Best case	Nominal	Worst case
1:9"	1.76	1.54	1.33
1:10"	1.42	1.24	1.08
1:11"	1.18	1.03	0.89
1:12"	0.99	0.86	0.75

Minimum twist required for stability is 1:10.2"
Recommended twist for optimal performance is 1:9.1" or faster

Berger 270 caliber 130 grain Classic Hunter

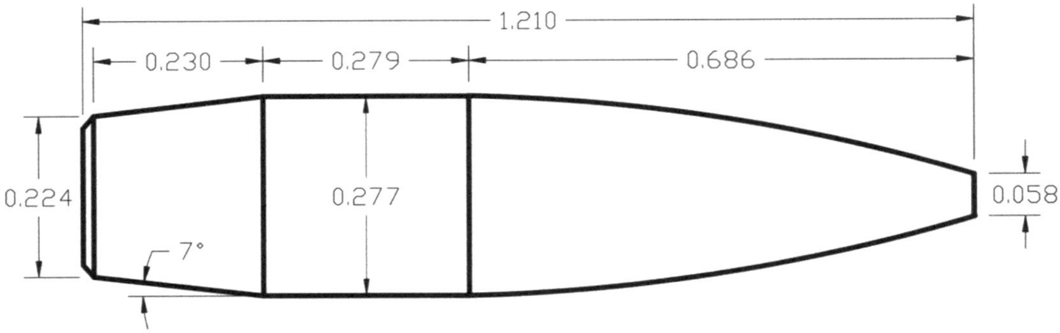

Lot #:	7850	Ogive Radius:	Hybrid
Sectional Density:	0.242 lb/in^2	Rt/R:	Hybrid

Ballistic Performance

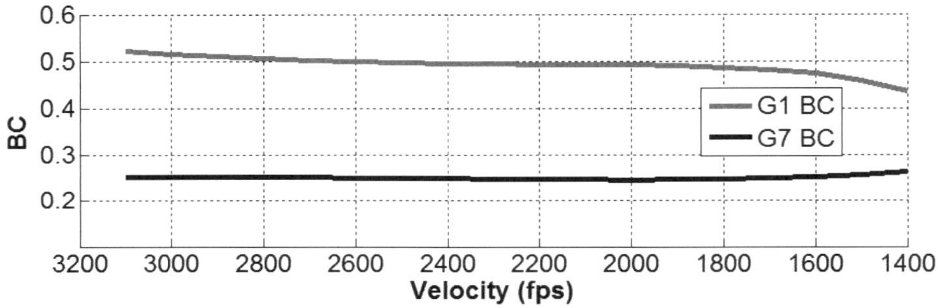

	Above 3000	3000-2500	2500-2000	2000-1500	Below 1500
G1 BC	0.531	0.507	0.494	0.475	0.458
G7 BC	0.252	0.251	0.249	0.252	0.258

Averages from 3000 fps to 1500 fps:
G1 BC: 0.491, variation of G1 BC: 0.058
G7 BC: 0.252, variation of G7 BC: 0.011
G7 form factor: 0.961

Rifling Twist Rate and Stability

Barrel Twist	Stability Factor Best case	Nominal	Worst case
1:9"	2.23	1.94	1.68
1:10"	1.80	1.57	1.36
1:11"	1.49	1.30	1.13
1:12"	1.25	1.09	0.95

Minimum twist required for stability is 1:11.5"
Recommended twist for optimal performance is 1:10.2" or faster

Berger 270 caliber 130 grain Hunting VLD

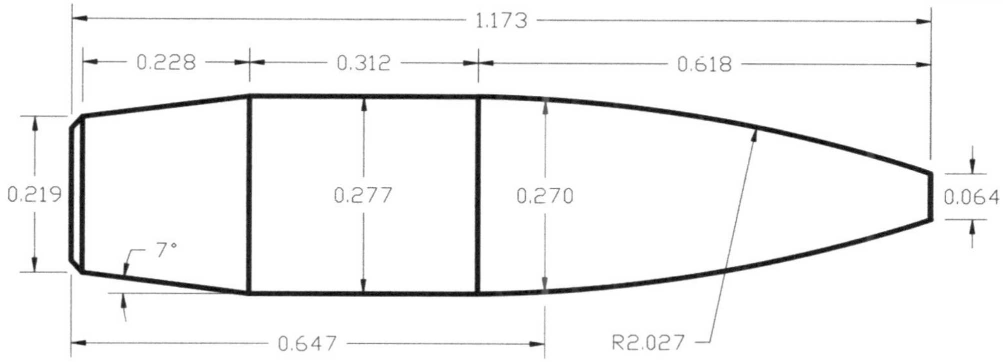

Lot #:	5917	Ogive Radius:	7.32 calibers
Sectional Density:	0.242 lb/in²	Rt/R:	0.91

Ballistic Performance

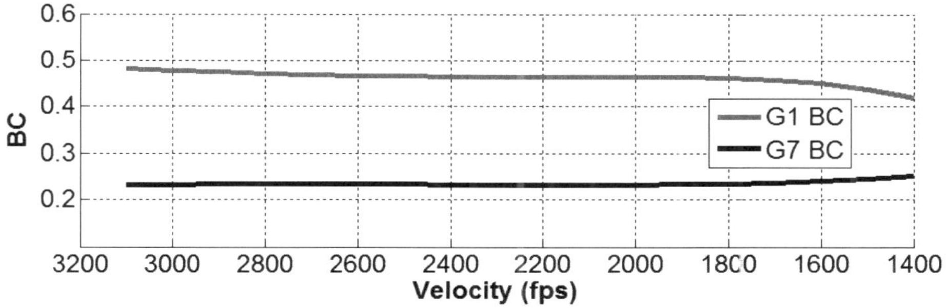

	Above 3000	3000-2500	2500-2000	2000-1500	Below 1500
G1 BC	0.489	0.471	0.464	0.449	0.435
G7 BC	0.232	0.234	0.233	0.239	0.245

Averages from 3000 fps to 1500 fps:
G1 BC: 0.460, variation of G1 BC: 0.042
G7 BC: 0.236, variation of G7 BC: 0.013
G7 form factor: 1.025

Rifling Twist Rate and Stability

Barrel Twist	Stability Factor Best case	Nominal	Worst case
1:9"	2.48	2.17	1.87
1:10"	2.01	1.76	1.52
1:11"	1.66	1.45	1.25
1:12"	1.40	1.22	1.05

Minimum twist required for stability is 1:12.1"
Recommended twist for optimal performance is 1:10.8" or faster

Berger 270 caliber 140 grain Hunting VLD

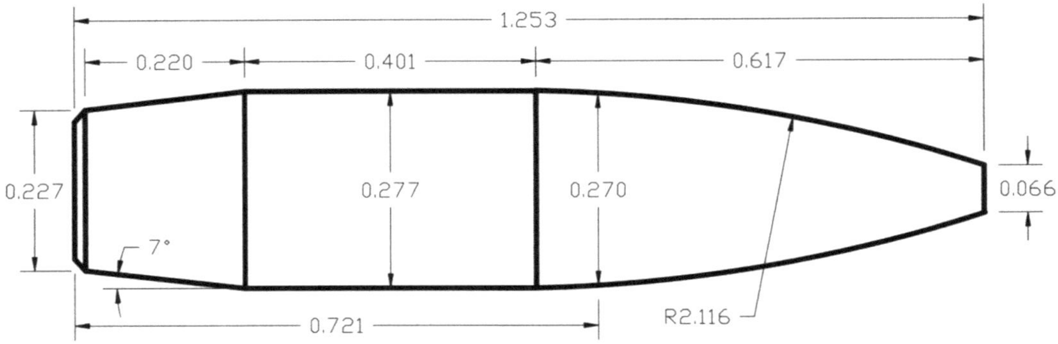

Lot #:	5904	Ogive Radius:	7.64 calibers
Sectional Density:	0.261 lb/in^2	Rt/R:	0.88

Ballistic Performance

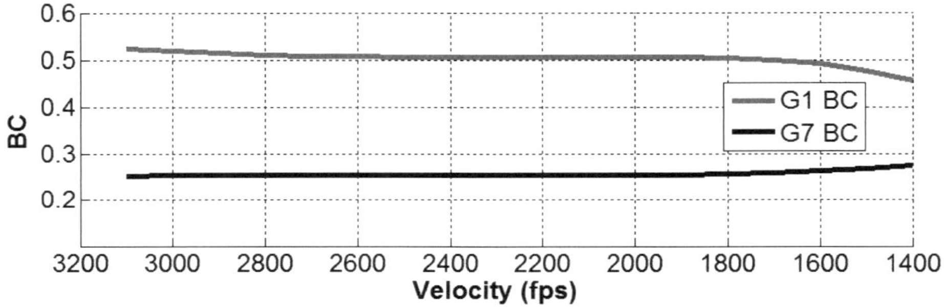

	Above 3000	3000-2500	2500-2000	2000-1500	Below 1500
G1 BC	0.532	0.513	0.506	0.492	0.477
G7 BC	0.252	0.254	0.255	0.262	0.269

Averages from 3000 fps to 1500 fps:
G1 BC: 0.502, variation of G1 BC: 0.042
G7 BC: 0.258, variation of G7 BC: 0.015
G7 form factor: 1.011

Rifling Twist Rate and Stability

Barrel Twist	Stability Factor Best case	Nominal	Worst case
1:9"	2.21	1.93	1.67
1:10"	1.79	1.56	1.35
1:11"	1.48	1.29	1.12
1:12"	1.24	1.08	0.94

Minimum twist required for stability is 1:11.4"
Recommended twist for optimal performance is 1:10.2" or faster

Berger 270 caliber 150 grain Hunting VLD

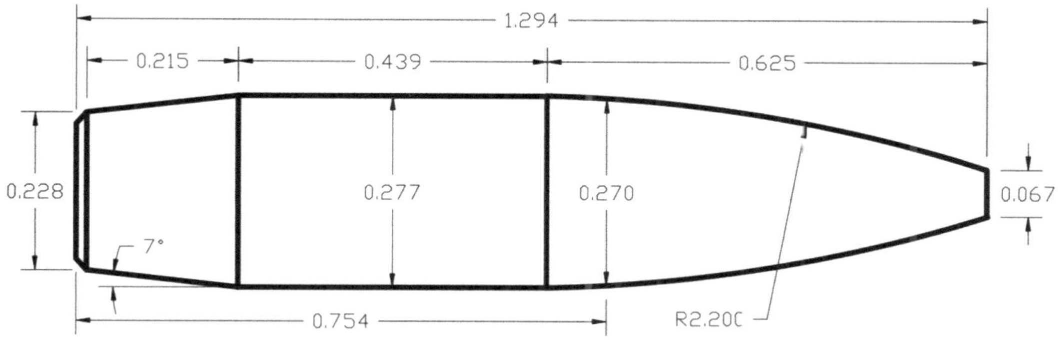

Lot #:	5931	Ogive Radius:	7.94 calibers
Sectional Density:	0.279 lb/in^2	Rt/R:	0.87

Ballistic Performance

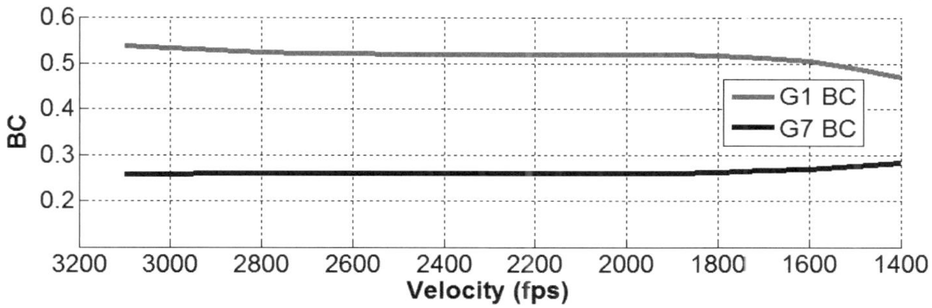

	Above 3000	3000-2500	2500-2000	2000-1500	Below 1500
G1 BC	0.545	0.526	0.520	0.506	0.491
G7 BC	0.258	0.261	0.261	0.269	0.277

Averages from 3000 fps to 1500 fps:
G1 BC: 0.516, variation of G1 BC: 0.042
G7 BC: 0.265, variation of G7 BC: 0.016
G7 form factor: 1.054

Rifling Twist Rate and Stability

Barrel Twist	Stability Factor Best case	Nominal	Worst case
1:9"	2.15	1.88	1.63
1:10"	1.74	1.52	1.32
1:11"	1.44	1.26	1.09
1:12"	1.21	1.06	0.91

Minimum twist required for stability is 1:11.3"
Recommended twist for optimal performance is 1:10.1" or faster

Cutting Edge 270 caliber 120 grain HPBT

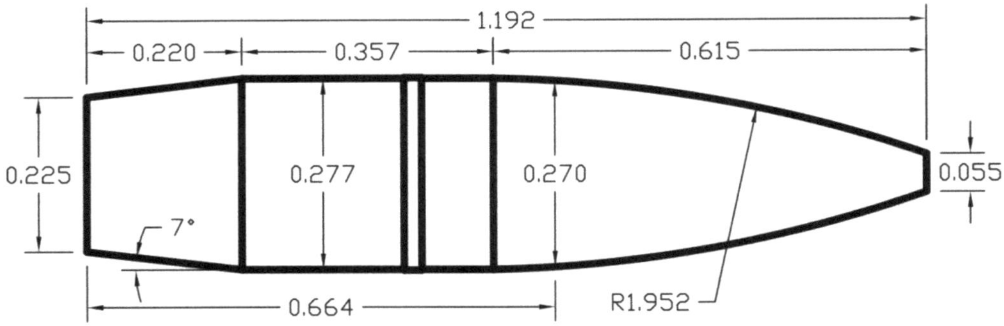

Lot #:	Unknown	Ogive Radius:	7.05 calibers
Sectional Density:	0.223 lb/in^2	Rt/R:	0.91

Ballistic Performance

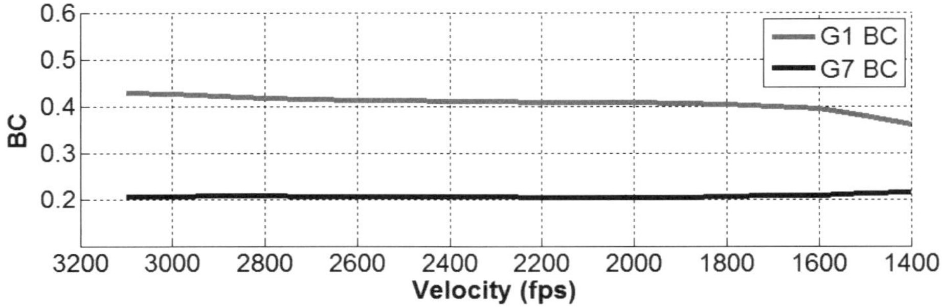

	Above 3000	3000-2500	2500-2000	2000-1500	Below 1500
G1 BC	0.438	0.419	0.411	0.394	0.380
G7 BC	0.207	0.208	0.206	0.210	0.214

Averages from 3000 fps to 1500 fps:
G1 BC: 0.407, variation of G1 BC: 0.046
G7 BC: 0.209, variation of G7 BC: 0.009
G7 form factor: 1.070

Rifling Twist Rate and Stability

Barrel Twist	Stability Factor Best case	Stability Factor Nominal	Stability Factor Worst case
1:9"	2.19	1.91	1.65
1:10"	1.77	1.55	1.34
1:11"	1.46	1.28	1.11
1:12"	1.23	1.07	0.93

Minimum twist required for stability is 1:11.4"
Recommended twist for optimal performance is 1:10.2" or faster

Cutting Edge 270 caliber 130 grain HPBT

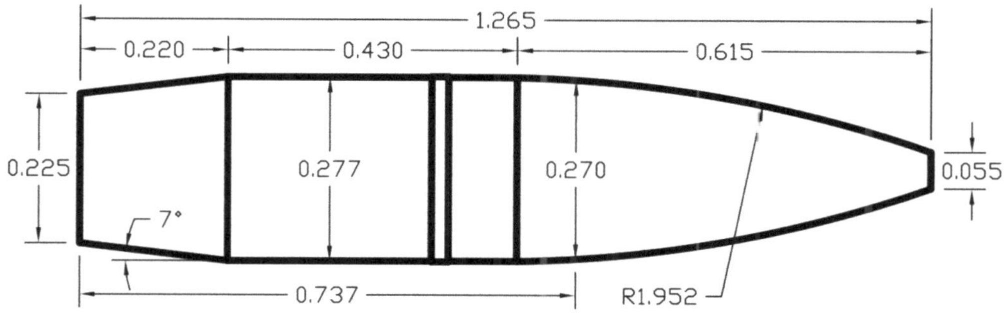

Lot #:	Unknown	Ogive Radius:	7.05 calibers
Sectional Density:	0.242 lb/in^2	Rt/R:	0.91

Ballistic Performance

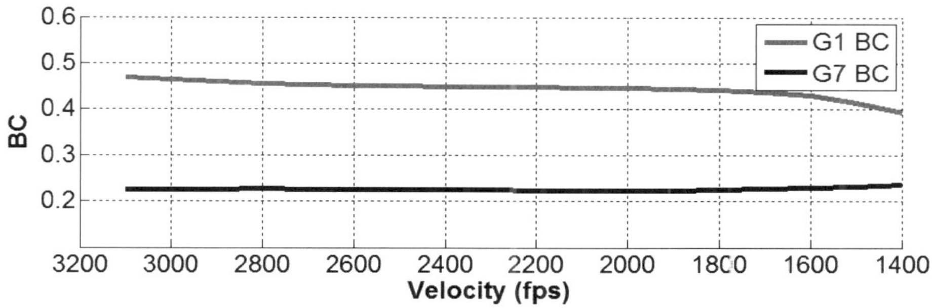

	Above 3000	3000-2500	2500-2000	2000-1500	Below 1500
G1 BC	0.475	0.456	0.447	0.430	0.414
G7 BC	0.225	0.226	0.225	0.228	0.233

Averages from 3000 fps to 1500 fps:
G1 BC: 0.443, variation of G1 BC: 0.049
G7 BC: 0.227, variation of G7 BC: 0.010
G7 form factor: 1.065

Rifling Twist Rate and Stability

Barrel Twist	Stability Factor Best case	Nominal	Worst case
1:9"	1.99	1.74	1.51
1:10"	1.61	1.41	1.22
1:11"	1.33	1.17	1.01
1:12"	1.12	0.98	0.85

Minimum twist required for stability is 1:10.8"
Recommended twist for optimal performance is 1:9.7" or faster

GS Custom 270 caliber 105 grain SP

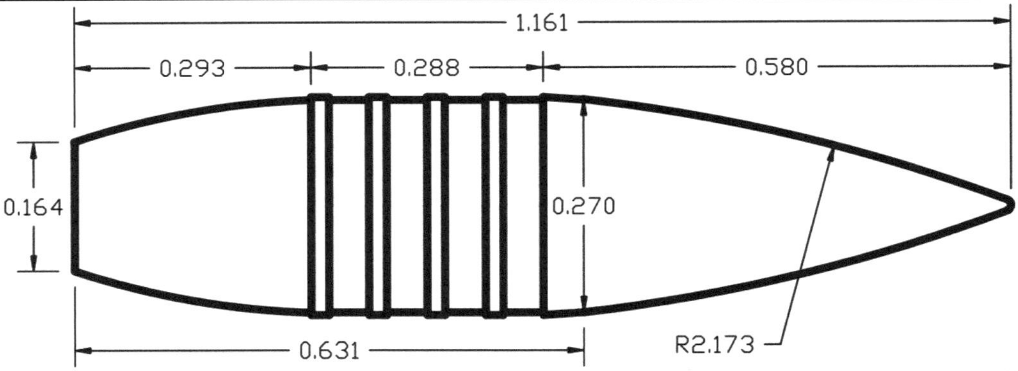

Lot #:	**Unknown**	**Ogive Radius:**	**7.85 calibers**
Sectional Density:	**0.196 lb/in²**	**Rt/R:**	**0.61**

Ballistic Performance

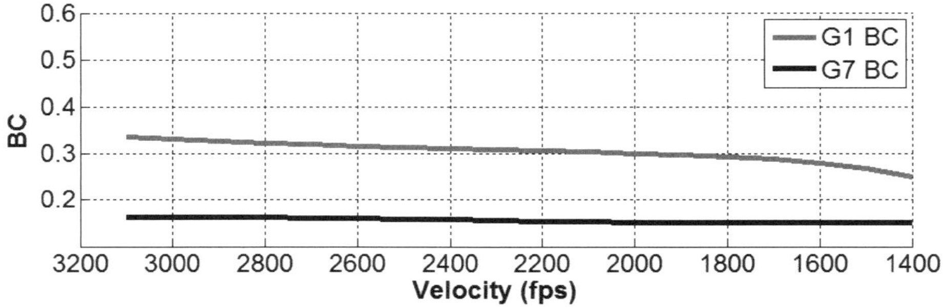

	Above 3000	3000-2500	2500-2000	2000-1500	Below 1500
G1 BC	0.344	0.323	0.308	0.284	0.267
G7 BC	0.163	0.160	0.155	0.151	0.150

Averages from 3000 fps to 1500 fps:
G1 BC: 0.304, variation of G1 BC: 0.065
G7 BC: 0.156, variation of G7 BC: 0.012
G7 form factor: 1.258

Rifling Twist Rate and Stability

Barrel Twist	**Stability Factor** Best case	Nominal	Worst case
1:9"	2.07	1.80	1.56
1:10"	1.67	1.46	1.26
1:11"	1.38	1.21	1.04
1:12"	1.16	1.01	0.88

Minimum twist required for stability is 1:11.0"
Recommended twist for optimal performance is 1:9.9" or faster

Hornady 270 caliber 110 grain BTHP

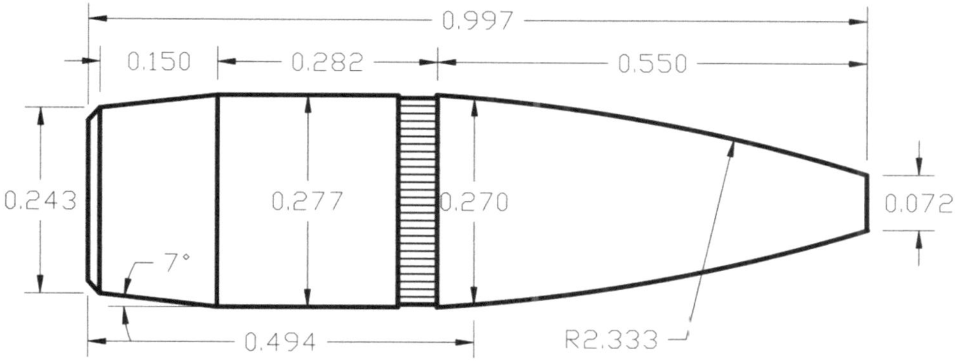

Lot #:	**2141471**	**Ogive Radius:**	**8.42 calibers**
Sectional Density:	**0.205 lb/in²**	**Rt/R:**	**0.65**

Ballistic Performance

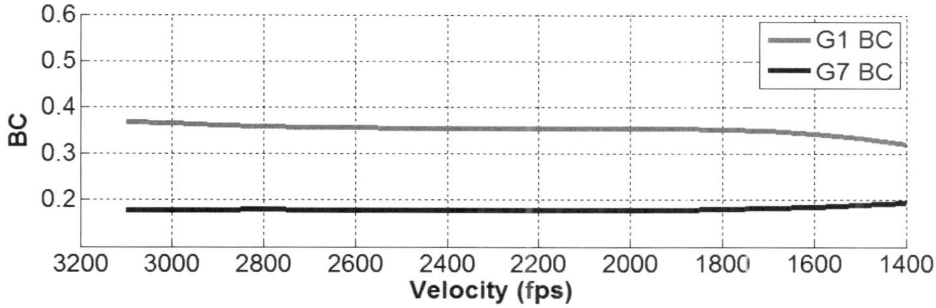

	Above 3000	3000-2500	2500-2000	2000-1500	Below 1500
G1 BC	0.375	0.360	0.355	0.344	0.334
G7 BC	0.178	0.179	0.178	0.183	0.189

Averages from 3000 fps to 1500 fps:
G1 BC: 0.352, variation of G1 BC: 0.031
G7 BC: 0.181, variation of G7 BC: 0.011
G7 form factor: 1.132

Rifling Twist Rate and Stability

Barrel Twist	Stability Factor Best case	Nominal	Worst case
1:11"	2.24	1.96	1.70
1:12"	1.89	1.65	1.42
1:13"	1.61	1.40	1.21
1:14"	1.39	1.21	1.05

Minimum twist required for stability is 1:14.1"
Recommended twist for optimal performance is 1:12.6" or faster

Hornady 270 caliber 130 grain GMX

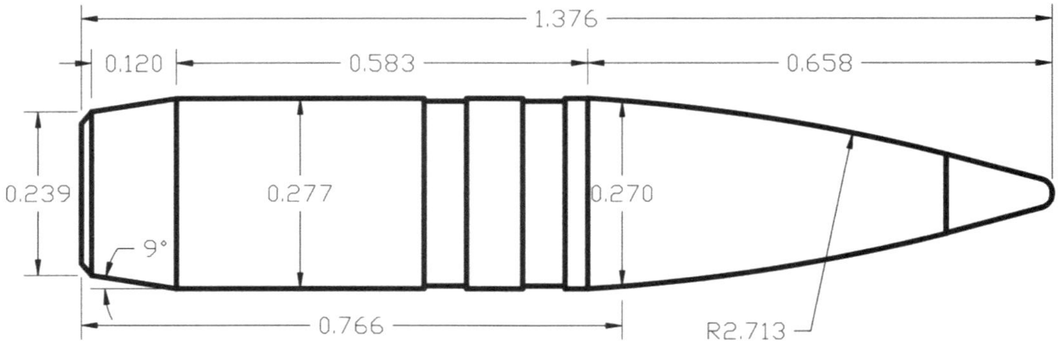

Lot #:	2150322	Ogive Radius:	9.79 calibers
Sectional Density:	0.242 lb/in²	Rt/R:	0.79

Ballistic Performance

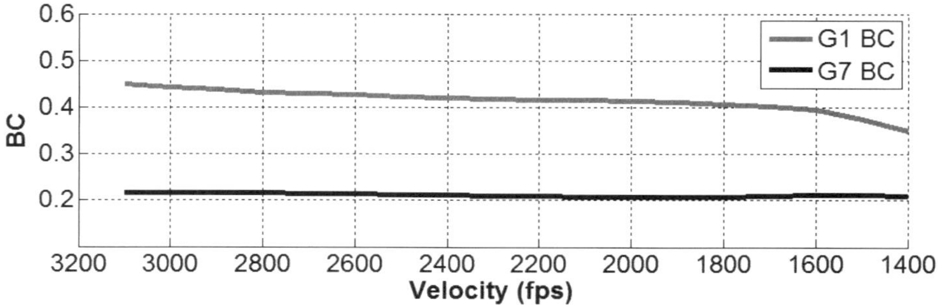

	Above 3000	3000-2500	2500-2000	2000-1500	Below 1500
G1 BC	0.459	0.433	0.418	0.394	0.375
G7 BC	0.217	0.215	0.210	0.209	0.211

Averages from 3000 fps to 1500 fps:
G1 BC: 0.414, variation of G1 BC: 0.068
G7 BC: 0.212, variation of G7 BC: 0.009
G7 form factor: 1.142

Rifling Twist Rate and Stability

Barrel Twist	Stability Factor Best case	Nominal	Worst case
1:7"	2.58	2.25	1.95
1:8"	1.97	1.72	1.49
1:9"	1.56	1.36	1.18
1:10"	1.26	1.10	0.95

Minimum twist required for stability is 1:9.6"
Recommended twist for optimal performance is 1:8.6" or faster

Hornady 270 caliber 130 grain Interbond

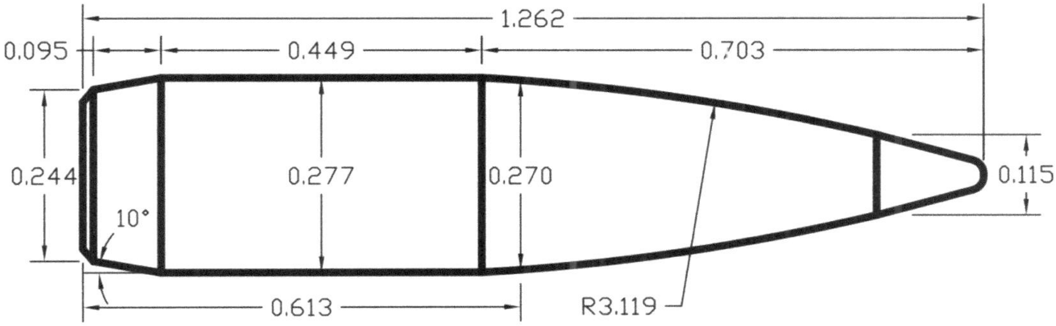

Lot #:	2070350	Ogive Radius:	11.26 calibers
Sectional Density:	0.242 lb/in^2	Rt/R:	0.69

Ballistic Performance

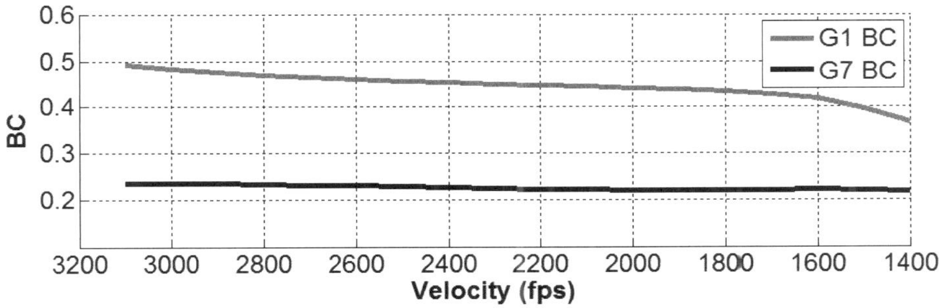

	Above 3000	3000-2500	2500-2000	2000-1500	Below 1500
G1 BC	0.507	0.473	0.451	0.420	0.397
G7 BC	0.240	0.235	0.227	0.223	0.223

Averages from 3000 fps to 1500 fps:
G1 BC: 0.447, variation of G1 BC: 0.090
G7 BC: 0.229, variation of G7 BC: 0.015
G7 form factor: 1.059

Rifling Twist Rate and Stability

Barrel Twist	Stability Factor Best case	Nominal	Worst case
1:9"	2.01	1.75	1.52
1:10"	1.63	1.42	1.23
1:11"	1.34	1.17	1.02
1:12"	1.13	0.99	0.85

Minimum twist required for stability is 1:10.9"
Recommended twist for optimal performance is 1:9.7" or faster

Hornady 270 caliber 140 grain BTSP Interlock

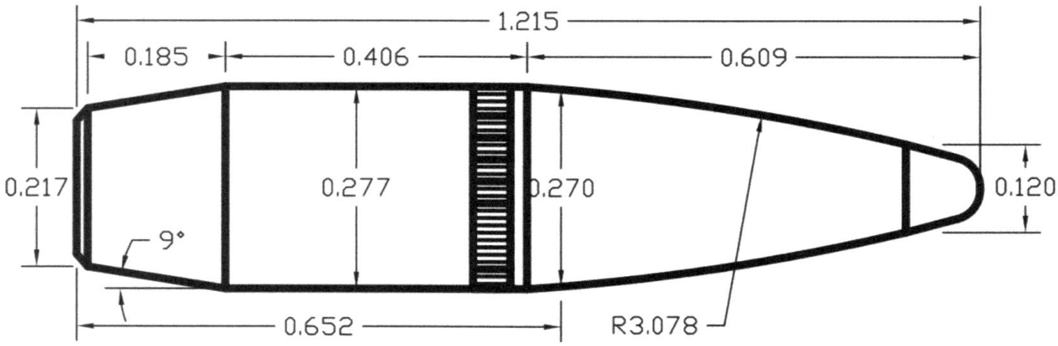

Lot #:	2070339	Ogive Radius:	11.11 calibers
Sectional Density:	0.261 lb/in²	Rt/R:	0.66

Ballistic Performance

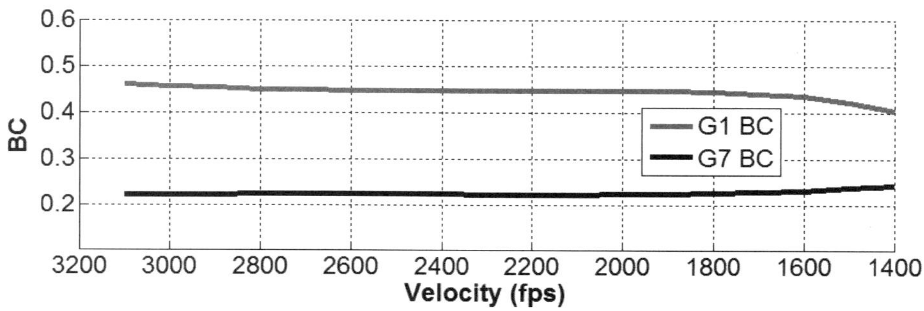

	Above 3000	3000-2500	2500-2000	2000-1500	Below 1500
G1 BC	0.466	0.452	0.447	0.435	0.423
G7 BC	0.221	0.224	0.225	0.231	0.238

Averages from 3000 fps to 1500 fps:
G1 BC: 0.443, variation of G1 BC: 0.034
G7 BC: 0.228, variation of G7 BC: 0.015
G7 form factor: 1.145

Rifling Twist Rate and Stability

Barrel Twist	Stability Factor Best case	Nominal	Worst case
1:9"	2.41	2.11	1.82
1:10"	1.96	1.71	1.48
1:11"	1.62	1.41	1.22
1:12"	1.36	1.19	1.03

Minimum twist required for stability is 1:11.9"
Recommended twist for optimal performance is 1:10.7" or faster

Hornady 270 caliber 150 grain SP Interlock

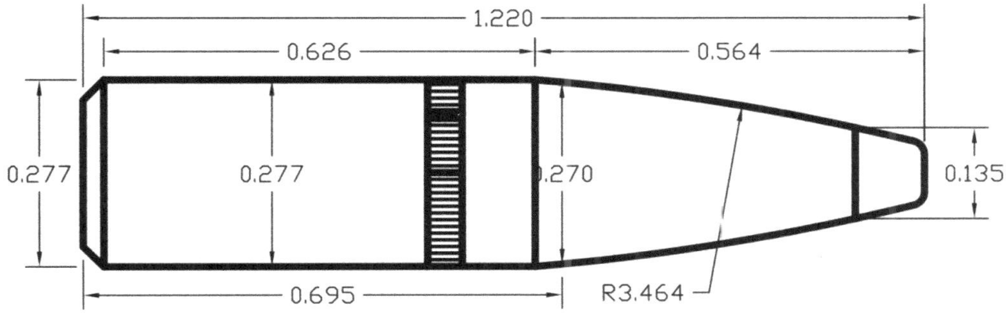

Lot #:	2080124	Ogive Radius:	12.50 calibers
Sectional Density:	0.279 lb/in^2	Rt/R:	0.53

Ballistic Performance

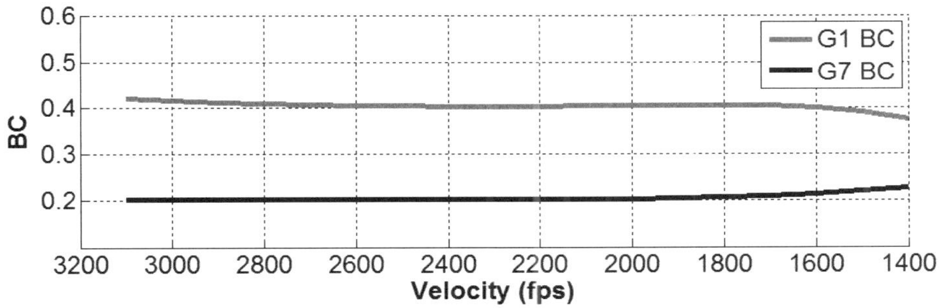

	Above 3000	3000-2500	2500-2000	2000-1500	Below 1500
G1 BC	0.426	0.409	0.404	0.397	0.390
G7 BC	0.202	0.203	0.203	0.212	0.220

Averages from 3000 fps to 1500 fps:
G1 BC: 0.403, variation of G1 BC: 0.025
G7 BC: 0.207, variation of G7 BC: 0.017
G7 form factor: 1.348

Rifling Twist Rate and Stability

Barrel Twist	Stability Factor Best case	Nominal	Worst case
1:9"	2.56	2.23	1.93
1:10"	2.07	1.81	1.56
1:11"	1.71	1.49	1.29
1:12"	1.44	1.26	1.09

Minimum twist required for stability is 1:12.3"
Recommended twist for optimal performance is 1:11.0" or faster

Hornady 270 caliber 120 grain SST BT

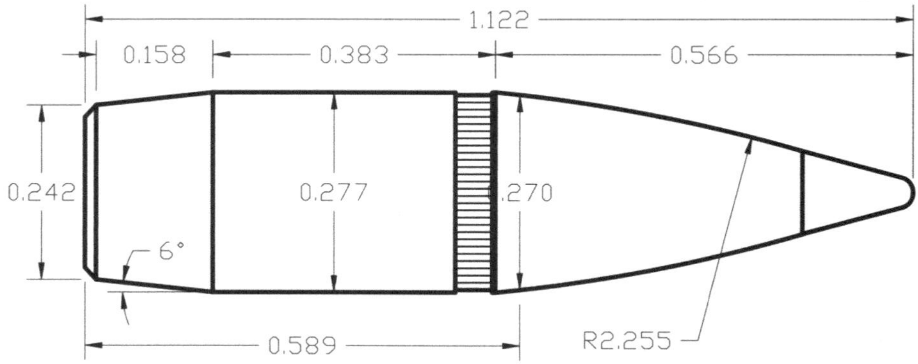

Lot #:	**2150478**	**Ogive Radius:**	**8.14 calibers**
Sectional Density:	**0.223 lb/in²**	**Rt/R:**	**0.64**

Ballistic Performance

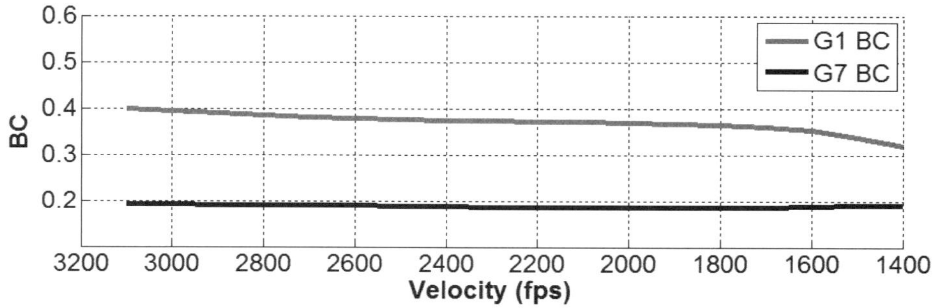

	Above 3000	3000-2500	2500-2000	2000-1500	Below 1500
G1 BC	0.407	0.386	0.374	0.355	0.339
G7 BC	0.193	0.191	0.188	0.189	0.191

Averages from 3000 fps to 1500 fps:
G1 BC: 0.370, variation of G1 BC: 0.055
G7 BC: 0.190, variation of G7 BC: 0.007
G7 form factor: 1.176

Rifling Twist Rate and Stability

Barrel Twist	Stability Factor Best case	Stability Factor Nominal	Stability Factor Worst case
1:10"	2.11	1.84	1.59
1:11"	1.74	1.52	1.32
1:12"	1.47	1.28	1.11
1:13"	1.25	1.09	0.94

Minimum twist required for stability is 1:12.4"
Recommended twist for optimal performance is 1:11.1" or faster

Hornady 270 caliber 130 grain SST BT

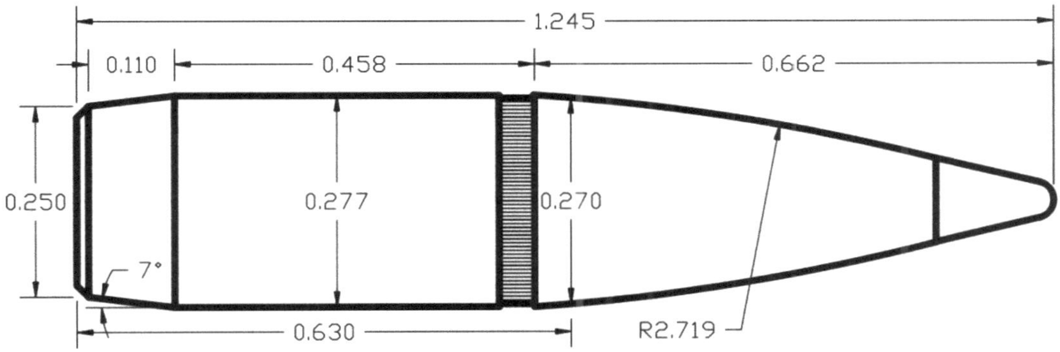

Lot #:	2140545	Ogive Radius:	9.82 calibers
Sectional Density:	0.242 lb/in^2	Rt/R:	0.70

Ballistic Performance

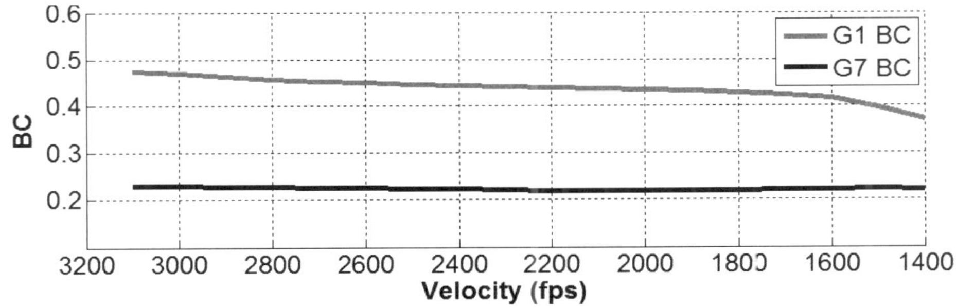

	Above 3000	3000-2500	2500-2000	2000-1500	Below 1500
G1 BC	0.486	0.457	0.440	0.415	0.395
G7 BC	0.230	0.227	0.221	0.220	0.222

Averages from 3000 fps to 1500 fps:
G1 BC: 0.436, variation of G1 BC: 0.074
G7 BC: 0.223, variation of G7 BC: 0.011
G7 form factor: 1.084

Rifling Twist Rate and Stability

Barrel Twist	Stability Factor Best case	Stability Factor Nominal	Stability Factor Worst case
1:9"	2.09	1.82	1.58
1:10"	1.69	1.48	1.28
1:11"	1.40	1.22	1.06
1:12"	1.17	1.03	0.89

Minimum twist required for stability is 1:11.1"
Recommended twist for optimal performance is 1:9.9" or faster

Hornady 270 caliber 140 grain SST BT

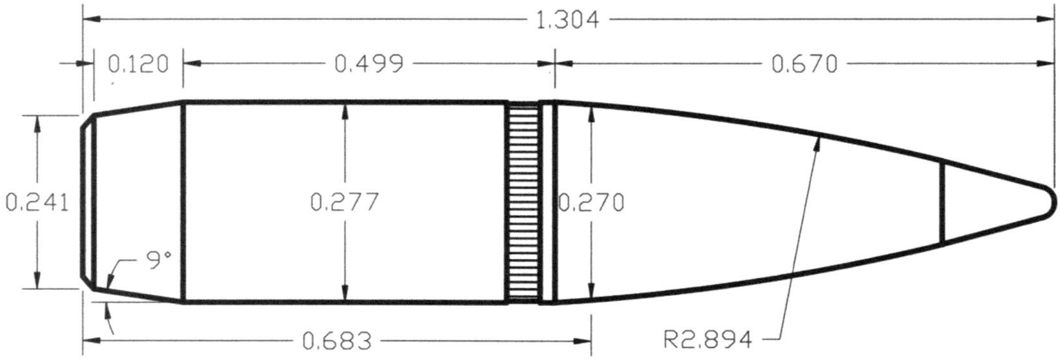

Lot #:	2142009	Ogive Radius:	10.45 calibers
Sectional Density:	0.261 lb/in^2	Rt/R:	0.57

Ballistic Performance

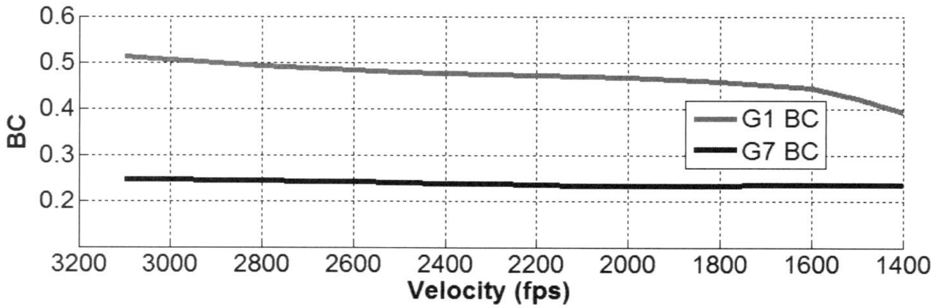

	Above 3000	3000-2500	2500-2000	2000-1500	Below 1500
G1 BC	0.524	0.493	0.473	0.444	0.422
G7 BC	0.248	0.244	0.238	0.236	0.237

Averages from 3000 fps to 1500 fps:
G1 BC: 0.469, variation of G1 BC: 0.084
G7 BC: 0.240, variation of G7 BC: 0.013
G7 form factor: 1.086

Rifling Twist Rate and Stability

Barrel Twist	Stability Factor Best case	Stability Factor Nominal	Stability Factor Worst case
1:8"	2.49	2.17	1.88
1:9"	1.97	1.72	1.48
1:10"	1.59	1.39	1.20
1:11"	1.32	1.15	0.99

Minimum twist required for stability is 1:10.8"
Recommended twist for optimal performance is 1:9.6" or faster

Hornady 270 caliber 110 grain Vmax

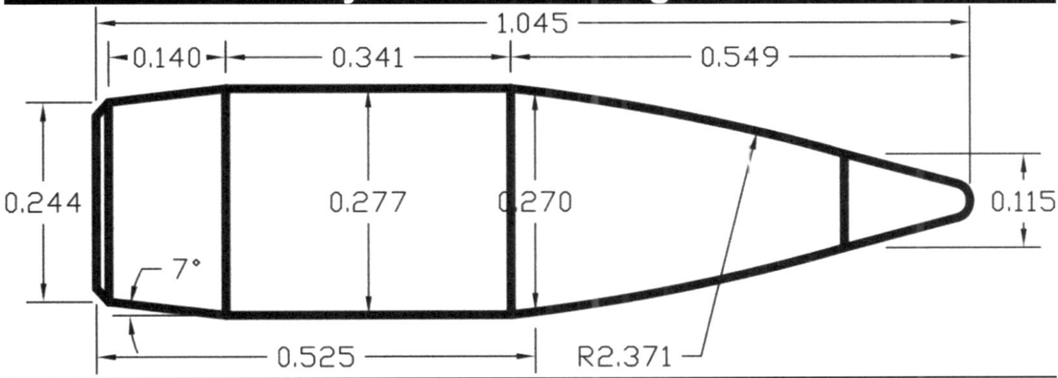

Lot #:	2080067	Ogive Radius:	8.56 calibers
Sectional Density:	0.205 lb/in^2	Rt/R:	0.53

Ballistic Performance

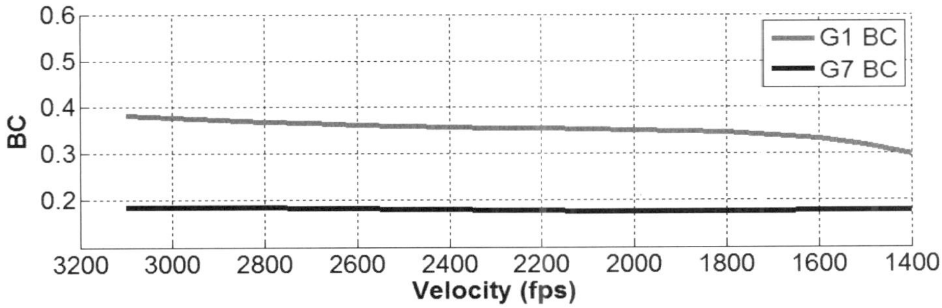

	Above 3000	3000-2500	2500-2000	2000-1500	Below 1500
G1 BC	0.391	0.369	0.355	0.334	0.317
G7 BC	0.185	0.183	0.178	0.177	0.179

Averages from 3000 fps to 1500 fps:
G1 BC: 0.351, variation of G1 BC: 0.060
G7 BC: 0.180, variation of G7 BC: 0.009
G7 form factor: 1.138

Rifling Twist Rate and Stability

Barrel Twist	Stability Factor Best case	Stability Factor Nominal	Stability Factor Worst case
1:9"	2.93	2.56	2.21
1:10"	2.37	2.07	1.79
1:11"	1.96	1.71	1.48
1:12"	1.65	1.44	1.24

Minimum twist required for stability is 1:13.1"
Recommended twist for optimal performance is 1:11.8" or faster

Matrix 270 caliber 150 grain RBT

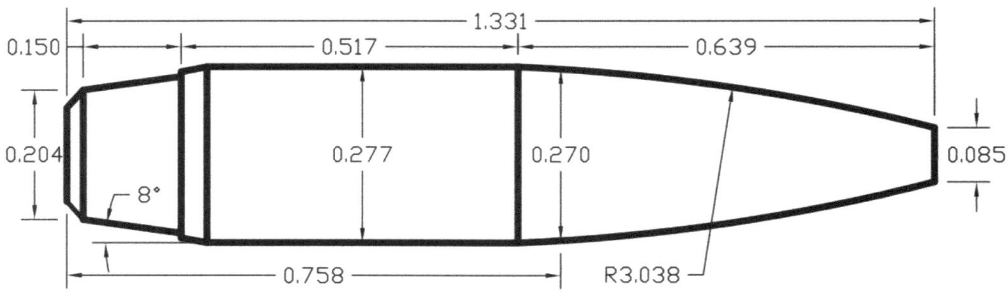

Lot #:	Unknown	Ogive Radius:	10.97 calibers
Sectional Density:	0.279 lb/in²	Rt/R:	0.79

Ballistic Performance

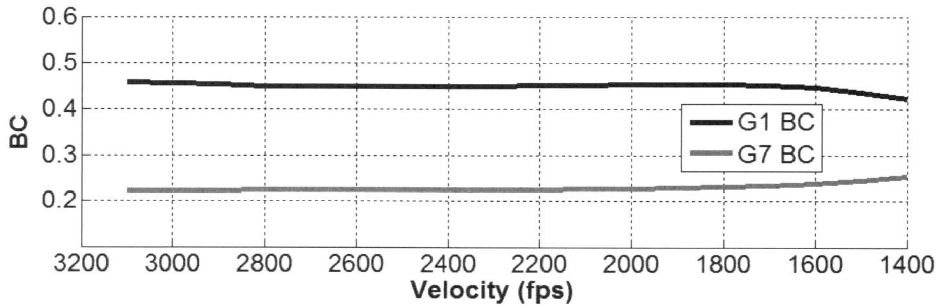

	Above 3000	3000-2500	2500-2000	2000-1500	Below 1500
G1 BC	0.465	0.453	0.451	0.445	0.437
G7 BC	0.221	0.225	0.227	0.237	0.246

Averages from 3000 fps to 1500 fps:
G1 BC: 0.449, variation of G1 BC: 0.020
G7 BC: 0.231, variation of G7 BC: 0.023
G7 form factor: 1.211

Rifling Twist Rate and Stability

Barrel Twist	Stability Factor Best case	Nominal	Worst case
1:9"	1.98	1.73	1.50
1:10"	1.61	1.40	1.21
1:11"	1.33	1.16	1.00
1:12"	1.12	0.97	0.84

Minimum twist required for stability is 1:10.8"
Recommended twist for optimal performance is 1:9.7" or faster

Nosler 270 caliber 100 grain Accubond

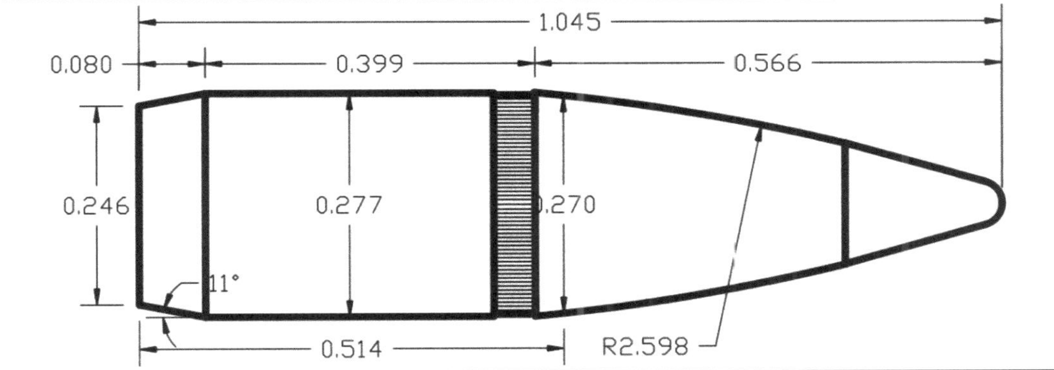

Lot #:	FJ20F6	Ogive Radius:	9.38 calibers
Sectional Density:	0.186 lb/in^2	Rt/R:	0.54

Ballistic Performance

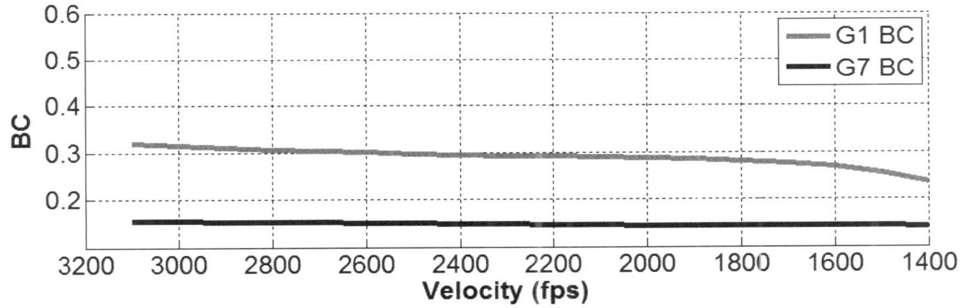

	Above 3000	3000-2500	2500-2000	2000-1500	Below 1500
G1 BC	0.329	0.308	0.294	0.272	0.257
G7 BC	0.156	0.153	0.148	0.144	0.144

Averages from 3000 fps to 1500 fps:
G1 BC: 0.290, variation of G1 BC: 0.060
G7 BC: 0.149, variation of G7 BC: 0.011
G7 form factor: 1.255

Rifling Twist Rate and Stability

Barrel Twist	Stability Factor Best case	Nominal	Worst case
1:9"	2.66	2.33	2.01
1:10"	2.16	1.88	1.63
1:11"	1.78	1.56	1.35
1:12"	1.50	1.31	1.13

Minimum twist required for stability is 1:12.5"
Recommended twist for optimal performance is 1:11.2" or faster

Nosler 270 caliber 130 grain Accubond

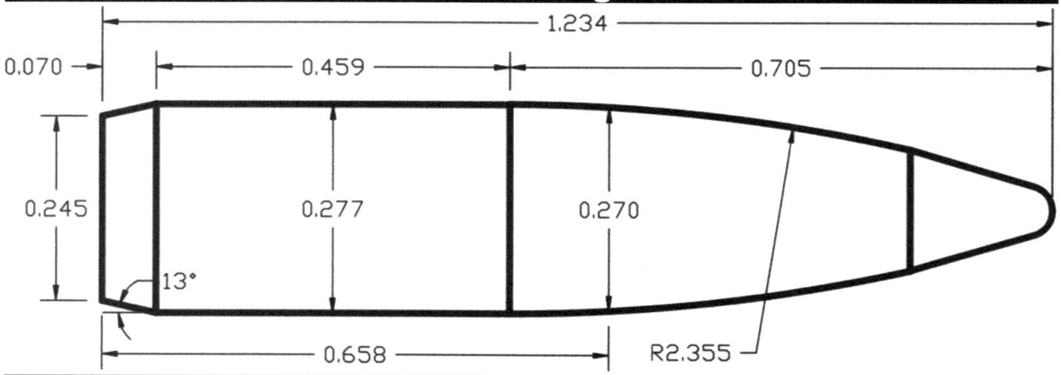

Lot #:	**Unknown**	**Ogive Radius:**	**8.82 calibers**
Sectional Density:	**0.242 lb/in²**	**Rt/R:**	**0.88**

Ballistic Performance

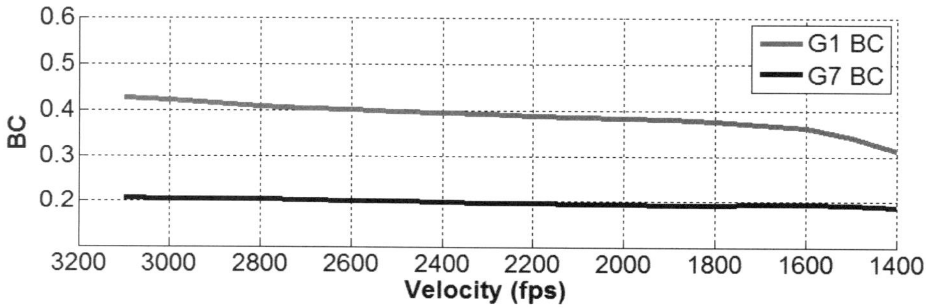

	Above 3000	3000-2500	2500-2000	2000-1500	Below 1500
G1 BC	0.438	0.410	0.391	0.364	0.343
G7 BC	0.207	0.203	0.197	0.193	0.192

Averages from 3000 fps to 1500 fps:

G1 BC: 0.387, variation of G1 BC: 0.079

G7 BC: 0.198, variation of G7 BC: 0.014

G7 form factor: 1.224

Rifling Twist Rate and Stability

Barrel Twist	Stability Factor Best case	Stability Factor Nominal	Stability Factor Worst case
1:9"	2.14	1.87	1.62
1:10"	1.74	1.52	1.31
1:11"	1.43	1.25	1.08
1:12"	1.21	1.05	0.91

Minimum twist required for stability is 1:11.2"

Recommended twist for optimal performance is 1:10.1" or faster

Nosler 270 caliber 140 grain Accubond

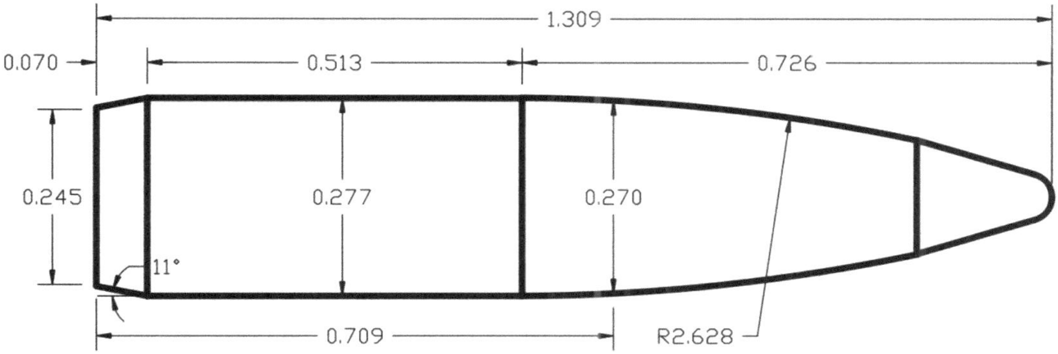

Lot #:	Unknown	Ogive Radius:	9.49 calibers
Sectional Density:	0.261 lb/in^2	Rt/R:	0.96

Ballistic Performance

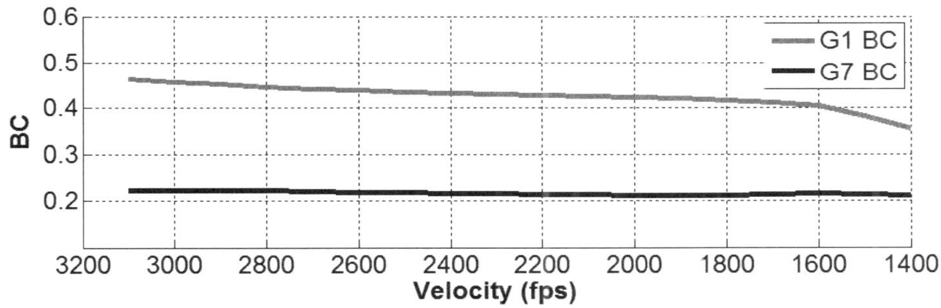

	Above 3000	3000-2500	2500-2000	2000-1500	Below 1500
G1 BC	0.474	0.446	0.429	0.403	0.383
G7 BC	0.224	0.221	0.216	0.214	0.215

Averages from 3000 fps to 1500 fps:
G1 BC: 0.424, variation of G1 BC: 0.074
G7 BC: 0.217, variation of G7 BC: 0.011
G7 form factor: 1.200

Rifling Twist Rate and Stability

Barrel Twist	Stability Factor Best case	Nominal	Worst case
1:9"	1.94	1.70	1.47
1:10"	1.57	1.38	1.19
1:11"	1.30	1.14	0.98
1:12"	1.09	0.96	0.83

Minimum twist required for stability is 1:10.7"
Recommended twist for optimal performance is 1:9.6" or faster

Nosler 270 caliber 150 grain LR Accubond

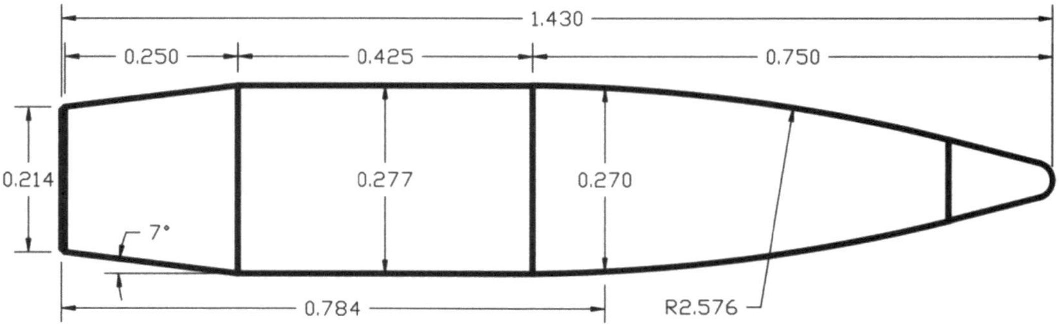

Lot #:	Unknown	Ogive Radius:	9.27 calibers
Sectional Density:	0.279 lb/in^2	Rt/R:	0.89

Ballistic Performance

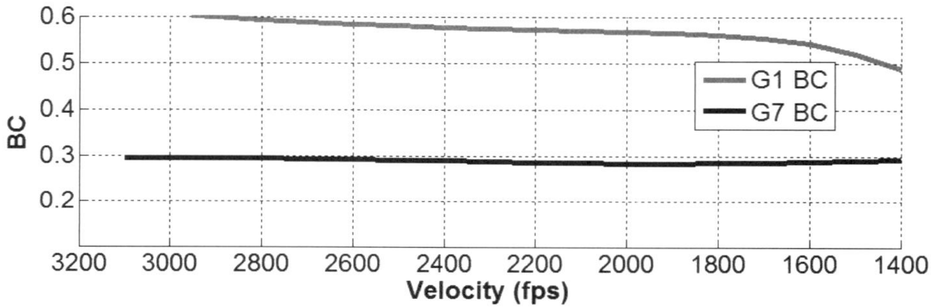

	Above 3000	3000-2500	2500-2000	2000-1500	Below 1500
G1 BC	0.625	0.593	0.574	0.543	0.519
G7 BC	0.296	0.294	0.288	0.289	0.292

Averages from 3000 fps to 1500 fps:
G1 BC: 0.568, variation of G1 BC: 0.087
G7 BC: 0.291, variation of G7 BC: 0.011
G7 form factor: 0.959

Rifling Twist Rate and Stability

Barrel Twist	Stability Factor Best case	Nominal	Worst case
1:9"	1.61	1.41	1.21
1:10"	1.30	1.14	0.98
1:11"	1.08	0.94	0.81
1:12"	0.90	0.79	0.68

Minimum twist required for stability is 1:9.7"
Recommended twist for optimal performance is 1:8.7" or faster

Nosler 270 caliber 130 grain Ballistic Silver Tip

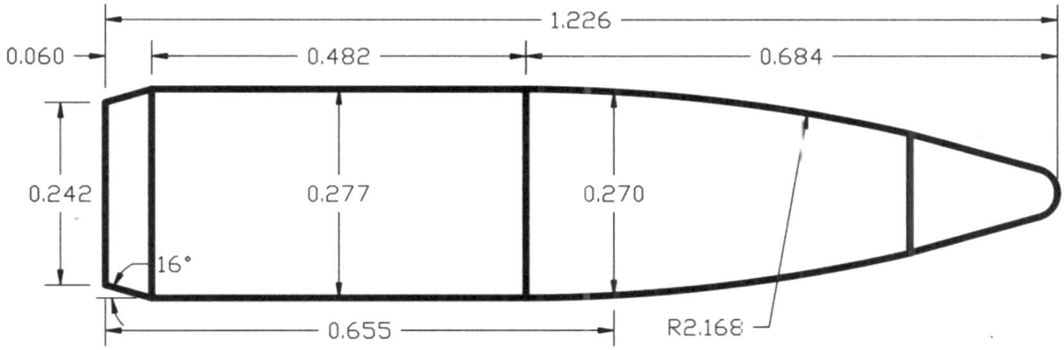

Lot #:	BL18F9	Ogive Radius:	7.83 calibers
Sectional Density:	0.242 lb/in²	Rt/R:	0.98

Ballistic Performance

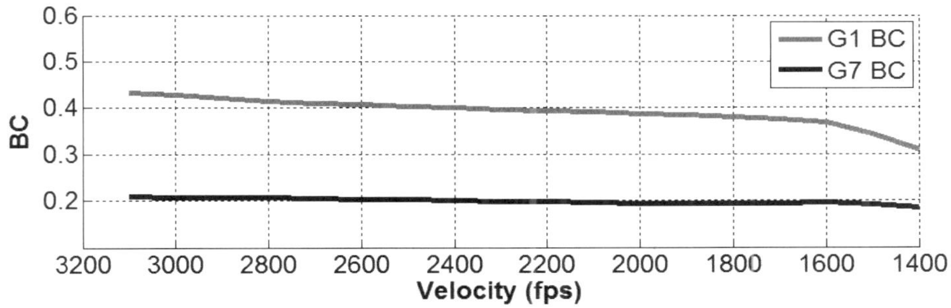

	Above 3000	3000-2500	2500-2000	2000-1500	Below 1500
G1 BC	0.443	0.414	0.395	0.366	0.344
G7 BC	0.210	0.205	0.198	0.193	0.192

Averages from 3000 fps to 1500 fps:
G1 BC: 0.390, variation of G1 BC: 0.082
G7 BC: 0.199, variation of G7 BC: 0.016
G7 form factor: 1.215

Rifling Twist Rate and Stability

Barrel Twist	Stability Factor Best case	Nominal	Worst case
1:9"	2.18	1.91	1.65
1:10"	1.77	1.55	1.34
1:11"	1.46	1.28	1.10
1:12"	1.23	1.07	0.93

Minimum twist required for stability is 1:11.3"
Recommended twist for optimal performance is 1:10.1" or faster

Nosler 270 caliber 140 grain Ballistic Tip

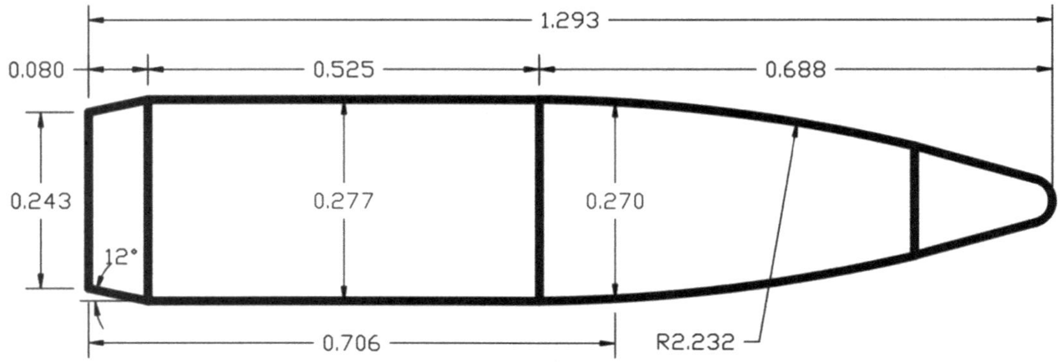

Lot #:	E10F9-1	Ogive Radius:	8.06 calibers
Sectional Density:	0.261 lb/in^2	Rt/R:	1.00

Ballistic Performance

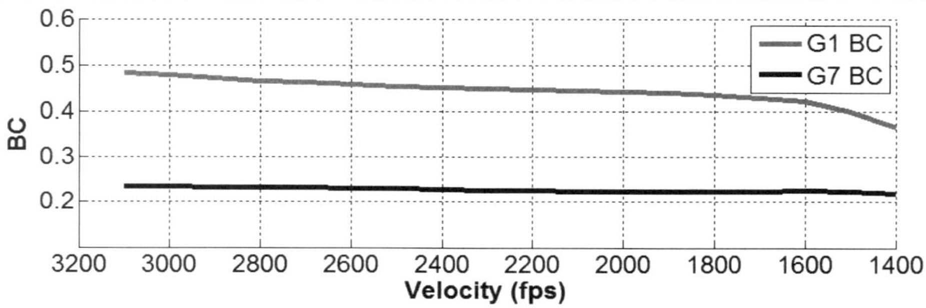

	Above 3000	3000-2500	2500-2000	2000-1500	Below 1500
G1 BC	0.495	0.467	0.449	0.421	0.399
G7 BC	0.235	0.231	0.226	0.223	0.224

Averages from 3000 fps to 1500 fps:
G1 BC: 0.444, variation of G1 BC: 0.079
G7 BC: 0.227, variation of G7 BC: 0.011
G7 form factor: 1.148

Rifling Twist Rate and Stability

Barrel Twist	Stability Factor Best case	Nominal	Worst case
1:9"	2.01	1.76	1.52
1:10"	1.63	1.43	1.23
1:11"	1.35	1.18	1.02
1:12"	1.13	0.99	0.86

Minimum twist required for stability is 1:10.9"
Recommended twist for optimal performance is 1:9.7" or faster

Nosler 270 caliber 115 grain Custom Competition

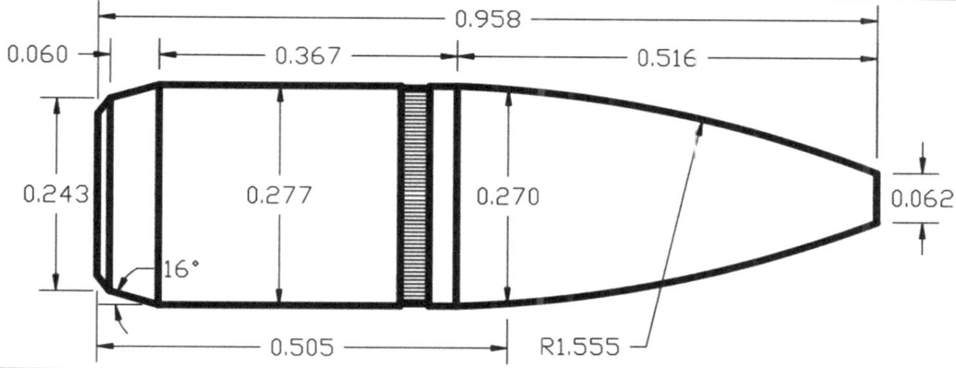

Lot #:	EL15F15	Ogive Radius:	5.61 calibers
Sectional Density:	0.214 lb/in^2	Rt/R:	0.83

Ballistic Performance

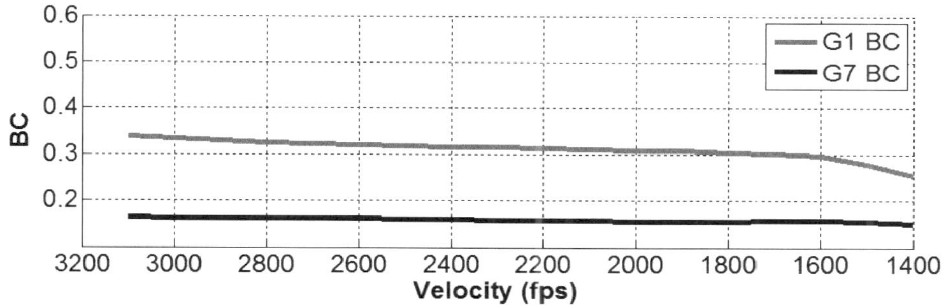

	Above 3000	3000-2500	2500-2000	2000-1500	Below 1500
G1 BC	0.345	0.326	0.315	0.295	0.279
G7 BC	0.163	0.162	0.158	0.156	0.157

Averages from 3000 fps to 1500 fps:
G1 BC: 0.311, variation of G1 BC: 0.055
G7 BC: 0.159, variation of G7 BC: 0.007
G7 form factor: 1.347

Rifling Twist Rate and Stability

Barrel Twist	Stability Factor Best case	Nominal	Worst case
1:9"	3.90	3.41	2.95
1:10"	3.16	2.76	2.39
1:11"	2.61	2.28	1.97
1:12"	2.20	1.92	1.66

Minimum twist required for stability is 1:15.2"
Recommended twist for optimal performance is 1:13.6" or faster

Nosler 270 caliber 130 grain E-Tip

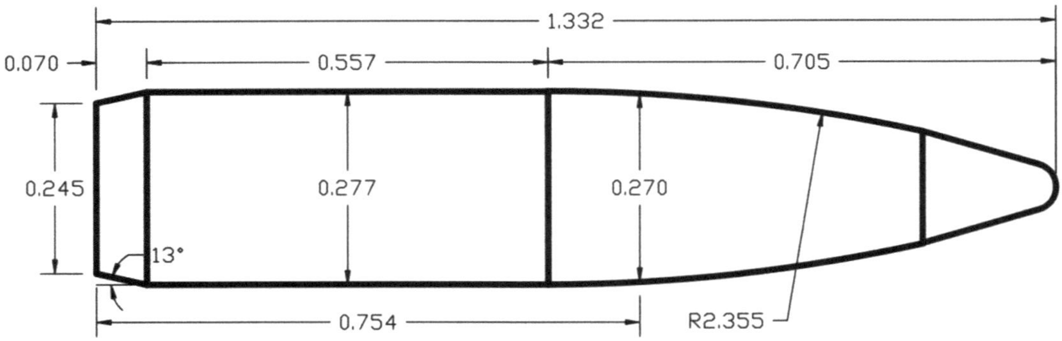

Lot #:	Unknown	Ogive Radius:	8.50 calibers
Sectional Density:	0.242 lb/in²	Rt/R:	0.99

Ballistic Performance

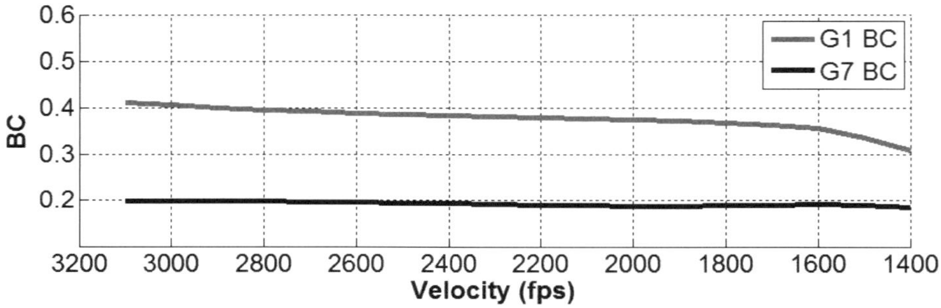

	Above 3000	3000-2500	2500-2000	2000-1500	Below 1500
G1 BC	0.421	0.396	0.380	0.356	0.337
G7 BC	0.199	0.196	0.191	0.189	0.189

Averages from 3000 fps to 1500 fps:
G1 BC: 0.376, variation of G1 BC: 0.069
G7 BC: 0.192, variation of G7 BC: 0.010
G7 form factor: 1.258

Rifling Twist Rate and Stability

Barrel Twist	Stability Factor Best case	Stability Factor Nominal	Stability Factor Worst case
1:9"	1.72	1.50	1.30
1:10"	1.39	1.21	1.05
1:11"	1.15	1.00	0.87
1:12"	0.96	0.84	0.73

Minimum twist required for stability is 1:10.1"
Recommended twist for optimal performance is 1:9.0" or faster

Nosler 270 caliber 140 grain Partition

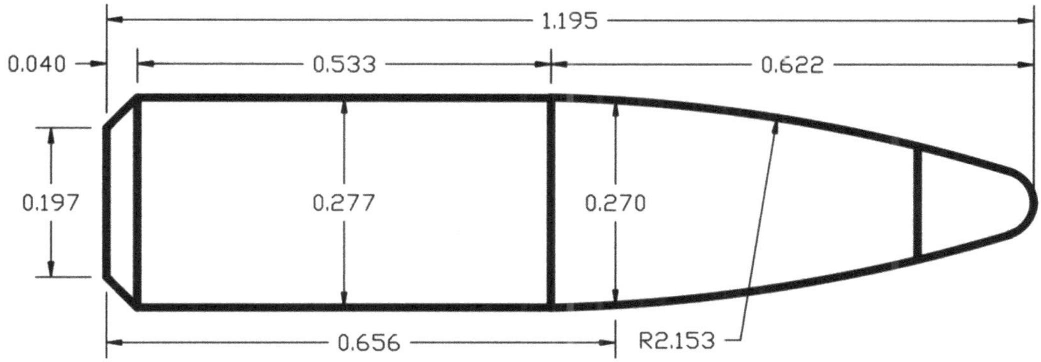

Lot #:	**GE16F1-2**	**Ogive Radius:**	**7.77 calibers**
Sectional Density:	**0.261 lb/in²**	**Rt/R:**	**0.88**

Ballistic Performance

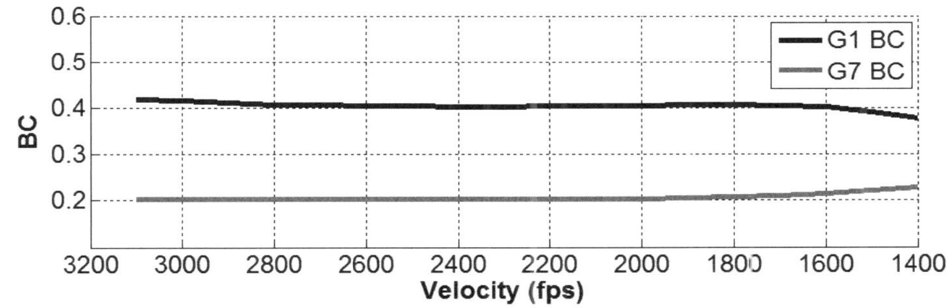

	Above 3000	3000-2500	2500-2000	2000-1500	Below 1500
G1 BC	0.426	0.409	0.404	0.398	0.392
G7 BC	0.202	0.203	0.203	0.212	0.221

Averages from 3000 fps to 1500 fps:
G1 BC: 0.404, variation of G1 BC: 0.023
G7 BC: 0.208, variation of G7 BC: 0.018
G7 form factor: 1.257

Rifling Twist Rate and Stability

	Stability Factor		
Barrel Twist	**Best case**	**Nominal**	**Worst case**
1:9"	2.53	2.21	1.91
1:10"	2.05	1.79	1.55
1:11"	1.70	1.48	1.28
1:12"	1.42	1.24	1.08

Minimum twist required for stability is 1:12.2"
Recommended twist for optimal performance is 1:10.9" or faster

Sierra 270 caliber 140 grain GameKing

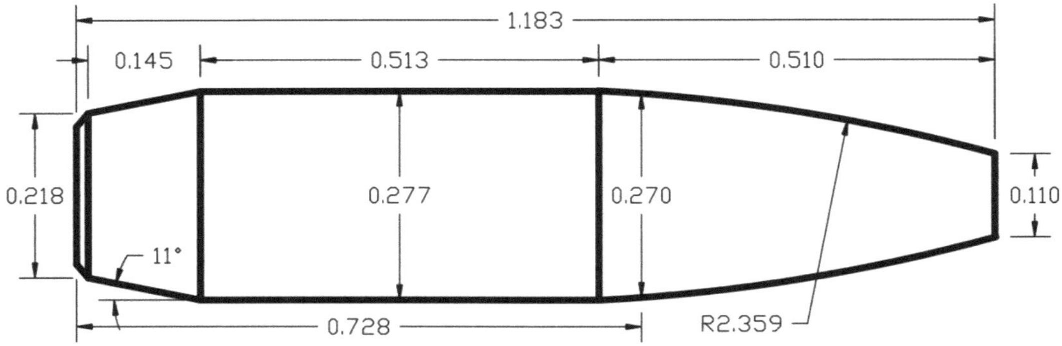

Lot #:	75685369561	**Ogive Radius:**	**8.52 calibers**
Sectional Density:	**0.261 lb/in²**	**Rt/R:**	**0.68**

Ballistic Performance

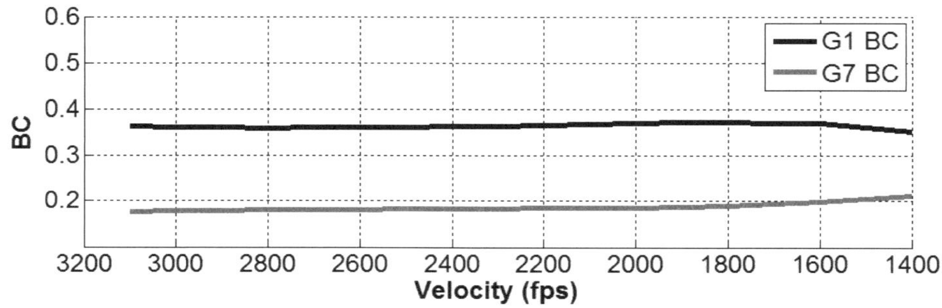

	Above 3000	3000-2500	2500-2000	2000-1500	Below 1500
G1 BC	0.367	0.362	0.366	0.366	0.362
G7 BC	0.174	0.180	0.184	0.195	0.205

Averages from 3000 fps to 1500 fps:
G1 BC: 0.364, variation of G1 BC: 0.008
G7 BC: 0.187, variation of G7 BC: 0.027
G7 form factor: 1.394

Rifling Twist Rate and Stability

Barrel Twist	Stability Factor Best case	Nominal	Worst case
1:9"	2.61	2.28	1.97
1:10"	2.11	1.85	1.60
1:11"	1.75	1.53	1.32
1:12"	1.47	1.28	1.11

Minimum twist required for stability is 1:12.4"
Recommended twist for optimal performance is 1:11.1" or faster

Sierra 270 caliber 150 grain GameKing

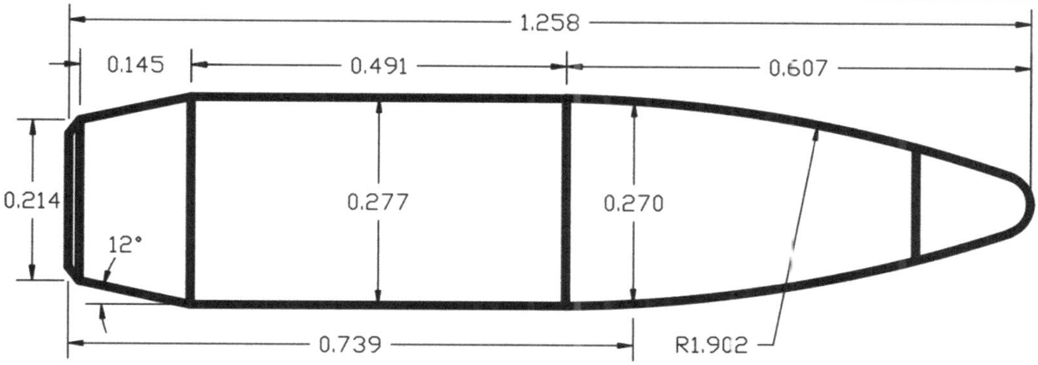

Lot #:	4032496	Ogive Radius:	6.87 calibers
Sectional Density:	0.279 lb/in²	Rt/R:	0.92

Ballistic Performance

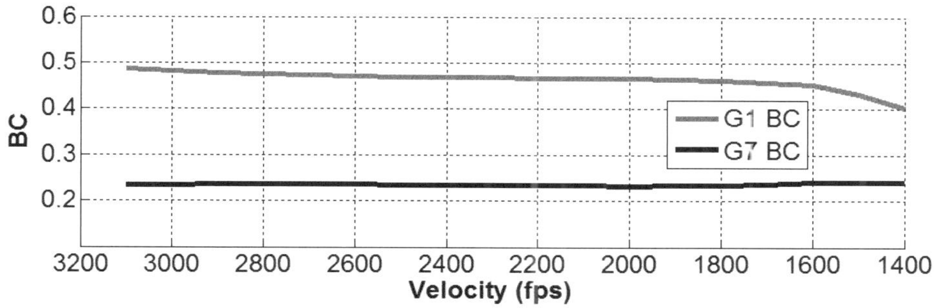

	Above 3000	3000-2500	2500-2000	2000-1500	Below 1500
G1 BC	0.493	0.475	0.467	0.448	0.431
G7 BC	0.233	0.236	0.235	0.238	0.242

Averages from 3000 fps to 1500 fps:
G1 BC: 0.461, variation of G1 BC: 0.051
G7 BC: 0.237, variation of G7 BC: 0.009
G7 form factor: 1.180

Rifling Twist Rate and Stability

Barrel Twist	Stability Factor: Best case	Nominal	Worst case
1:9"	2.34	2.04	1.77
1:10"	1.89	1.65	1.43
1:11"	1.56	1.37	1.18
1:12"	1.31	1.15	0.99

Minimum twist required for stability is 1:11.7"
Recommended twist for optimal performance is 1:10.5" or faster

Sierra 270 caliber 115 grain MatchKing

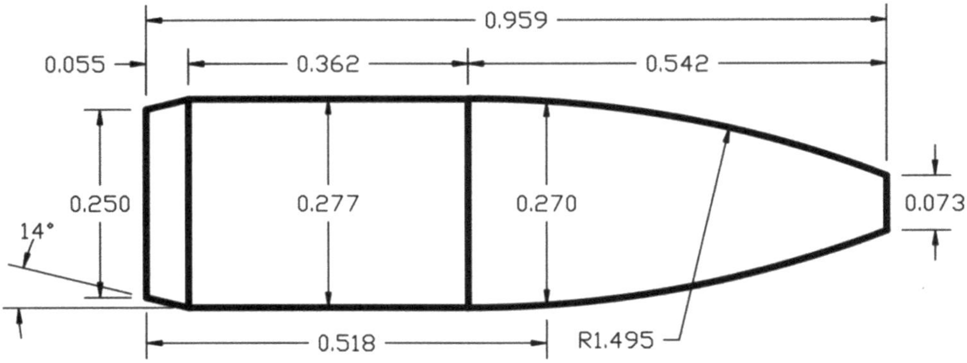

Lot #:	3949857	Ogive Radius:	5.40 calibers
Sectional Density:	0.214 lb/in^2	Rt/R:	1.00

Ballistic Performance

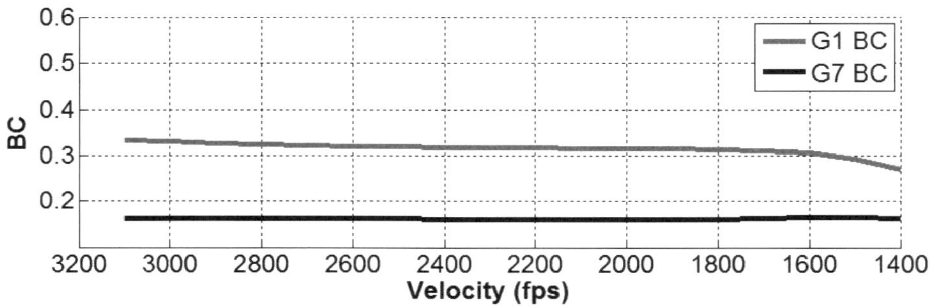

	Above 3000	3000-2500	2500-2000	2000-1500	Below 1500
G1 BC	0.340	0.326	0.318	0.304	0.292
G7 BC	0.161	0.162	0.160	0.162	0.164

Averages from 3000 fps to 1500 fps:
G1 BC: 0.315, variation of G1 BC: 0.039
G7 BC: 0.162, variation of G7 BC: 0.005
G7 form factor: 1.325

Rifling Twist Rate and Stability

Barrel Twist	Stability Factor Best case	Nominal	Worst case
1:9"	3.91	3.42	2.96
1:10"	3.17	2.77	2.40
1:11"	2.62	2.29	1.98
1:12"	2.20	1.92	1.66

Minimum twist required for stability is 1:15.2"
Recommended twist for optimal performance is 1:13.6" or faster

Sierra 270 caliber 135 grain MatchKing

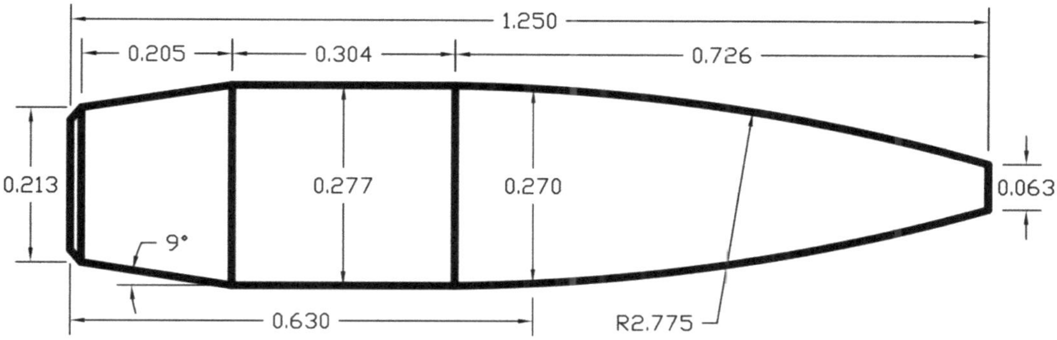

Lot #:	4029059	Ogive Radius:	10.02 calibers
Sectional Density:	0.251 lb/in^2	Rt/R:	0.91

Ballistic Performance

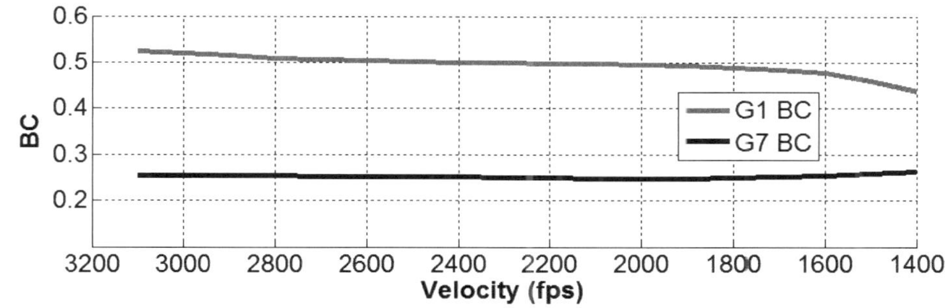

	Above 3000	3000-2500	2500-2000	2000-1500	Below 1500
G1 BC	0.534	0.511	0.498	0.477	0.459
G7 BC	0.253	0.253	0.250	0.253	0.258

Averages from 3000 fps to 1500 fps:
G1 BC: 0.494, variation of G1 BC: 0.061
G7 BC: 0.253, variation of G7 BC: 0.010
G7 form factor: 0.992

Rifling Twist Rate and Stability

Barrel Twist	Stability Factor Best case	Nominal	Worst case
1:9"	2.14	1.87	1.62
1:10"	1.74	1.52	1.31
1:11"	1.43	1.25	1.08
1:12"	1.21	1.05	0.91

Minimum twist required for stability is 1:11.2"
Recommended twist for optimal performance is 1:10.1" or faster

Swift 270 caliber 140 grain A-Frame

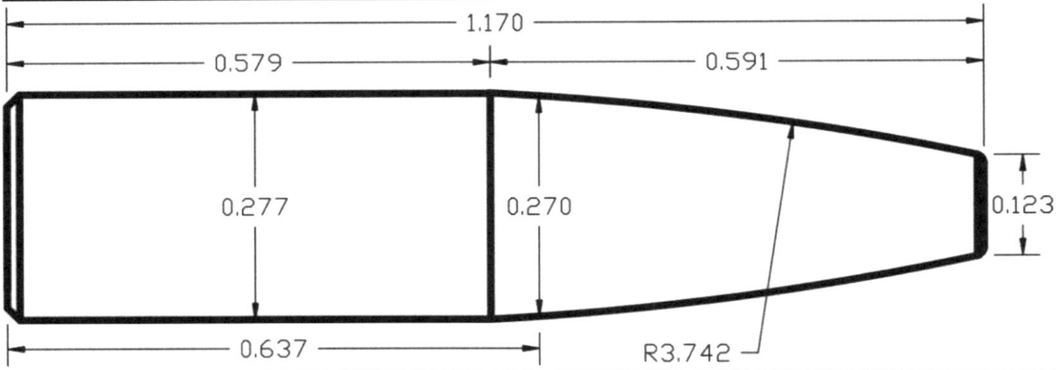

Lot #:	862	Ogive Radius:	13.51 calibers
Sectional Density:	0.261 lb/in^2	Rt/R:	0.64

Ballistic Performance

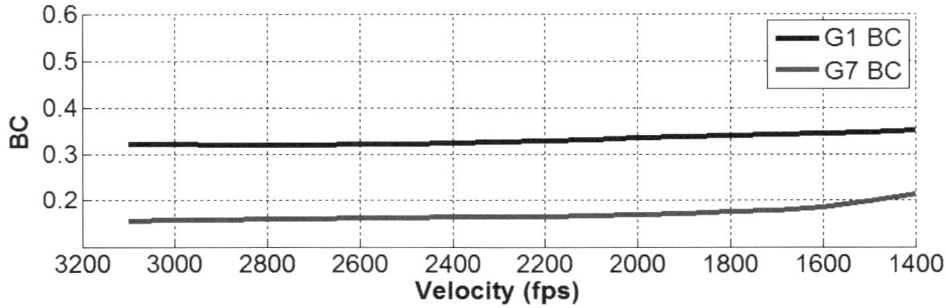

	Above 3000	3000-2500	2500-2000	2000-1500	Below 1500
G1 BC	0.326	0.323	0.330	0.342	0.348
G7 BC	0.155	0.160	0.166	0.183	0.197

Averages from 3000 fps to 1500 fps:
G1 BC: 0.332, variation of G1 BC: 0.026
G7 BC: 0.172, variation of G7 BC: 0.040
G7 form factor: 1.531

Rifling Twist Rate and Stability

Barrel Twist	Stability Factor Best case	Nominal	Worst case
1:9"	2.69	2.35	2.03
1:10"	2.18	1.91	1.65
1:11"	1.80	1.57	1.36
1:12"	1.51	1.32	1.14

Minimum twist required for stability is 1:12.6"
Recommended twist for optimal performance is 1:11.3" or faster

Swift 270 caliber 150 grain A-Frame

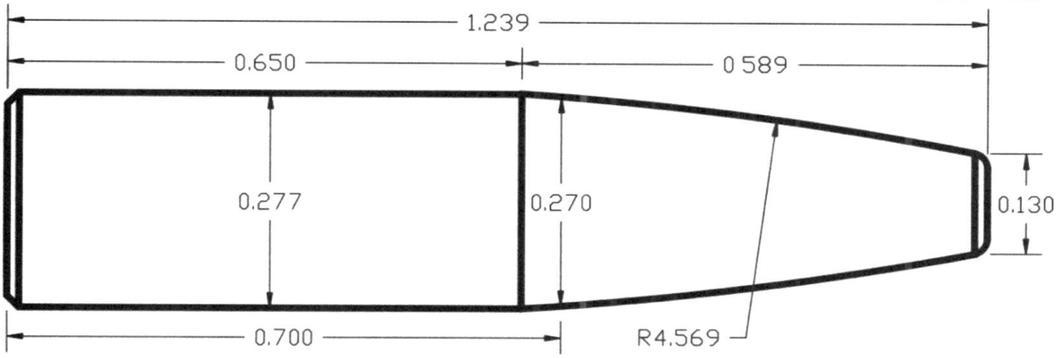

Lot #:	861	Ogive Radius:	16.50 calibers
Sectional Density:	0.279 lb/in^2	Rt/R:	0.54

Ballistic Performance

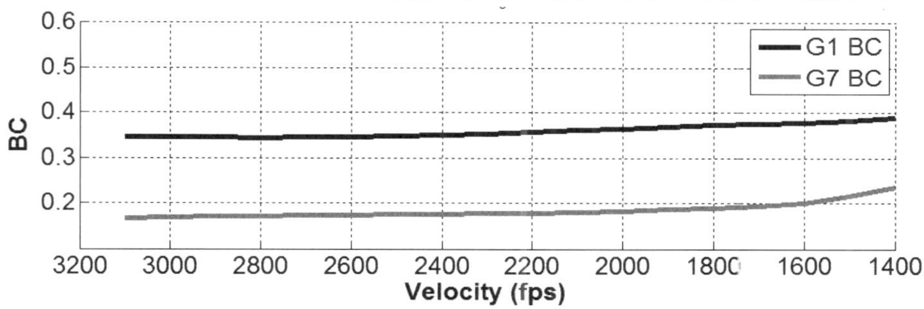

	Above 3000	3000-2500	2500-2000	2000-1500	Below 1500
G1 BC	0.348	0.347	0.356	0.373	0.381
G7 BC	0.165	0.172	0.179	0.200	0.217

Averages from 3000 fps to 1500 fps:
G1 BC: 0.360, variation of G1 BC: 0.036
G7 BC: 0.186, variation of G7 BC: 0 048
G7 form factor: 1.516

Rifling Twist Rate and Stability

Barrel Twist	Stability Factor Best case	Nominal	Worst case
1:9"	2.44	2.13	1.85
1:10"	1.98	1.73	1.50
1:11"	1.64	1.43	1.24
1:12"	1.37	1.20	1.04

Minimum twist required for stability is 1:12.0"
Recommended twist for optimal performance is 1:10.7" or faster

.284 Caliber (7mm)

.284 caliber bullets typically range from 110 to 180 grains. The image below relates bullet weight to sectional density for .284 caliber bullets.

Range of Weight and Sectional Density for 0.284 Caliber

Bullet Weight (grains)	110	122	133	145	157	169	180
Sectional Density (lb/in^2)	0.194	0.215	0.236	0.257	0.278	0.299	0.319

Sectional Density (SD) is a measure of how much mass is behind the bullets frontal area. Higher SD means a greater ability for the bullet to penetrate targets on impact, and also retain velocity in flight.

Spectrum of G7 form factors

0.80 0.85 0.90 0.95 1.00 1.05 1.10 1.15 1.20

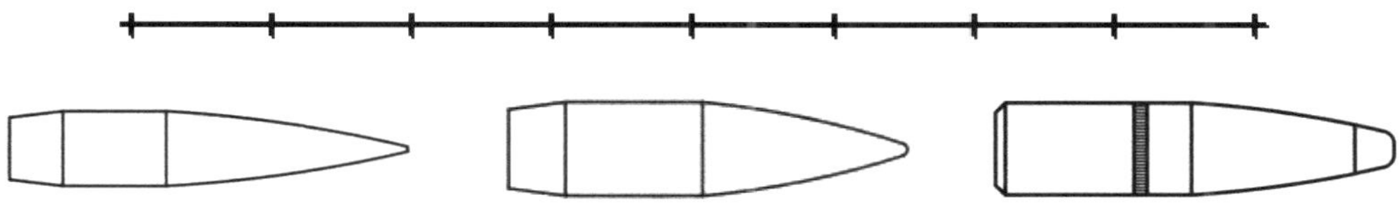

G7 Form factors range from 0.80 for the lowest drag shapes, up to 1.20+ for blunter shapes.

G7 Ballistic Coefficients are simply Sectional Density (SD) divided by G7 form factor.

You can determine a bullets' suitability for long range performance by seeing where it lies on the SD and form factor lines. Ideally, you want a bullet with a high SD and a low G7 form factor.

As an example, a 180 grain .284 caliber bullet has an SD of 0.319. If that bullet had a G7 form factor of 0.95, it's G7 BC would be:

G7 BC = SD/FF = 0.319/0.95 = 0.336.

Barnes .284 caliber 145 LRXBT

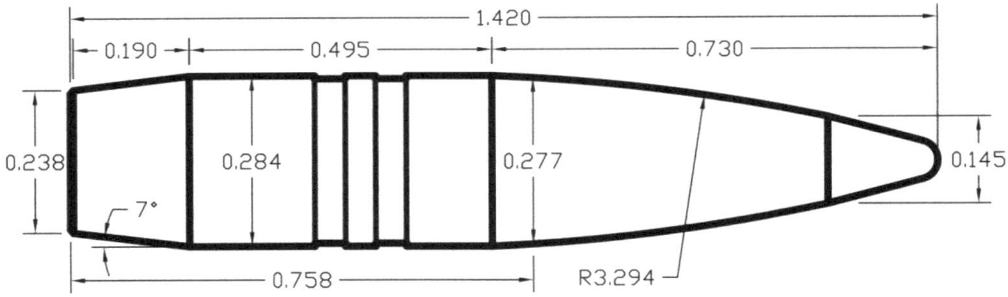

	Lot #:	Unknown	Ogive Radius:	11.60 calibers
	Sectional Density:	0.257 lb/in^2	Rt/R:	0.67

Ballistic Performance

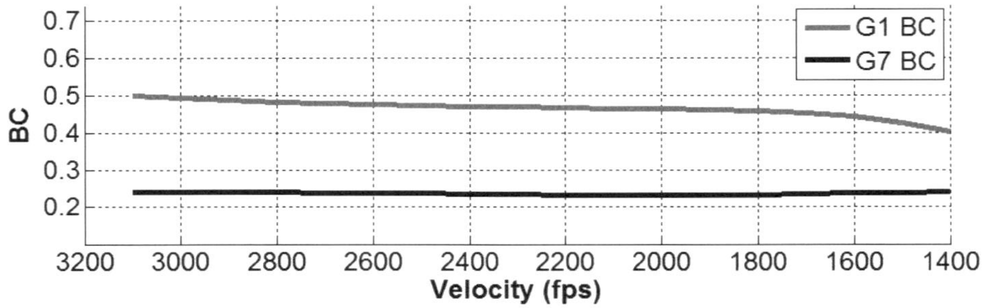

	Above 3000	3000-2500	2500-2000	2000-1500	Below 1500
G1 BC	0.510	0.484	0.469	0.445	0.427
G7 BC	0.242	0.240	0.236	0.237	0.240

Averages from 3000 fps to 1500 fps:
G1 BC: 0.465, variation of G1 BC: 0.067
G7 BC: 0.238, variation of G7 BC: 0.009
G7 form factor: 1.078

Rifling Twist Rate and Stability

Barrel Twist	Stability Factor Best case	Nominal	Worst case
1:8"	2.09	1.83	1.58
1:9"	1.65	1.45	1.25
1:10"	1.34	1.17	1.01
1:11"	1.11	0.97	0.84

Minimum twist required for stability is 1:9.9"
Recommended twist for optimal performance is 1:8.9" or faster

Barnes .284 caliber 171 grain Match Burner

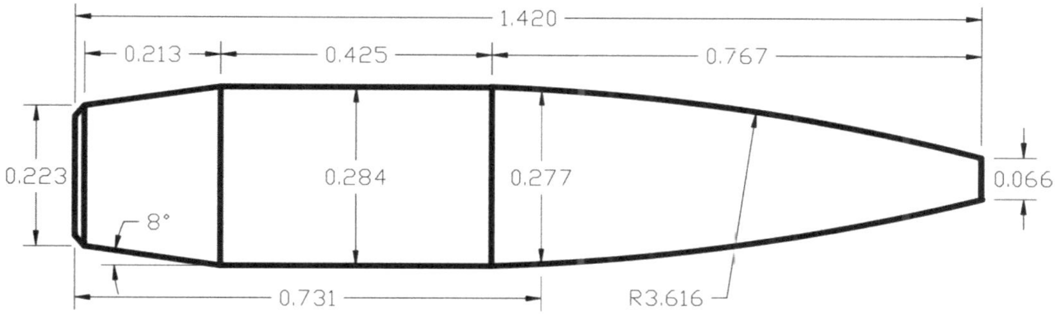

Lot #:	6398296-1	Ogive Radius:	12.73 calibers
Sectional Density:	0.303 lb/in^2	Rt/R:	0.76

Ballistic Performance

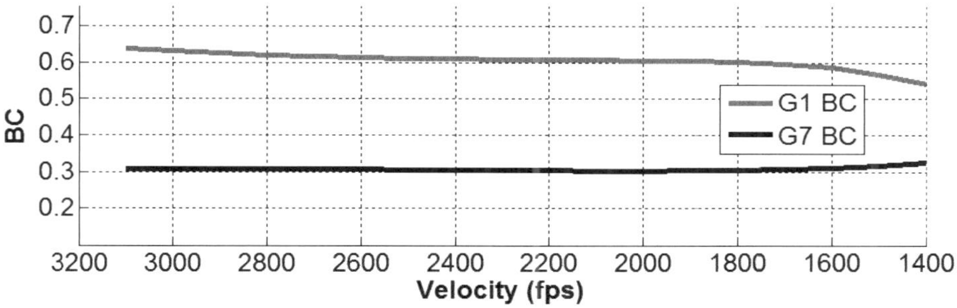

	Above 3000	3000-2500	2500-2000	2000-1500	Below 1500
G1 BC	0.648	0.621	0.609	0.586	0.567
G7 BC	0.307	0.308	0.306	0.312	0.319

Averages from 3000 fps to 1500 fps:
G1 BC: 0.604, variation of G1 BC: 0.064
G7 BC: 0.310, variation of G7 BC: 0.015
G7 form factor: 0.977

Rifling Twist Rate and Stability

Barrel Twist	Stability Factor Best case	Nominal	Worst case
1:8"	2.49	2.17	1.88
1:9"	1.96	1.72	1.48
1:10"	1.59	1.39	1.20
1:11"	1.31	1.15	0.99

Minimum twist required for stability is 1:10.8"
Recommended twist for optimal performance is 1:9.6" or faster

Barnes .284 caliber 175 TSX FB

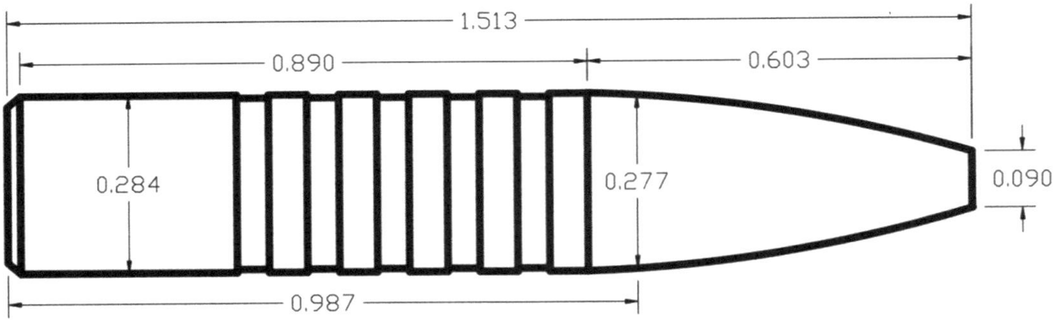

Lot #:	**Unknown**	**Ogive Radius:**	**8.25 calibers**
Sectional Density:	**0.310 lb/in^2**	**Rt/R:**	**0.92**

Ballistic Performance

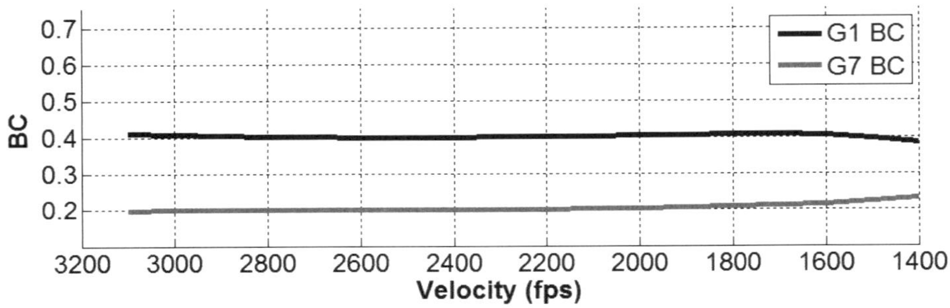

	Above 3000	3000-2500	2500-2000	2000-1500	Below 1500
G1 BC	0.418	0.405	0.403	0.402	0.398
G7 BC	0.198	0.201	0.203	0.214	0.225

Averages from 3000 fps to 1500 fps:
G1 BC: 0.403, variation of G1 BC: 0.011
G7 BC: 0.207, variation of G7 BC: 0.025
G7 form factor: 1.497

Rifling Twist Rate and Stability

	Stability Factor		
Barrel Twist	**Best case**	**Nominal**	**Worst case**
1:8"	2.11	1.85	1.60
1:9"	1.67	1.46	1.26
1:10"	1.35	1.18	1.02
1:11"	1.12	0.98	0.84

Minimum twist required for stability is 1:9.9"
Recommended twist for optimal performance is 1:8.9" or faster

Barnes .284 caliber 110 grain TTSX FB

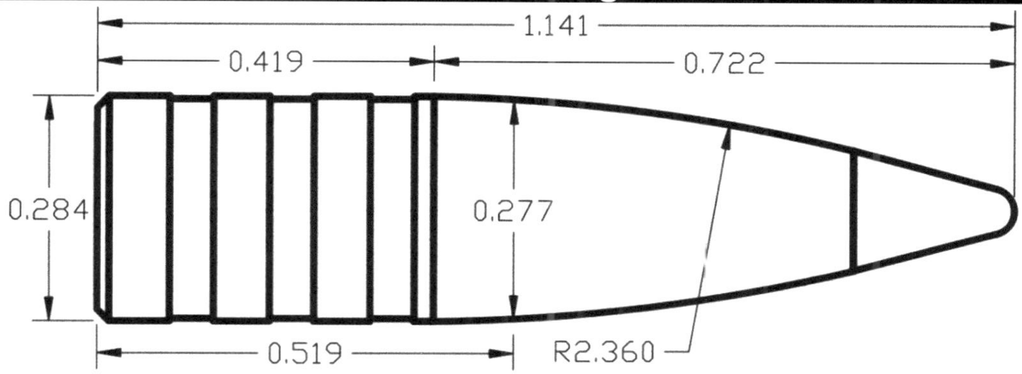

Lot #:	3984-1 028M TMS-M	Ogive Radius:	8.31 calibers
Sectional Density:	0.195 lb/in^2	Rt/R:	0.87

Ballistic Performance

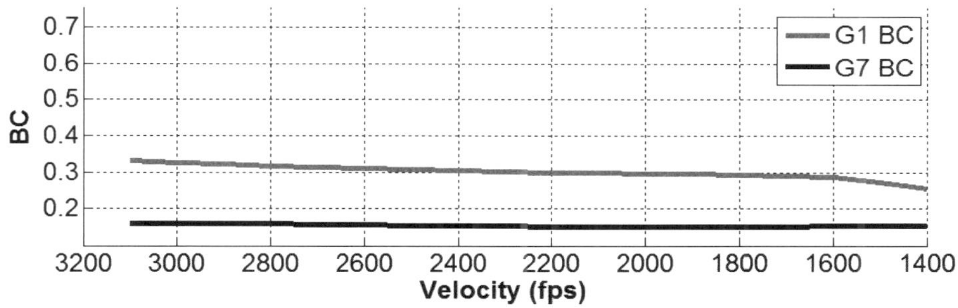

	Above 3000	3000-2500	2500-2000	2000-1500	Below 1500
G1 BC	0.343	0.318	0.304	0.287	0.275
G7 BC	0.162	0.158	0.153	0.152	0.155

Averages from 3000 fps to 1500 fps:
G1 BC: 0.303, variation of G1 BC: 0.053
G7 BC: 0.155, variation of G7 BC: 0.010
G7 form factor: 1.257

Rifling Twist Rate and Stability

	Stability Factor		
Barrel Twist	**Best case**	**Nominal**	**Worst case**
1:8"	3.02	2.64	2.28
1:9"	2.38	2.08	1.80
1:10"	1.93	1.69	1.46
1:11"	1.60	1.39	1.21

Minimum twist required for stability is 1:11.9"
Recommended twist for optimal performance is 1:10.6" or faster

Barnes .284 caliber 120 TTSXBT

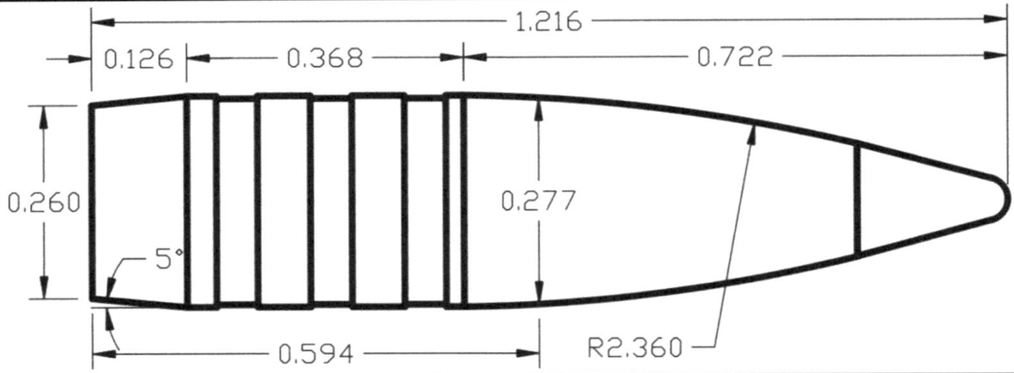

Lot #:	5706-2 132N JB-M	**Ogive Radius:**	**8.31 calibers**
Sectional Density:	**0.213 lb/in²**	**Rt/R:**	**0.87**

Ballistic Performance

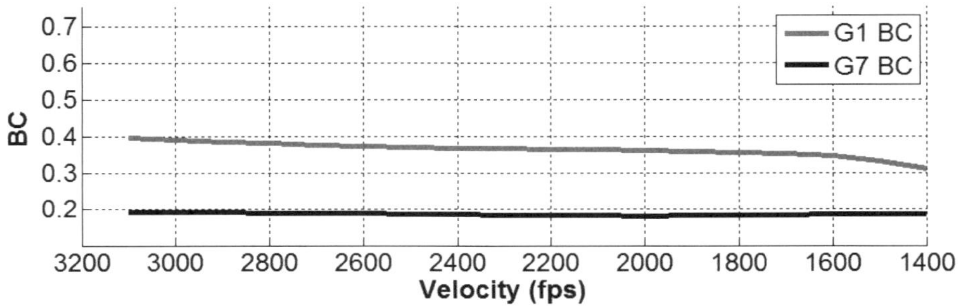

	Above 3000	3000-2500	2500-2000	2000-1500	Below 1500
G1 BC	0.407	0.382	0.367	0.347	0.332
G7 BC	0.193	0.189	0.184	0.184	0.187

Averages from 3000 fps to 1500 fps:
G1 BC: 0.364, variation of G1 BC: 0.060
G7 BC: 0.187, variation of G7 BC: 0.010
G7 form factor: 1.138

Rifling Twist Rate and Stability

Barrel Twist	Stability Factor Best case	Nominal	Worst case
1:8"	2.75	2.40	2.07
1:9"	2.17	1.90	1.64
1:10"	1.76	1.54	1.33
1:11"	1.45	1.27	1.10

Minimum twist required for stability is 1:11.3"
Recommended twist for optimal performance is 1:10.1" or faster

Barnes .284 caliber 140 TTSXBT

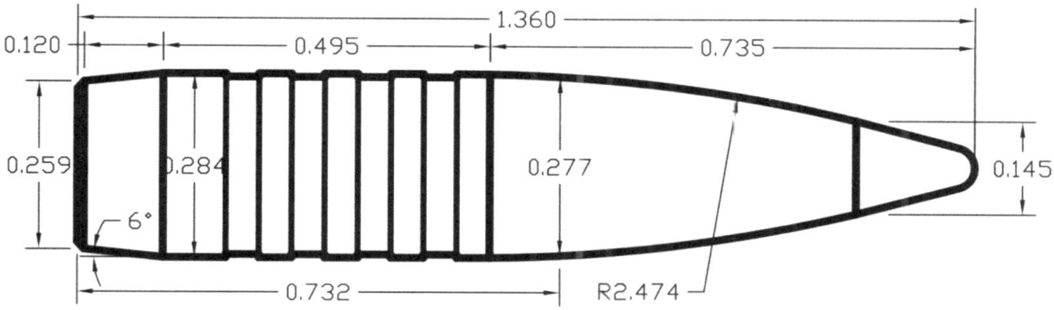

Lot #:	**Unknown**	**Ogive Radius:**	**8.71 calibers**
Sectional Density:	**0.248 lb/in²**	**Rt/R:**	**0.91**

Ballistic Performance

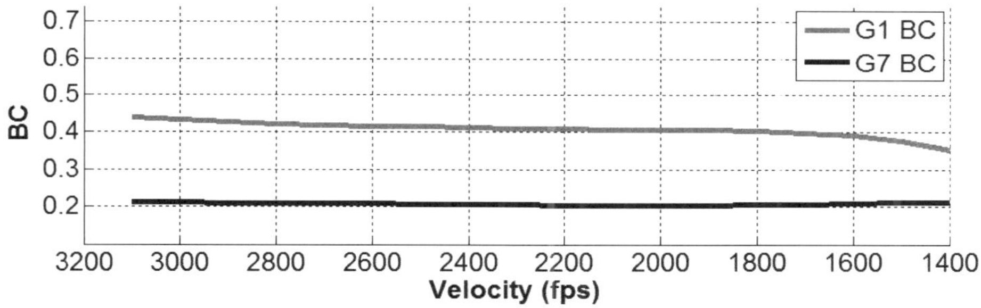

	Above 3000	3000-2500	2500-2000	2000-1500	Below 1500
G1 BC	0.447	0.423	0.410	0.391	0.376
G7 BC	0.212	0.210	0.206	0.208	0.211

Averages from 3000 fps to 1500 fps:
G1 BC: 0.407, variation of G1 BC: 0.057
G7 BC: 0.209, variation of G7 BC: 0.007
G7 form factor: 1.188

Rifling Twist Rate and Stability

Barrel Twist	Stability Factor Best case	Nominal	Worst case
1:8"	2.29	2.00	1.73
1:9"	1.81	1.58	1.37
1:10"	1.46	1.28	1.11
1:11"	1.21	1.06	0.91

Minimum twist required for stability is 1:10.4"
Recommended twist for optimal performance is 1:9.3" or faster

Berger .284 caliber 140 grain Hunting VLD

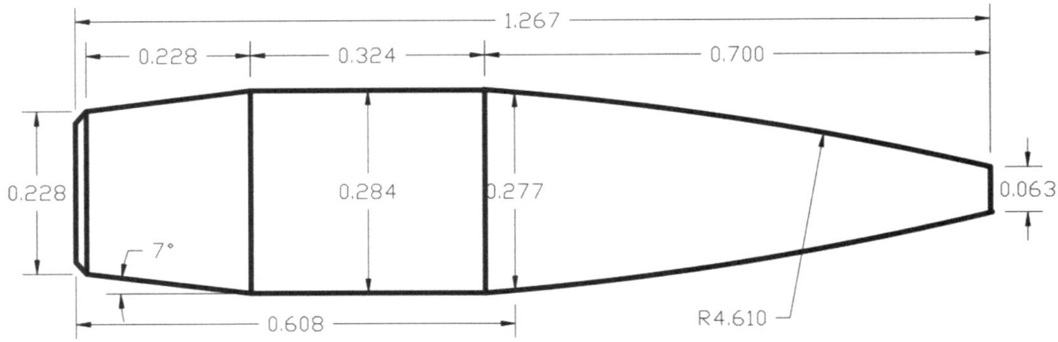

Lot #:	**7544**	**Ogive Radius:**	**16.23 calibers**
Sectional Density:	**0.248 lb/in^2**	**Rt/R:**	**0.47**

Ballistic Performance

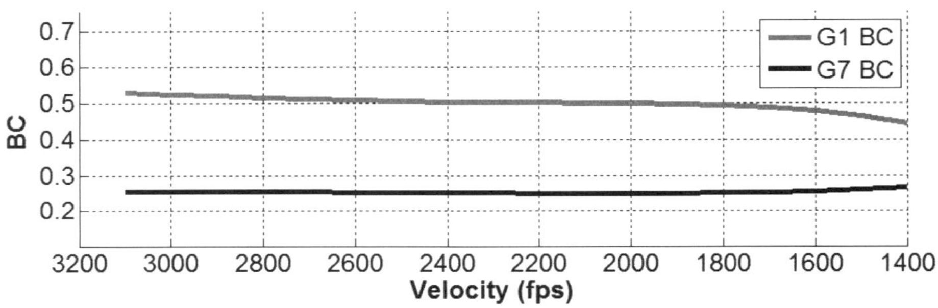

	Above 3000	3000-2500	2500-2000	2000-1500	Below 1500
G1 BC	0.539	0.515	0.502	0.482	0.465
G7 BC	0.255	0.255	0.252	0.256	0.262

Averages from 3000 fps to 1500 fps:
G1 BC: 0.498, variation of G1 BC: 0.059
G7 BC: 0.256, variation of G7 BC: 0.011
G7 form factor: 0.970

Rifling Twist Rate and Stability

Barrel Twist	Stability Factor Best case	Nominal	Worst case
1:9"	2.24	1.96	1.69
1:10"	1.82	1.59	1.37
1:11"	1.50	1.31	1.13
1:12"	1.26	1.10	0.95

Minimum twist required for stability is 1:11.5"
Recommended twist for optimal performance is 1:10.3" or faster

Berger .284 caliber 168 grain Classic Hunter

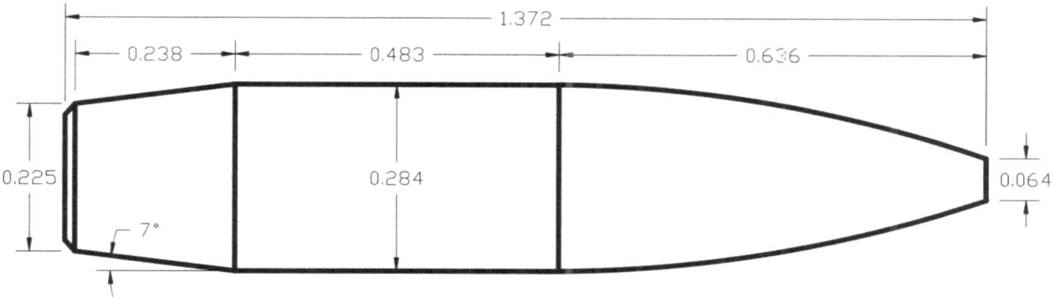

Lot #:	7627	Ogive Radius:	Hybrid
Sectional Density:	0.298 lb/in²	Rt/R:	Hybrid

Ballistic Performance

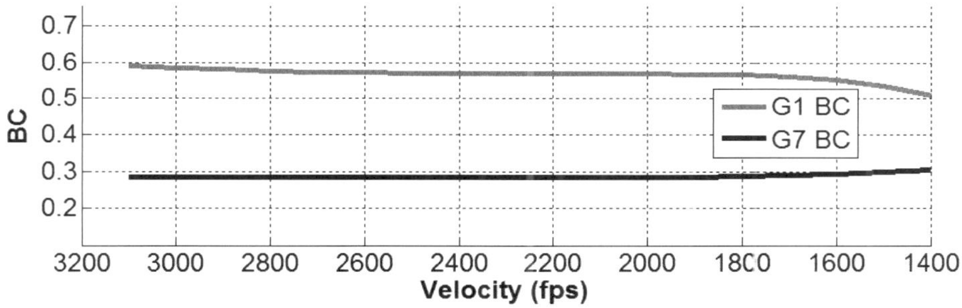

	Above 3000	3000-2500	2500-2000	2000-1500	Below 1500
G1 BC	0.599	0.578	0.570	0.552	0.535
G7 BC	0.284	0.287	0.286	0.293	0.301

Averages from 3000 fps to 1500 fps:
G1 BC: 0.565, variation of G1 BC: 0.051
G7 BC: 0.290, variation of G7 BC: 0.015
G7 form factor: 1.026

Rifling Twist Rate and Stability

Barrel Twist	Stability Factor Best case	Nominal	Worst case
1:9"	2.13	1.86	1.61
1:10"	1.73	1.51	1.31
1:11"	1.43	1.25	1.08
1:12"	1.20	1.05	0.91

Minimum twist required for stability is 1:11.2"
Recommended twist for optimal performance is 1:10.0" or faster

Berger .284 caliber 168 grain Hunting VLD

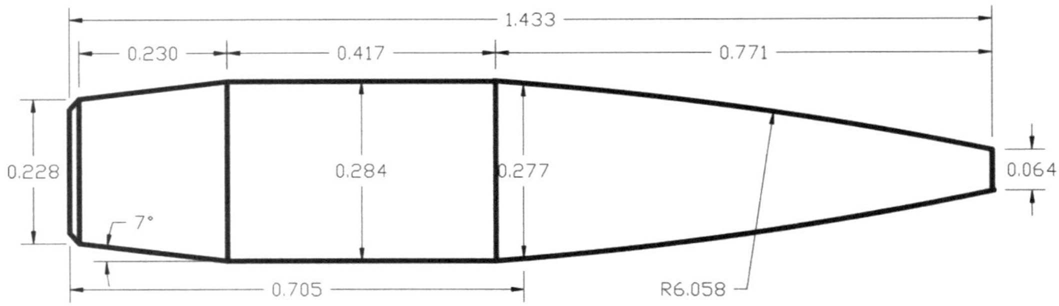

Lot #:	7654	Ogive Radius:	21.33 calibers
Sectional Density:	0.298 lb/in^2	Rt/R:	0.45

Ballistic Performance

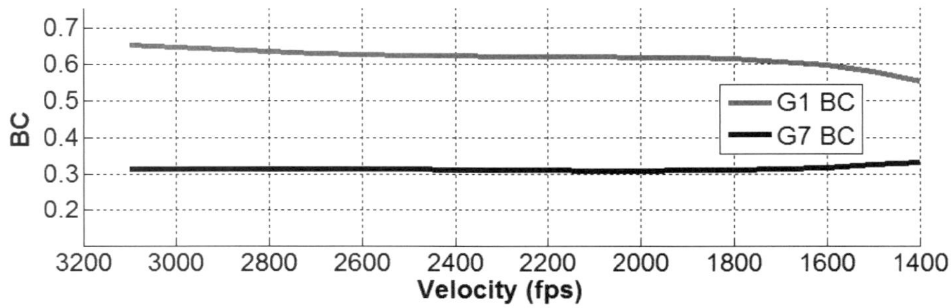

	Above 3000	3000-2500	2500-2000	2000-1500	Below 1500
G1 BC	0.664	0.635	0.621	0.598	0.578
G7 BC	0.315	0.315	0.312	0.318	0.326

Averages from 3000 fps to 1500 fps:
G1 BC: 0.616, variation of G1 BC: 0.068
G7 BC: 0.316, variation of G7 BC: 0.015
G7 form factor: 0.941

Rifling Twist Rate and Stability

Barrel Twist	Stability Factor Best case	Nominal	Worst case
1:8"	2.38	2.08	1.80
1:9"	1.88	1.64	1.42
1:10"	1.52	1.33	1.15
1:11"	1.26	1.10	0.95

Minimum twist required for stability is 1:10.5"
Recommended twist for optimal performance is 1:9.4" or faster

Berger .284 caliber 180 grain Hunting VLD

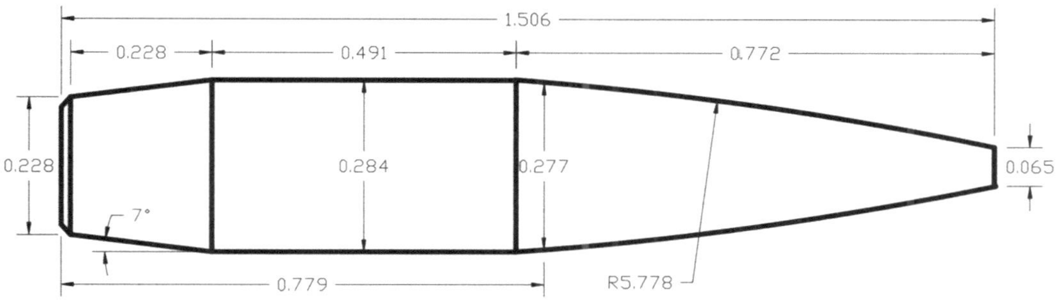

Lot #:	7509	Ogive Radius:	20.35 calibers
Sectional Density:	0.319 lb/in^2	Rt/R:	0.48

Ballistic Performance

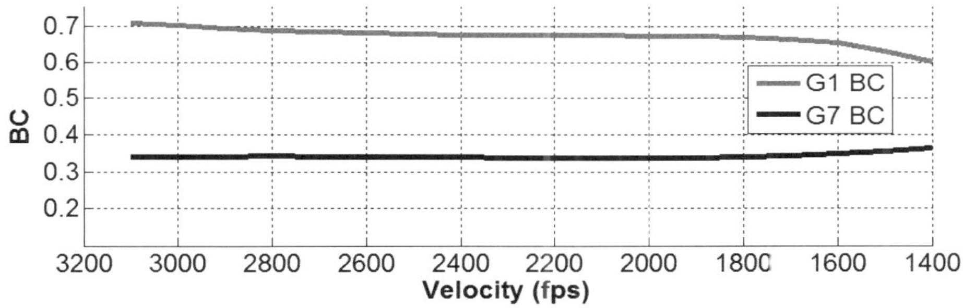

	Above 3000	3000-2500	2500-2000	2000-1500	Below 1500
G1 BC	0.720	0.690	0.676	0.653	0.632
G7 BC	0.341	0.342	0.340	0.347	0.356

Averages from 3000 fps to 1500 fps:

G1 BC: 0.671, variation of G1 BC: 0.069

G7 BC: 0.345, variation of G7 BC: 0.018

G7 form factor: 0.925

Rifling Twist Rate and Stability

Barrel Twist	Stability Factor Best case	Nominal	Worst case
1:8"	2.20	1.92	1.66
1:9"	1.74	1.52	1.31
1:10"	1.41	1.23	1.06
1:11"	1.16	1.02	0.88

Minimum twist required for stability is 1:10.1"

Recommended twist for optimal performance is 1:9.1" or faster

Berger .284 caliber 195 grain Hunting Hybrid

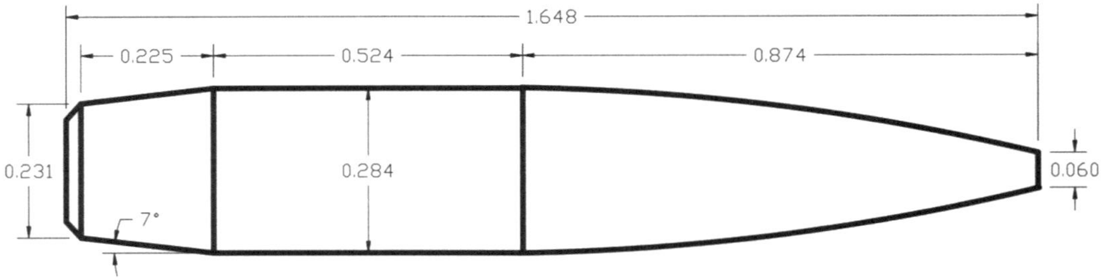

Lot #:	0000	Ogive Radius:	Hybrid
Sectional Density:	0.345 lb/in^2	Rt/R:	Hybrid

Ballistic Performance

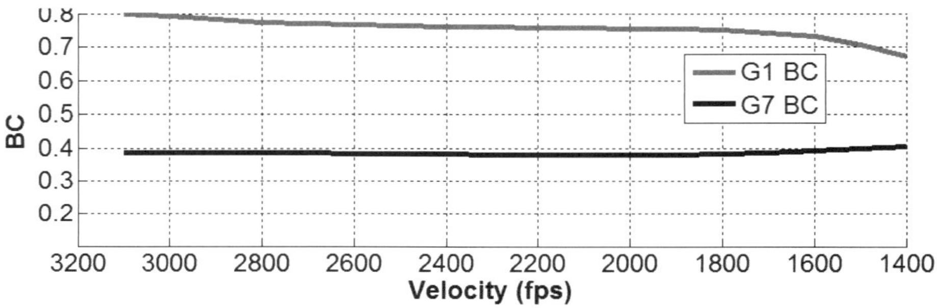

	Above 3000	3000-2500	2500-2000	2000-1500	Below 1500
G1 BC	0.814	0.777	0.759	0.731	0.706
G7 BC	0.385	0.385	0.382	0.389	0.398

Averages from 3000 fps to 1500 fps:
G1 BC: 0.754, variation of G1 BC: 0.084
G7 BC: 0.387, variation of G7 BC: 0.018
G7 form factor: 0.893

Rifling Twist Rate and Stability

Barrel Twist	Stability Factor Best case	Nominal	Worst case
1:7"	2.39	2.09	1.81
1:8"	1.83	1.60	1.38
1:9"	1.45	1.26	1.09
1:10"	1.17	1.02	0.89

Minimum twist required for stability is 1:9.2"
Recommended twist for optimal performance is 1:8.3" or faster

Berger .284 caliber 168 grain Target VLD

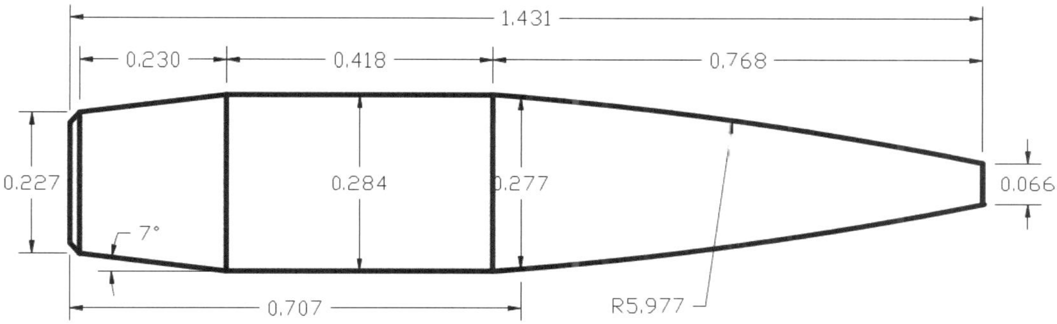

Lot #:	**2918**	**Ogive Radius:**	**21.05 calibers**
Sectional Density:	**0.298 lb/in^2**	**Rt/R:**	**0.46**

Ballistic Performance

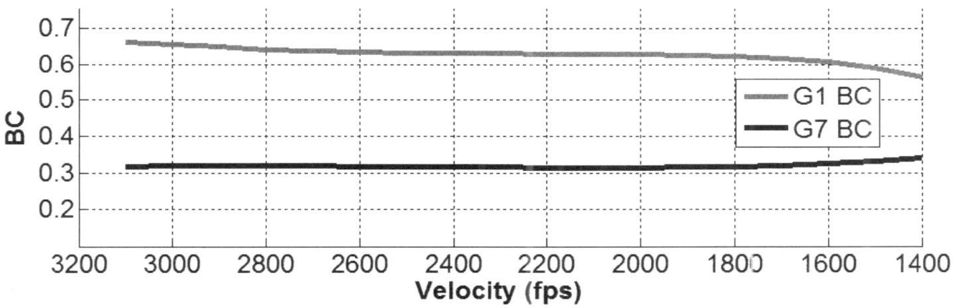

	Above 3000	3000-2500	2500-2000	2000-1500	Below 1500
G1 BC	0.672	0.643	0.630	0.609	0.590
G7 BC	0.318	0.319	0.317	0.324	0.332

Averages from 3000 fps to 1500 fps:
G1 BC: 0.626, variation of G1 BC: 0.064
G7 BC: 0.322, variation of G7 BC: 0.017
G7 form factor: 0.926

Rifling Twist Rate and Stability

	Stability Factor		
Barrel Twist	**Best case**	**Nominal**	**Worst case**
1:8"	2.39	2.09	1.80
1:9"	1.89	1.65	1.42
1:10"	1.53	1.33	1.15
1:11"	1.26	1.10	0.95

Minimum twist required for stability is 1:10.5"
Recommended twist for optimal performance is 1:9.4" or faster

Berger .284 caliber 180 grain Target Hybrid

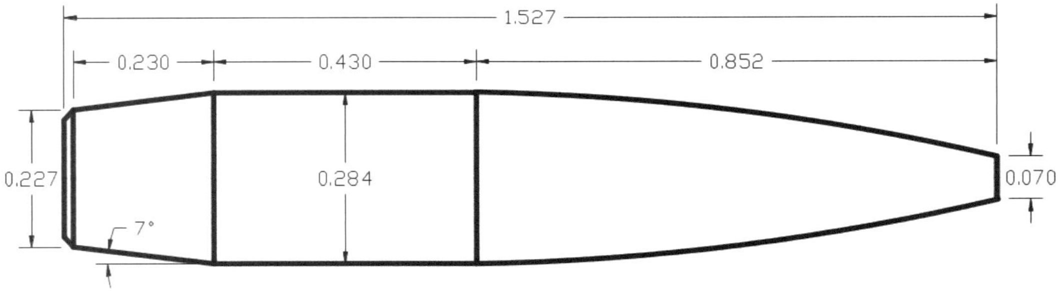

Lot #:	6771	Ogive Radius:	Hybrid
Sectional Density:	0.319 lb/in^2	Rt/R:	Hybrid

Ballistic Performance

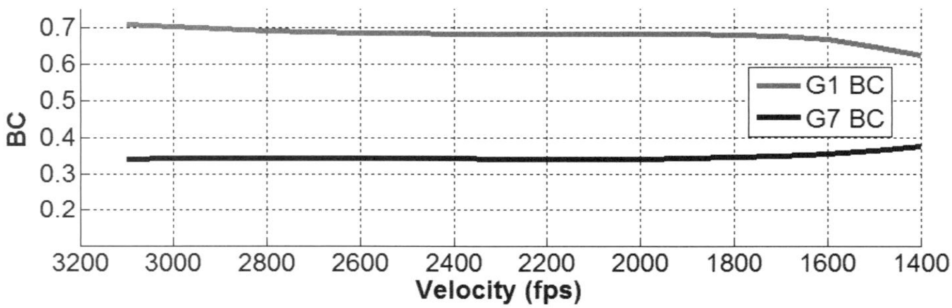

	Above 3000	3000-2500	2500-2000	2000-1500	Below 1500
G1 BC	0.720	0.693	0.682	0.664	0.647
G7 BC	0.341	0.344	0.343	0.354	0.365

Averages from 3000 fps to 1500 fps:
G1 BC: 0.679, variation of G1 BC: 0.055
G7 BC: 0.349, variation of G7 BC: 0.023
G7 form factor: 0.915

Rifling Twist Rate and Stability

Barrel Twist	Stability Factor Best case	Nominal	Worst case
1:7"	2.76	2.41	2.09
1:8"	2.11	1.85	1.60
1:9"	1.67	1.46	1.26
1:10"	1.35	1.18	1.02

Minimum twist required for stability is 1:9.9"
Recommended twist for optimal performance is 1:8.9" or faster

Berger .284 caliber 180 grain Target VLD

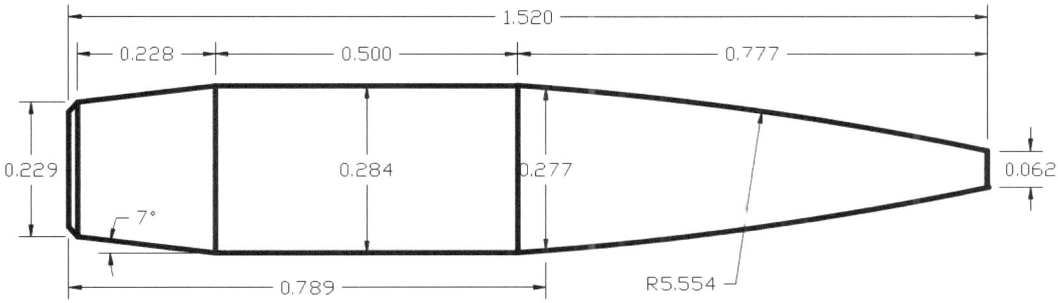

Lot #:	6769	Ogive Radius:	19.56 calibers
Sectional Density:	0.319 lb/in^2	Rt/R:	0.50

Ballistic Performance

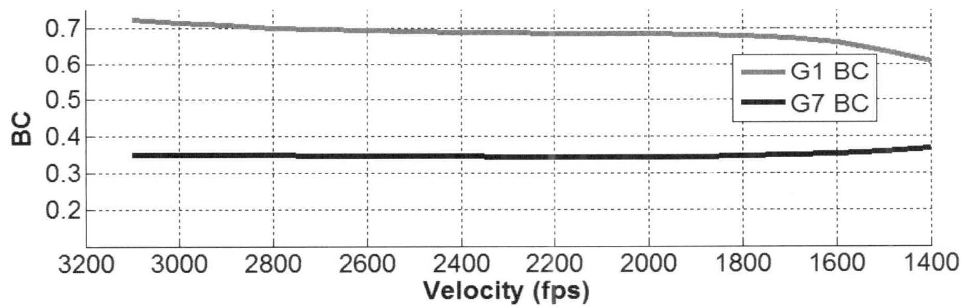

	Above 3000	3000-2500	2500-2000	2000-1500	Below 1500
G1 BC	0.734	0.702	0.687	0.661	0.638
G7 BC	0.348	0.348	0.345	0.351	0.360

Averages from 3000 fps to 1500 fps:
G1 BC: 0.681, variation of G1 BC: 0.076
G7 BC: 0.350, variation of G7 BC: 0.017
G7 form factor: 0.912

Rifling Twist Rate and Stability

Barrel Twist	Stability Factor Best case	Nominal	Worst case
1:7"	2.80	2.45	2.11
1:8"	2.14	1.87	1.62
1:9"	1.69	1.48	1.28
1:10"	1.37	1.20	1.04

Minimum twist required for stability is 1:10.0"
Recommended twist for optimal performance is 1:8.9" or faster

Cauterucio .284 caliber 177 grain VLD

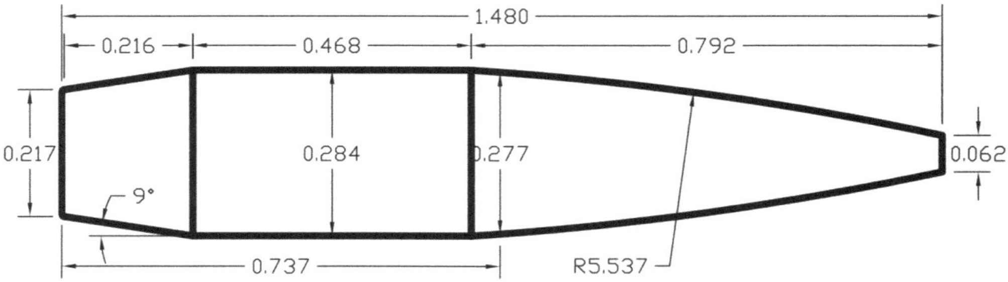

Lot #:	**Unknown**		**Ogive Radius:**	**19.50 calibers**
Sectional Density:	**0.314 lb/in^2**		**Rt/R:**	**0.56**

Ballistic Performance

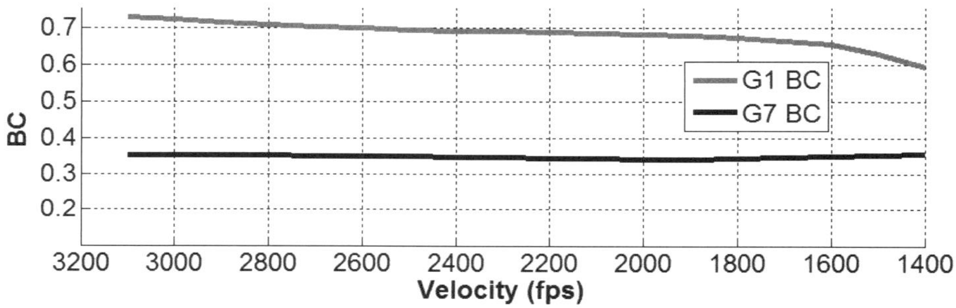

	Above 3000	3000-2500	2500-2000	2000-1500	Below 1500
G1 BC	0.743	0.708	0.688	0.655	0.628
G7 BC	0.352	0.351	0.346	0.348	0.353

Averages from 3000 fps to 1500 fps:
G1 BC: 0.681, variation of G1 BC: 0.094
G7 BC: 0.349, variation of G7 BC: 0.011
G7 form factor: 0.897

Rifling Twist Rate and Stability

Barrel Twist	Stability Factor Best case	Stability Factor Nominal	Stability Factor Worst case
1:8"	2.28	1.99	1.72
1:9"	1.80	1.57	1.36
1:10"	1.46	1.27	1.10
1:11"	1.21	1.05	0.91

Minimum twist required for stability is 1:10.3"
Recommended twist for optimal performance is 1:9.2" or faster

Cauterucio .284 caliber 189 grain VLD

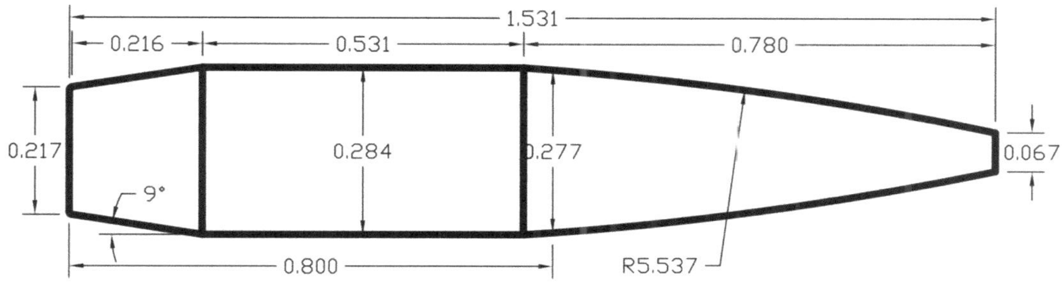

Lot #:	Unknown	Ogive Radius:	19.50 calibers
Sectional Density:	0.335 lb/in^2	Rt/R:	0.56

Ballistic Performance

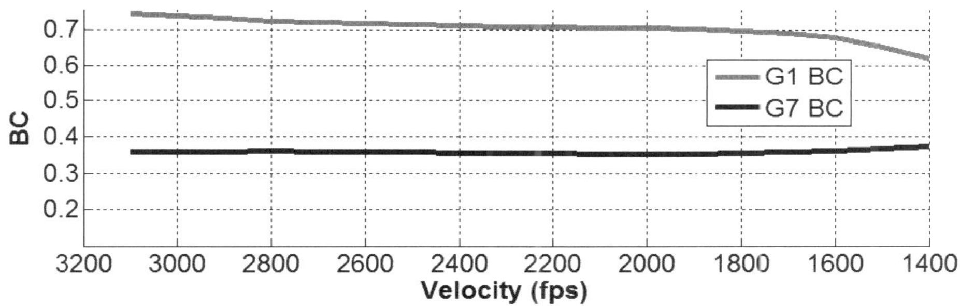

	Above 3000	3000-2500	2500-2000	2000-1500	Below 1500
G1 BC	0.757	0.724	0.708	0.678	0.653
G7 BC	0.359	0.359	0.356	0.361	0.368

Averages from 3000 fps to 1500 fps:
G1 BC: 0.701, variation of G1 BC: 0.084
G7 BC: 0.360, variation of G7 BC: 0.014
G7 form factor: 0.930

Rifling Twist Rate and Stability

Barrel Twist	Stability Factor Best case	Nominal	Worst case
1:8"	2.20	1.92	1.66
1:9"	1.74	1.52	1.32
1:10"	1.41	1.23	1.07
1:11"	1.17	1.02	0.88

Minimum twist required for stability is 1:10.1"
Recommended twist for optimal performance is 1:9.1" or faster

GS Custom .284 Caliber 125 grain SP

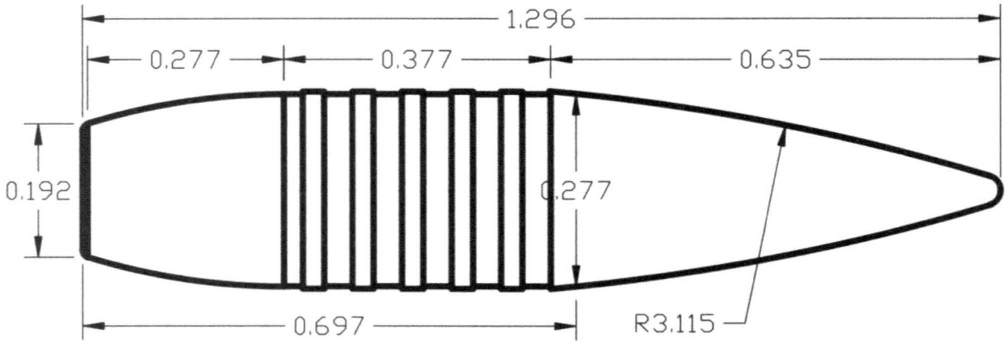

Lot #:	**Unknown**	**Ogive Radius:**	**10.97 calibers**
Sectional Density:	**0.221 lb/in^2**	**Rt/R:**	**0.49**

Ballistic Performance

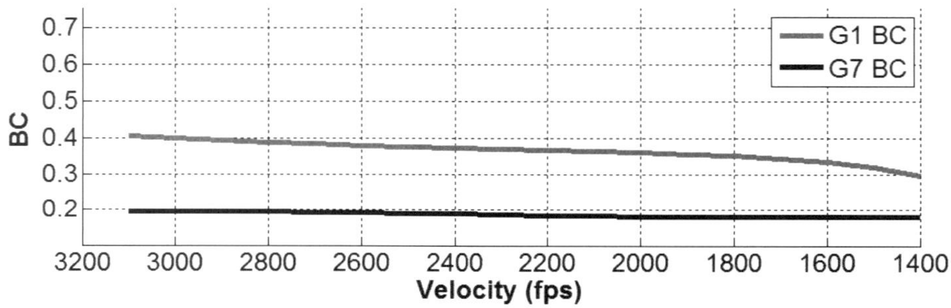

	Above 3000	3000-2500	2500-2000	2000-1500	Below 1500
G1 BC	0.415	0.388	0.369	0.341	0.320
G7 BC	0.196	0.193	0.186	0.181	0.180

Averages from 3000 fps to 1500 fps:
G1 BC: 0.364, variation of G1 BC: 0.079
G7 BC: 0.187, variation of G7 BC: 0.015
G7 form factor: 1.188

Rifling Twist Rate and Stability

Barrel Twist	Stability Factor Best case	Nominal	Worst case
1:8"	2.37	2.07	1.79
1:9"	1.87	1.64	1.42
1:10"	1.52	1.33	1.15
1:11"	1.25	1.10	0.95

Minimum twist required for stability is 1:10.5"
Recommended twist for optimal performance is 1:9.4" or faster

Hornady .284 caliber 162 grain Amax

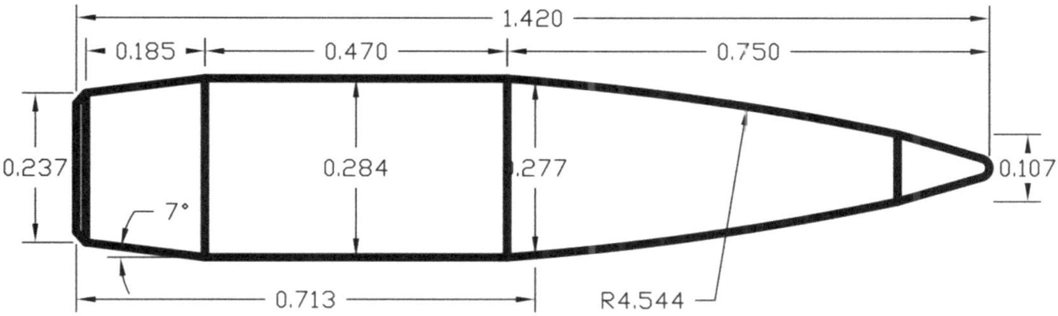

Lot #:	28402	Ogive Radius:	16.0 calibers
Sectional Density:	0.287 lb/in^2	Rt/R:	0.55

Ballistic Performance

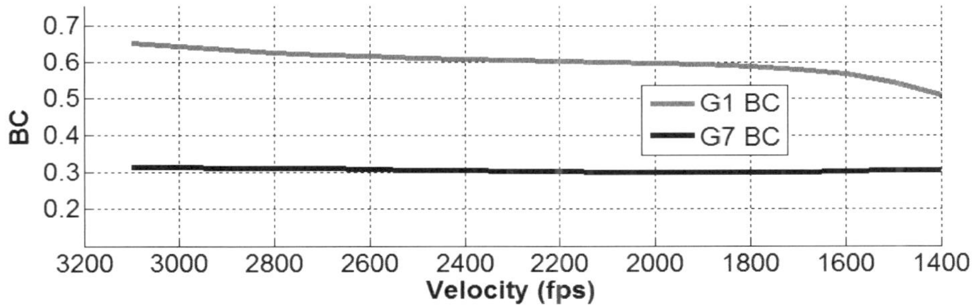

	Above 3000	3000-2500	2500-2000	2000-1500	Below 1500
G1 BC	0.666	0.627	0.604	0.569	0.543
G7 BC	0.315	0.311	0.303	0.302	0.305

Averages from 3000 fps to 1500 fps:

G1 BC: 0.598, variation of G1 BC: 0.100

G7 BC: 0.307, variation of G7 BC: 0.015

G7 form factor: 0.936

Rifling Twist Rate and Stability

Barrel Twist	Stability Factor Best case	Nominal	Worst case
1:8"	2.35	2.06	1.78
1:9"	1.86	1.63	1.41
1:10"	1.51	1.32	1.14
1:11"	1.25	1.09	0.94

Minimum twist required for stability is 1:10.5"

Recommended twist for optimal performance is 1:9.4" or faster

Hornady .284 caliber 162 grain BTHP Match

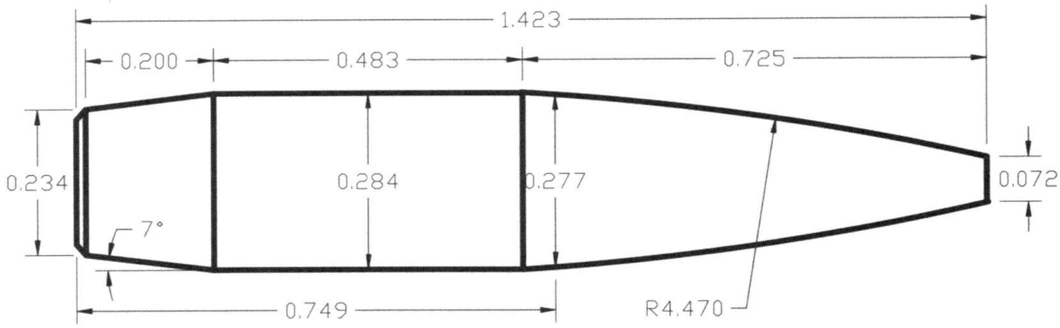

Lot #:	**2150249**	**Ogive Radius:**	**15.74 calibers**
Sectional Density:	**0.287 lb/in²**	**Rt/R:**	**0.57**

Ballistic Performance

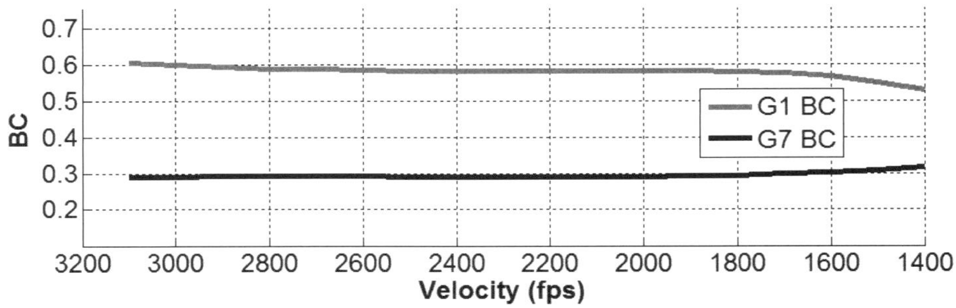

	Above 3000	3000-2500	2500-2000	2000-1500	Below 1500
G1 BC	0.614	0.591	0.582	0.566	0.550
G7 BC	0.291	0.293	0.293	0.301	0.310

Averages from 3000 fps to 1500 fps:
G1 BC: 0.578, variation of G1 BC: 0.048
G7 BC: 0.297, variation of G7 BC: 0.018
G7 form factor: 0.966

Rifling Twist Rate and Stability

	Stability Factor		
Barrel Twist	**Best case**	**Nominal**	**Worst case**
1:8"	2.34	2.04	1.77
1:9"	1.85	1.62	1.40
1:10"	1.50	1.31	1.13
1:11"	1.24	1.08	0.93

Minimum twist required for stability is 1:10.4"
Recommended twist for optimal performance is 1:9.3" or faster

Hornady .284 caliber 139 grain GMX

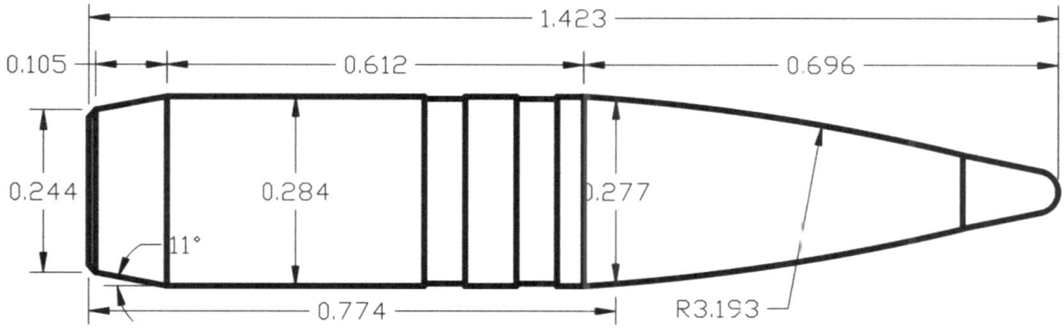

Lot #:	2140672	**Ogive Radius:**	**11.24 calibers**
Sectional Density:	**0.246 lb/in^2**	**Rt/R:**	**0.57**

Ballistic Performance

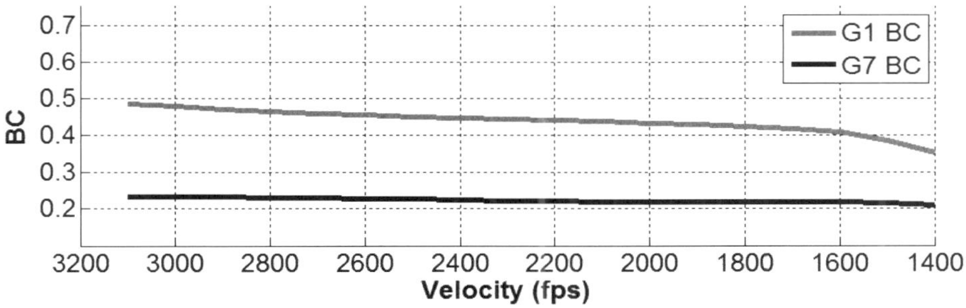

	Above 3000	3000-2500	2500-2000	2000-1500	Below 1500
G1 BC	0.497	0.464	0.442	0.410	0.385
G7 BC	0.235	0.230	0.222	0.217	0.216

Averages from 3000 fps to 1500 fps:
G1 BC: 0.437, variation of G1 BC: 0.093
G7 BC: 0.224, variation of G7 BC: 0.018
G7 form factor: 1.102

Rifling Twist Rate and Stability

Barrel Twist	Stability Factor Best case	Nominal	Worst case
1:8"	2.01	1.75	1.52
1:9"	1.59	1.39	1.20
1:10"	1.28	1.12	0.97
1:11"	1.06	0.93	0.80

Minimum twist required for stability is 1:9.7"
Recommended twist for optimal performance is 1:8.7" or faster

Hornady .284 caliber 139 grain Interbond

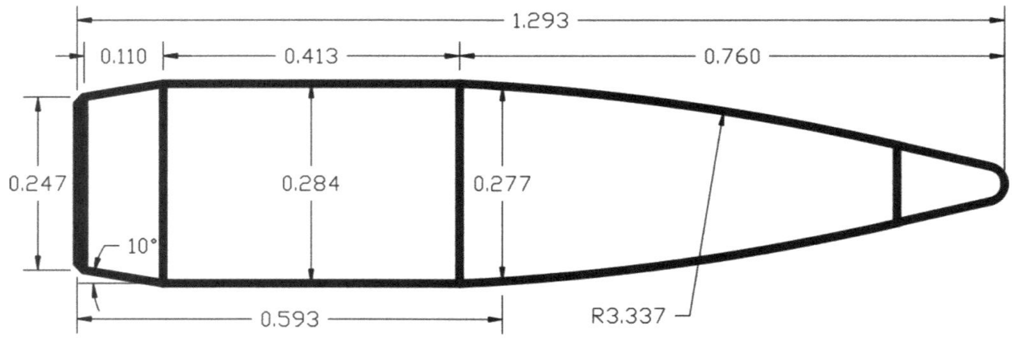

Lot #:	Unknown	Ogive Radius:	11.75 calibers
Sectional Density:	0.246 lb/in²	Rt/R:	0.65

Ballistic Performance

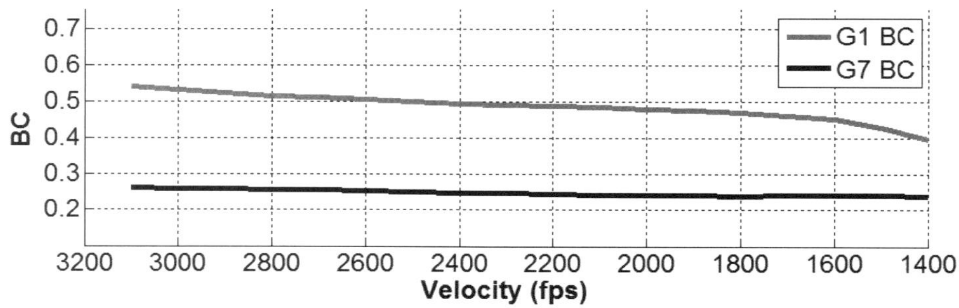

	Above 3000	3000-2500	2500-2000	2000-1500	Below 1500
G1 BC	0.554	0.516	0.490	0.454	0.428
G7 BC	0.262	0.256	0.246	0.241	0.241

Averages from 3000 fps to 1500 fps:
G1 BC: 0.485, variation of G1 BC: 0.103
G7 BC: 0.248, variation of G7 BC: 0.019
G7 form factor: 0.993

Rifling Twist Rate and Stability

Barrel Twist	Stability Factor Best case	Nominal	Worst case
1:8"	2.65	2.32	2.01
1:9"	2.10	1.83	1.58
1:10"	1.70	1.48	1.28
1:11"	1.40	1.23	1.06

Minimum twist required for stability is 1:11.1"
Recommended twist for optimal performance is 1:9.9" or faster

Hornady .284 caliber 154 grain Interbond

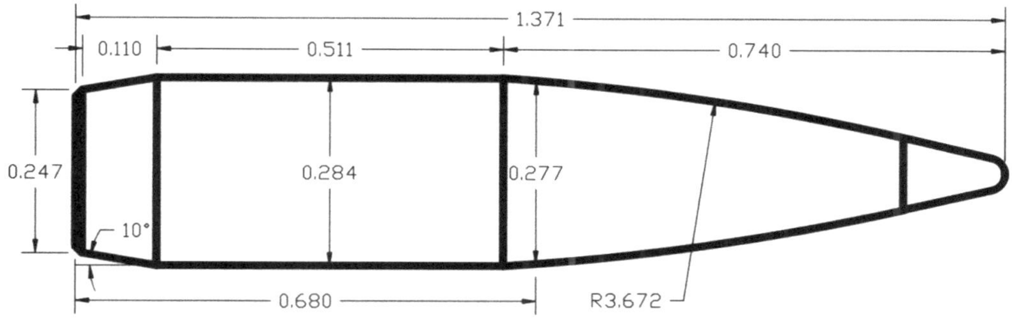

Lot #:	Unknown	Ogive Radius:	12.93 calibers
Sectional Density:	0.273 lb/in²	Rt/R:	0.56

Ballistic Performance

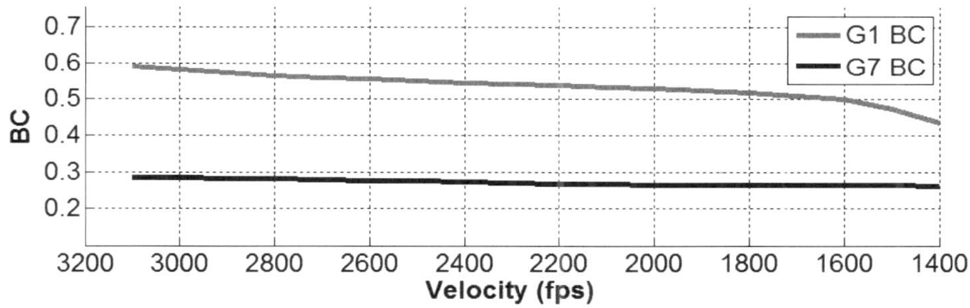

	Above 3000	3000-2500	2500-2000	2000-1500	Below 1500
G1 BC	0.606	0.566	0.539	0.500	0.472
G7 BC	0.287	0.280	0.271	0.265	0.265

Averages from 3000 fps to 1500 fps:
G1 BC: 0.533, variation of G1 BC: 0.111
G7 BC: 0.273, variation of G7 BC: 0.020
G7 form factor: 1.001

Rifling Twist Rate and Stability

Barrel Twist	Stability Factor Best case	Nominal	Worst case
1:8"	2.48	2.17	1.87
1:9"	1.96	1.71	1.48
1:10"	1.59	1.39	1.20
1:11"	1.31	1.15	0.99

Minimum twist required for stability is 1:10.7"
Recommended twist for optimal performance is 1:9.6" or faster

Hornady .284 caliber 175 Interlock

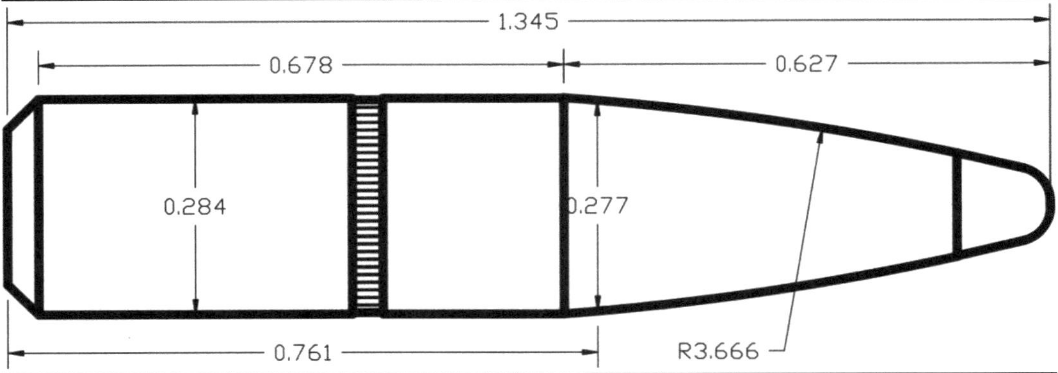

Lot #:	Unknown	Ogive Radius:	12.91 calibers
Sectional Density:	0.310 lb/in^2	Rt/R:	0.60

Ballistic Performance

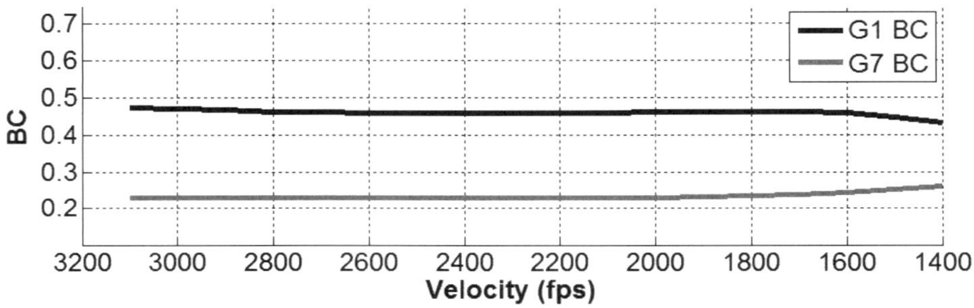

	Above 3000	3000-2500	2500-2000	2000-1500	Below 1500
G1 BC	0.483	0.464	0.459	0.454	0.448
G7 BC	0.229	0.230	0.231	0.242	0.253

Averages from 3000 fps to 1500 fps:
G1 BC: 0.459, variation of G1 BC: 0.023
G7 BC: 0.236, variation of G7 BC: 0.023
G7 form factor: 1.314

Rifling Twist Rate and Stability

Barrel Twist	Stability Factor Best case	Nominal	Worst case
1:8"	2.98	2.60	2.25
1:9"	2.35	2.06	1.78
1:10"	1.91	1.67	1.44
1:11"	1.58	1.38	1.19

Minimum twist required for stability is 1:11.8"
Recommended twist for optimal performance is 1:10.5" or faster

Hornady .284 caliber 139 grain BTSP Interlock

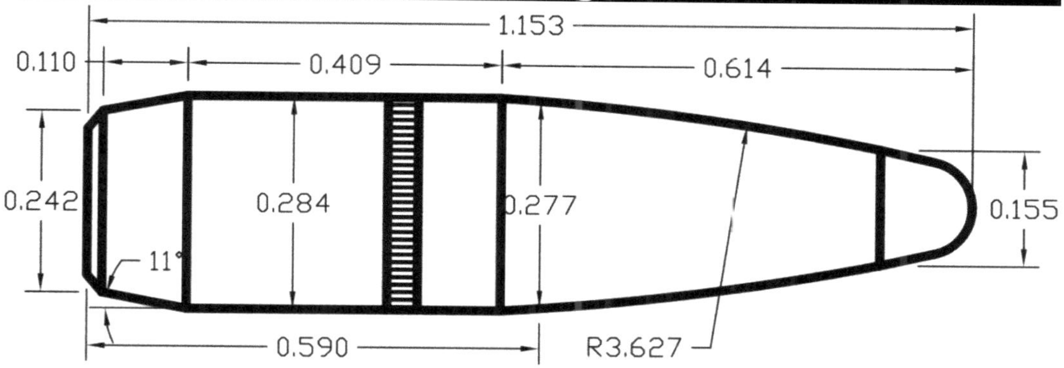

Lot #:	Unknown	Ogive Radius:	12.77 calibers
Sectional Density:	0.246 lb/in²	Rt/R:	0.65

Ballistic Performance

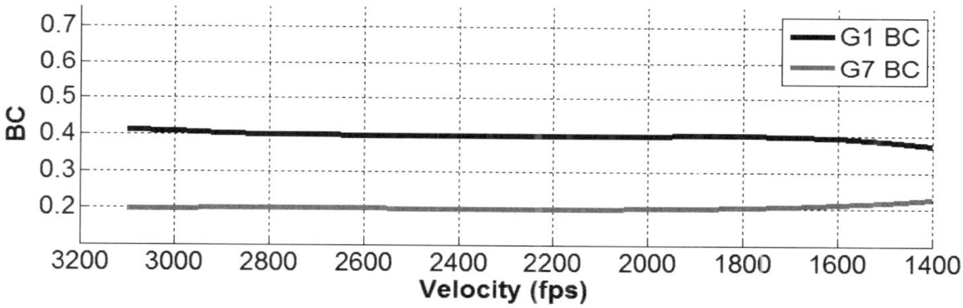

	Above 3000	3000-2500	2500-2000	2000-1500	Below 1500
G1 BC	0.419	0.402	0.397	0.392	0.386
G7 BC	0.198	0.200	0.200	0.209	0.218

Averages from 3000 fps to 1500 fps:

G1 BC: 0.397, variation of G1 BC: 0.022

G7 BC: 0.204, variation of G7 BC: 0.018

G7 form factor: 1.207

Rifling Twist Rate and Stability

Barrel Twist	Stability Factor Best case	Nominal	Worst case
1:8"	3.70	3.23	2.80
1:9"	2.92	2.55	2.21
1:10"	2.37	2.07	1.79
1:11"	1.96	1.71	1.48

Minimum twist required for stability is 1:13.1"

Recommended twist for optimal performance is 1:11.7" or faster

Hornady .284 caliber 162 grain BTSP Interlock

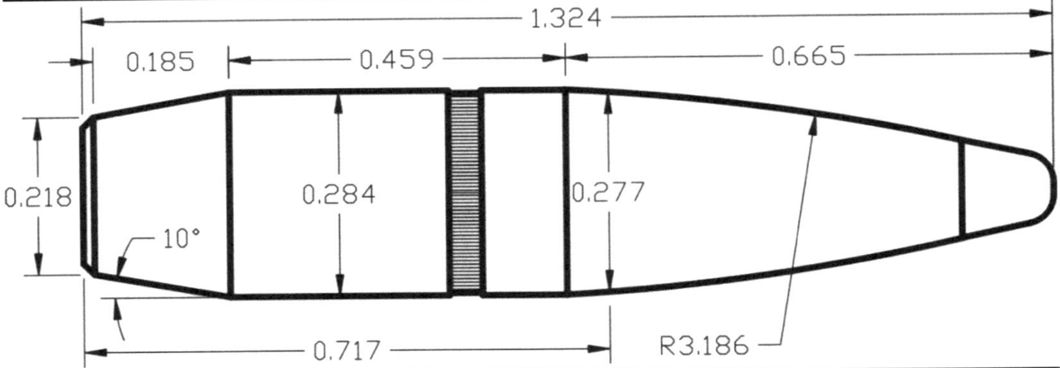

Lot #:	2140810	Ogive Radius:	11.22 calibers
Sectional Density:	0.287 lb/in²	Rt/R:	0.63

Ballistic Performance

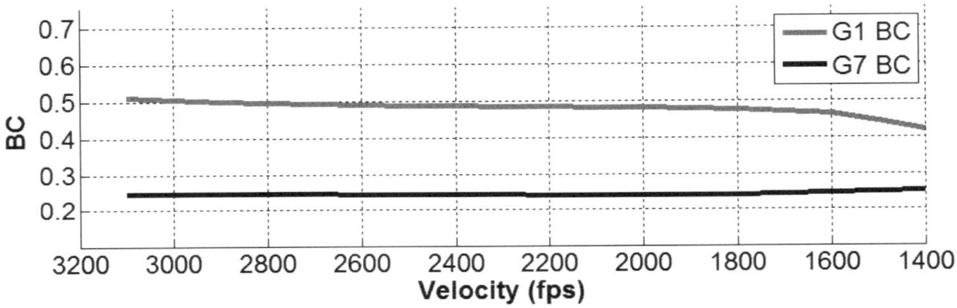

	Above 3000	3000-2500	2500-2000	2000-1500	Below 1500
G1 BC	0.519	0.497	0.485	0.463	0.445
G7 BC	0.246	0.246	0.244	0.246	0.250

Averages from 3000 fps to 1500 fps:
G1 BC: 0.480, variation of G1 BC: 0.061
G7 BC: 0.246, variation of G7 BC: 0.008
G7 form factor: 1.165

Rifling Twist Rate and Stability

Barrel Twist	Stability Factor Best case	Nominal	Worst case
1:8"	2.89	2.52	2.18
1:9"	2.28	1.99	1.72
1:10"	1.85	1.61	1.40
1:11"	1.53	1.33	1.15

Minimum twist required for stability is 1:11.6"
Recommended twist for optimal performance is 1:10.4" or faster

Hornady .284 caliber 139 grain SP Interlock

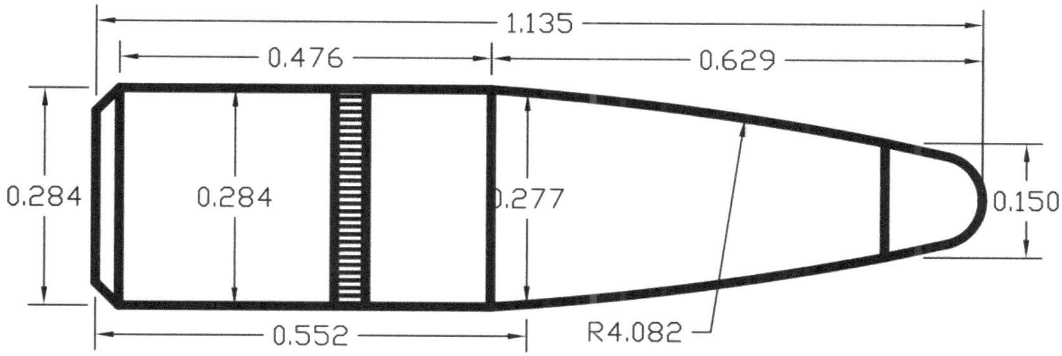

Lot #:	Unknown	Ogive Radius:	4.08 calibers
Sectional Density:	0.246 lb/in²	Rt/R:	0.59

Ballistic Performance

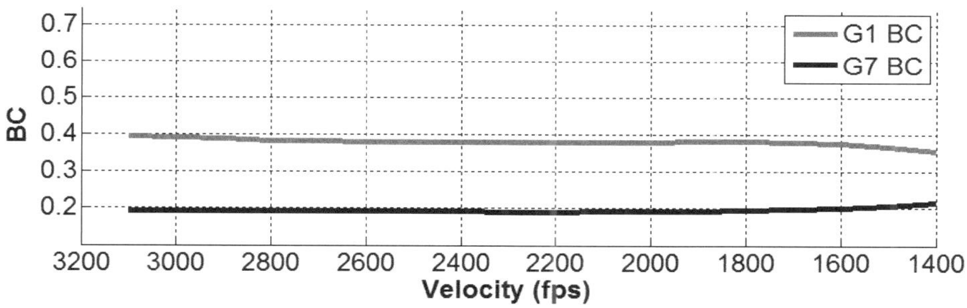

	Above 3000	3000-2500	2500-2000	2000-1500	Below 1500
G1 BC	0.402	0.385	0.380	0.375	0.369
G7 BC	0.190	0.191	0.191	0.200	0.208

Averages from 3000 fps to 1500 fps:
G1 BC: 0.380, variation of G1 BC: 0.022
G7 BC: 0.195, variation of G7 BC: 0.017
G7 form factor: 1.261

Rifling Twist Rate and Stability

	Stability Factor		
Barrel Twist	**Best case**	**Nominal**	**Worst case**
1:8"	3.87	3.38	2.92
1:9"	3.06	2.67	2.31
1:10"	2.48	2.16	1.87
1:11"	2.05	1.79	1.55

Minimum twist required for stability is 1:13.4'
Recommended twist for optimal performance is 1:12.0" or faster

Hornady .284 caliber 154 grain SP Interlock

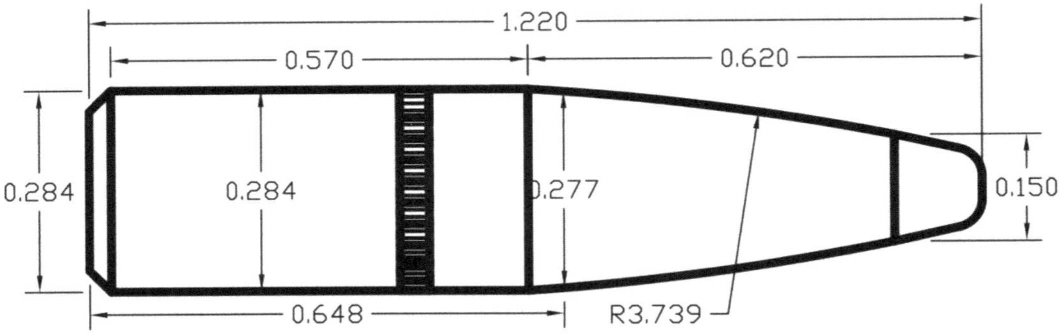

Lot #:	**Unknown**	**Ogive Radius:**	**13.17 calibers**
Sectional Density:	**0.273 lb/in²**	**Rt/R:**	**0.62**

Ballistic Performance

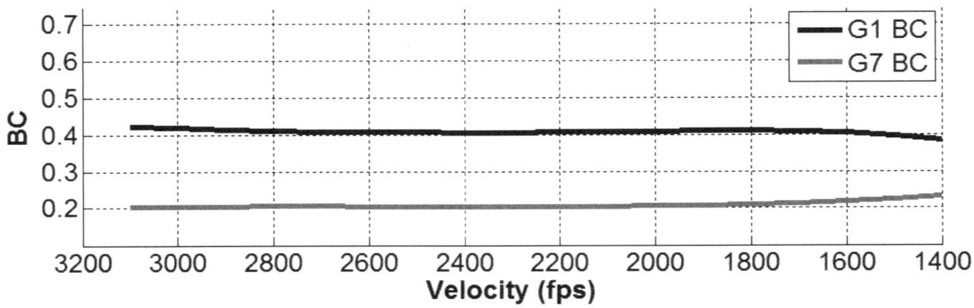

	Above 3000	3000-2500	2500-2000	2000-1500	Below 1500
G1 BC	0.430	0.413	0.409	0.404	0.398
G7 BC	0.204	0.205	0.205	0.215	0.224

Averages from 3000 fps to 1500 fps:
G1 BC: 0.409, variation of G1 BC: 0.022
G7 BC: 0.210, variation of G7 BC: 0.019
G7 form factor: 1.300

Rifling Twist Rate and Stability

	Stability Factor		
Barrel Twist	**Best case**	**Nominal**	**Worst case**
1:8"	3.48	3.04	2.63
1:9"	2.75	2.40	2.08
1:10"	2.23	1.95	1.68
1:11"	1.84	1.61	1.39

Minimum twist required for stability is 1:12.7"
Recommended twist for optimal performance is 1:11.4" or faster

Hornady .284 caliber 139 grain SST BT

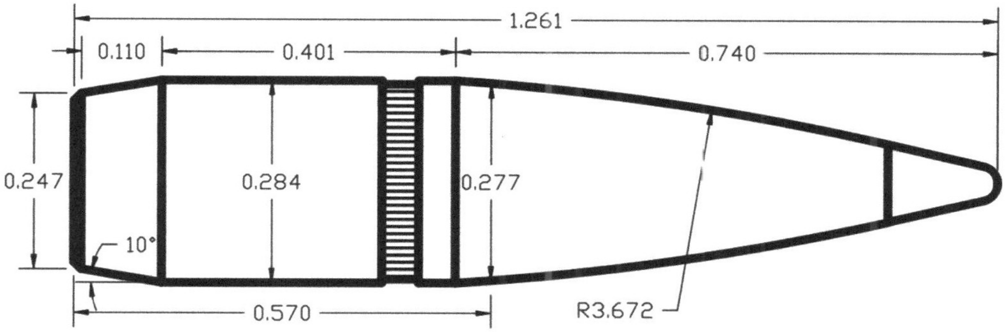

Lot #:	Unknown	Ogive Radius:	12.93 calibers
Sectional Density:	0.246 lb/in^2	Rt/R:	0.56

Ballistic Performance

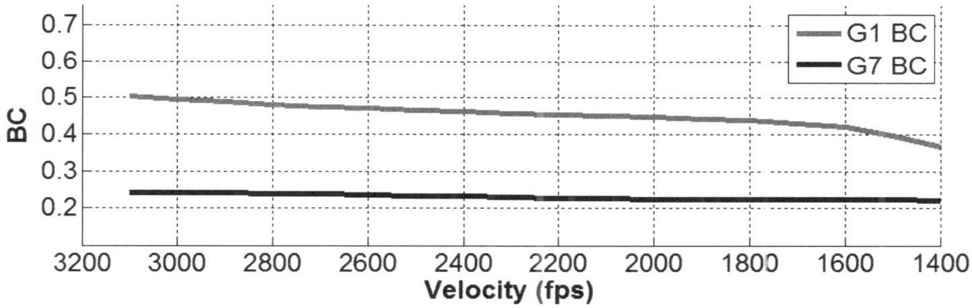

	Above 3000	3000-2500	2500-2000	2000-1500	Below 1500
G1 BC	0.516	0.480	0.456	0.422	0.398
G7 BC	0.244	0.238	0.229	0.224	0.223

Averages from 3000 fps to 1500 fps:
G1 BC: 0.451, variation of G1 BC: 0.097
G7 BC: 0.231, variation of G7 BC: 0.019
G7 form factor: 1.068

Rifling Twist Rate and Stability

Barrel Twist	Stability Factor Best case	Nominal	Worst case
1:8"	2.86	2.49	2.16
1:9"	2.26	1.97	1.70
1:10"	1.83	1.60	1.38
1:11"	1.51	1.32	1.14

Minimum twist required for stability is 1:11.5"
Recommended twist for optimal performance is 1:10.3" or faster

Hornady .284 caliber 154 grain SST BT

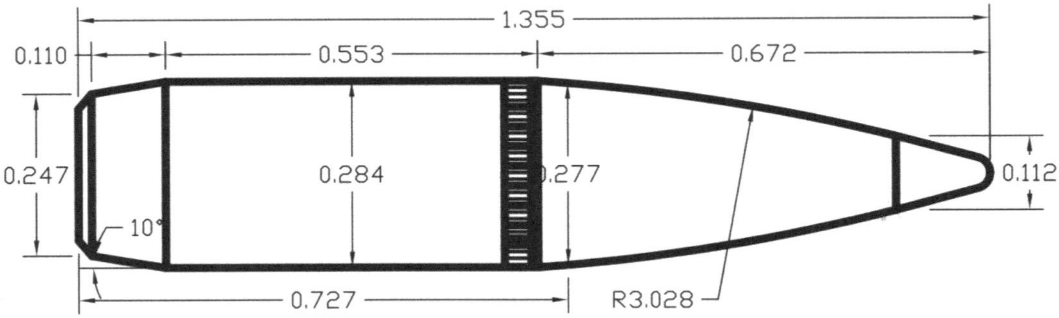

Lot #:	2070270	Ogive Radius:	10.66 calibers
Sectional Density:	0.273 lb/in^2	Rt/R:	0.70

Ballistic Performance

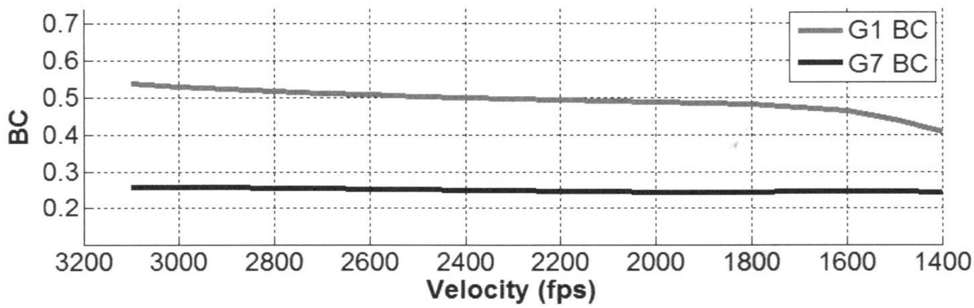

	Above 3000	3000-2500	2500-2000	2000-1500	Below 1500
G1 BC	0.549	0.516	0.496	0.465	0.441
G7 BC	0.260	0.256	0.249	0.247	0.247

Averages from 3000 fps to 1500 fps:
G1 BC: 0.491, variation of G1 BC: 0.089
G7 BC: 0.251, variation of G7 BC: 0.013
G7 form factor: 1.086

Rifling Twist Rate and Stability

Barrel Twist	Stability Factor Best case	Nominal	Worst case
1:8"	2.57	2.24	1.94
1:9"	2.03	1.77	1.53
1:10"	1.64	1.43	1.24
1:11"	1.36	1.19	1.03

Minimum twist required for stability is 1:10.9"
Recommended twist for optimal performance is 1:9.8" or faster

Hornady .284 caliber 162 grain SST BT

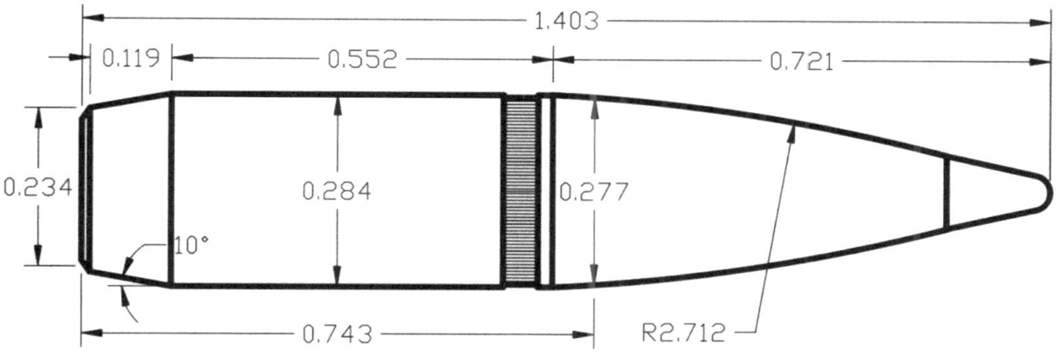

Lot #:	2140586	**Ogive Radius:**	**9.55 calibers**
Sectional Density:	**0.287 lb/in²**	**Rt/R:**	**0.70**

Ballistic Performance

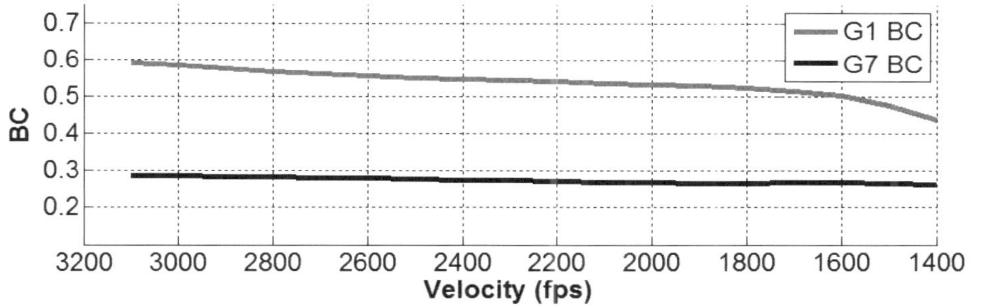

	Above 3000	3000-2500	2500-2000	2000-1500	Below 1500
G1 BC	0.606	0.567	0.541	0.504	0.475
G7 BC	0.287	0.281	0.272	0.267	0.267

Averages from 3000 fps to 1500 fps:
G1 BC: 0.535, variation of G1 BC: 0.108
G7 BC: 0.274, variation of G7 BC: 0.018
G7 form factor: 1.048

Rifling Twist Rate and Stability

Barrel Twist	Stability Factor Best case	Nominal	Worst case
1:8"	2.44	2.13	1.85
1:9"	1.93	1.69	1.46
1:10"	1.56	1.37	1.18
1:11"	1.29	1.13	0.98

Minimum twist required for stability is 1:10.7"
Recommended twist for optimal performance is 1:9.5" or faster

Hornady .284 caliber 120 grain Vmax

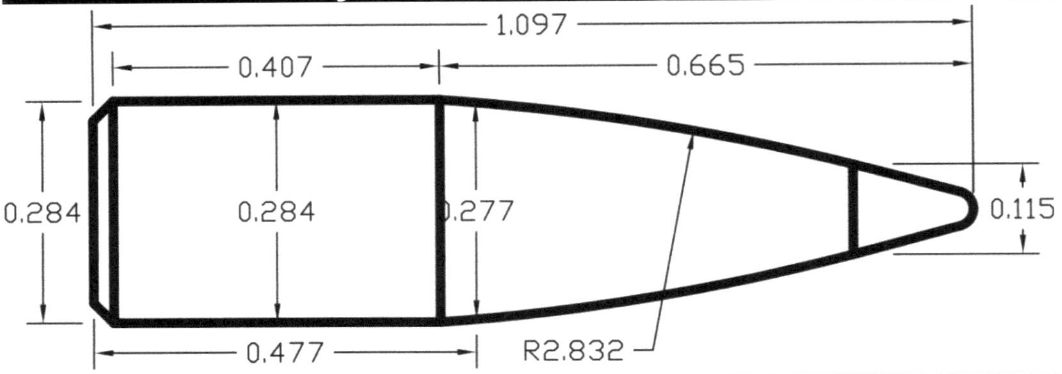

Lot #:	Unknown	Ogive Radius:	9.97 calibers
Sectional Density:	0.213 lb/in^2	Rt/R:	0.65

Ballistic Performance

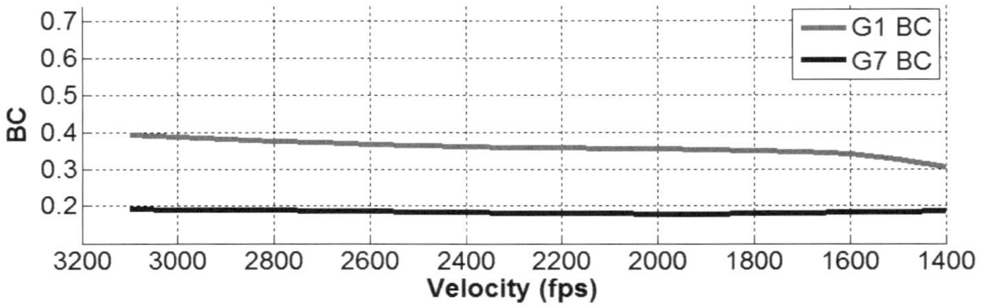

	Above 3000	3000-2500	2500-2000	2000-1500	Below 1500
G1 BC	0.405	0.377	0.360	0.341	0.327
G7 BC	0.192	0.187	0.181	0.181	0.184

Averages from 3000 fps to 1500 fps:
G1 BC: 0.359, variation of G1 BC: 0.061
G7 BC: 0.184, variation of G7 BC: 0.011
G7 form factor: 1.155

Rifling Twist Rate and Stability

Barrel Twist	Stability Factor Best case	Nominal	Worst case
1:8"	3.69	3.22	2.79
1:9"	2.91	2.54	2.20
1:10"	2.36	2.06	1.78
1:11"	1.95	1.70	1.47

Minimum twist required for stability is 1:13.1"
Recommended twist for optimal performance is 1:11.7" or faster

JLK .284 caliber 180 grain BTHP

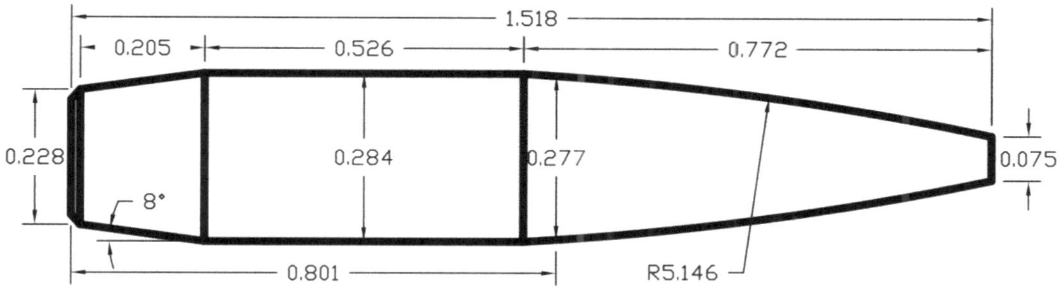

Lot #:	**Unknown**	**Ogive Radius:**	**18.12 calibers**
Sectional Density:	**0.319 lb/in^2**	**Rt/R:**	**0.61**

Ballistic Performance

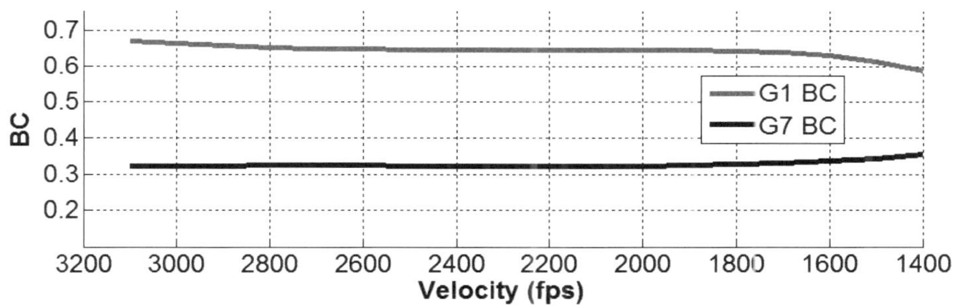

	Above 3000	3000-2500	2500-2000	2000-1500	Below 1500
G1 BC	0.679	0.655	0.647	0.630	0.613
G7 BC	0.322	0.325	0.325	0.335	0.346

Averages from 3000 fps to 1500 fps:
G1 BC: 0.643, variation of G1 BC: 0.050
G7 BC: 0.330, variation of G7 BC: 0.022
G7 form factor: 0.966

Rifling Twist Rate and Stability

Barrel Twist	**Stability Factor** Best case	Nominal	Worst case
1:8"	2.15	1.88	1.63
1:9"	1.70	1.49	1.28
1:10"	1.38	1.20	1.04
1:11"	1.14	0.99	0.86

Minimum twist required for stability is 1:10.0"
Recommended twist for optimal performance is 1:9.0" or faster

Lapua .284 caliber 150 grain Scenar L

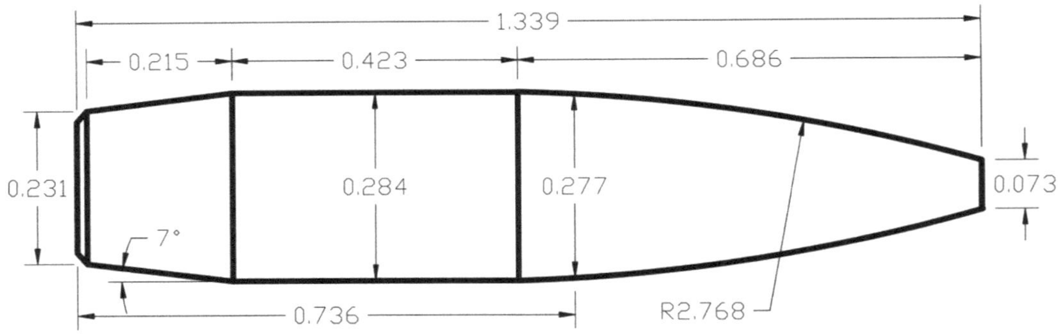

Lot #:	**3124513**	**Ogive Radius:**	**9.75 calibers**
Sectional Density:	**0.266 lb/in^2**	**Rt/R:**	**0.83**

Ballistic Performance

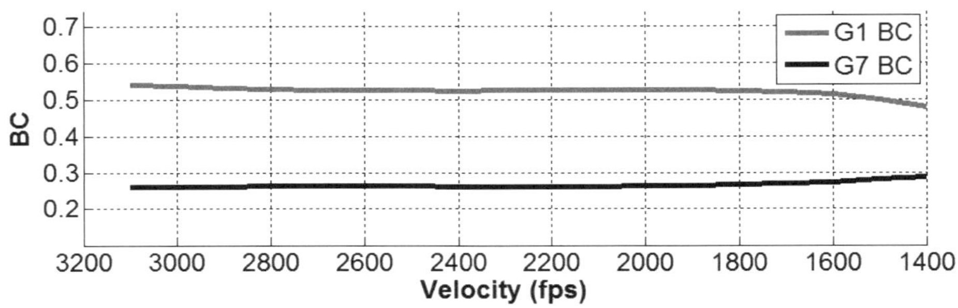

	Above 3000	3000-2500	2500-2000	2000-1500	Below 1500
G1 BC	0.550	0.531	0.525	0.513	0.500
G7 BC	0.261	0.263	0.264	0.273	0.282

Averages from 3000 fps to 1500 fps:
G1 BC: 0.522, variation of G1 BC: 0.038
G7 BC: 0.268, variation of G7 BC: 0.019
G7 form factor: 0.991

Rifling Twist Rate and Stability

Barrel Twist	Stability Factor Best case	Nominal	Worst case
1:8"	2.59	2.26	1.95
1:9"	2.04	1.79	1.54
1:10"	1.66	1.45	1.25
1:11"	1.37	1.20	1.03

Minimum twist required for stability is 1:11.0"
Recommended twist for optimal performance is 1:9.8" or faster

Lapua .284 caliber 180 grain Scenar L

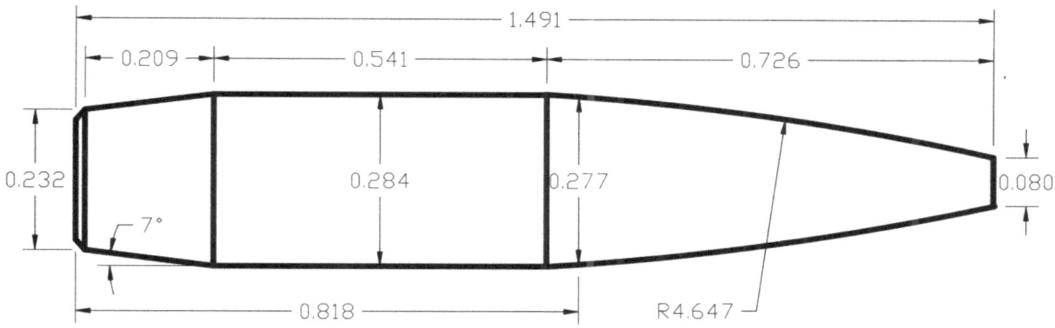

Lot #:	1422114	Ogive Radius:	16.36 calibers
Sectional Density:	0.319 lb/in^2	Rt/R:	0.57

Ballistic Performance

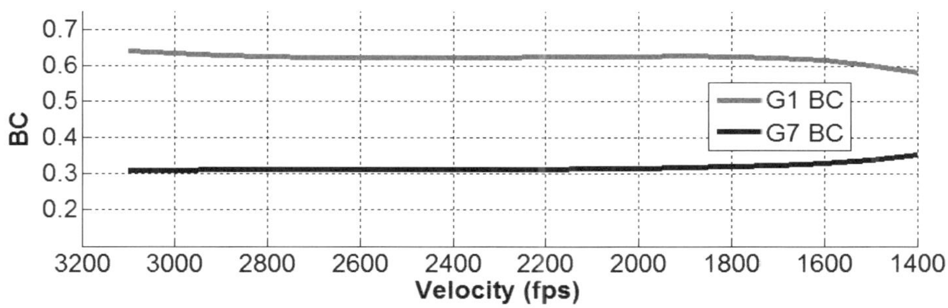

	Above 3000	3000-2500	2500-2000	2000-1500	Below 1500
G1 BC	0.648	0.628	0.624	0.614	0.602
G7 BC	0.307	0.312	0.314	0.327	0.340

Averages from 3000 fps to 1500 fps:
G1 BC: 0.621, variation of G1 BC: 0.033
G7 BC: 0.319, variation of G7 BC: 0.029
G7 form factor: 0.999

Rifling Twist Rate and Stability

Barrel Twist	Stability Factor Best case	Nominal	Worst case
1:8"	2.27	1.98	1.71
1:9"	1.79	1.57	1.35
1:10"	1.45	1.27	1.10
1:11"	1.20	1.05	0.91

Minimum twist required for stability is 1:10.3"
Recommended twist for optimal performance is 1:9.2" or faster

Matrix .284 caliber 175 grain RBT

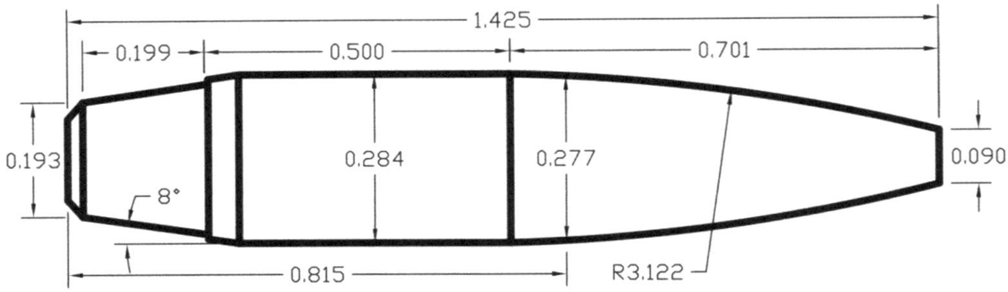

Lot #:	Unknown	Ogive Radius:	10.99 calibers
Sectional Density:	0.310 lb/in^2	Rt/R:	0.89

Ballistic Performance

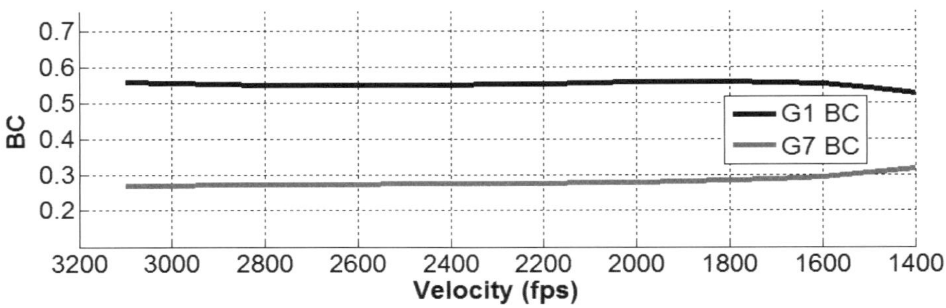

	Above 3000	3000-2500	2500-2000	2000-1500	Below 1500
G1 BC	0.564	0.552	0.553	0.549	0.541
G7 BC	0.268	0.274	0.278	0.293	0.305

Averages from 3000 fps to 1500 fps:

G1 BC: 0.551, variation of G1 BC: 0.017

G7 BC: 0.283, variation of G7 BC: 0.034

G7 form factor: 1.096

Rifling Twist Rate and Stability

Barrel Twist	Stability Factor Best case	Nominal	Worst case
1:8"	2.52	2.20	1.90
1:9"	1.99	1.74	1.50
1:10"	1.61	1.41	1.22
1:11"	1.33	1.16	1.01

Minimum twist required for stability is 1:10.8"

Recommended twist for optimal performance is 1:9.7" or faster

Matrix .284 caliber 168 grain VLD

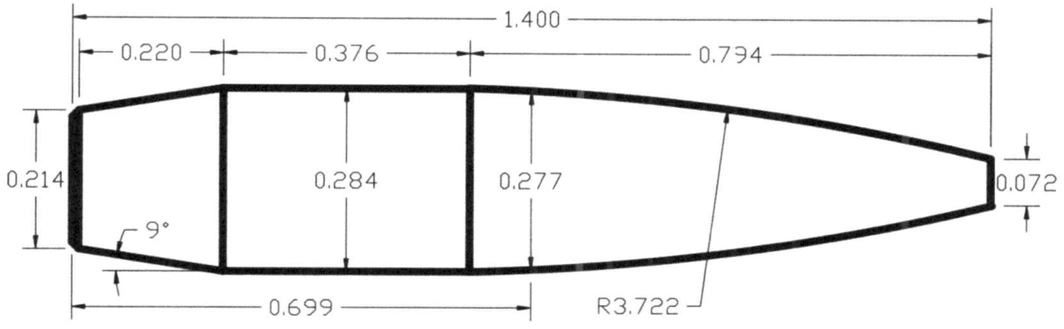

Lot #:	**Unknown**	**Ogive Radius:**	**13.11 calibers**
Sectional Density:	**0.298 lb/in^2**	**Rt/R:**	**0.82**

Ballistic Performance

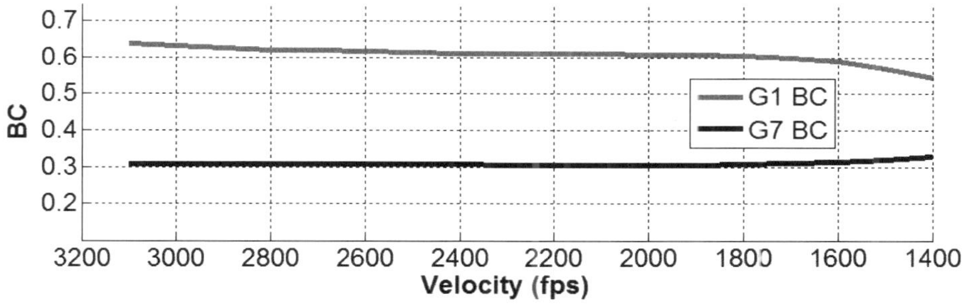

	Above 3000	3000-2500	2500-2000	2000-1500	Below 1500
G1 BC	0.648	0.623	0.611	0.589	0.570
G7 BC	0.307	0.309	0.307	0.314	0.321

Averages from 3000 fps to 1500 fps:
G1 BC: 0.606, variation of G1 BC: 0.062
G7 BC: 0.311, variation of G7 BC: 0.016
G7 form factor: 0.956

Rifling Twist Rate and Stability

Barrel Twist	Stability Factor Best case	Nominal	Worst case
1:8"	2.54	2.22	1.92
1:9"	2.01	1.76	1.52
1:10"	1.63	1.42	1.23
1:11"	1.35	1.18	1.02

Minimum twist required for stability is 1:10.9"
Recommended twist for optimal performance is 1:9.7" or faster

Matrix .284 caliber 190 grain VLD

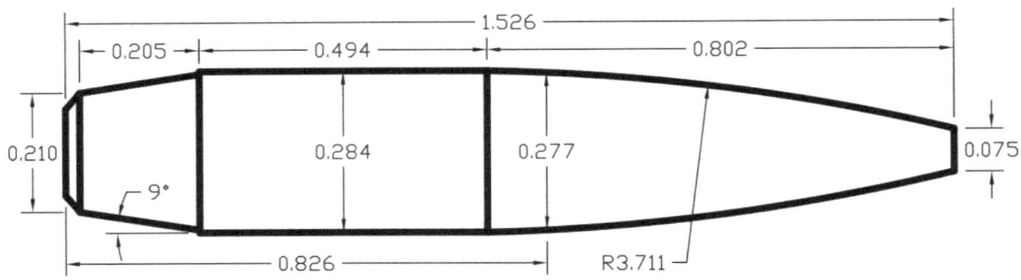

Lot #:	**Unknown**	**Ogive Radius:**	**13.07 calibers**
Sectional Density:	**0.337 lb/in²**	**Rt/R:**	**0.78**

Ballistic Performance

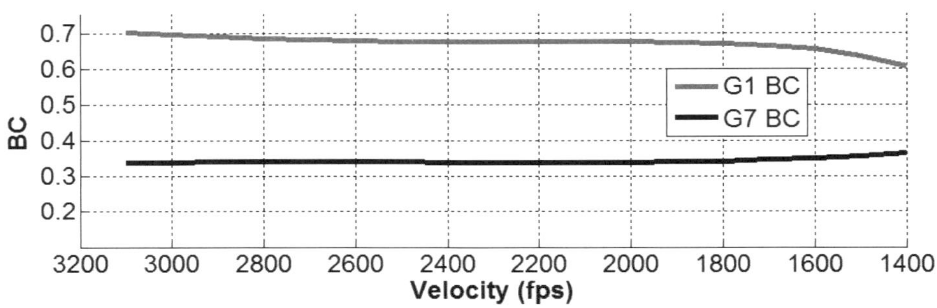

	Above 3000	3000-2500	2500-2000	2000-1500	Below 1500
G1 BC	0.711	0.686	0.675	0.654	0.634
G7 BC	0.337	0.340	0.340	0.348	0.358

Averages from 3000 fps to 1500 fps:
G1 BC: 0.670, variation of G1 BC: 0.060
G7 BC: 0.344, variation of G7 BC: 0.019
G7 form factor: 0.978

Rifling Twist Rate and Stability

Barrel Twist	Best case	Stability Factor Nominal	Worst case
1:8"	2.24	1.95	1.69
1:9"	1.77	1.54	1.33
1:10"	1.43	1.25	1.08
1:11"	1.18	1.03	0.89

Minimum twist required for stability is 1:10.2"
Recommended twist for optimal performance is 1:9.1" or faster

Nosler .284 caliber 150 grain LR Accubond

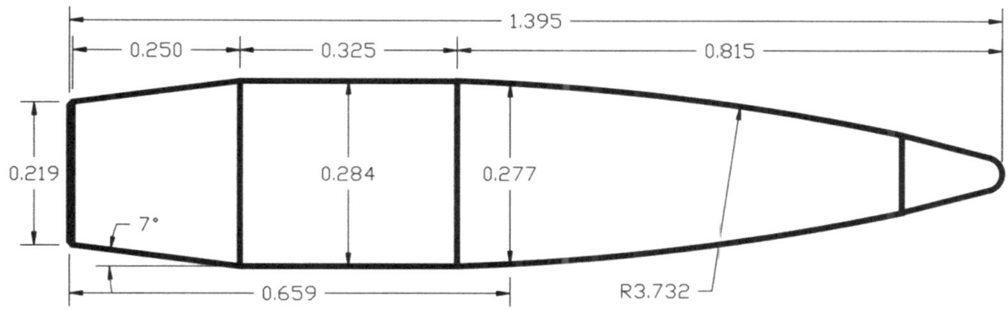

Lot #:	Unknown	Ogive Radius:	13.14 calibers
Sectional Density:	0.266 lb/in^2	Rt/R:	0.72

Ballistic Performance

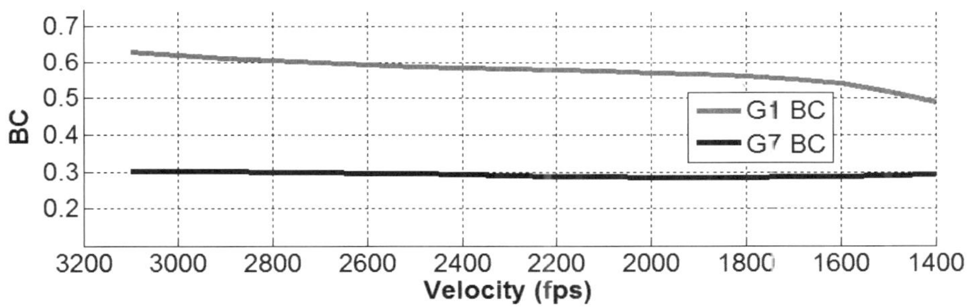

	Above 3000	3000-2500	2500-2000	2000-1500	Below 1500
G1 BC	0.643	0.605	0.580	0.545	0.519
G7 BC	0.305	0.300	0.292	0.289	0.292

Averages from 3000 fps to 1500 fps:
G1 BC: 0.575, variation of G1 BC: 0.102
G7 BC: 0.295, variation of G7 BC: 0.016
G7 form factor: 0.902

Rifling Twist Rate and Stability

Barrel Twist	Stability Factor Best case	Nominal	Worst case
1:8"	2.30	2.01	1.73
1:9"	1.81	1.58	1.37
1:10"	1.47	1.28	1.11
1:11"	1.21	1.06	0.92

Minimum twist required for stability is 1:10.3"
Recommended twist for optimal performance is 1:9.3" or faster

Nosler .284 caliber 168 grain LR Accubond

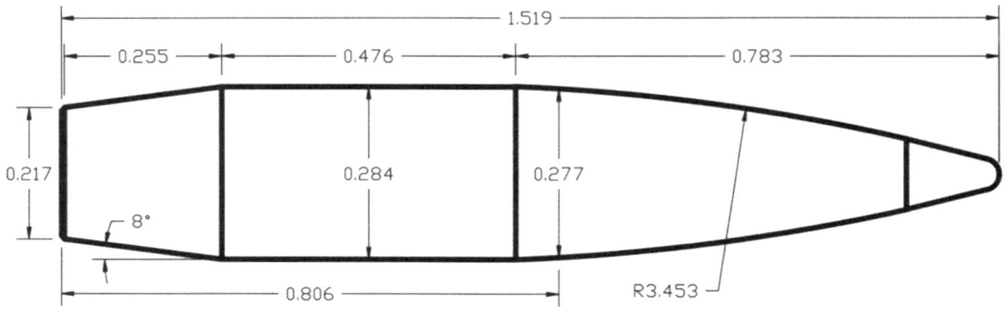

Lot #:	Unknown	Ogive Radius:	12.16 calibers
Sectional Density:	0.298 lb/in^2	Rt/R:	0.70

Ballistic Performance

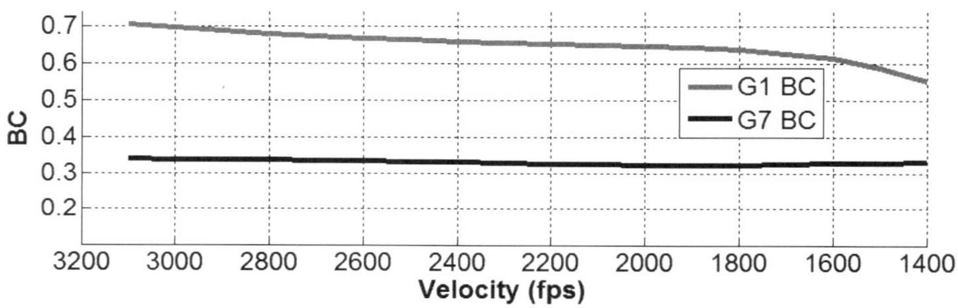

	Above 3000	3000-2500	2500-2000	2000-1500	Below 1500
G1 BC	0.719	0.679	0.654	0.617	0.588
G7 BC	0.341	0.337	0.329	0.327	0.330

Averages from 3000 fps to 1500 fps:
G1 BC: 0.648, variation of G1 BC: 0.108
G7 BC: 0.332, variation of G7 BC: 0.015
G7 form factor: 0.896

Rifling Twist Rate and Stability

Barrel Twist	Stability Factor Best case	Stability Factor Nominal	Stability Factor Worst case
1:8"	2.00	1.75	1.51
1:9"	1.58	1.38	1.20
1:10"	1.28	1.12	0.97
1:11"	1.06	0.93	0.80

Minimum twist required for stability is 1:9.7"
Recommended twist for optimal performance is 1:8.6" or faster

Nosler .284 caliber 175 grain Long Range Accubond

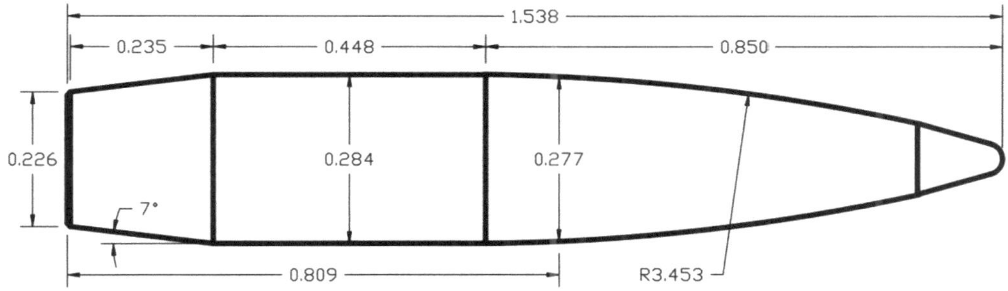

Lot #:	Unknown	Ogive Radius:	12.16 calibers
Sectional Density:	0.310 lb/in^2	Rt/R:	0.90

Ballistic Performance

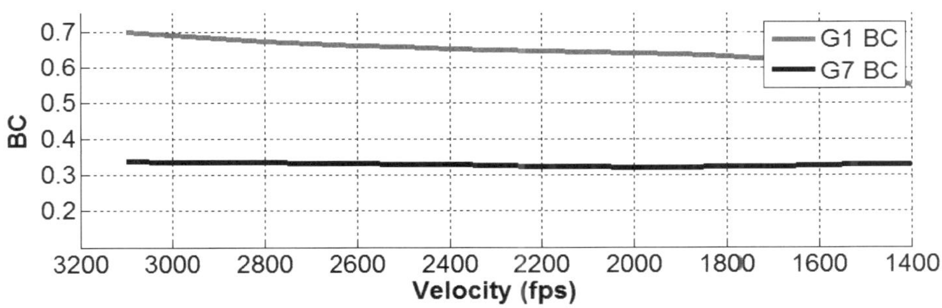

	Above 3000	3000-2500	2500-2000	2000-1500	Below 1500
G1 BC	0.714	0.673	0.649	0.613	0.585
G7 BC	0.338	0.334	0.326	0.326	0.329

Averages from 3000 fps to 1500 fps:
G1 BC: 0.643, variation of G1 BC: 0.104
G7 BC: 0.330, variation of G7 BC: 0.015
G7 form factor: 0.940

Rifling Twist Rate and Stability

Barrel Twist	Stability Factor Best case	Nominal	Worst case
1:8"	2.01	1.76	1.52
1:9"	1.59	1.39	1.20
1:10"	1.29	1.13	0.97
1:11"	1.06	0.93	0.80

Minimum twist required for stability is 1:9.7"
Recommended twist for optimal performance is 1:8.7" or faster

Nosler .284 caliber 140 grain Ballistic Silver Tip

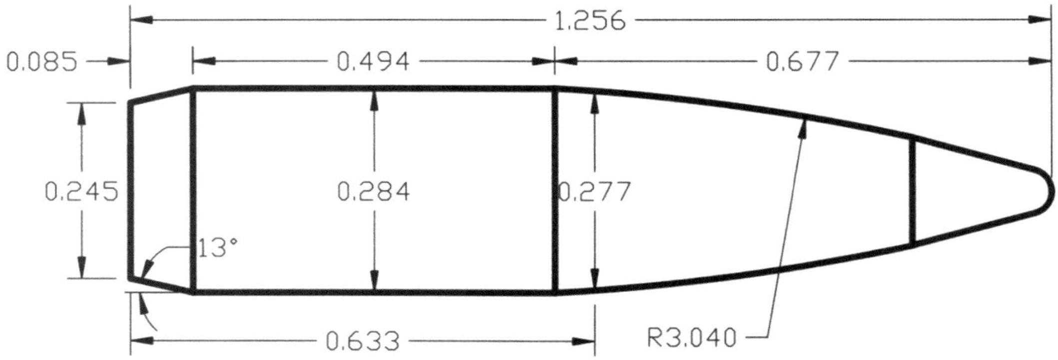

Lot #:	DL14F9	**Ogive Radius:**	**10.70 calibers**
Sectional Density:	**0.248 lb/in^2**	**Rt/R:**	**0.59**

Ballistic Performance

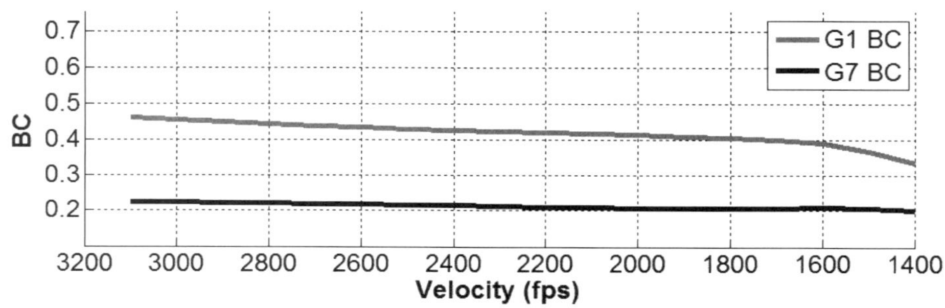

	Above 3000	3000-2500	2500-2000	2000-1500	Below 1500
G1 BC	0.475	0.443	0.422	0.391	0.368
G7 BC	0.225	0.220	0.212	0.207	0.206

Averages from 3000 fps to 1500 fps:
G1 BC: 0.417, variation of G1 BC: 0.089
G7 BC: 0.213, variation of G7 BC: 0.017
G7 form factor: 1.163

Rifling Twist Rate and Stability

Barrel Twist	Stability Factor Best case	Stability Factor Nominal	Stability Factor Worst case
1:8"	2.92	2.55	2.20
1:9"	2.30	2.01	1.74
1:10"	1.87	1.63	1.41
1:11"	1.54	1.35	1.17

Minimum twist required for stability is 1:11.6"
Recommended twist for optimal performance is 1:10.4" or faster

Nosler .284 caliber 120 grain Ballistic Tip

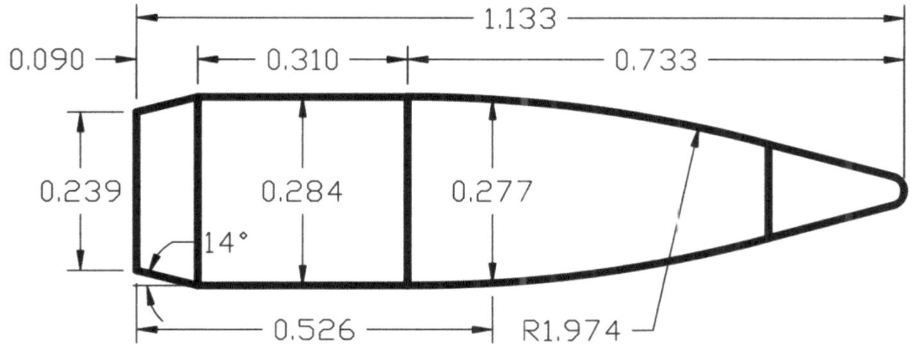

Lot #:	**1119 958 343**	**Ogive Radius:**	**7.0 calibers**
Sectional Density:	**0.213 lb/in^2**	**Rt/R:**	**1.0**

Ballistic Performance

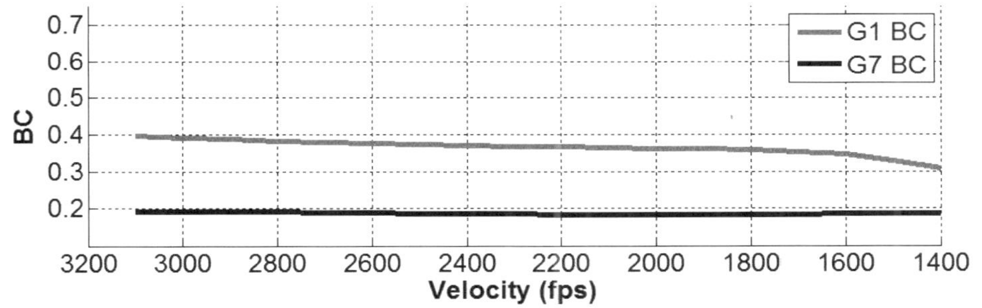

	Above 3000	3000-2500	2500-2000	2000-1500	Below 1500
G1 BC	0.406	0.383	0.368	0.347	0.330
G7 BC	0.192	0.190	0.185	0.184	0.186

Averages from 3000 fps to 1500 fps:
G1 BC: 0.365, variation of G1 BC: 0.062
G7 BC: 0.187, variation of G7 BC: 0.009
G7 form factor: 1.137

Rifling Twist Rate and Stability

Barrel Twist	Stability Factor Best case	Stability Factor Nominal	Stability Factor Worst case
1:8"	3.36	2.93	2.54
1:9"	2.65	2.32	2.01
1:10"	2.15	1.88	1.62
1:11"	1.78	1.55	1.34

Minimum twist required for stability is 1:12.5"
Recommended twist for optimal performance is 1:11.2" or faster

Nosler .284 caliber 150 grain Ballistic Tip

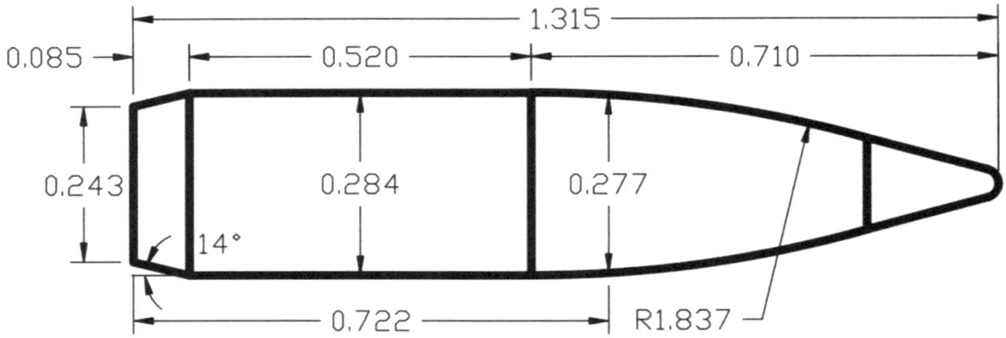

Lot #:	1141 9581165	Ogive Radius:	6.5 calibers
Sectional Density:	0.266 lb/in^2	Rt/R:	0.95

Ballistic Performance

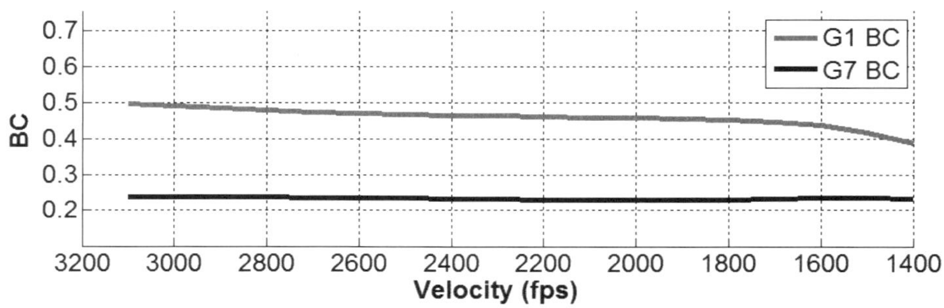

	Above 3000	3000-2500	2500-2000	2000-1500	Below 1500
G1 BC	0.506	0.479	0.463	0.438	0.418
G7 BC	0.240	0.237	0.233	0.233	0.235

Averages from 3000 fps to 1500 fps:
G1 BC: 0.458, variation of G1 BC: 0.071
G7 BC: 0.235, variation of G7 BC: 0.009
G7 form factor: 1.131

Rifling Twist Rate and Stability

Barrel Twist	Stability Factor Best case	Nominal	Worst case
1:8"	2.73	2.38	2.06
1:9"	2.16	1.88	1.63
1:10"	1.75	1.53	1.32
1:11"	1.44	1.26	1.09

Minimum twist required for stability is 1:11.3"
Recommended twist for optimal performance is 1:10.1" or faster

Nosler .284 caliber 168 grain Custom Competition

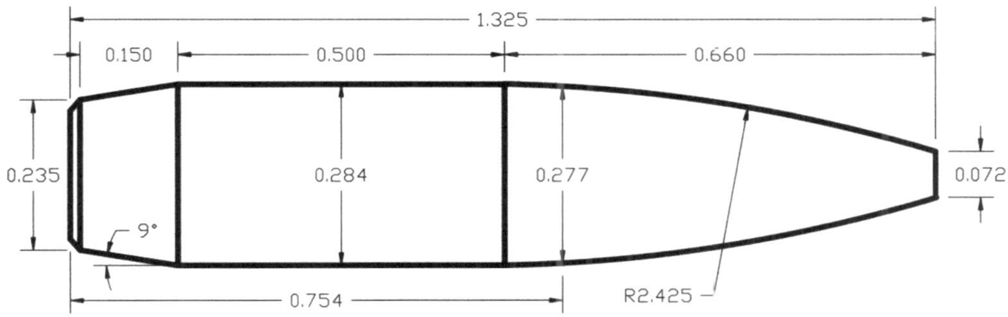

Lot #:	**Unknown**	**Ogive Radius:**	**8.54 calibers**
Sectional Density:	**0.298 lb/in^2**	**Rt/R:**	**0.87**

Ballistic Performance

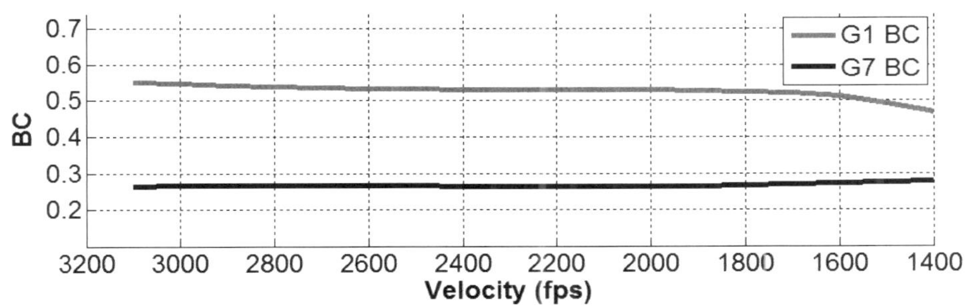

	Above 3000	3000-2500	2500-2000	2000-1500	Below 1500
G1 BC	0.559	0.538	0.529	0.510	0.492
G7 BC	0.265	0.267	0.266	0.271	0.277

Averages from 3000 fps to 1500 fps:
G1 BC: 0.524, variation of G1 BC: 0.054
G7 BC: 0.269, variation of G7 BC: 0.012
G7 form factor: 1.106

Rifling Twist Rate and Stability

Barrel Twist	Stability Factor Best case	Nominal	Worst case
1:8"	2.99	2.61	2.26
1:9"	2.36	2.06	1.78
1:10"	1.91	1.67	1.44
1:11"	1.58	1.38	1.19

Minimum twist required for stability is 1:11.8"
Recommended twist for optimal performance is 1:10.6" or faster

Nosler .284 caliber 140 grain E-Tip

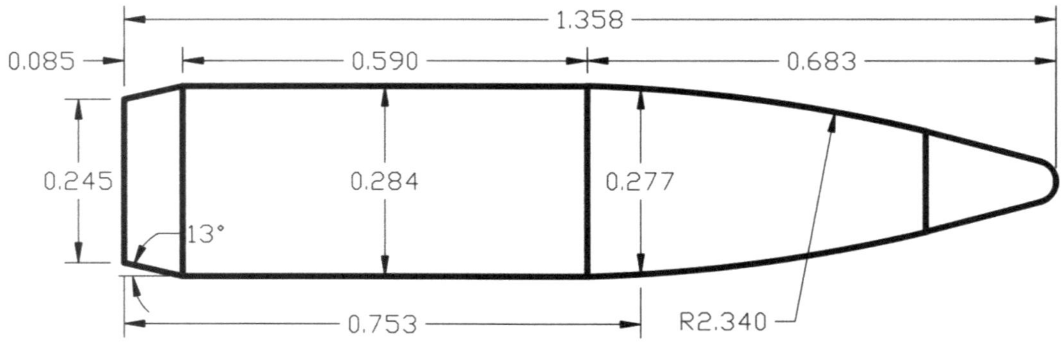

Lot #:	LH02F14	**Ogive Radius:**	**8.24 calibers**
Sectional Density:	**0.248 lb/in²**	**Rt/R:**	**0.79**

Ballistic Performance

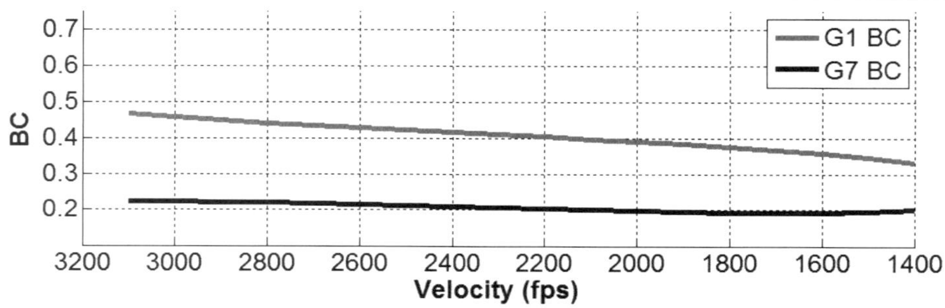

	Above 3000	3000-2500	2500-2000	2000-1500	Below 1500
G1 BC	0.480	0.441	0.407	0.369	0.347
G7 BC	0.227	0.218	0.205	0.196	0.196

Averages from 3000 fps to 1500 fps:
G1 BC: 0.405, variation of G1 BC: 0.111
G7 BC: 0.207, variation of G7 BC: 0.028
G7 form factor: 1.200

Rifling Twist Rate and Stability

Barrel Twist	Stability Factor Best case	Nominal	Worst case
1:8"	2.32	2.03	1.75
1:9"	1.83	1.60	1.38
1:10"	1.48	1.30	1.12
1:11"	1.23	1.07	0.93

Minimum twist required for stability is 1:10.4"
Recommended twist for optimal performance is 1:9.3" or faster

Nosler .284 caliber 150 grain E-Tip

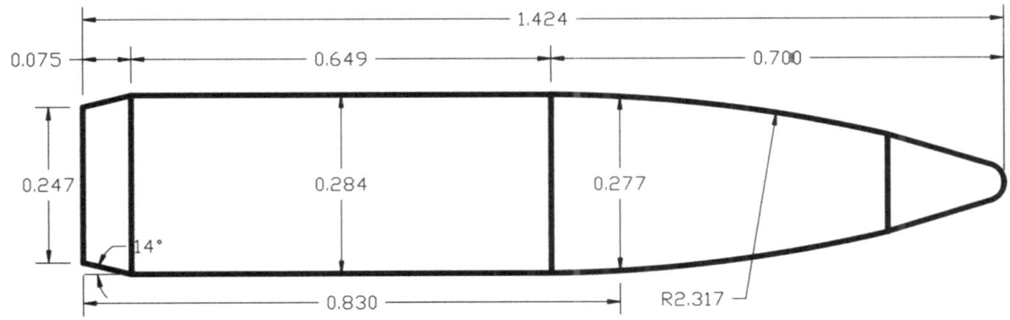

Lot #:	Unknown	Ogive Radius:	8.16 calibers
Sectional Density:	0.266 lb/in^2	Rt/R:	0.92

Ballistic Performance

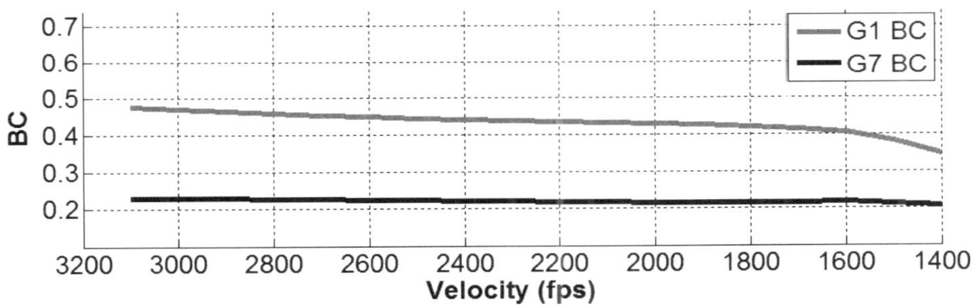

	Above 3000	3000-2500	2500-2000	2000-1500	Below 1500
G1 BC	0.489	0.458	0.437	0.406	0.382
G7 BC	0.231	0.227	0.220	0.215	0.214

Averages from 3000 fps to 1500 fps:
G1 BC: 0.432, variation of G1 BC: 0.089
G7 BC: 0.221, variation of G7 BC: 0.016
G7 form factor: 1.204

Rifling Twist Rate and Stability

Barrel Twist	Stability Factor Best case	Nominal	Worst case
1:8"	2.16	1.89	1.63
1:9"	1.71	1.49	1.29
1:10"	1.38	1.21	1.05
1:11"	1.14	1.00	0.86

Minimum twist required for stability is 1:10.0"
Recommended twist for optimal performance is 1:9.0" or faster

Nosler .284 caliber 150 grain Partition

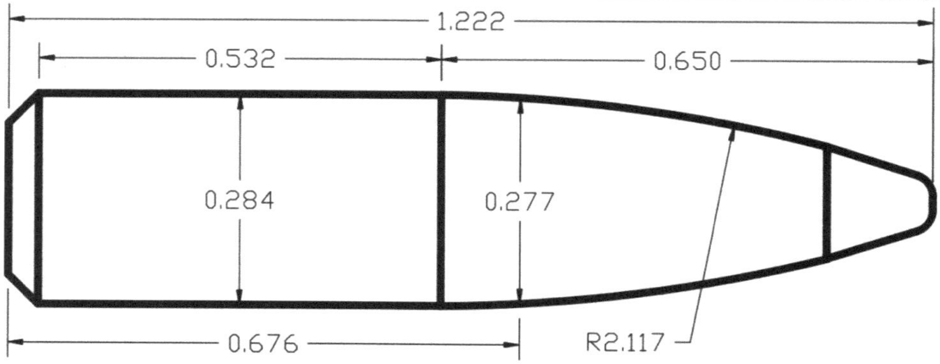

Lot #:	**HC04F11**	**Ogive Radius:**	**7.43 calibers**
Sectional Density:	**0.266 lb/in²**	**Rt/R:**	**0.95**

Ballistic Performance

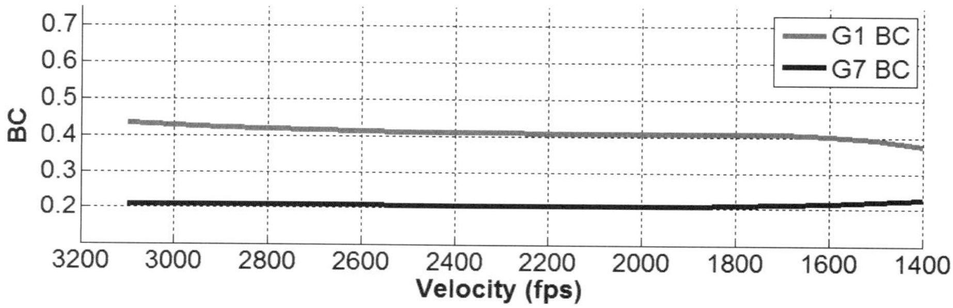

	Above 3000	3000-2500	2500-2000	2000-1500	Below 1500
G1 BC	0.443	0.421	0.411	0.401	0.393
G7 BC	0.210	0.209	0.207	0.214	0.222

Averages from 3000 fps to 1500 fps:
G1 BC: 0.411, variation of G1 BC: 0.037
G7 BC: 0.211, variation of G7 BC: 0.016
G7 form factor: 1.258

Rifling Twist Rate and Stability

Barrel Twist	Stability Factor Best case	Nominal	Worst case
1:8"	3.38	2.95	2.55
1:9"	2.67	2.33	2.01
1:10"	2.16	1.89	1.63
1:11"	1.79	1.56	1.35

Minimum twist required for stability is 1:12.5"
Recommended twist for optimal performance is 1:11.2" or faster

Nosler .284 caliber 175 grain Partition

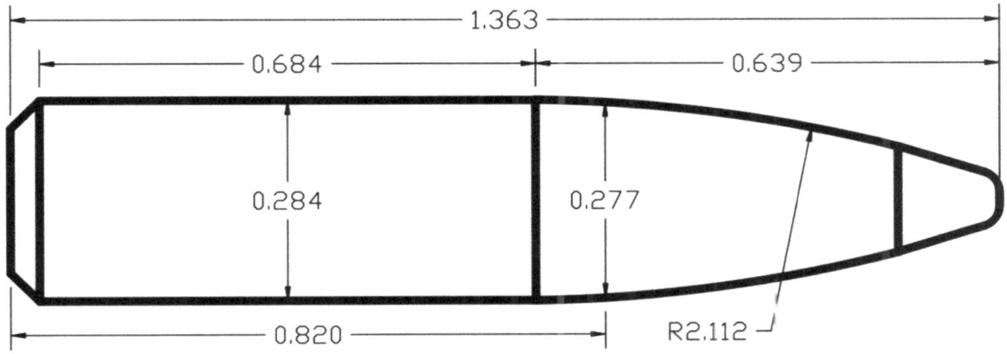

Lot #:	**LC07F1-1**	**Ogive Radius:**	**7.44 calibers**
Sectional Density:	**0.310 lb/in^2**	**Rt/R:**	**0.93**

Ballistic Performance

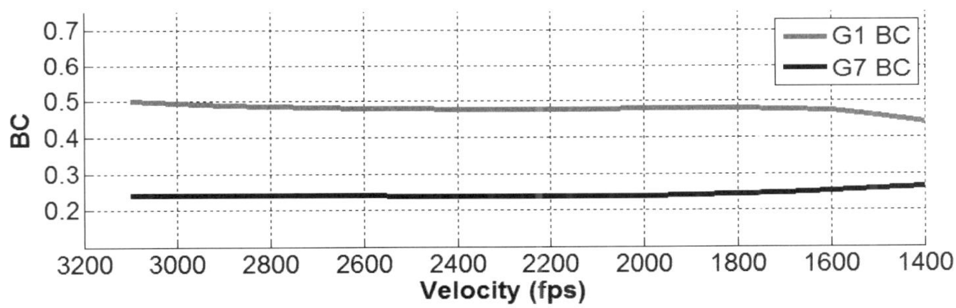

	Above 3000	3000-2500	2500-2000	2000-1500	Below 1500
G1 BC	0.507	0.486	0.478	0.469	0.459
G7 BC	0.240	0.241	0.240	0.250	0.259

Averages from 3000 fps to 1500 fps:
G1 BC: 0.477, variation of G1 BC: 0.034
G7 BC: 0.245, variation of G7 BC: 0.019
G7 form factor: 1.265

Rifling Twist Rate and Stability

Barrel Twist	Stability Factor Best case	Nominal	Worst case
1:8"	2.87	2.50	2.17
1:9"	2.26	1.98	1.71
1:10"	1.83	1.60	1.39
1:11"	1.52	1.32	1.15

Minimum twist required for stability is 1:11.6"
Recommended twist for optimal performance is 1:10.3" or faster

Sierra .284 caliber 150 grain GameKing

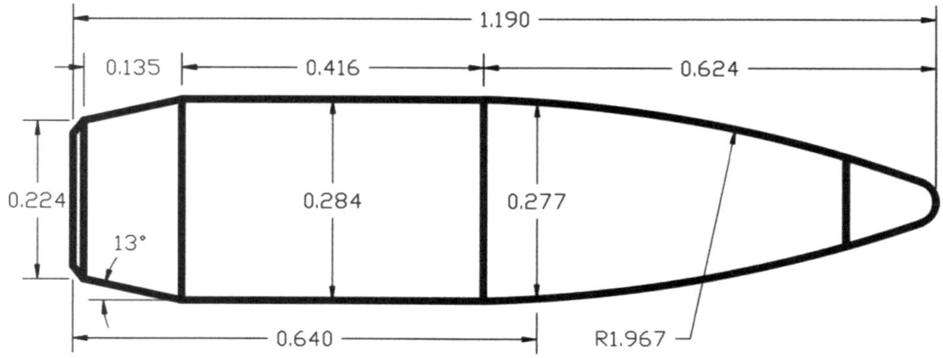

Lot #:	003787195	Ogive Radius:	6.926 calibers
Sectional Density:	0.266 lb/in^2	Rt/R:	0.86

Ballistic Performance

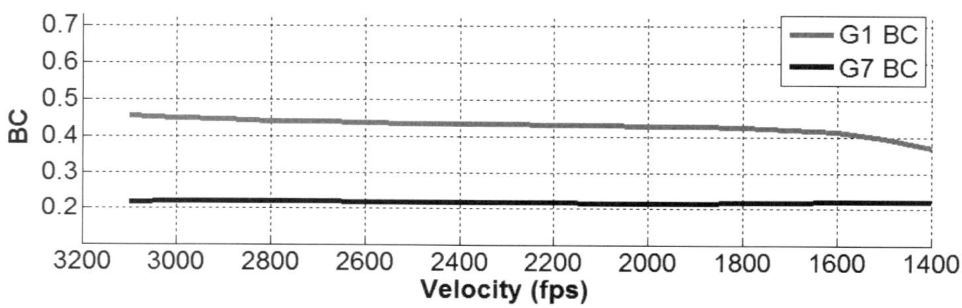

	Above 3000	3000-2500	2500-2000	2000-1500	Below 1500
G1 BC	0.462	0.443	0.433	0.414	0.397
G7 BC	0.219	0.220	0.218	0.220	0.223

Averages from 3000 fps to 1500 fps:
G1 BC: 0.428, variation of G1 BC: 0.054
G7 BC: 0.220, variation of G7 BC: 0.007
G7 form factor: 1.209

Rifling Twist Rate and Stability

Barrel Twist	Stability Factor Best case	Nominal	Worst case
1:8"	3.64	3.18	2.75
1:9"	2.88	2.52	2.18
1:10"	2.33	2.04	1.76
1:11"	1.93	1.68	1.46

Minimum twist required for stability is 1:13.0"
Recommended twist for optimal performance is 1:11.7" or faster

Sierra .284 caliber 175 grain GameKing

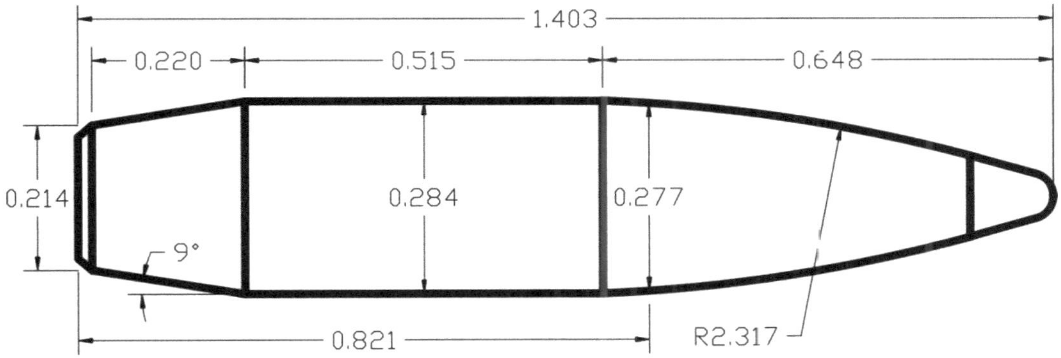

Lot #:	**004437465**	**Ogive Radius:**	**8.16 calibers**
Sectional Density:	**0.310 lb/in²**	**Rt/R:**	**0.90**

Ballistic Performance

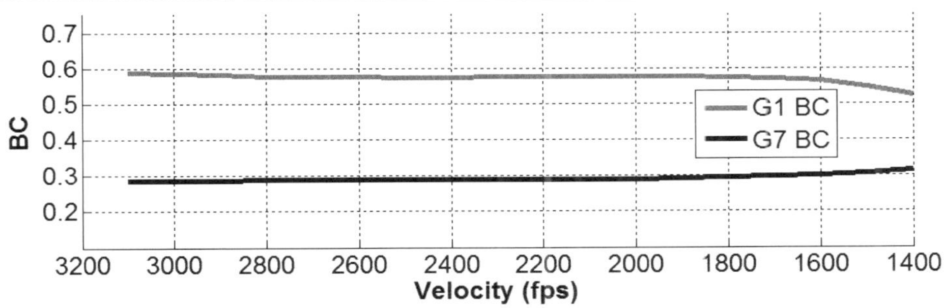

	Above 3000	3000-2500	2500-2000	2000-1500	Below 1500
G1 BC	0.597	0.580	0.575	0.561	0.546
G7 BC	0.283	0.288	0.289	0.299	0.308

Averages from 3000 fps to 1500 fps:
G1 BC: 0.570, variation of G1 BC: 0.039
G7 BC: 0.293, variation of G7 BC: 0.022
G7 form factor: 1.058

Rifling Twist Rate and Stability

Barrel Twist	Stability Factor Best case	Stability Factor Nominal	Stability Factor Worst case
1:8"	2.63	2.30	1.99
1:9"	2.08	1.82	1.57
1:10"	1.69	1.47	1.27
1:11"	1.39	1.22	1.05

Minimum twist required for stability is 1:11.1"
Recommended twist for optimal performance is 1:9.9" or faster

Sierra .284 caliber 140 grain HPBT GameKing

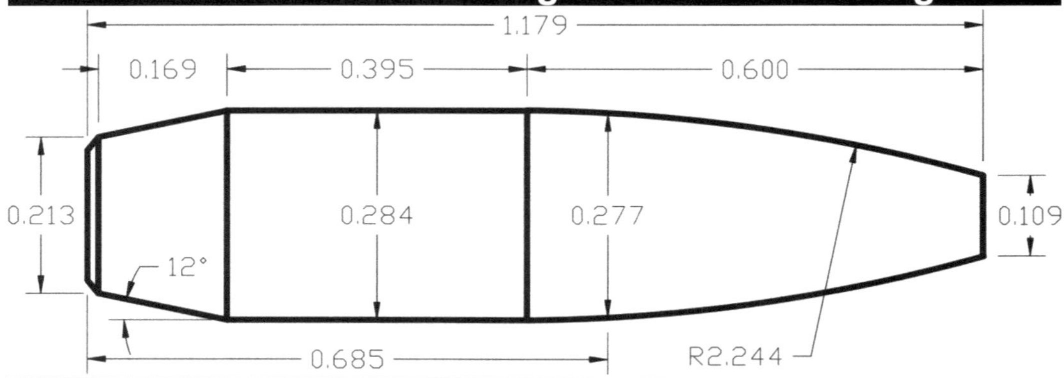

Lot #:	969203158	Ogive Radius:	7.90 calibers
Sectional Density:	0.248 lb/in^2	Rt/R:	0.94

Ballistic Performance

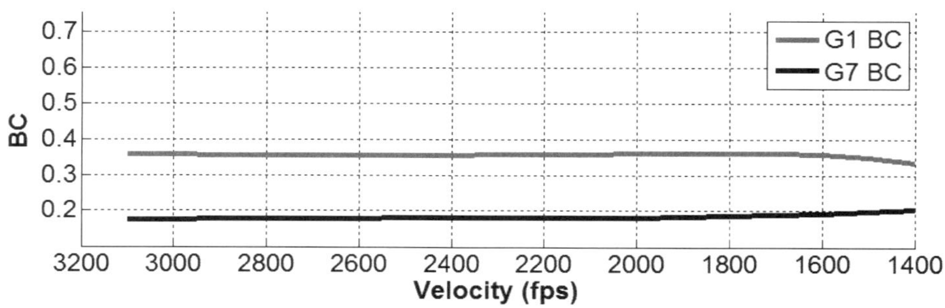

	Above 3000	3000-2500	2500-2000	2000-1500	Below 1500
G1 BC	0.363	0.357	0.359	0.356	0.350
G7 BC	0.172	0.177	0.180	0.189	0.197

Averages from 3000 fps to 1500 fps:
G1 BC: 0.356, variation of G1 BC: 0.012
G7 BC: 0.183, variation of G7 BC: 0.022
G7 form factor: 1.356

Rifling Twist Rate and Stability

Barrel Twist	Stability Factor Best case	Nominal	Worst case
1:10"	2.24	1.95	1.69
1:11"	1.85	1.61	1.40
1:12"	1.55	1.36	1.17
1:13"	1.32	1.16	1.00

Minimum twist required for stability is 1:12.8"
Recommended twist for optimal performance is 1:11.4" or faster

Sierra .284 caliber 160 grain HPBT GameKing

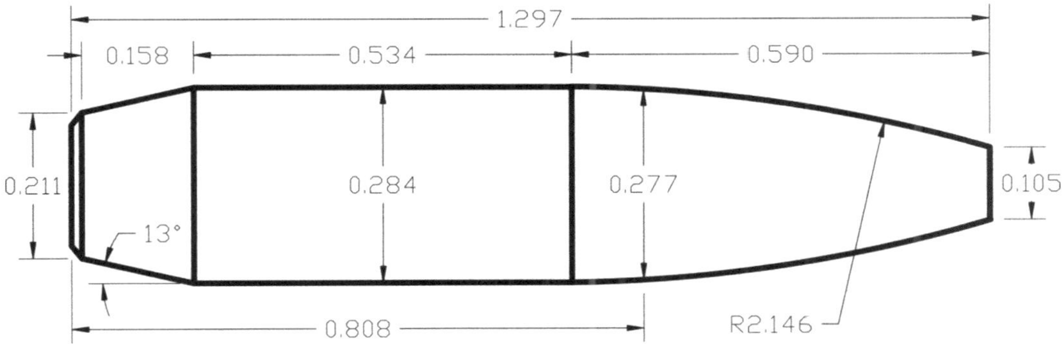

Lot #:	966715233		Ogive Radius:	7.56 calibers
Sectional Density:	0.283 lb/in^2		Rt/R:	0.93

Ballistic Performance

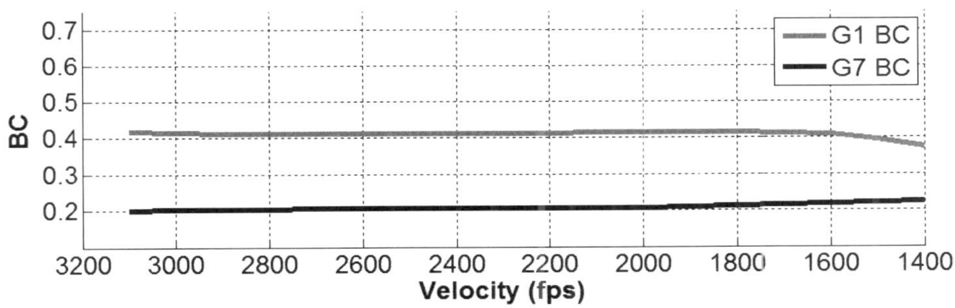

	Above 3000	3000-2500	2500-2000	2000-1500	Below 1500
G1 BC	0.422	0.413	0.413	0.405	0.396
G7 BC	0.200	0.205	0.208	0.216	0.223

Averages from 3000 fps to 1500 fps:
G1 BC: 0.409, variation of G1 BC: 0.020
G7 BC: 0.210, variation of G7 BC: 0.020
G7 form factor: 1.350

Rifling Twist Rate and Stability

Barrel Twist	Best case	Stability Factor Nominal	Worst case
1:9"	2.39	2.09	1.81
1:10"	1.94	1.69	1.46
1:11"	1.60	1.40	1.21
1:12"	1.35	1.18	1.02

Minimum twist required for stability is 1:11.9"
Recommended twist for optimal performance is 1:10.6" or faster

Sierra .284 caliber 140 grain SBT GameKing

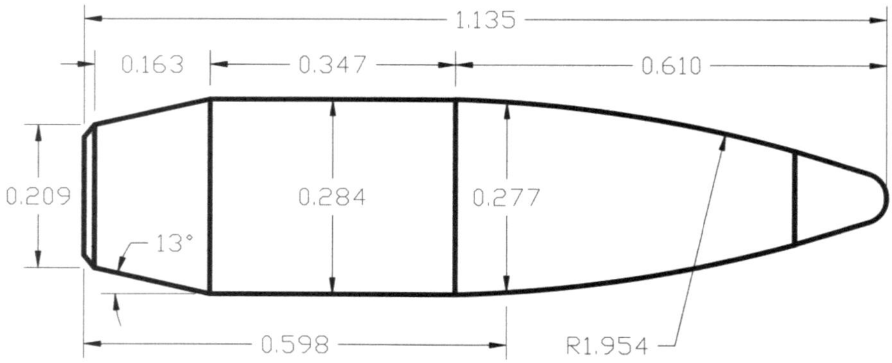

Lot #:	97522958	Ogive Radius:	6.88 calibers
Sectional Density:	0.248 lb/in^2	Rt/R:	0.81

Ballistic Performance

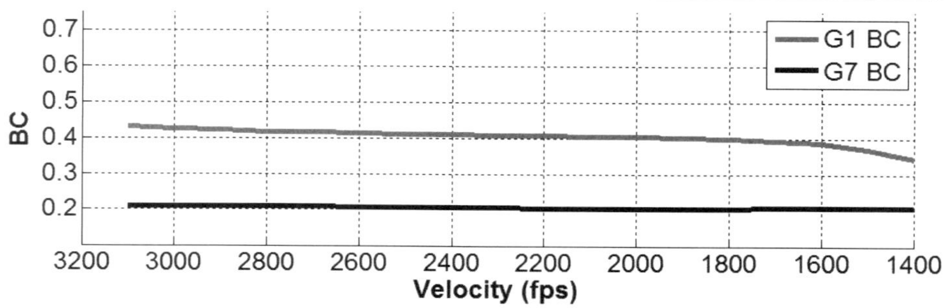

	Above 3000	3000-2500	2500-2000	2000-1500	Below 1500
G1 BC	0.438	0.420	0.409	0.388	0.371
G7 BC	0.208	0.208	0.205	0.206	0.208

Averages from 3000 fps to 1500 fps:
G1 BC: 0.404, variation of G1 BC: 0.056
G7 BC: 0.207, variation of G7 BC: 0.005
G7 form factor: 1.198

Rifling Twist Rate and Stability

Barrel Twist	Stability Factor Best case	Nominal	Worst case
1:11"	2.06	1.80	1.56
1:12"	1.73	1.51	1.31
1:13"	1.48	1.29	1.12
1:14"	1.27	1.11	0.96

Minimum twist required for stability is 1:13.5"
Recommended twist for optimal performance is 1:12.1" or faster

Sierra .284 caliber 130 grain MatchKing

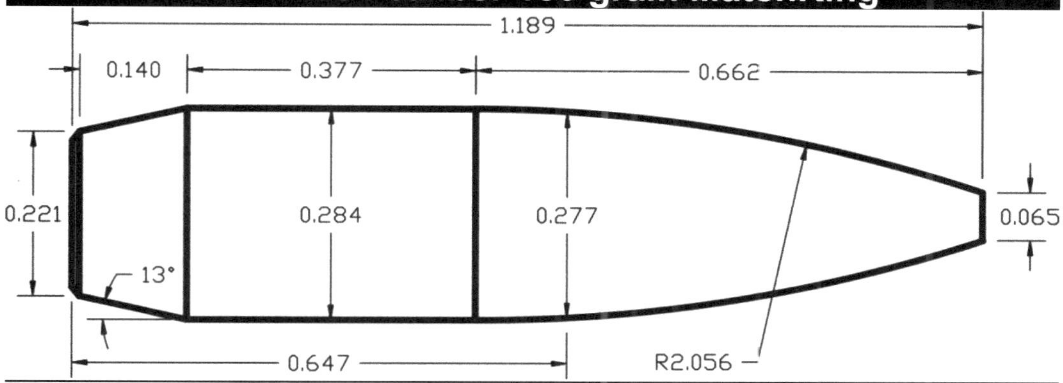

Lot #:	Unknown	Ogive Radius:	7.24 calibers
Sectional Density:	0.230 lb/in^2	Rt/R:	1.00

Ballistic Performance

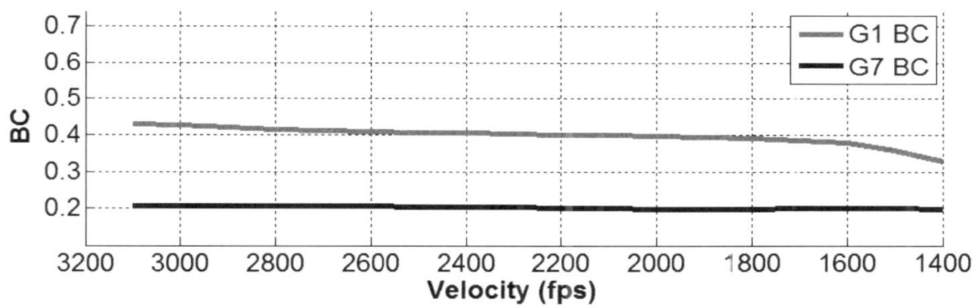

	Above 3000	3000-2500	2500-2000	2000-1500	Below 1500
G1 BC	0.438	0.416	0.402	0.378	0.358
G7 BC	0.208	0.206	0.202	0.200	0.201

Averages from 3000 fps to 1500 fps:
G1 BC: 0.397, variation of G1 BC: 0.067
G7 BC: 0.203, variation of G7 BC: 0.008
G7 form factor: 1.133

Rifling Twist Rate and Stability

Barrel Twist	Stability Factor Best case	Nominal	Worst case
1:8"	3.17	2.77	2.39
1:9"	2.50	2.19	1.89
1:10"	2.03	1.77	1.53
1:11"	1.67	1.46	1.27

Minimum twist required for stability is 1:12.1"
Recommended twist for optimal performance is 1:10.9" or faster

Sierra .284 caliber 150 grain MatchKing

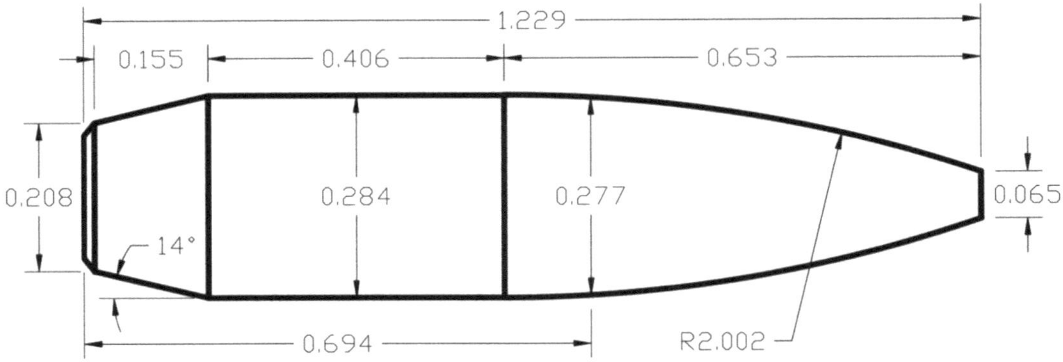

Lot #:	994759956	Ogive Radius:	7.05 calibers
Sectional Density:	0.266 lb/in^2	Rt/R:	1.00

Ballistic Performance

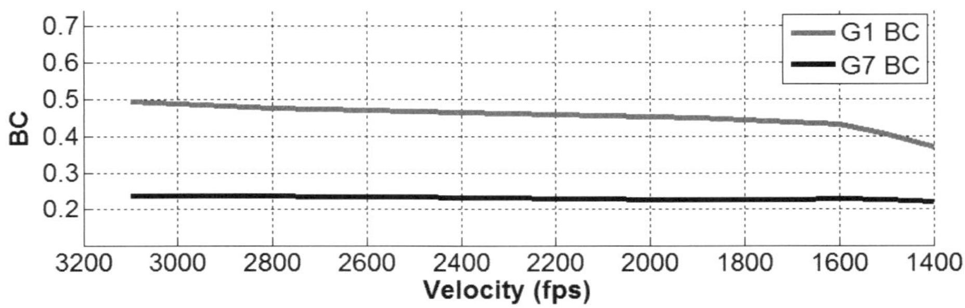

	Above 3000	3000-2500	2500-2000	2000-1500	Below 1500
G1 BC	0.504	0.478	0.460	0.430	0.406
G7 BC	0.239	0.237	0.231	0.228	0.228

Averages from 3000 fps to 1500 fps:
G1 BC: 0.454, variation of G1 BC: 0.082
G7 BC: 0.232, variation of G7 BC: 0.011
G7 form factor: 1.144

Rifling Twist Rate and Stability

Barrel Twist	Stability Factor Best case	Nominal	Worst case
1:10"	2.12	1.86	1.60
1:11"	1.76	1.53	1.33
1:12"	1.48	1.29	1.11
1:13"	1.26	1.10	0.95

Minimum twist required for stability is 1:12.4"
Recommended twist for optimal performance is 1:11.1" or faster

Sierra .284 caliber 168 grain MatchKing

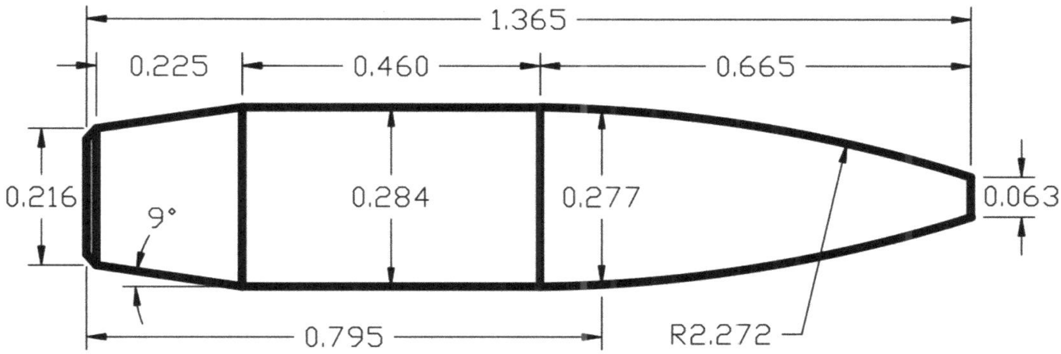

Lot #:	003734142	Ogive Radius:	8.00 calibers
Sectional Density:	0.298 lb/in^2	Rt/R:	0.91

Ballistic Performance

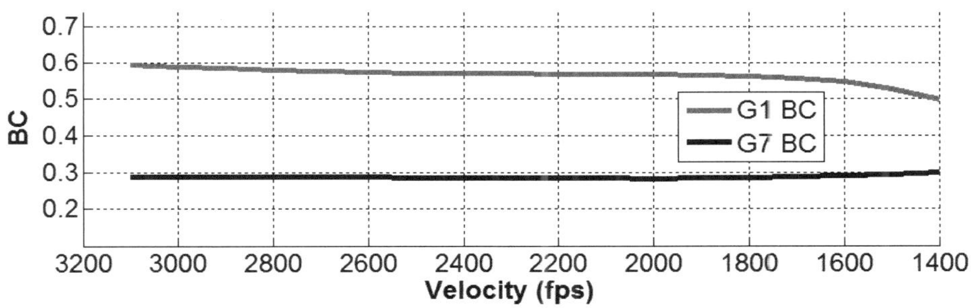

	Above 3000	3000-2500	2500-2000	2000-1500	Below 1500
G1 BC	0.604	0.580	0.569	0.546	0.525
G7 BC	0.286	0.288	0.286	0.290	0.296

Averages from 3000 fps to 1500 fps:
G1 BC: 0.563, variation of G1 BC: 0.063
G7 BC: 0.289, variation of G7 BC: 0.012
G7 form factor: 1.030

Rifling Twist Rate and Stability

Barrel Twist	Stability Factor Best case	Nominal	Worst case
1:8"	2.74	2.39	2.07
1:9"	2.16	1.89	1.64
1:10"	1.75	1.53	1.32
1:11"	1.45	1.27	1.09

Minimum twist required for stability is 1:11.3"
Recommended twist for optimal performance is 1:10.1" or faster

Sierra .284 caliber 175 grain MatchKing

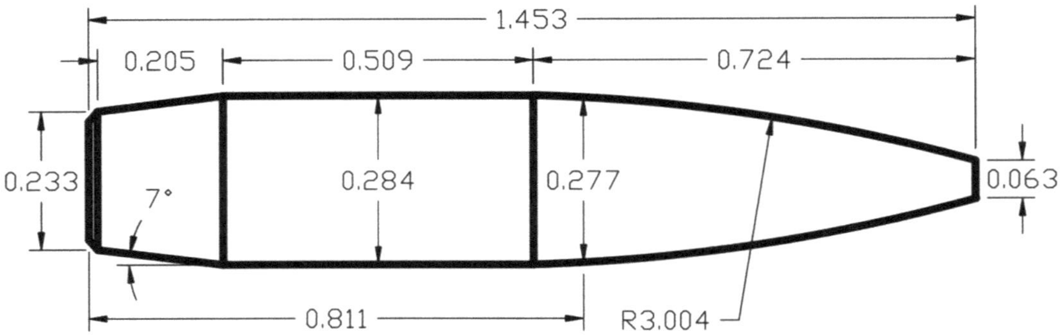

Lot #:	**002906329**	**Ogive Radius:**	**10.58 calibers**
Sectional Density:	**0.310 lb/in²**	**Rt/R:**	**0.83**

Ballistic Performance

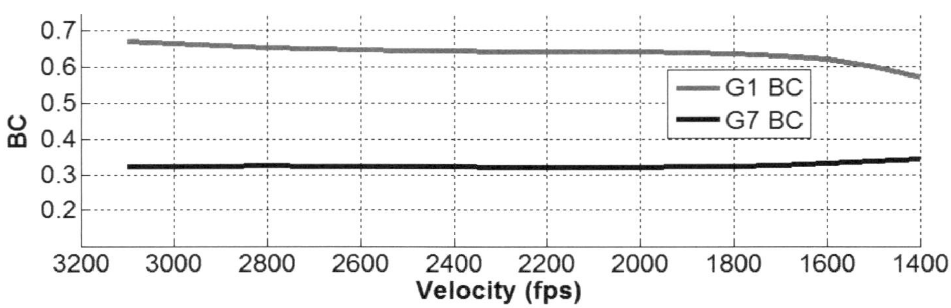

	Above 3000	3000-2500	2500-2000	2000-1500	Below 1500
G1 BC	0.682	0.654	0.642	0.619	0.599
G7 BC	0.323	0.325	0.323	0.329	0.337

Averages from 3000 fps to 1500 fps:
G1 BC: 0.637, variation of G1 BC: 0.066
G7 BC: 0.327, variation of G7 BC: 0.016
G7 form factor: 0.948

Rifling Twist Rate and Stability

Barrel Twist	Stability Factor Best case	Nominal	Worst case
1:8"	2.38	2.08	1.80
1:9"	1.88	1.64	1.42
1:10"	1.52	1.33	1.15
1:11"	1.26	1.10	0.95

Minimum twist required for stability is 1:10.5"
Recommended twist for optimal performance is 1:9.4" or faster

Sierra .284 caliber 180 grain MatchKing

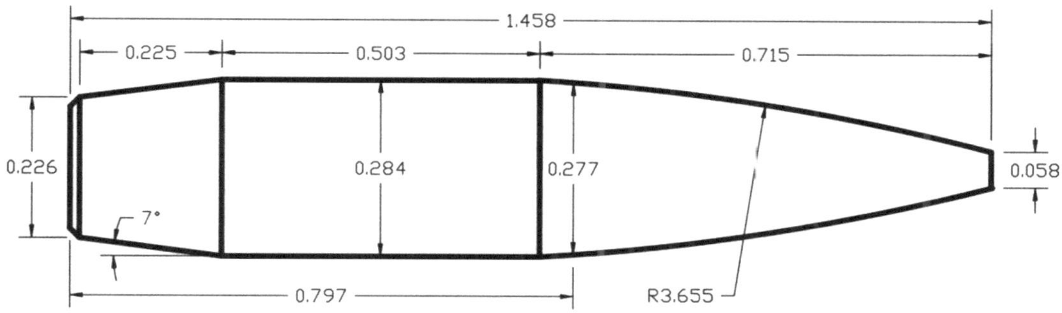

Lot #:	Unknown	Ogive Radius:	12.87 calibers
Sectional Density:	0.319 lb/in^2	Rt/R:	0.63

Ballistic Performance

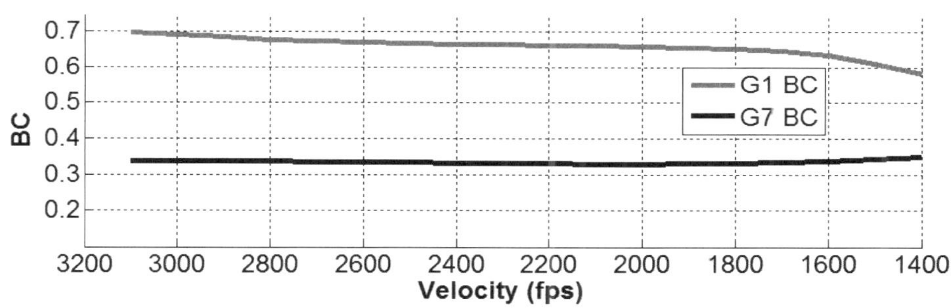

	Above 3000	3000-2500	2500-2000	2000-1500	Below 1500
G1 BC	0.710	0.678	0.662	0.635	0.612
G7 BC	0.336	0.336	0.333	0.338	0.344

Averages from 3000 fps to 1500 fps:
G1 BC: 0.657, variation of G1 BC: 0.079
G7 BC: 0.337, variation of G7 BC: 0.014
G7 form factor: 0.946

Rifling Twist Rate and Stability

Barrel Twist	Stability Factor Best case	Nominal	Worst case
1:8"	2.42	2.12	1.83
1:9"	1.91	1.67	1.45
1:10"	1.55	1.35	1.17
1:11"	1.28	1.12	0.97

Minimum twist required for stability is 1:10.6"
Recommended twist for optimal performance is 1:9.5" or faster

Swift .284 caliber 150 grain Scirocco II

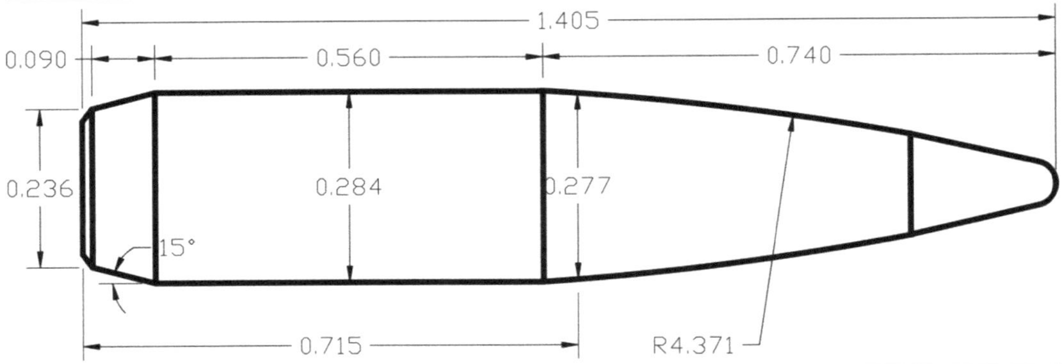

Lot #:	920W	Ogive Radius:	15.39 calibers
Sectional Density:	0.266 lb/in^2	Rt/R:	0.49

Ballistic Performance

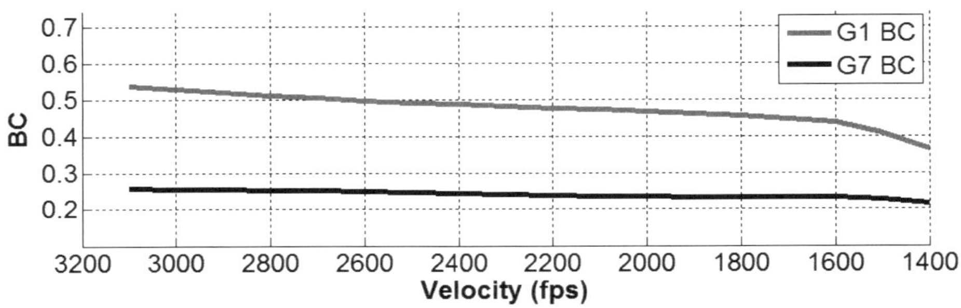

	Above 3000	3000-2500	2500-2000	2000-1500	Below 1500
G1 BC	0.552	0.510	0.480	0.438	0.408
G7 BC	0.261	0.253	0.241	0.231	0.228

Averages from 3000 fps to 1500 fps:
G1 BC: 0.474, variation of G1 BC: 0.120
G7 BC: 0.242, variation of G7 BC: 0.030
G7 form factor: 1.100

Rifling Twist Rate and Stability

Barrel Twist	Stability Factor Best case	Stability Factor Nominal	Stability Factor Worst case
1:8"	2.25	1.96	1.70
1:9"	1.78	1.55	1.34
1:10"	1.44	1.26	1.09
1:11"	1.19	1.04	0.90

Minimum twist required for stability is 1:10.2"
Recommended twist for optimal performance is 1:9.2" or faster

Wildcat .284 caliber 200 grain ULD

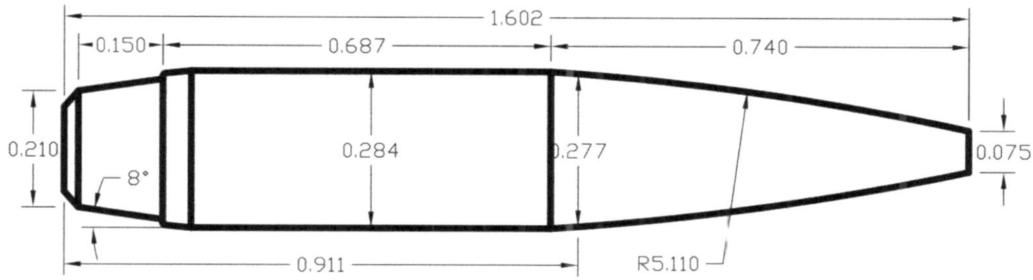

Lot #:	Unknown	Ogive Radius:	17.99 calibers
Sectional Density:	0.354 lb/in^2	Rt/R:	0.58

Ballistic Performance

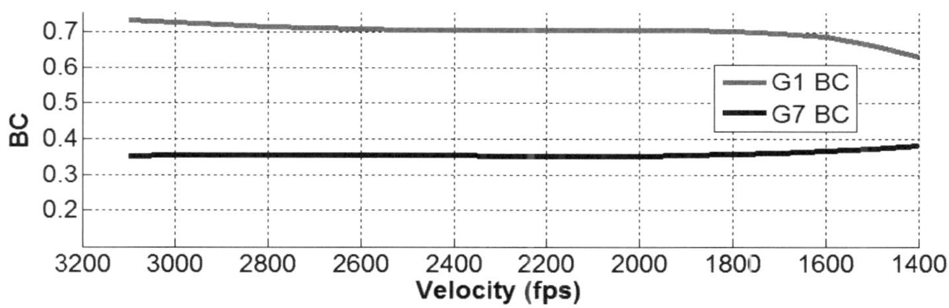

	Above 3000	3000-2500	2500-2000	2000-1500	Below 1500
G1 BC	0.744	0.716	0.705	0.685	0.665
G7 BC	0.353	0.355	0.355	0.364	0.374

Averages from 3000 fps to 1500 fps:
G1 BC: 0.700, variation of G1 BC: 0.061
G7 BC: 0.360, variation of G7 BC: 0.020
G7 form factor: 0.985

Rifling Twist Rate and Stability

Barrel Twist	Stability Factor Best case	Nominal	Worst case
1:8"	2.04	1.78	1.54
1:9"	1.61	1.41	1.22
1:10"	1.31	1.14	0.99
1:11"	1.08	0.94	0.82

Minimum twist required for stability is 1:9.8"
Recommended twist for optimal performance is 1:8.7" or faster

.308 Caliber (7.62mm)

.308 caliber bullets typically range from 140 to 230 grains. The image below relates bullet weight to sectional density for .308 caliber bullets.

Range of Weight and Sectional Density for 0.308 Caliber

Bullet Weight (grains)	140	155	170	185	200	215	230
Sectional Density (lb/in^2)	0.211	0.233	0.256	0.279	0.301	0.324	0.346

Sectional Density (SD) is a measure of how much mass is behind the bullets frontal area. Higher SD means a greater ability for the bullet to penetrate targets on impact, and also retain velocity in flight.

Spectrum of G7 form factors

0.80 0.85 0.90 0.95 1.00 1.05 1.10 1.15 1.20

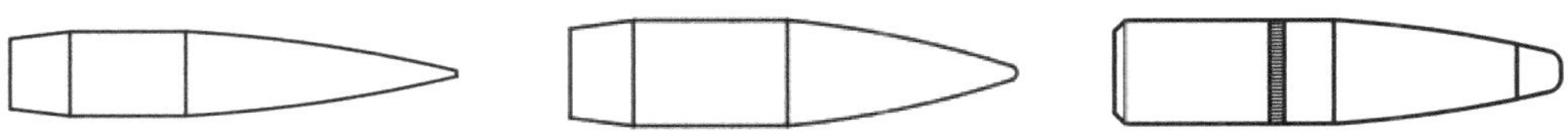

G7 Form factors range from 0.80 for the lowest drag shapes, up to 1.20+ for blunter shapes.

G7 Ballistic Coefficients are simply Sectional Density (SD) divided by G7 form factor.

You can determine a bullets' suitability for long range performance by seeing where it lies on the SD and form factor lines. Ideally, you want a bullet with a high SD and a low G7 form factor.

As an example, a 185 grain .308 caliber bullet has an SD of 0.279. If that bullet had a G7 form factor of 0.95, it's G7 BC would be:

G7 BC = SD/FF = 0.279/0.95 = 0.294.

Alco .308 caliber 220 grain Tipped RBBT

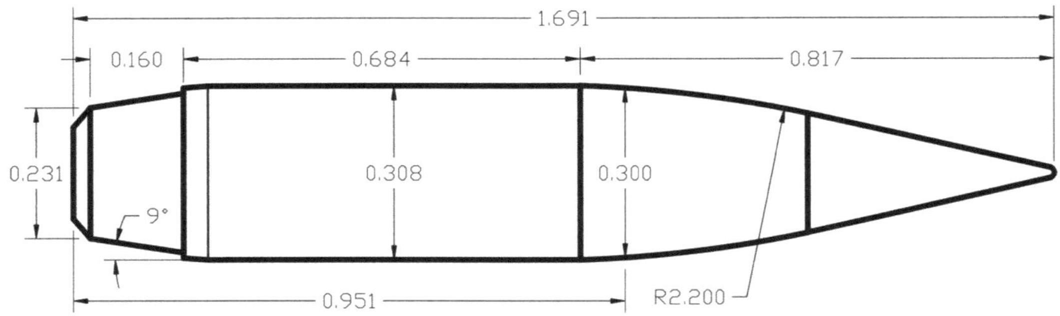

Lot #:	unknown	Ogive Radius:	7.14 calibers
Sectional Density:	0.331 lb/in²	Rt/R:	1.00

Ballistic Performance

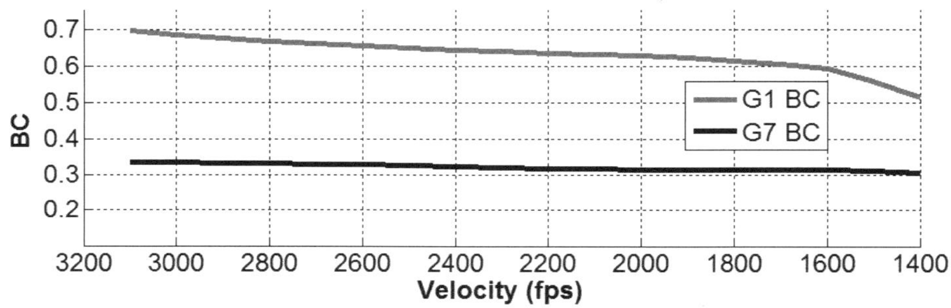

	Above 3000	3000-2500	2500-2000	2000-1500	Below 1500
G1 BC	0.711	0.667	0.638	0.593	0.558
G7 BC	0.337	0.331	0.321	0.314	0.313

Averages from 3000 fps to 1500 fps:
G1 BC: 0.630, variation of G1 BC: 0.127
G7 BC: 0.322, variation of G7 BC: 0.022
G7 form factor: 1.029

Rifling Twist Rate and Stability

Barrel Twist	Stability Factor Best case	Stability Factor Nominal	Stability Factor Worst case
1:8"	2.24	1.96	1.69
1:9"	1.77	1.55	1.34
1:10"	1.43	1.25	1.08
1:11"	1.19	1.04	0.90

Minimum twist required for stability is 1:10.2"
Recommended twist for optimal performance is 1:9.1" or faster

Barnes .308 caliber 175 grain LRXBT

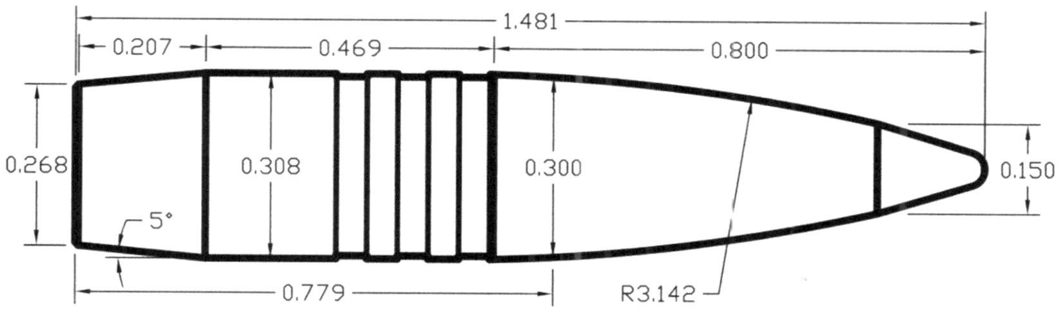

Lot #:	5293-1 104N AH-M	Ogive Radius:	10.20 calibers
Sectional Density:	**0.264 lb/in²**	**Rt/R:**	**0.80**

Ballistic Performance

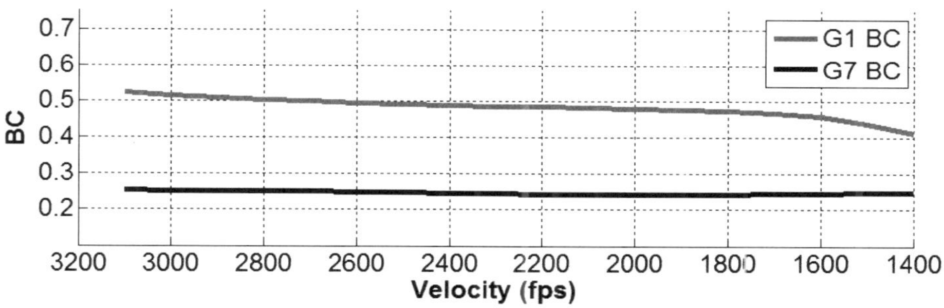

	Above 3000	3000-2500	2500-2000	2000-1500	Below 1500
G1 BC	0.534	0.503	0.486	0.460	0.439
G7 BC	0.253	0.250	0.244	0.244	0.247

Averages from 3000 fps to 1500 fps:
G1 BC: 0.481, variation of G1 BC: 0.077
G7 BC: 0.247, variation of G7 BC: 0.011
G7 form factor: 1.068

Rifling Twist Rate and Stability

Barrel Twist	Stability Factor Best case	Nominal	Worst case
1:10"	1.68	1.47	1.27
1:11"	1.39	1.21	1.05
1:12"	1.17	1.02	0.88
1:13"	1.00	0.87	0.75

Minimum twist required for stability is 1:11.1"
Recommended twist for optimal performance is 1:9.9" or faster

Barnes .308 caliber 200 grain LRXBT

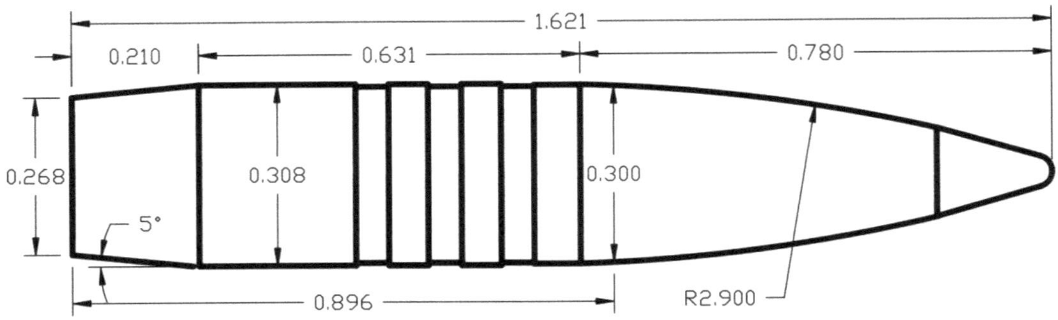

Lot #: 5141D-1 149M	**Ogive Radius:** **9.42 calibers**
Sectional Density: **0.301 lb/in^2**	**Rt/R:** **0.80**

Ballistic Performance

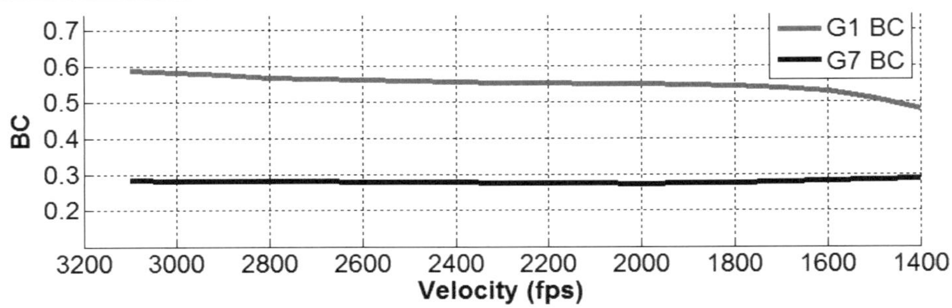

	Above 3000	3000-2500	2500-2000	2000-1500	Below 1500
G1 BC	0.601	0.569	0.553	0.528	0.508
G7 BC	0.284	0.282	0.278	0.281	0.286

Averages from 3000 fps to 1500 fps:
G1 BC: 0.549, variation of G1 BC: 0.074
G7 BC: 0.281, variation of G7 BC: 0.010
G7 form factor: 1.070

Rifling Twist Rate and Stability

	Stability Factor		
Barrel Twist	**Best case**	**Nominal**	**Worst case**
1:8"	2.31	2.02	1.74
1:9"	1.82	1.59	1.38
1:10"	1.48	1.29	1.12
1:11"	1.22	1.07	0.92

Minimum twist required for stability is 1:10.4"
Recommended twist for optimal performance is 1:9.3" or faster

Barnes .308 caliber 155 grain Match Burner

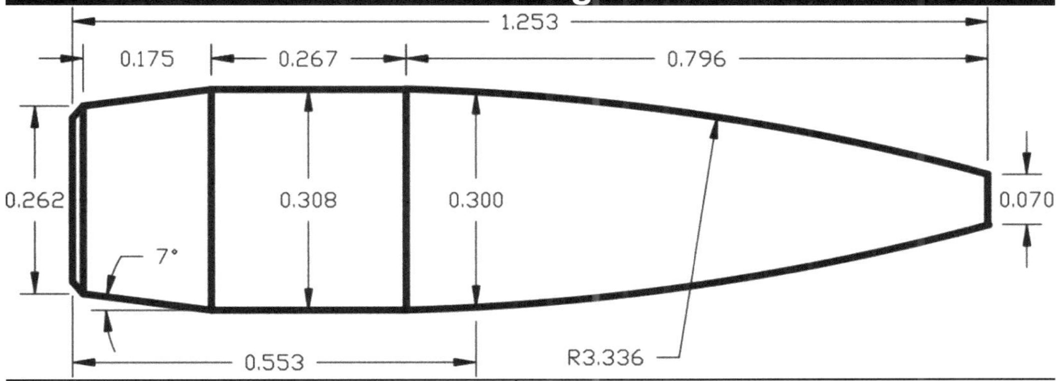

Lot #: 5206A-1 280-M CDW-M		**Ogive Radius:**	**10.83 calibers**
Sectional Density:	**0.233 lb/in²**	**Rt/R:**	**0.82**

Ballistic Performance

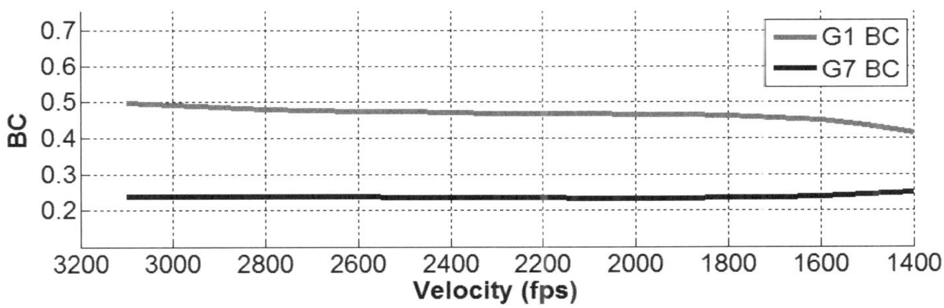

	Above 3000	3000-2500	2500-2000	2000-1500	Below 1500
G1 BC	0.506	0.481	0.469	0.450	0.434
G7 BC	0.240	0.239	0.236	0.239	0.244

Averages from 3000 fps to 1500 fps:
G1 BC: 0.465, variation of G1 BC: 0.057
G7 BC: 0.239, variation of G7 BC: 0.011
G7 form factor: 0.977

Rifling Twist Rate and Stability

Barrel Twist	Stability Factor Best case	Nominal	Worst case
1:10"	2.42	2.11	1.83
1:11"	2.00	1.75	1.51
1:12"	1.68	1.47	1.27
1:13"	1.43	1.25	1.08

Minimum twist required for stability is 1:13.3'
Recommended twist for optimal performance is 1:11.9" or faster

Barnes .308 caliber 175 grain Match Burner

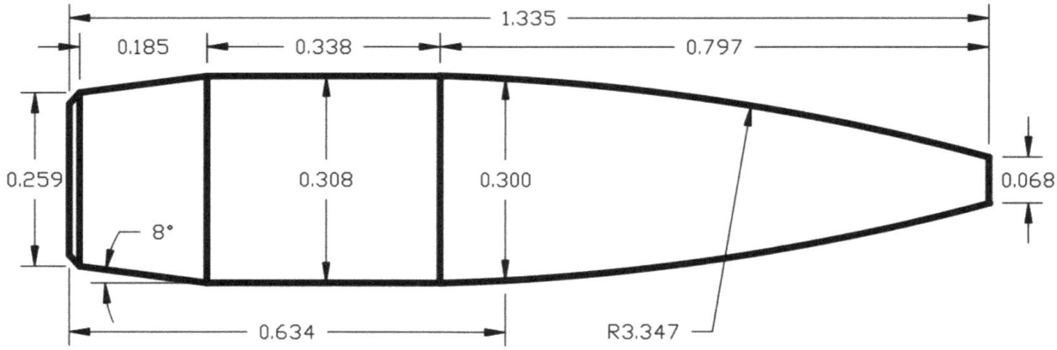

Lot #: 5236A-4 140N JH-M		Ogive Radius:	10.87 calibers
Sectional Density:	0.264 lb/in^2	Rt/R:	0.81

Ballistic Performance

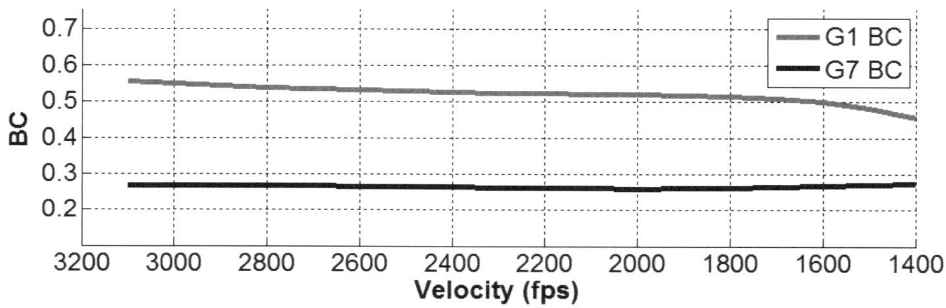

	Above 3000	3000-2500	2500-2000	2000-1500	Below 1500
G1 BC	0.565	0.538	0.524	0.500	0.481
G7 BC	0.268	0.267	0.263	0.266	0.271

Averages from 3000 fps to 1500 fps:
G1 BC: 0.519, variation of G1 BC: 0.068
G7 BC: 0.266, variation of G7 BC: 0.010
G7 form factor: 0.989

Rifling Twist Rate and Stability

Barrel Twist	Stability Factor Best case	Stability Factor Nominal	Stability Factor Worst case
1:10"	2.27	1.99	1.72
1:11"	1.88	1.64	1.42
1:12"	1.58	1.38	1.19
1:13"	1.35	1.18	1.02

Minimum twist required for stability is 1:12.9"
Recommended twist for optimal performance is 1:11.5" or faster

Barnes .308 caliber 150 grain TAC RRLP FB

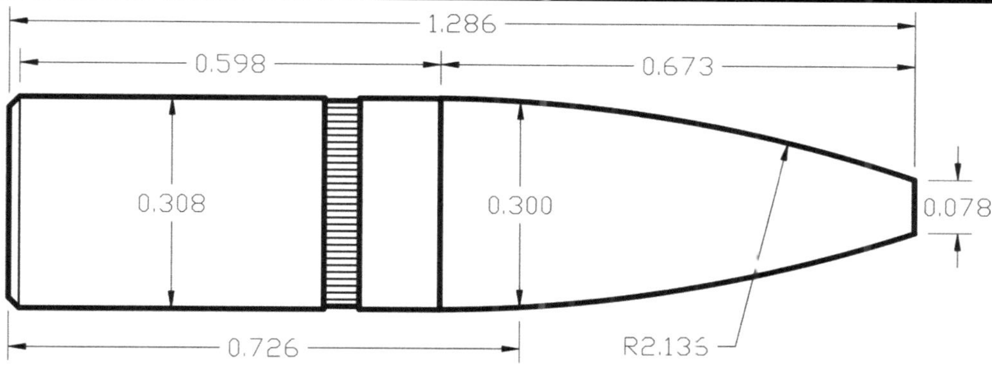

Lot #:	6421176		Ogive Radius:	6.94 calibers
Sectional Density:	0.226 lb/in^2		Rt/R:	0.97

Ballistic Performance

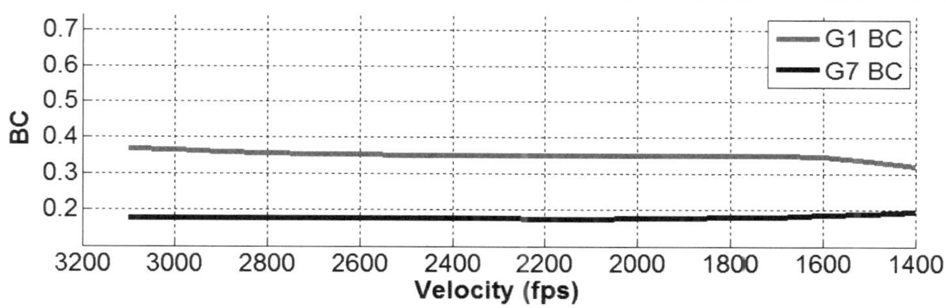

	Above 3000	3000-2500	2500-2000	2000-1500	Below 1500
G1 BC	0.374	0.358	0.351	0.343	0.336
G7 BC	0.177	0.177	0.177	0.183	0.189

Averages from 3000 fps to 1500 fps:
G1 BC: 0.351, variation of G1 BC: 0.028
G7 BC: 0.180, variation of G7 BC: 0.013
G7 form factor: 1.255

Rifling Twist Rate and Stability

Barrel Twist	Best case	Stability Factor Nominal	Worst case
1:10"	2.16	1.89	1.63
1:11"	1.79	1.56	1.35
1:12"	1.50	1.31	1.13
1:13"	1.28	1.12	0.97

Minimum twist required for stability is 1:12.5"
Recommended twist for optimal performance is 1:11.2" or faster

Barnes .308 caliber 150 grain TAC-X BT

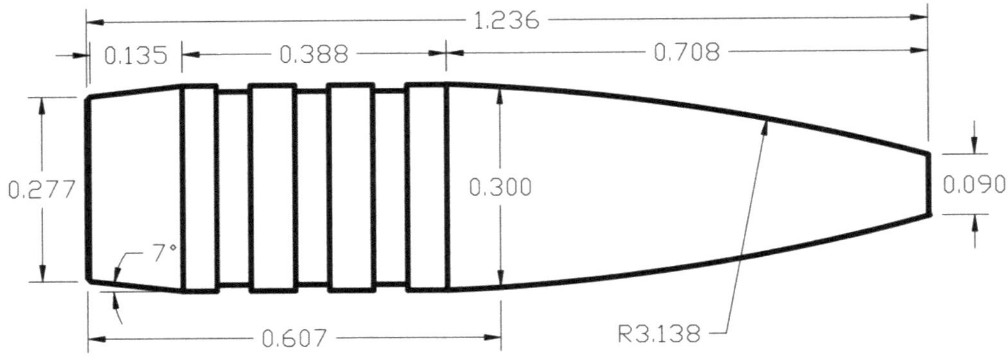

Lot #:	**64211 99-2 260N**	**Ogive Radius:**	**10.19 calibers**
Sectional Density:	**0.226 lb/in^2**	**Rt/R:**	**0.75**

Ballistic Performance

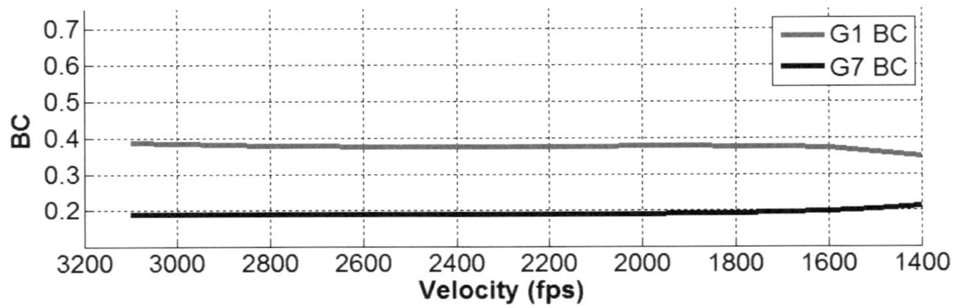

	Above 3000	3000-2500	2500-2000	2000-1500	Below 1500
G1 BC	0.394	0.381	0.377	0.371	0.363
G7 BC	0.187	0.189	0.190	0.198	0.205

Averages from 3000 fps to 1500 fps:
G1 BC: 0.376, variation of G1 BC: 0.022
G7 BC: 0.193, variation of G7 BC: 0.017
G7 form factor: 1.170

Rifling Twist Rate and Stability

Barrel Twist	Stability Factor Best case	Nominal	Worst case
1:10"	2.44	2.13	1.84
1:11"	2.01	1.76	1.52
1:12"	1.69	1.48	1.28
1:13"	1.44	1.26	1.09

Minimum twist required for stability is 1:13.3"
Recommended twist for optimal performance is 1:11.9" or faster

Barnes .308 caliber 110 grain TacX FB M/LE

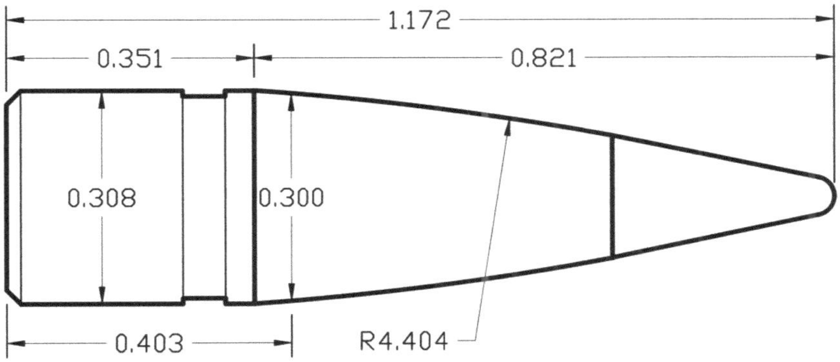

Lot #:	**6469337**	**Ogive Radius:**	**14.3 calibers**
Sectional Density:	**0.166 lb/in^2**	**Rt/R:**	**0.45**

Ballistic Performance

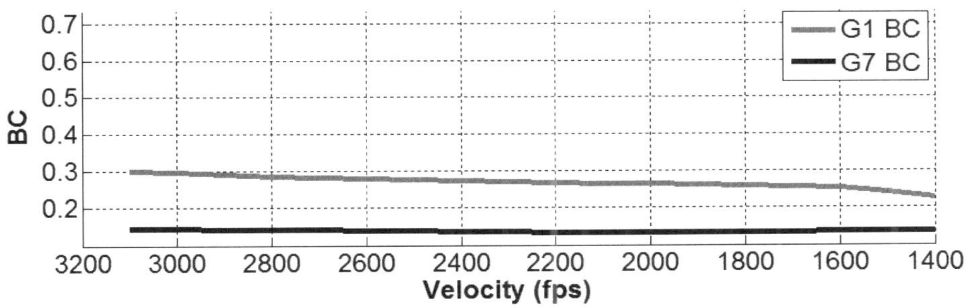

	Above 3000	3000-2500	2500-2000	2000-1500	Below 1500
G1 BC	0.311	0.286	0.270	0.253	0.242
G7 BC	0.147	0.142	0.136	0.135	0.136

Averages from 3000 fps to 1500 fps:
G1 BC: 0.270, variation of G1 BC: 0.055
G7 BC: 0.138, variation of G7 BC: 0.012
G7 form factor: 1.200

Rifling Twist Rate and Stability

Barrel Twist	Stability Factor Best case	Nominal	Worst case
1:10"	2.08	1.82	1.57
1:11"	1.72	1.50	1.30
1:12"	1.45	1.26	1.09
1:13"	1.23	1.08	0.93

Minimum twist required for stability is 1:12.3"
Recommended twist for optimal performance is 1:11.0" or faster

Barnes .308 caliber 168 grain Tac-X M/LE

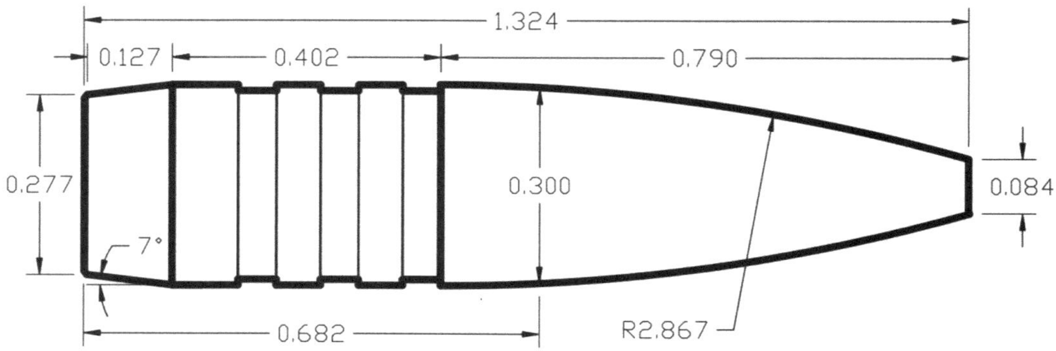

Lot #:	6368440-4 259N	Ogive Radius:	9.31 calibers
Sectional Density:	0.253 lb/in²	Rt/R:	0.99

Ballistic Performance

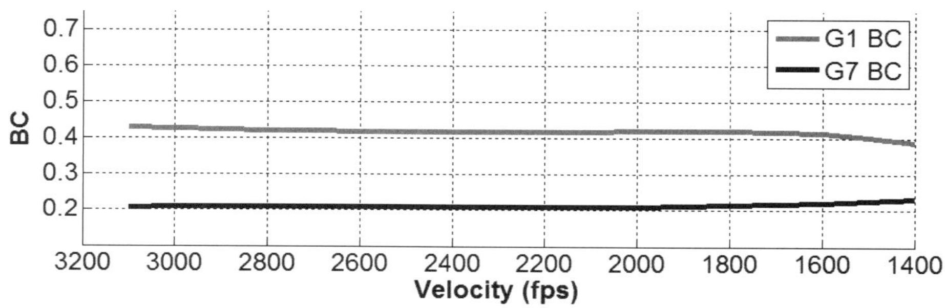

	Above 3000	3000-2500	2500-2000	2000-1500	Below 1500
G1 BC	0.437	0.422	0.419	0.411	0.403
G7 BC	0.207	0.209	0.210	0.219	0.227

Averages from 3000 fps to 1500 fps:
G1 BC: 0.417, variation of G1 BC: 0.024
G7 BC: 0.214, variation of G7 BC: 0.018
G7 form factor: 1.182

Rifling Twist Rate and Stability

Barrel Twist	Stability Factor Best case	Nominal	Worst case
1:10"	2.24	1.96	1.69
1:11"	1.85	1.62	1.40
1:12"	1.56	1.36	1.18
1:13"	1.33	1.16	1.00

Minimum twist required for stability is 1:12.8"
Recommended twist for optimal performance is 1:11.4" or faster

Barnes .308 caliber 165 grain TSX BT

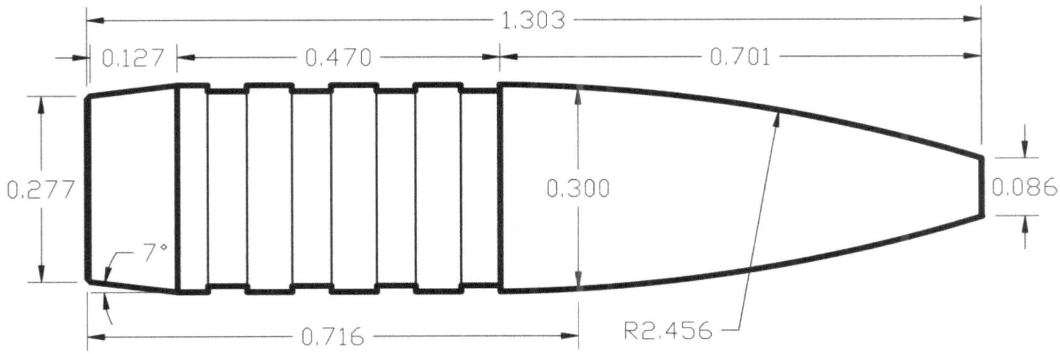

Lot #:	6470081-1	Ogive Radius:	7.97 calibers
Sectional Density:	0.249 lb/in^2	Rt/R:	0.92

Ballistic Performance

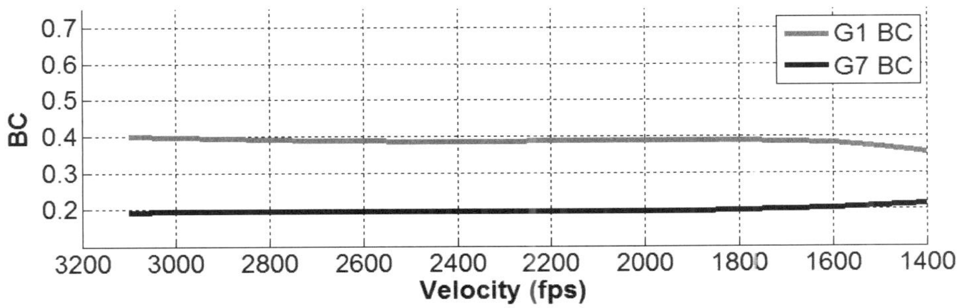

	Above 3000	3000-2500	2500-2000	2000-1500	Below 1500
G1 BC	0.406	0.392	0.388	0.379	0.371
G7 BC	0.192	0.194	0.195	0.202	0.209

Averages from 3000 fps to 1500 fps:
G1 BC: 0.385, variation of G1 BC: 0.026
G7 BC: 0.198, variation of G7 BC: 0.015
G7 form factor: 1.255

Rifling Twist Rate and Stability

Barrel Twist	Stability Factor Best case	Nominal	Worst case
1:10"	2.30	2.01	1.74
1:11"	1.90	1.66	1.44
1:12"	1.60	1.40	1.21
1:13"	1.36	1.19	1.03

Minimum twist required for stability is 1:12.9"
Recommended twist for optimal performance is 1:11.6" or faster

Barnes .308 caliber 168 grain TSX BT

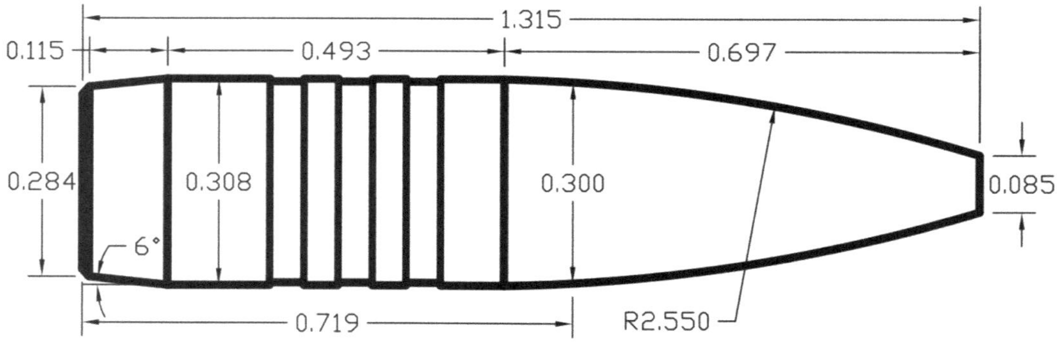

Lot #:	3302-2 325H BS	Ogive Radius:	8.28 calibers
Sectional Density:	0.253 lb/in²	Rt/R:	0.90

Ballistic Performance

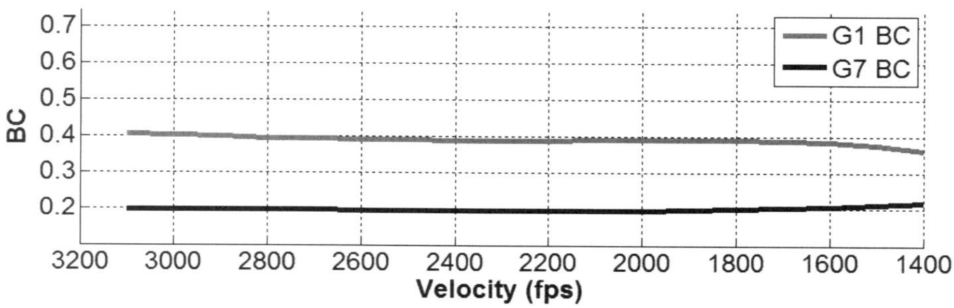

	Above 3000	3000-2500	2500-2000	2000-1500	Below 1500
G1 BC	0.414	0.396	0.390	0.383	0.376
G7 BC	0.196	0.197	0.196	0.204	0.212

Averages from 3000 fps to 1500 fps:
G1 BC: 0.390, variation of G1 BC: 0.027
G7 BC: 0.200, variation of G7 BC: 0.016
G7 form factor: 1.264

Rifling Twist Rate and Stability

Barrel Twist	Stability Factor Best case	Nominal	Worst case
1:10"	2.28	1.99	1.72
1:11"	1.89	1.65	1.42
1:12"	1.58	1.38	1.20
1:13"	1.35	1.18	1.02

Minimum twist required for stability is 1:12.9"
Recommended twist for optimal performance is 1:11.5" or faster

Barnes .308 caliber 180 grain TSX BT

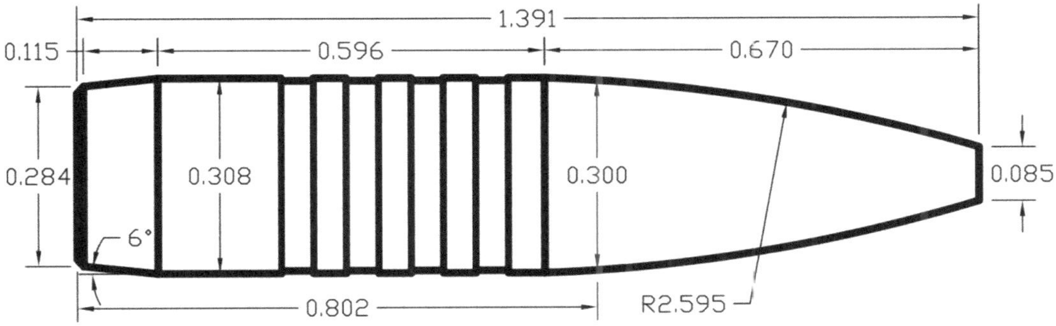

Lot #:	3443A-1 005I LO	Ogive Radius:	8.43 calibers
Sectional Density:	0.271 lb/in^2	Rt/R:	0.85

Ballistic Performance

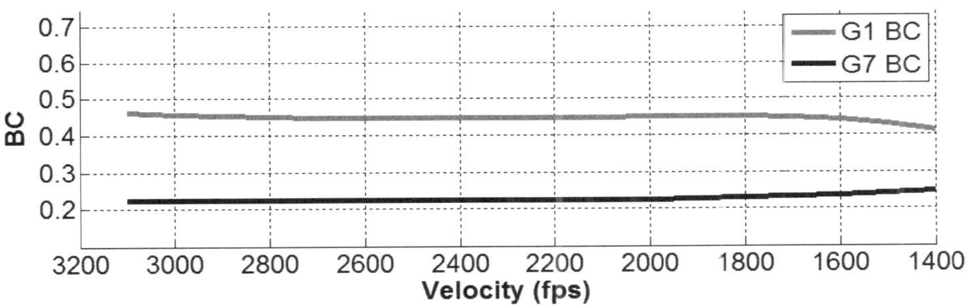

	Above 3000	3000-2500	2500-2000	2000-1500	Below 1500
G1 BC	0.463	0.447	0.443	0.435	0.425
G7 BC	0.220	0.222	0.223	0.231	0.240

Averages from 3000 fps to 1500 fps:
G1 BC: 0.441, variation of G1 BC: 0.027
G7 BC: 0.227, variation of G7 BC: 0.019
G7 form factor: 1.197

Rifling Twist Rate and Stability

Barrel Twist	Stability Factor Best case	Nominal	Worst case
1:10"	2.08	1.81	1.57
1:11"	1.72	1.50	1.30
1:12"	1.44	1.26	1.09
1:13"	1.23	1.07	0.93

Minimum twist required for stability is 1:12.3"
Recommended twist for optimal performance is 1:11.0" or faster

Barnes .308 caliber 130 grain TTSX BT

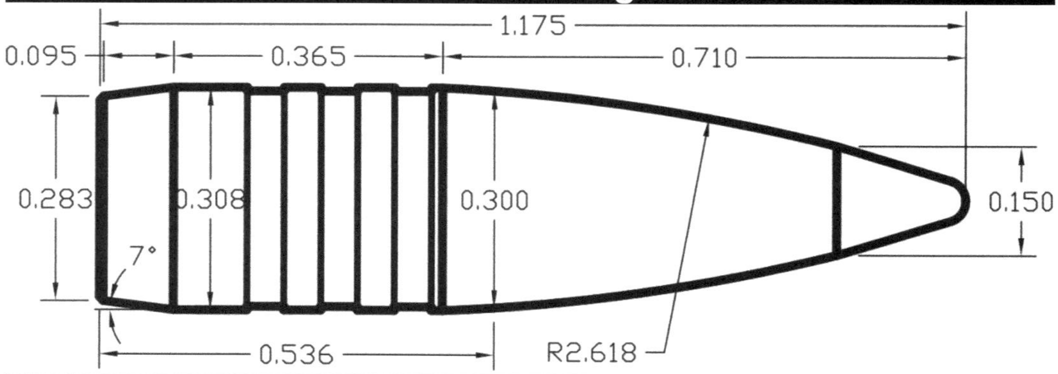

Lot #:	Unknown	Ogive Radius:	8.39 calibers
Sectional Density:	0.196 lb/in^2	Rt/R:	0.60

Ballistic Performance

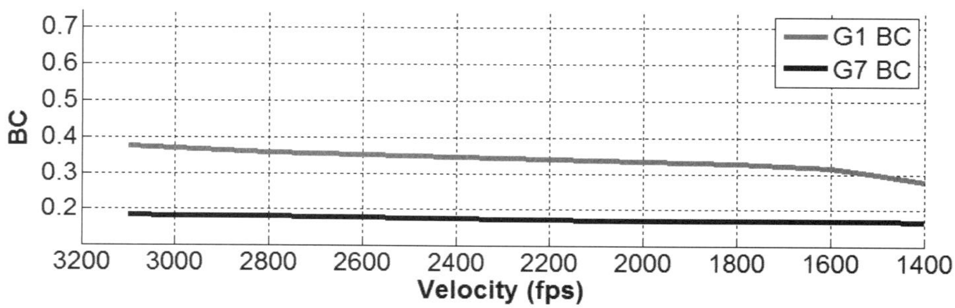

	Above 3000	3000-2500	2500-2000	2000-1500	Below 1500
G1 BC	0.387	0.361	0.343	0.318	0.300
G7 BC	0.183	0.179	0.172	0.169	0.169

Averages from 3000 fps to 1500 fps:
G1 BC: 0.339, variation of G1 BC: 0.071
G7 BC: 0.174, variation of G7 BC: 0.013
G7 form factor: 1.128

Rifling Twist Rate and Stability

Barrel Twist	Stability Factor Best case	Nominal	Worst case
1:10"	2.44	2.13	1.85
1:11"	2.02	1.76	1.53
1:12"	1.70	1.48	1.28
1:13"	1.45	1.26	1.09

Minimum twist required for stability is 1:13.3"
Recommended twist for optimal performance is 1:11.9" or faster

Barnes .308 caliber 150 grain TTSX BT

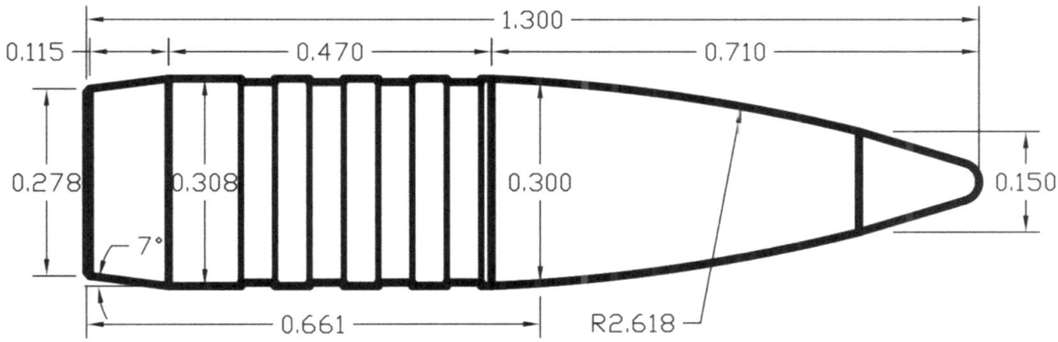

Lot #:	**Unknown**	**Ogive Radius:**	**8.73 calibers**
Sectional Density:	**0.226 lb/in²**	**Rt/R:**	**0.66**

Ballistic Performance

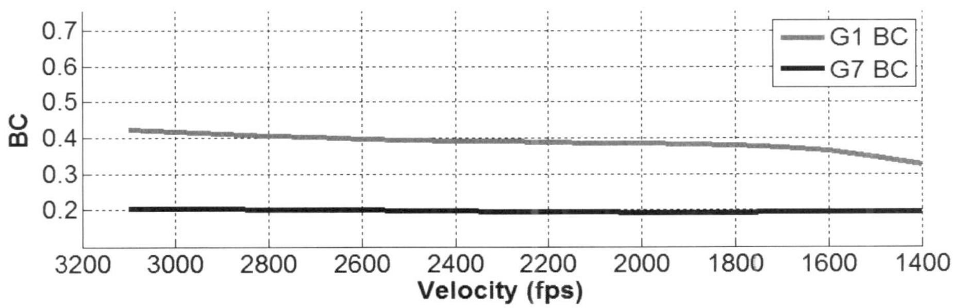

	Above 3000	3000-2500	2500-2000	2000-1500	Below 1500
G1 BC	0.433	0.406	0.390	0.366	0.349
G7 BC	0.205	0.201	0.196	0.194	0.196

Averages from 3000 fps to 1500 fps:
G1 BC: 0.386, variation of G1 BC: 0.068
G7 BC: 0.198, variation of G7 BC: 0.011
G7 form factor: 1.142

Rifling Twist Rate and Stability

Barrel Twist	Stability Factor Best case	Nominal	Worst case
1:10"	2.11	1.84	1.59
1:11"	1.74	1.52	1.31
1:12"	1.46	1.28	1.10
1:13"	1.25	1.09	0.94

Minimum twist required for stability is 1:12.4"
Recommended twist for optimal performance is 1:11.1" or faster

Barnes .308 caliber 168 grain TTSX BT

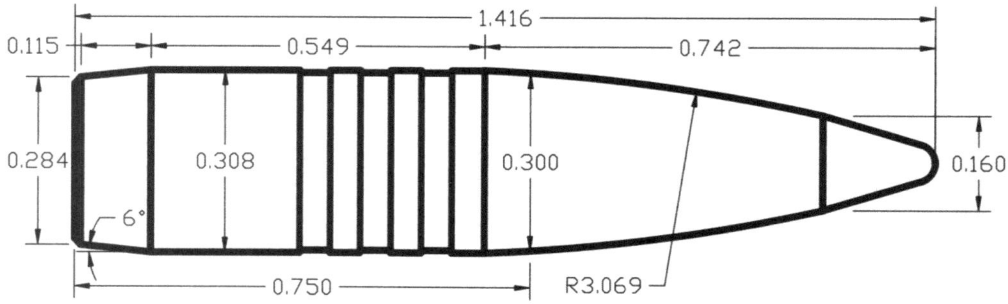

Lot #:	3447-1 353H CW	Ogive Radius:	9.96 calibers
Sectional Density:	0.253 lb/in²	Rt/R:	0.78

Ballistic Performance

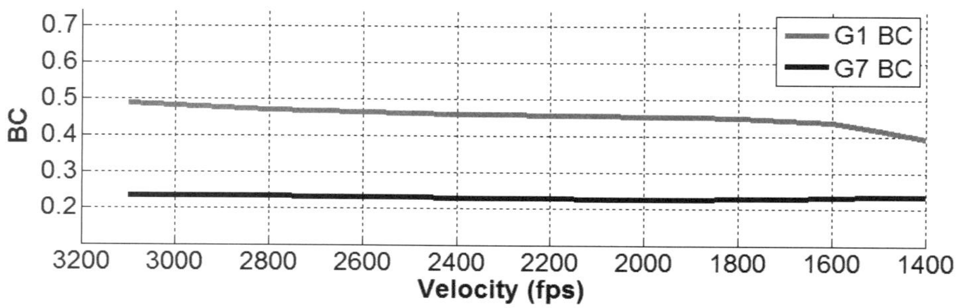

	Above 3000	3000-2500	2500-2000	2000-1500	Below 1500
G1 BC	0.500	0.472	0.457	0.436	0.419
G7 BC	0.237	0.234	0.230	0.232	0.236

Averages from 3000 fps to 1500 fps:
G1 BC: 0.454, variation of G1 BC: 0.064
G7 BC: 0.233, variation of G7 BC: 0.009
G7 form factor: 1.086

Rifling Twist Rate and Stability

Barrel Twist	Stability Factor Best case	Stability Factor Nominal	Stability Factor Worst case
1:10"	1.84	1.61	1.39
1:11"	1.52	1.33	1.15
1:12"	1.28	1.12	0.97
1:13"	1.09	0.95	0.82

Minimum twist required for stability is 1:11.6"
Recommended twist for optimal performance is 1:10.4" or faster

Berger .308 caliber 155 grain Hunting VLD

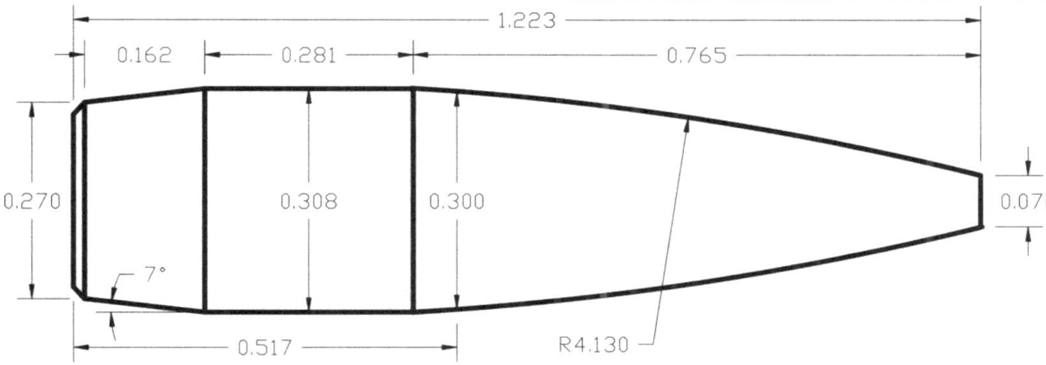

Lot #:	**7518**	**Ogive Radius:**	**13.41 calibers**
Sectional Density:	**0.233 lb/in^2**	**Rt/R:**	**0.61**

Ballistic Performance

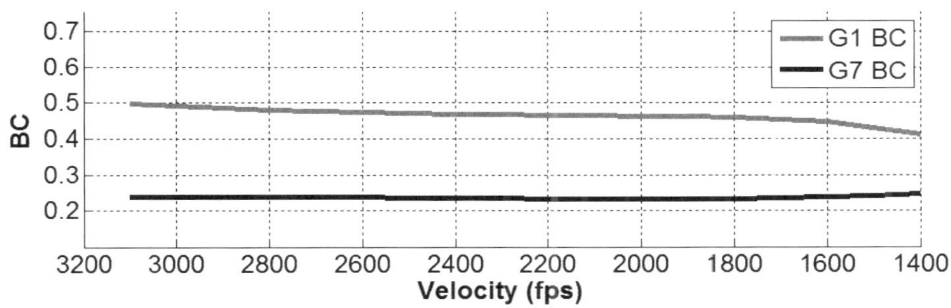

	Above 3000	3000-2500	2500-2000	2000-1500	Below 1500
G1 BC	0.506	0.481	0.467	0.447	0.430
G7 BC	0.240	0.238	0.235	0.238	0.243

Averages from 3000 fps to 1500 fps:
G1 BC: 0.464, variation of G1 BC: 0.060
G7 BC: 0.238, variation of G7 BC: 0.010
G7 form factor: 0.981

Rifling Twist Rate and Stability

Barrel Twist	Stability Factor Best case	Stability Factor Nominal	Stability Factor Worst case
1:11"	2.15	1.87	1.62
1:12"	1.80	1.57	1.36
1:13"	1.54	1.34	1.16
1:14"	1.32	1.16	1.00

Minimum twist required for stability is 1:13.7"
Recommended twist for optimal performance is 1:12.3" or faster

Berger .308 caliber 168 grain Classic Hunter

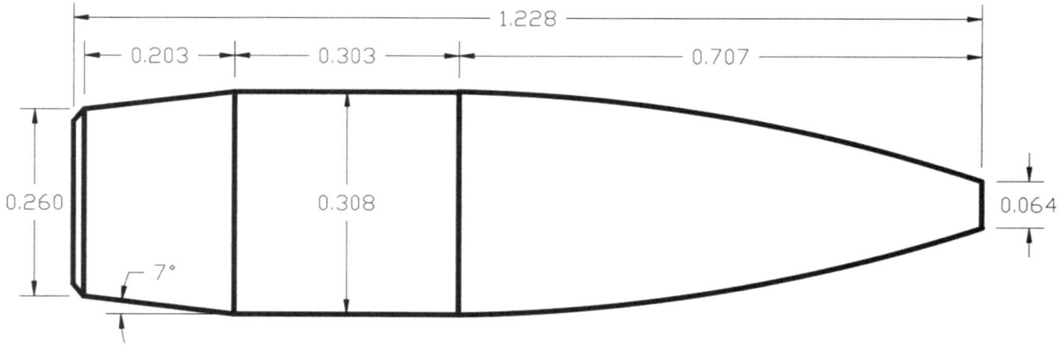

Lot #:	7488	Ogive Radius:	Hybrid
Sectional Density:	0.253 lb/in^2	Rt/R:	Hybrid

Ballistic Performance

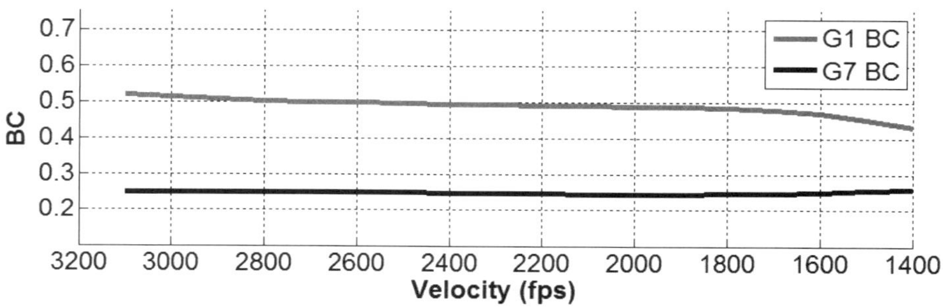

	Above 3000	3000-2500	2500-2000	2000-1500	Below 1500
G1 BC	0.530	0.505	0.492	0.471	0.454
G7 BC	0.251	0.251	0.248	0.251	0.256

Averages from 3000 fps to 1500 fps:
G1 BC: 0.488, variation of G1 BC: 0.061
G7 BC: 0.251, variation of G7 BC: 0.010
G7 form factor: 1.010

Rifling Twist Rate and Stability

Barrel Twist	Stability Factor Best case	Nominal	Worst case
1:11"	2.30	2.01	1.74
1:12"	1.93	1.69	1.46
1:13"	1.65	1.44	1.24
1:14"	1.42	1.24	1.07

Minimum twist required for stability is 1:14.2"
Recommended twist for optimal performance is 1:12.7" or faster

Berger .308 caliber 168 grain Hunting VLD

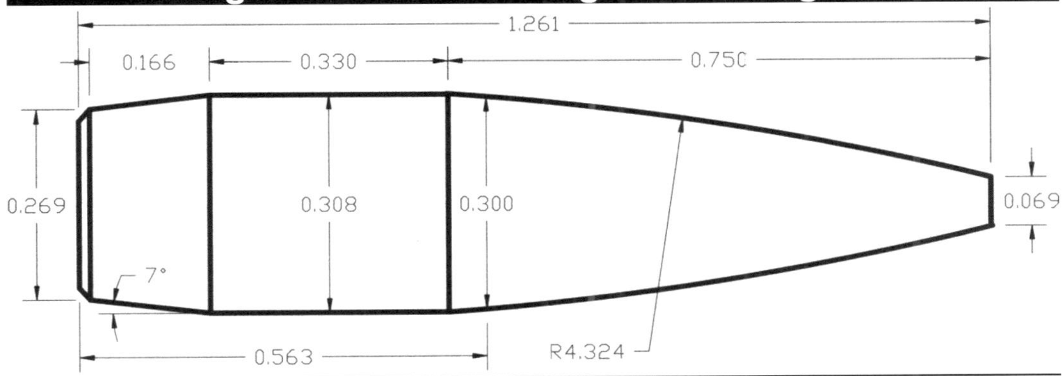

Lot #:	7528	Ogive Radius:	14.04 calibers
Sectional Density:	0.253 lb/in^2	Rt/R:	0.56

Ballistic Performance

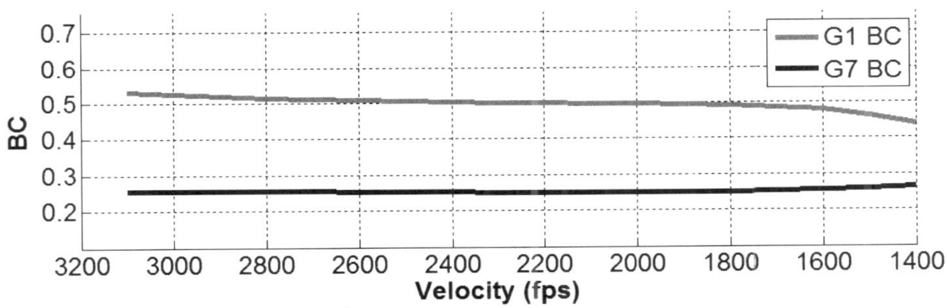

	Above 3000	3000-2500	2500-2000	2000-1500	Below 1500
G1 BC	0.542	0.515	0.500	0.479	0.461
G7 BC	0.257	0.255	0.252	0.255	0.260

Averages from 3000 fps to 1500 fps:
G1 BC: 0.497, variation of G1 BC: 0.064
G7 BC: 0.255, variation of G7 BC: 0.010
G7 form factor: 0.992

Rifling Twist Rate and Stability

Barrel Twist	Stability Factor Best case	Nominal	Worst case
1:11"	2.13	1.86	1.61
1:12"	1.79	1.56	1.35
1:13"	1.52	1.33	1.15
1:14"	1.31	1.15	0.99

Minimum twist required for stability is 1:13.7"
Recommended twist for optimal performance is 1:12.2" or faster

Berger .308 caliber 175 grain Hunting VLD

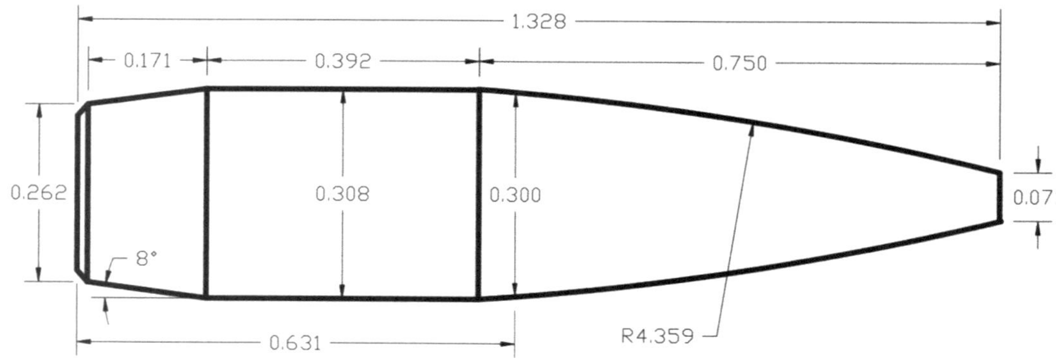

Lot #:	5747	Ogive Radius:	14.15 calibers
Sectional Density:	0.264 lb/in^2	Rt/R:	0.56

Ballistic Performance

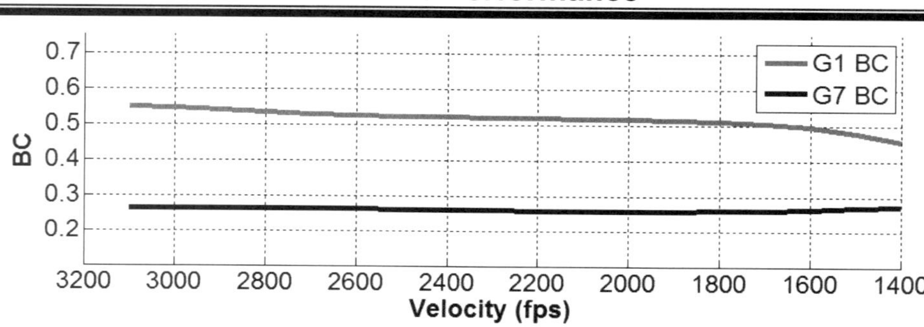

	Above 3000	3000-2500	2500-2000	2000-1500	Below 1500
G1 BC	0.561	0.535	0.521	0.498	0.479
G7 BC	0.266	0.265	0.262	0.265	0.270

Averages from 3000 fps to 1500 fps:
G1 BC: 0.516, variation of G1 BC: 0.066
G7 BC: 0.265, variation of G7 BC: 0.010
G7 form factor: 0.995

Rifling Twist Rate and Stability

Barrel Twist	Stability Factor Best case	Stability Factor Nominal	Stability Factor Worst case
1:10"	2.31	2.02	1.74
1:11"	1.91	1.67	1.44
1:12"	1.60	1.40	1.21
1:13"	1.37	1.19	1.03

Minimum twist required for stability is 1:13.0"
Recommended twist for optimal performance is 1:11.6" or faster

Berger .308 caliber 185 grain Classic Hunter

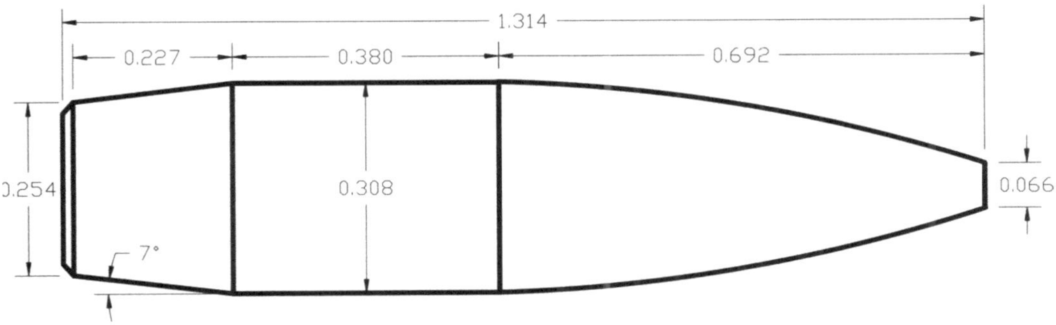

Lot #:	**7435**	**Ogive Radius:**	**Hybrid**
Sectional Density:	**0.279 lb/in²**	**Rt/R:**	**Hybrid**

Ballistic Performance

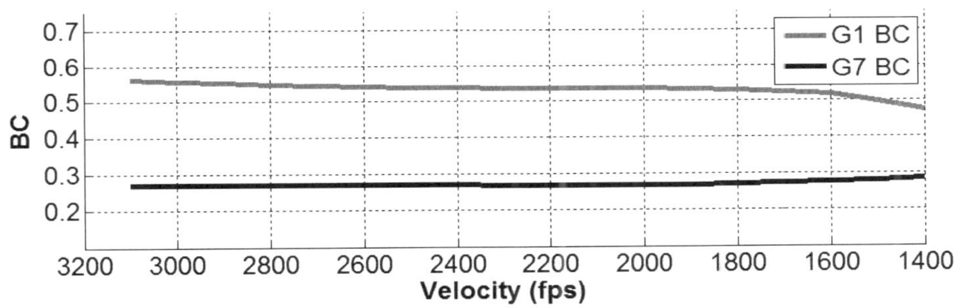

	Above 3000	3000-2500	2500-2000	2000-1500	Below 1500
G1 BC	0.571	0.547	0.536	0.513	0.498
G7 BC	0.271	0.272	0.270	0.274	0.281

Averages from 3000 fps to 1500 fps:
G1 BC: 0.532, variation of G1 BC: 0.058
G7 BC: 0.273, variation of G7 BC: 0.012
G7 form factor: 1.021

Rifling Twist Rate and Stability

Barrel Twist	**Stability Factor** Best case	Nominal	Worst case
1:11"	2.08	1.82	1.57
1:12"	1.75	1.53	1.32
1:13"	1.49	1.30	1.13
1:14"	1.28	1.12	0.97

Minimum twist required for stability is 1:13.5"
Recommended twist for optimal performance is 1:12.1" or faster

Berger .308 caliber 185 grain Hunting VLD

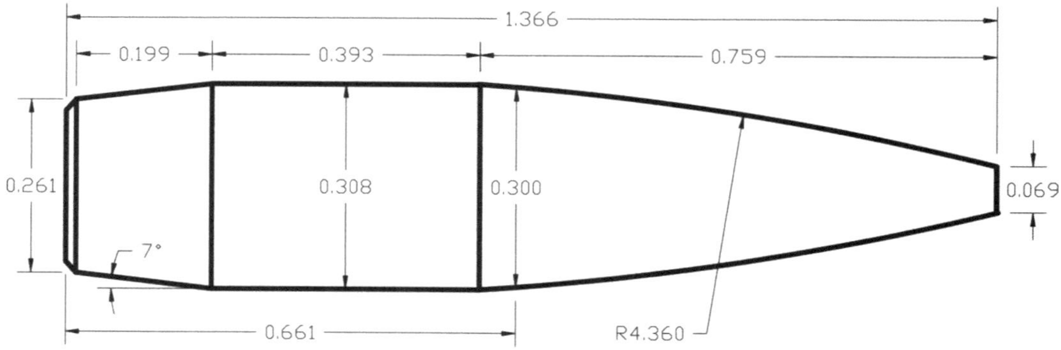

Lot #:	7576	Ogive Radius:	14.15 calibers
Sectional Density:	0.279 lb/in²	Rt/R:	0.57

Ballistic Performance

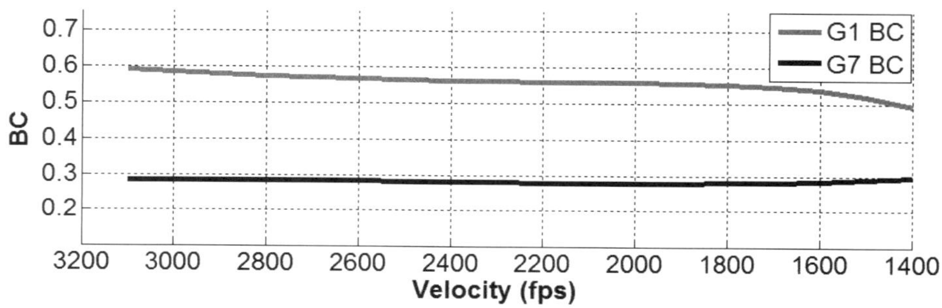

	Above 3000	3000-2500	2500-2000	2000-1500	Below 1500
G1 BC	0.602	0.574	0.560	0.538	0.519
G7 BC	0.285	0.285	0.282	0.286	0.292

Averages from 3000 fps to 1500 fps:
G1 BC: 0.556, variation of G1 BC: 0.066
G7 BC: 0.285, variation of G7 BC: 0.013
G7 form factor: 0.976

Rifling Twist Rate and Stability

Barrel Twist	Stability Factor Best case	Stability Factor Nominal	Stability Factor Worst case
1:10"	2.25	1.97	1.70
1:11"	1.86	1.62	1.40
1:12"	1.56	1.37	1.18
1:13"	1.33	1.16	1.01

Minimum twist required for stability is 1:12.8"
Recommended twist for optimal performance is 1:11.4" or faster

Berger .308 caliber 190 grain Hunting VLD

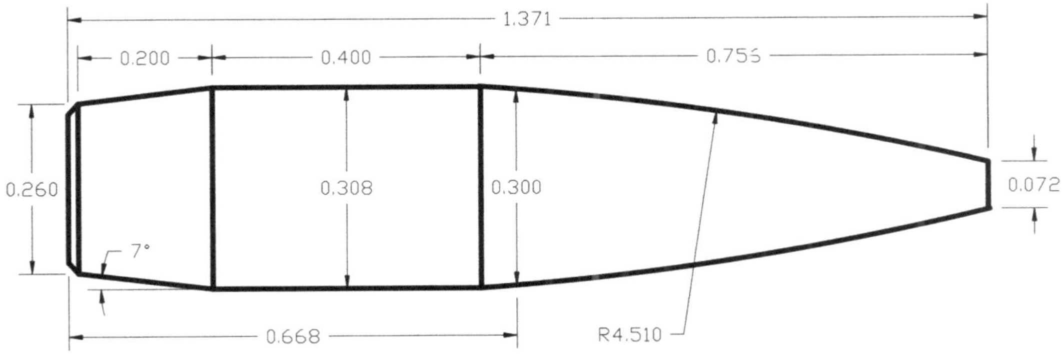

Lot #:	6778	Ogive Radius:	14.64 calibers
Sectional Density:	0.286 lb/in^2	Rt/R:	0.55

Ballistic Performance

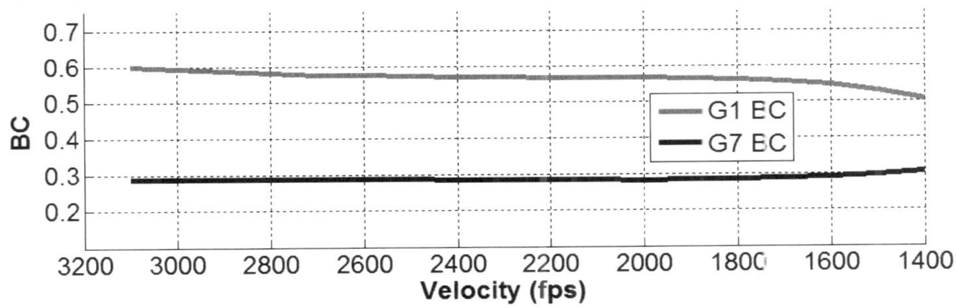

	Above 3000	3000-2500	2500-2000	2000-1500	Below 1500
G1 BC	0.609	0.582	0.569	0.548	0.530
G7 BC	0.288	0.289	0.286	0.291	0.299

Averages from 3000 fps to 1500 fps:

G1 BC: 0.565, variation of G1 BC: 0.062

G7 BC: 0.290, variation of G7 BC: 0.014

G7 form factor: 0.987

Rifling Twist Rate and Stability

Barrel Twist	Stability Factor Best case	Nominal	Worst case
1:10"	2.29	2.00	1.73
1:11"	1.89	1.65	1.43
1:12"	1.59	1.39	1.20
1:13"	1.35	1.18	1.02

Minimum twist required for stability is 1:12.9"

Recommended twist for optimal performance is 1:11.5" or faster

Berger .308 caliber 210 grain Hunting VLD

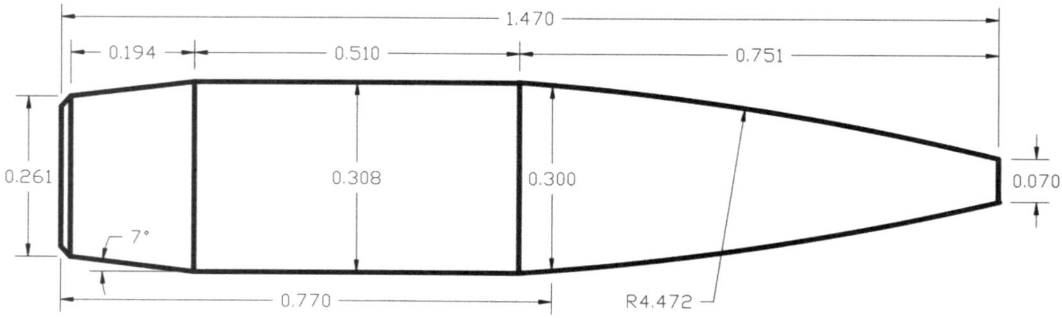

Lot #:	7577	Ogive Radius:	14.52 calibers
Sectional Density:	0.316 lb/in^2	Rt/R:	0.54

Ballistic Performance

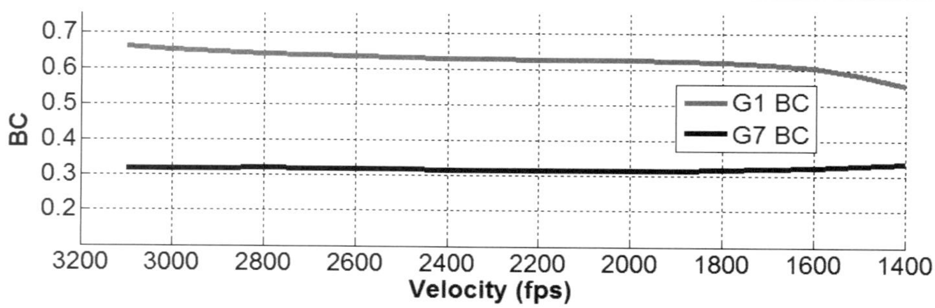

	Above 3000	3000-2500	2500-2000	2000-1500	Below 1500
G1 BC	0.672	0.642	0.628	0.605	0.584
G7 BC	0.318	0.319	0.316	0.321	0.329

Averages from 3000 fps to 1500 fps:
G1 BC: 0.623, variation of G1 BC: 0.069
G7 BC: 0.320, variation of G7 BC: 0.015
G7 form factor: 0.989

Rifling Twist Rate and Stability

Barrel Twist	Best case	Stability Factor Nominal	Worst case
1:9"	2.55	2.23	1.92
1:10"	2.06	1.80	1.56
1:11"	1.70	1.49	1.29
1:12"	1.43	1.25	1.08

Minimum twist required for stability is 1:12.3"
Recommended twist for optimal performance is 1:11.0" or faster

Berger .308 caliber 175 grain Tactical OTM

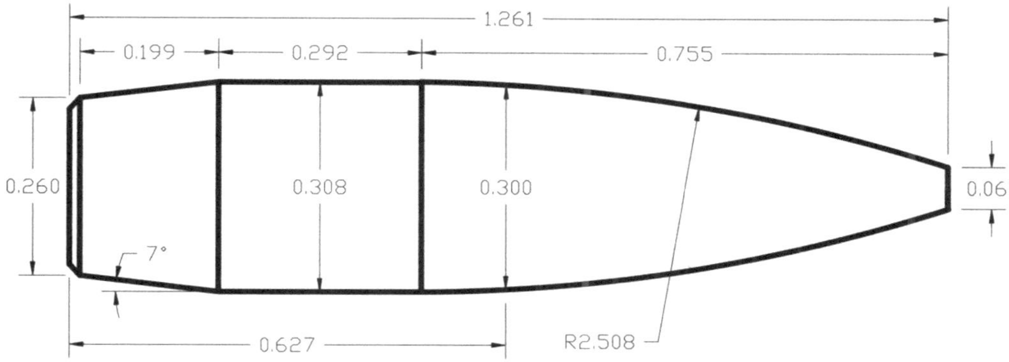

Lot #:	**5555**	**Ogive Radius:**	**8.14 calibers**
Sectional Density:	**0.264 lb/in^2**	**Rt/R:**	**0.95**

Ballistic Performance

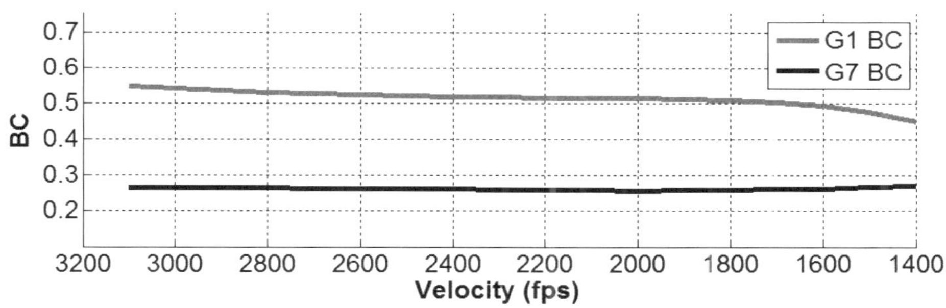

	Above 3000	3000-2500	2500-2000	2000-1500	Below 1500
G1 BC	0.558	0.531	0.517	0.494	0.475
G7 BC	0.264	0.263	0.260	0.263	0.267

Averages from 3000 fps to 1500 fps:
G1 BC: 0.512, variation of G1 BC: 0.067
G7 BC: 0.263, variation of G7 BC: 0.010
G7 form factor: 1.002

Rifling Twist Rate and Stability

Barrel Twist	Stability Factor Best case	Nominal	Worst case
1:11"	2.22	1.94	1.68
1:12"	1.86	1.63	1.41
1:13"	1.59	1.39	1.20
1:14"	1.37	1.20	1.03

Minimum twist required for stability is 1:14.0"
Recommended twist for optimal performance is 1:12.5" or faster

Berger .308 caliber 185 grain Tactical OTM Juggernaut

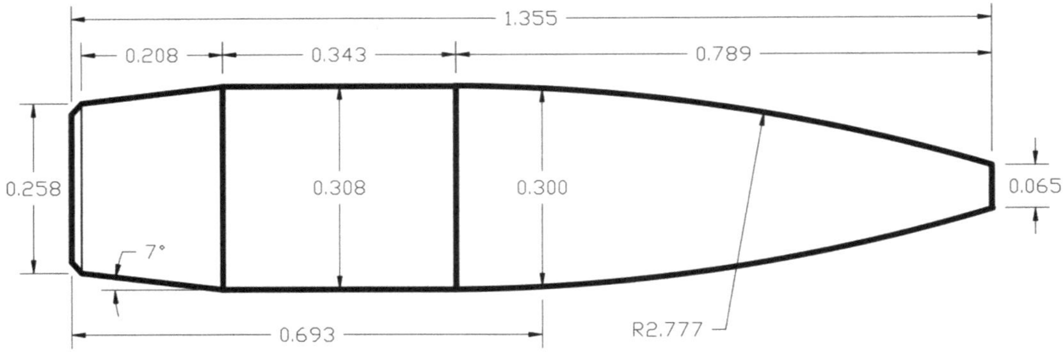

Lot #:	**7561**	**Ogive Radius:**	**9.02 calibers**
Sectional Density:	**0.279 lb/in^2**	**Rt/R:**	**0.94**

Ballistic Performance

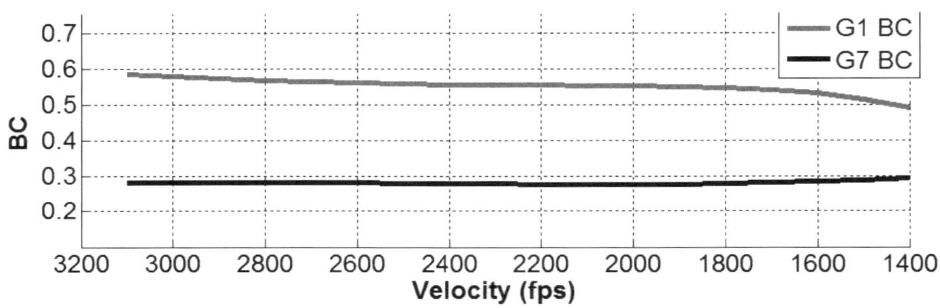

	Above 3000	3000-2500	2500-2000	2000-1500	Below 1500
G1 BC	0.595	0.568	0.555	0.533	0.514
G7 BC	0.282	0.282	0.279	0.283	0.290

Averages from 3000 fps to 1500 fps:

G1 BC: 0.551, variation of G1 BC: 0.064

G7 BC: 0.283, variation of G7 BC: 0.012

G7 form factor: 0.986

Rifling Twist Rate and Stability

Barrel Twist	Stability Factor Best case	Nominal	Worst case
1:10"	2.30	2.01	1.74
1:11"	1.90	1.66	1.44
1:12"	1.60	1.40	1.21
1:13"	1.36	1.19	1.03

Minimum twist required for stability is 1:12.9"

Recommended twist for optimal performance is 1:11.6" or faster

Berger .308 caliber 230 grain Tactical OTM Hybrid

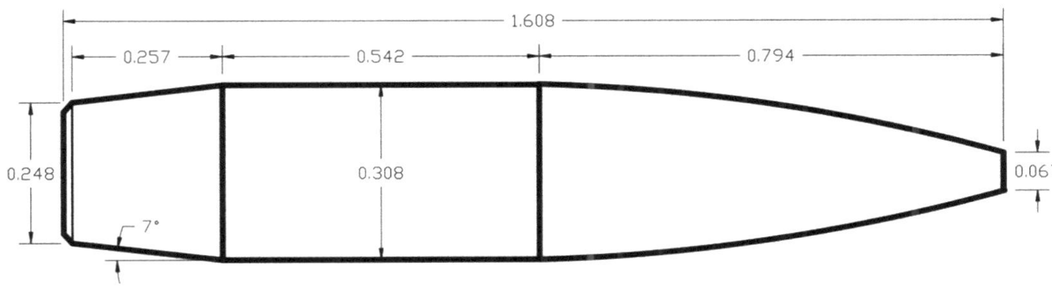

Lot #:	6241	Ogive Radius:	Hybrid
Sectional Density:	0.346 lb/in²	Rt/R:	Hybrid

Ballistic Performance

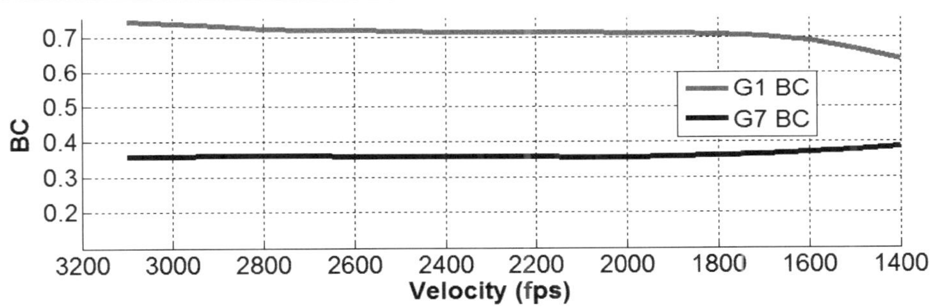

	Above 3000	3000-2500	2500-2000	2000-1500	Below 1500
G1 BC	0.755	0.726	0.714	0.690	0.668
G7 BC	0.358	0.360	0.359	0.367	0.376

Averages from 3000 fps to 1500 fps:

G1 BC: 0.708, variation of G1 BC: 0.069

G7 BC: 0.364, variation of G7 BC: 0.019

G7 form factor: 0.953

Rifling Twist Rate and Stability

Barrel Twist	Stability Factor Best case	Nominal	Worst case
1:9"	2.15	1.87	1.62
1:10"	1.74	1.52	1.31
1:11"	1.44	1.26	1.09
1:12"	1.21	1.05	0.91

Minimum twist required for stability is 1:11.2"

Recommended twist for optimal performance is 1:10.1" or faster

Berger .308 caliber 115 grain Target FB

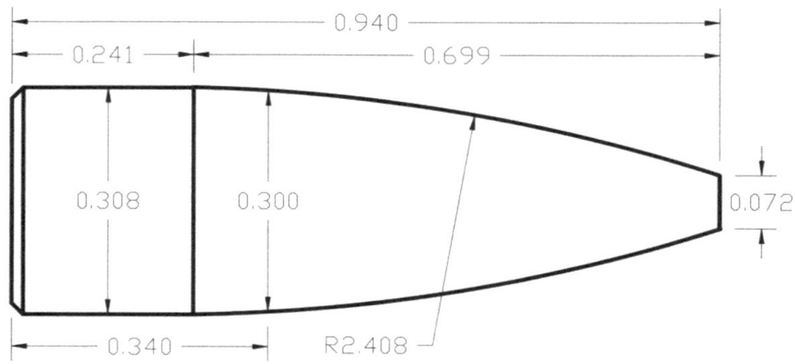

Lot #:	7601	Ogive Radius:	7.82 calibers
Sectional Density:	0.173 lb/in^2	Rt/R:	0.88

Ballistic Performance

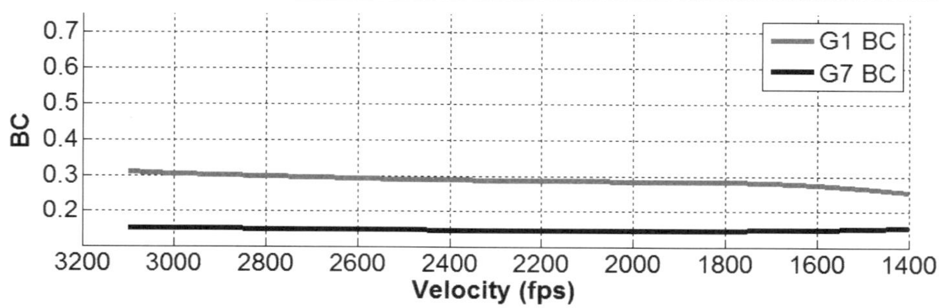

	Above 3000	3000-2500	2500-2000	2000-1500	Below 1500
G1 BC	0.319	0.300	0.290	0.278	0.269
G7 BC	0.151	0.149	0.146	0.148	0.151

Averages from 3000 fps to 1500 fps:
G1 BC: 0.289, variation of G1 BC: 0.039
G7 BC: 0.148, variation of G7 BC: 0.007
G7 form factor: 1.168

Rifling Twist Rate and Stability

Barrel Twist	Stability Factor Best case	Nominal	Worst case
1:14"	2.08	1.82	1.57
1:15"	1.81	1.58	1.37
1:16"	1.59	1.39	1.20
1:17"	1.41	1.23	1.06

Minimum twist required for stability is 1:17.2"
Recommended twist for optimal performance is 1:15.4" or faster

Berger .308 caliber 150 grain Target FB

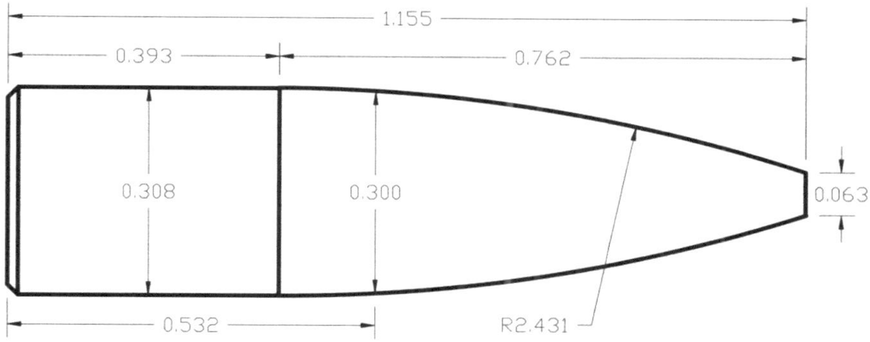

Lot #:	**7568**	**Ogive Radius:**	**7.89 calibers**
Sectional Density:	**0.226 lb/in²**	**Rt/R:**	**1.00**

Ballistic Performance

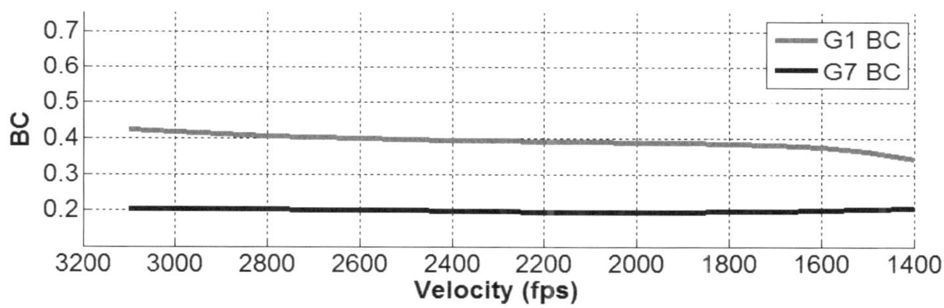

	Above 3000	3000-2500	2500-2000	2000-1500	Below 1500
G1 BC	0.433	0.407	0.393	0.377	0.364
G7 BC	0.205	0.202	0.198	0.200	0.205

Averages from 3000 fps to 1500 fps:
G1 BC: 0.392, variation of G1 BC: 0.054
G7 BC: 0.201, variation of G7 BC: 0.009
G7 form factor: 1.124

Rifling Twist Rate and Stability

Barrel Twist	Stability Factor Best case	 Nominal	 Worst case
1:12"	2.06	1.80	1.55
1:13"	1.75	1.53	1.32
1:14"	1.51	1.32	1.14
1:15"	1.32	1.15	0.99

Minimum twist required for stability is 1:14.7"
Recommended twist for optimal performance is 1:13.[illegible]" or faster

Berger .308 caliber 155 grain Target Hybrid

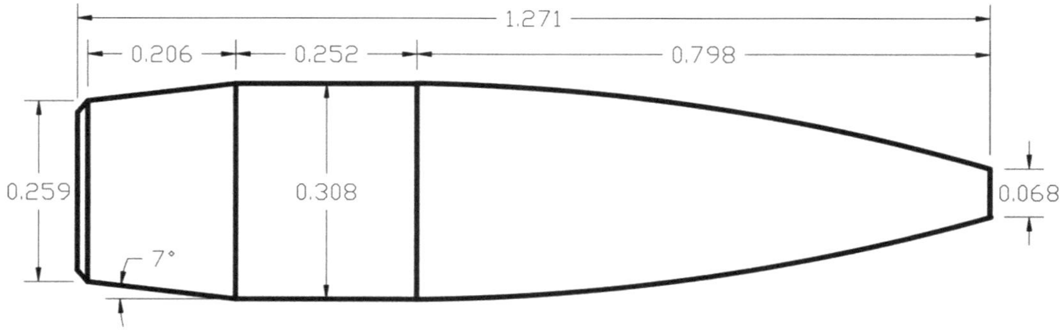

Lot #:	7459	Ogive Radius:	Hybrid
Sectional Density:	0.233 lb/in^2	Rt/R:	Hybrid

Ballistic Performance

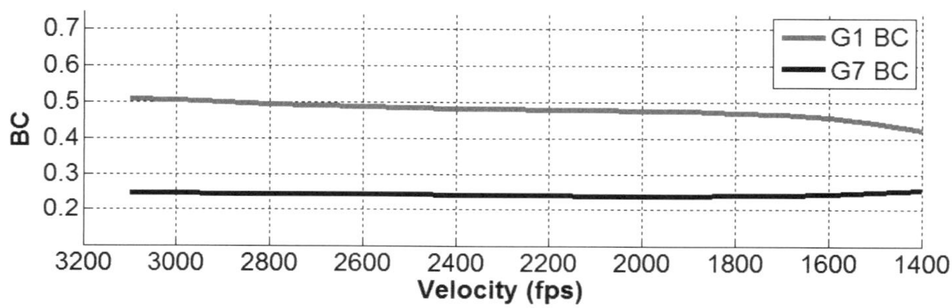

	Above 3000	3000-2500	2500-2000	2000-1500	Below 1500
G1 BC	0.520	0.494	0.481	0.460	0.444
G7 BC	0.246	0.245	0.242	0.245	0.250

Averages from 3000 fps to 1500 fps:
G1 BC: 0.477, variation of G1 BC: 0.060
G7 BC: 0.245, variation of G7 BC: 0.011
G7 form factor: 0.953

Rifling Twist Rate and Stability

Barrel Twist	Stability Factor Best case	Nominal	Worst case
1:10"	2.32	2.03	1.75
1:11"	1.92	1.68	1.45
1:12"	1.61	1.41	1.22
1:13"	1.37	1.20	1.04

Minimum twist required for stability is 1:13.0"
Recommended twist for optimal performance is 1:11.6" or faster

Berger .308 caliber 155 grain Target VLD

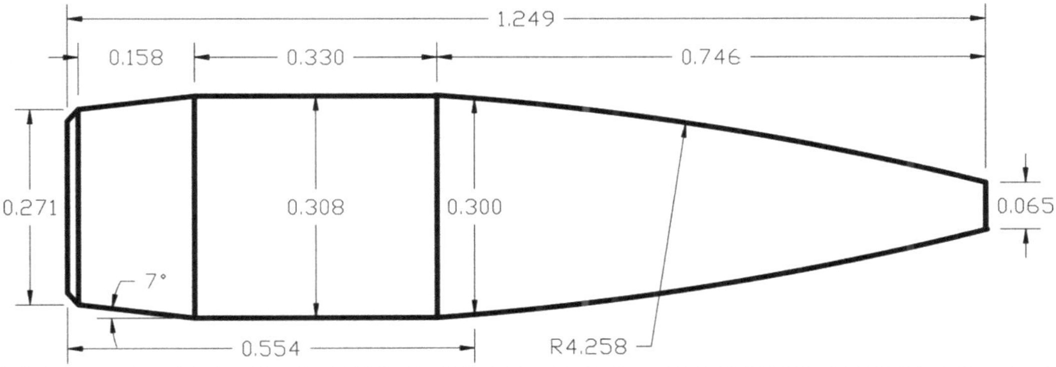

Lot #:	7751	Ogive Radius:	13.82 calibers
Sectional Density:	0.233 lb/in^2	Rt/R:	0.55

Ballistic Performance

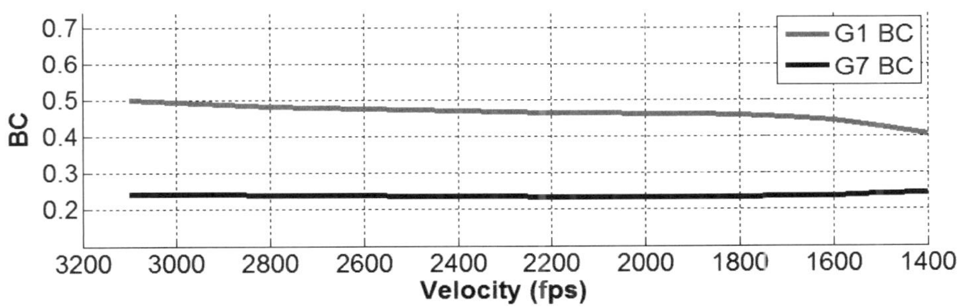

	Above 3000	3000-2500	2500-2000	2000-1500	Below 1500
G1 BC	0.510	0.483	0.467	0.445	0.428
G7 BC	0.241	0.239	0.235	0.237	0.241

Averages from 3000 fps to 1500 fps:

G1 BC: 0.464, variation of G1 BC: 0.066

G7 BC: 0.238, variation of G7 BC: 0.009

G7 form factor: 0.981

Rifling Twist Rate and Stability

Barrel Twist	Stability Factor Best case	Nominal	Worst case
1:10"	2.44	2.13	1.85
1:11"	2.02	1.76	1.53
1:12"	1.70	1.48	1.28
1:13"	1.45	1.26	1.09

Minimum twist required for stability is 1:13.3"

Recommended twist for optimal performance is 1:11.9" or faster

Berger 155.5 grain Target FULLBORE

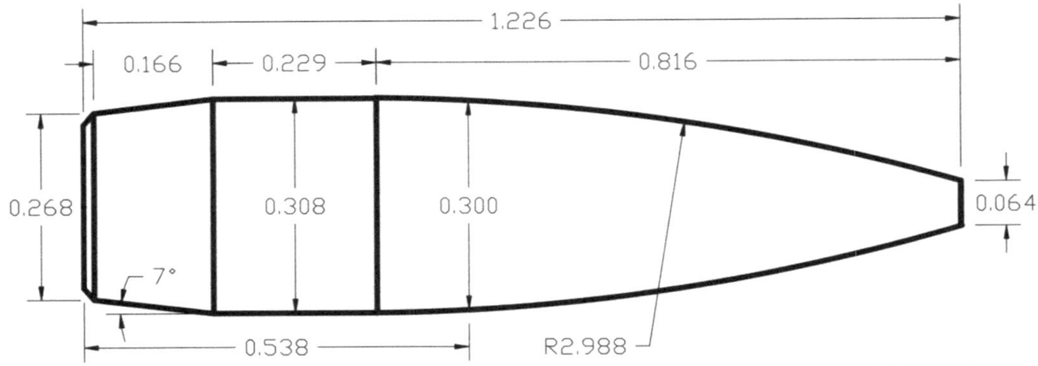

Lot #:	7461	Ogive Radius:	9.70 calibers
Sectional Density:	0.234 lb/in^2	Rt/R:	0.93

Ballistic Performance

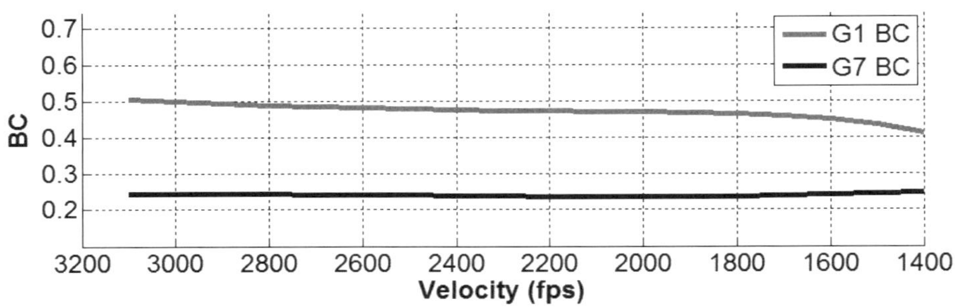

	Above 3000	3000-2500	2500-2000	2000-1500	Below 1500
G1 BC	0.516	0.489	0.474	0.451	0.434
G7 BC	0.244	0.242	0.238	0.240	0.244

Averages from 3000 fps to 1500 fps:
G1 BC: 0.470, variation of G1 BC: 0.066
G7 BC: 0.241, variation of G7 BC: 0.009
G7 form factor: 0.968

Rifling Twist Rate and Stability

Barrel Twist	Stability Factor Best case	Nominal	Worst case
1:11"	2.13	1.86	1.61
1:12"	1.79	1.56	1.35
1:13"	1.53	1.33	1.15
1:14"	1.31	1.15	0.99

Minimum twist required for stability is 1:13.7"
Recommended twist for optimal performance is 1:12.3" or faster

Berger .308 caliber 168 grain Target Hybrid

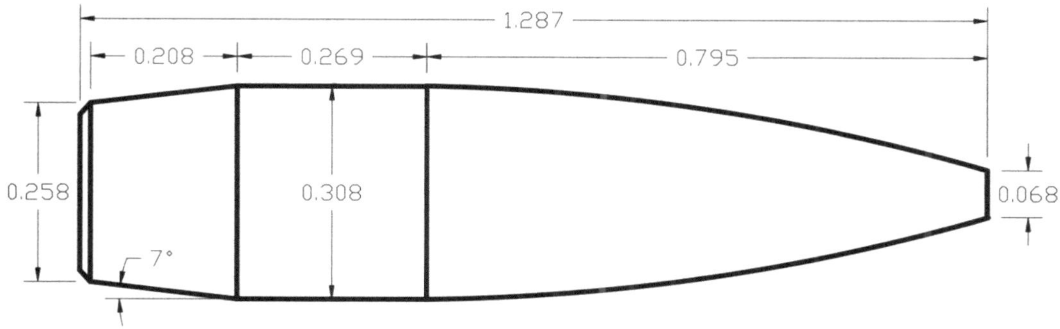

Lot #:	7469	Ogive Radius:	Hybrid
Sectional Density:	0.253 lb/in^2	Rt/R:	Hybrid

Ballistic Performance

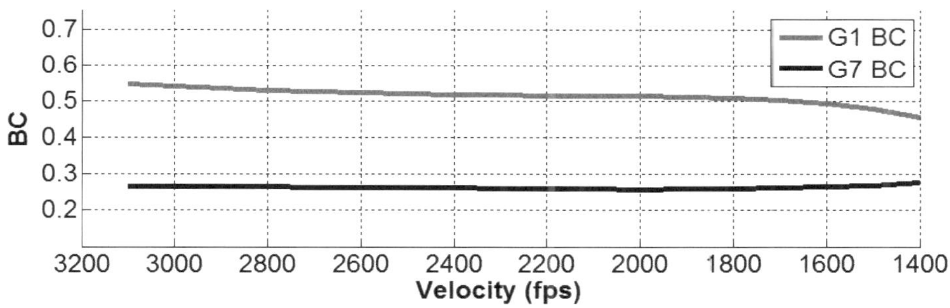

	Above 3000	3000-2500	2500-2000	2000-1500	Below 1500
G1 BC	0.558	0.531	0.517	0.495	0.478
G7 BC	0.264	0.263	0.260	0.264	0.269

Averages from 3000 fps to 1500 fps:
G1 BC: 0.513, variation of G1 BC: 0.064
G7 BC: 0.263, variation of G7 BC: 0.012
G7 form factor: 0.960

Rifling Twist Rate and Stability

Barrel Twist	Stability Factor Best case	Nominal	Worst case
1:10"	2.43	2.12	1.83
1:11"	2.01	1.75	1.52
1:12"	1.69	1.47	1.27
1:13"	1.44	1.26	1.09

Minimum twist required for stability is 1:13.3"
Recommended twist for optimal performance is 1:11.9" or faster

Berger .308 caliber 168 grain Target VLD

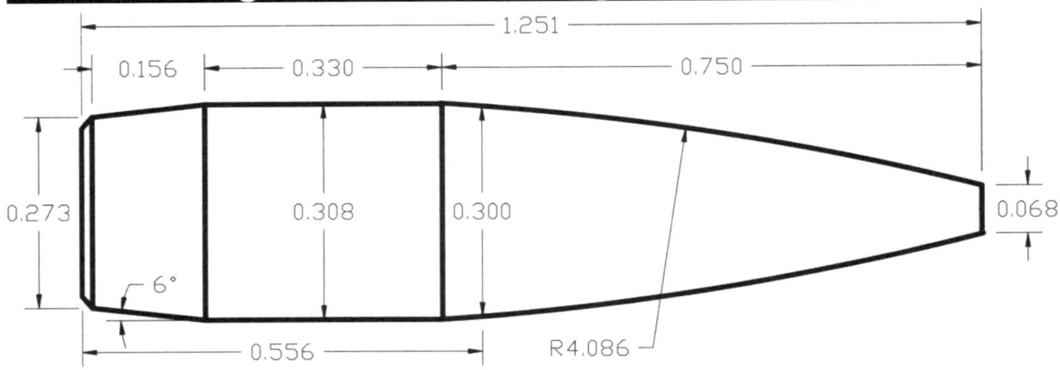

Lot #:	5641	Ogive Radius:	13.27 calibers
Sectional Density:	0.253 lb/in^2	Rt/R:	0.59

Ballistic Performance

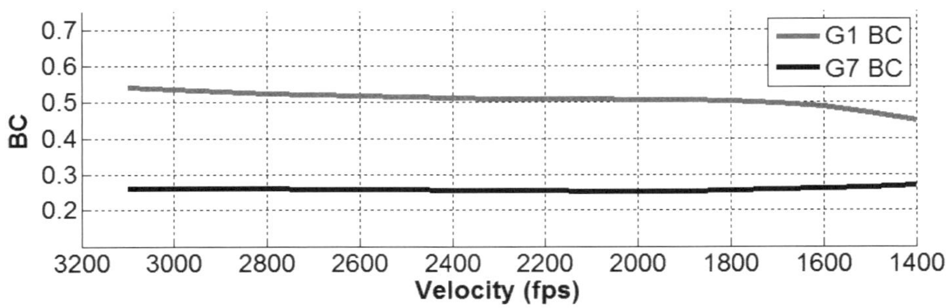

	Above 3000	3000-2500	2500-2000	2000-1500	Below 1500
G1 BC	0.553	0.524	0.510	0.489	0.472
G7 BC	0.262	0.260	0.256	0.260	0.266

Averages from 3000 fps to 1500 fps:
G1 BC: 0.507, variation of G1 BC: 0.064
G7 BC: 0.260, variation of G7 BC: 0.012
G7 form factor: 0.973

Rifling Twist Rate and Stability

Barrel Twist	Stability Factor Best case	Nominal	Worst case
1:11"	2.18	1.90	1.65
1:12"	1.83	1.60	1.38
1:13"	1.56	1.36	1.18
1:14"	1.34	1.17	1.02

Minimum twist required for stability is 1:13.9"
Recommended twist for optimal performance is 1:12.4" or faster

Berger .308 caliber 175 grain Target Long Range BT

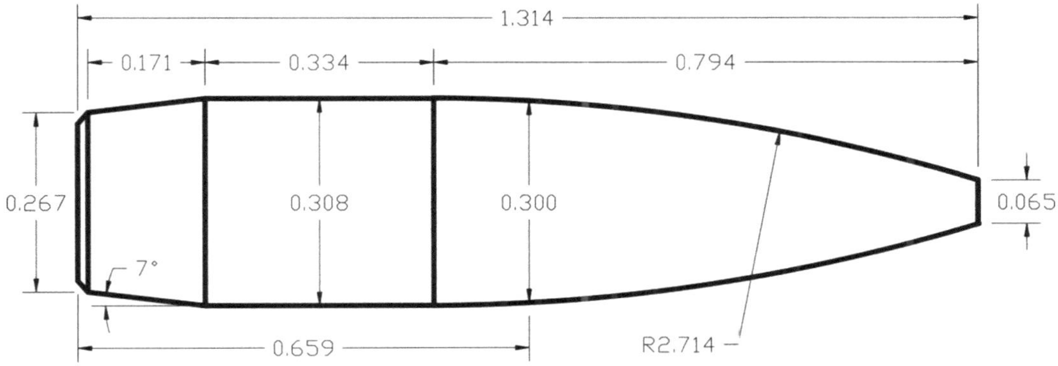

Lot #:	**5838**	**Ogive Radius:**	**8.81 calibers**
Sectional Density:	**0.264 lb/in^2**	**Rt/R:**	**0.98**

Ballistic Performance

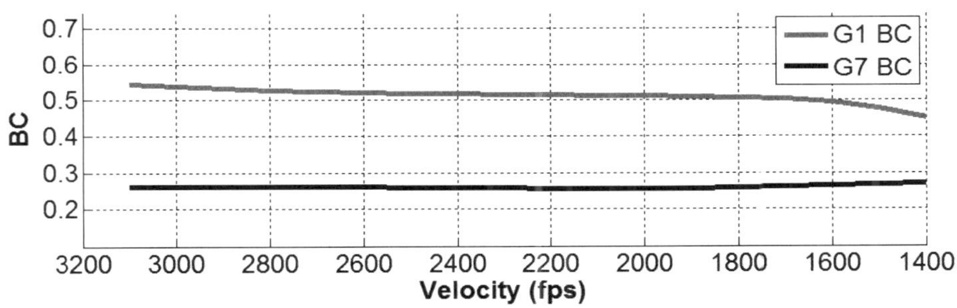

	Above 3000	3000-2500	2500-2000	2000-1500	Below 1500
G1 BC	0.555	0.529	0.515	0.493	0.475
G7 BC	0.263	0.262	0.259	0.262	0.268

Averages from 3000 fps to 1500 fps:
G1 BC: 0.511, variation of G1 BC: 0.064
G7 BC: 0.262, variation of G7 BC: 0.011
G7 form factor: 1.005

Rifling Twist Rate and Stability

Barrel Twist	Stability Factor Best case	Nominal	Worst case
1:10"	2.38	2.08	1.80
1:11"	1.97	1.72	1.49
1:12"	1.65	1.45	1.25
1:13"	1.41	1.23	1.06

Minimum twist required for stability is 1:13.2"
Recommended twist for optimal performance is 1:11.8" or faster

Berger .308 caliber 175 grain Target VLD

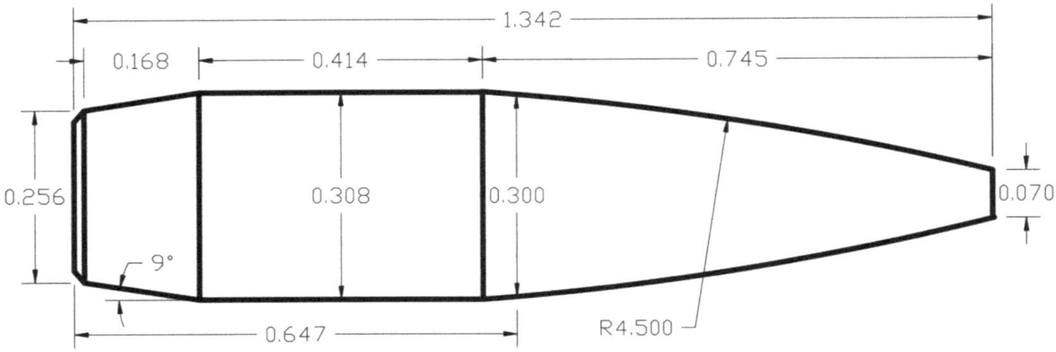

Lot #:	5471	Ogive Radius:	14.61 calibers
Sectional Density:	0.264 lb/in^2	Rt/R:	0.53

Ballistic Performance

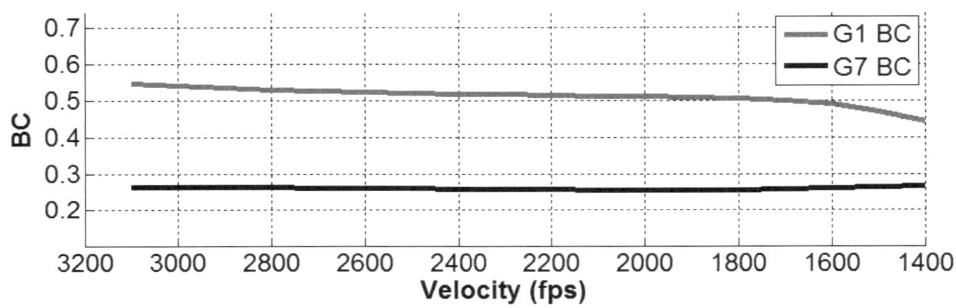

	Above 3000	3000-2500	2500-2000	2000-1500	Below 1500
G1 BC	0.559	0.531	0.516	0.490	0.470
G7 BC	0.265	0.263	0.259	0.261	0.264

Averages from 3000 fps to 1500 fps:
G1 BC: 0.511, variation of G1 BC: 0.072
G7 BC: 0.262, variation of G7 BC: 0.008
G7 form factor: 1.006

Rifling Twist Rate and Stability

Barrel Twist	Stability Factor Best case	Stability Factor Nominal	Stability Factor Worst case
1:10"	2.24	1.96	1.69
1:11"	1.85	1.62	1.40
1:12"	1.56	1.36	1.18
1:13"	1.33	1.16	1.00

Minimum twist required for stability is 1:12.8"
Recommended twist for optimal performance is 1:11.4" or faster

Berger .308 caliber 185 grain Target Hybrid

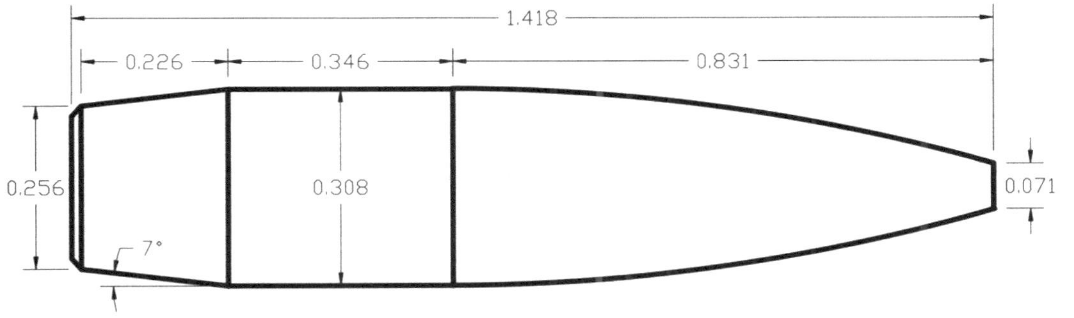

Lot #:	**6773**		**Ogive Radius:**	**Hybrid**
Sectional Density:	**0.279 lb/in^2**		**Rt/R:**	**Hybrid**

Ballistic Performance

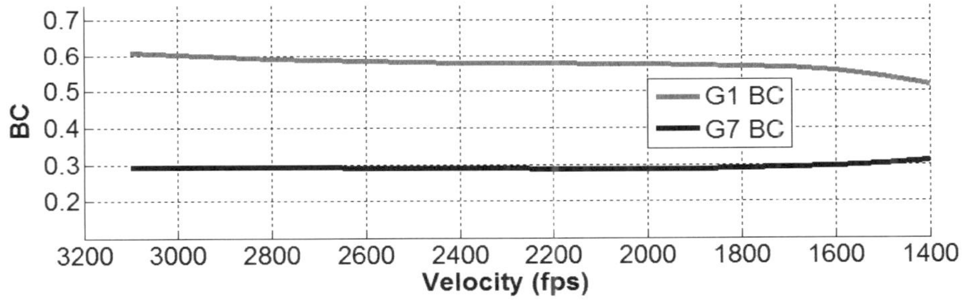

	Above 3000	3000-2500	2500-2000	2000-1500	Below 1500
G1 BC	0.618	0.591	0.578	0.558	0.540
G7 BC	0.293	0.293	0.291	0.297	0.304

Averages from 3000 fps to 1500 fps:

G1 BC: 0.575, variation of G1 BC: 0.061

G7 BC: 0.295, variation of G7 BC: 0.015

G7 form factor: 0.945

Rifling Twist Rate and Stability

Barrel Twist	**Stability Factor** **Best case**	**Nominal**	**Worst case**
1:9"	2.49	2.18	1.88
1:10"	2.02	1.76	1.52
1:11"	1.67	1.46	1.26
1:12"	1.40	1.22	1.06

Minimum twist required for stability is 1:12.1"

Recommended twist for optimal performance is 1:10.8" or faster

Berger .308 caliber 185 grain Target LRBT- Juggernaut

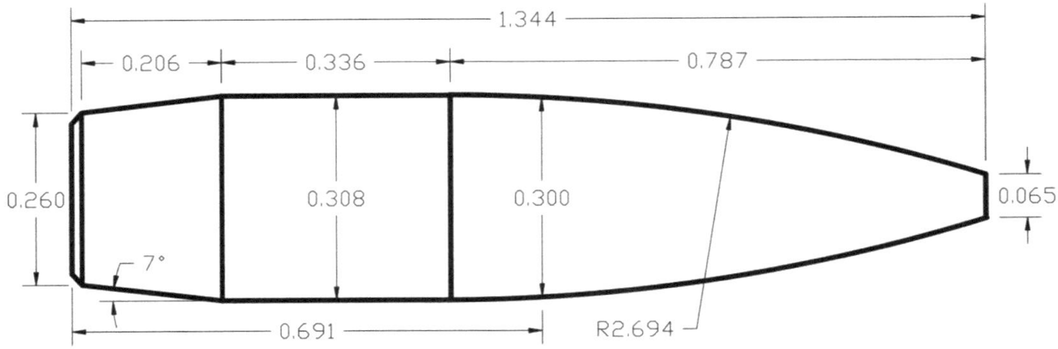

Lot #:	7555	Ogive Radius:	8.75 calibers
Sectional Density:	0.279 lb/in^2	Rt/R:	0.97

Ballistic Performance

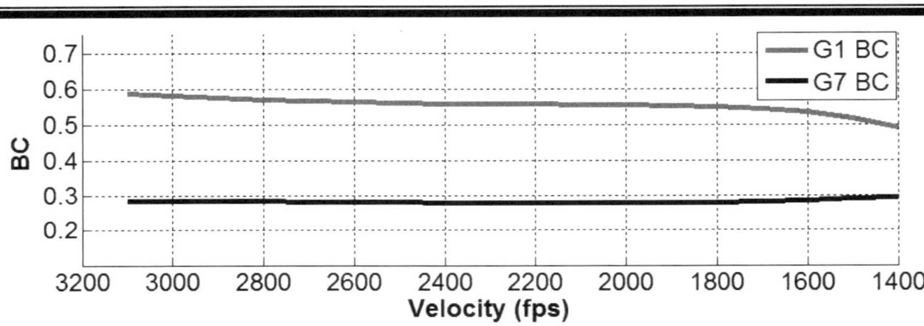

	Above 3000	3000-2500	2500-2000	2000-1500	Below 1500
G1 BC	0.599	0.572	0.558	0.536	0.516
G7 BC	0.284	0.284	0.281	0.285	0.291

Averages from 3000 fps to 1500 fps:
G1 BC: 0.554, variation of G1 BC: 0.065
G7 BC: 0.284, variation of G7 BC: 0.012
G7 form factor: 0.981

Rifling Twist Rate and Stability

Barrel Twist	Stability Factor Best case	Nominal	Worst case
1:10"	2.36	2.06	1.78
1:11"	1.95	1.70	1.47
1:12"	1.64	1.43	1.24
1:13"	1.40	1.22	1.05

Minimum twist required for stability is 1:13.1"
Recommended twist for optimal performance is 1:11.7" or faster

Berger .308 caliber 185 grain Target VLD

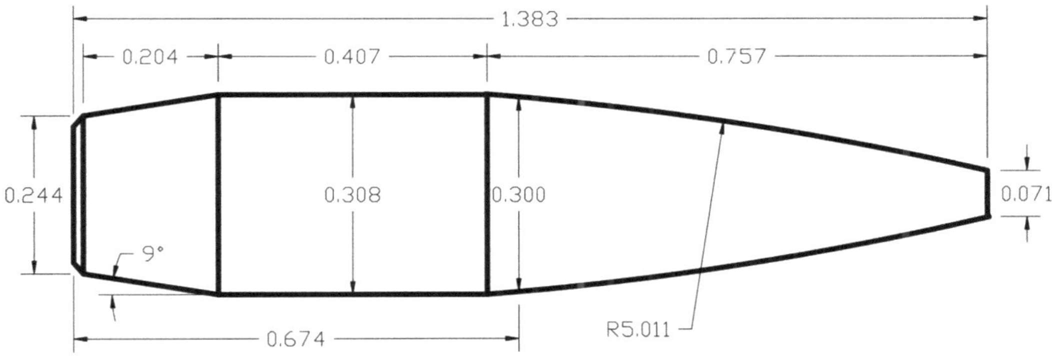

Lot #:	2959	Ogive Radius:	14.32 calibers
Sectional Density:	0.279 lb/in²	Rt/R:	0.56

Ballistic Performance

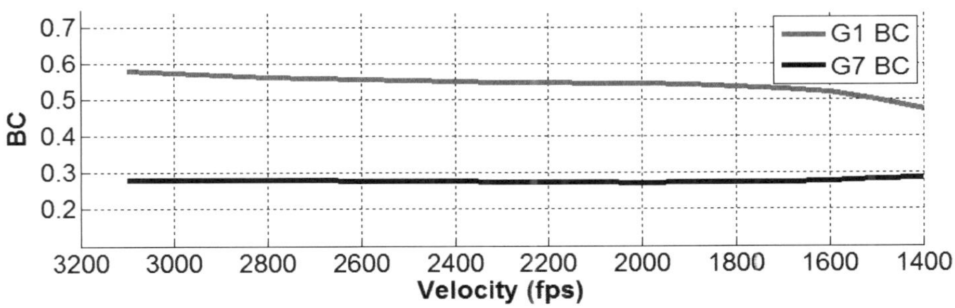

	Above 3000	3000-2500	2500-2000	2000-1500	Below 1500
G1 BC	0.590	0.562	0.547	0.521	0.500
G7 BC	0.279	0.279	0.275	0.277	0.282

Averages from 3000 fps to 1500 fps:
G1 BC: 0.542, variation of G1 BC: 0.073
G7 BC: 0.278, variation of G7 BC: 0.009
G7 form factor: 1.002

Rifling Twist Rate and Stability

Barrel Twist	Stability Factor Best case	Nominal	Worst case
1:10"	2.17	1.90	1.64
1:11"	1.79	1.57	1.36
1:12"	1.51	1.32	1.14
1:13"	1.28	1.12	0.97

Minimum twist required for stability is 1:12.6"
Recommended twist for optimal performance is 1:11.2" or faster

Berger .308 caliber 200 grain Target Hybrid

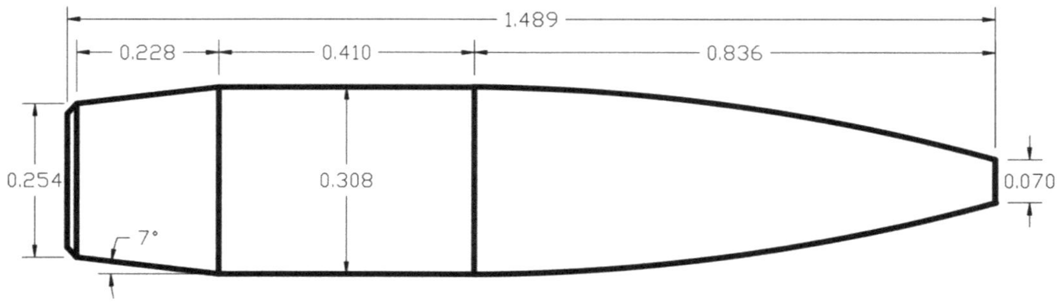

Lot #:	7669	Ogive Radius:	Hybrid
Sectional Density:	0.301 lb/in^2	Rt/R:	Hybrid

Ballistic Performance

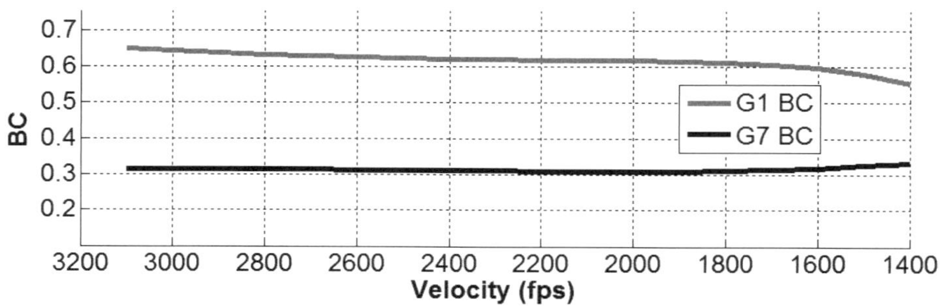

	Above 3000	3000-2500	2500-2000	2000-1500	Below 1500
G1 BC	0.661	0.632	0.619	0.597	0.578
G7 BC	0.313	0.314	0.311	0.318	0.326

Averages from 3000 fps to 1500 fps:
G1 BC: 0.615, variation of G1 BC: 0.065
G7 BC: 0.316, variation of G7 BC: 0.016
G7 form factor: 0.955

Rifling Twist Rate and Stability

Barrel Twist	Stability Factor Best case	Nominal	Worst case
1:9"	2.34	2.04	1.77
1:10"	1.89	1.65	1.43
1:11"	1.56	1.37	1.18
1:12"	1.31	1.15	0.99

Minimum twist required for stability is 1:11.7"
Recommended twist for optimal performance is 1:10.5" or faster

Berger .308 caliber 210 grain Target Long Range BT

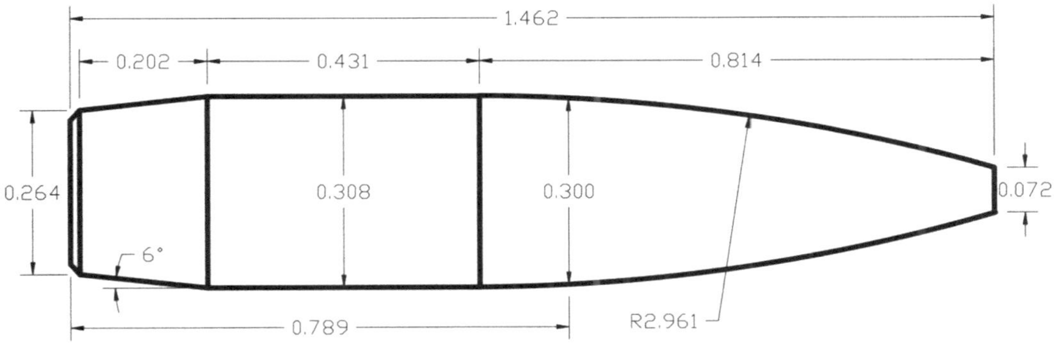

Lot #:	6304	Ogive Radius:	9.61 calibers
Sectional Density:	0.316 lb/in^2	Rt/R:	0.97

Ballistic Performance

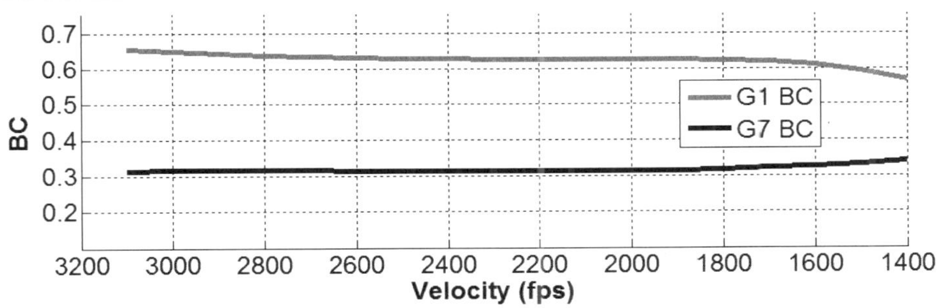

	Above 3000	3000-2500	2500-2000	2000-1500	Below 1500
G1 BC	0.665	0.638	0.627	0.609	0.592
G7 BC	0.315	0.317	0.315	0.324	0.334

Averages from 3000 fps to 1500 fps:
G1 BC: 0.624, variation of G1 BC: 0.056
G7 BC: 0.320, variation of G7 BC: 0.019
G7 form factor: 0.988

Rifling Twist Rate and Stability

Barrel Twist	Stability Factor Best case	Nominal	Worst case
1:10"	2.10	1.83	1.58
1:11"	1.73	1.51	1.31
1:12"	1.46	1.27	1.10
1:13"	1.24	1.08	0.94

Minimum twist required for stability is 1:12.4"
Recommended twist for optimal performance is 1:11.0" or faster

Berger .308 caliber 210 grain Target VLD

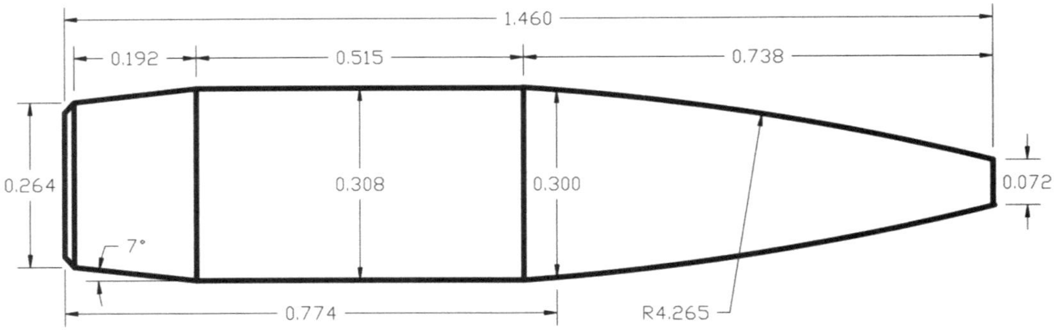

Lot #:	7397	Ogive Radius:	13.85 calibers
Sectional Density:	0.316 lb/in^2	Rt/R:	0.55

Ballistic Performance

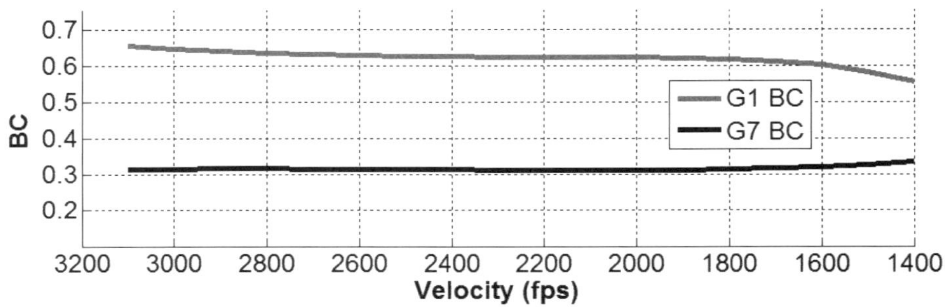

	Above 3000	3000-2500	2500-2000	2000-1500	Below 1500
G1 BC	0.665	0.637	0.624	0.602	0.582
G7 BC	0.315	0.316	0.314	0.320	0.328

Averages from 3000 fps to 1500 fps:
G1 BC: 0.619, variation of G1 BC: 0.065
G7 BC: 0.318, variation of G7 BC: 0.016
G7 form factor: 0.995

Rifling Twist Rate and Stability

Barrel Twist	Stability Factor Best case	Nominal	Worst case
1:10"	2.10	1.84	1.59
1:11"	1.74	1.52	1.31
1:12"	1.46	1.28	1.10
1:13"	1.25	1.09	0.94

Minimum twist required for stability is 1:12.4"
Recommended twist for optimal performance is 1:11.1" or faster

Berger .308 caliber 215 grain Target Hybrid

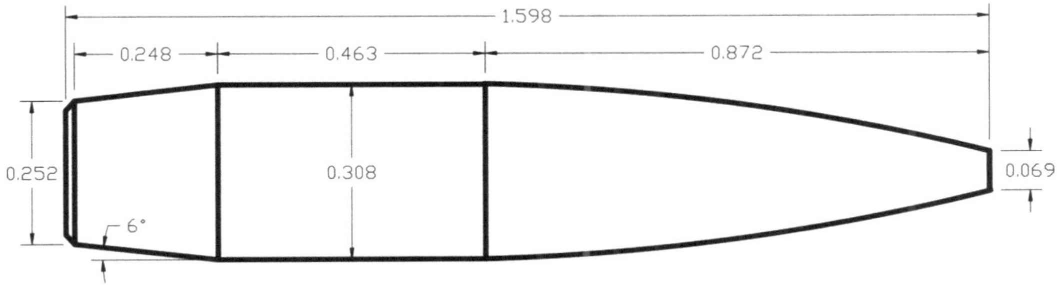

Lot #:	**6326**	**Ogive Radius:**	**Hybrid**
Sectional Density:	**0.324 lb/in^2**	**Rt/R:**	**Hybrid**

Ballistic Performance

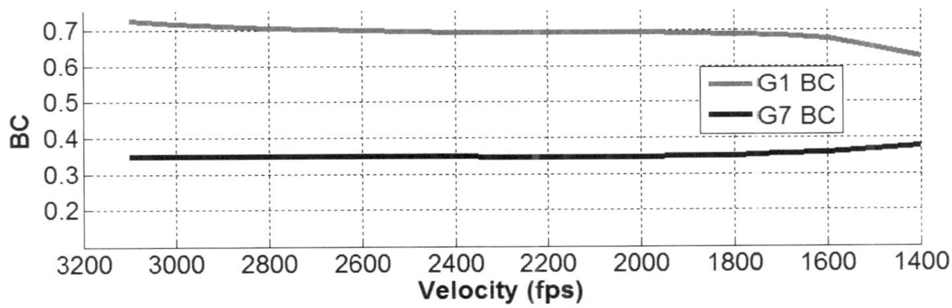

	Above 3000	3000-2500	2500-2000	2000-1500	Below 1500
G1 BC	0.738	0.706	0.693	0.672	0.653
G7 BC	0.349	0.350	0.349	0.358	0.368

Averages from 3000 fps to 1500 fps:
G1 BC: 0.689, variation of G1 BC: 0.065
G7 BC: 0.354, variation of G7 BC: 0.021
G7 form factor: 0.915

Rifling Twist Rate and Stability

Barrel Twist	Stability Factor Best case	Nominal	Worst case
1:8"	2.59	2.26	1.95
1:9"	2.04	1.78	1.54
1:10"	1.65	1.45	1.25
1:11"	1.37	1.19	1.03

Minimum twist required for stability is 1:11.0"
Recommended twist for optimal performance is 1:9 8" or faster

Berger .308 caliber 230 grain Target Hybrid

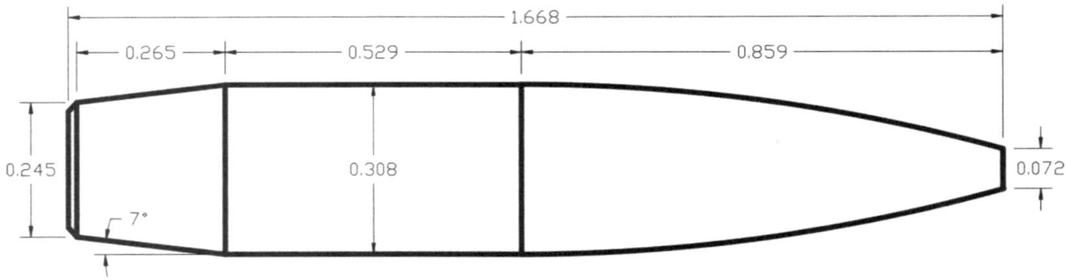

Lot #:	5341	Ogive Radius:	Hybrid
Sectional Density:	0.346 lb/in^2	Rt/R:	Hybrid

Ballistic Performance

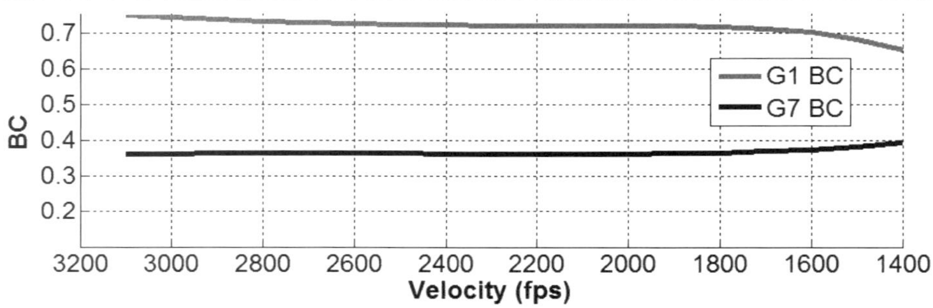

	Above 3000	3000-2500	2500-2000	2000-1500	Below 1500
G1 BC	0.762	0.733	0.721	0.700	0.680
G7 BC	0.361	0.363	0.363	0.373	0.384

Averages from 3000 fps to 1500 fps:
G1 BC: 0.717, variation of G1 BC: 0.063
G7 BC: 0.368, variation of G7 BC: 0.022
G7 form factor: 0.942

Rifling Twist Rate and Stability

Barrel Twist	Stability Factor Best case	Nominal	Worst case
1:8"	2.44	2.13	1.84
1:9"	1.93	1.68	1.46
1:10"	1.56	1.36	1.18
1:11"	1.29	1.13	0.97

Minimum twist required for stability is 1:10.7"
Recommended twist for optimal performance is 1:9.5" or faster

Cutting Edge .308 caliber 180 grain HPBT

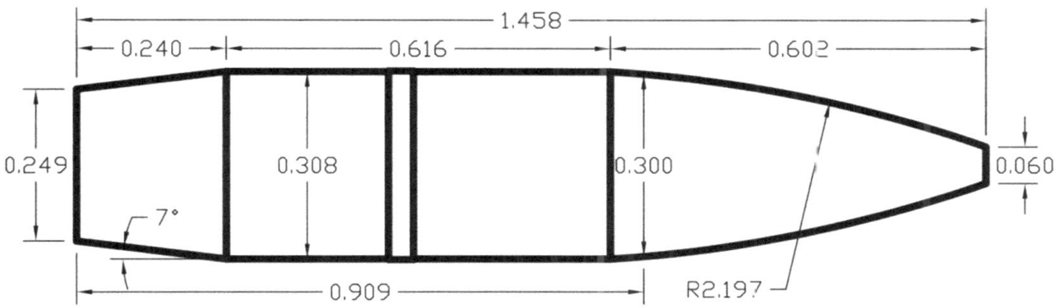

Lot #:	**Unknown**	**Ogive Radius:**	**7.13 calibers**
Sectional Density:	**0.271 lb/in^2**	**Rt/R:**	**0.70**

Ballistic Performance

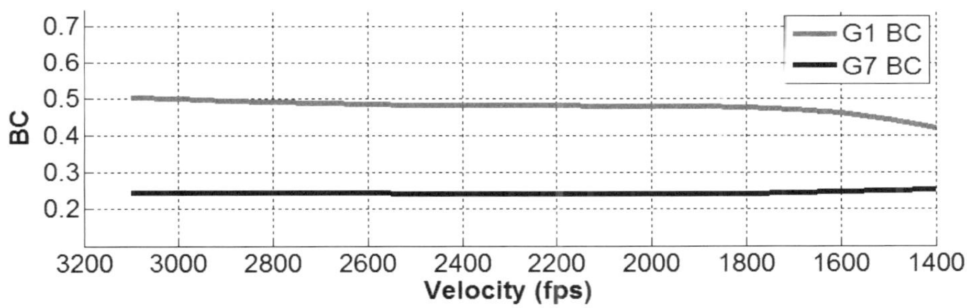

	Above 3000	3000-2500	2500-2000	2000-1500	Below 1500
G1 BC	0.512	0.491	0.482	0.462	0.445
G7 BC	0.242	0.244	0.242	0.246	0.250

Averages from 3000 fps to 1500 fps:
G1 BC: 0.477, variation of G1 BC: 0.054
G7 BC: 0.245, variation of G7 BC: 0.010
G7 form factor: 1.108

Rifling Twist Rate and Stability

Barrel Twist	Stability Factor Best case	Nominal	Worst case
1:10"	1.81	1.58	1.37
1:11"	1.50	1.31	1.13
1:12"	1.26	1.10	0.95
1:13"	1.07	0.94	0.81

Minimum twist required for stability is 1:11.5"
Recommended twist for optimal performance is 1:10.3" or faster

Cutting Edge .308 caliber 165 grain Lazer

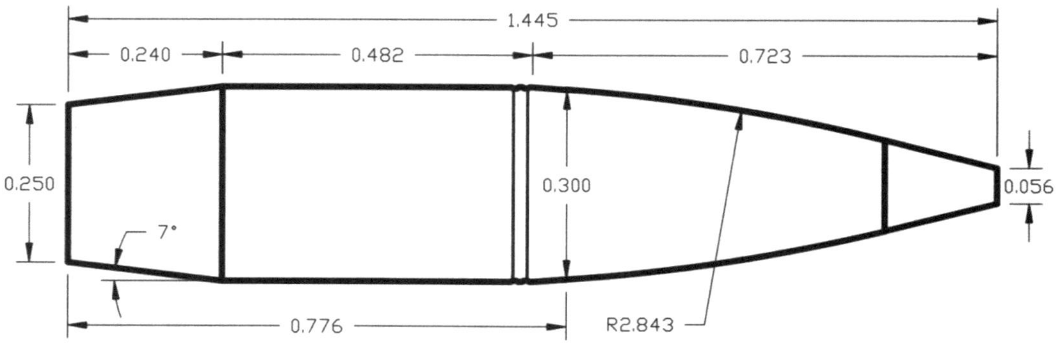

Lot #:	Unknown	Ogive Radius:	9.23 calibers
Sectional Density:	0.249 lb/in²	Rt/R:	0.67

Ballistic Performance

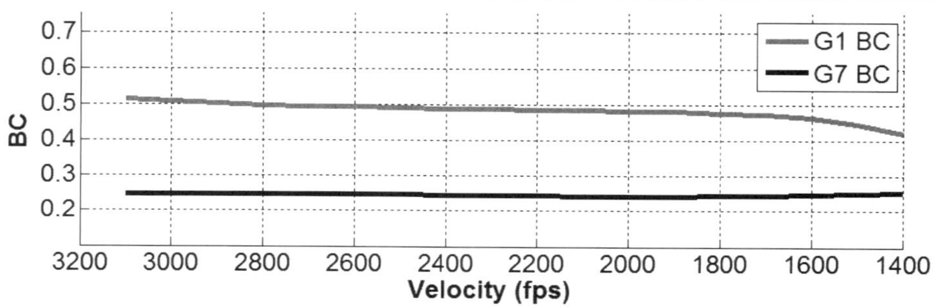

	Above 3000	3000-2500	2500-2000	2000-1500	Below 1500
G1 BC	0.524	0.499	0.486	0.464	0.446
G7 BC	0.248	0.248	0.245	0.247	0.251

Averages from 3000 fps to 1500 fps:
G1 BC: 0.482, variation of G1 BC: 0.063
G7 BC: 0.247, variation of G7 BC: 0.008
G7 form factor: 1.005

Rifling Twist Rate and Stability

Barrel Twist	Stability Factor Best case	Nominal	Worst case
1:9"	2.10	1.84	1.59
1:10"	1.70	1.49	1.29
1:11"	1.41	1.23	1.06
1:12"	1.18	1.03	0.89

Minimum twist required for stability is 1:11.1"
Recommended twist for optimal performance is 1:10.0" or faster

Cutting Edge .308 caliber 180 grain LRHPBT-Max

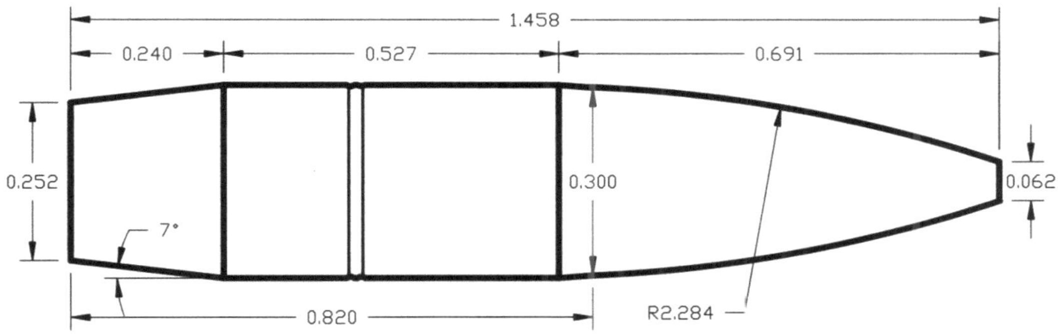

Lot #:	**Unknown**	**Ogive Radius:**	**7.42 calibers**
Sectional Density:	**0.271 lb/in^2**	**Rt/R:**	**0.88**

Ballistic Performance

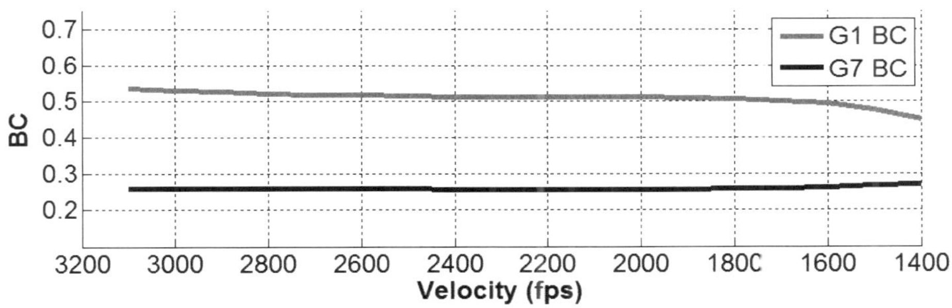

	Above 3000	3000-2500	2500-2000	2000-1500	Below 1500
G1 BC	0.544	0.522	0.512	0.492	0.475
G7 BC	0.258	0.259	0.257	0.262	0.267

Averages from 3000 fps to 1500 fps:
G1 BC: 0.507, variation of G1 BC: 0.055
G7 BC: 0.260, variation of G7 BC: 0.011
G7 form factor: 1.041

Rifling Twist Rate and Stability

	Stability Factor		
Barrel Twist	**Best case**	**Nominal**	**Worst case**
1:9"	2.24	1.95	1.69
1:10"	1.81	1.58	1.37
1:11"	1.50	1.31	1.13
1:12"	1.26	1.10	0.95

Minimum twist required for stability is 1:11.5"
Recommended twist for optimal performance is 1:10.3" or faster

Cutting Edge .308 caliber 180 grain LZR 180 MAX

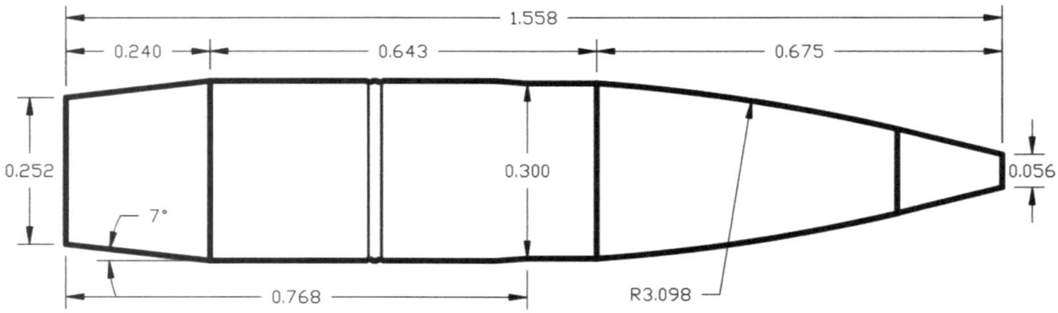

Lot #:	Unknown	Ogive Radius:	10.06 calibers
Sectional Density:	0.271 lb/in^2	Rt/R:	0.51

Ballistic Performance

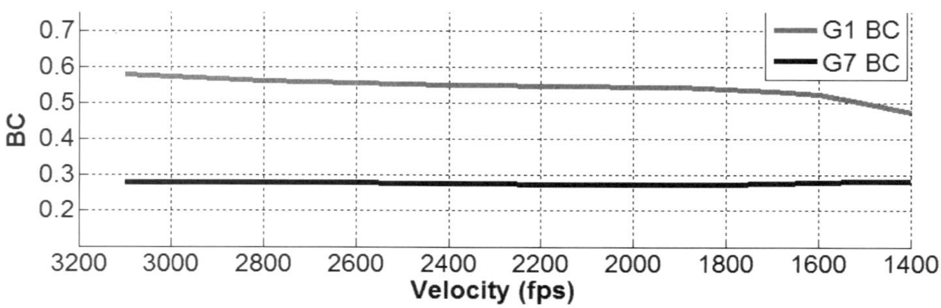

	Above 3000	3000-2500	2500-2000	2000-1500	Below 1500
G1 BC	0.591	0.563	0.549	0.523	0.501
G7 BC	0.280	0.279	0.276	0.278	0.282

Averages from 3000 fps to 1500 fps:
G1 BC: 0.543, variation of G1 BC: 0.073
G7 BC: 0.279, variation of G7 BC: 0.008
G7 form factor: 0.973

Rifling Twist Rate and Stability

Barrel Twist	Stability Factor Best case	Stability Factor Nominal	Stability Factor Worst case
1:9"	1.84	1.61	1.39
1:10"	1.49	1.30	1.13
1:11"	1.23	1.08	0.93
1:12"	1.04	0.91	0.78

Minimum twist required for stability is 1:10.4"
Recommended twist for optimal performance is 1:9.3" or faster

Cutting Edge .308 caliber 180 grain LZR 180 STD

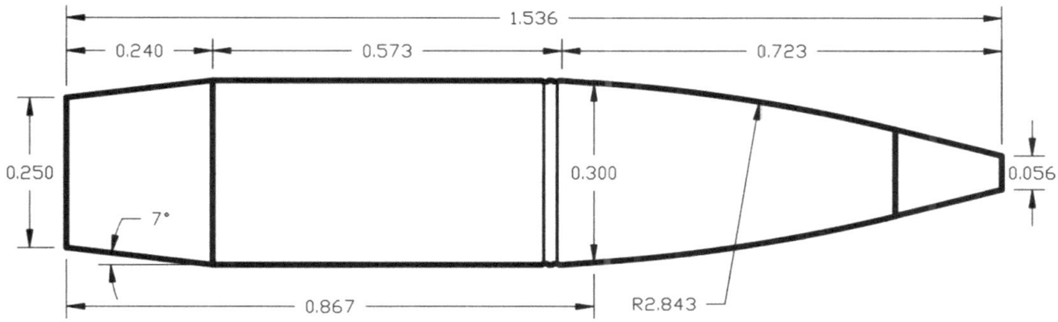

Lot #:	Unknown	Ogive Radius:	9.23 calibers
Sectional Density:	0.271 lb/in^2	Rt/R:	0.67

Ballistic Performance

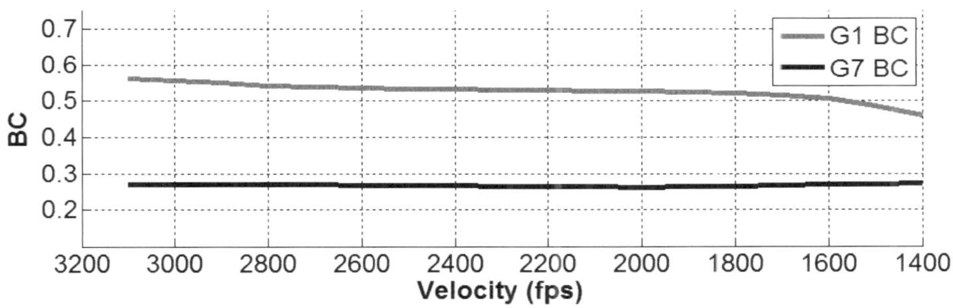

	Above 3000	3000-2500	2500-2000	2000-1500	Below 1500
G1 BC	0.571	0.543	0.529	0.505	0.484
G7 BC	0.270	0.270	0.266	0.268	0.273

Averages from 3000 fps to 1500 fps:

G1 BC: 0.524, variation of G1 BC: 0.069

G7 BC: 0.269, variation of G7 BC: 0.009

G7 form factor: 1.008

Rifling Twist Rate and Stability

Barrel Twist	Stability Factor Best case	Nominal	Worst case
1:9"	1.92	1.68	1.45
1:10"	1.56	1.36	1.18
1:11"	1.29	1.12	0.97
1:12"	1.08	0.94	0.82

Minimum twist required for stability is 1:10.6"

Recommended twist for optimal performance is 1:9.5" or faster

Cutting Edge .308 caliber 165 grain MTH

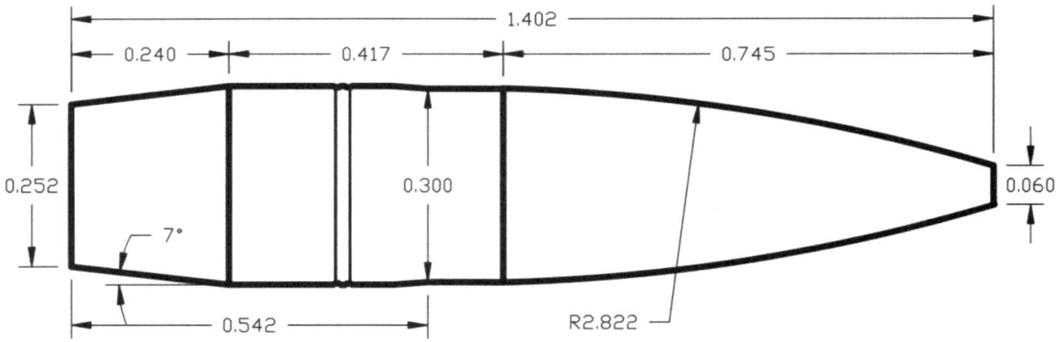

Lot #:	Unknown	Ogive Radius:	9.16 calibers
Sectional Density:	0.249 lb/in^2	Rt/R:	0.82

Ballistic Performance

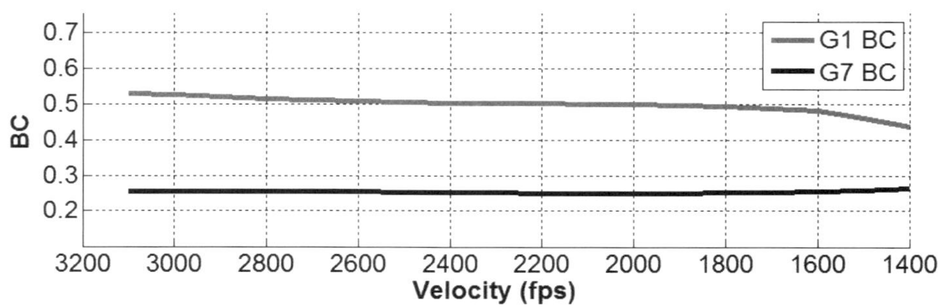

	Above 3000	3000-2500	2500-2000	2000-1500	Below 1500
G1 BC	0.540	0.515	0.502	0.481	0.463
G7 BC	0.256	0.256	0.253	0.256	0.261

Averages from 3000 fps to 1500 fps:
G1 BC: 0.498, variation of G1 BC: 0.062
G7 BC: 0.256, variation of G7 BC: 0.010
G7 form factor: 0.972

Rifling Twist Rate and Stability

Barrel Twist	Stability Factor Best case	Nominal	Worst case
1:10"	1.86	1.63	1.41
1:11"	1.54	1.34	1.16
1:12"	1.29	1.13	0.98
1:13"	1.10	0.96	0.83

Minimum twist required for stability is 1:11.6"
Recommended twist for optimal performance is 1:10.4" or faster

Cutting Edge .308 caliber 180 grain MTH C21

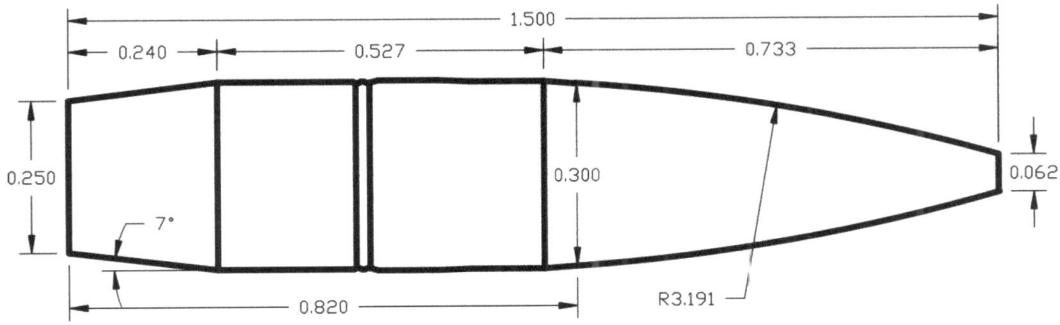

Lot #:	**Unknown**	**Ogive Radius:**	**10.36 calibers**
Sectional Density:	**0.271 lb/in²**	**Rt/R:**	**0.70**

Ballistic Performance

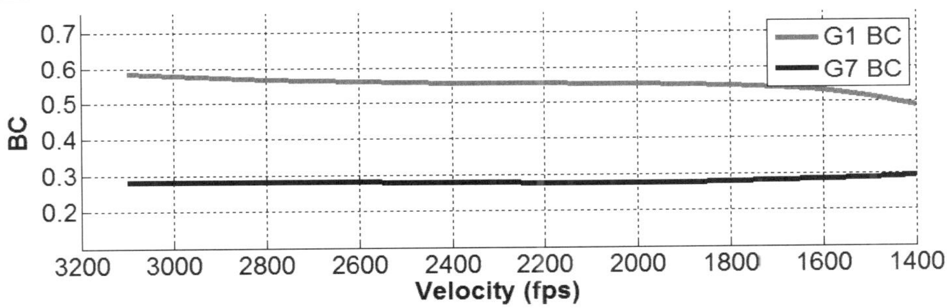

	Above 3000	3000-2500	2500-2000	2000-1500	Below 1500
G1 BC	0.595	0.569	0.556	0.533	0.513
G7 BC	0.282	0.282	0.279	0.283	0.289

Averages from 3000 fps to 1500 fps:

G1 BC: 0.551, variation of G1 BC: 0.066

G7 BC: 0.283, variation of G7 BC: 0.011

G7 form factor: 0.959

Rifling Twist Rate and Stability

Barrel Twist	Stability Factor Best case	Nominal	Worst case
1:9"	2.06	1.80	1.55
1:10"	1.67	1.46	1.26
1:11"	1.38	1.20	1.04
1:12"	1.16	1.01	0.87

Minimum twist required for stability is 1:11.0"

Recommended twist for optimal performance is 1:9.9" or faster

DAG .308 caliber 147 grain FMJ BT

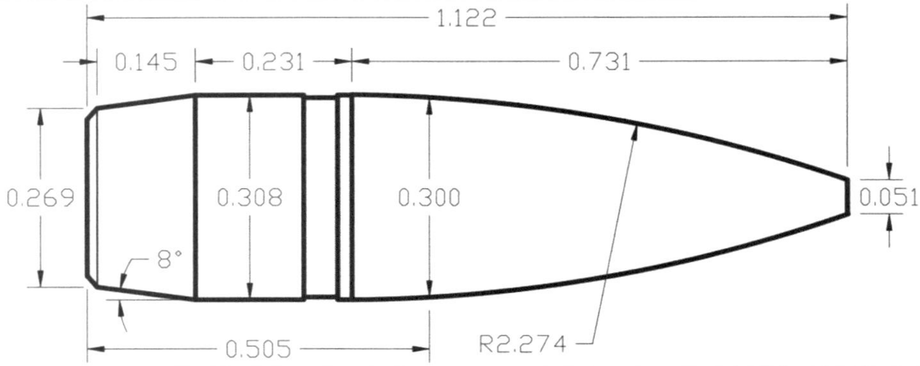

Lot #:	**Unknown**	**Ogive Radius:**	**7.38 calibers**
Sectional Density:	**0.221 lb/in^2**	**Rt/R:**	**0.94**

Ballistic Performance

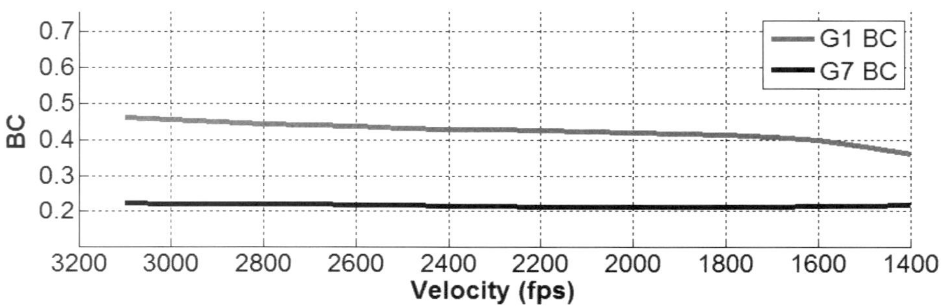

	Above 3000	3000-2500	2500-2000	2000-1500	Below 1500
G1 BC	0.472	0.444	0.427	0.403	0.383
G7 BC	0.223	0.220	0.215	0.214	0.216

Averages from 3000 fps to 1500 fps:
G1 BC: 0.423, variation of G1 BC: 0.072
G7 BC: 0.217, variation of G7 BC: 0.011
G7 form factor: 1.020

Rifling Twist Rate and Stability

Barrel Twist	Stability Factor Best case	Stability Factor Nominal	Stability Factor Worst case
1:12"	2.19	1.91	1.65
1:13"	1.87	1.63	1.41
1:14"	1.61	1.41	1.22
1:15"	1.40	1.22	1.06

Minimum twist required for stability is 1:15.2"
Recommended twist for optimal performance is 1:13.6" or faster

Dietlein .308 caliber 155.5 grain ULD

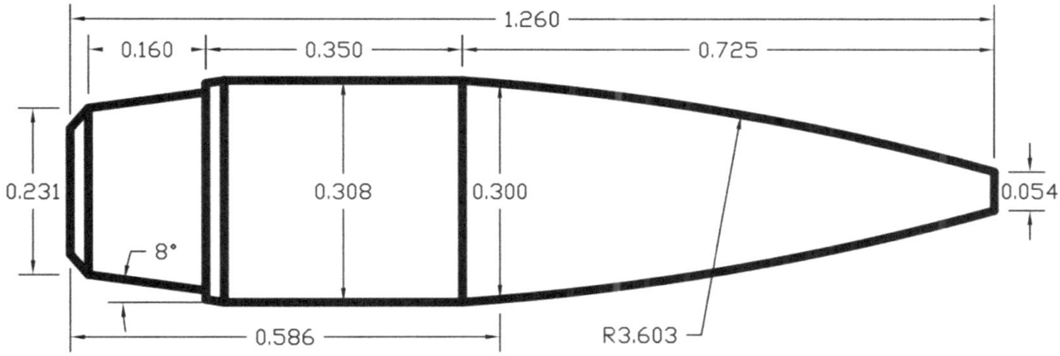

Lot #:	**21**	**Ogive Radius:**	**13.42 calibers**
Sectional Density:	**0.234 lb/in^2**	**Rt/R:**	**0.57**

Ballistic Performance

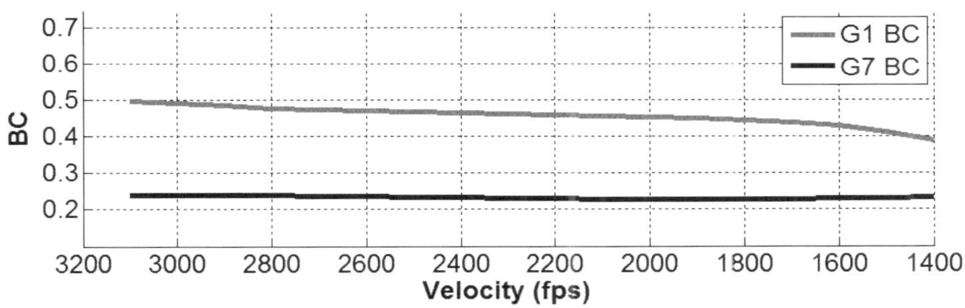

	Above 3000	3000-2500	2500-2000	2000-1500	Below 1500
G1 BC	0.507	0.478	0.460	0.432	0.411
G7 BC	0.240	0.237	0.231	0.229	0.231

Averages from 3000 fps to 1500 fps:
G1 BC: 0.455, variation of G1 BC: 0.079
G7 BC: 0.233, variation of G7 BC: 0.012
G7 form factor: 1.004

Rifling Twist Rate and Stability

Barrel Twist	**Stability Factor** Best case	**Stability Factor** Nominal	**Stability Factor** Worst case
1:10"	2.39	2.09	1.81
1:11"	1.97	1.73	1.49
1:12"	1.66	1.45	1.25
1:13"	1.41	1.24	1.07

Minimum twist required for stability is 1:13.2"
Recommended twist for optimal performance is 1:11.8" or faster

Dietlein .308 caliber 168 grain ULD

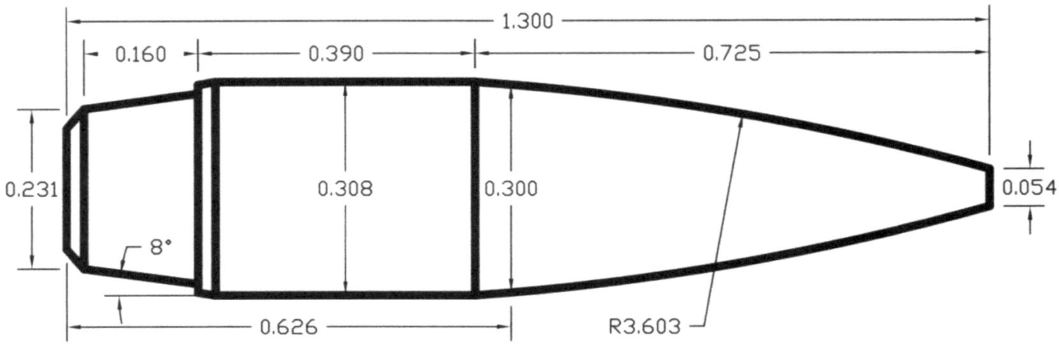

Lot #:	14	Ogive Radius:	13.42 calibers
Sectional Density:	0.253 lb/in^2	Rt/R:	0.57

Ballistic Performance

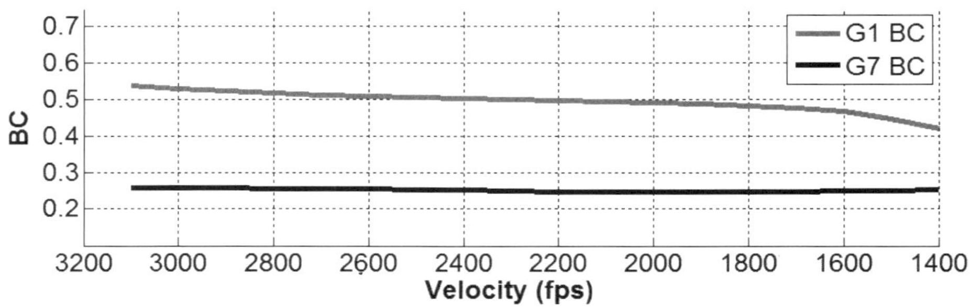

	Above 3000	3000-2500	2500-2000	2000-1500	Below 1500
G1 BC	0.548	0.517	0.498	0.469	0.447
G7 BC	0.260	0.256	0.250	0.249	0.251

Averages from 3000 fps to 1500 fps:
G1 BC: 0.493, variation of G1 BC: 0.083
G7 BC: 0.253, variation of G7 BC: 0.012
G7 form factor: 1.001

Rifling Twist Rate and Stability

Barrel Twist	Stability Factor Best case	Nominal	Worst case
1:10"	2.36	2.06	1.78
1:11"	1.95	1.70	1.47
1:12"	1.64	1.43	1.24
1:13"	1.40	1.22	1.05

Minimum twist required for stability is 1:13.1"
Recommended twist for optimal performance is 1:11.7" or faster

Dietlein .308 caliber 173 grain ULD

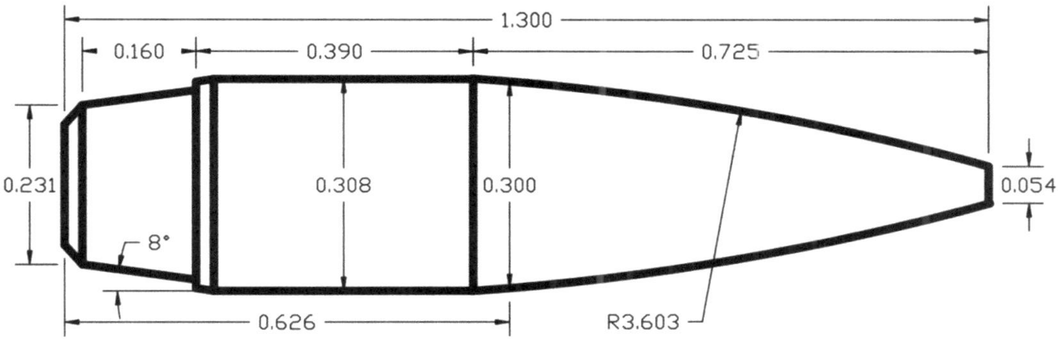

Lot #:	19	Ogive Radius:	13.42 calibers
Sectional Density:	0.261 lb/in^2	Rt/R:	0.57

Ballistic Performance

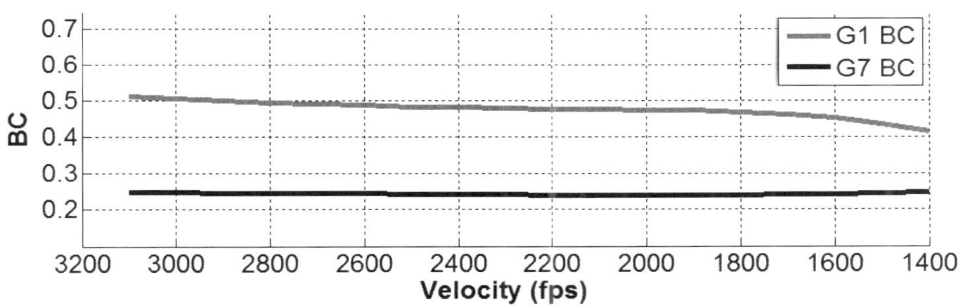

	Above 3000	3000-2500	2500-2000	2000-1500	Below 1500
G1 BC	0.520	0.494	0.479	0.455	0.436
G7 BC	0.246	0.245	0.241	0.242	0.246

Averages from 3000 fps to 1500 fps:
G1 BC: 0.475, variation of G1 BC: 0.068
G7 BC: 0.243, variation of G7 BC: 0.008
G7 form factor: 1.070

Rifling Twist Rate and Stability

Barrel Twist	Stability Factor Best case	Nominal	Worst case
1:10"	2.43	2.12	1.83
1:11"	2.01	1.75	1.52
1:12"	1.69	1.47	1.27
1:13"	1.44	1.26	1.09

Minimum twist required for stability is 1:13.3"
Recommended twist for optimal performance is 1:11.9" or faster

Dietlein .308 caliber 175 grain ULD

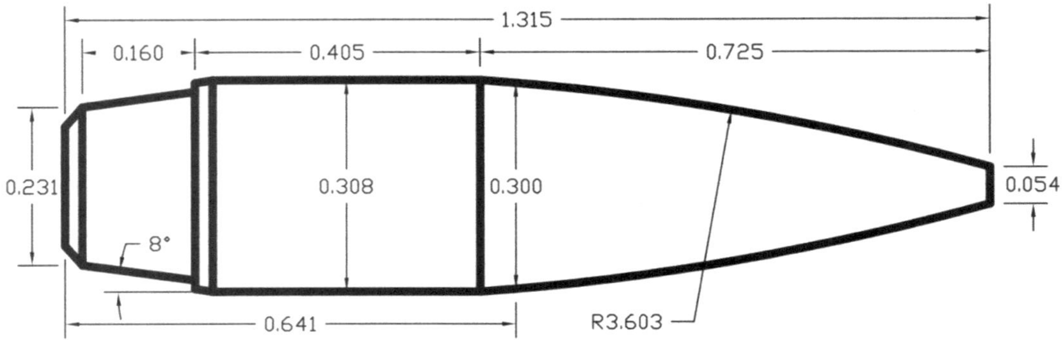

Lot #:	16PW	Ogive Radius:	13.42 calibers
Sectional Density:	0.264 lb/in^2	Rt/R:	0.57

Ballistic Performance

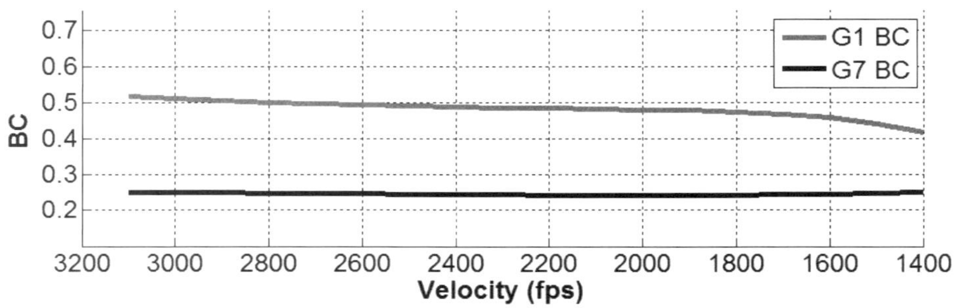

	Above 3000	3000-2500	2500-2000	2000-1500	Below 1500
G1 BC	0.527	0.500	0.485	0.461	0.442
G7 BC	0.250	0.248	0.244	0.245	0.249

Averages from 3000 fps to 1500 fps:
G1 BC: 0.481, variation of G1 BC: 0.069
G7 BC: 0.246, variation of G7 BC: 0.009
G7 form factor: 1.069

Rifling Twist Rate and Stability

	Stability Factor		
Barrel Twist	Best case	Nominal	Worst case
1:10"	2.43	2.12	1.83
1:11"	2.01	1.75	1.52
1:12"	1.69	1.47	1.27
1:13"	1.44	1.26	1.09

Minimum twist required for stability is 1:13.4"
Recommended twist for optimal performance is 1:12.0" or faster

Dietlein .308 caliber 185 grain ULD

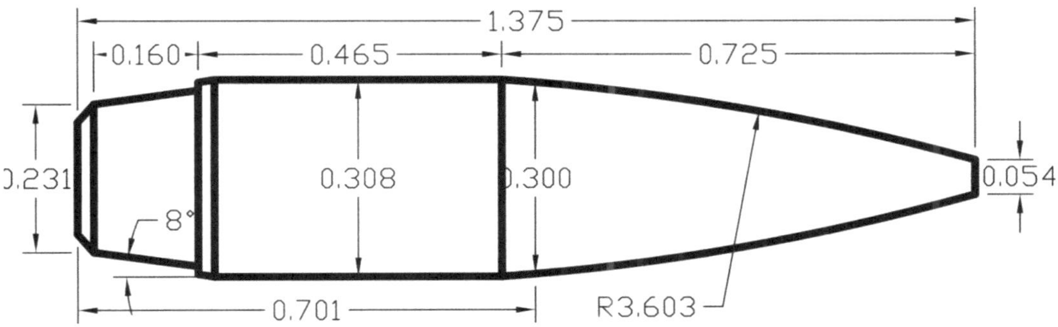

Lot #:	18	Ogive Radius:	13.42 calibers
Sectional Density:	0.279 lb/in^2	Rt/R:	0.57

Ballistic Performance

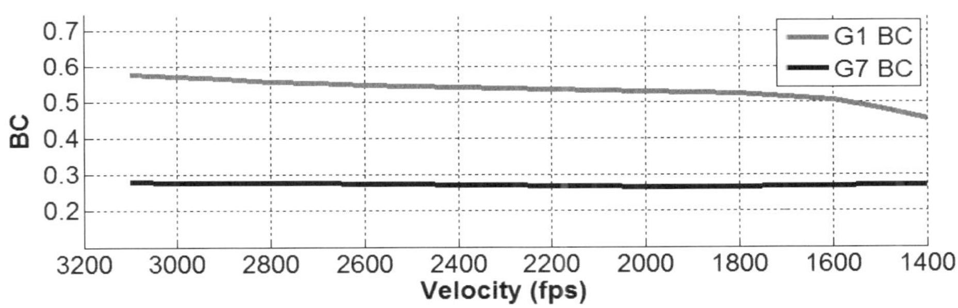

	Above 3000	3000-2500	2500-2000	2000-1500	Below 1500
G1 BC	0.589	0.556	0.537	0.506	0.482
G7 BC	0.279	0.276	0.270	0.269	0.271

Averages from 3000 fps to 1500 fps:
G1 BC: 0.531, variation of G1 BC: 0.087
G7 BC: 0.272, variation of G7 BC: 0.012
G7 form factor: 1.023

Rifling Twist Rate and Stability

Barrel Twist	Stability Factor Best case	Nominal	Worst case
1:10"	2.21	1.93	1.67
1:11"	1.82	1.59	1.38
1:12"	1.53	1.34	1.16
1:13"	1.31	1.14	0.99

Minimum twist required for stability is 1:12.7"
Recommended twist for optimal performance is 1:11.3" or faster

Dietlein .308 caliber 190 grain ULD

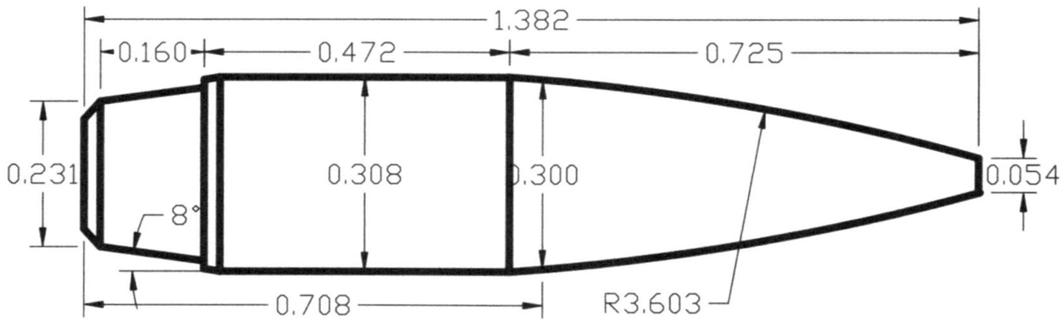

Lot #:	12-1	Ogive Radius:	13.42 calibers
Sectional Density:	0.286 lb/in²	Rt/R:	0.57

Ballistic Performance

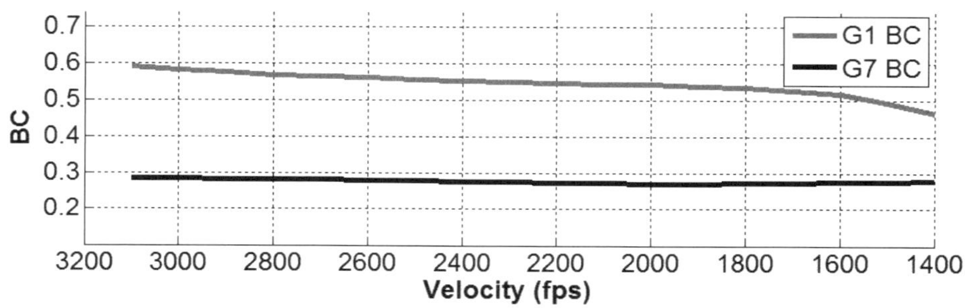

	Above 3000	3000-2500	2500-2000	2000-1500	Below 1500
G1 BC	0.603	0.569	0.549	0.518	0.493
G7 BC	0.285	0.282	0.276	0.275	0.278

Averages from 3000 fps to 1500 fps:
G1 BC: 0.544, variation of G1 BC: 0.089
G7 BC: 0.279, variation of G7 BC: 0.013
G7 form factor: 1.027

Rifling Twist Rate and Stability

Barrel Twist	Stability Factor Best case	Nominal	Worst case
1:10"	2.23	1.95	1.69
1:11"	1.85	1.61	1.39
1:12"	1.55	1.36	1.17
1:13"	1.32	1.15	1.00

Minimum twist required for stability is 1:12.8"
Recommended twist for optimal performance is 1:11.4" or faster

DRT .308 caliber 175 grain Frangible

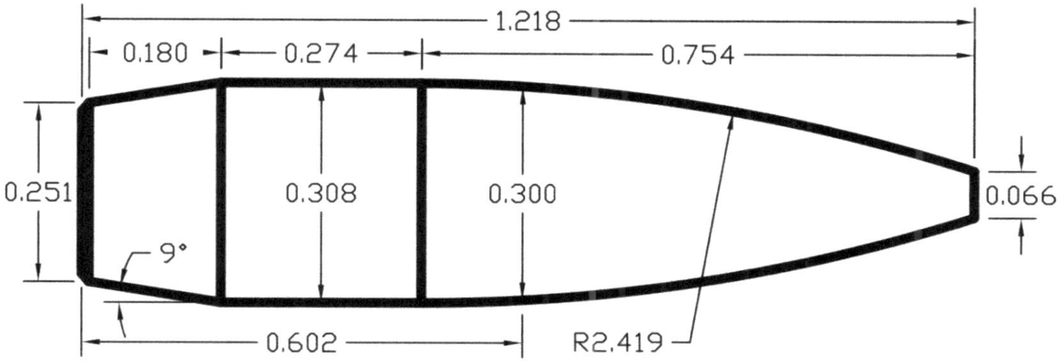

Lot #:	Unknown	Ogive Radius:	7.86 calibers
Sectional Density:	0.264 lb/in^2	Rt/R:	0.99

Ballistic Performance

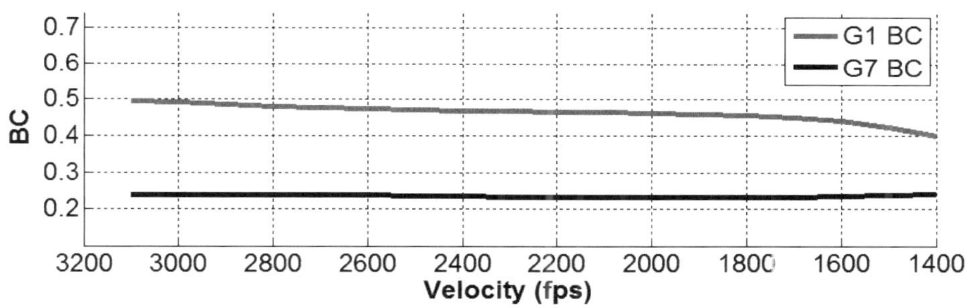

	Above 3000	3000-2500	2500-2000	2000-1500	Below 1500
G1 BC	0.505	0.480	0.466	0.443	0.424
G7 BC	0.239	0.238	0.234	0.235	0.239

Averages from 3000 fps to 1500 fps:
G1 BC: 0.462, variation of G1 BC: 0.066
G7 BC: 0.237, variation of G7 BC: 0.008
G7 form factor: 1.113

Rifling Twist Rate and Stability

Barrel Twist	Stability Factor Best case	Nominal	Worst case
1:10"	2.97	2.59	2.24
1:11"	2.45	2.14	1.85
1:12"	2.06	1.80	1.56
1:13"	1.75	1.53	1.33

Minimum twist required for stability is 1:14.7"
Recommended twist for optimal performance is 1:13.1" or faster

DRT .308 caliber 200 grain Frangible

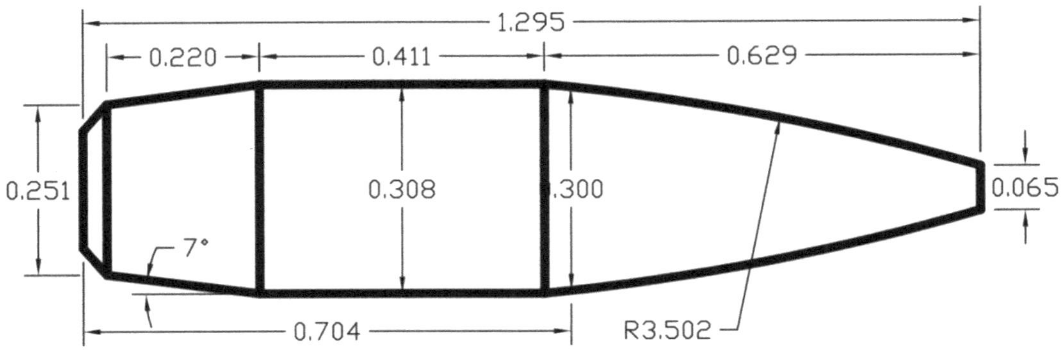

Lot #:	**Unknown**	**Ogive Radius:**	**11.37 calibers**
Sectional Density:	**0.301 lb/in^2**	**Rt/R:**	**0.52**

Ballistic Performance

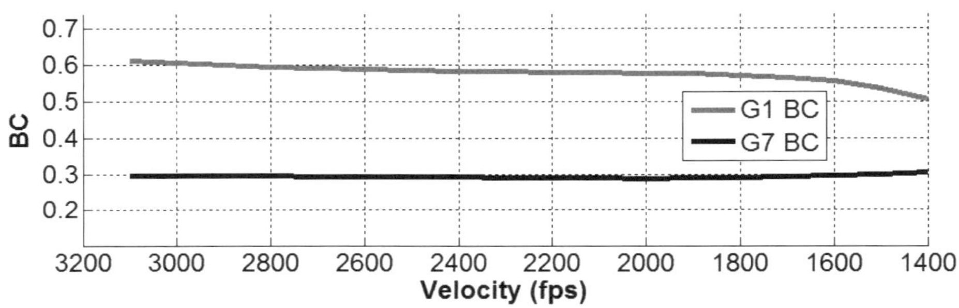

	Above 3000	3000-2500	2500-2000	2000-1500	Below 1500
G1 BC	0.623	0.595	0.581	0.555	0.533
G7 BC	0.295	0.295	0.292	0.295	0.300

Averages from 3000 fps to 1500 fps:
G1 BC: 0.575, variation of G1 BC: 0.073
G7 BC: 0.295, variation of G7 BC: 0.011
G7 form factor: 1.020

Rifling Twist Rate and Stability

	Stability Factor		
Barrel Twist	**Best case**	**Nominal**	**Worst case**
1:10"	2.84	2.48	2.14
1:11"	2.35	2.05	1.77
1:12"	1.97	1.72	1.49
1:13"	1.68	1.47	1.27

Minimum twist required for stability is 1:14.4"
Recommended twist for optimal performance is 1:12.9" or faster

GS Custom .308 caliber 137 grain SP

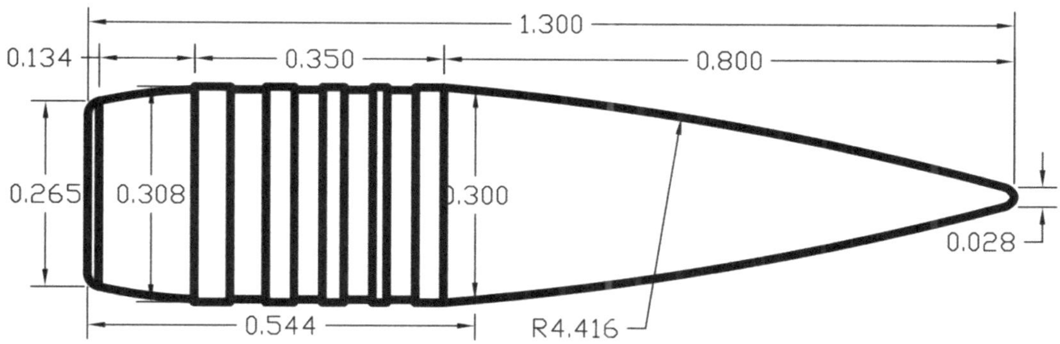

Lot #:	Unknown	Ogive Radius:	14.44 calibers
Sectional Density:	0.206 lb/in^2	Rt/R:	0.62

Ballistic Performance

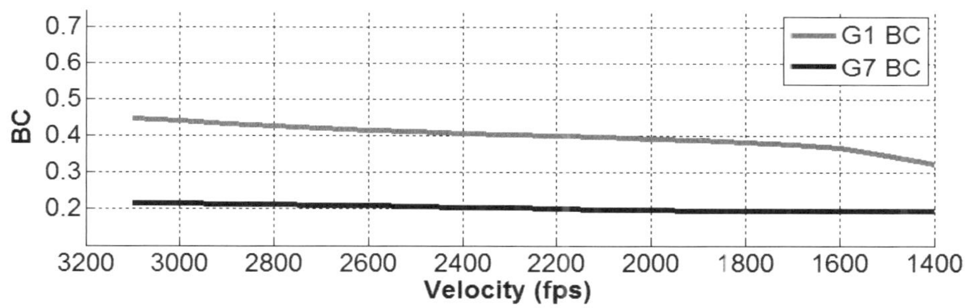

	Above 3000	3000-2500	2500-2000	2000-1500	Below 1500
G1 BC	0.457	0.422	0.398	0.368	0.346
G7 BC	0.216	0.209	0.200	0.195	0.194

Averages from 3000 fps to 1500 fps:

G1 BC: 0.395, variation of G1 BC: 0.091

G7 BC: 0.202, variation of G7 BC: 0.019

G7 form factor: 1.023

Rifling Twist Rate and Stability

Barrel Twist	Stability Factor: Best case	Stability Factor: Nominal	Stability Factor: Worst case
1:10"	1.92	1.68	1.45
1:11"	1.59	1.39	1.20
1:12"	1.34	1.17	1.01
1:13"	1.14	0.99	0.86

Minimum twist required for stability is 1:11.8"

Recommended twist for optimal performance is 1:10.6" or faster

GS Custom .308 caliber 158 grain SP

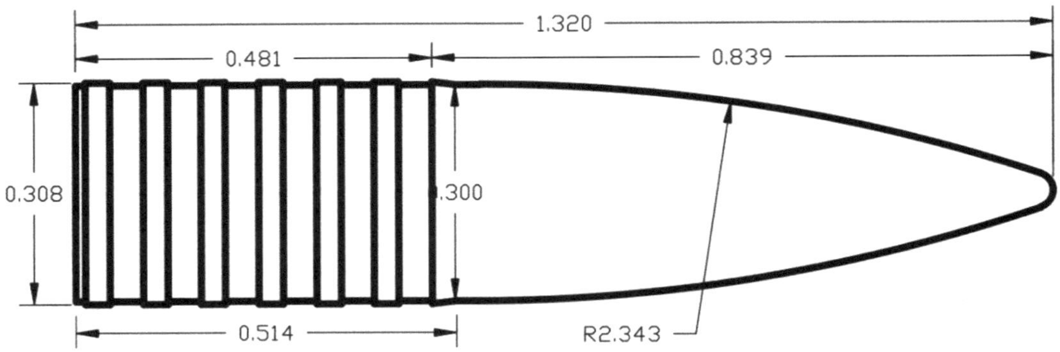

Lot #:	**Unknown**	**Ogive Radius:**	**7.61 calibers**
Sectional Density:	**0.238 lb/in²**	**Rt/R:**	**1.00**

Ballistic Performance

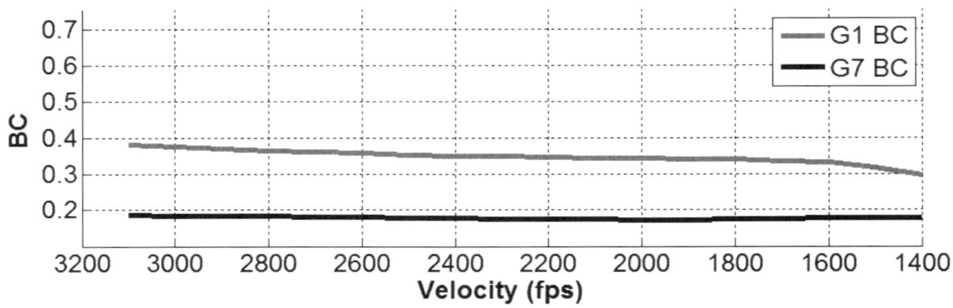

	Above 3000	3000-2500	2500-2000	2000-1500	Below 1500
G1 BC	0.393	0.365	0.349	0.330	0.317
G7 BC	0.186	0.181	0.176	0.175	0.178

Averages from 3000 fps to 1500 fps:
G1 BC: 0.348, variation of G1 BC: 0.060
G7 BC: 0.178, variation of G7 BC: 0.011
G7 form factor: 1.335

Rifling Twist Rate and Stability

Barrel Twist	Stability Factor Best case	Nominal	Worst case
1:10"	2.12	1.85	1.60
1:11"	1.75	1.53	1.33
1:12"	1.47	1.29	1.11
1:13"	1.26	1.10	0.95

Minimum twist required for stability is 1:12.4"
Recommended twist for optimal performance is 1:11.1" or faster

GS Custom .308 caliber 167 grain SP

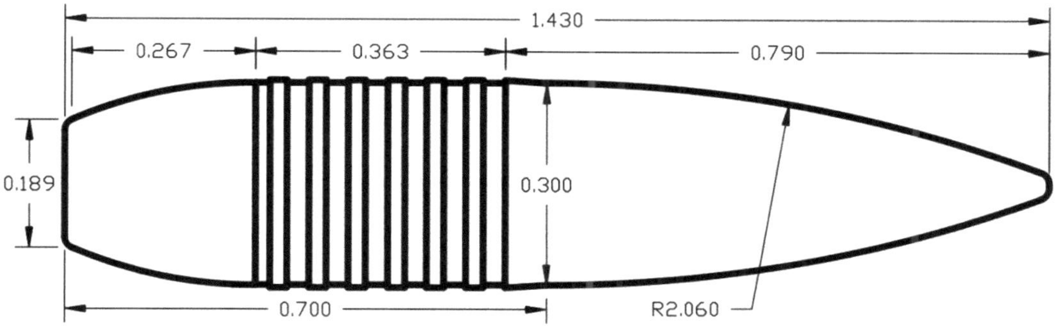

Lot #:	**Unknown**	**Ogive Radius:**	**6.69 calibers**
Sectional Density:	**0.252 lb/in^2**	**Rt/R:**	**1.00**

Ballistic Performance

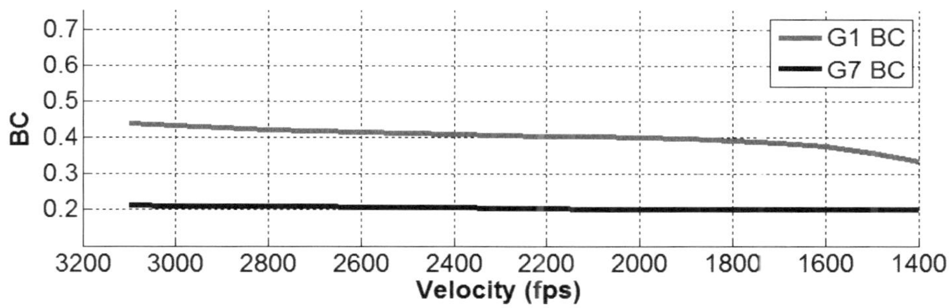

	Above 3000	3000-2500	2500-2000	2000-1500	Below 1500
G1 BC	0.446	0.421	0.405	0.380	0.360
G7 BC	0.211	0.209	0.204	0.201	0.202

Averages from 3000 fps to 1500 fps:
G1 BC: 0.400, variation of G1 BC: 0.072
G7 BC: 0.205, variation of G7 BC: 0.010
G7 form factor: 1.226

Rifling Twist Rate and Stability

Barrel Twist	Stability Factor Best case	 Nominal	 Worst case
1:10"	1.78	1.55	1.34
1:11"	1.47	1.28	1.11
1:12"	1.23	1.08	0.93
1:13"	1.05	0.92	0.79

Minimum twist required for stability is 1:11.4"
Recommended twist for optimal performance is 1:10.2" or faster

GS Custom .308 caliber 197 grain SP

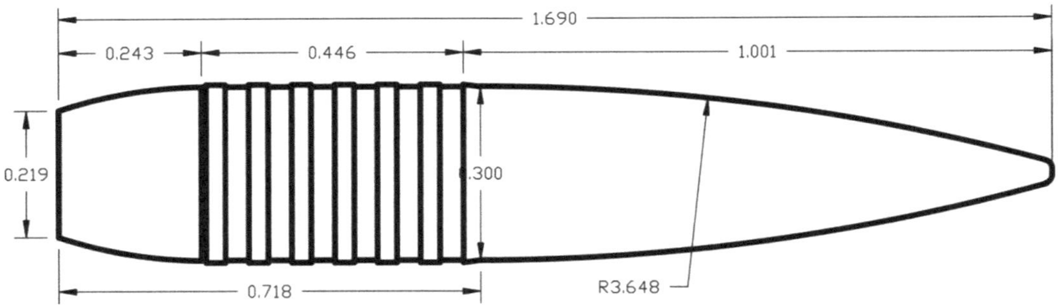

Lot #:	**Unknown**	**Ogive Radius:**	**11.84 calibers**
Sectional Density:	**0.297 lb/in²**	**Rt/R:**	**1.00**

Ballistic Performance

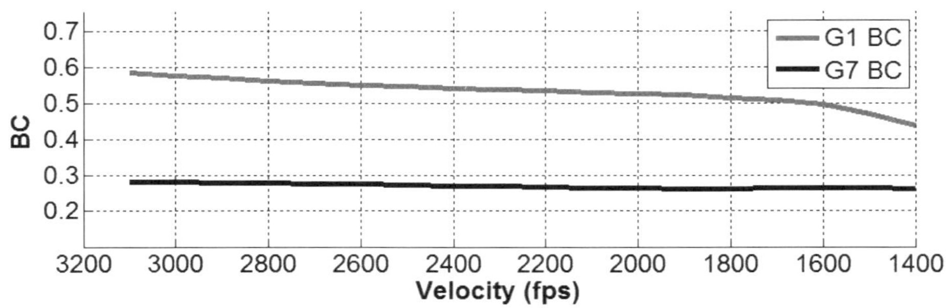

	Above 3000	3000-2500	2500-2000	2000-1500	Below 1500
G1 BC	0.599	0.561	0.536	0.499	0.472
G7 BC	0.283	0.278	0.269	0.265	0.265

Averages from 3000 fps to 1500 fps:
G1 BC: 0.530, variation of G1 BC: 0.105
G7 BC: 0.271, variation of G7 BC: 0.017
G7 form factor: 1.094

Rifling Twist Rate and Stability

Barrel Twist	Best case	Stability Factor Nominal	Worst case
1:8"	2.01	1.76	1.52
1:9"	1.59	1.39	1.20
1:10"	1.29	1.12	0.97
1:11"	1.06	0.93	0.80

Minimum twist required for stability is 1:9.7"
Recommended twist for optimal performance is 1:8.7" or faster

.308 caliber 155 grain HBC (Australia)

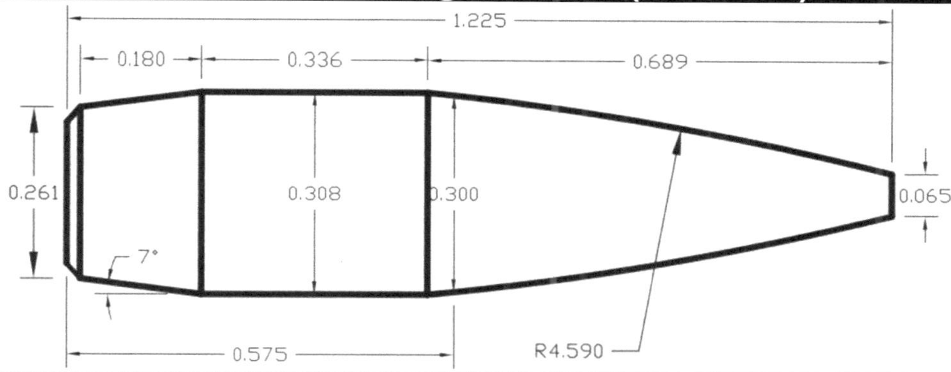

Lot #:	**Unknown**	**Ogive Radius:**	**14.90 calibers**
Sectional Density:	**0.233 lb/in²**	**Rt/R:**	**0.46**

Ballistic Performance

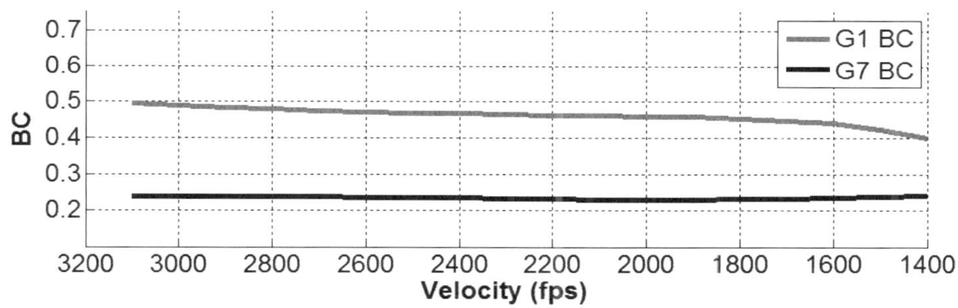

	Above 3000	3000-2500	2500-2000	2000-1500	Below 1500
G1 BC	0.504	0.479	0.464	0.441	0.423
G7 BC	0.239	0.237	0.233	0.235	0.238

Averages from 3000 fps to 1500 fps:
G1 BC: 0.460, variation of G1 BC: 0.066
G7 BC: 0.236, variation of G7 BC: 0.008
G7 form factor: 0.989

Rifling Twist Rate and Stability

Barrel Twist	Best case (Stability Factor)	Nominal (Stability Factor)	Worst case (Stability Factor)
1:10"	2.58	2.26	1.95
1:11"	2.14	1.87	1.61
1:12"	1.79	1.57	1.36
1:13"	1.53	1.34	1.15

Minimum twist required for stability is 1:13.7"
Recommended twist for optimal performance is 1:12.3" or faster

Hornady .308 caliber 155 grain Amax

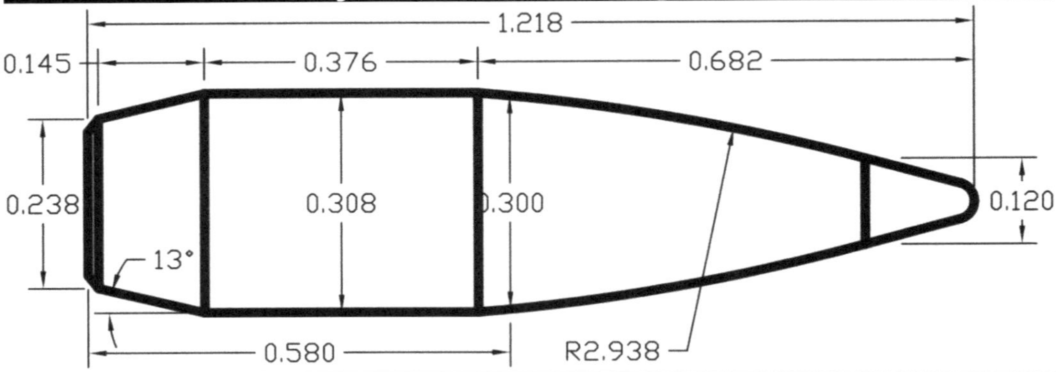

Lot #:	30312	Ogive Radius:	9.54 calibers
Sectional Density:	0.233 lb/in^2	Rt/R:	0.57

Ballistic Performance

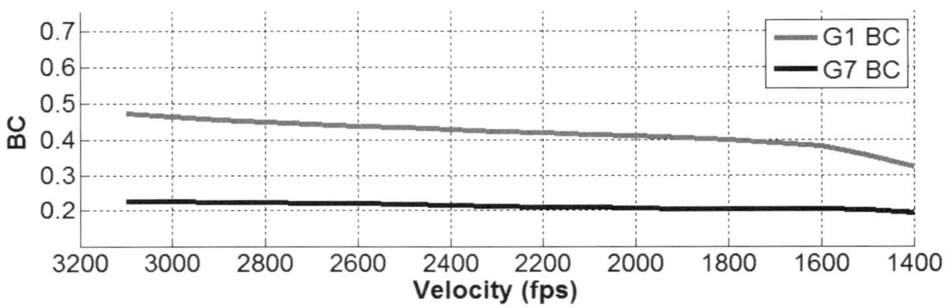

	Above 3000	3000-2500	2500-2000	2000-1500	Below 1500
G1 BC	0.485	0.449	0.422	0.384	0.357
G7 BC	0.229	0.223	0.212	0.203	0.200

Averages from 3000 fps to 1500 fps:
G1 BC: 0.417, variation of G1 BC: 0.107
G7 BC: 0.213, variation of G7 BC: 0.027
G7 form factor: 1.099

Rifling Twist Rate and Stability

Barrel Twist	Stability Factor Best case	Nominal	Worst case
1:10"	2.63	2.29	1.98
1:11"	2.17	1.90	1.64
1:12"	1.82	1.59	1.38
1:13"	1.55	1.36	1.17

Minimum twist required for stability is 1:13.8"
Recommended twist for optimal performance is 1:12.4" or faster

Hornady .308 caliber 168 grain Amax

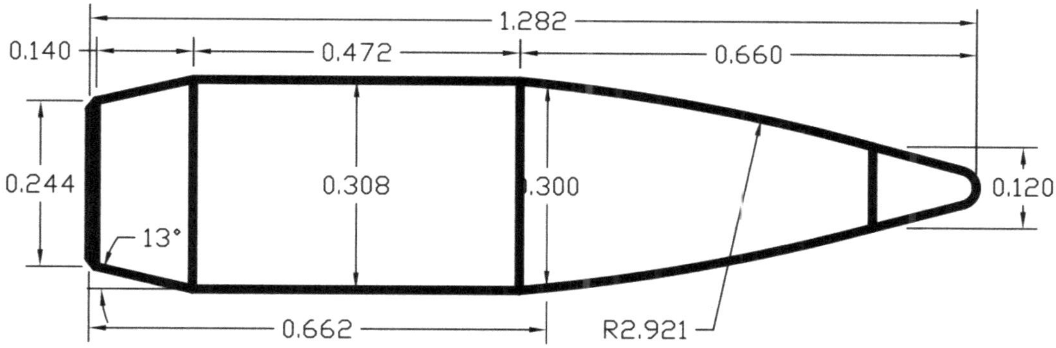

Lot #:	2080203	Ogive Radius:	9.48 calibers
Sectional Density:	0.253 lb/in^2	Rt/R:	0.58

Ballistic Performance

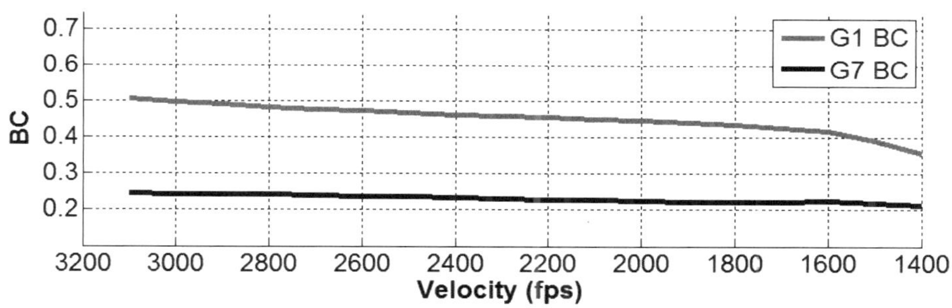

	Above 3000	3000-2500	2500-2000	2000-1500	Below 1500
G1 BC	0.518	0.483	0.457	0.419	0.391
G7 BC	0.245	0.239	0.230	0.222	0.219

Averages from 3000 fps to 1500 fps:
G1 BC: 0.451, variation of G1 BC: 0.107
G7 BC: 0.230, variation of G7 BC: 0.024
G7 form factor: 1.100

Rifling Twist Rate and Stability

Barrel Twist	Stability Factor Best case	Nominal	Worst case
1:10"	2.46	2.15	1.86
1:11"	2.03	1.77	1.53
1:12"	1.71	1.49	1.29
1:13"	1.45	1.27	1.10

Minimum twist required for stability is 1:13.4"
Recommended twist for optimal performance is 1:12.0" or faster

Hornady .308 caliber 178 grain Amax

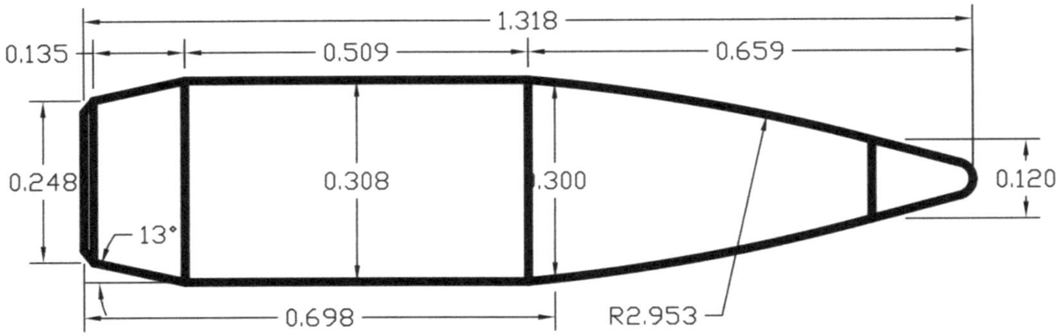

Lot #:	**2080527**	**Ogive Radius:**	**9.59 calibers**
Sectional Density:	**0.268 lb/in^2**	**Rt/R:**	**0.56**

Ballistic Performance

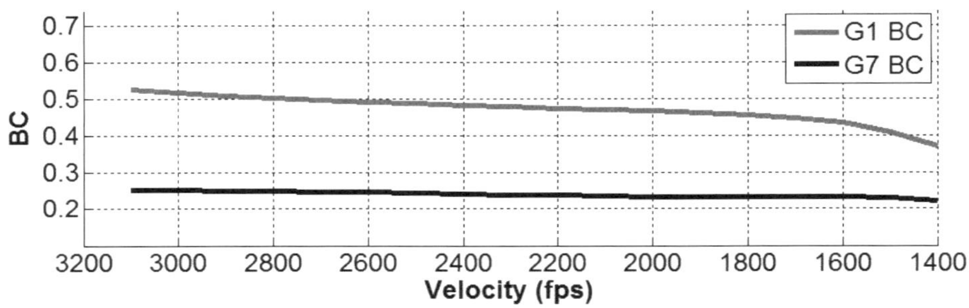

	Above 3000	3000-2500	2500-2000	2000-1500	Below 1500
G1 BC	0.538	0.502	0.476	0.438	0.409
G7 BC	0.255	0.249	0.240	0.232	0.229

Averages from 3000 fps to 1500 fps:
G1 BC: 0.470, variation of G1 BC: 0.108
G7 BC: 0.240, variation of G7 BC: 0.024
G7 form factor: 1.117

Rifling Twist Rate and Stability

Barrel Twist	Stability Factor Best case	Nominal	Worst case
1:10"	2.40	2.10	1.81
1:11"	1.98	1.73	1.50
1:12"	1.67	1.46	1.26
1:13"	1.42	1.24	1.07

Minimum twist required for stability is 1:13.2"
Recommended twist for optimal performance is 1:11.8" or faster

Hornady .308 caliber 208 grain Amax

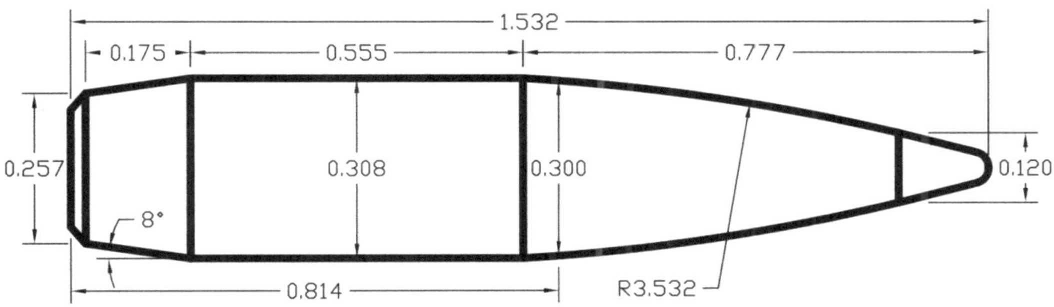

Lot #:	2080070	Ogive Radius:	11.47 calibers
Sectional Density:	0.313 lb/in²	Rt/R:	0.68

Ballistic Performance

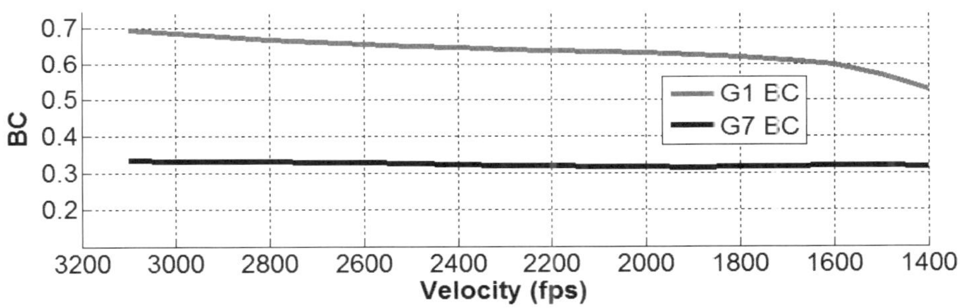

	Above 3000	3000-2500	2500-2000	2000-1500	Below 1500
G1 BC	0.709	0.666	0.640	0.600	0.569
G7 BC	0.336	0.330	0.322	0.318	0.320

Averages from 3000 fps to 1500 fps:
G1 BC: 0.633, variation of G1 BC: 0.114
G7 BC: 0.324, variation of G7 BC: 0.017
G7 form factor: 0.966

Rifling Twist Rate and Stability

Barrel Twist	Stability Factor Best case	Nominal	Worst case
1:10"	1.81	1.58	1.37
1:11"	1.50	1.31	1.13
1:12"	1.26	1.10	0.95
1:13"	1.07	0.94	0.81

Minimum twist required for stability is 1:11.5"
Recommended twist for optimal performance is 1:10.3" or faster

Hornady .308 caliber 155 grain BTHP Match

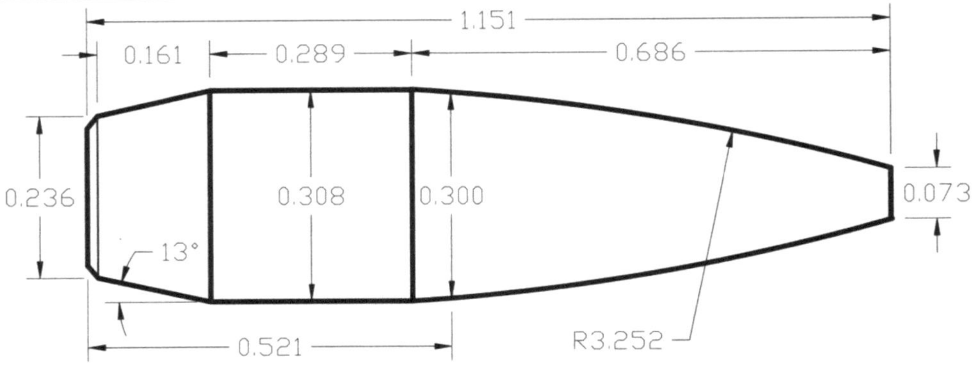

Lot #:	2142023	Ogive Radius:	10.56 calibers
Sectional Density:	0.233 lb/in^2	Rt/R:	0.63

Ballistic Performance

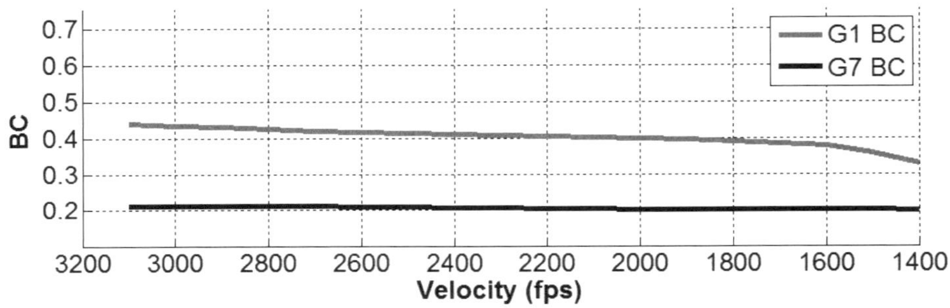

	Above 3000	3000-2500	2500-2000	2000-1500	Below 1500
G1 BC	0.451	0.426	0.408	0.380	0.359
G7 BC	0.214	0.211	0.205	0.201	0.201

Averages from 3000 fps to 1500 fps:
G1 BC: 0.403, variation of G1 BC: 0.078
G7 BC: 0.206, variation of G7 BC: 0.012
G7 form factor: 1.133

Rifling Twist Rate and Stability

Barrel Twist	Stability Factor Best case	Nominal	Worst case
1:12"	2.15	1.87	1.62
1:13"	1.83	1.60	1.38
1:14"	1.58	1.38	1.19
1:15"	1.37	1.20	1.04

Minimum twist required for stability is 1:15.0"
Recommended twist for optimal performance is 1:13.4" or faster

Hornady .308 caliber 168 grain HPBT Match

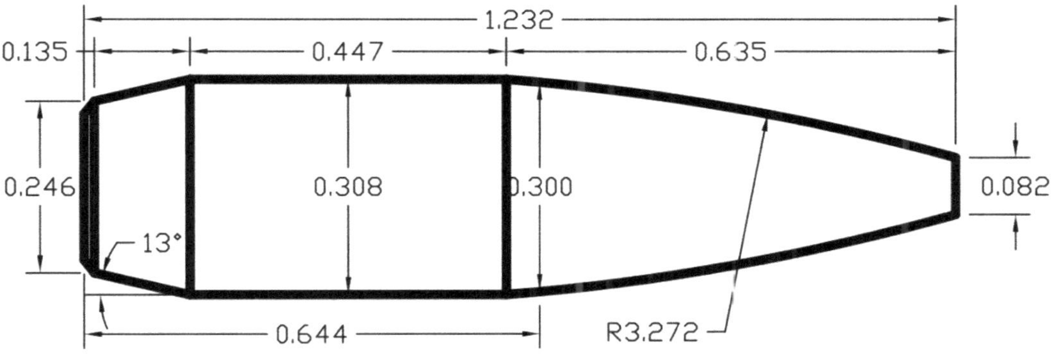

Lot #:	2080097	Ogive Radius:	10.62 calibers
Sectional Density:	0.253 lb/in²	Rt/R:	0.58

Ballistic Performance

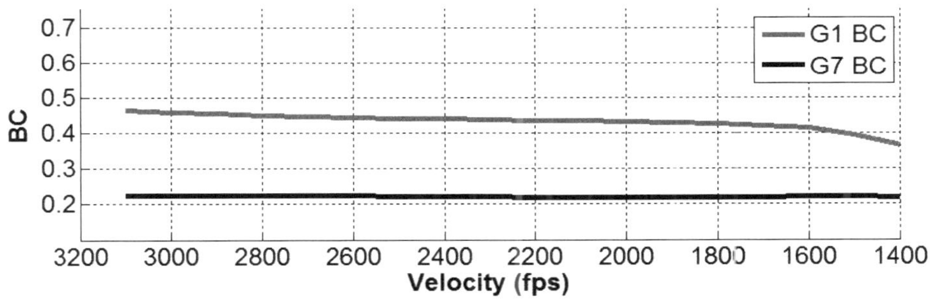

	Above 3000	3000-2500	2500-2000	2000-1500	Below 1500
G1 BC	0.473	0.451	0.437	0.413	0.394
G7 BC	0.224	0.223	0.220	0.219	0.221

Averages from 3000 fps to 1500 fps:
G1 BC: 0.432, variation of G1 BC: 0.065
G7 BC: 0.221, variation of G7 BC: 0.007
G7 form factor: 1.143

Rifling Twist Rate and Stability

Barrel Twist	Stability Factor Best case	Nominal	Worst case
1:10"	2.75	2.41	2.08
1:11"	2.28	1.99	1.72
1:12"	1.91	1.67	1.45
1:13"	1.63	1.42	1.23

Minimum twist required for stability is 1:14.2"
Recommended twist for optimal performance is 1:12.7" or faster

Hornady .308 caliber 178 grain HPBT Match

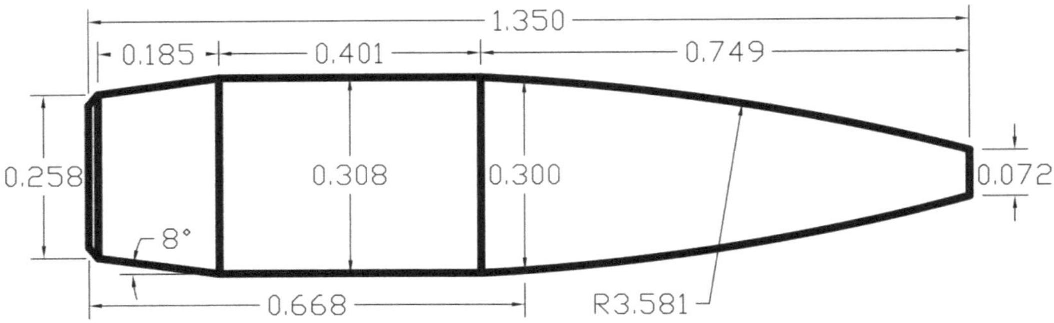

Lot #:	2100474	Ogive Radius:	11.63 calibers
Sectional Density:	0.268 lb/in²	Rt/R:	0.70

Ballistic Performance

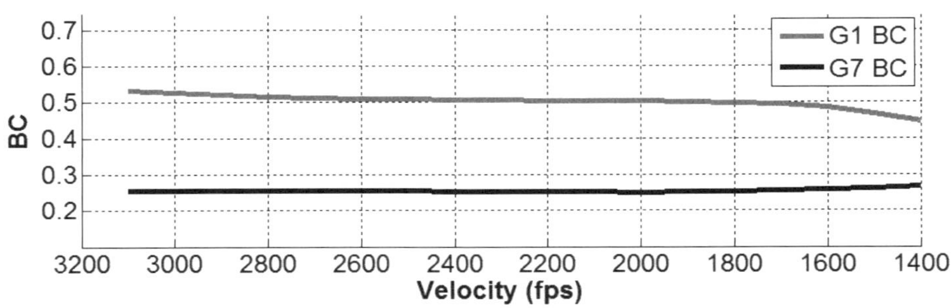

	Above 3000	3000-2500	2500-2000	2000-1500	Below 1500
G1 BC	0.540	0.516	0.505	0.485	0.468
G7 BC	0.256	0.256	0.254	0.258	0.264

Averages from 3000 fps to 1500 fps:
G1 BC: 0.501, variation of G1 BC: 0.057
G7 BC: 0.257, variation of G7 BC: 0.012
G7 form factor: 1.043

Rifling Twist Rate and Stability

Barrel Twist	Stability Factor Best case	Nominal	Worst case
1:10"	2.24	1.96	1.69
1:11"	1.85	1.62	1.40
1:12"	1.56	1.36	1.18
1:13"	1.33	1.16	1.00

Minimum twist required for stability is 1:12.8"
Recommended twist for optimal performance is 1:11.4" or faster

Hornady .308 caliber 195 grain BTHP Match

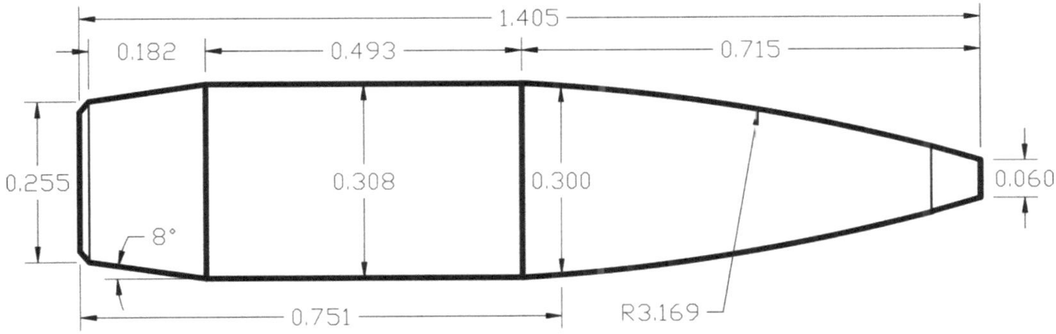

Lot #:	**2150419**	**Ogive Radius:**	**10.29 calibers**
Sectional Density:	**0.294 lb/in²**	**Rt/R:**	**0.57**

Ballistic Performance

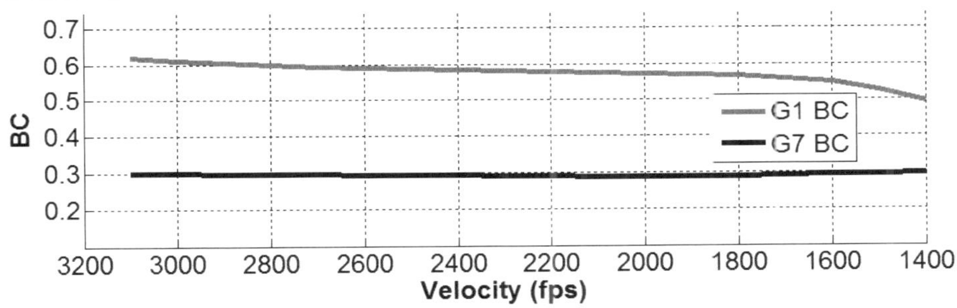

	Above 3000	3000-2500	2500-2000	2000-1500	Below 1500
G1 BC	0.632	0.599	0.580	0.550	0.525
G7 BC	0.299	0.297	0.292	0.292	0.295

Averages from 3000 fps to 1500 fps:
G1 BC: 0.574, variation of G1 BC: 0.087
G7 BC: 0.295, variation of G7 BC: 0.011
G7 form factor: 0.997

Rifling Twist Rate and Stability

Barrel Twist	Stability Factor Best case	Nominal	Worst case
1:10"	2.19	1.91	1.65
1:11"	1.81	1.58	1.36
1:12"	1.52	1.33	1.15
1:13"	1.29	1.13	0.98

Minimum twist required for stability is 1:12.6"
Recommended twist for optimal performance is 1:11.3" or faster

Hornady .308 caliber 208 grain HPBT Match

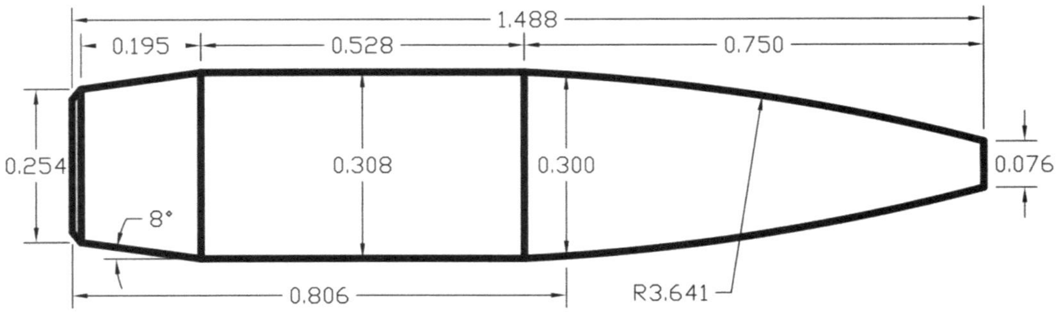

Lot #:	Unknown	Ogive Radius:	11.82 calibers
Sectional Density:	0.313 lb/in^2	Rt/R:	0.68

Ballistic Performance

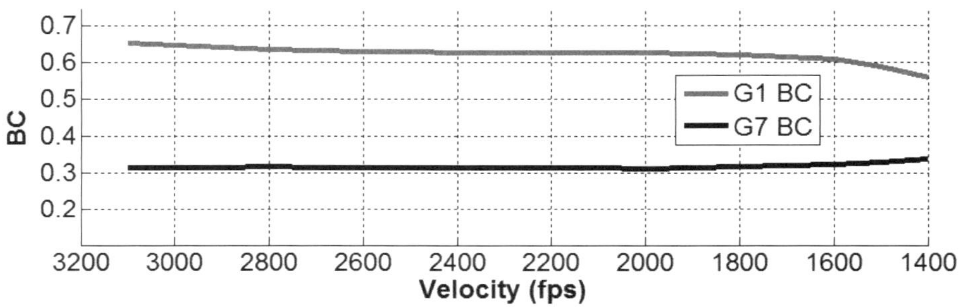

	Above 3000	3000-2500	2500-2000	2000-1500	Below 1500
G1 BC	0.663	0.637	0.626	0.605	0.586
G7 BC	0.314	0.316	0.315	0.322	0.330

Averages from 3000 fps to 1500 fps:
G1 BC: 0.621, variation of G1 BC: 0.060
G7 BC: 0.319, variation of G7 BC: 0.017
G7 form factor: 0.982

Rifling Twist Rate and Stability

Barrel Twist	Stability Factor Best case	Nominal	Worst case
1:10"	1.97	1.72	1.49
1:11"	1.63	1.42	1.23
1:12"	1.37	1.20	1.03
1:13"	1.17	1.02	0.88

Minimum twist required for stability is 1:12.0"
Recommended twist for optimal performance is 1:10.7" or faster

Hornady .308 caliber 225 grain HPBT Match

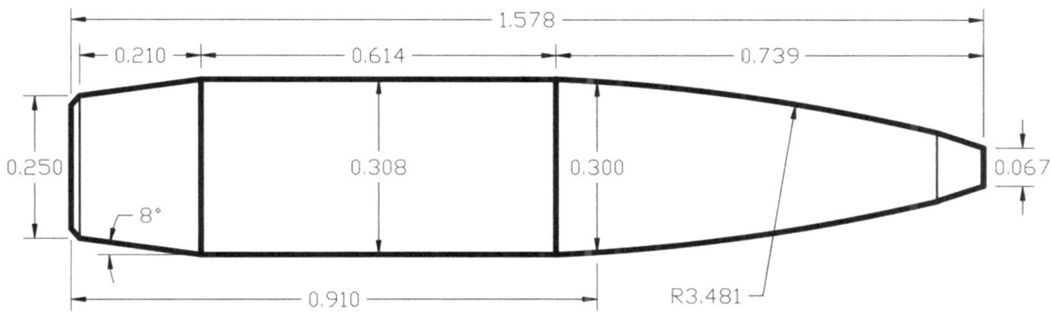

Lot #:	2150349	Ogive Radius:	11.30 calibers
Sectional Density:	0.339 lb/in^2	Rt/R:	0.67

Ballistic Performance

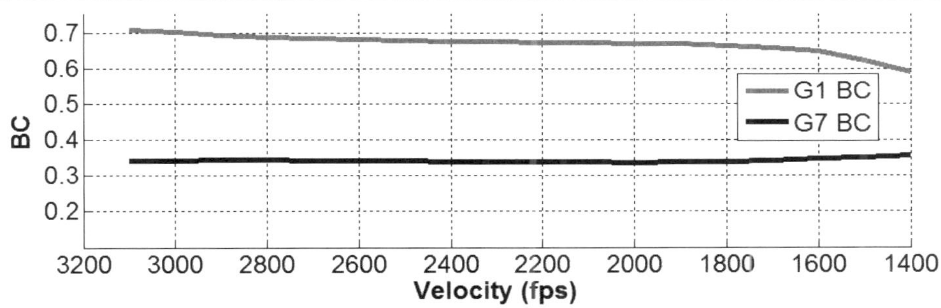

	Above 3000	3000-2500	2500-2000	2000-1500	Below 1500
G1 BC	0.721	0.689	0.674	0.647	0.623
G7 BC	0.341	0.342	0.339	0.344	0.351

Averages from 3000 fps to 1500 fps:
G1 BC: 0.668, variation of G1 BC: 0.078
G7 BC: 0.343, variation of G7 BC: 0.014
G7 form factor: 0.988

Rifling Twist Rate and Stability

Barrel Twist	Stability Factor Best case	Nominal	Worst case
1:9"	2.22	1.94	1.68
1:10"	1.80	1.57	1.36
1:11"	1.49	1.30	1.12
1:12"	1.25	1.09	0.94

Minimum twist required for stability is 1:11.4"
Recommended twist for optimal performance is 1:10.2" or faster

Hornady .308 caliber 165 grain BTSP Interlock

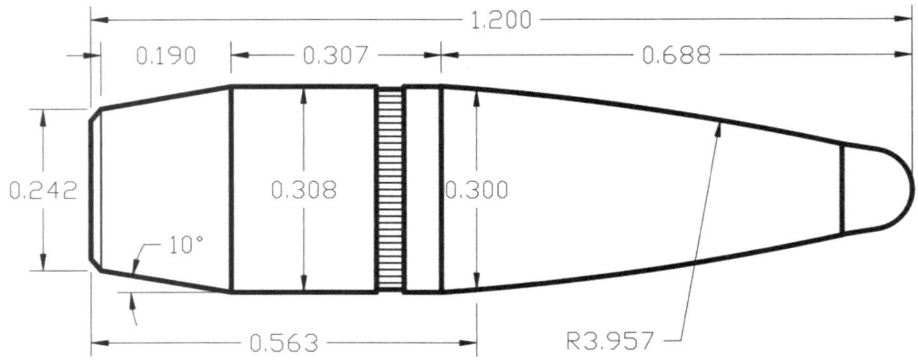

Lot #:	2141429	Ogive Radius:	12.85 calibers
Sectional Density:	0.249 lb/in^2	Rt/R:	0.51

Ballistic Performance

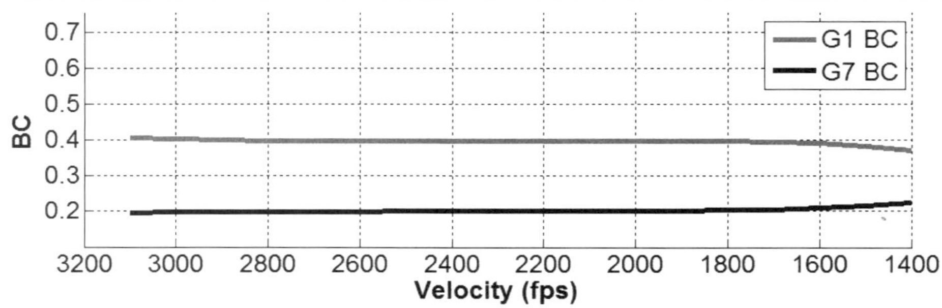

	Above 3000	3000-2500	2500-2000	2000-1500	Below 1500
G1 BC	0.411	0.400	0.398	0.390	0.382
G7 BC	0.195	0.199	0.200	0.208	0.216

Averages from 3000 fps to 1500 fps:
G1 BC: 0.395, variation of G1 BC: 0.022
G7 BC: 0.203, variation of G7 BC: 0.018
G7 form factor: 1.224

Rifling Twist Rate and Stability

Barrel Twist	Stability Factor Best case	Nominal	Worst case
1:12"	2.03	1.77	1.53
1:13"	1.73	1.51	1.30
1:14"	1.49	1.30	1.12
1:15"	1.30	1.13	0.98

Minimum twist required for stability is 1:14.6"
Recommended twist for optimal performance is 1:13.0" or faster

Hornady .308 caliber 150 grain BTSP

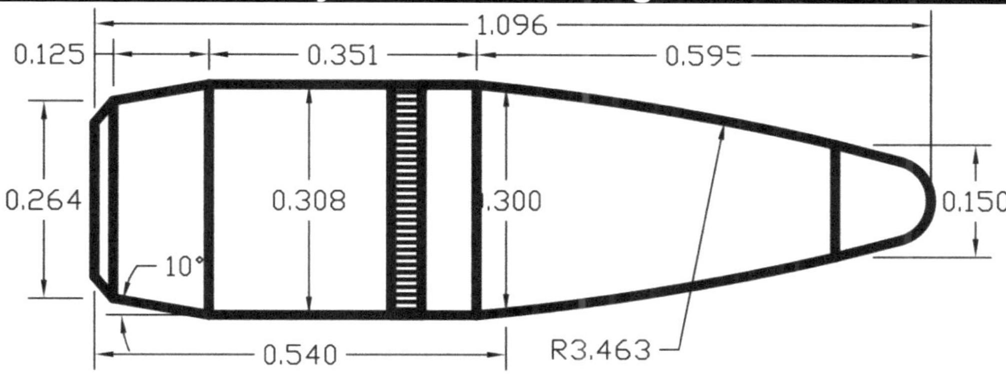

Lot #:	96-020	Ogive Radius:	11.24 calibers
Sectional Density:	0.226 lb/in^2	Rt/R:	0.53

Ballistic Performance

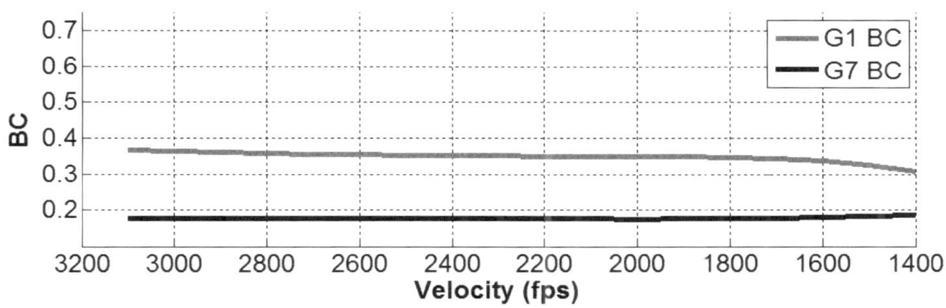

	Above 3000	3000-2500	2500-2000	2000-1500	Below 1500
G1 BC	0.374	0.359	0.352	0.338	0.326
G7 BC	0.177	0.178	0.177	0.180	0.184

Averages from 3000 fps to 1500 fps:
G1 BC: 0.349, variation of G1 BC: 0.039
G7 BC: 0.179, variation of G7 BC: 0.008
G7 form factor: 1.262

Rifling Twist Rate and Stability

Barrel Twist	Stability Factor Best case	Nominal	Worst case
1:10"	3.44	3.01	2.60
1:11"	2.84	2.48	2.15
1:12"	2.39	2.09	1.80
1:13"	2.04	1.78	1.54

Minimum twist required for stability is 1:15.8"
Recommended twist for optimal performance is 1:14.2" or faster

Hornady .308 caliber 190 grain BTSP

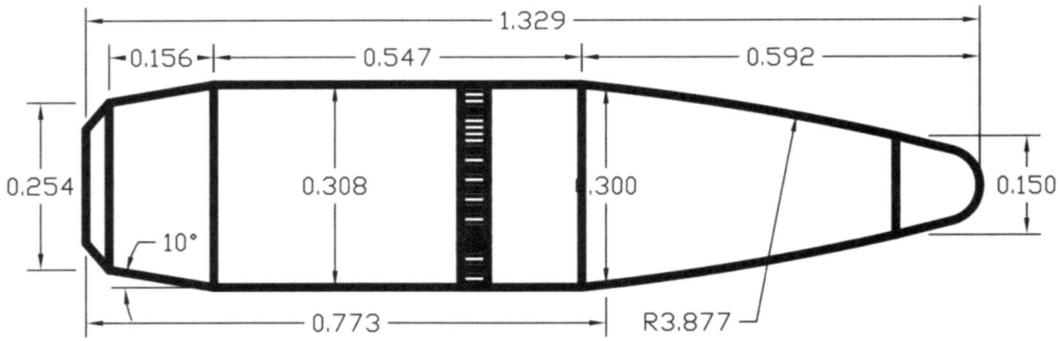

Lot #:	2060308	Ogive Radius:	12.59 calibers
Sectional Density:	0.286 lb/in^2	Rt/R:	0.49

Ballistic Performance

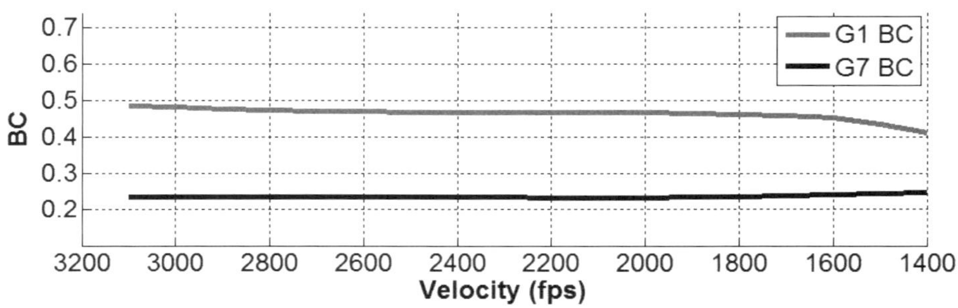

	Above 3000	3000-2500	2500-2000	2000-1500	Below 1500
G1 BC	0.492	0.475	0.467	0.450	0.435
G7 BC	0.233	0.235	0.235	0.239	0.245

Averages from 3000 fps to 1500 fps:
G1 BC: 0.463, variation of G1 BC: 0.046
G7 BC: 0.237, variation of G7 BC: 0.011
G7 form factor: 1.205

Rifling Twist Rate and Stability

Barrel Twist	Stability Factor Best case	Nominal	Worst case
1:10"	2.50	2.19	1.89
1:11"	2.07	1.81	1.56
1:12"	1.74	1.52	1.31
1:13"	1.48	1.29	1.12

Minimum twist required for stability is 1:13.5"
Recommended twist for optimal performance is 1:12.1" or faster

Hornady .308 caliber 150 grain FMJBT

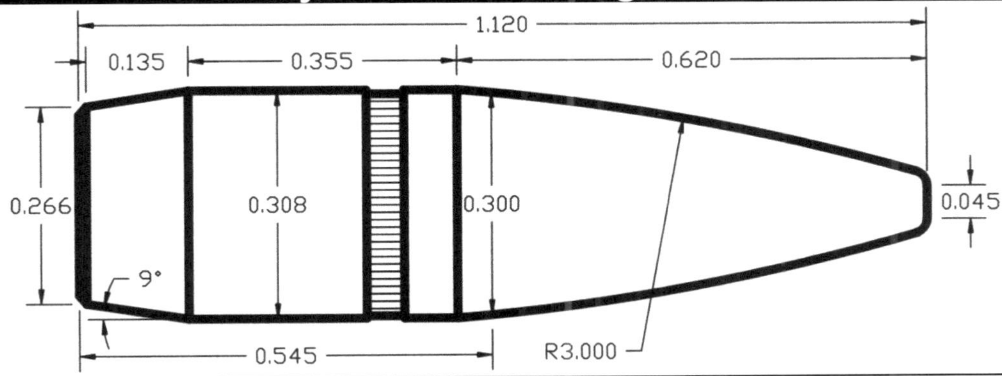

Lot #:	**Unknown**	**Ogive Radius:**	**9.74 calibers**
Sectional Density:	**0.226 lb/in^2**	**Rt/R:**	**0.51**

Ballistic Performance

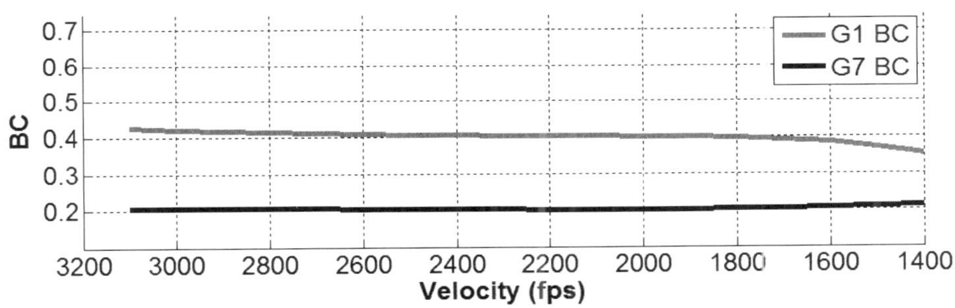

	Above 3000	3000-2500	2500-2000	2000-1500	Below 1500
G1 BC	0.434	0.414	0.404	0.386	0.371
G7 BC	0.206	0.205	0.203	0.205	0.209

Averages from 3000 fps to 1500 fps:
G1 BC: 0.400, variation of G1 BC: 0.051
G7 BC: 0.205, variation of G7 BC: 0.007
G7 form factor: 1.100

Rifling Twist Rate and Stability

Barrel Twist	**Stability Factor** Best case	Nominal	Worst case
1:10"	3.23	2.82	2.44
1:11"	2.67	2.33	2.02
1:12"	2.25	1.96	1.70
1:13"	1.91	1.67	1.45

Minimum twist required for stability is 1:15.3"
Recommended twist for optimal performance is 1:13.7" or faster

Hornady .308 caliber 165 grain GMX

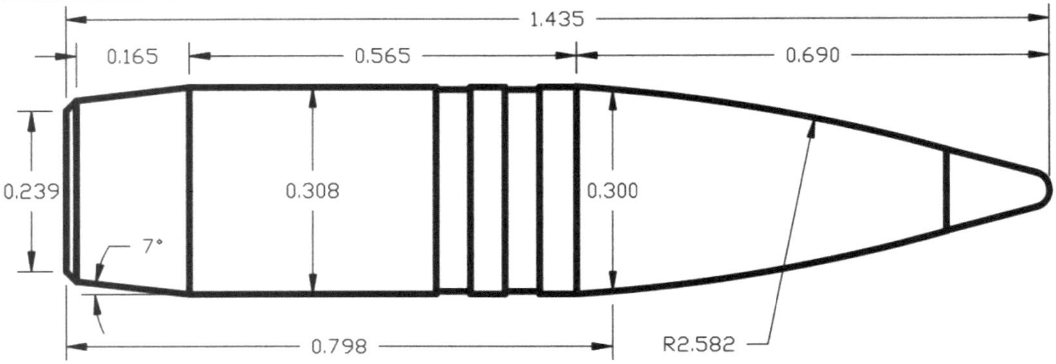

Lot #:	2140504	**Ogive Radius:**	**8.38 calibers**
Sectional Density:	**0.316 lb/in^2**	**Rt/R:**	**0.62**

Ballistic Performance

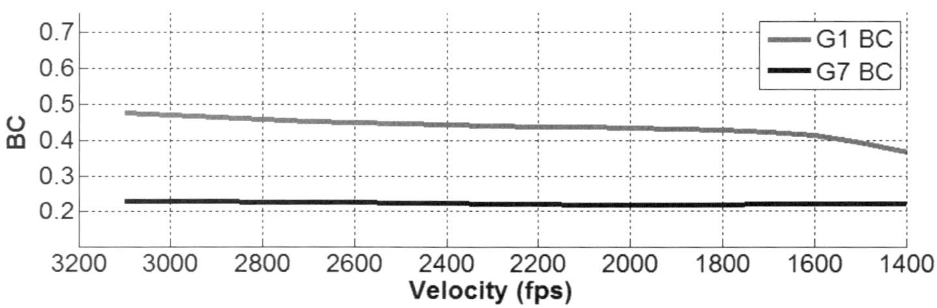

	Above 3000	3000-2500	2500-2000	2000-1500	Below 1500
G1 BC	0.487	0.458	0.441	0.415	0.395
G7 BC	0.231	0.227	0.222	0.220	0.222

Averages from 3000 fps to 1500 fps:
G1 BC: 0.437, variation of G1 BC: 0.075
G7 BC: 0.224, variation of G7 BC: 0.011
G7 form factor: 1.111

Rifling Twist Rate and Stability

Barrel Twist	Stability Factor Best case	Stability Factor Nominal	Stability Factor Worst case
1:10"	1.74	1.52	1.31
1:11"	1.44	1.26	1.09
1:12"	1.21	1.05	0.91
1:13"	1.03	0.90	0.78

Minimum twist required for stability is 1:11.3"
Recommended twist for optimal performance is 1:10.1" or faster

Hornady .308 caliber 150 grain SST BT

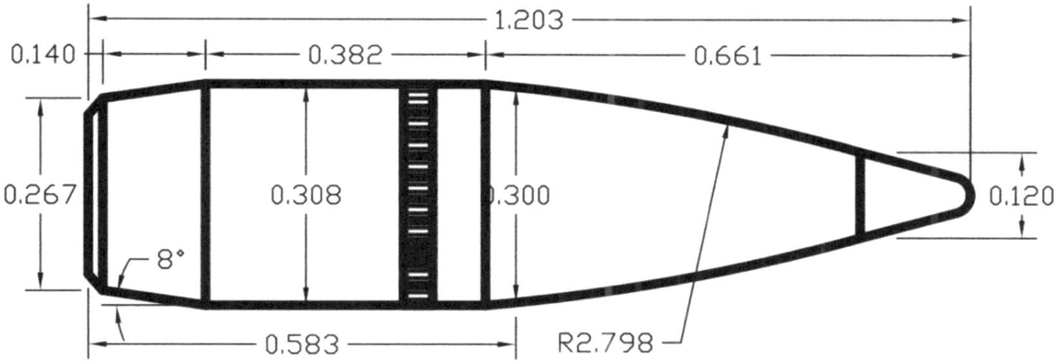

Lot #:	2071086	Ogive Radius:	9.09 calibers
Sectional Density:	0.226 lb/in^2	Rt/R:	0.58

Ballistic Performance

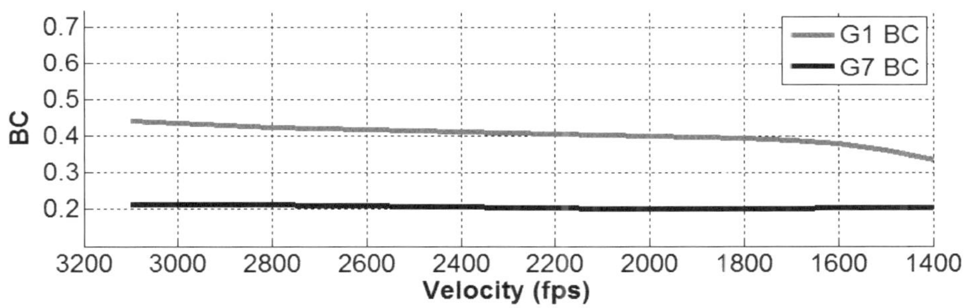

	Above 3000	3000-2500	2500-2000	2000-1500	Below 1500
G1 BC	0.456	0.428	0.410	0.383	0.363
G7 BC	0.216	0.212	0.206	0.203	0.204

Averages from 3000 fps to 1500 fps:
G1 BC: 0.405, variation of G1 BC: 0.076
G7 BC: 0.208, variation of G7 BC: 0.012
G7 form factor: 1.088

Rifling Twist Rate and Stability

Barrel Twist	Stability Factor Best case	Nominal	Worst case
1:10"	2.63	2.30	1.99
1:11"	2.18	1.90	1.64
1:12"	1.83	1.60	1.38
1:13"	1.56	1.36	1.18

Minimum twist required for stability is 1:13.8'
Recommended twist for optimal performance is 1:12.4" or faster

Hornady .308 caliber 165 grain SST BT

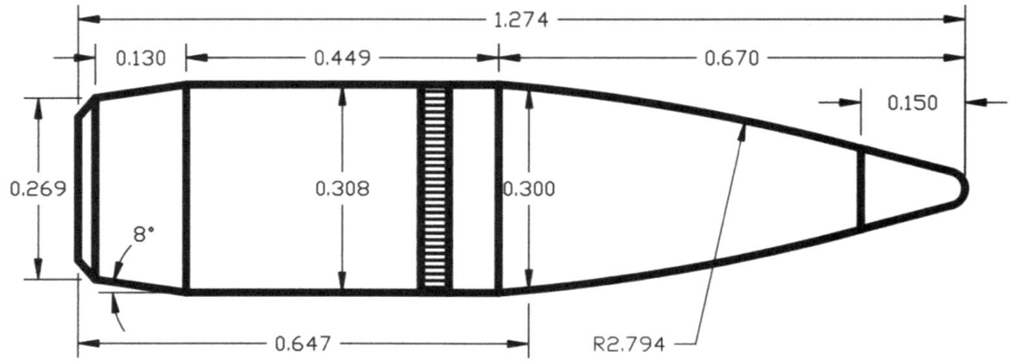

Lot #:	2071017	Ogive Radius:	9.07 calibers
Sectional Density:	0.249 lb/in^2	Rt/R:	0.61

Ballistic Performance

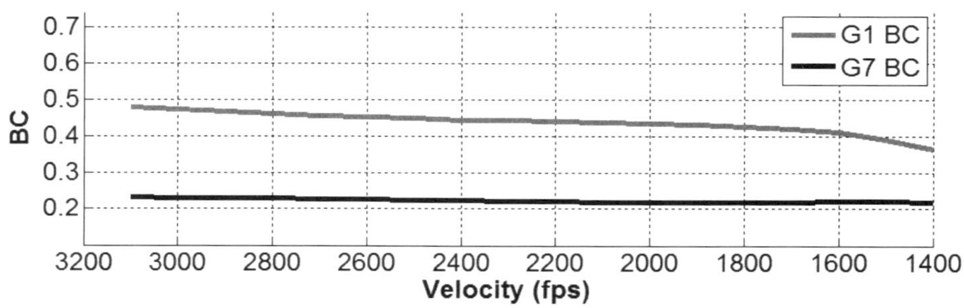

	Above 3000	3000-2500	2500-2000	2000-1500	Below 1500
G1 BC	0.490	0.461	0.442	0.414	0.392
G7 BC	0.232	0.228	0.222	0.220	0.220

Averages from 3000 fps to 1500 fps:
G1 BC: 0.437, variation of G1 BC: 0.081
G7 BC: 0.224, variation of G7 BC: 0.013
G7 form factor: 1.110

Rifling Twist Rate and Stability

Barrel Twist	Stability Factor Best case	Nominal	Worst case
1:10"	2.46	2.15	1.86
1:11"	2.03	1.77	1.53
1:12"	1.71	1.49	1.29
1:13"	1.45	1.27	1.10

Minimum twist required for stability is 1:13.4"
Recommended twist for optimal performance is 1:12.0" or faster

Hornady .308 caliber 180 grain SST BT

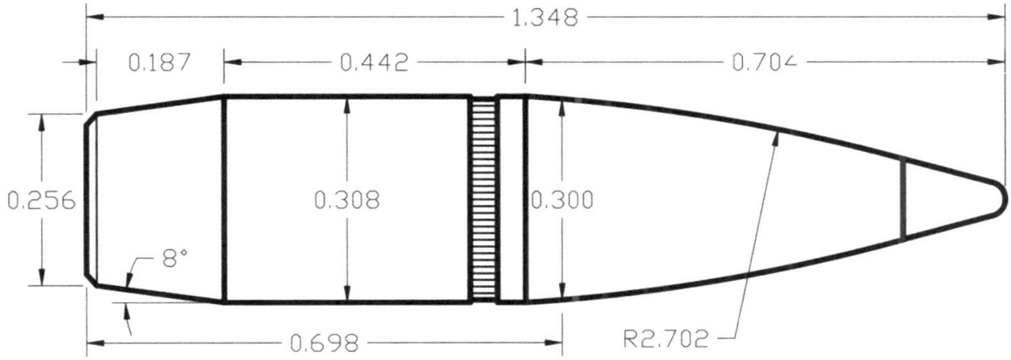

Lot #:	**2141308**	**Ogive Radius:**	**8.77 calibers**
Sectional Density:	**0.271 lb/in²**	**Rt/R:**	**0.61**

Ballistic Performance

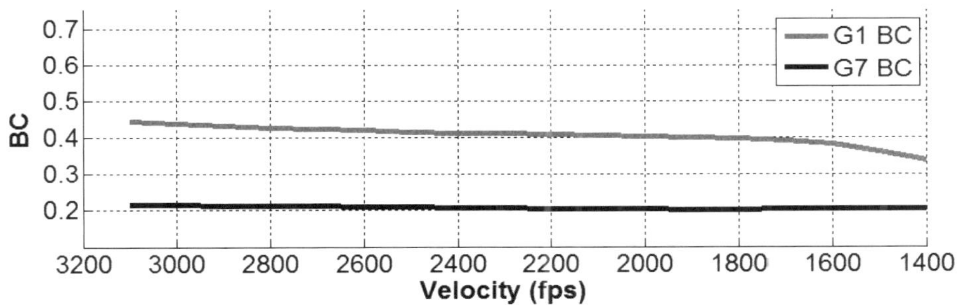

	Above 3000	3000-2500	2500-2000	2000-1500	Below 1500
G1 BC	0.555	0.521	0.500	0.467	0.443
G7 BC	0.263	0.258	0.251	0.248	0.249

Averages from 3000 fps to 1500 fps:
G1 BC: 0.494, variation of G1 BC: 0.092
G7 BC: 0.253, variation of G7 BC: 0.014
G7 form factor: 1.071

Rifling Twist Rate and Stability

Barrel Twist	Stability Factor Best case	Stability Factor Nominal	Stability Factor Worst case
1:10"	2.28	1.99	1.72
1:11"	1.88	1.64	1.42
1:12"	1.58	1.38	1.19
1:13"	1.35	1.18	1.02

Minimum twist required for stability is 1:12.9"
Recommended twist for optimal performance is 1:11.5" or faster

JLK .308 caliber 155 grain VLD

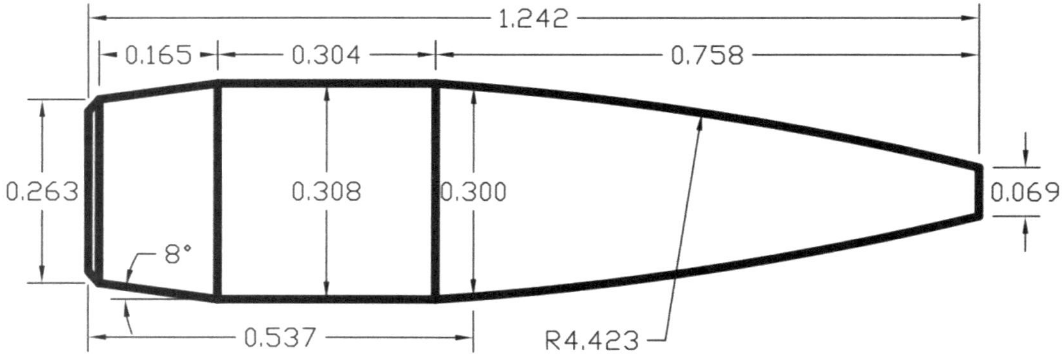

Lot #:	**Unknown**	**Ogive Radius:**	**14.36 calibers**
Sectional Density:	**0.233 lb/in^2**	**Rt/R:**	**0.60**

Ballistic Performance

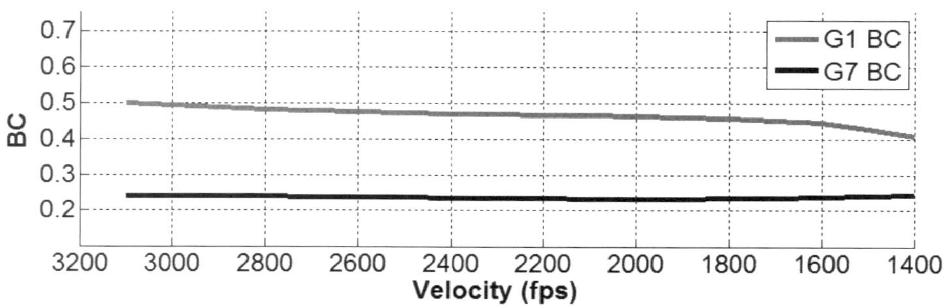

	Above 3000	3000-2500	2500-2000	2000-1500	Below 1500
G1 BC	0.508	0.482	0.468	0.445	0.428
G7 BC	0.241	0.239	0.235	0.237	0.241

Averages from 3000 fps to 1500 fps:
G1 BC: 0.464, variation of G1 BC: 0.065
G7 BC: 0.238, variation of G7 BC: 0.008
G7 form factor: 0.981

Rifling Twist Rate and Stability

Barrel Twist	**Stability Factor** Best case	Nominal	Worst case
1:10"	2.48	2.17	1.88
1:11"	2.05	1.79	1.55
1:12"	1.72	1.51	1.30
1:13"	1.47	1.28	1.11

Minimum twist required for stability is 1:13.4"
Recommended twist for optimal performance is 1:12.0" or faster

JLK .308 caliber 210 grain VLD

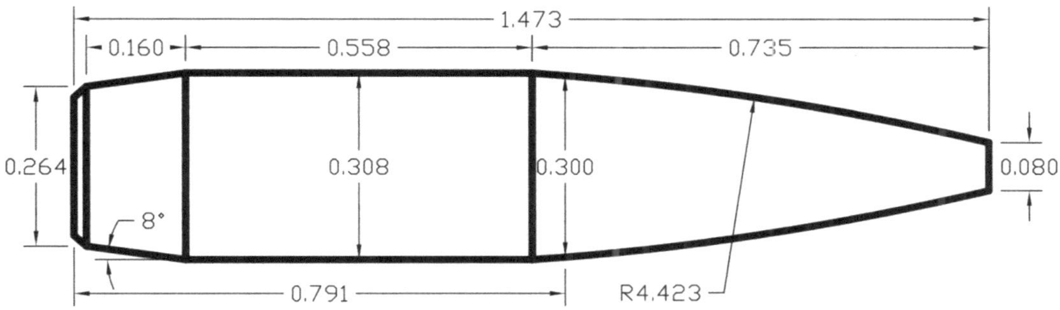

Lot #:	Unknown	Ogive Radius:	14.36 calibers
Sectional Density:	0.316 lb/in^2	Rt/R:	0.60

Ballistic Performance

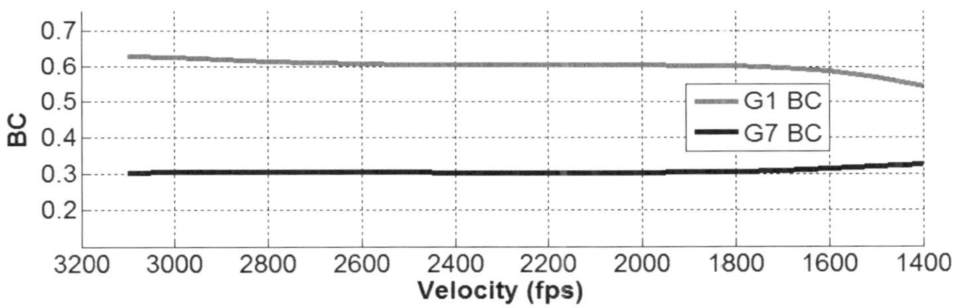

	Above 3000	3000-2500	2500-2000	2000-1500	Below 1500
G1 BC	0.640	0.615	0.605	0.586	0.569
G7 BC	0.303	0.305	0.304	0.312	0.320

Averages from 3000 fps to 1500 fps:
G1 BC: 0.601, variation of G1 BC: 0.055
G7 BC: 0.309, variation of G7 BC: 0.017
G7 form factor: 1.025

Rifling Twist Rate and Stability

Barrel Twist	Stability Factor Best case	Nominal	Worst case
1:10"	2.05	1.79	1.55
1:11"	1.69	1.48	1.28
1:12"	1.42	1.24	1.08
1:13"	1.21	1.06	0.92

Minimum twist required for stability is 1:12.2"
Recommended twist for optimal performance is 1:10.9" or faster

JLK .308 caliber 210 grain VLD Long BT

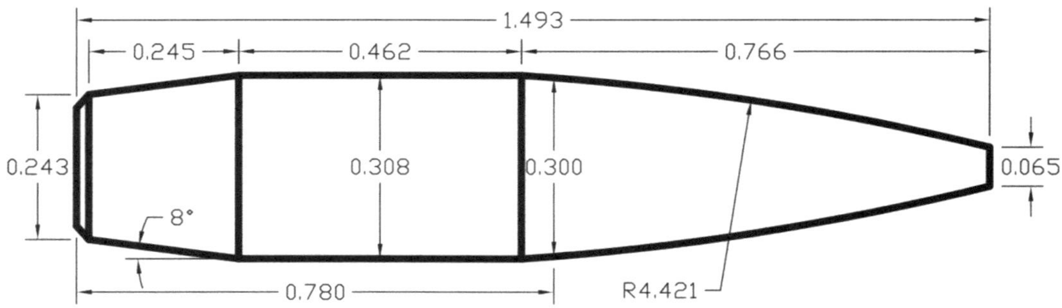

Lot #:	Unknown	Ogive Radius:	14.36 calibers
Sectional Density:	0.316 lb/in^2	Rt/R:	0.60

Ballistic Performance

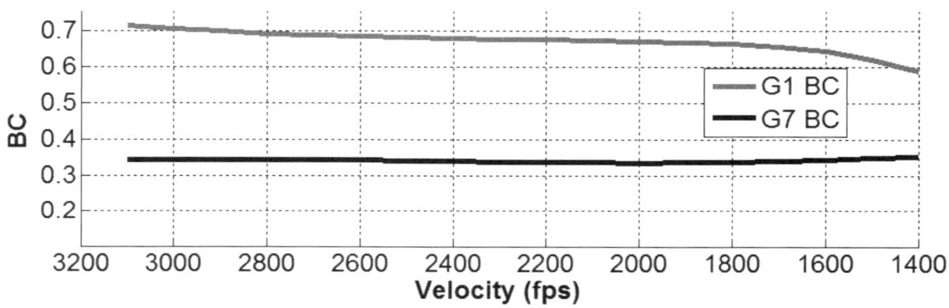

	Above 3000	3000-2500	2500-2000	2000-1500	Below 1500
G1 BC	0.726	0.693	0.675	0.645	0.620
G7 BC	0.344	0.344	0.339	0.343	0.349

Averages from 3000 fps to 1500 fps:
G1 BC: 0.669, variation of G1 BC: 0.086
G7 BC: 0.343, variation of G7 BC: 0.013
G7 form factor: 0.921

Rifling Twist Rate and Stability

Barrel Twist	Stability Factor Best case	Nominal	Worst case
1:10"	1.97	1.72	1.49
1:11"	1.63	1.42	1.23
1:12"	1.37	1.20	1.03
1:13"	1.17	1.02	0.88

Minimum twist required for stability is 1:12.0"
Recommended twist for optimal performance is 1:10.7" or faster

Lapua .308 caliber 185 grain FMJBT (D46)

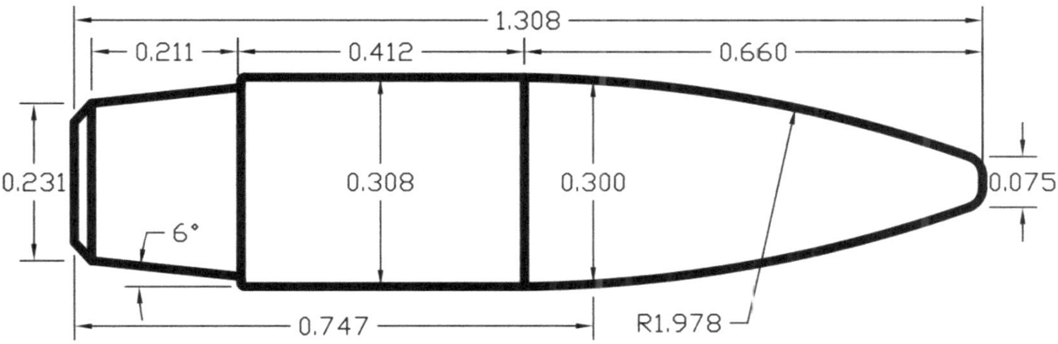

Lot #:	**P00389601**	**Ogive Radius:**	**6.42 calibers**
Sectional Density:	**0.279 lb/in²**	**Rt/R:**	**0.98**

Ballistic Performance

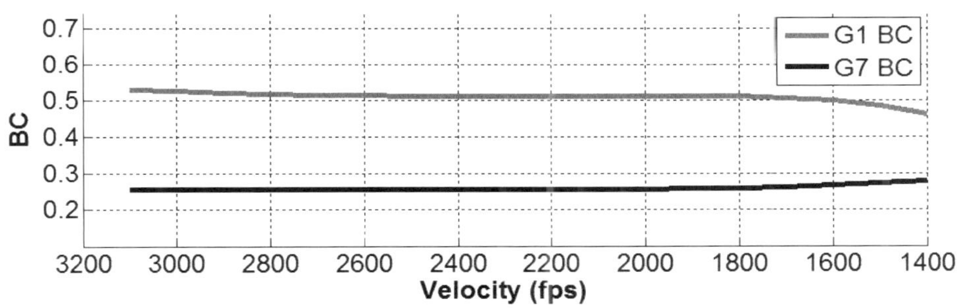

	Above 3000	3000-2500	2500-2000	2000-1500	Below 1500
G1 BC	0.537	0.518	0.512	0.498	0.484
G7 BC	0.255	0.257	0.258	0.265	0.273

Averages from 3000 fps to 1500 fps:
G1 BC: 0.508, variation of G1 BC: 0.041
G7 BC: 0.261, variation of G7 BC: 0.016
G7 form factor: 1.067

Rifling Twist Rate and Stability

Barrel Twist	Stability Factor Best case	Stability Factor Nominal	Stability Factor Worst case
1:10"	2.55	2.23	1.93
1:11"	2.11	1.84	1.59
1:12"	1.77	1.55	1.34
1:13"	1.51	1.32	1.14

Minimum twist required for stability is 1:13.6"
Recommended twist for optimal performance is 1:12.2" or faster

Lapua .308 caliber 200 grain FMJBT

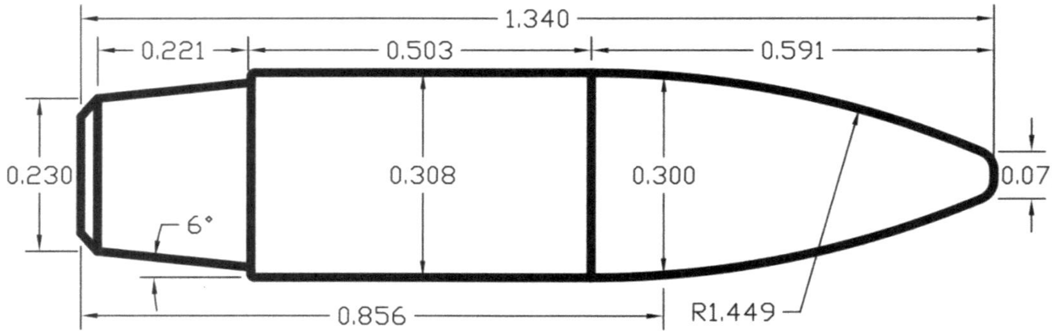

Lot #:	4PL7060	Ogive Radius:	4.70 calibers
Sectional Density:	0.301 lb/in^2	Rt/R:	1.00

Ballistic Performance

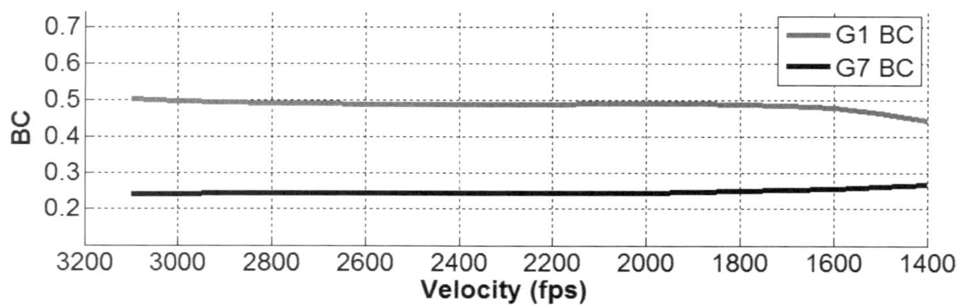

	Above 3000	3000-2500	2500-2000	2000-1500	Below 1500
G1 BC	0.508	0.493	0.489	0.477	0.465
G7 BC	0.241	0.244	0.246	0.254	0.262

Averages from 3000 fps to 1500 fps:
G1 BC: 0.485, variation of G1 BC: 0.033
G7 BC: 0.249, variation of G7 BC: 0.018
G7 form factor: 1.210

Rifling Twist Rate and Stability

Barrel Twist	Stability Factor Best case	Stability Factor Nominal	Stability Factor Worst case
1:10"	2.57	2.25	1.94
1:11"	2.13	1.86	1.61
1:12"	1.79	1.56	1.35
1:13"	1.52	1.33	1.15

Minimum twist required for stability is 1:13.7"
Recommended twist for optimal performance is 1:12.2" or faster

Lapua .308 caliber 150 grain FMJBT

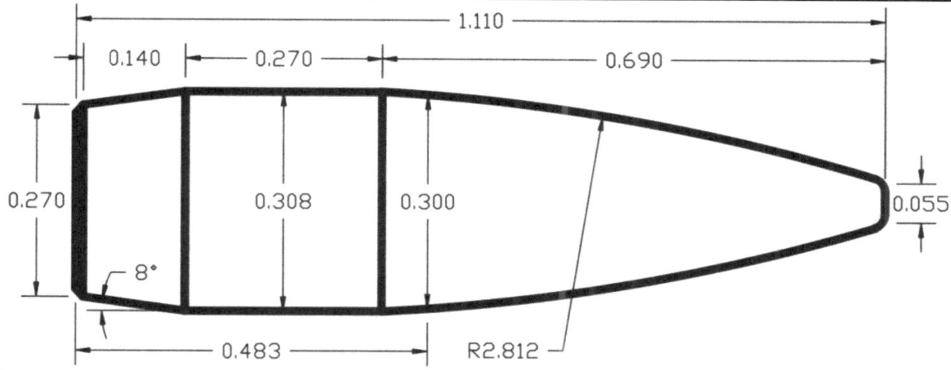

Lot #:	Unknown	Ogive Radius:	9.13 calibers
Sectional Density:	0.226 lb/in^2	Rt/R:	0.71

Ballistic Performance

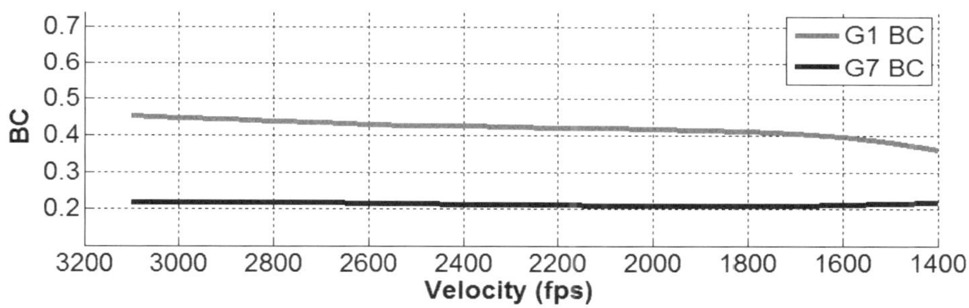

	Above 3000	3000-2500	2500-2000	2000-1500	Below 1500
G1 BC	0.463	0.438	0.423	0.400	0.382
G7 BC	0.219	0.217	0.212	0.212	0.215

Averages from 3000 fps to 1500 fps:
G1 BC: 0.419, variation of G1 BC: 0.066
G7 BC: 0.215, variation of G7 BC: 0.009
G7 form factor: 1.052

Rifling Twist Rate and Stability

	Stability Factor		
Barrel Twist	**Best case**	**Nominal**	**Worst case**
1:10"	3.32	2.90	2.51
1:11"	2.74	2.40	2.07
1:12"	2.30	2.01	1.74
1:13"	1.96	1.71	1.48

Minimum twist required for stability is 1:15.5"
Recommended twist for optimal performance is 1:13.9" or faster

Lapua .308 caliber 170 grain FMJBT

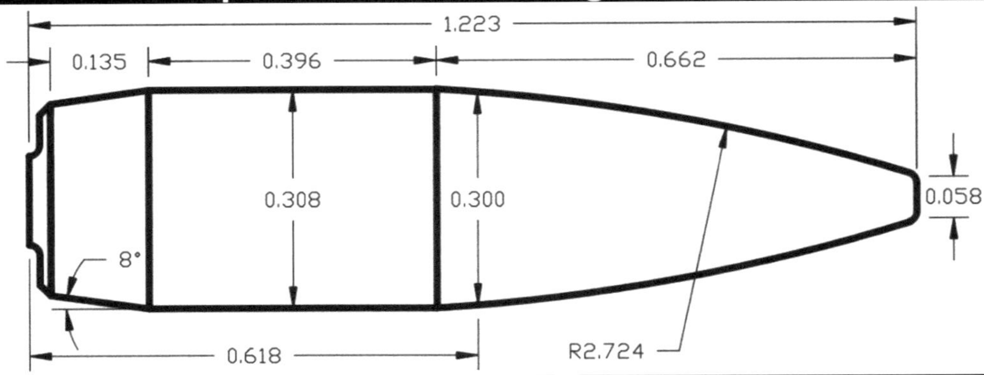

Lot #:	P00648003/1612413	**Ogive Radius:**	**8.84 calibers**
Sectional Density:	**0.256 lb/in²**	**Rt/R:**	**0.67**

Ballistic Performance

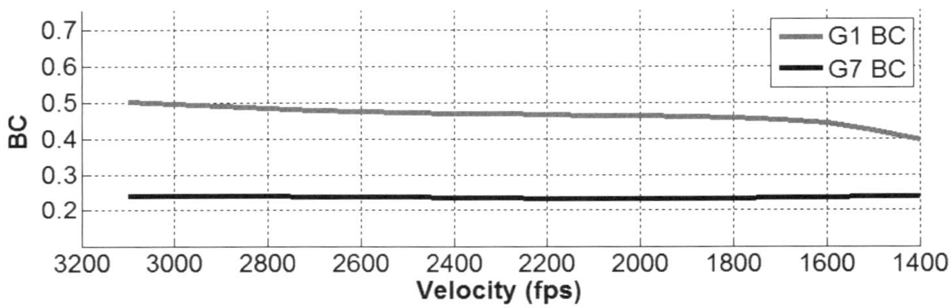

	Above 3000	3000-2500	2500-2000	2000-1500	Below 1500
G1 BC	0.512	0.484	0.468	0.444	0.424
G7 BC	0.242	0.240	0.236	0.236	0.238

Averages from 3000 fps to 1500 fps:
G1 BC: 0.464, variation of G1 BC: 0.071
G7 BC: 0.238, variation of G7 BC: 0.009
G7 form factor: 1.076

Rifling Twist Rate and Stability

Barrel Twist	Stability Factor Best case	Nominal	Worst case
1:10"	2.85	2.49	2.15
1:11"	2.35	2.06	1.78
1:12"	1.98	1.73	1.49
1:13"	1.68	1.47	1.27

Minimum twist required for stability is 1:14.4"
Recommended twist for optimal performance is 1:12.9" or faster

Lapua .308 caliber 155 grain Scenar

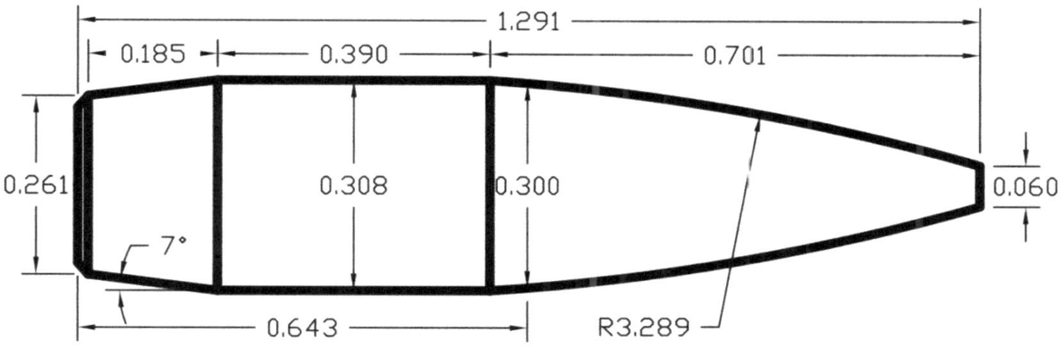

Lot #:	4PL7073	Ogive Radius:	10.68 calibers
Sectional Density:	0.233 lb/in^2	Rt/R:	0.64

Ballistic Performance

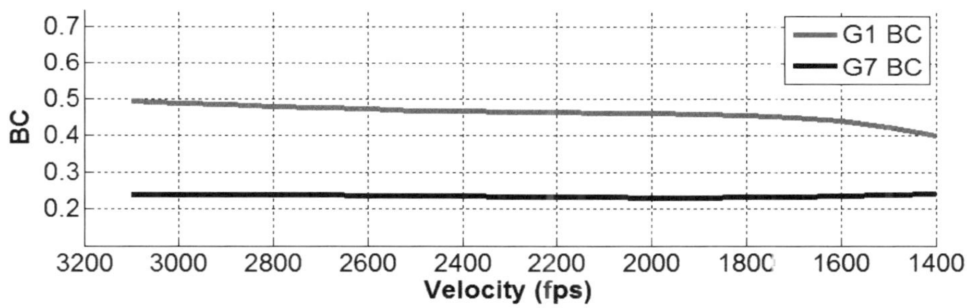

	Above 3000	3000-2500	2500-2000	2000-1500	Below 1500
G1 BC	0.505	0.479	0.465	0.442	0.424
G7 BC	0.239	0.238	0.234	0.235	0.239

Averages from 3000 fps to 1500 fps:
G1 BC: 0.461, variation of G1 BC: 0.066
G7 BC: 0.236, variation of G7 BC: 0.008
G7 form factor: 0.988

Rifling Twist Rate and Stability

	Stability Factor		
Barrel Twist	**Best case**	**Nominal**	**Worst case**
1:10"	2.22	1.94	1.68
1:11"	1.83	1.60	1.39
1:12"	1.54	1.35	1.16
1:13"	1.31	1.15	0.99

Minimum twist required for stability is 1:12.7"
Recommended twist for optimal performance is 1:11.4" or faster

Lapua .308 caliber 167 grain Scenar

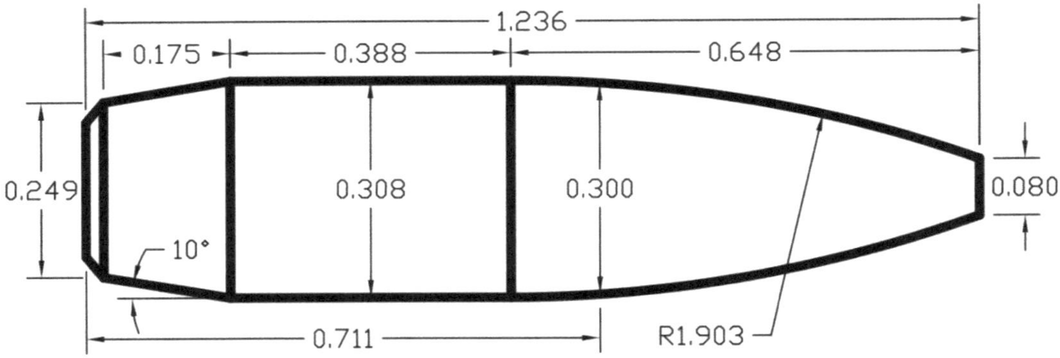

Lot #:	**4PL7069**	**Ogive Radius:**	**6.18 calibers**
Sectional Density:	**0.252 lb/in^2**	**Rt/R:**	**1.00**

Ballistic Performance

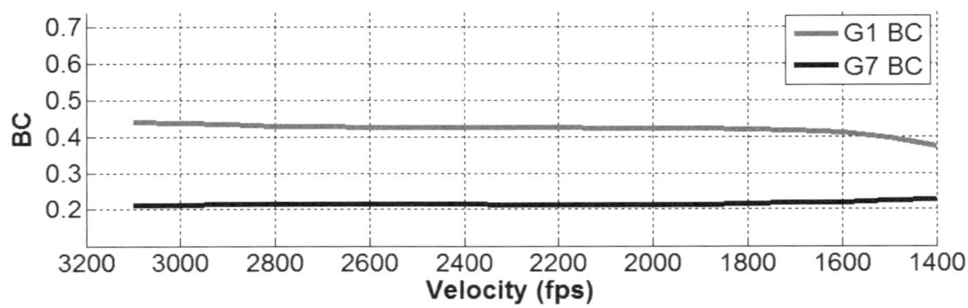

	Above 3000	3000-2500	2500-2000	2000-1500	Below 1500
G1 BC	0.448	0.432	0.425	0.410	0.396
G7 BC	0.212	0.214	0.214	0.218	0.223

Averages from 3000 fps to 1500 fps:

G1 BC: 0.421, variation of G1 BC: 0.042

G7 BC: 0.216, variation of G7 BC: 0.010

G7 form factor: 1.163

Rifling Twist Rate and Stability

Barrel Twist	**Stability Factor** Best case	Nominal	Worst case
1:10"	2.71	2.37	2.05
1:11"	2.24	1.96	1.69
1:12"	1.88	1.65	1.42
1:13"	1.61	1.40	1.21

Minimum twist required for stability is 1:14.1"

Recommended twist for optimal performance is 1:12.6" or faster

Lapua .308 caliber 185 grain Scenar

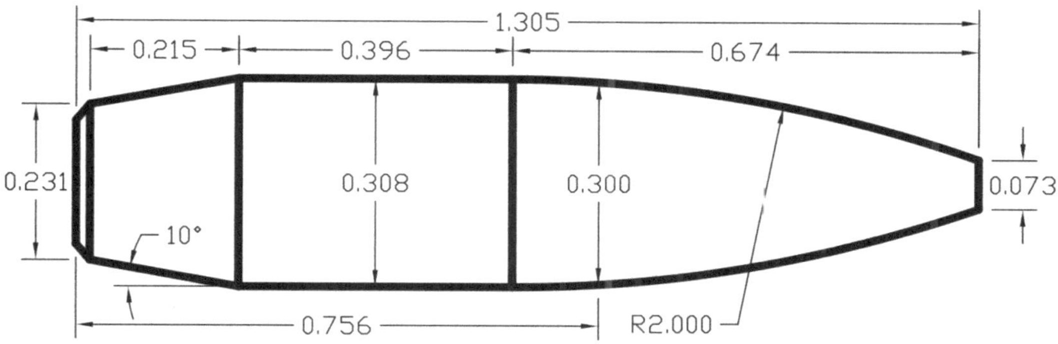

Lot #:	Unknown	Ogive Radius:	6.49 calibers
Sectional Density:	0.279 lb/in^2	Rt/R:	1.00

Ballistic Performance

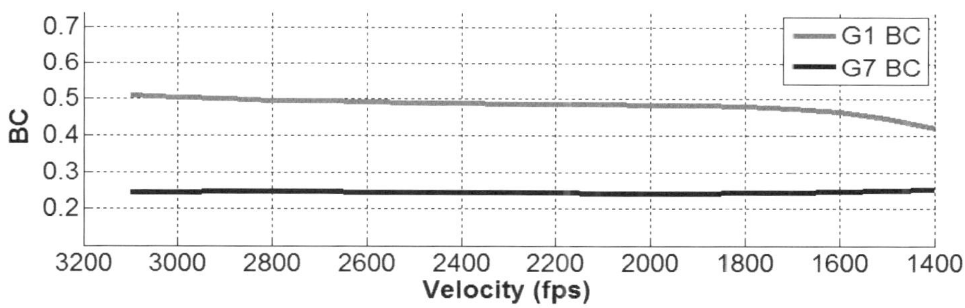

	Above 3000	3000-2500	2500-2000	2000-1500	Below 1500
G1 BC	0.516	0.496	0.486	0.465	0.446
G7 BC	0.245	0.246	0.244	0.247	0.251

Averages from 3000 fps to 1500 fps:
G1 BC: 0.480, variation of G1 BC: 0.057
G7 BC: 0.246, variation of G7 BC: 0.008
G7 form factor: 1.130

Rifling Twist Rate and Stability

Barrel Twist	Stability Factor Best case	Nominal	Worst case
1:10"	2.57	2.24	1.94
1:11"	2.12	1.85	1.60
1:12"	1.78	1.56	1.35
1:13"	1.52	1.33	1.15

Minimum twist required for stability is 1:13.7"
Recommended twist for optimal performance is 1:12.2" or faster

Lapua .308 caliber 155 grain Scenar L

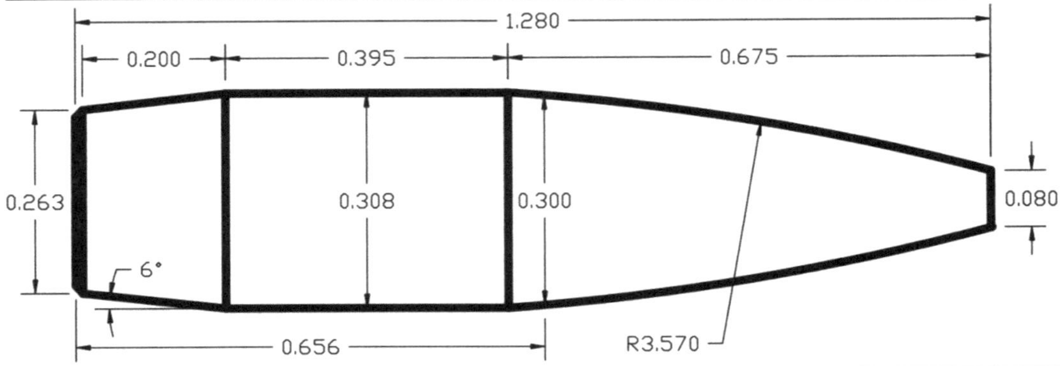

	Lot #:	Unknown	Ogive Radius:	11.59 calibers
	Sectional Density:	0.233 lb/in^2	Rt/R:	0.57

Ballistic Performance

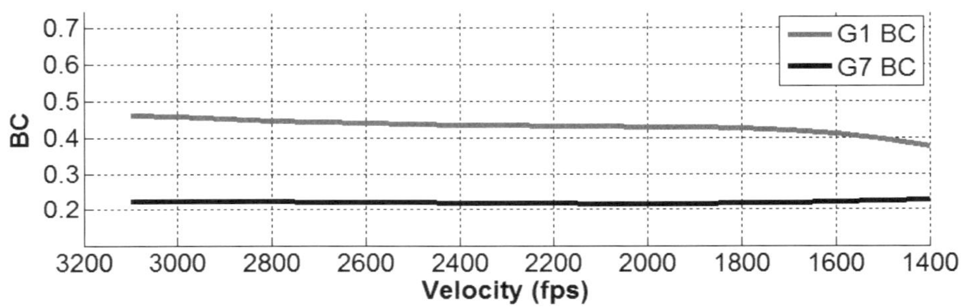

	Above 3000	3000-2500	2500-2000	2000-1500	Below 1500
G1 BC	0.472	0.448	0.434	0.414	0.397
G7 BC	0.224	0.222	0.218	0.220	0.223

Averages from 3000 fps to 1500 fps:
G1 BC: 0.431, variation of G1 BC: 0.061
G7 BC: 0.221, variation of G7 BC: 0.008
G7 form factor: 1.056

Rifling Twist Rate and Stability

	Stability Factor		
Barrel Twist	**Best case**	**Nominal**	**Worst case**
1:10"	2.28	1.99	1.72
1:11"	1.88	1.64	1.42
1:12"	1.58	1.38	1.19
1:13"	1.35	1.18	1.02

Minimum twist required for stability is 1:12.7"
Recommended twist for optimal performance is 1:11.4" or faster

Lapua .308 caliber 175 grain Scenar-L

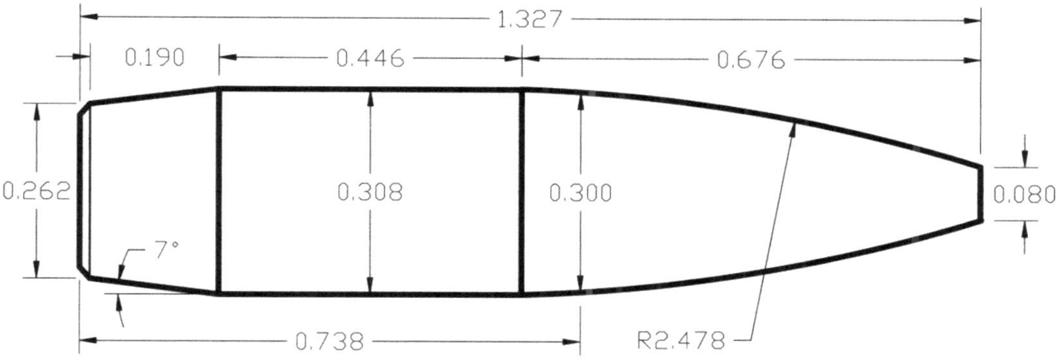

Lot #:	P00841301	Ogive Radius:	8.05 calibers
Sectional Density:	0.264 lb/in^2	Rt/R:	0.83

Ballistic Performance

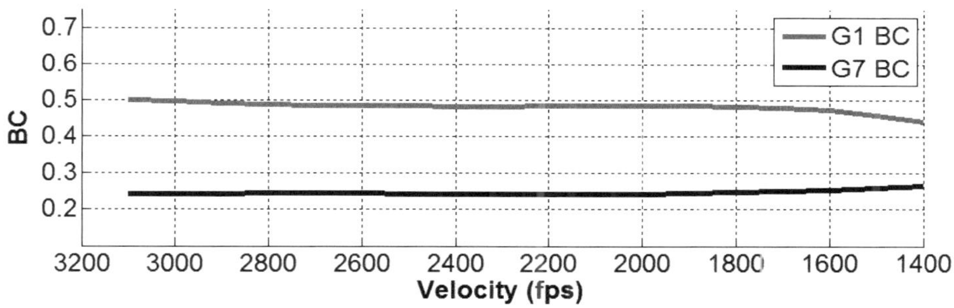

	Above 3000	3000-2500	2500-2000	2000-1500	Below 1500
G1 BC	0.507	0.490	0.484	0.472	0.459
G7 BC	0.240	0.243	0.243	0.251	0.259

Averages from 3000 fps to 1500 fps:
G1 BC: 0.481, variation of G1 BC: 0.037
G7 BC: 0.247, variation of G7 BC: 0.017
G7 form factor: 1.067

Rifling Twist Rate and Stability

Barrel Twist	Stability Factor Best case	Nominal	Worst case
1:10"	2.31	2.02	1.75
1:11"	1.91	1.67	1.45
1:12"	1.61	1.40	1.21
1:13"	1.37	1.20	1.03

Minimum twist required for stability is 1:13.0"
Recommended twist for optimal performance is 1:11.6" or faster

Lapua .308 caliber 220 grain Scenar L

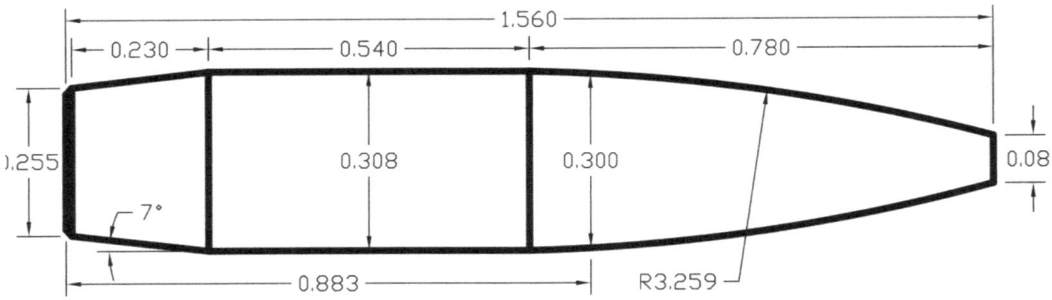

Lot #:	Unknown	Ogive Radius:	10.58 calibers
Sectional Density:	0.331 lb/in^2	Rt/R:	0.84

Ballistic Performance

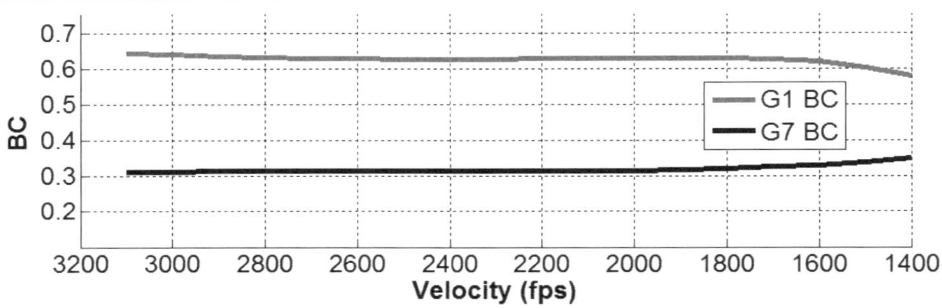

	Above 3000	3000-2500	2500-2000	2000-1500	Below 1500
G1 BC	0.654	0.633	0.628	0.616	0.602
G7 BC	0.310	0.314	0.316	0.328	0.340

Averages from 3000 fps to 1500 fps:
G1 BC: 0.624, variation of G1 BC: 0.038
G7 BC: 0.321, variation of G7 BC: 0.027
G7 form factor: 1.033

Rifling Twist Rate and Stability

Barrel Twist	Stability Factor Best case	Nominal	Worst case
1:10"	1.82	1.59	1.37
1:11"	1.50	1.31	1.13
1:12"	1.26	1.10	0.95
1:13"	1.08	0.94	0.81

Minimum twist required for stability is 1:11.5"
Recommended twist for optimal performance is 1:10.3" or faster

Matrix .308 caliber 168 grain RBT

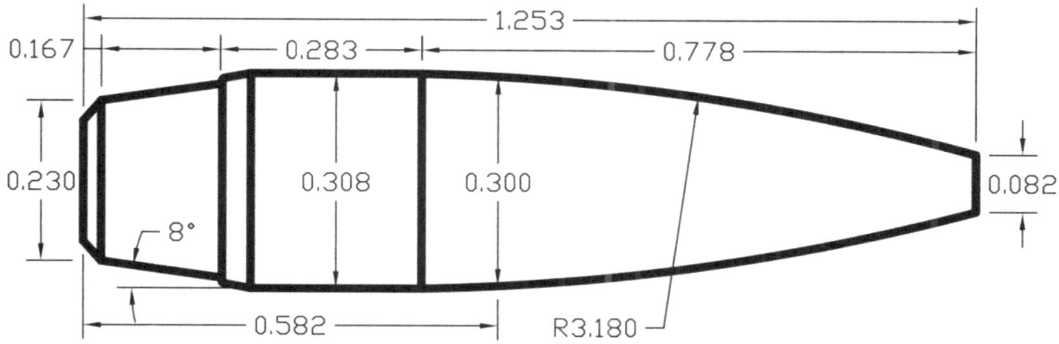

Lot #:	Unknown	Ogive Radius:	10.32 calibers
Sectional Density:	0.253 lb/in²	Rt/R:	0.85

Ballistic Performance

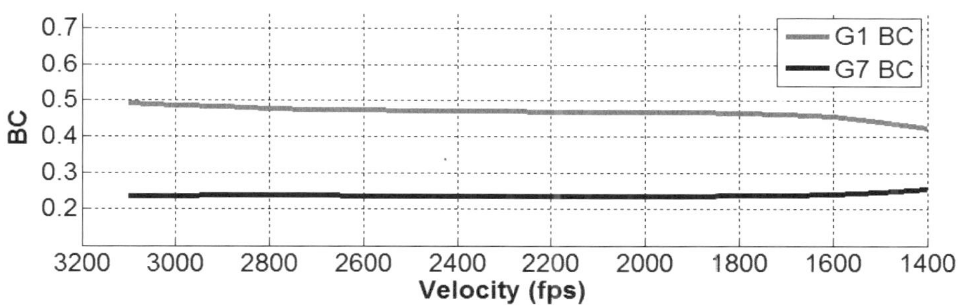

	Above 3000	3000-2500	2500-2000	2000-1500	Below 1500
G1 BC	0.498	0.478	0.470	0.455	0.441
G7 BC	0.236	0.237	0.236	0.242	0.249

Averages from 3000 fps to 1500 fps:
G1 BC: 0.466, variation of G1 BC: 0.044
G7 BC: 0.240, variation of G7 BC: 0.014
G7 form factor: 1.056

Rifling Twist Rate and Stability

Barrel Twist	Stability Factor Best case	Nominal	Worst case
1:10"	2.62	2.29	1.98
1:11"	2.17	1.89	1.64
1:12"	1.82	1.59	1.38
1:13"	1.55	1.36	1.17

Minimum twist required for stability is 1:13.8"
Recommended twist for optimal performance is 1:12.4" or faster

Matrix .308 caliber 210 grain VLD

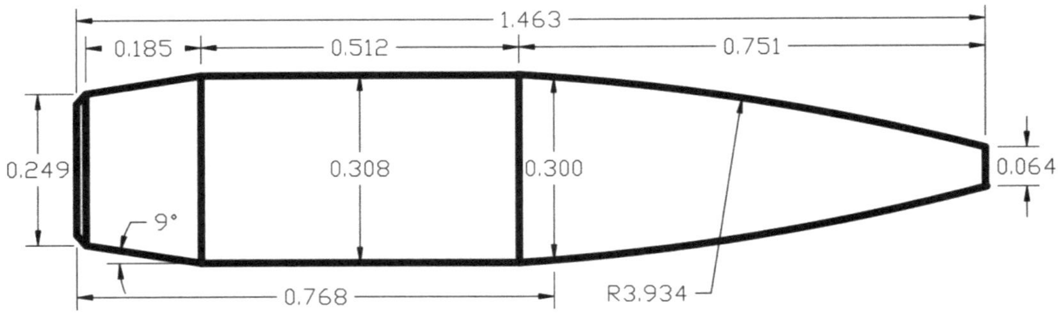

Lot #:	**Unknown**	**Ogive Radius:**	**12.77 calibers**
Sectional Density:	**0.316 lb/in²**	**Rt/R:**	**0.65**

Ballistic Performance

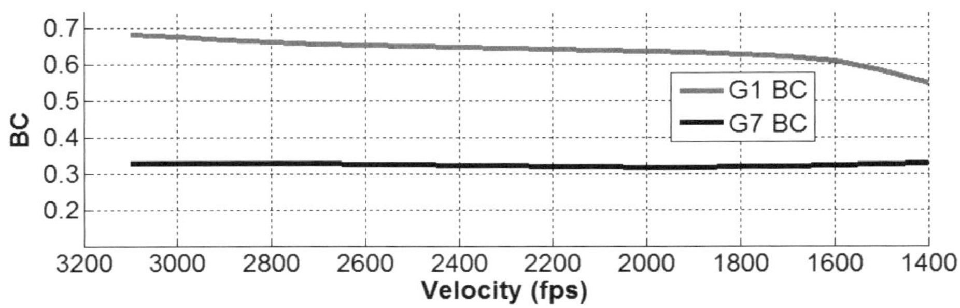

	Above 3000	3000-2500	2500-2000	2000-1500	Below 1500
G1 BC	0.696	0.661	0.642	0.609	0.582
G7 BC	0.330	0.328	0.323	0.323	0.327

Averages from 3000 fps to 1500 fps:
G1 BC: 0.635, variation of G1 BC: 0.093
G7 BC: 0.326, variation of G7 BC: 0.011
G7 form factor: 0.971

Rifling Twist Rate and Stability

Barrel Twist	Stability Factor Best case	Nominal	Worst case
1:10"	2.09	1.83	1.58
1:11"	1.73	1.51	1.31
1:12"	1.45	1.27	1.10
1:13"	1.24	1.08	0.94

Minimum twist required for stability is 1:12.3"
Recommended twist for optimal performance is 1:11.0" or faster

Nosler .308 caliber 150 grain Accubond

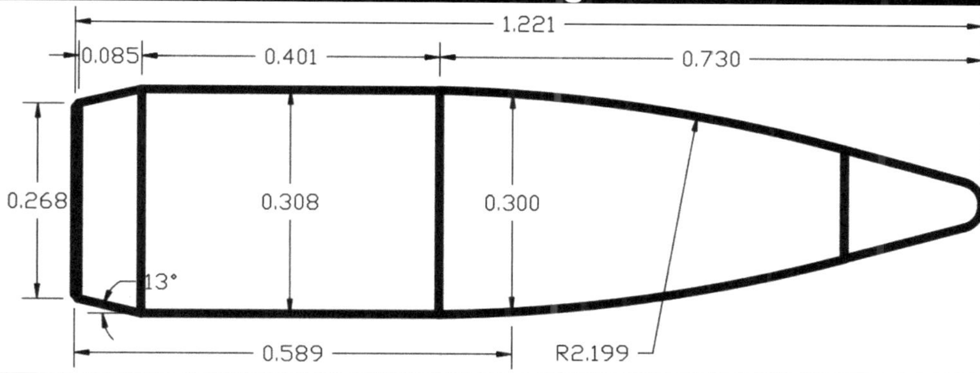

Lot #:	Unknown	Ogive Radius:	7.14 calibers
Sectional Density:	0.226 lb/in^2	Rt/R:	0.87

Ballistic Performance

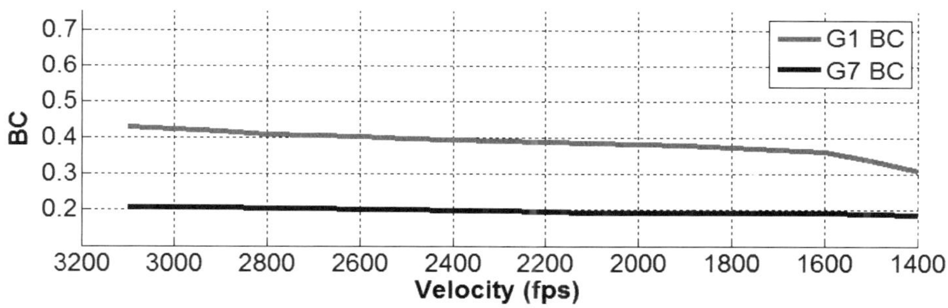

	Above 3000	3000-2500	2500-2000	2000-1500	Below 1500
G1 BC	0.439	0.410	0.390	0.361	0.339
G7 BC	0.208	0.203	0.196	0.191	0.190

Averages from 3000 fps to 1500 fps:
G1 BC: 0.386, variation of G1 BC: 0.083
G7 BC: 0.197, variation of G7 BC: 0.016
G7 form factor: 1.146

Rifling Twist Rate and Stability

Barrel Twist	Best case	Stability Factor Nominal	Worst case
1:10"	2.52	2.20	1.91
1:11"	2.09	1.82	1.58
1:12"	1.75	1.53	1.32
1:13"	1.49	1.30	1.13

Minimum twist required for stability is 1:13.6"
Recommended twist for optimal performance is 1:12.1" or faster

Nosler .308 caliber 165 grain Accubond

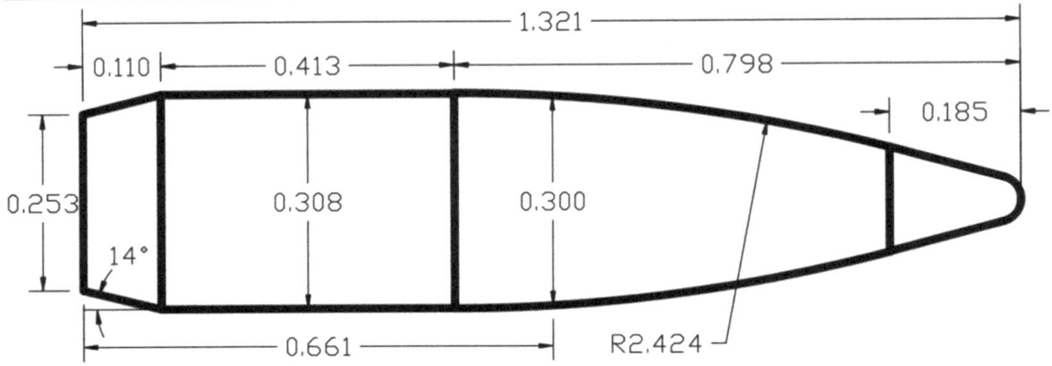

Lot #:	Unknown	Ogive Radius:	7.87 calibers
Sectional Density:	0.249 lb/in^2	Rt/R:	1.00

Ballistic Performance

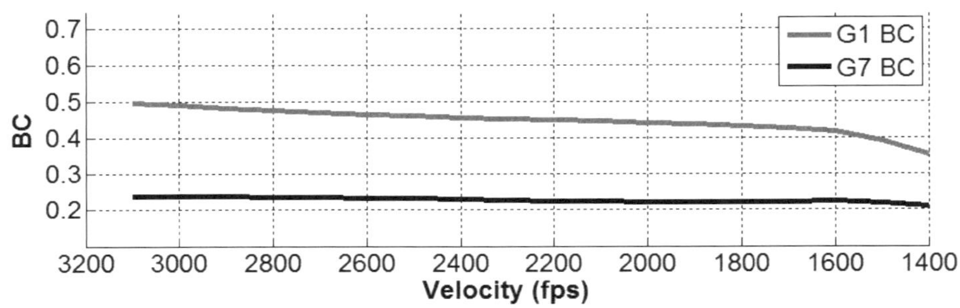

	Above 3000	3000-2500	2500-2000	2000-1500	Below 1500
G1 BC	0.509	0.475	0.452	0.417	0.391
G7 BC	0.241	0.236	0.227	0.221	0.219

Averages from 3000 fps to 1500 fps:
G1 BC: 0.446, variation of G1 BC: 0.099
G7 BC: 0.228, variation of G7 BC: 0.020
G7 form factor: 1.091

Rifling Twist Rate and Stability

Barrel Twist	Stability Factor Best case	Stability Factor Nominal	Stability Factor Worst case
1:10"	2.21	1.93	1.67
1:11"	1.83	1.60	1.38
1:12"	1.54	1.34	1.16
1:13"	1.31	1.14	0.99

Minimum twist required for stability is 1:12.7"
Recommended twist for optimal performance is 1:11.3" or faster

Nosler .308 caliber 180 grain Accubond

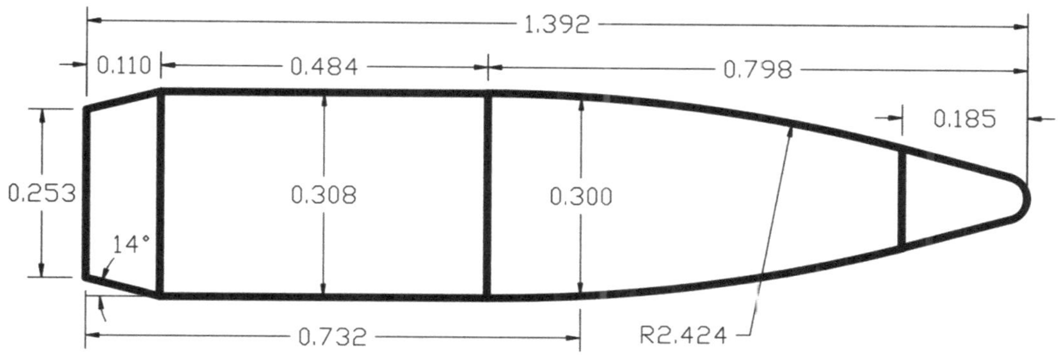

Lot #:	Unknown	Ogive Radius:	7.87 calibers
Sectional Density:	0.271 lb/in^2	Rt/R:	1.00

Ballistic Performance

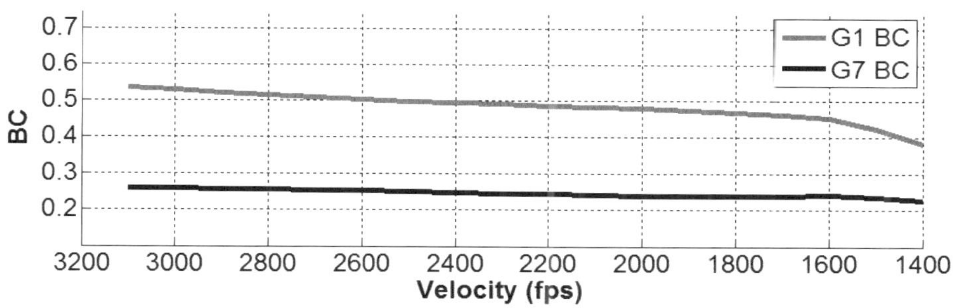

	Above 3000	3000-2500	2500-2000	2000-1500	Below 1500
G1 BC	0.549	0.513	0.488	0.451	0.423
G7 BC	0.260	0.254	0.246	0.239	0.237

Averages from 3000 fps to 1500 fps:
G1 BC: 0.482, variation of G1 BC: 0.105
G7 BC: 0.247, variation of G7 BC: 0.021
G7 form factor: 1.101

Rifling Twist Rate and Stability

Barrel Twist	Stability Factor Best case	Nominal	Worst case
1:10"	2.07	1.81	1.57
1:11"	1.71	1.50	1.29
1:12"	1.44	1.26	1.09
1:13"	1.23	1.07	0.93

Minimum twist required for stability is 1:12.3"
Recommended twist for optimal performance is 1:11.0" or faster

Nosler .308 caliber 200 grain Accubond

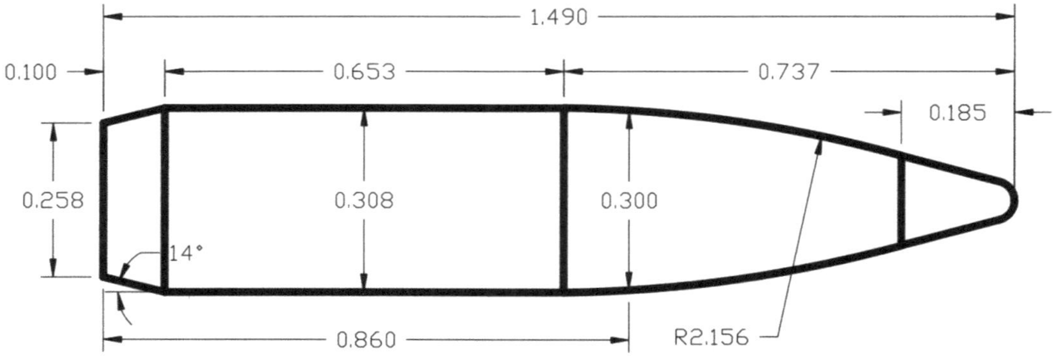

Lot #:	Unknown	Ogive Radius:	7.00 calibers
Sectional Density:	0.301 lb/in^2	Rt/R:	1.00

Ballistic Performance

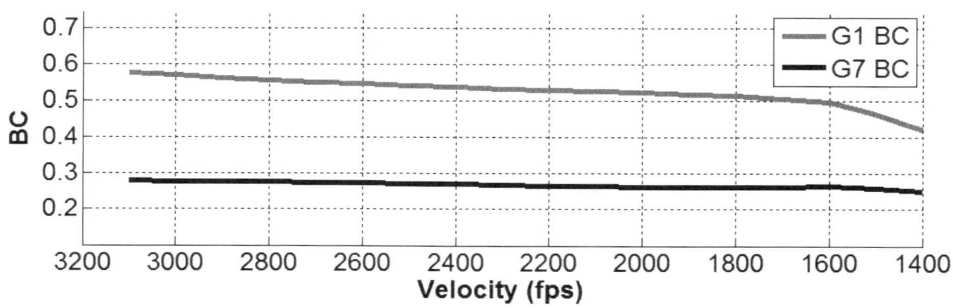

	Above 3000	3000-2500	2500-2000	2000-1500	Below 1500
G1 BC	0.590	0.555	0.532	0.494	0.465
G7 BC	0.280	0.275	0.267	0.262	0.261

Averages from 3000 fps to 1500 fps:
G1 BC: 0.525, variation of G1 BC: 0.104
G7 BC: 0.268, variation of G7 BC: 0.018
G7 form factor: 1.123

Rifling Twist Rate and Stability

Barrel Twist	Stability Factor Best case	Nominal	Worst case
1:10"	1.89	1.65	1.43
1:11"	1.56	1.36	1.18
1:12"	1.31	1.15	0.99
1:13"	1.12	0.98	0.84

Minimum twist required for stability is 1:11.7"
Recommended twist for optimal performance is 1:10.5" or faster

Nosler .308 caliber 190 grain Long Range Accubond

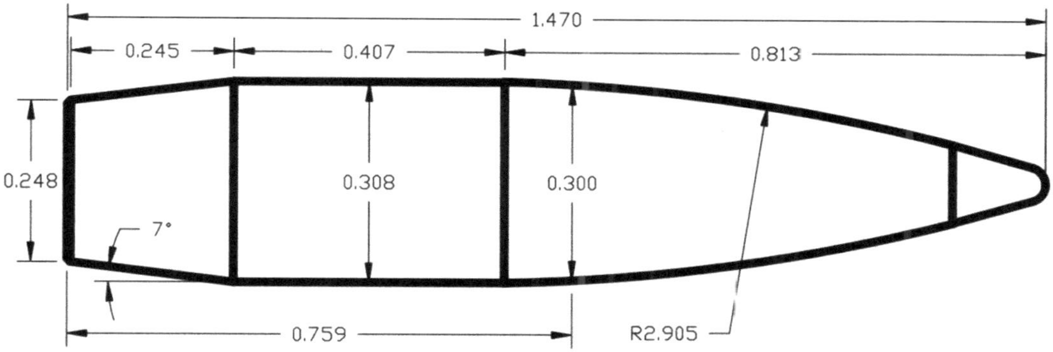

Lot #:	FL26F10	Ogive Radius:	9.43 calibers
Sectional Density:	0.286 lb/in^2	Rt/R:	0.85

Ballistic Performance

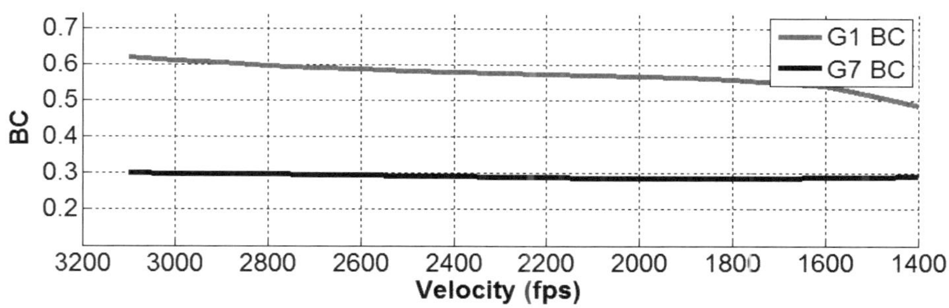

	Above 3000	3000-2500	2500-2000	2000-1500	Below 1500
G1 BC	0.633	0.597	0.575	0.541	0.516
G7 BC	0.300	0.296	0.289	0.287	0.290

Averages from 3000 fps to 1500 fps:
G1 BC: 0.569, variation of G1 BC: 0.096
G7 BC: 0.292, variation of G7 BC: 0.014
G7 form factor: 0.981

Rifling Twist Rate and Stability

Barrel Twist	Stability Factor Best case	Stability Factor Nominal	Stability Factor Worst case
1:10"	1.87	1.63	1.41
1:11"	1.54	1.35	1.17
1:12"	1.30	1.13	0.98
1:13"	1.10	0.96	0.83

Minimum twist required for stability is 1:11.7"
Recommended twist for optimal performance is 1:10.4" or faster

Nosler .308 caliber 210 grain Long Range Accubond

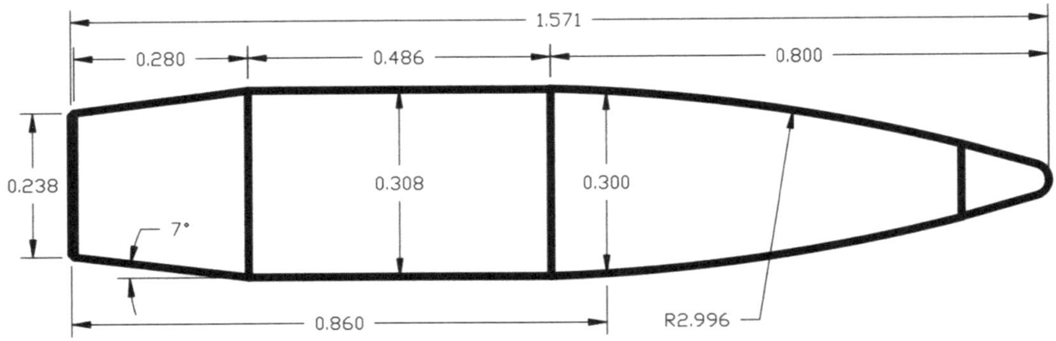

Lot #:	Unknown	Ogive Radius:	9.73 calibers
Sectional Density:	0.316 lb/in^2	Rt/R:	0.79

Ballistic Performance

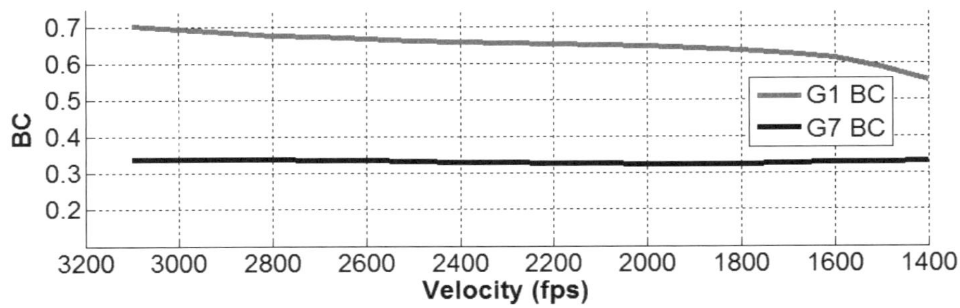

	Above 3000	3000-2500	2500-2000	2000-1500	Below 1500
G1 BC	0.716	0.677	0.653	0.616	0.587
G7 BC	0.339	0.336	0.329	0.327	0.330

Averages from 3000 fps to 1500 fps:
G1 BC: 0.647, variation of G1 BC: 0.106
G7 BC: 0.332, variation of G7 BC: 0.015
G7 form factor: 0.954

Rifling Twist Rate and Stability

	Stability Factor		
Barrel Twist	**Best case**	**Nominal**	**Worst case**
1:10"	1.88	1.64	1.42
1:11"	1.56	1.36	1.17
1:12"	1.31	1.14	0.99
1:13"	1.11	0.97	0.84

Minimum twist required for stability is 1:11.1"
Recommended twist for optimal performance is 1:9.9" or faster

Nosler .308 caliber 125 grain Ballistic Tip

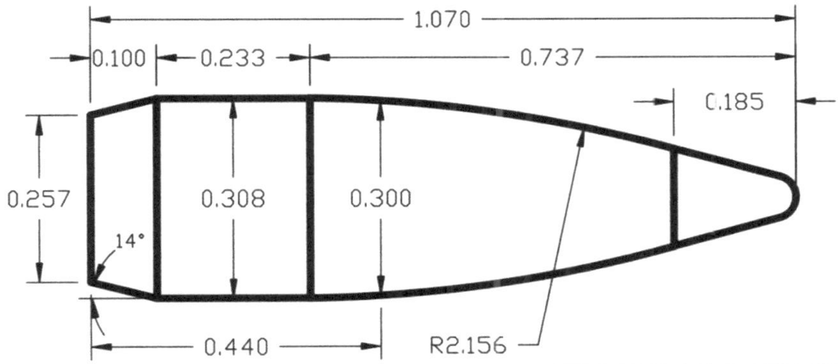

Lot #:	**Unknown**	**Ogive Radius:**	**7.0 calibers**
Sectional Density:	**0.188 lb/in²**	**Rt/R:**	**1.00**

Ballistic Performance

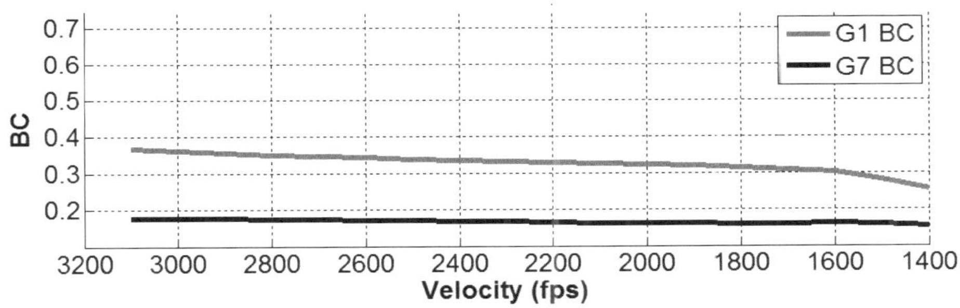

	Above 3000	3000-2500	2500-2000	2000-1500	Below 1500
G1 BC	0.378	0.351	0.332	0.304	0.284
G7 BC	0.179	0.174	0.167	0.161	0.159

Averages from 3000 fps to 1500 fps:
G1 BC: 0.327, variation of G1 BC: 0.079
G7 BC: 0.167, variation of G7 BC: 0.018
G7 form factor: 1.127

Rifling Twist Rate and Stability

Barrel Twist	Stability Factor Best case	Nominal	Worst case
1:10"	3.07	2.68	2.32
1:11"	2.54	2.22	1.92
1:12"	2.13	1.86	1.61
1:13"	1.82	1.59	1.37

Minimum twist required for stability is 1:14.9"
Recommended twist for optimal performance is 1:13.4" or faster

Nosler .308 caliber 150 grain Ballistic Tip

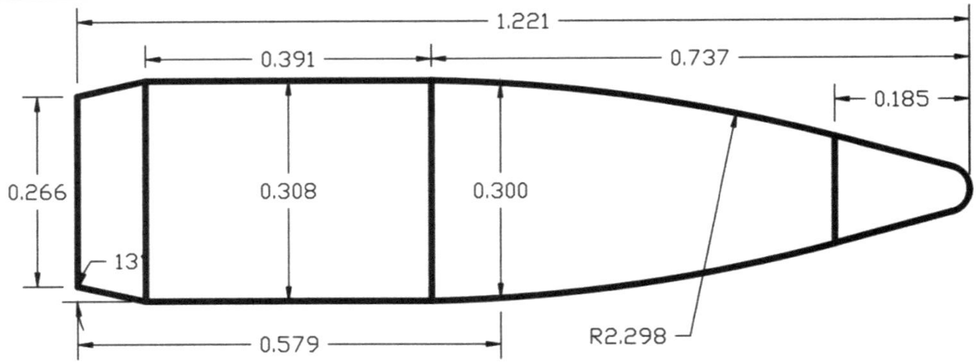

Lot #:	**1210 3431211**	**Ogive Radius:**	**7.0 calibers**
Sectional Density:	**0.226 lb/in^2**	**Rt/R:**	**1.00**

Ballistic Performance

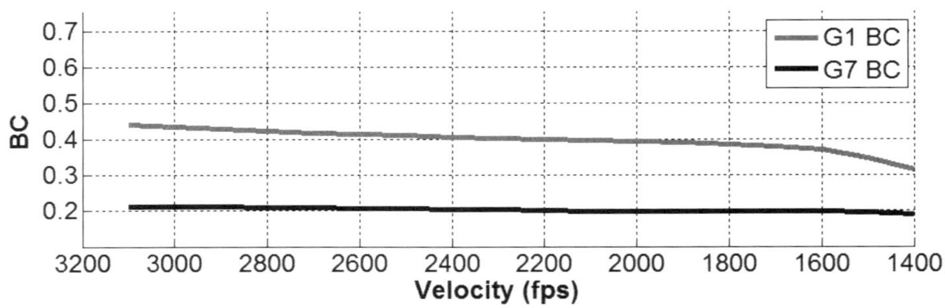

	Above 3000	3000-2500	2500-2000	2000-1500	Below 1500
G1 BC	0.453	0.423	0.402	0.371	0.348
G7 BC	0.214	0.210	0.202	0.196	0.195

Averages from 3000 fps to 1500 fps:
G1 BC: 0.397, variation of G1 BC: 0.088
G7 BC: 0.203, variation of G7 BC: 0.018
G7 form factor: 1.114

Rifling Twist Rate and Stability

Barrel Twist	**Stability Factor** Best case	Nominal	Worst case
1:10"	2.48	2.17	1.88
1:11"	2.05	1.79	1.55
1:12"	1.72	1.51	1.30
1:13"	1.47	1.28	1.11

Minimum twist required for stability is 1:13.4"
Recommended twist for optimal performance is 1:12.0" or faster

Nosler .308 caliber 165 grain Ballistic Tip

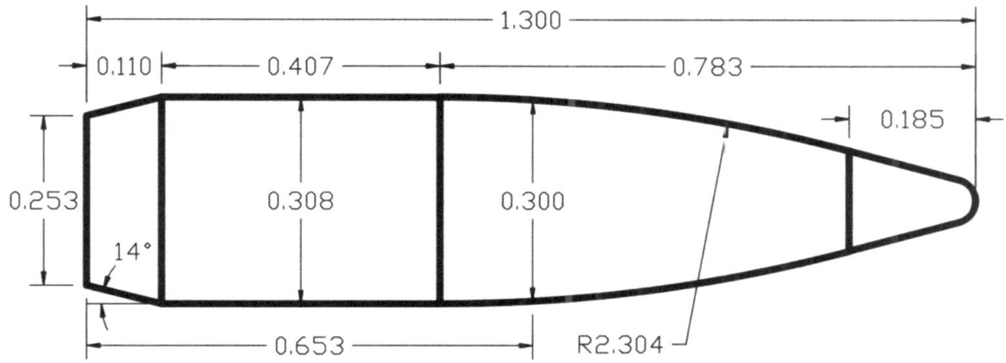

Lot #:	**Unknown**	**Ogive Radius:**	**7.48 calibers**
Sectional Density:	**0.249 lb/in^2**	**Rt/R:**	**1.00**

Ballistic Performance

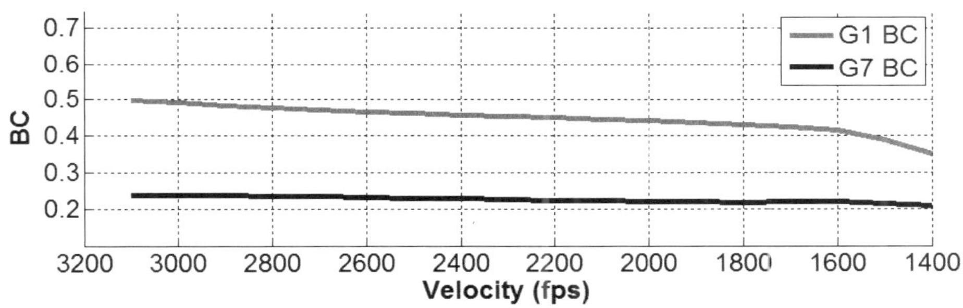

	Above 3000	3000-2500	2500-2000	2000-1500	Below 1500
G1 BC	0.510	0.475	0.450	0.414	0.388
G7 BC	0.241	0.236	0.226	0.219	0.217

Averages from 3000 fps to 1500 fps:
G1 BC: 0.445, variation of G1 BC: 0.102
G7 BC: 0.227, variation of G7 BC: 0.022
G7 form factor: 1.094

Rifling Twist Rate and Stability

	Stability Factor		
Barrel Twist	**Best case**	**Nominal**	**Worst case**
1:10"	2.32	2.02	1.75
1:11"	1.91	1.67	1.45
1:12"	1.61	1.41	1.22
1:13"	1.37	1.20	1.04

Minimum twist required for stability is 1:13.0"
Recommended twist for optimal performance is 1:11.6" or faster

Nosler .308 caliber 168 grain Ballistic Tip

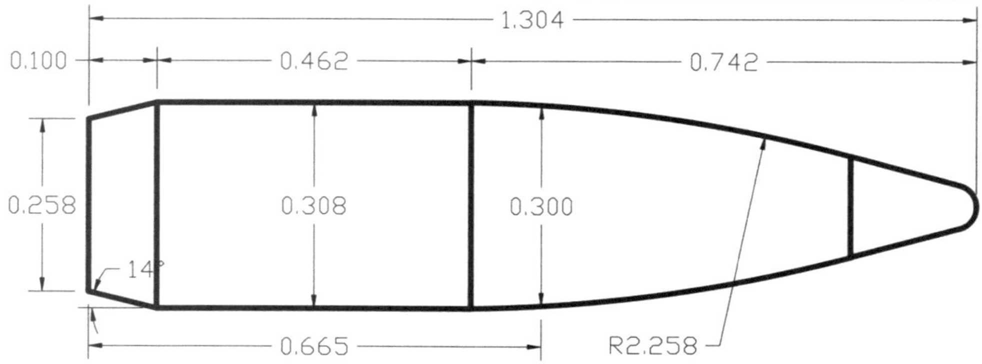

Lot #:	HL19F3-1/822	Ogive Radius:	7.33 calibers
Sectional Density:	0.253 lb/in²	Rt/R:	0.87

Ballistic Performance

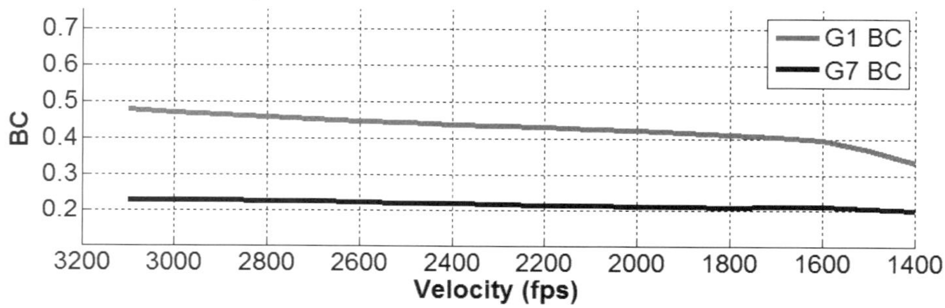

	Above 3000	3000-2500	2500-2000	2000-1500	Below 1500
G1 BC	0.491	0.457	0.433	0.398	0.372
G7 BC	0.233	0.227	0.218	0.210	0.208

Averages from 3000 fps to 1500 fps:
G1 BC: 0.427, variation of G1 BC: 0.100
G7 BC: 0.219, variation of G7 BC: 0.022
G7 form factor: 1.159

Rifling Twist Rate and Stability

Barrel Twist	Stability Factor Best case	Stability Factor Nominal	Stability Factor Worst case
1:10"	2.34	2.04	1.77
1:11"	1.93	1.69	1.46
1:12"	1.62	1.42	1.23
1:13"	1.38	1.21	1.04

Minimum twist required for stability is 1:13.0"
Recommended twist for optimal performance is 1:11.7" or faster

Nosler .308 caliber 180 grain Ballistic Tip

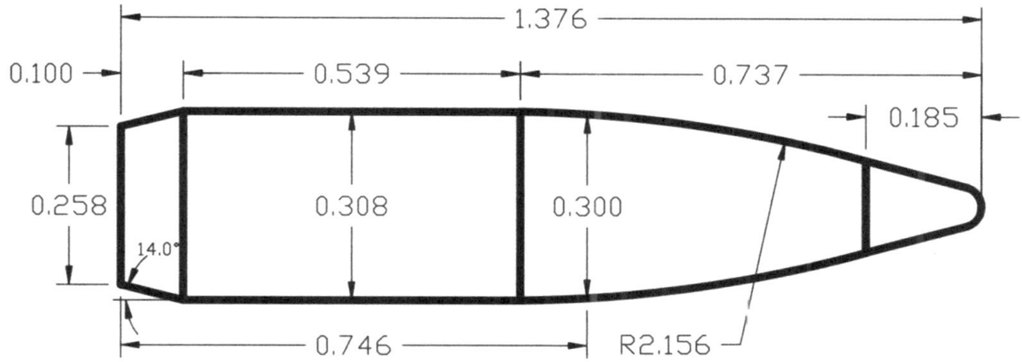

Lot #:	Unknown	Ogive Radius:	7.0 calibers
Sectional Density:	0.271 lb/in^2	Rt/R:	1.00

Ballistic Performance

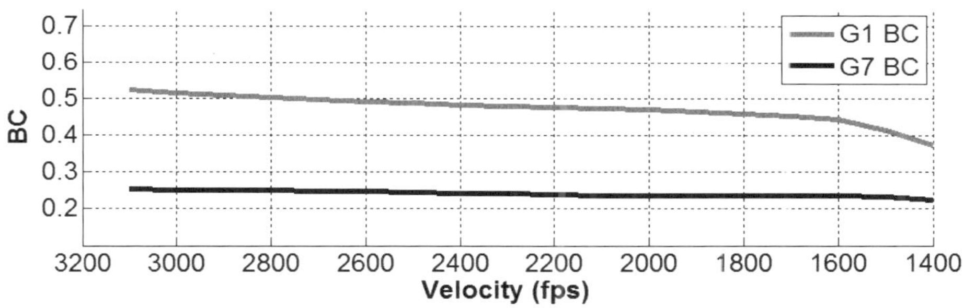

	Above 3000	3000-2500	2500-2000	2000-1500	Below 1500
G1 BC	0.535	0.501	0.478	0.442	0.415
G7 BC	0.253	0.249	0.240	0.234	0.232

Averages from 3000 fps to 1500 fps:
G1 BC: 0.472, variation of G1 BC: 0.100
G7 BC: 0.241, variation of G7 BC: 0.019
G7 form factor: 1.125

Rifling Twist Rate and Stability

Barrel Twist	Stability Factor Best case	Nominal	Worst case
1:10"	2.14	1.87	1.62
1:11"	1.77	1.55	1.34
1:12"	1.49	1.30	1.12
1:13"	1.27	1.11	0.96

Minimum twist required for stability is 1:12.6"
Recommended twist for optimal performance is 1:11.3" or faster

Nosler .308 caliber 140 grain Custom Competition

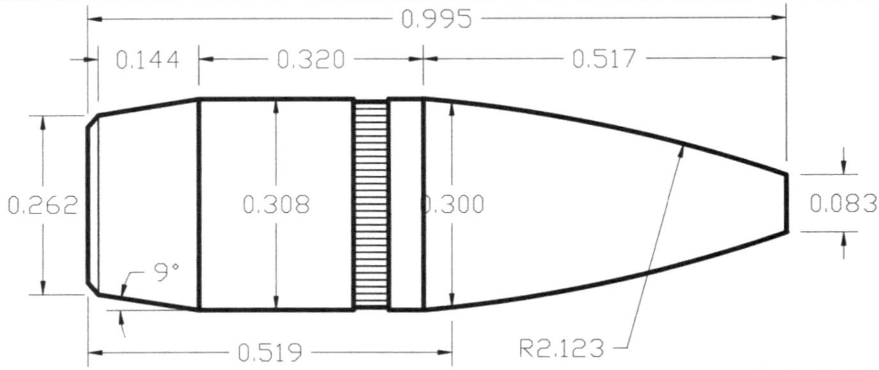

	Lot #:	KL07F15	Ogive Radius:	6.89 calibers
	Sectional Density:	0.211 lb/in^2	Rt/R:	0.59

Ballistic Performance

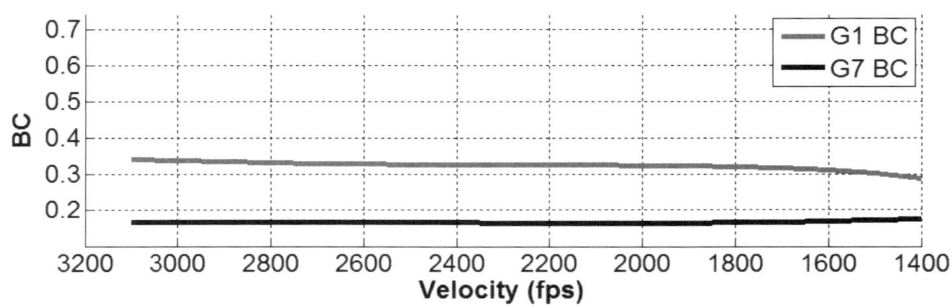

	Above 3000	3000-2500	2500-2000	2000-1500	Below 1500
G1 BC	0.345	0.333	0.327	0.314	0.303
G7 BC	0.164	0.165	0.164	0.167	0.171

Averages from 3000 fps to 1500 fps:
G1 BC: 0.323, variation of G1 BC: 0.034
G7 BC: 0.166, variation of G7 BC: 0.007
G7 form factor: 1.271

Rifling Twist Rate and Stability

Barrel Twist	Best case	Stability Factor Nominal	Worst case
1:14"	2.14	1.87	1.62
1:15"	1.87	1.63	1.41
1:16"	1.64	1.43	1.24
1:17"	1.45	1.27	1.10

Minimum twist required for stability is 1:17.5"
Recommended twist for optimal performance is 1:15.6" or faster

Nosler .308 caliber 155 grain Custom Competition

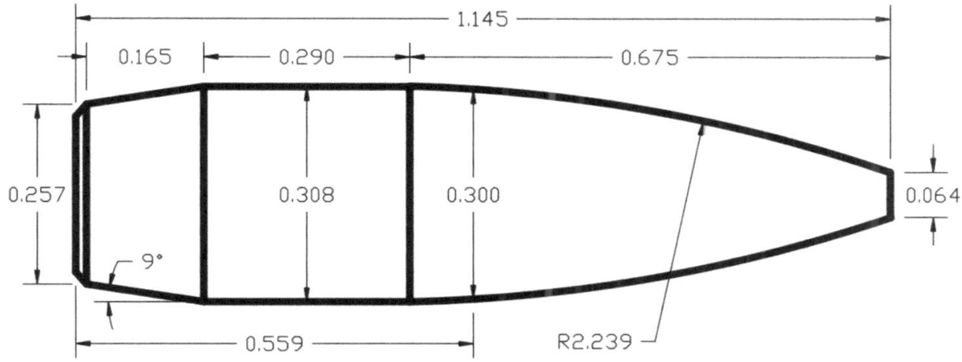

Lot #:	Unknown	Ogive Radius:	7.27 calibers
Sectional Density:	0.233 lb/in^2	Rt/R:	0.88

Ballistic Performance

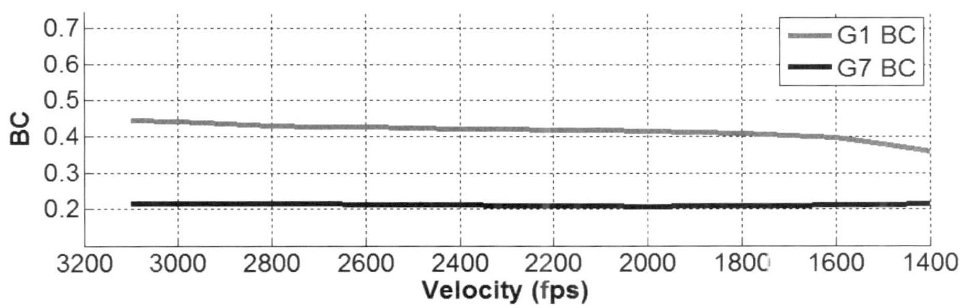

	Above 3000	3000-2500	2500-2000	2000-1500	Below 1500
G1 BC	0.454	0.432	0.419	0.397	0.380
G7 BC	0.215	0.214	0.210	0.211	0.214

Averages from 3000 fps to 1500 fps:
G1 BC: 0.414, variation of G1 BC: 0.060
G7 BC: 0.213, variation of G7 BC: 0.007
G7 form factor: 1.098

Rifling Twist Rate and Stability

Barrel Twist	Stability Factor Best case	Nominal	Worst case
1:10"	3.14	2.74	2.37
1:11"	2.59	2.26	1.96
1:12"	2.18	1.90	1.65
1:13"	1.86	1.62	1.40

Minimum twist required for stability is 1:15.1"
Recommended twist for optimal performance is 1:13.5" or faster

Nosler .308 caliber 168 grain Custom Competition

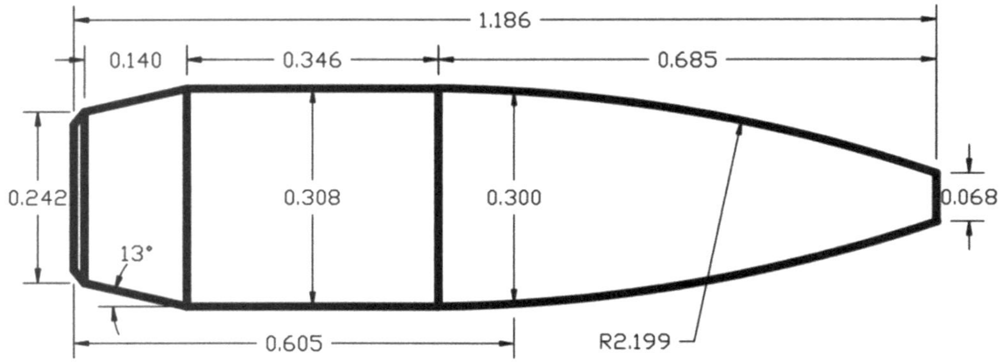

Lot #:	JE29F15-2	Ogive Radius:	7.14 calibers
Sectional Density:	0.253 lb/in^2	Rt/R:	0.92

Ballistic Performance

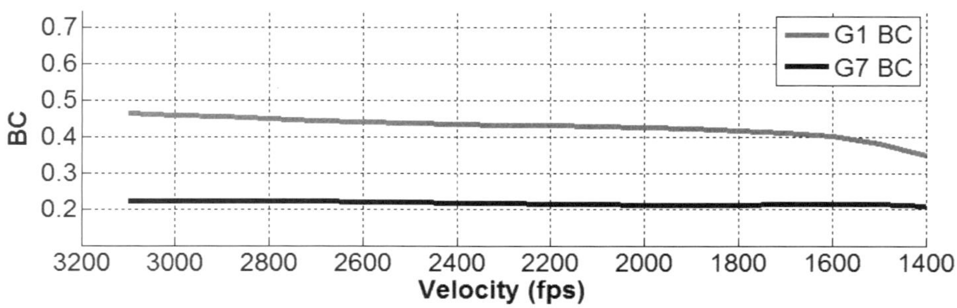

	Above 3000	3000-2500	2500-2000	2000-1500	Below 1500
G1 BC	0.475	0.449	0.432	0.404	0.382
G7 BC	0.225	0.223	0.217	0.214	0.214

Averages from 3000 fps to 1500 fps:
G1 BC: 0.427, variation of G1 BC: 0.078
G7 BC: 0.218, variation of G7 BC: 0.011
G7 form factor: 1.159

Rifling Twist Rate and Stability

Barrel Twist	Stability Factor Best case	Stability Factor Nominal	Stability Factor Worst case
1:10"	3.07	2.68	2.32
1:11"	2.54	2.22	1.92
1:12"	2.13	1.86	1.61
1:13"	1.82	1.59	1.37

Minimum twist required for stability is 1:15.0"
Recommended twist for optimal performance is 1:13.4" or faster

Nosler .308 caliber 175 grain Custom Competition

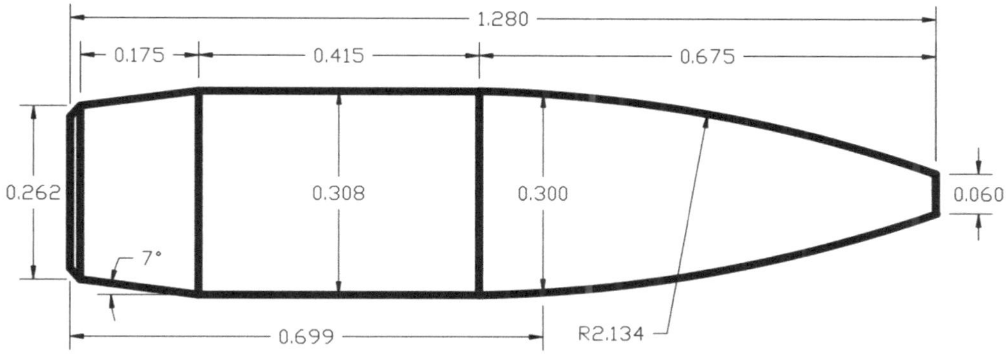

Lot #:	Unknown	Ogive Radius:	6.93 calibers
Sectional Density:	0.264 lb/in^2	Rt/R:	0.89

Ballistic Performance

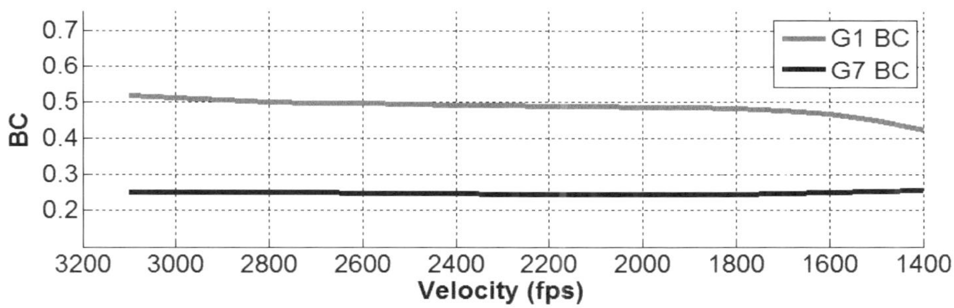

	Above 3000	3000-2500	2500-2000	2000-1500	Below 1500
G1 BC	0.526	0.502	0.490	0.468	0.449
G7 BC	0.249	0.249	0.246	0.249	0.253

Averages from 3000 fps to 1500 fps:
G1 BC: 0.485, variation of G1 BC: 0.062
G7 BC: 0.249, variation of G7 BC: 0.009
G7 form factor: 1.059

Rifling Twist Rate and Stability

Barrel Twist	Stability Factor Best case	Nominal	Worst case
1:10"	2.57	2.24	1.94
1:11"	2.12	1.86	1.60
1:12"	1.78	1.56	1.35
1:13"	1.52	1.33	1.15

Minimum twist required for stability is 1:13.7"
Recommended twist for optimal performance is 1:12.2" or faster

Nosler .308 caliber 190 grain Custom Competition

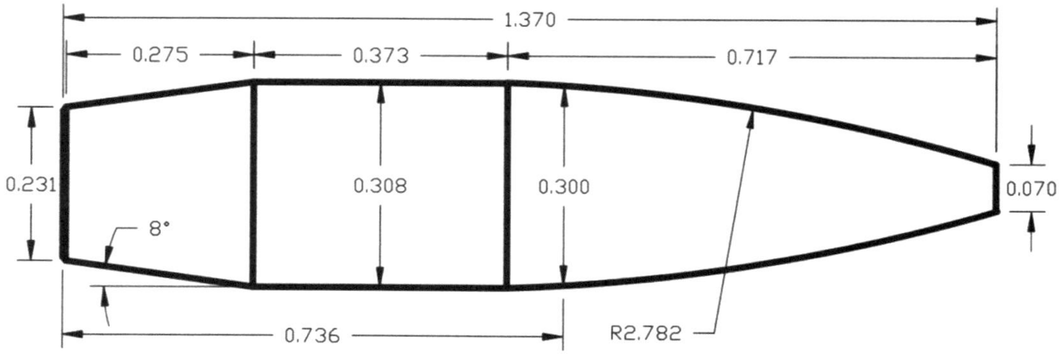

Lot #:	DL07F20	Ogive Radius:	9.03 calibers
Sectional Density:	0.286 lb/in^2	Rt/R:	0.80

Ballistic Performance

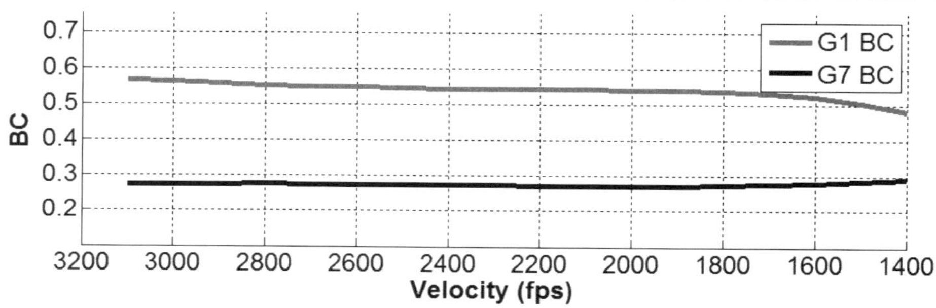

	Above 3000	3000-2500	2500-2000	2000-1500	Below 1500
G1 BC	0.579	0.557	0.546	0.525	0.508
G7 BC	0.274	0.276	0.274	0.279	0.286

Averages from 3000 fps to 1500 fps:
G1 BC: 0.541, variation of G1 BC: 0.057
G7 BC: 0.278, variation of G7 BC: 0.013
G7 form factor: 1.030

Rifling Twist Rate and Stability

Barrel Twist	Stability Factor Best case	Nominal	Worst case
1:10"	2.29	2.00	1.73
1:11"	1.89	1.65	1.43
1:12"	1.59	1.39	1.20
1:13"	1.36	1.18	1.02

Minimum twist required for stability is 1:12.9"
Recommended twist for optimal performance is 1:11.6" or faster

Nosler .308 caliber 150 grain E-Tip

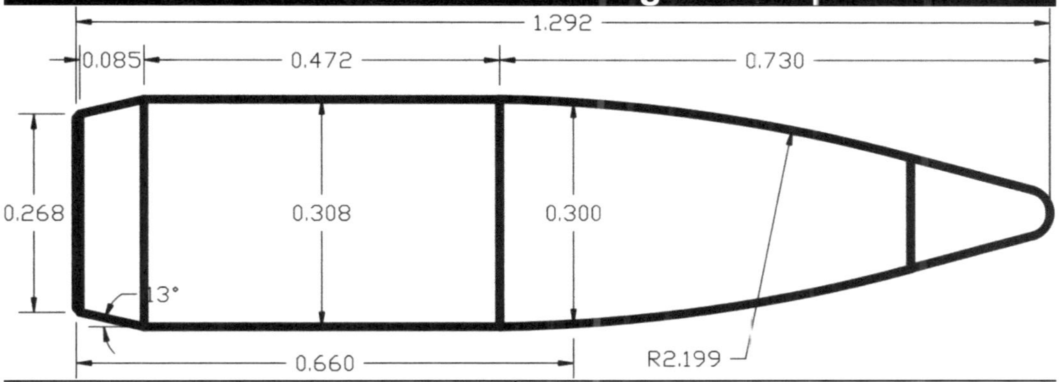

Lot #:	Unknown	Ogive Radius:	7.52 calibers
Sectional Density:	0.226 lb/in^2	Rt/R:	0.83

Ballistic Performance

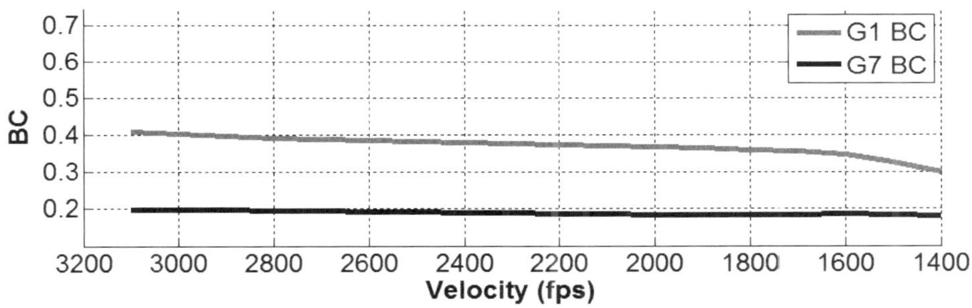

	Above 3000	3000-2500	2500-2000	2000-1500	Below 1500
G1 BC	0.420	0.393	0.375	0.348	0.327
G7 BC	0.199	0.195	0.188	0.184	0.184

Averages from 3000 fps to 1500 fps:
G1 BC: 0.370, variation of G1 BC: 0.077
G7 BC: 0.189, variation of G7 BC: 0.014
G7 form factor: 1.193

Rifling Twist Rate and Stability

Barrel Twist	Stability Factor Best case	Nominal	Worst case
1:10"	2.14	1.87	1.62
1:11"	1.77	1.55	1.34
1:12"	1.49	1.30	1.12
1:13"	1.27	1.11	0.96

Minimum twist required for stability is 1:12.5"
Recommended twist for optimal performance is 1:11.2" or faster

Nosler .308 caliber 168 grain E-Tip

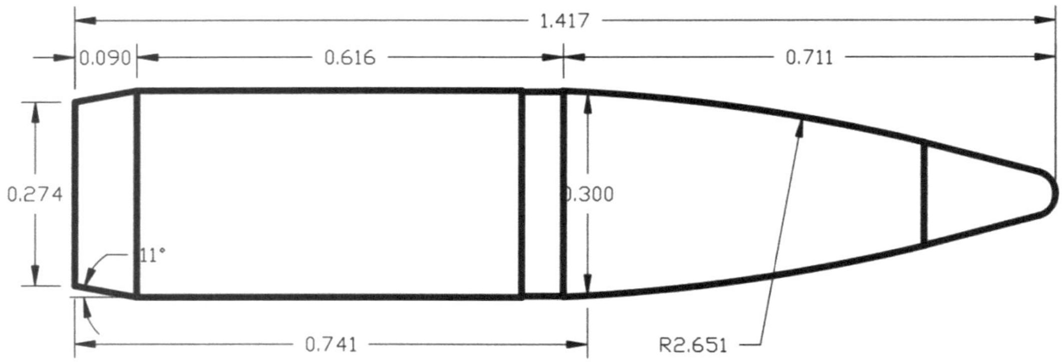

Lot #:	FL19F14	Ogive Radius:	8.61 calibers
Sectional Density:	0.253 lb/in²	Rt/R:	0.62

Ballistic Performance

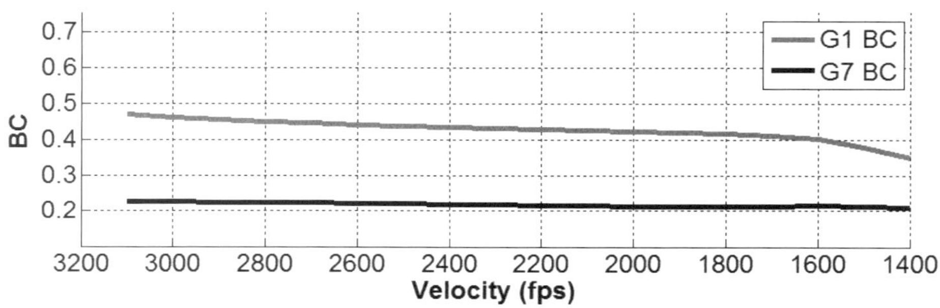

	Above 3000	3000-2500	2500-2000	2000-1500	Below 1500
G1 BC	0.481	0.451	0.431	0.403	0.381
G7 BC	0.228	0.223	0.217	0.214	0.214

Averages from 3000 fps to 1500 fps:
G1 BC: 0.427, variation of G1 BC: 0.082
G7 BC: 0.218, variation of G7 BC: 0.013
G7 form factor: 1.159

Rifling Twist Rate and Stability

Barrel Twist	Stability Factor Best case	Nominal	Worst case
1:10"	1.84	1.60	1.39
1:11"	1.52	1.33	1.15
1:12"	1.28	1.11	0.96
1:13"	1.09	0.95	0.82

Minimum twist required for stability is 1:11.6"
Recommended twist for optimal performance is 1:10.3" or faster

Nosler .308 caliber 180 grain E-Tip

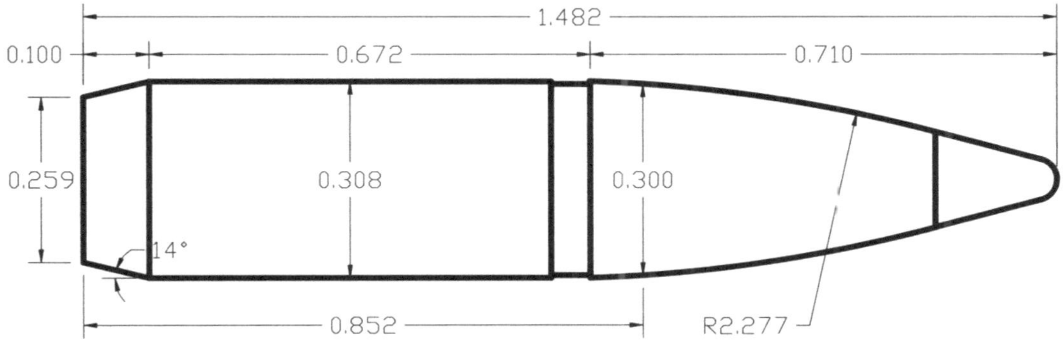

Lot #:	GL02F14	Ogive Radius:	7.39 calibers
Sectional Density:	0.271 lb/in^2	Rt/R:	0.76

Ballistic Performance

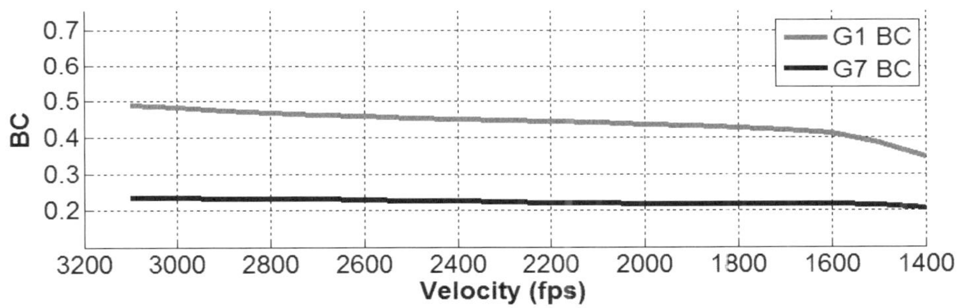

	Above 3000	3000-2500	2500-2000	2000-1500	Below 1500
G1 BC	0.501	0.468	0.445	0.411	0.385
G7 BC	0.237	0.232	0.224	0.217	0.216

Averages from 3000 fps to 1500 fps:
G1 BC: 0.439, variation of G1 BC: 0.096
G7 BC: 0.225, variation of G7 BC: 0.020
G7 form factor: 1.208

Rifling Twist Rate and Stability

Barrel Twist	Stability Factor Best case	Nominal	Worst case
1:9"	2.13	1.86	1.61
1:10"	1.73	1.51	1.30
1:11"	1.43	1.25	1.08
1:12"	1.20	1.05	0.91

Minimum twist required for stability is 1:11.2"
Recommended twist for optimal performance is 1:10.0" or faster

Nosler .308 caliber 165 grain Partition

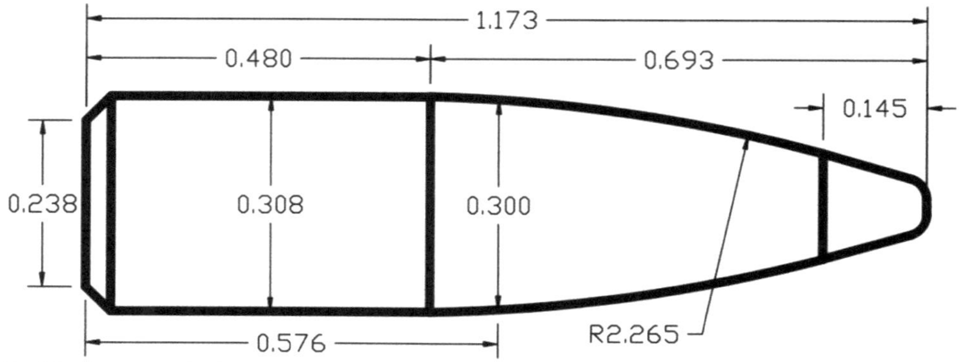

Lot #:	EE09F1-2	Ogive Radius:	7.34 calibers
Sectional Density:	0.249 lb/in^2	Rt/R:	0.90

Ballistic Performance

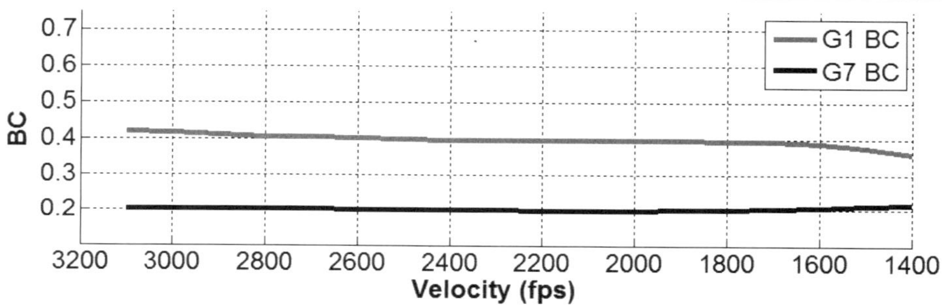

	Above 3000	3000-2500	2500-2000	2000-1500	Below 1500
G1 BC	0.430	0.408	0.398	0.386	0.376
G7 BC	0.204	0.202	0.200	0.206	0.212

Averages from 3000 fps to 1500 fps:
G1 BC: 0.397, variation of G1 BC: 0.040
G7 BC: 0.204, variation of G7 BC: 0.013
G7 form factor: 1.218

Rifling Twist Rate and Stability

Barrel Twist	Stability Factor Best case	Nominal	Worst case
1:10"	3.12	2.72	2.35
1:11"	2.57	2.25	1.95
1:12"	2.16	1.89	1.63
1:13"	1.84	1.61	1.39

Minimum twist required for stability is 1:15.1"
Recommended twist for optimal performance is 1:13.5" or faster

Nosler .308 caliber 180 grain Partition

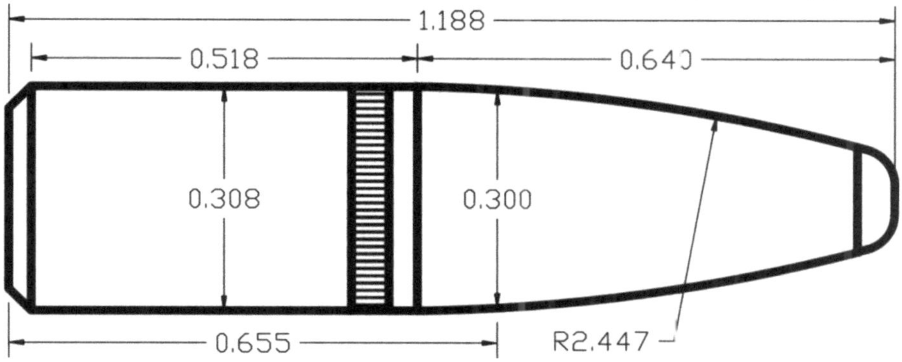

Lot #:	Unknown	Ogive Radius:	7.95 calibers
Sectional Density:	0.271 lb/in²	Rt/R:	0.91

Ballistic Performance

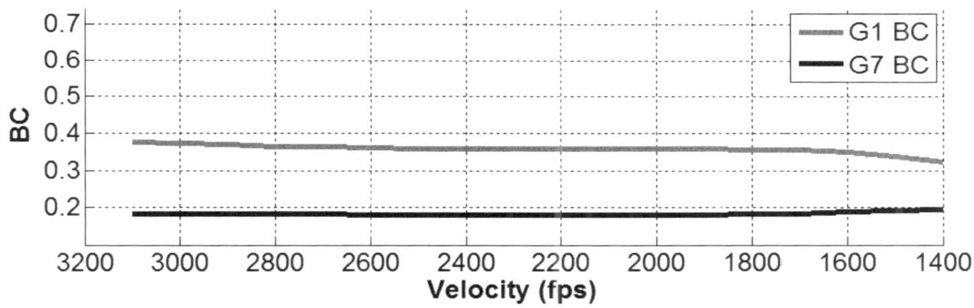

	Above 3000	3000-2500	2500-2000	2000-1500	Below 1500
G1 BC	0.385	0.367	0.359	0.349	0.340
G7 BC	0.182	0.182	0.181	0.186	0.192

Averages from 3000 fps to 1500 fps:
G1 BC: 0.358, variation of G1 BC: 0.034
G7 BC: 0.184, variation of G7 BC: 0.012
G7 form factor: 1.475

Rifling Twist Rate and Stability

Barrel Twist	Stability Factor Best case	Nominal	Worst case
1:10"	3.28	2.86	2.48
1:11"	2.71	2.37	2.05
1:12"	2.28	1.99	1.72
1:13"	1.94	1.69	1.46

Minimum twist required for stability is 1:15.4"
Recommended twist for optimal performance is 1:13.8" or faster

Nosler .308 caliber 200 grain Partition

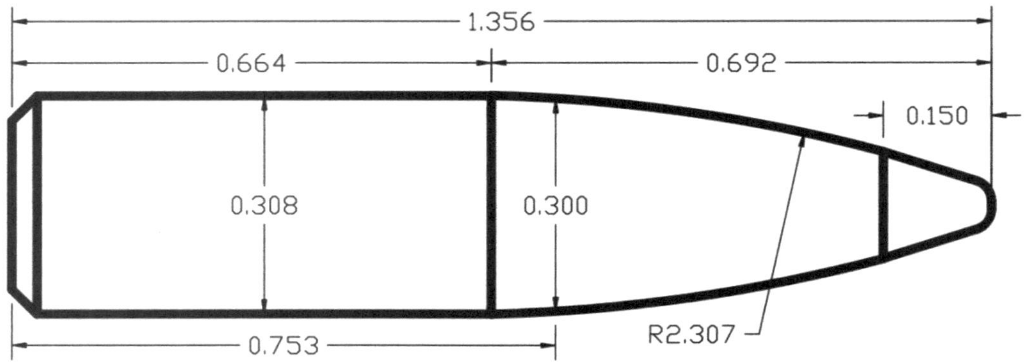

Lot #:	**HD07F2-1**	**Ogive Radius:**	**7.49 calibers**
Sectional Density:	**0.301 lb/in²**	**Rt/R:**	**0.85**

Ballistic Performance

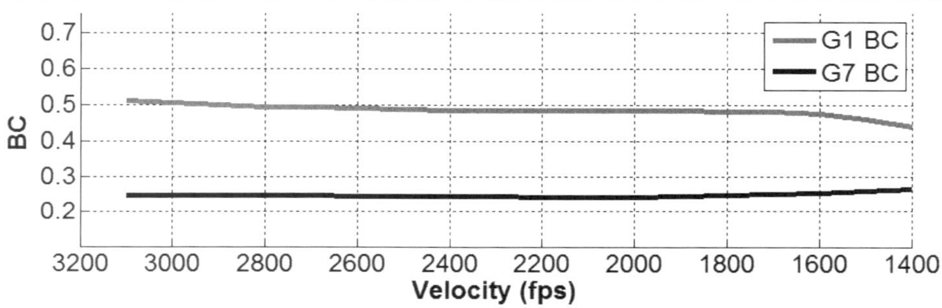

	Above 3000	3000-2500	2500-2000	2000-1500	Below 1500
G1 BC	0.522	0.497	0.486	0.473	0.461
G7 BC	0.247	0.246	0.244	0.252	0.260

Averages from 3000 fps to 1500 fps:
G1 BC: 0.485, variation of G1 BC: 0.045
G7 BC: 0.249, variation of G7 BC: 0.017
G7 form factor: 1.210

Rifling Twist Rate and Stability

Barrel Twist	Stability Factor Best case	Stability Factor Nominal	Stability Factor Worst case
1:10"	2.48	2.17	1.88
1:11"	2.05	1.79	1.55
1:12"	1.73	1.51	1.30
1:13"	1.47	1.28	1.11

Minimum twist required for stability is 1:13.5"
Recommended twist for optimal performance is 1:12.0" or faster

PMP .308 caliber 155 grain HPBT (S. Africa)

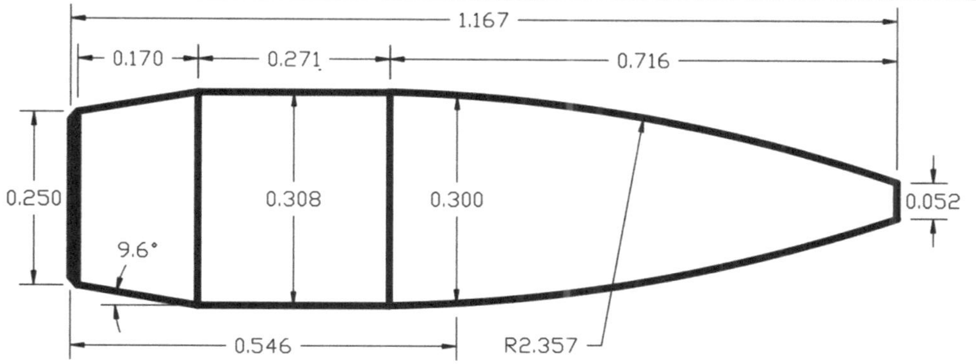

Lot #:	66	Ogive Radius:	7.65 calibers
Sectional Density:	0.233 lb/in^2	Rt/R:	0.90

Ballistic Performance

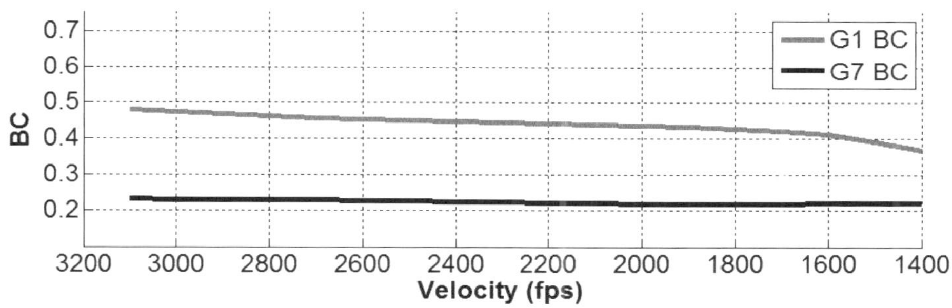

	Above 3000	3000-2500	2500-2000	2000-1500	Below 1500
G1 BC	0.490	0.461	0.443	0.414	0.393
G7 BC	0.232	0.229	0.223	0.220	0.221

Averages from 3000 fps to 1500 fps:
G1 BC: 0.438, variation of G1 BC: 0.081
G7 BC: 0.224, variation of G7 BC: 0.012
G7 form factor: 1.041

Rifling Twist Rate and Stability

Barrel Twist	Stability Factor Best case	Nominal	Worst case
1:10"	2.97	2.59	2.24
1:11"	2.45	2.14	1.85
1:12"	2.06	1.80	1.56
1:13"	1.76	1.54	1.33

Minimum twist required for stability is 1:14.7"
Recommended twist for optimal performance is 1:13.2" or faster

Pri Partizan .308 caliber 150 grain FMJ

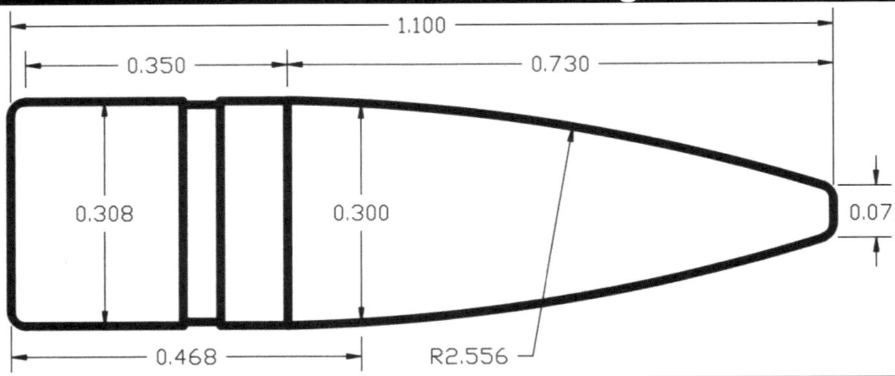

Lot #:	Unknown	Ogive Radius:	8.30 calibers
Sectional Density:	0.226 lb/in^2	Rt/R:	0.87

Ballistic Performance

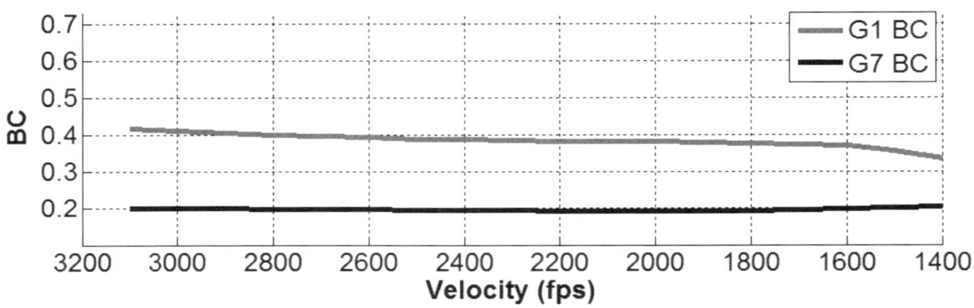

	Above 3000	3000-2500	2500-2000	2000-1500	Below 1500
G1 BC	0.427	0.400	0.385	0.368	0.355
G7 BC	0.202	0.199	0.194	0.196	0.200

Averages from 3000 fps to 1500 fps:
G1 BC: 0.384, variation of G1 BC: 0.055
G7 BC: 0.197, variation of G7 BC: 0.009
G7 form factor: 1.146

Rifling Twist Rate and Stability

Barrel Twist	Stability Factor Best case	Nominal	Worst case
1:10"	3.40	2.97	2.57
1:11"	2.81	2.46	2.13
1:12"	2.36	2.07	1.79
1:13"	2.01	1.76	1.52

Minimum twist required for stability is 1:15.7"
Recommended twist for optimal performance is 1:14.1" or faster

RUAG Swiss Match .308 caliber 175 HPBT

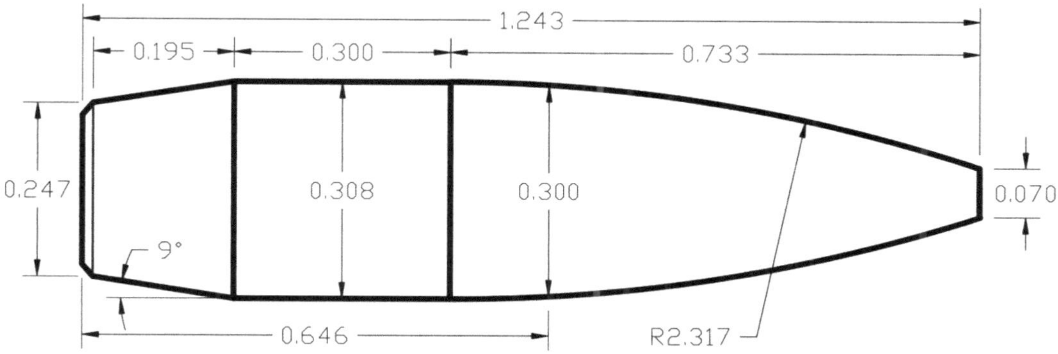

Lot #:	**unknown**	**Ogive Radius:**	**7.52 calibers**
Sectional Density:	**0.264 lb/in^2**	**Rt/R:**	**1.00**

Ballistic Performance

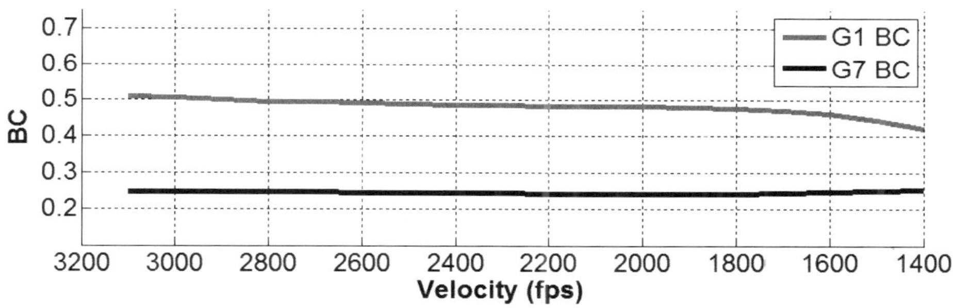

	Above 3000	3000-2500	2500-2000	2000-1500	Below 1500
G1 BC	0.519	0.496	0.484	0.463	0.444
G7 BC	0.246	0.246	0.243	0.246	0.250

Averages from 3000 fps to 1500 fps:
G1 BC: 0.479, variation of G1 BC: 0.061
G7 BC: 0.246, variation of G7 BC: 0.009
G7 form factor: 1.071

Rifling Twist Rate and Stability

	Stability Factor		
Barrel Twist	**Best case**	**Nominal**	**Worst case**
1:11"	2.31	2.02	1.75
1:12"	1.94	1.70	1.47
1:13"	1.65	1.45	1.25
1:14"	1.43	1.25	1.08

Minimum twist required for stability is 1:14.3"
Recommended twist for optimal performance is 1:12 8" or faster

RUAG Swiss Match .308 caliber 200 HPBT

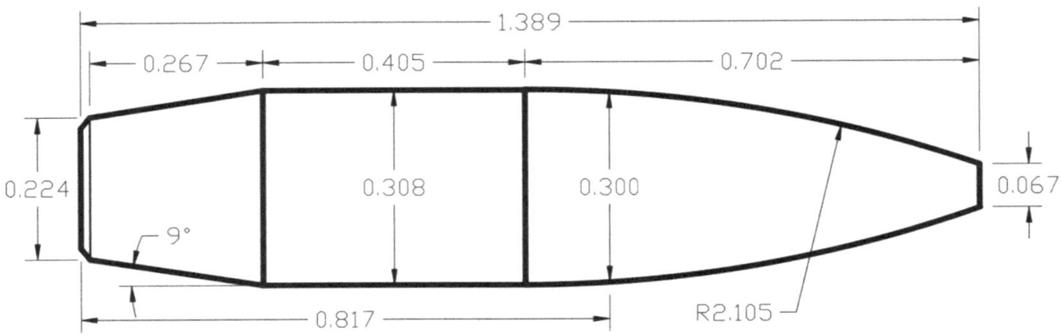

Lot #:	10-HE	Ogive Radius:	6.83 calibers
Sectional Density:	0.301 lb/in^2	Rt/R:	1.00

Ballistic Performance

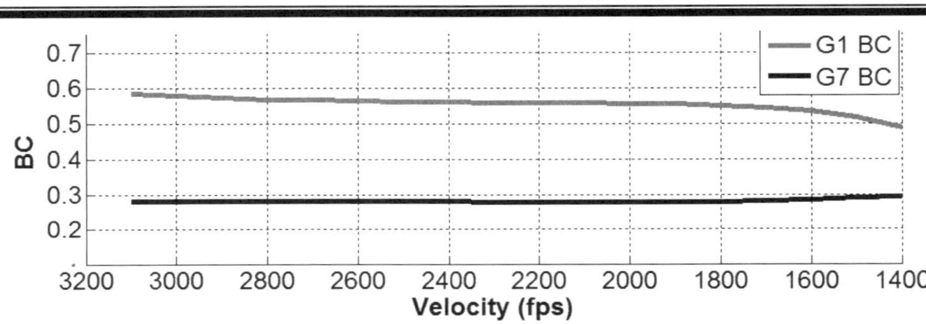

	Above 3000	3000-2500	2500-2000	2000-1500	Below 1500
G1 BC	0.592	0.570	0.559	0.536	0.516
G7 BC	0.281	0.283	0.281	0.285	0.291

Averages from 3000 fps to 1500 fps:
G1 BC: 0.553, variation of G1 BC: 0.062
G7 BC: 0.284, variation of G7 BC: 0.011
G7 form factor: 1.061

Rifling Twist Rate and Stability

Barrel Twist	Stability Factor Best case	Nominal	Worst case
1:10"	2.32	2.02	1.75
1:11"	1.92	1.67	1.45
1:12"	1.61	1.41	1.22
1:13"	1.37	1.20	1.04

Minimum twist required for stability is 1:13.0"
Recommended twist for optimal performance is 1:11.6" or faster

RUAG Swiss Match .308 caliber 220 HPBT

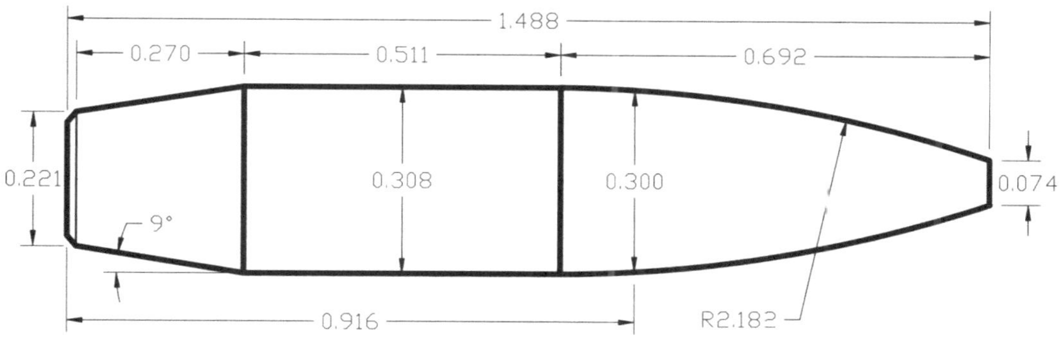

Lot #:	5001554	Ogive Radius:	7.08 calibers
Sectional Density:	0.331 lb/in^2	Rt/R:	0.97

Ballistic Performance

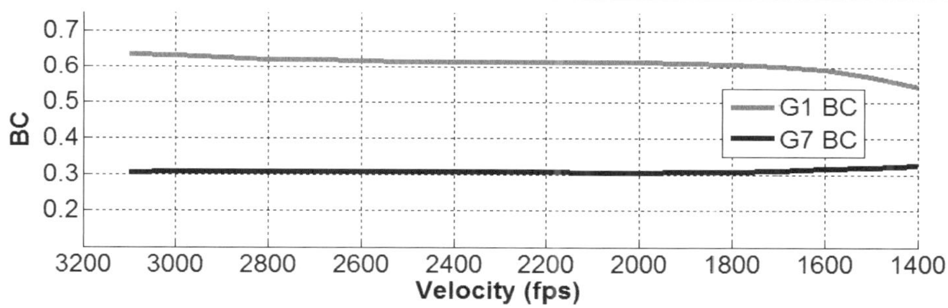

	Above 3000	3000-2500	2500-2000	2000-1500	Below 1500
G1 BC	0.644	0.623	0.614	0.593	0.573
G7 BC	0.305	0.309	0.309	0.315	0.322

Averages from 3000 fps to 1500 fps:
G1 BC: 0.608, variation of G1 BC: 0.058
G7 BC: 0.312, variation of G7 BC: 0.015
G7 form factor: 1.062

Rifling Twist Rate and Stability

Barrel Twist	Stability Factor Best case	Nominal	Worst case
1:10"	2.09	1.82	1.58
1:11"	1.72	1.51	1.30
1:12"	1.45	1.27	1.09
1:13"	1.23	1.08	0.93

Minimum twist required for stability is 1:12.3"
Recommended twist for optimal performance is 1:11.0" or faster

RUAG Swiss P Target .308 caliber 168 HPBT

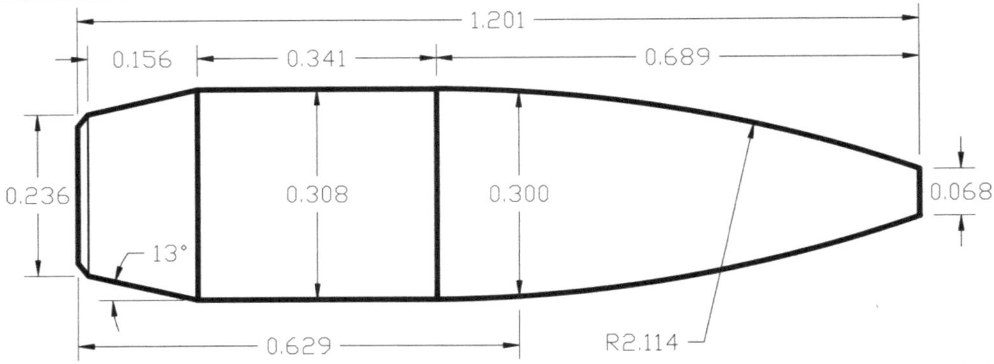

Lot #:	unknown	Ogive Radius:	6.86 calibers
Sectional Density:	0.253 lb/in^2	Rt/R:	0.96

Ballistic Performance

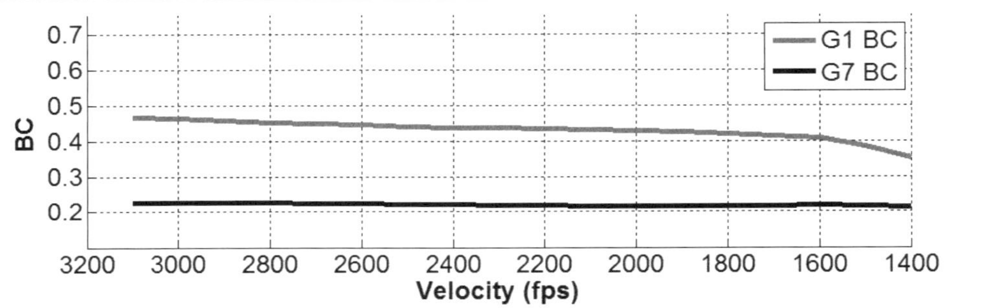

	Above 3000	3000-2500	2500-2000	2000-1500	Below 1500
G1 BC	0.478	0.453	0.436	0.407	0.385
G7 BC	0.227	0.225	0.219	0.216	0.216

Averages from 3000 fps to 1500 fps:
G1 BC: 0.430, variation of G1 BC: 0.078
G7 BC: 0.220, variation of G7 BC: 0.011
G7 form factor: 1.150

Rifling Twist Rate and Stability

Barrel Twist	Best case	Stability Factor Nominal	Worst case
1:12"	2.06	1.80	1.56
1:13"	1.75	1.53	1.32
1:14"	1.51	1.32	1.14
1:15"	1.32	1.15	1.00

Minimum twist required for stability is 1:14.7"
Recommended twist for optimal performance is 1:13.1" or faster

Sierra .308 caliber 150 grain FMJBT

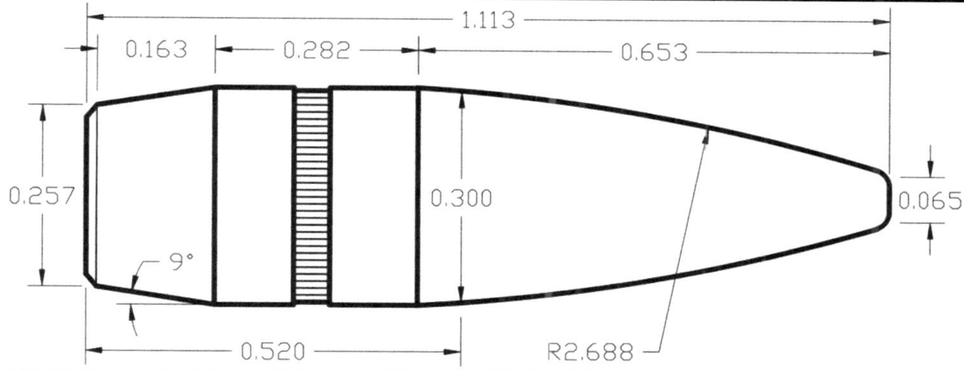

Lot #:	**9433303**	**Ogive Radius:**	**8.73 calibers**
Sectional Density:	**0.226 lb/in^2**	**Rt/R:**	**0.69**

Ballistic Performance

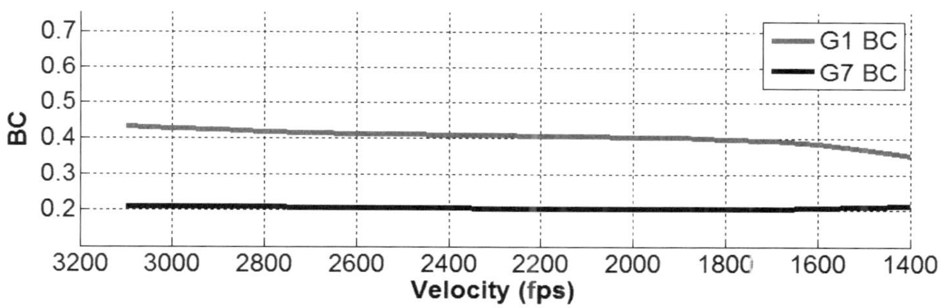

	Above 3000	3000-2500	2500-2000	2000-1500	Below 1500
G1 BC	0.439	0.419	0.407	0.388	0.372
G7 BC	0.208	0.208	0.205	0.206	0.210

Averages from 3000 fps to 1500 fps:
G1 BC: 0.403, variation of G1 BC: 0.055
G7 BC: 0.207, variation of G7 BC: 0.007
G7 form factor: 1.091

Rifling Twist Rate and Stability

	Stability Factor		
Barrel Twist	**Best case**	**Nominal**	**Worst case**
1:12"	2.29	2.00	1.73
1:13"	1.95	1.70	1.47
1:14"	1.68	1.47	1.27
1:15"	1.46	1.28	1.11

Minimum twist required for stability is 1:15.5"
Recommended twist for optimal performance is 1:13.8" or faster

Sierra .308 caliber 173 grain FMJBT

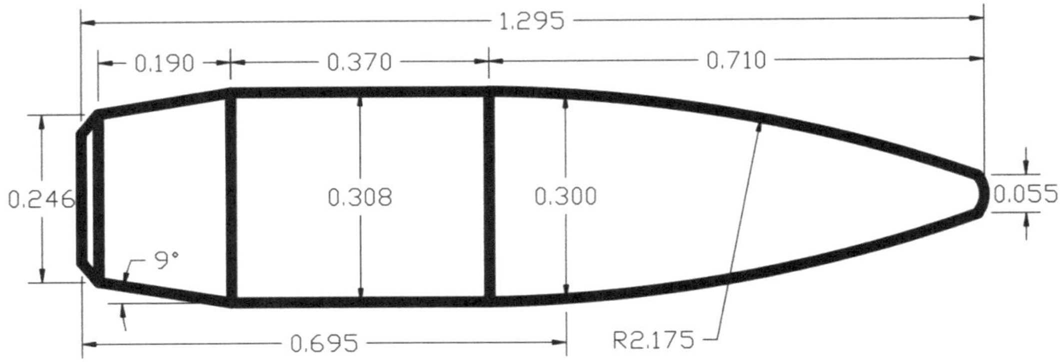

Lot #:	Unknown	Ogive Radius:	7.06 calibers
Sectional Density:	0.261 lb/in^2	Rt/R:	0.94

Ballistic Performance

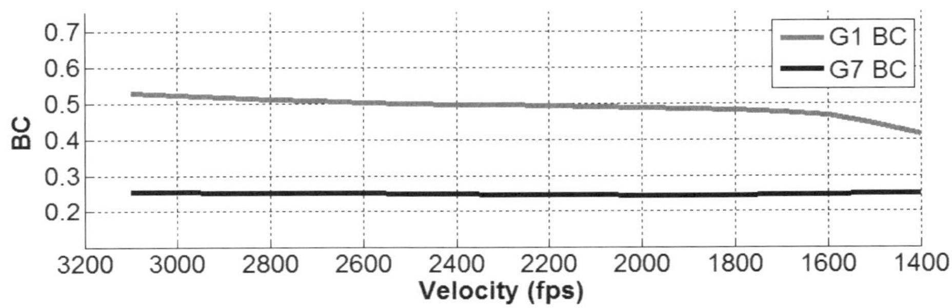

	Above 3000	3000-2500	2500-2000	2000-1500	Below 1500
G1 BC	0.539	0.511	0.495	0.467	0.445
G7 BC	0.255	0.254	0.249	0.248	0.250

Averages from 3000 fps to 1500 fps:
G1 BC: 0.489, variation of G1 BC: 0.078
G7 BC: 0.251, variation of G7 BC: 0.010
G7 form factor: 1.039

Rifling Twist Rate and Stability

Barrel Twist	Stability Factor Best case	Nominal	Worst case
1:10"	2.46	2.15	1.86
1:11"	2.03	1.77	1.53
1:12"	1.71	1.49	1.29
1:13"	1.45	1.27	1.10

Minimum twist required for stability is 1:13.4"
Recommended twist for optimal performance is 1:12.0" or faster

Sierra .308 caliber 150 grain GameKing

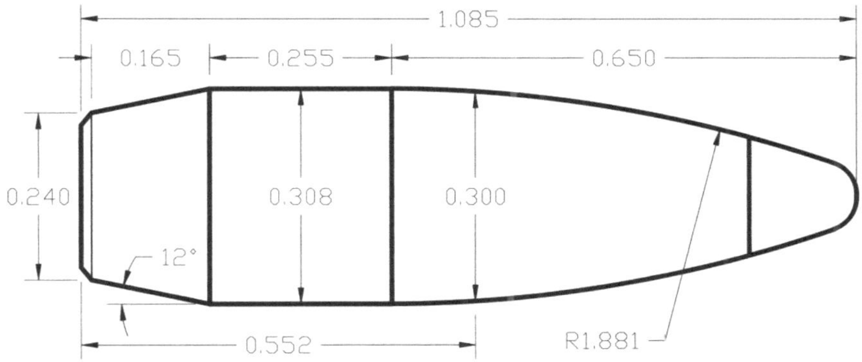

Lot #:	981430260	Ogive Radius:	6.11 calibers
Sectional Density:	0.226 lb/in^2	Rt/R:	0.98

Ballistic Performance

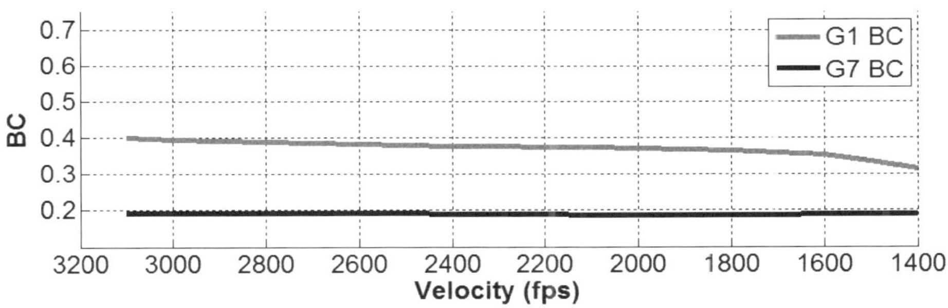

	Above 3000	3000-2500	2500-2000	2000-1500	Below 1500
G1 BC	0.407	0.388	0.375	0.354	0.337
G7 BC	0.193	0.192	0.189	0.188	0.189

Averages from 3000 fps to 1500 fps:
G1 BC: 0.371, variation of G1 BC: 0.059
G7 BC: 0.190, variation of G7 BC: 0.007
G7 form factor: 1.189

Rifling Twist Rate and Stability

Barrel Twist	Stability Factor Best case	Nominal	Worst case
1:13"	2.09	1.83	1.58
1:14"	1.81	1.58	1.36
1:15"	1.57	1.37	1.19
1:16"	1.38	1.21	1.04

Minimum twist required for stability is 1:16.1"
Recommended twist for optimal performance is 1:14.4" or faster

Sierra .308 caliber 165 grain GameKing

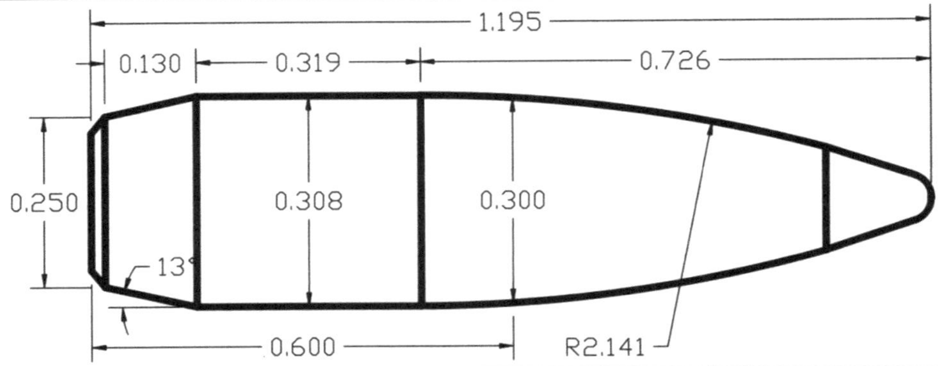

Lot #:	**004680245**	**Ogive Radius:**	**6.95 calibers**
Sectional Density:	**0.249 lb/in²**	**Rt/R:**	**1.00**

Ballistic Performance

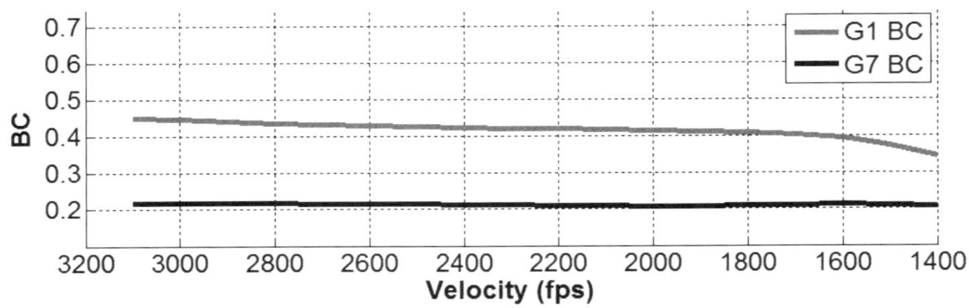

	Above 3000	3000-2500	2500-2000	2000-1500	Below 1500
G1 BC	0.460	0.436	0.420	0.395	0.374
G7 BC	0.218	0.216	0.211	0.209	0.210

Averages from 3000 fps to 1500 fps:
G1 BC: 0.415, variation of G1 BC: 0.071
G7 BC: 0.213, variation of G7 BC: 0.010
G7 form factor: 1.169

Rifling Twist Rate and Stability

Barrel Twist	Stability Factor Best case	Nominal	Worst case
1:10"	2.95	2.58	2.23
1:11"	2.44	2.13	1.84
1:12"	2.05	1.79	1.55
1:13"	1.75	1.53	1.32

Minimum twist required for stability is 1:14.7"
Recommended twist for optimal performance is 1:13.1" or faster

Sierra .308 caliber 180 grain GameKing

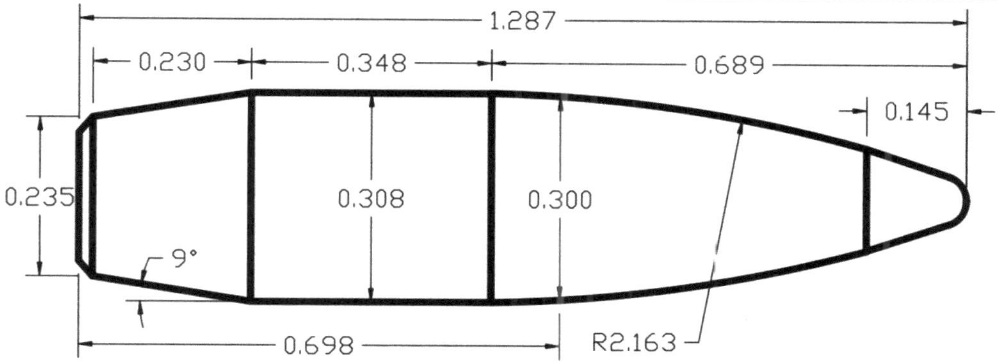

Lot #:	**0036195825817**	**Ogive Radius:**	**7.02 calibers**
Sectional Density:	**0.271 lb/in²**	**Rt/R:**	**0.90**

Ballistic Performance

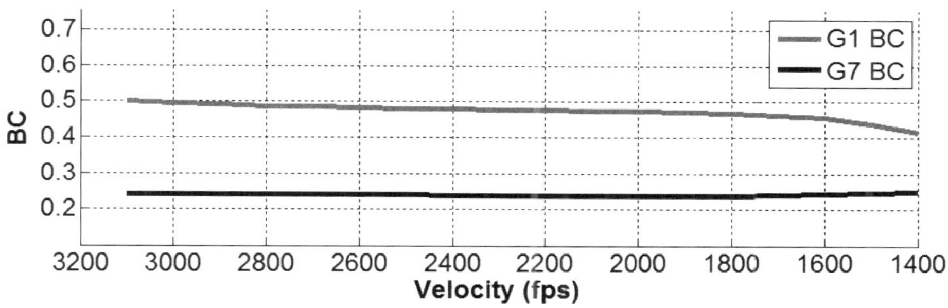

	Above 3000	3000-2500	2500-2000	2000-1500	Below 1500
G1 BC	0.508	0.487	0.476	0.456	0.438
G7 BC	0.241	0.242	0.239	0.242	0.247

Averages from 3000 fps to 1500 fps:
G1 BC: 0.472, variation of G1 BC: 0.057
G7 BC: 0.242, variation of G7 BC: 0.009
G7 form factor: 1.120

Rifling Twist Rate and Stability

Barrel Twist	Stability Factor Best case	 Nominal	 Worst case
1:10"	2.60	2.27	1.97
1:11"	2.15	1.88	1.62
1:12"	1.81	1.58	1.36
1:13"	1.54	1.34	1.16

Minimum twist required for stability is 1:13.8"
Recommended twist for optimal performance is 1:12.3" or faster

Sierra .308 caliber 200 grain GameKing

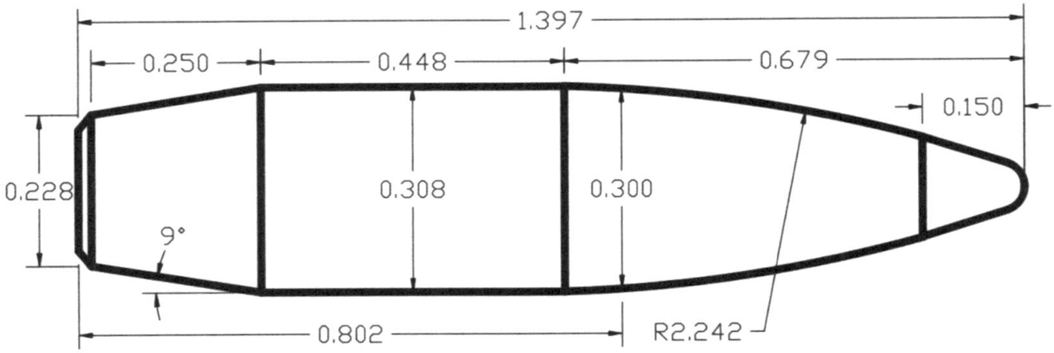

Lot #:	**Unknown**	**Ogive Radius:**	**7.28 calibers**
Sectional Density:	**0.301 lb/in^2**	**Rt/R:**	**0.85**

Ballistic Performance

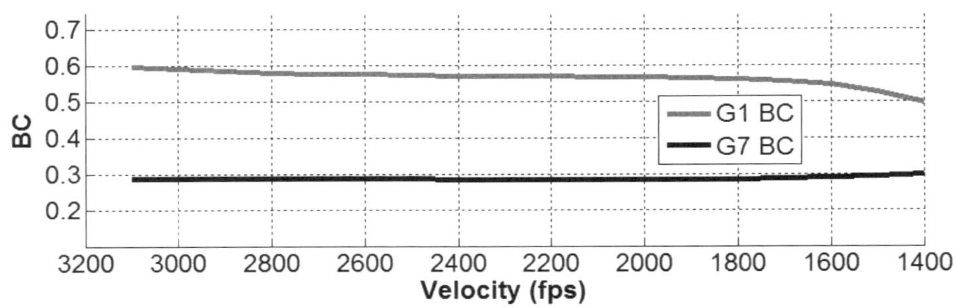

	Above 3000	3000-2500	2500-2000	2000-1500	Below 1500
G1 BC	0.605	0.581	0.570	0.546	0.525
G7 BC	0.287	0.288	0.286	0.290	0.296

Averages from 3000 fps to 1500 fps:
G1 BC: 0.564, variation of G1 BC: 0.064
G7 BC: 0.289, variation of G7 BC: 0.011
G7 form factor: 1.041

Rifling Twist Rate and Stability

Barrel Twist	Stability Factor Best case	Nominal	Worst case
1:10"	2.28	1.99	1.72
1:11"	1.88	1.65	1.42
1:12"	1.58	1.38	1.20
1:13"	1.35	1.18	1.02

Minimum twist required for stability is 1:12.9"
Recommended twist for optimal performance is 1:11.5" or faster

Sierra .308 caliber 165 grain HPBT GameKing

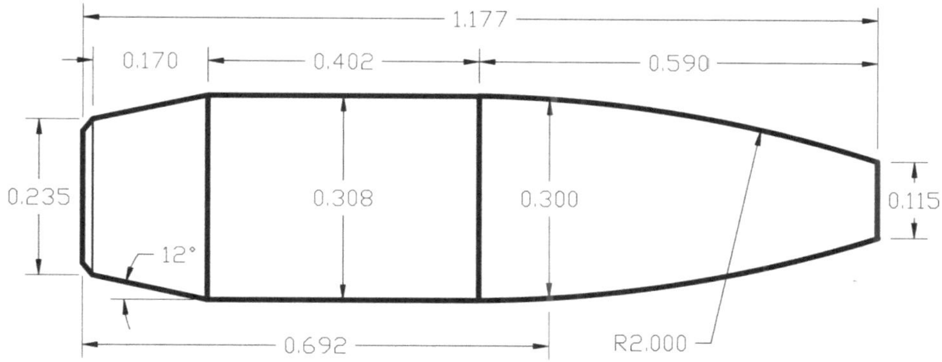

Lot #:	9527576KN	Ogive Radius:	6.49 calibers
Sectional Density:	0.249 lb/in²	Rt/R:	0.93

Ballistic Performance

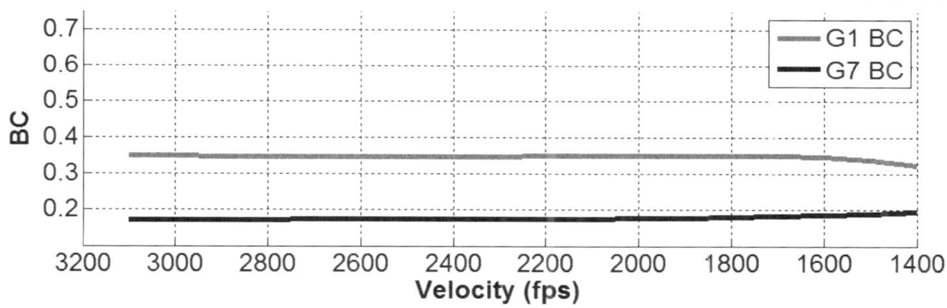

	Above 3000	3000-2500	2500-2000	2000-1500	Below 1500
G1 BC	0.355	0.349	0.349	0.345	0.338
G7 BC	0.168	0.173	0.176	0.183	0.190

Averages from 3000 fps to 1500 fps:
G1 BC: 0.347, variation of G1 BC: 0.014
G7 BC: 0.178, variation of G7 BC: 0.019
G7 form factor: 1.397

Rifling Twist Rate and Stability

Barrel Twist	Stability Factor Best case	Nominal	Worst case
1:12"	2.14	1.87	1.62
1:13"	1.83	1.59	1.38
1:14"	1.57	1.38	1.19
1:15"	1.37	1.20	1.04

Minimum twist required for stability is 1:15.0"
Recommended twist for optimal performance is 1:13.4" or faster

Sierra .308 caliber 125 grain MatchKing

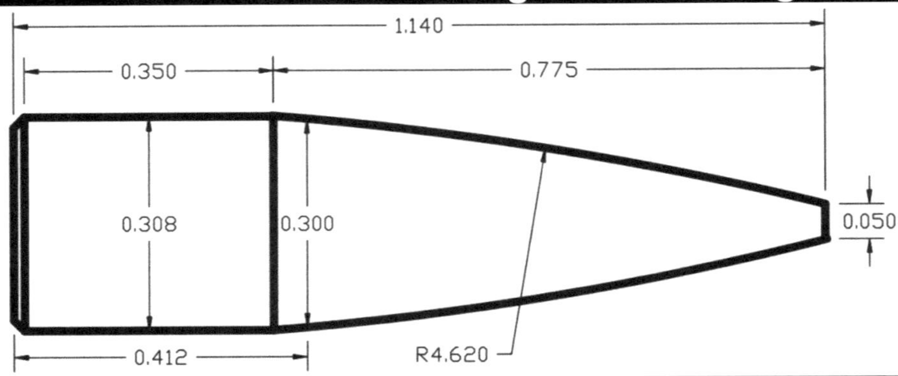

Lot #:	Unknown	Ogive Radius:	15.00 calibers
Sectional Density:	0.188 lb/in^2	Rt/R:	0.52

Ballistic Performance

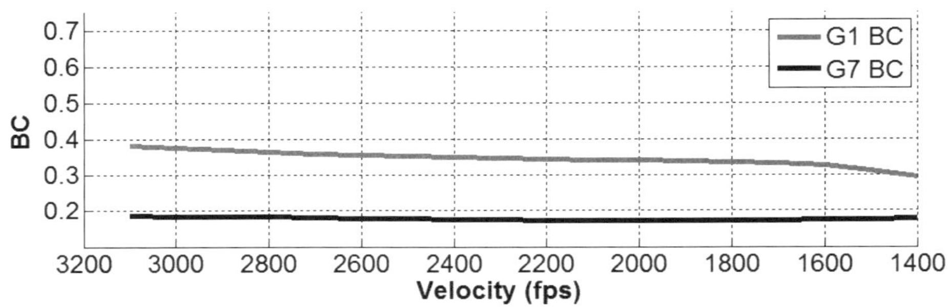

	Above 3000	3000-2500	2500-2000	2000-1500	Below 1500
G1 BC	0.394	0.365	0.347	0.327	0.313
G7 BC	0.187	0.181	0.174	0.174	0.176

Averages from 3000 fps to 1500 fps:
G1 BC: 0.346, variation of G1 BC: 0.064
G7 BC: 0.177, variation of G7 BC: 0.013
G7 form factor: 1.063

Rifling Twist Rate and Stability

Barrel Twist	Stability Factor Best case	Nominal	Worst case
1:10"	2.56	2.24	1.94
1:11"	2.12	1.85	1.60
1:12"	1.78	1.55	1.34
1:13"	1.52	1.32	1.15

Minimum twist required for stability is 1:13.7"
Recommended twist for optimal performance is 1:12.2" or faster

Sierra .308 caliber 135 grain MatchKing

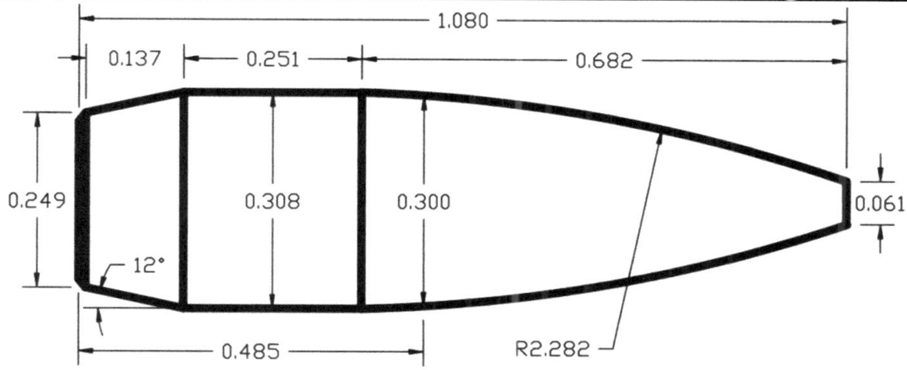

Lot #:	**Unknown**	**Ogive Radius:**	**7.41 calibers**
Sectional Density:	**0.203 lb/in^2**	**Rt/R:**	**0.85**

Ballistic Performance

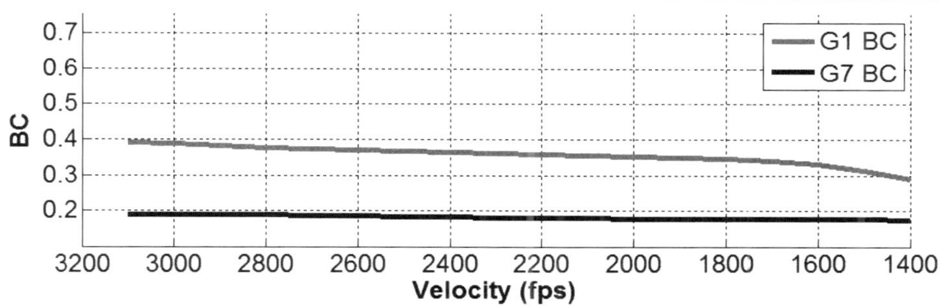

	Above 3000	3000-2500	2500-2000	2000-1500	Below 1500
G1 BC	0.401	0.377	0.361	0.335	0.316
G7 BC	0.190	0.187	0.181	0.178	0.177

Averages from 3000 fps to 1500 fps:
G1 BC: 0.356, variation of G1 BC: 0.071
G7 BC: 0.182, variation of G7 BC: 0.012
G7 form factor: 1.116

Rifling Twist Rate and Stability

	Stability Factor		
Barrel Twist	**Best case**	**Nominal**	**Worst case**
1:10"	3.23	2.82	2.44
1:11"	2.67	2.33	2.02
1:12"	2.24	1.96	1.69
1:13"	1.91	1.67	1.44

Minimum twist required for stability is 1:15.3"
Recommended twist for optimal performance is 1:13.7" or faster

Sierra .308 caliber 150 grain MatchKing

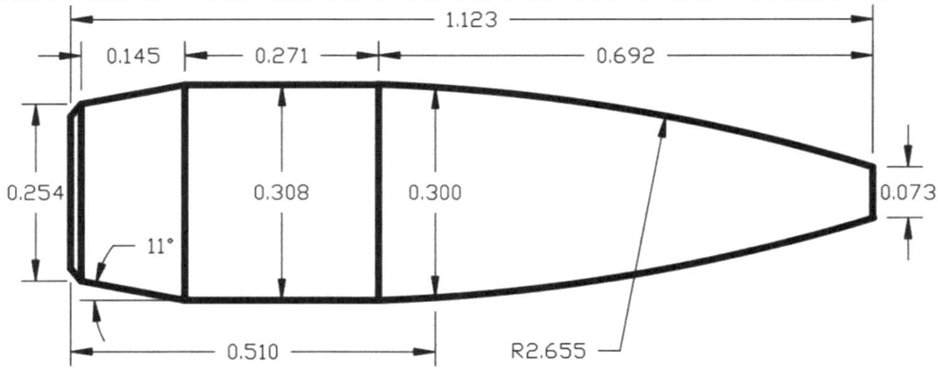

Lot #:	0090526825658	**Ogive Radius:**	**8.62 calibers**
Sectional Density:	**0.226 lb/in^2**	**Rt/R:**	**0.79**

Ballistic Performance

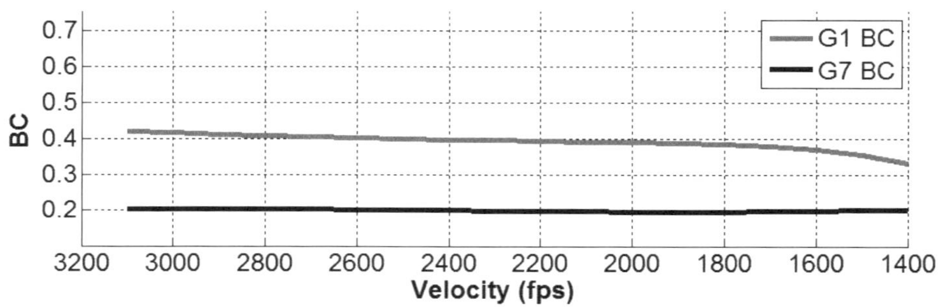

	Above 3000	3000-2500	2500-2000	2000-1500	Below 1500
G1 BC	0.430	0.408	0.395	0.373	0.355
G7 BC	0.204	0.203	0.199	0.198	0.200

Averages from 3000 fps to 1500 fps:
G1 BC: 0.391, variation of G1 BC: 0.062
G7 BC: 0.200, variation of G7 BC: 0.008
G7 form factor: 1.128

Rifling Twist Rate and Stability

Barrel Twist	Stability Factor Best case	Nominal	Worst case
1:10"	3.21	2.80	2.42
1:11"	2.65	2.32	2.00
1:12"	2.23	1.95	1.68
1:13"	1.90	1.66	1.43

Minimum twist required for stability is 1:15.3"
Recommended twist for optimal performance is 1:13.7" or faster

Sierra .308 caliber 155 grain MatchKing (2155)

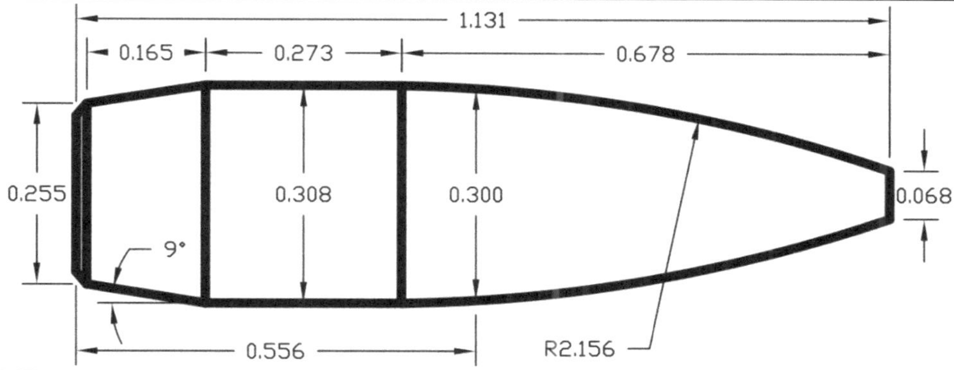

Lot #:	003125976	Ogive Radius:	7.0 calibers
Sectional Density:	0.233 lb/in²	Rt/R:	0.91

Ballistic Performance

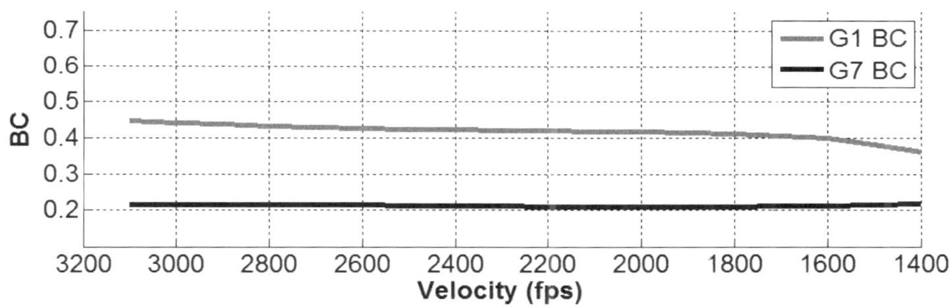

	Above 3000	3000-2500	2500-2000	2000-1500	Below 1500
G1 BC	0.454	0.433	0.421	0.400	0.383
G7 BC	0.215	0.215	0.212	0.213	0.216

Averages from 3000 fps to 1500 fps:
G1 BC: 0.417, variation of G1 BC: 0.058
G7 BC: 0.214, variation of G7 BC: 0.006
G7 form factor: 1.092

Rifling Twist Rate and Stability

Barrel Twist	Stability Factor Best case	Nominal	Worst case
1:10"	3.25	2.84	2.45
1:11"	2.69	2.35	2.03
1:12"	2.26	1.97	1.70
1:13"	1.92	1.68	1.45

Minimum twist required for stability is 1:15.4"
Recommended twist for optimal performance is 1:13.8" or faster

Sierra .308 caliber 155 grain Palma (2156)

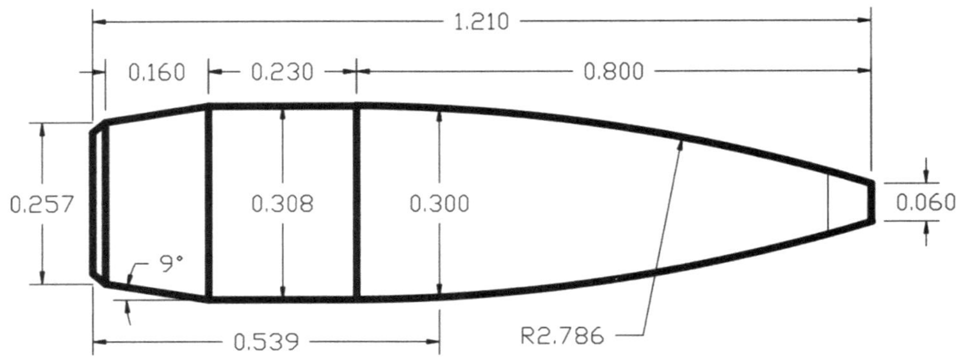

Lot #:	Unknown	Ogive Radius:	9.05 calibers
Sectional Density:	0.233 lb/in^2	Rt/R:	0.94

Ballistic Performance

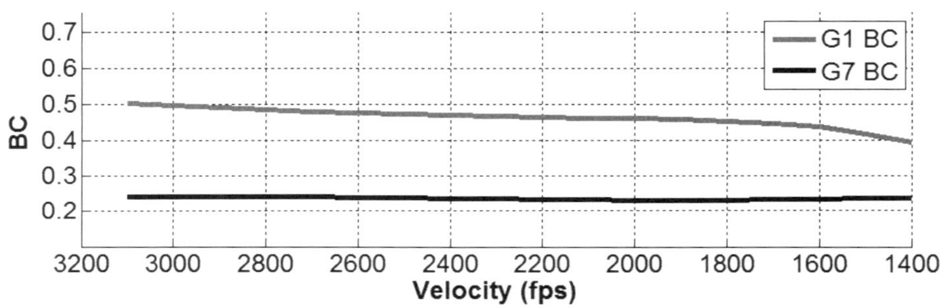

	Above 3000	3000-2500	2500-2000	2000-1500	Below 1500
G1 BC	0.514	0.485	0.467	0.440	0.419
G7 BC	0.243	0.241	0.235	0.233	0.236

Averages from 3000 fps to 1500 fps:
G1 BC: 0.462, variation of G1 BC: 0.078
G7 BC: 0.237, variation of G7 BC: 0.012
G7 form factor: 0.985

Rifling Twist Rate and Stability

Barrel Twist	Stability Factor Best case	Nominal	Worst case
1:10"	2.68	2.34	2.02
1:11"	2.21	1.93	1.67
1:12"	1.86	1.62	1.40
1:13"	1.58	1.38	1.20

Minimum twist required for stability is 1:14.0"
Recommended twist for optimal performance is 1:12.5" or faster

Sierra .308 caliber 168 grain MatchKing

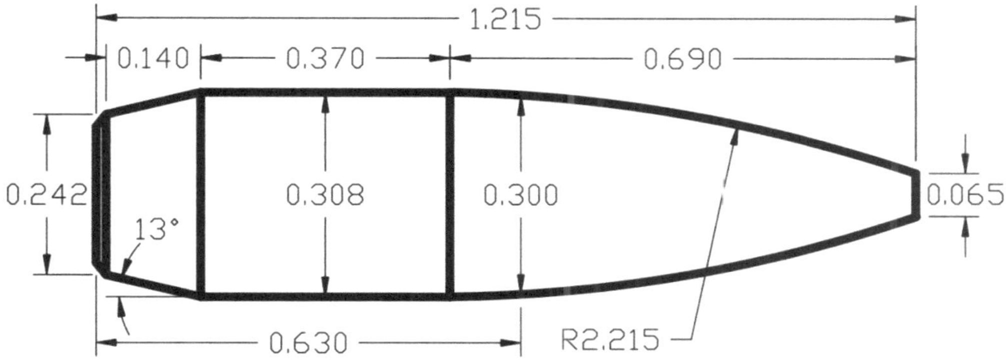

Lot #:	29414	Ogive Radius:	7.19 calibers
Sectional Density:	0.253 lb/in²	Rt/R:	0.90

Ballistic Performance

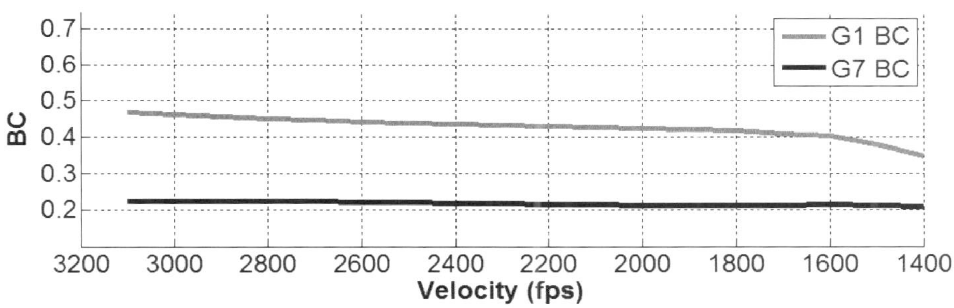

	Above 3000	3000-2500	2500-2000	2000-1500	Below 1500
G1 BC	0.476	0.450	0.432	0.402	0.380
G7 BC	0.226	0.223	0.217	0.213	0.213

Averages from 3000 fps to 1500 fps:
G1 BC: 0.426, variation of G1 BC: 0.081
G7 BC: 0.218, variation of G7 BC: 0.012
G7 form factor: 1.160

Rifling Twist Rate and Stability

Barrel Twist	Stability Factor Best case	Nominal	Worst case
1:10"	2.87	2.50	2.17
1:11"	2.37	2.07	1.79
1:12"	1.99	1.74	1.50
1:13"	1.70	1.48	1.28

Minimum twist required for stability is 1:14.4"
Recommended twist for optimal performance is 1:12.9" or faster

Sierra .308 caliber 175 grain MatchKing

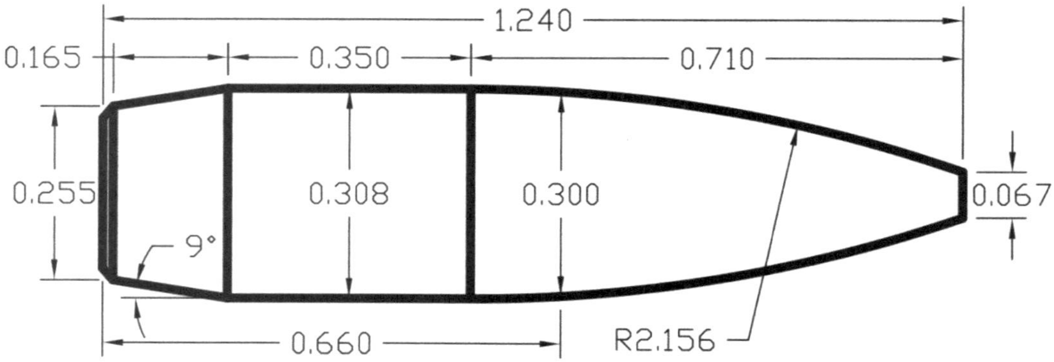

Lot #:	003843112	Ogive Radius:	7.0 calibers
Sectional Density:	0.264 lb/in^2	Rt/R:	1.00

Ballistic Performance

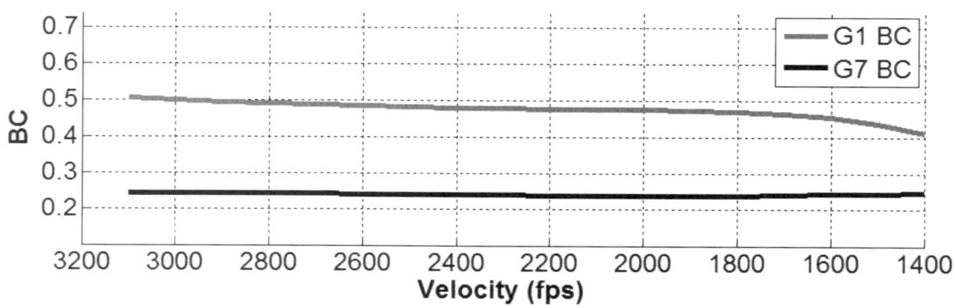

	Above 3000	3000-2500	2500-2000	2000-1500	Below 1500
G1 BC	0.514	0.491	0.478	0.456	0.437
G7 BC	0.244	0.243	0.240	0.242	0.246

Averages from 3000 fps to 1500 fps:
G1 BC: 0.473, variation of G1 BC: 0.063
G7 BC: 0.243, variation of G7 BC: 0.008
G7 form factor: 1.085

Rifling Twist Rate and Stability

Barrel Twist	Stability Factor Best case	Nominal	Worst case
1:10"	2.82	2.46	2.13
1:11"	2.33	2.03	1.76
1:12"	1.96	1.71	1.48
1:13"	1.67	1.46	1.26

Minimum twist required for stability is 1:14.3"
Recommended twist for optimal performance is 1:12.8" or faster

Sierra .308 caliber 180 grain MatchKing

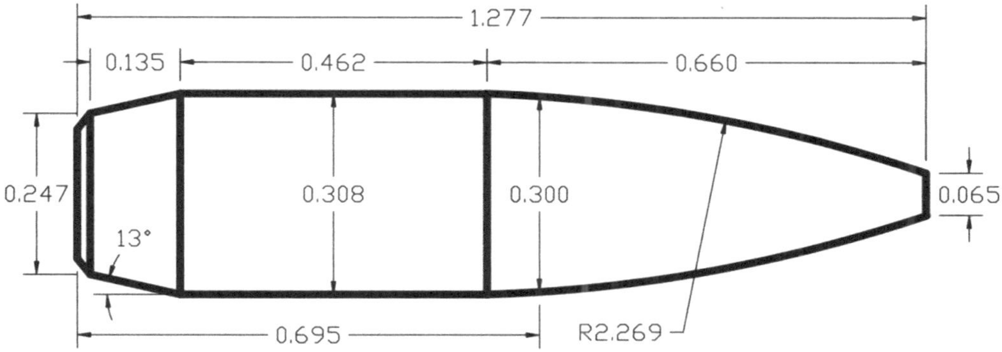

Lot #:	**0035486206245**	**Ogive Radius:**	**7.37 calibers**
Sectional Density:	**0.271 lb/in²**	**Rt/R:**	**0.87**

Ballistic Performance

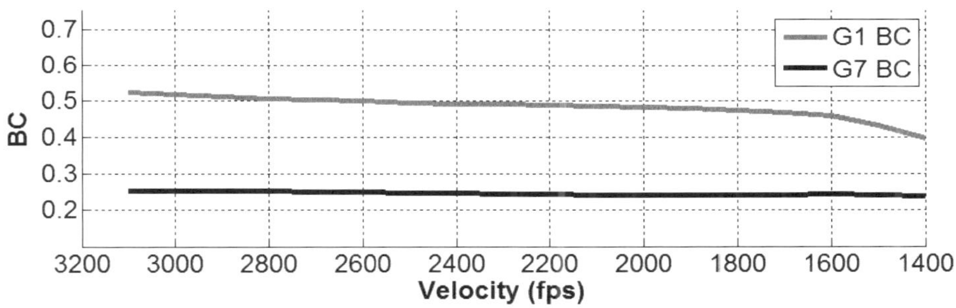

	Above 3000	3000-2500	2500-2000	2000-1500	Below 1500
G1 BC	0.535	0.507	0.488	0.457	0.433
G7 BC	0.253	0.251	0.246	0.242	0.243

Averages from 3000 fps to 1500 fps:
G1 BC: 0.482, variation of G1 BC: 0.085
G7 BC: 0.247, variation of G7 BC: 0.011
G7 form factor: 1.099

Rifling Twist Rate and Stability

Barrel Twist	**Stability Factor** Best case	Nominal	Worst case
1:10"	2.66	2.32	2.01
1:11"	2.20	1.92	1.66
1:12"	1.85	1.61	1.40
1:13"	1.57	1.38	1.19

Minimum twist required for stability is 1:13.9"
Recommended twist for optimal performance is 1:12.4" or faster

Sierra .308 caliber 190 grain MatchKing

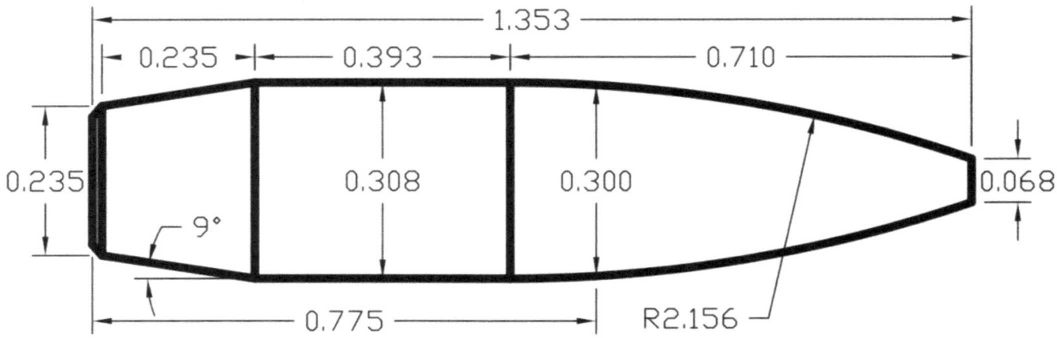

Lot #:	002323479	Ogive Radius:	7.0 calibers
Sectional Density:	0.286 lb/in^2	Rt/R:	1.00

Ballistic Performance

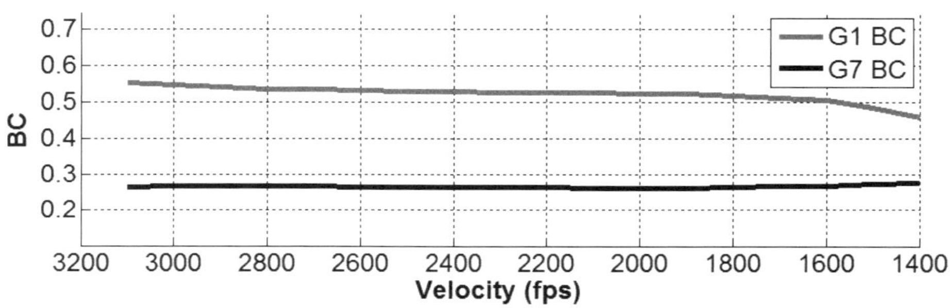

	Above 3000	3000-2500	2500-2000	2000-1500	Below 1500
G1 BC	0.560	0.538	0.526	0.504	0.485
G7 BC	0.265	0.267	0.265	0.268	0.273

Averages from 3000 fps to 1500 fps:
G1 BC: 0.521, variation of G1 BC: 0.061
G7 BC: 0.267, variation of G7 BC: 0.010
G7 form factor: 1.070

Rifling Twist Rate and Stability

Barrel Twist	Stability Factor Best case	Nominal	Worst case
1:10"	2.38	2.08	1.79
1:11"	1.96	1.72	1.48
1:12"	1.65	1.44	1.25
1:13"	1.41	1.23	1.06

Minimum twist required for stability is 1:13.2"
Recommended twist for optimal performance is 1:11.8" or faster

Sierra .308 caliber 200 grain MatchKing

Drawing not available
Bullet length of 1.393" used for stability calculations.

Lot #:	Unknown	Ogive Radius:	Unavailable
Sectional Density:	0.301 lb/in^2	Rt/R:	Unavailable

Ballistic Performance

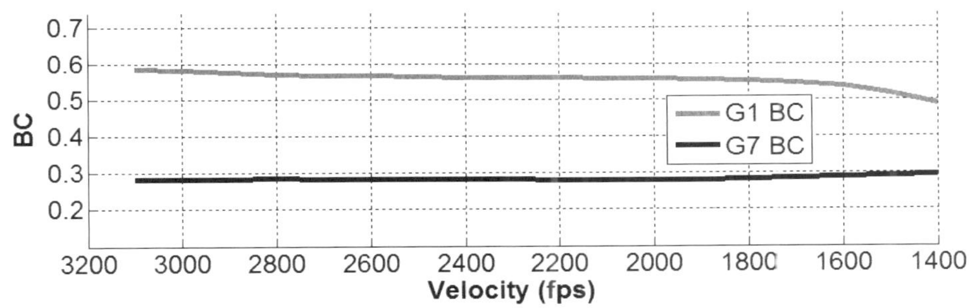

	Above 3000	3000-2500	2500-2000	2000-1500	Below 1500
G1 BC	0.596	0.572	0.561	0.537	0.517
G7 BC	0.282	0.284	0.282	0.286	0.291

Averages from 3000 fps to 1500 fps:
G1 BC: 0.555, variation of G1 BC: 0.064
G7 BC: 0.285, variation of G7 BC: 0.011
G7 form factor: 1.058

Rifling Twist Rate and Stability

Barrel Twist	Stability Factor Best case	Nominal	Worst case
1:10"	2.30	2.01	1.74
1:11"	1.90	1.66	1.43
1:12"	1.60	1.39	1.21
1:13"	1.36	1.19	1.03

Minimum twist required for stability is 1:12.9"
Recommended twist for optimal performance is 1:11.6" or faster

Sierra .308 caliber 210 grain MatchKing

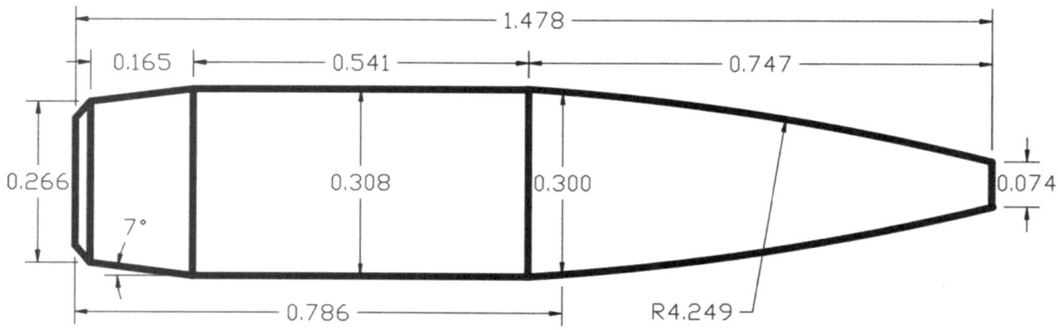

Lot #:	**Unknown**		**Ogive Radius:**	**13.80 calibers**
Sectional Density:	**0.316 lb/in^2**		**Rt/R:**	**0.62**

Ballistic Performance

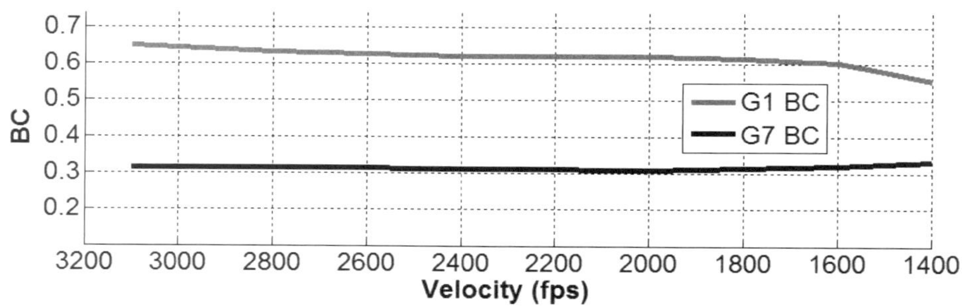

	Above 3000	3000-2500	2500-2000	2000-1500	Below 1500
G1 BC	0.661	0.633	0.620	0.599	0.580
G7 BC	0.313	0.314	0.312	0.319	0.327

Averages from 3000 fps to 1500 fps:
G1 BC: 0.616, variation of G1 BC: 0.063
G7 BC: 0.316, variation of G7 BC: 0.016
G7 form factor: 1.000

Rifling Twist Rate and Stability

Barrel Twist	Stability Factor Best case	Stability Factor Nominal	Stability Factor Worst case
1:10"	2.03	1.77	1.53
1:11"	1.68	1.47	1.27
1:12"	1.41	1.23	1.07
1:13"	1.20	1.05	0.91

Minimum twist required for stability is 1:12.2"
Recommended twist for optimal performance is 1:10.9" or faster

Sierra .308 caliber 220 grain MatchKing

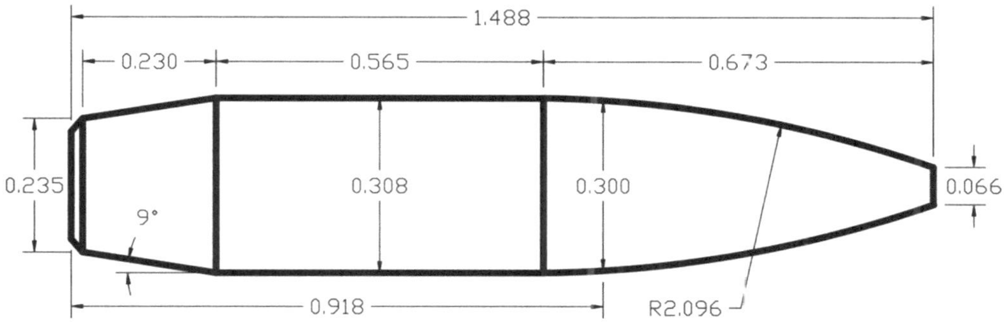

Lot #:	**0039477615817**	**Ogive Radius:**	**6.82 calibers**
Sectional Density:	**0.331 lb/in^2**	**Rt/R:**	**0.95**

Ballistic Performance

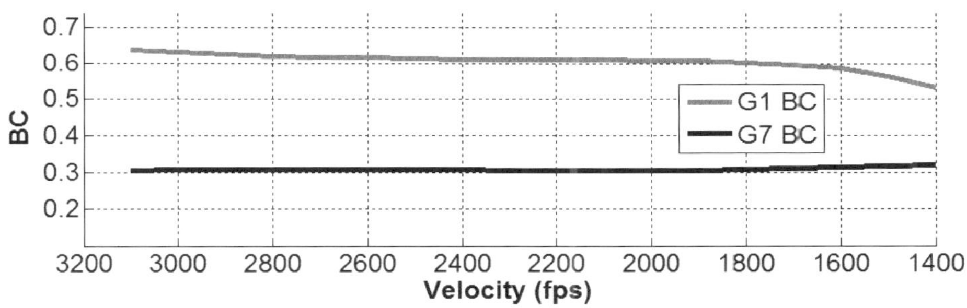

	Above 3000	3000-2500	2500-2000	2000-1500	Below 1500
G1 BC	0.646	0.622	0.611	0.587	0.565
G7 BC	0.306	0.309	0.307	0.312	0.318

Averages from 3000 fps to 1500 fps:
G1 BC: 0.604, variation of G1 BC: 0.066
G7 BC: 0.310, variation of G7 BC: 0.012
G7 form factor: 1.068

Rifling Twist Rate and Stability

Barrel Twist	Stability Factor: Best case	Stability Factor: Nominal	Stability Factor: Worst case
1:10"	2.08	1.82	1.57
1:11"	1.72	1.50	1.30
1:12"	1.45	1.26	1.09
1:13"	1.23	1.08	0.93

Minimum twist required for stability is 1:12.3"
Recommended twist for optimal performance is 1:11.0" or faster

Sierra .308 caliber 240 grain MatchKing

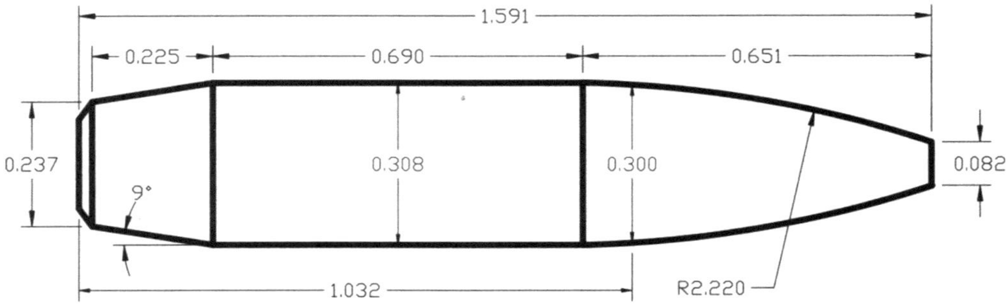

Lot #:	0035667661313		Ogive Radius:	7.2 calibers
Sectional Density:	0.361 lb/in^2		Rt/R:	0.89

Ballistic Performance

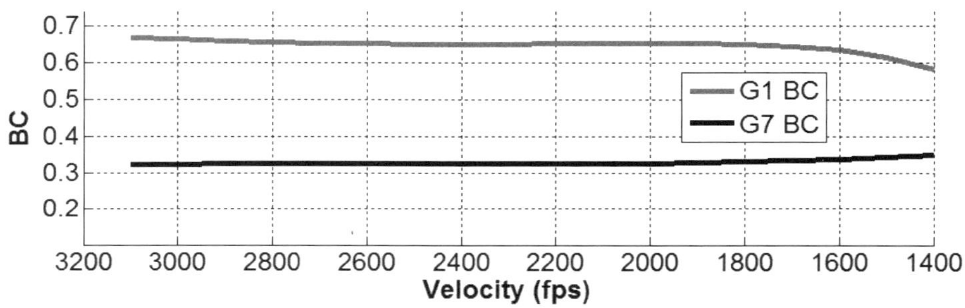

	Above 3000	3000-2500	2500-2000	2000-1500	Below 1500
G1 BC	0.677	0.657	0.651	0.632	0.613
G7 BC	0.321	0.326	0.327	0.336	0.345

Averages from 3000 fps to 1500 fps:
G1 BC: 0.644, variation of G1 BC: 0.050
G7 BC: 0.331, variation of G7 BC: 0.021
G7 form factor: 1.093

Rifling Twist Rate and Stability

Barrel Twist	Stability Factor Best case	Stability Factor Nominal	Stability Factor Worst case
1:10"	1.87	1.63	1.41
1:11"	1.55	1.35	1.17
1:12"	1.30	1.14	0.98
1:13"	1.11	0.97	0.84

Minimum twist required for stability is 1:11.7"
Recommended twist for optimal performance is 1:10.4" or faster

Sierra .308 caliber 125 grain Tipped MatchKing

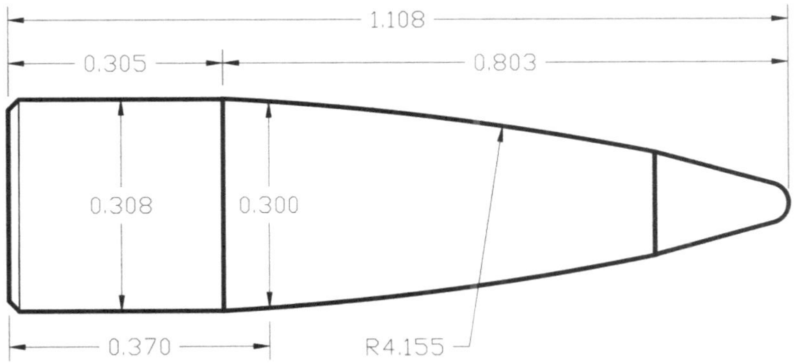

Lot #:	**97152252**	**Ogive Radius:**	**13.49 calibers**
Sectional Density:	**0.188 lb/in²**	**Rt/R:**	**0.54**

Ballistic Performance

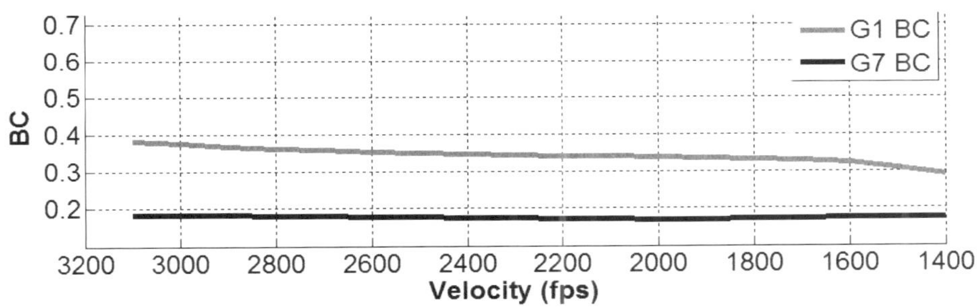

	Above 3000	3000-2500	2500-2000	2000-1500	Below 1500
G1 BC	0.393	0.363	0.344	0.324	0.311
G7 BC	0.186	0.180	0.173	0.172	0.175

Averages from 3000 fps to 1500 fps:
G1 BC: 0.344, variation of G1 BC: 0.065
G7 BC: 0.176, variation of G7 BC: 0.014
G7 form factor: 1.070

Rifling Twist Rate and Stability

Barrel Twist	Stability Factor Best case	Nominal	Worst case
1:11"	2.30	2.01	1.73
1:12"	1.93	1.69	1.46
1:13"	1.64	1.44	1.24
1:14"	1.42	1.24	1.07

Minimum twist required for stability is 1:14.2"
Recommended twist for optimal performance is 1:12.7" or faster

Sierra .308 caliber 155 grain Tipped MatchKing

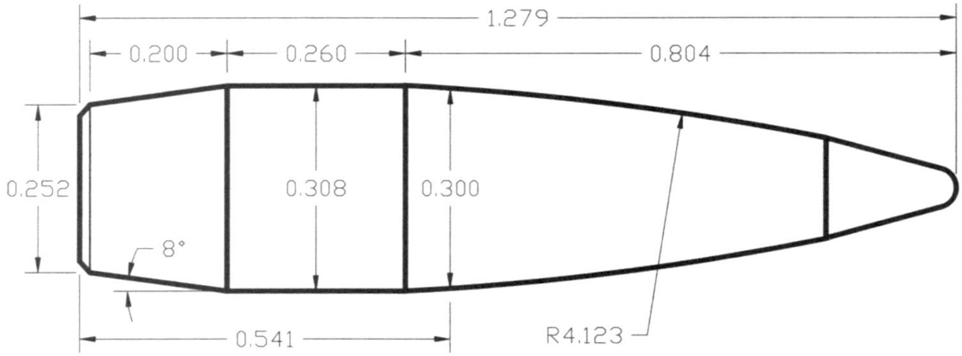

Lot #:	9466948	Ogive Radius:	13.39 calibers
Sectional Density:	0.233 lb/in^2	Rt/R:	0.55

Ballistic Performance

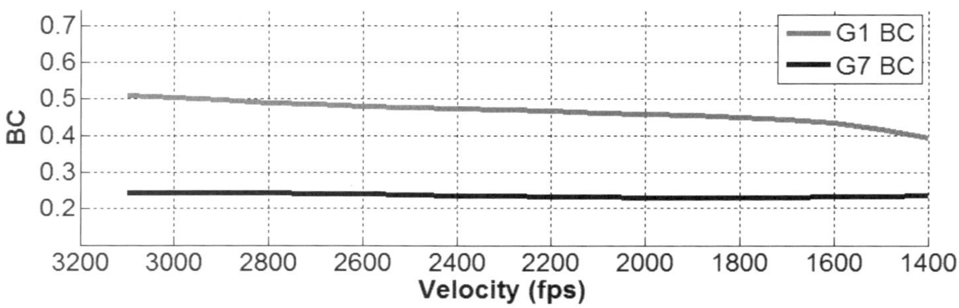

	Above 3000	3000-2500	2500-2000	2000-1500	Below 1500
G1 BC	0.521	0.489	0.468	0.438	0.416
G7 BC	0.247	0.242	0.235	0.233	0.234

Averages from 3000 fps to 1500 fps:
G1 BC: 0.464, variation of G1 BC: 0.086
G7 BC: 0.238, variation of G7 BC: 0.015
G7 form factor: 0.983

Rifling Twist Rate and Stability

	Stability Factor		
Barrel Twist	**Best case**	**Nominal**	**Worst case**
1:10"	2.28	1.99	1.72
1:11"	1.89	1.65	1.42
1:12"	1.58	1.38	1.20
1:13"	1.35	1.18	1.02

Minimum twist required for stability is 1:12.9"
Recommended twist for optimal performance is 1:11.5" or faster

Sierra .308 caliber 168 grain Tipped MatchKing

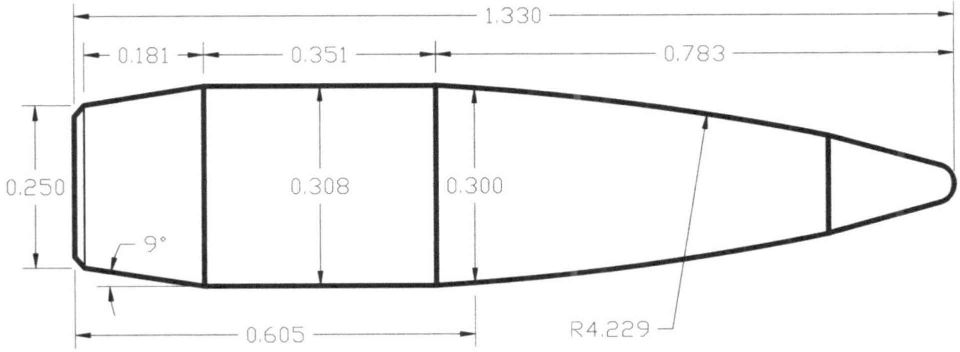

Lot #:	10062701485	Ogive Radius:	13.73 calibers
Sectional Density:	0.253 lb/in^2	Rt/R:	0.54

Ballistic Performance

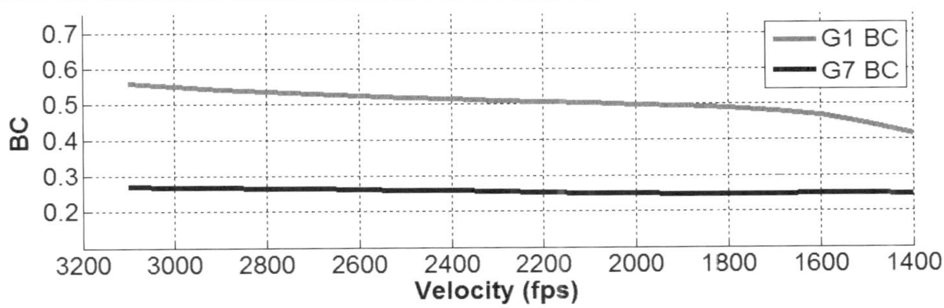

	Above 3000	3000-2500	2500-2000	2000-1500	Below 1500
G1 BC	0.572	0.534	0.508	0.471	0.444
G7 BC	0.271	0.265	0.255	0.250	0.250

Averages from 3000 fps to 1500 fps:
G1 BC: 0.503, variation of G1 BC: 0.106
G7 BC: 0.257, variation of G7 BC: 0.019
G7 form factor: 0.984

Rifling Twist Rate and Stability

Barrel Twist	Stability Factor Best case	Nominal	Worst case
1:10"	2.21	1.93	1.67
1:11"	1.82	1.59	1.38
1:12"	1.53	1.34	1.16
1:13"	1.31	1.14	0.99

Minimum twist required for stability is 1:12.7"
Recommended twist for optimal performance is 1:11.3" or faster

Sierra .308 caliber 175 grain Tipped MatchKing

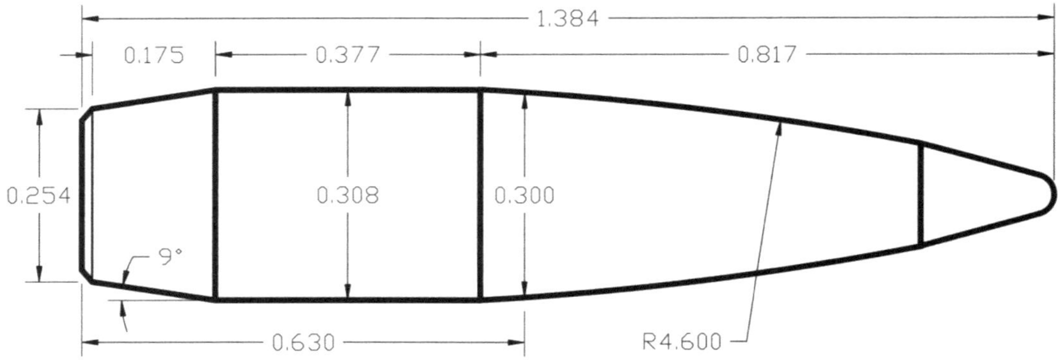

Lot #:	1009388796	Ogive Radius:	14.94 calibers
Sectional Density:	0.264 lb/in^2	Rt/R:	0.55

Ballistic Performance

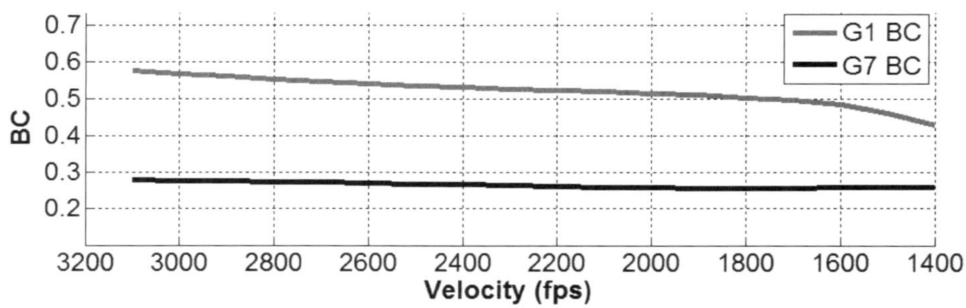

	Above 3000	3000-2500	2500-2000	2000-1500	Below 1500
G1 BC	0.591	0.552	0.525	0.488	0.461
G7 BC	0.280	0.274	0.264	0.259	0.259

Averages from 3000 fps to 1500 fps:
G1 BC: 0.520, variation of G1 BC: 0.107
G7 BC: 0.266, variation of G7 BC: 0.019
G7 form factor: 0.991

Rifling Twist Rate and Stability

Barrel Twist	Stability Factor Best case	Stability Factor Nominal	Stability Factor Worst case
1:9"	2.53	2.21	1.91
1:10"	2.05	1.79	1.55
1:11"	1.69	1.48	1.28
1:12"	1.42	1.24	1.07

Minimum twist required for stability is 1:12.2"
Recommended twist for optimal performance is 1:10.9" or faster

Swift .308 caliber 150 grain BT Spitzer

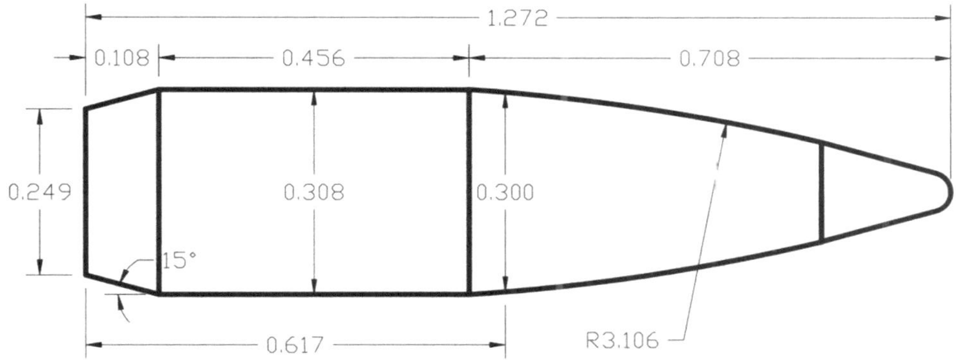

Lot #:	914W	Ogive Radius:	10.08 calibers
Sectional Density:	0.226 lb/in^2	Rt/R:	0.42

Ballistic Performance

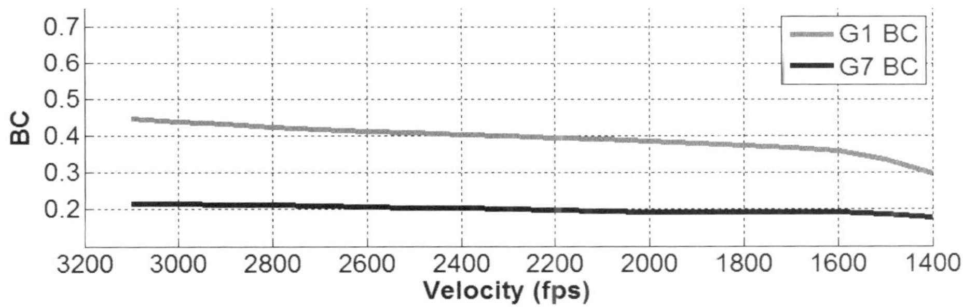

	Above 3000	3000-2500	2500-2000	2000-1500	Below 1500
G1 BC	0.458	0.423	0.397	0.360	0.334
G7 BC	0.217	0.210	0.199	0.190	0.187

Averages from 3000 fps to 1500 fps:
G1 BC: 0.392, variation of G1 BC: 0.104
G7 BC: 0.200, variation of G7 BC: 0.027
G7 form factor: 1.132

Rifling Twist Rate and Stability

	Stability Factor		
Barrel Twist	**Best case**	**Nominal**	**Worst case**
1:10"	2.24	1.96	1.69
1:11"	1.85	1.62	1.40
1:12"	1.56	1.36	1.18
1:13"	1.33	1.16	1.00

Minimum twist required for stability is 1:12.3"
Recommended twist for optimal performance is 1:11.4" or faster

Swift .308 caliber 165 grain BT Spitzer

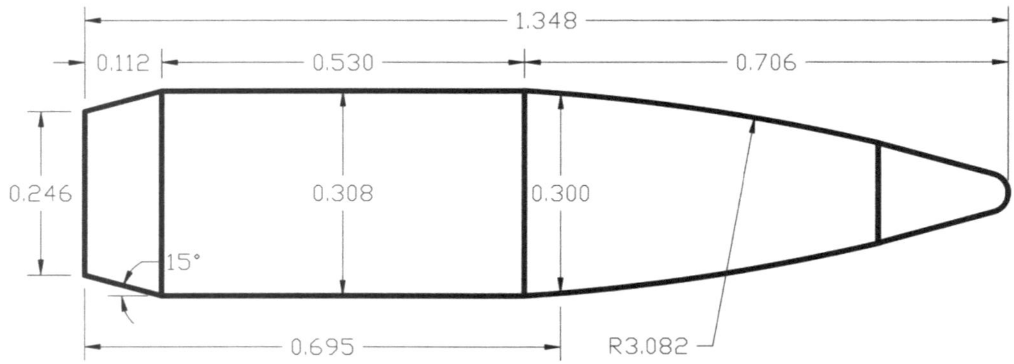

Lot #:	**919W**	**Ogive Radius:**	**10.01 calibers**
Sectional Density:	**0.249 lb/in²**	**Rt/R:**	**0.43**

Ballistic Performance

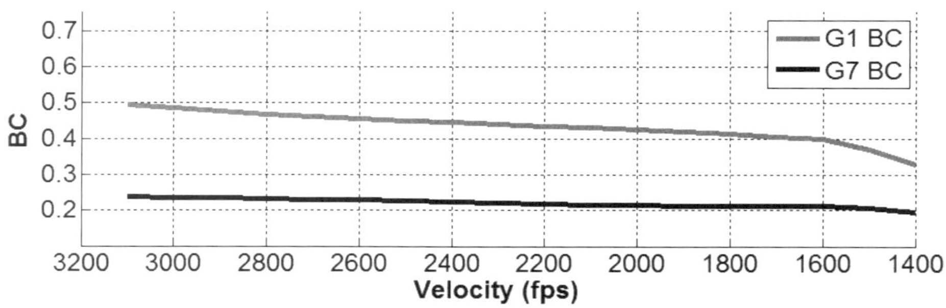

	Above 3000	3000-2500	2500-2000	2000-1500	Below 1500
G1 BC	0.506	0.468	0.439	0.398	0.370
G7 BC	0.240	0.232	0.221	0.211	0.207

Averages from 3000 fps to 1500 fps:
G1 BC: 0.433, variation of G1 BC: 0.114
G7 BC: 0.221, variation of G7 BC: 0.030
G7 form factor: 1.126

Rifling Twist Rate and Stability

Barrel Twist	**Stability Factor** **Best case**	**Nominal**	**Worst case**
1:10"	2.09	1.82	1.58
1:11"	1.72	1.51	1.30
1:12"	1.45	1.27	1.09
1:13"	1.23	1.08	0.93

Minimum twist required for stability is 1:12.3"
Recommended twist for optimal performance is 1:11.0" or faster

Swift .308 caliber 180 grain BT Spitzer

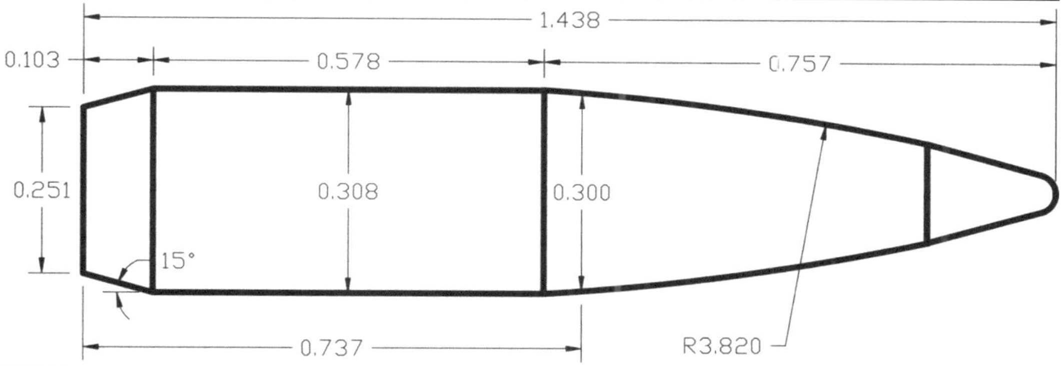

Lot #:	917W	Ogive Radius:	12.40 calibers
Sectional Density:	0.271 lb/in^2	Rt/R:	0.48

Ballistic Performance

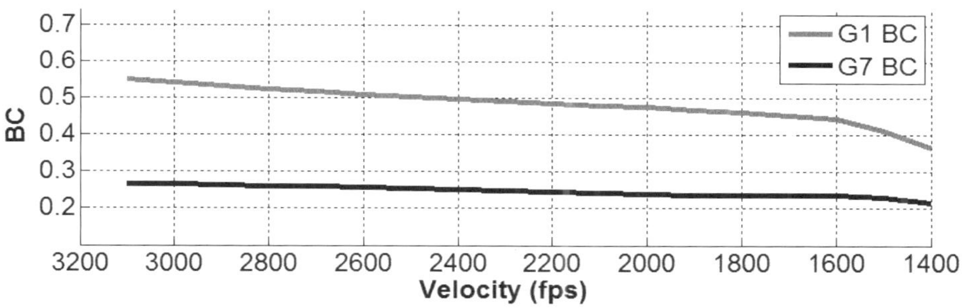

	Above 3000	3000-2500	2500-2000	2000-1500	Below 1500
G1 BC	0.567	0.522	0.489	0.443	0.411
G7 BC	0.268	0.259	0.246	0.234	0.230

Averages from 3000 fps to 1500 fps:
G1 BC: 0.483, variation of G1 BC: 0.130
G7 BC: 0.246, variation of G7 BC: 0.035
G7 form factor: 1.103

Rifling Twist Rate and Stability

	Stability Factor		
Barrel Twist	**Best case**	**Nominal**	**Worst case**
1:9"	2.33	2.03	1.76
1:10"	1.89	1.65	1.42
1:11"	1.56	1.36	1.18
1:12"	1.31	1.14	0.99

Minimum twist required for stability is 1:11.7"
Recommended twist for optimal performance is 1:10.5" or faster

Vapor Trail .308 caliber 207 grain BTHP

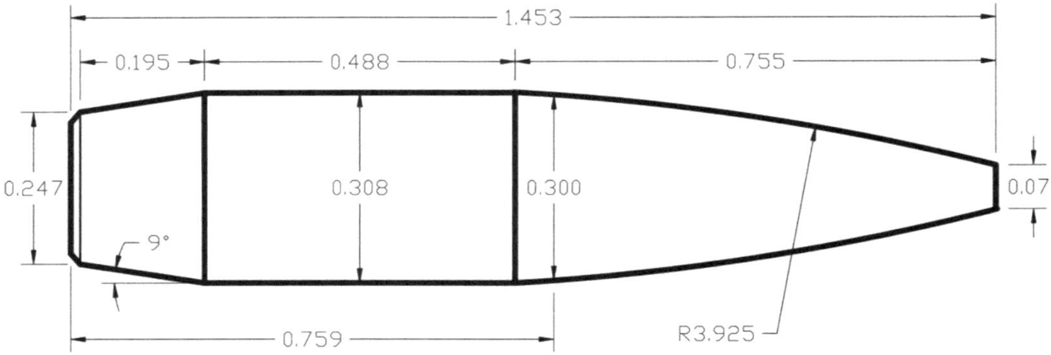

Lot #:	Unknown	Ogive Radius:	12.74 calibers
Sectional Density:	0.312 lb/in^2	Rt/R:	0.63

Ballistic Performance

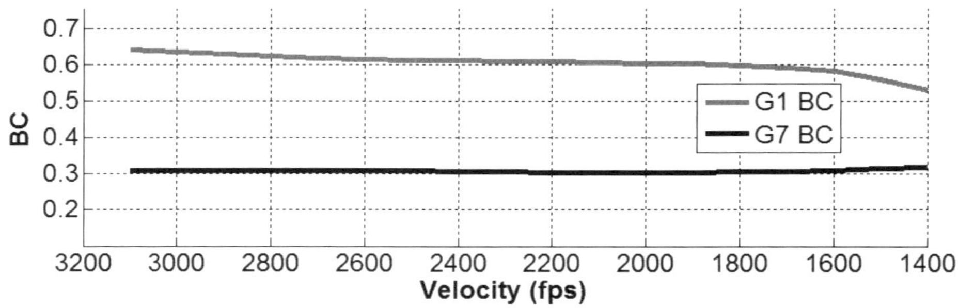

	Above 3000	3000-2500	2500-2000	2000-1500	Below 1500
G1 BC	0.652	0.623	0.608	0.581	0.558
G7 BC	0.309	0.309	0.306	0.309	0.314

Averages from 3000 fps to 1500 fps:
G1 BC: 0.602, variation of G1 BC: 0.076
G7 BC: 0.309, variation of G7 BC: 0.011
G7 form factor: 1.009

Rifling Twist Rate and Stability

Barrel Twist	Stability Factor Best case	Nominal	Worst case
1:10"	2.10	1.84	1.59
1:11"	1.74	1.52	1.31
1:12"	1.46	1.28	1.10
1:13"	1.24	1.09	0.94

Minimum twist required for stability is 1:12.4"
Recommended twist for optimal performance is 1:11.1" or faster

.338 Caliber

.338 caliber bullets typically range from 185 to 300 grains. The image below relates bullet weight to sectional density for .338 caliber bullets.

Range of Weight and Sectional Density for 0.338 Caliber

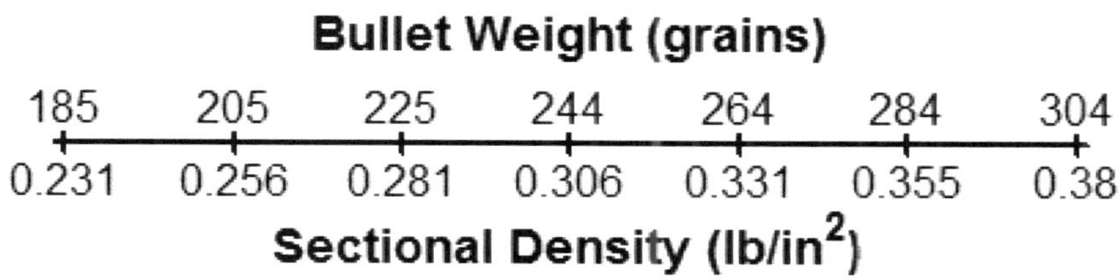

Sectional Density (SD) is a measure of how much mass is behind the bullets frontal area. Higher SD means a greater ability for the bullet to penetrate targets on impact, and also retain velocity in flight.

Spectrum of G7 form factors

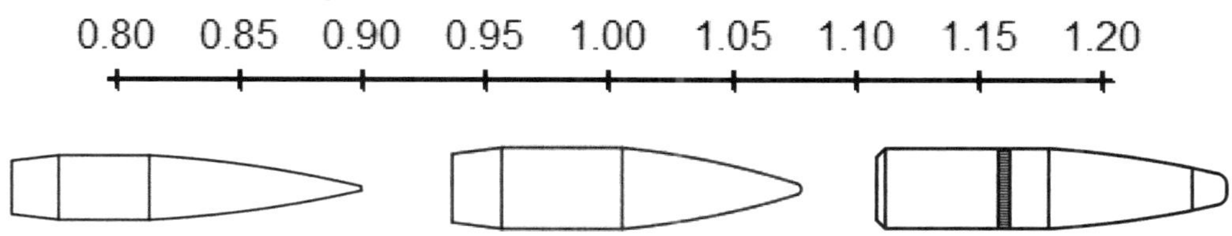

G7 Form factors range from 0.80 for the lowest drag shapes, up to 1.20+ for blunter shapes.

G7 Ballistic Coefficients are simply Sectional Density (SD) divided by G7 form factor.

You can determine a bullets' suitability for long range performance by seeing where it lies on the SD and form factor lines. Ideally, you want a bullet with a high SD and a low G7 form factor.

As an example, a 300 grain .338 caliber bullet has an SD of 0.375. If that bullet had a G7 form factor of 0.95, it's G7 BC would be:

G7 BC = SD/FF = 0.375/0.95 = 0.395.

Barnes 338 caliber 265 grain LRX BT

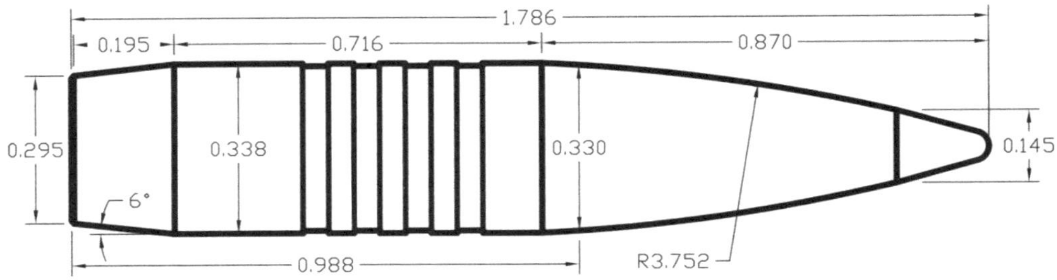

Lot #:	Unknown	Ogive Radius:	11.10 calibers
Sectional Density:	0.331 lb/in^2	Rt/R:	0.67

Ballistic Performance

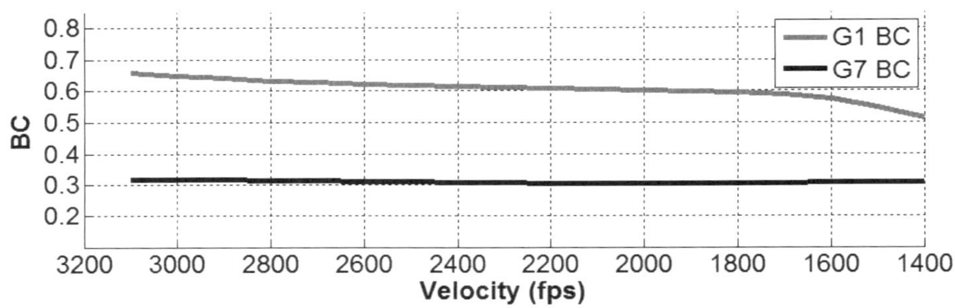

	Above 3000	3000-2500	2500-2000	2000-1500	Below 1500
G1 BC	0.674	0.634	0.610	0.576	0.549
G7 BC	0.319	0.314	0.307	0.306	0.309

Averages from 3000 fps to 1500 fps:
G1 BC: 0.605, variation of G1 BC: 0.101
G7 BC: 0.310, variation of G7 BC: 0.015
G7 form factor: 1.069

Rifling Twist Rate and Stability

	Stability Factor		
Barrel Twist	Best case	Nominal	Worst case
1:9"	2.18	1.90	1.64
1:10"	1.76	1.54	1.33
1:11"	1.46	1.27	1.10
1:12"	1.22	1.07	0.92

Minimum twist required for stability is 1:11.3"
Recommended twist for optimal performance is 1:10.1" or faster

Barnes 338 caliber 280 grain LRX BT

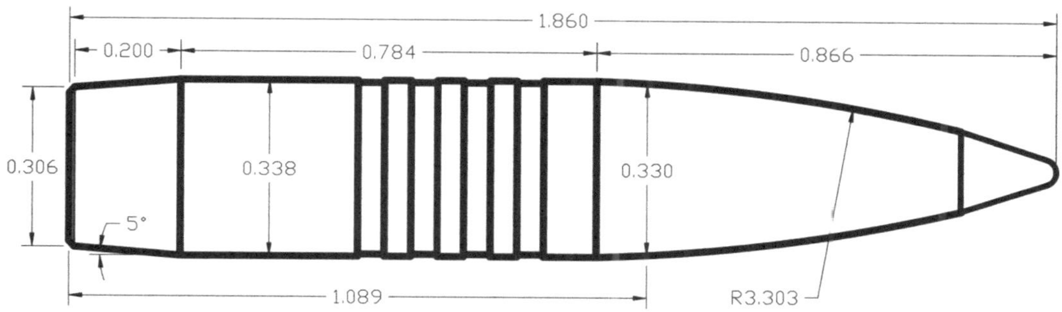

Lot #:	6480853	Ogive Radius:	9.77 calibers
Sectional Density:	0.350 lb/in^2	Rt/R:	0.80

Ballistic Performance

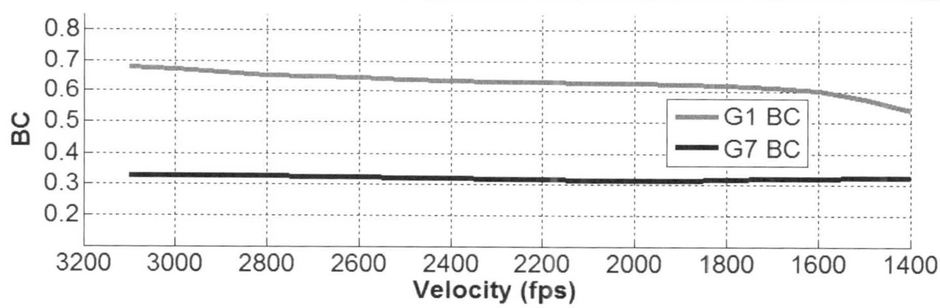

	Above 3000	3000-2500	2500-2000	2000-1500	Below 1500
G1 BC	0.695	0.655	0.633	0.602	0.577
G7 BC	0.329	0.325	0.318	0.320	0.325

Averages from 3000 fps to 1500 fps:
G1 BC: 0.629, variation of G1 BC: 0.094
G7 BC: 0.322, variation of G7 BC: 0.013
G7 form factor: 1.086

Rifling Twist Rate and Stability

Barrel Twist	Stability Factor Best case	Nominal	Worst case
1:8"	2.58	2.26	1.95
1:9"	2.04	1.78	1.54
1:10"	1.65	1.44	1.25
1:11"	1.37	1.19	1.03

Minimum twist required for stability is 1:11.0"
Recommended twist for optimal performance is 1:9.8" or faster

Barnes 338 caliber 225 grain TAC-TX BT

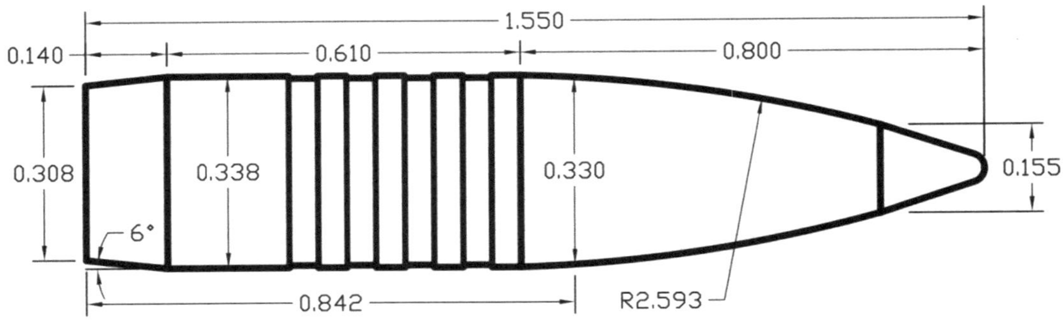

Lot #:	6421202	Ogive Radius:	7.67 calibers
Sectional Density:	0.281 lb/in^2	Rt/R:	0.75

Ballistic Performance

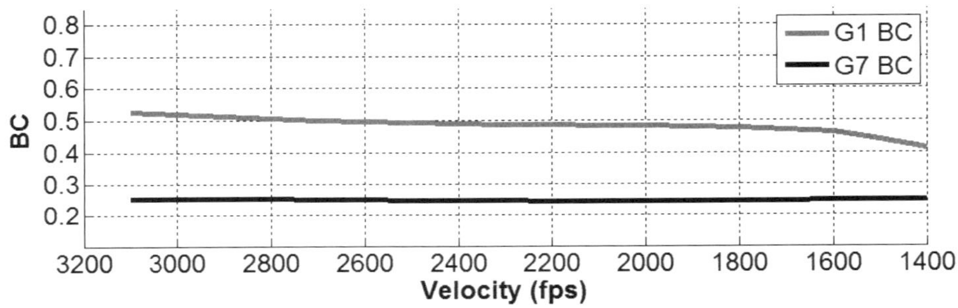

	Above 3000	3000-2500	2500-2000	2000-1500	Below 1500
G1 BC	0.538	0.506	0.488	0.462	0.441
G7 BC	0.255	0.251	0.245	0.245	0.248

Averages from 3000 fps to 1500 fps:
G1 BC: 0.484, variation of G1 BC: 0.078
G7 BC: 0.248, variation of G7 BC: 0.012
G7 form factor: 1.134

Rifling Twist Rate and Stability

Barrel Twist	Stability Factor Best case	Stability Factor Nominal	Stability Factor Worst case
1:10"	2.26	1.98	1.71
1:11"	1.87	1.63	1.41
1:12"	1.57	1.37	1.19
1:13"	1.34	1.17	1.01

Minimum twist required for stability is 1:12.8"
Recommended twist for optimal performance is 1:11.5" or faster

Barnes 338 caliber 265 grain TAC-TX BT

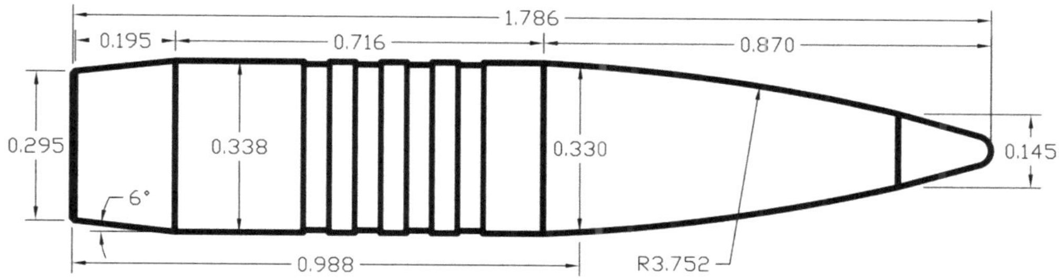

Lot #:	6470069	Ogive Radius:	11.10 calibers
Sectional Density:	0.331 lb/in²	Rt/R:	0.67

Ballistic Performance

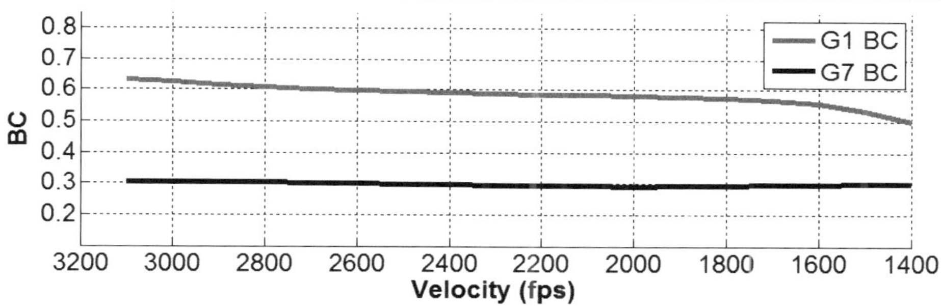

	Above 3000	3000-2500	2500-2000	2000-1500	Below 1500
G1 BC	0.647	0.609	0.588	0.558	0.535
G7 BC	0.306	0.302	0.295	0.296	0.301

Averages from 3000 fps to 1500 fps:
G1 BC: 0.583, variation of G1 BC: 0.089
G7 BC: 0.299, variation of G7 BC: 0.013
G7 form factor: 1.108

Rifling Twist Rate and Stability

Barrel Twist	Stability Factor Best case	Nominal	Worst case
1:9"	2.17	1.89	1.64
1:10"	1.76	1.53	1.33
1:11"	1.45	1.27	1.10
1:12"	1.22	1.07	0.92

Minimum twist required for stability is 1:11 3"
Recommended twist for optimal performance is 1:10.1" or faster

Barnes 338 caliber 285 grain TAC-X BT

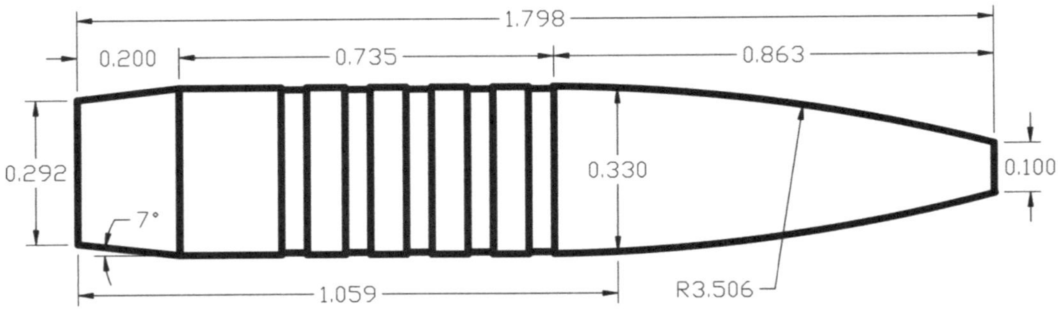

Lot #:	6398305	Ogive Radius:	10.37 calibers
Sectional Density:	0.356 lb/in²	Rt/R:	0.91

Ballistic Performance

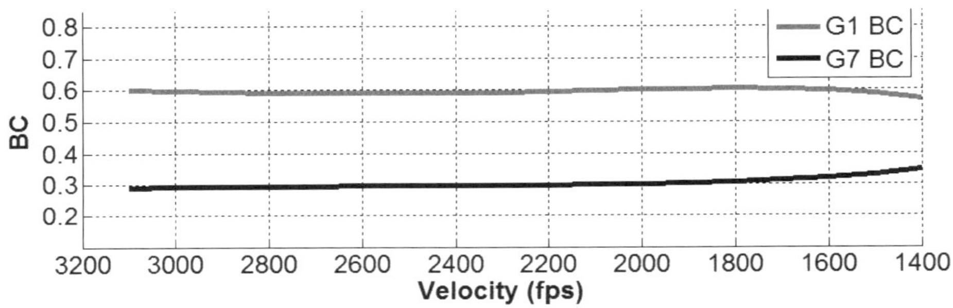

	Above 3000	3000-2500	2500-2000	2000-1500	Below 1500
G1 BC	0.610	0.594	0.595	0.594	0.588
G7 BC	0.289	0.295	0.299	0.317	0.332

Averages from 3000 fps to 1500 fps:
G1 BC: 0.594, variation of G1 BC: 0.012
G7 BC: 0.306, variation of G7 BC: 0.039
G7 form factor: 1.168

Rifling Twist Rate and Stability

Barrel Twist	Stability Factor Best case	Stability Factor Nominal	Stability Factor Worst case
1:9"	2.30	2.01	1.74
1:10"	1.87	1.63	1.41
1:11"	1.54	1.35	1.17
1:12"	1.30	1.13	0.98

Minimum twist required for stability is 1:11.7"
Recommended twist for optimal performance is 1:10.4" or faster

Barnes 338 caliber 285 grain TSX BT

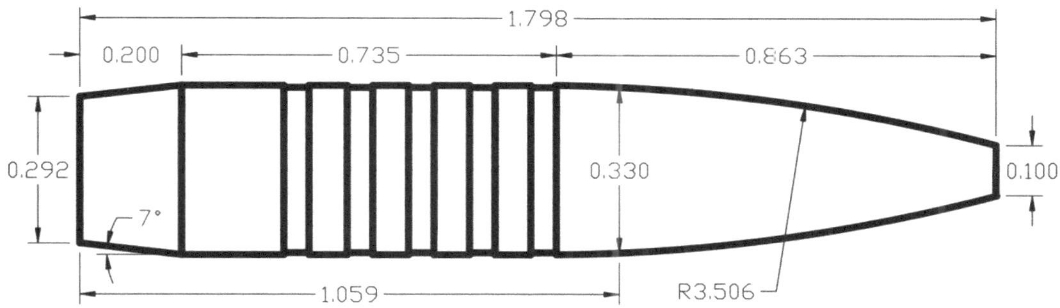

Lot #:	5058-1 092M PKG	**Ogive Radius:**	**10.37 calibers**
Sectional Density:	**0.356 lb/in^2**	**Rt/R:**	**0.91**

Ballistic Performance

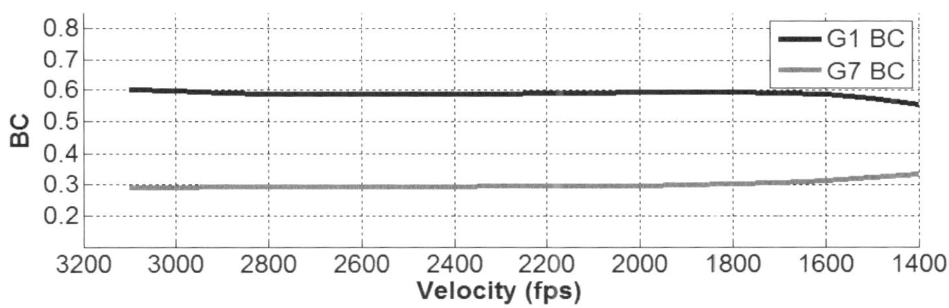

	Above 3000	3000-2500	2500-2000	2000-1500	Below 1500
G1 BC	0.609	0.592	0.591	0.584	0.574
G7 BC	0.288	0.294	0.297	0.311	0.324

Averages from 3000 fps to 1500 fps:
G1 BC: 0.588, variation of G1 BC: 0.023
G7 BC: 0.302, variation of G7 BC: 0.032
G7 form factor: 1.180

Rifling Twist Rate and Stability

Barrel Twist	**Stability Factor** Best case	Nominal	Worst case
1:9"	2.29	2.00	1.73
1:10"	1.86	1.62	1.40
1:11"	1.54	1.34	1.16
1:12"	1.29	1.13	0.97

Minimum twist required for stability is 1:11.6"
Recommended twist for optimal performance is 1:10.4" or faster

Barnes 338 caliber 185 grain TTSX BT

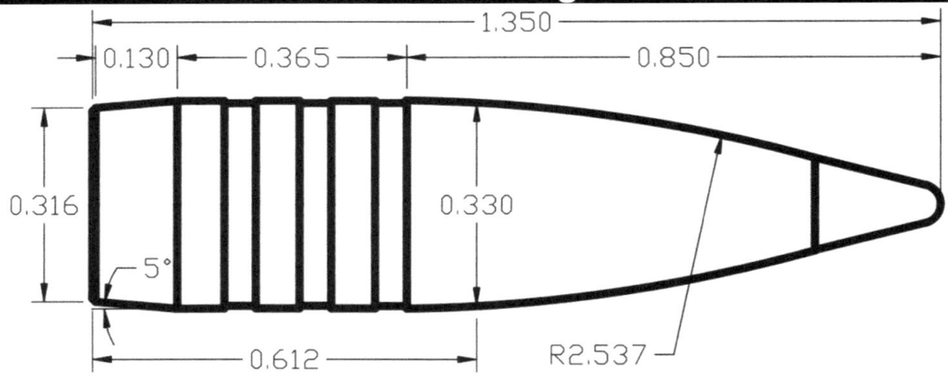

Lot #:	6298-1 113N AH-M	Ogive Radius:	7.51 calibers
Sectional Density:	0.231 lb/in^2	Rt/R:	0.90

Ballistic Performance

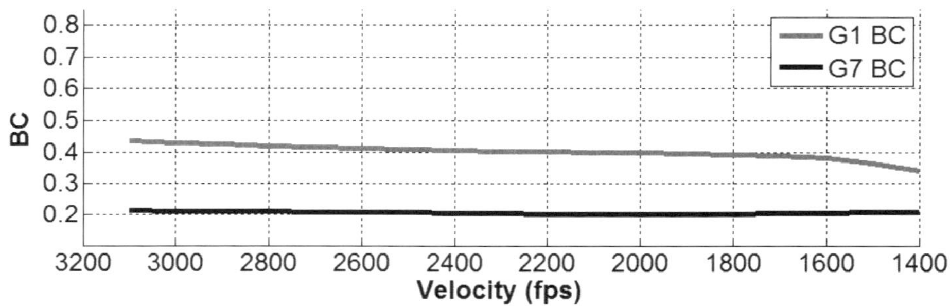

	Above 3000	3000-2500	2500-2000	2000-1500	Below 1500
G1 BC	0.448	0.420	0.403	0.381	0.364
G7 BC	0.212	0.208	0.202	0.202	0.205

Averages from 3000 fps to 1500 fps:
G1 BC: 0.400, variation of G1 BC: 0.067
G7 BC: 0.205, variation of G7 BC: 0.011
G7 form factor: 1.128

Rifling Twist Rate and Stability

Barrel Twist	Stability Factor Best case	Nominal	Worst case
1:9"	3.43	2.99	2.59
1:10"	2.78	2.43	2.10
1:11"	2.29	2.00	1.73
1:12"	1.93	1.68	1.46

Minimum twist required for stability is 1:14.2"
Recommended twist for optimal performance is 1:12.7" or faster

Barnes 338 caliber 210 grain TTSX BT

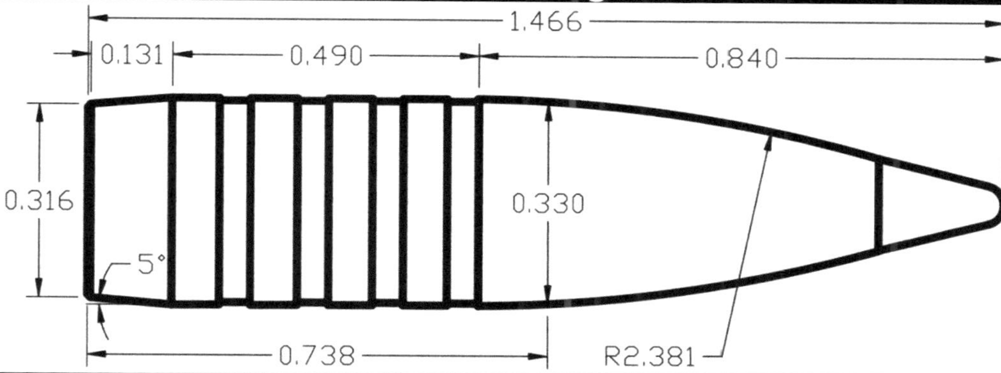

Lot #:	5299-1 051-n kdj-m	**Ogive Radius:**	**7.04 calibers**
Sectional Density:	**0.263 lb/in²**	**Rt/R:**	**0.82**

Ballistic Performance

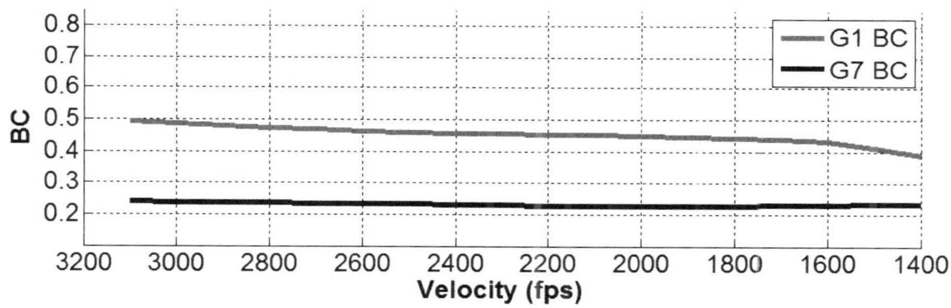

	Above 3000	3000-2500	2500-2000	2000-1500	Below 1500
G1 BC	0.505	0.474	0.455	0.432	0.413
G7 BC	0.239	0.235	0.229	0.229	0.232

Averages from 3000 fps to 1500 fps:
G1 BC: 0.453, variation of G1 BC: 0.073
G7 BC: 0.232, variation of G7 BC: 0.012
G7 form factor: 1.132

Rifling Twist Rate and Stability

	Stability Factor		
Barrel Twist	**Best case**	**Nominal**	**Worst case**
1:9"	3.07	2.68	2.32
1:10"	2.48	2.17	1.88
1:11"	2.05	1.79	1.55
1:12"	1.72	1.51	1.30

Minimum twist required for stability is 1:13.4"
Recommended twist for optimal performance is 1:12.0" or faster

Barnes 338 caliber 225 grain TTSX BT

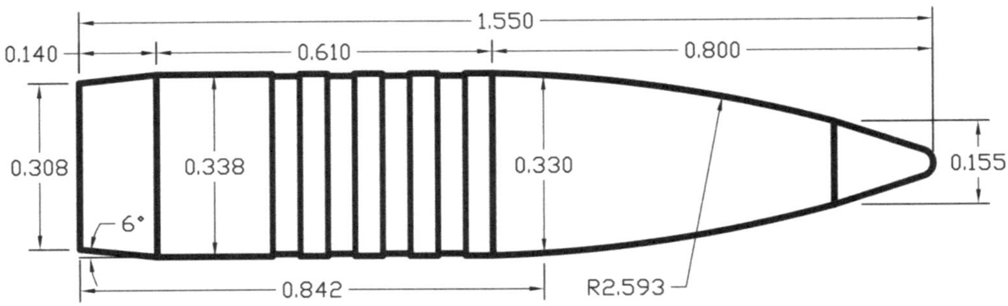

Lot #:	3155-1 009H CW	Ogive Radius:	7.67 calibers
Sectional Density:	0.281 lb/in^2	Rt/R:	0.75

Ballistic Performance

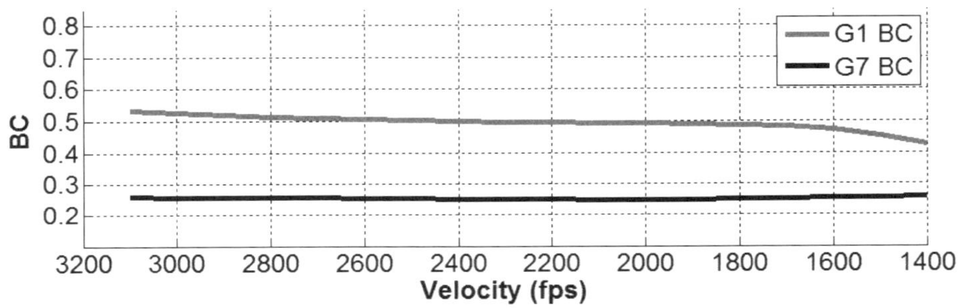

	Above 3000	3000-2500	2500-2000	2000-1500	Below 1500
G1 BC	0.547	0.516	0.498	0.473	0.453
G7 BC	0.259	0.256	0.250	0.251	0.255

Averages from 3000 fps to 1500 fps:
G1 BC: 0.494, variation of G1 BC: 0.075
G7 BC: 0.253, variation of G7 BC: 0.011
G7 form factor: 1.110

Rifling Twist Rate and Stability

Barrel Twist	Stability Factor Best case	Nominal	Worst case
1:9"	2.79	2.44	2.11
1:10"	2.26	1.98	1.71
1:11"	1.87	1.63	1.41
1:12"	1.57	1.37	1.19

Minimum twist required for stability is 1:12.8"
Recommended twist for optimal performance is 1:11.5" or faster

Berger 338 caliber 250 grain Elite Hunter

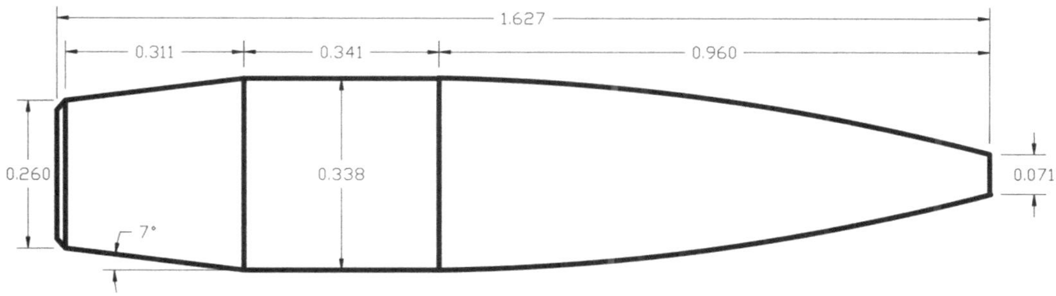

Lot #:	7476	Ogive Radius:	Hybrid
Sectional Density:	0.313 lb/in^2	Rt/R:	Hybrid

Ballistic Performance

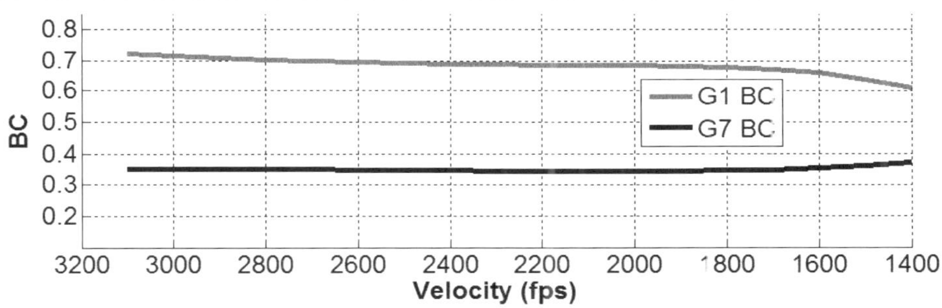

	Above 3000	3000-2500	2500-2000	2000-1500	Below 1500
G1 BC	0.738	0.705	0.689	0.662	0.640
G7 BC	0.350	0.350	0.346	0.352	0.361

Averages from 3000 fps to 1500 fps:
G1 BC: 0.684, variation of G1 BC: 0.077
G7 BC: 0.351, variation of G7 BC: 0.017
G7 form factor: 0.891

Rifling Twist Rate and Stability

Barrel Twist	Stability Factor Best case	Nominal	Worst case
1:10"	2.18	1.91	1.65
1:11"	1.80	1.58	1.36
1:12"	1.52	1.32	1.15
1:13"	1.29	1.13	0.98

Minimum twist required for stability is 1:12.6"
Recommended twist for optimal performance is 1:11.3" or faster

Berger 338 caliber 300 grain Elite Hunter

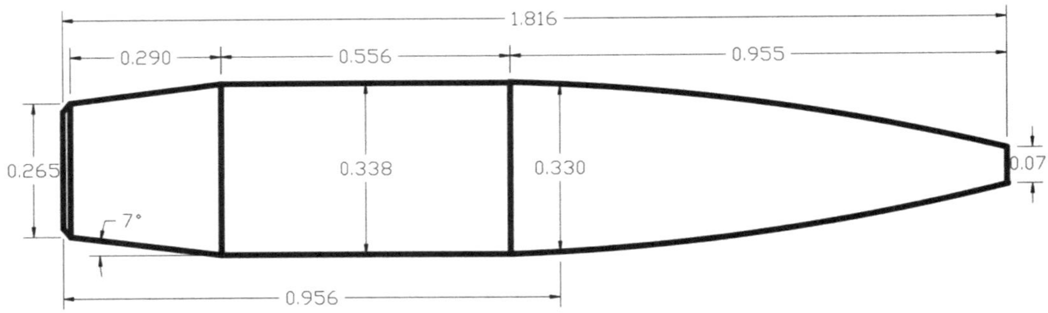

Lot #:	6293	Ogive Radius:	Hybrid
Sectional Density:	0.375 lb/in^2	Rt/R:	Hybrid

Ballistic Performance

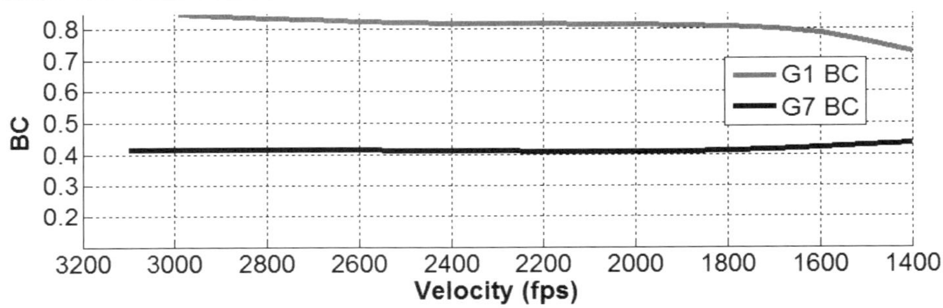

	Above 3000	3000-2500	2500-2000	2000-1500	Below 1500
G1 BC	0.875	0.836	0.818	0.788	0.762
G7 BC	0.414	0.415	0.411	0.419	0.429

Averages from 3000 fps to 1500 fps:
G1 BC: 0.812, variation of G1 BC: 0.089
G7 BC: 0.417, variation of G7 BC: 0.020
G7 form factor: 0.900

Rifling Twist Rate and Stability

Barrel Twist	Stability Factor Best case	Nominal	Worst case
1:9"	2.34	2.05	1.77
1:10"	1.90	1.66	1.43
1:11"	1.57	1.37	1.19
1:12"	1.32	1.15	1.00

Minimum twist required for stability is 1:11.8"
Recommended twist for optimal performance is 1:10.5" or faster

Berger 338 caliber 250 grain Tactical OTM

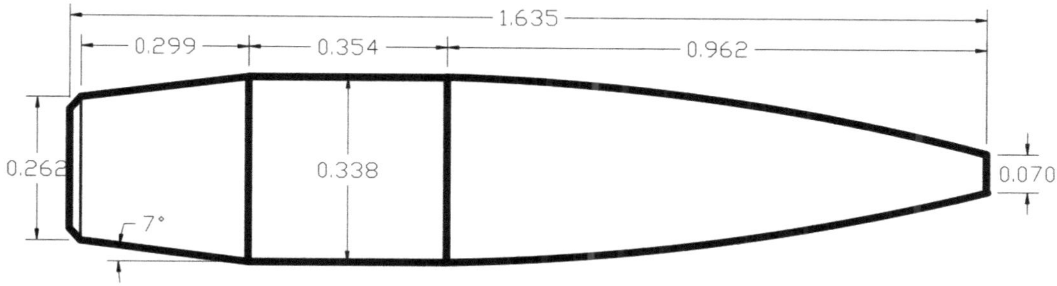

Lot #:	7468	Ogive Radius:	Hybrid
Sectional Density:	0.313 lb/in^2	Rt/R:	Hybrid

Ballistic Performance

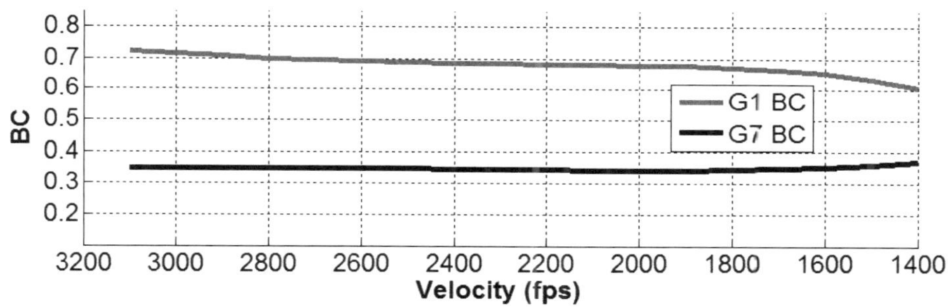

	Above 3000	3000-2500	2500-2000	2000-1500	Below 1500
G1 BC	0.737	0.703	0.685	0.658	0.636
G7 BC	0.349	0.349	0.344	0.350	0.358

Averages from 3000 fps to 1500 fps:
G1 BC: 0.680, variation of G1 BC: 0.080
G7 BC: 0.349, variation of G7 BC: 0.016
G7 form factor: 0.895

Rifling Twist Rate and Stability

Barrel Twist	Stability Factor Best case	Nominal	Worst case
1:10"	2.15	1.88	1.63
1:11"	1.78	1.55	1.34
1:12"	1.49	1.31	1.13
1:13"	1.27	1.11	0.96

Minimum twist required for stability is 1:12.5"
Recommended twist for optimal performance is 1:11.2" or faster

Berger 338 caliber 300 grain Tactical OTM

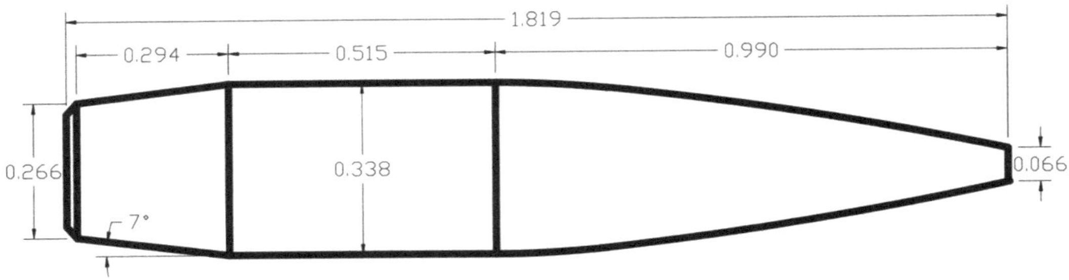

Lot #:	6070	Ogive Radius:	Hybrid
Sectional Density:	0.375 lb/in²	Rt/R:	Hybrid

Ballistic Performance

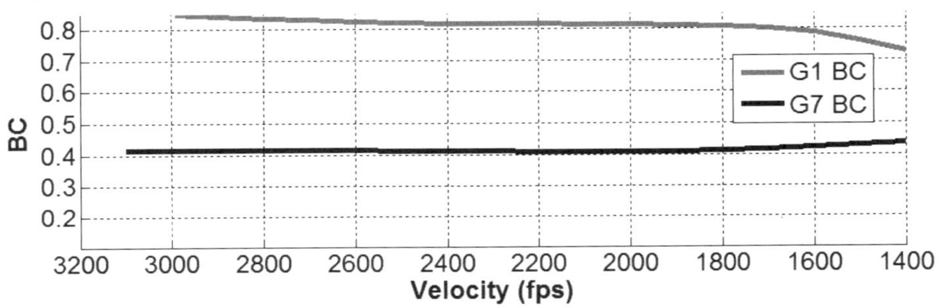

	Above 3000	3000-2500	2500-2000	2000-1500	Below 1500
G1 BC	0.875	0.836	0.818	0.788	0.762
G7 BC	0.414	0.415	0.411	0.419	0.429

Averages from 3000 fps to 1500 fps:
G1 BC: 0.812, variation of G1 BC: 0.089
G7 BC: 0.417, variation of G7 BC: 0.020
G7 form factor: 0.900

Rifling Twist Rate and Stability

Barrel Twist	Stability Factor Best case	Nominal	Worst case
1:9"	2.34	2.05	1.77
1:10"	1.90	1.66	1.43
1:11"	1.57	1.37	1.19
1:12"	1.32	1.15	1.00

Minimum twist required for stability is 1:11.8"
Recommended twist for optimal performance is 1:10.5" or faster

Cutting Edge 338 caliber 300 grain HPBT

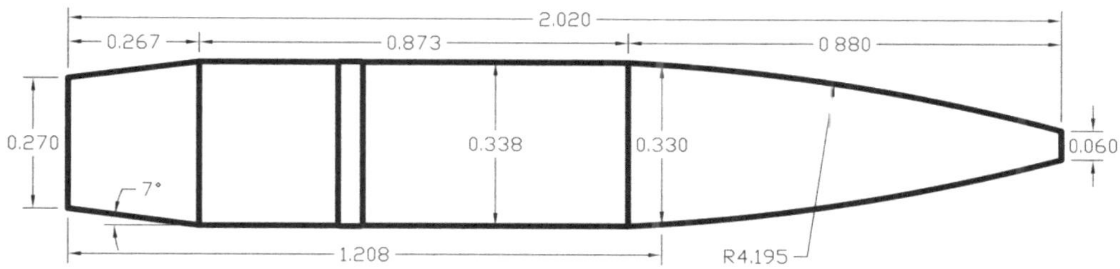

Lot #:	Unknown	Ogive Radius:	12.41 calibers
Sectional Density:	0.375 lb/in^2	Rt/R:	0.68

Ballistic Performance

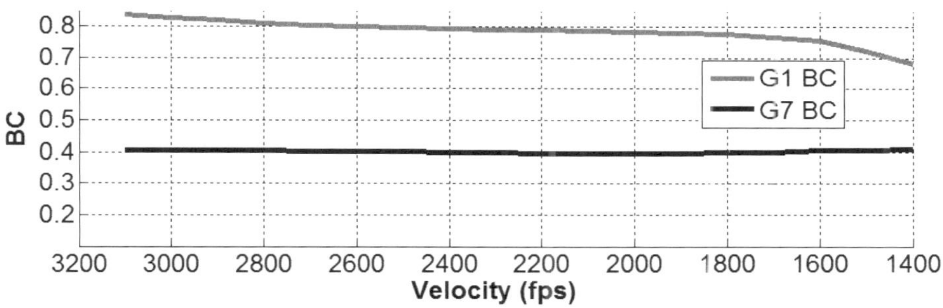

	Above 3000	3000-2500	2500-2000	2000-1500	Below 1500
G1 BC	0.860	0.819	0.797	0.759	0.727
G7 BC	0.408	0.406	0.401	0.403	0.409

Averages from 3000 fps to 1500 fps:
G1 BC: 0.789, variation of G1 BC: 0.107
G7 BC: 0.405, variation of G7 BC: 0.012
G7 form factor: 0.927

Rifling Twist Rate and Stability

Barrel Twist	Stability Factor Best case	Nominal	Worst case
1:8"	2.17	1.90	1.64
1:9"	1.71	1.50	1.30
1:10"	1.39	1.21	1.05
1:11"	1.15	1.00	0.87

Minimum twist required for stability is 1:10.1"
Recommended twist for optimal performance is 1:9.0" or faster

Cutting Edge 338 caliber 225 grain Lazer

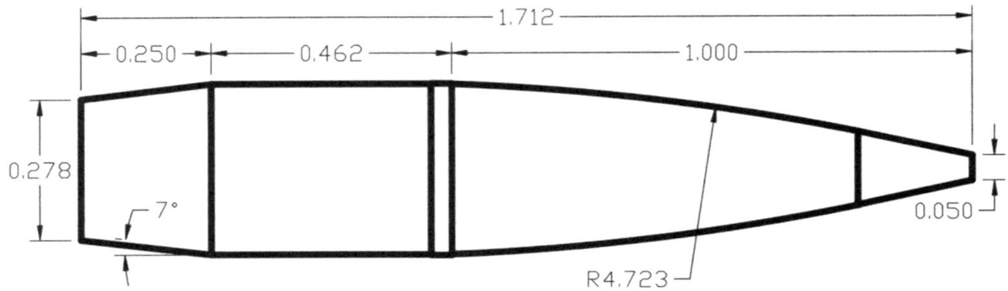

Lot #:	Unknown	Ogive Radius:	13.97 calibers
Sectional Density:	0.281 lb/in^2	Rt/R:	0.69

Ballistic Performance

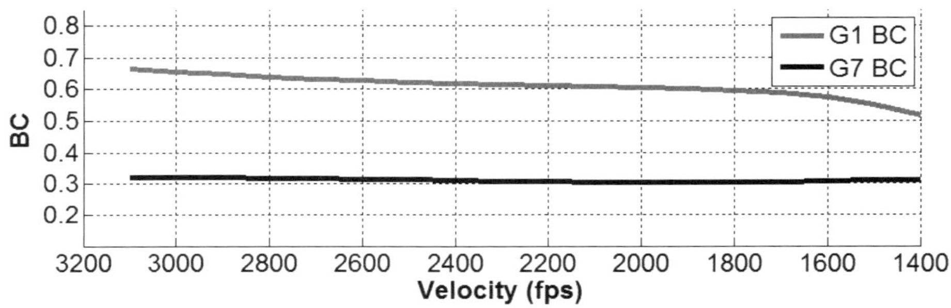

	Above 3000	3000-2500	2500-2000	2000-1500	Below 1500
G1 BC	0.681	0.640	0.614	0.578	0.551
G7 BC	0.323	0.317	0.309	0.307	0.310

Averages from 3000 fps to 1500 fps:
G1 BC: 0.609, variation of G1 BC: 0.106
G7 BC: 0.312, variation of G7 BC: 0.017
G7 form factor: 0.902

Rifling Twist Rate and Stability

Barrel Twist	Stability Factor Best case	Nominal	Worst case
1:9"	2.09	1.83	1.58
1:10"	1.69	1.48	1.28
1:11"	1.40	1.22	1.06
1:12"	1.18	1.03	0.89

Minimum twist required for stability is 1:11.1"
Recommended twist for optimal performance is 1:9.9" or faster

Cutting Edge 338 caliber 250 grain Lazer

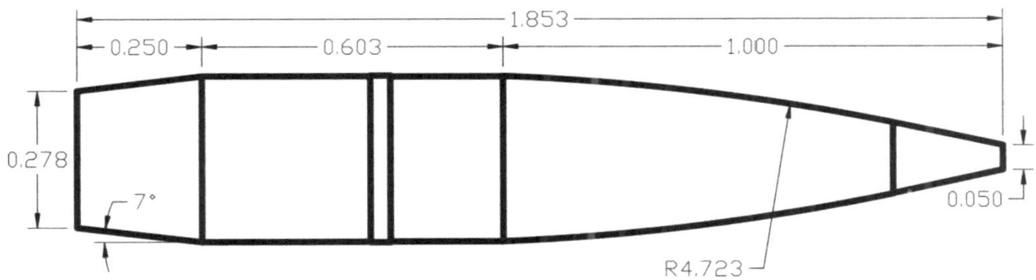

Lot #:	Unknown	Ogive Radius:	13.97 calibers
Sectional Density:	0.313 lb/in^2	Rt/R:	0.69

Ballistic Performance

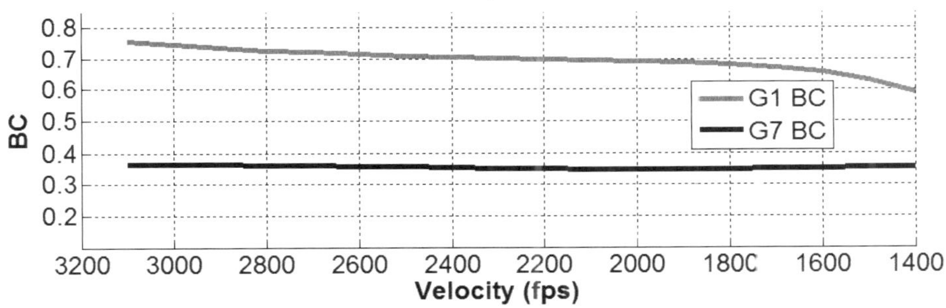

	Above 3000	3000-2500	2500-2000	2000-1500	Below 1500
G1 BC	0.775	0.729	0.702	0.662	0.631
G7 BC	0.367	0.362	0.353	0.351	0.355

Averages from 3000 fps to 1500 fps:
G1 BC: 0.696, variation of G1 BC: 0.116
G7 BC: 0.357, variation of G7 BC: 0.018
G7 form factor: 0.877

Rifling Twist Rate and Stability

Barrel Twist	Stability Factor Best case	Stability Factor Nominal	Stability Factor Worst case
1:9"	1.84	1.61	1.39
1:10"	1.49	1.30	1.13
1:11"	1.23	1.08	0.93
1:12"	1.04	0.91	0.78

Minimum twist required for stability is 1:10.4"
Recommended twist for optimal performance is 1:9.3" or faster

Cutting Edge 338 caliber 275 grain Lazer

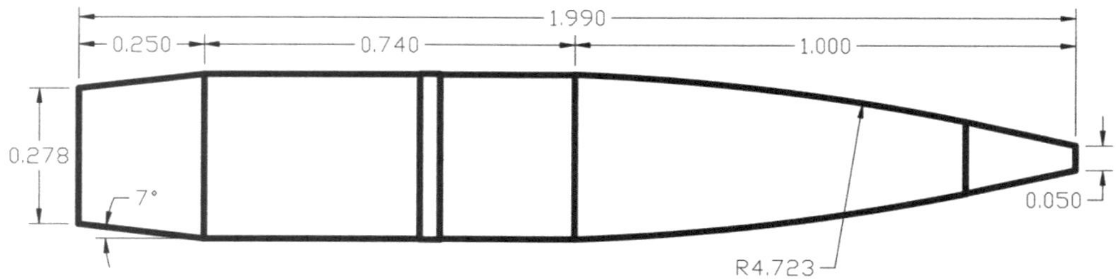

Lot #:	Unknown	Ogive Radius:	13.97 calibers
Sectional Density:	0.344 lb/in^2	Rt/R:	0.70

Ballistic Performance

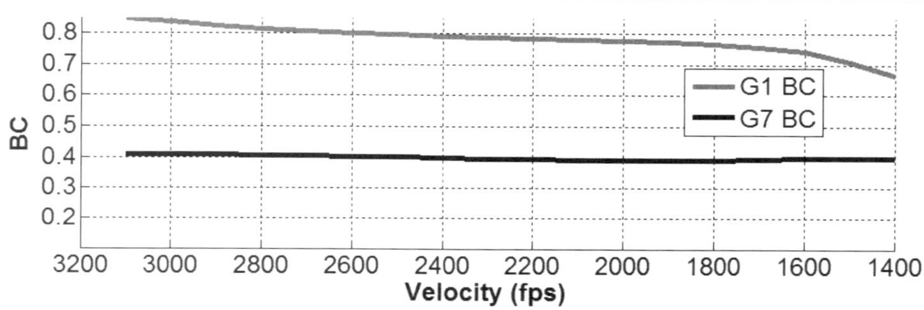

	Above 3000	3000-2500	2500-2000	2000-1500	Below 1500
G1 BC	0.867	0.817	0.787	0.744	0.710
G7 BC	0.410	0.405	0.396	0.395	0.399

Averages from 3000 fps to 1500 fps:
G1 BC: 0.780, variation of G1 BC: 0.127
G7 BC: 0.400, variation of G7 BC: 0.018
G7 form factor: 0.860

Rifling Twist Rate and Stability

Barrel Twist	Stability Factor Best case	Nominal	Worst case
1:8"	2.08	1.82	1.57
1:9"	1.64	1.44	1.24
1:10"	1.33	1.16	1.01
1:11"	1.10	0.96	0.83

Minimum twist required for stability is 1:9.8"
Recommended twist for optimal performance is 1:8.8" or faster

Cutting Edge 338 caliber 254 grain MTAC

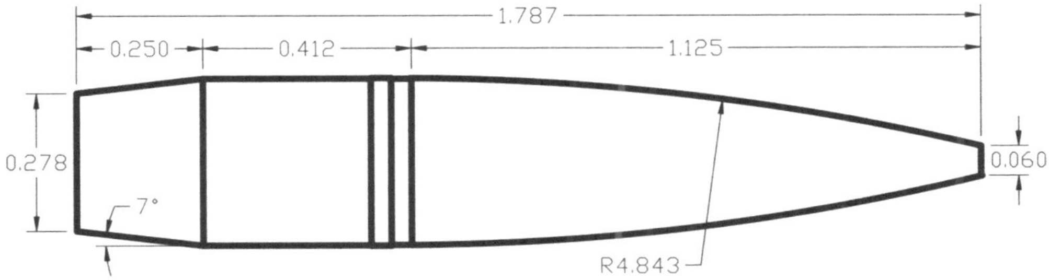

Lot #:	**Unknown**	**Ogive Radius:**	**14.33 calibers**
Sectional Density:	**0.318 lb/in²**	**Rt/R:**	**0.96**

Ballistic Performance

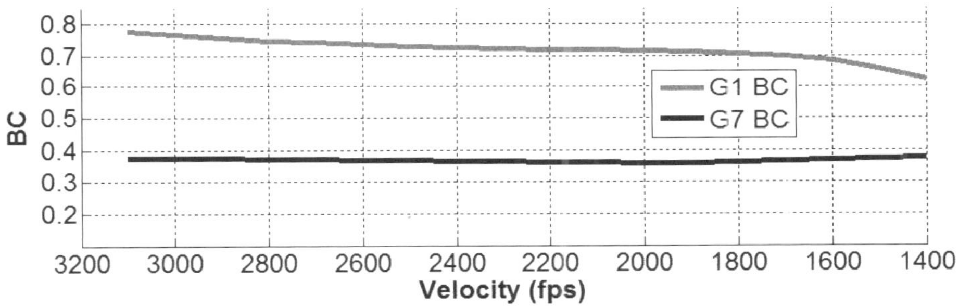

	Above 3000	3000-2500	2500-2000	2000-1500	Below 1500
G1 BC	0.793	0.749	0.724	0.683	0.660
G7 BC	0.376	0.371	0.364	0.365	0.371

Averages from 3000 fps to 1500 fps:
G1 BC: 0.718, variation of G1 BC: 0.107
G7 BC: 0.368, variation of G7 BC: 0.015
G7 form factor: 0.862

Rifling Twist Rate and Stability

Barrel Twist	Stability Factor Best case	Nominal	Worst case
1:9"	2.08	1.82	1.57
1:10"	1.69	1.47	1.27
1:11"	1.39	1.22	1.05
1:12"	1.17	1.02	0.88

Minimum twist required for stability is 1:11.1"
Recommended twist for optimal performance is 1:9.9" or faster

Cutting Edge 338 caliber 275 grain MTAC

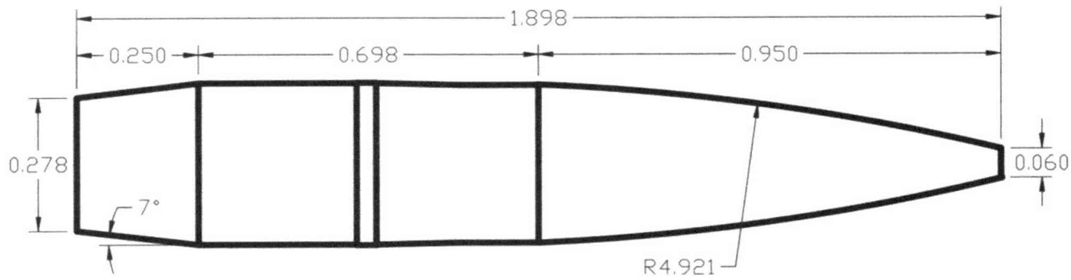

Lot #:	Unknown	Ogive Radius:	14.56 calibers
Sectional Density:	0.344 lb/in^2	Rt/R:	0.70

Ballistic Performance

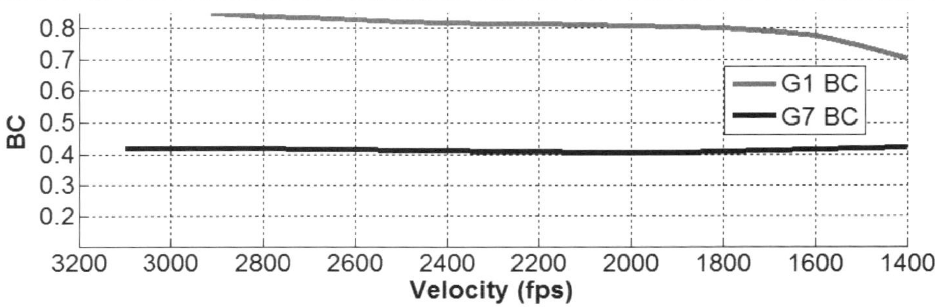

	Above 3000	3000-2500	2500-2000	2000-1500	Below 1500
G1 BC	0.887	0.841	0.816	0.777	0.745
G7 BC	0.420	0.417	0.410	0.413	0.419

Averages from 3000 fps to 1500 fps:
G1 BC: 0.809, variation of G1 BC: 0.114
G7 BC: 0.415, variation of G7 BC: 0.014
G7 form factor: 0.829

Rifling Twist Rate and Stability

Barrel Twist	Stability Factor Best case	Stability Factor Nominal	Stability Factor Worst case
1:9"	1.89	1.65	1.43
1:10"	1.53	1.34	1.16
1:11"	1.26	1.10	0.95
1:12"	1.06	0.93	0.80

Minimum twist required for stability is 1:10.6"
Recommended twist for optimal performance is 1:9.4" or faster

Cutting Edge 338 caliber 252 grain MTH

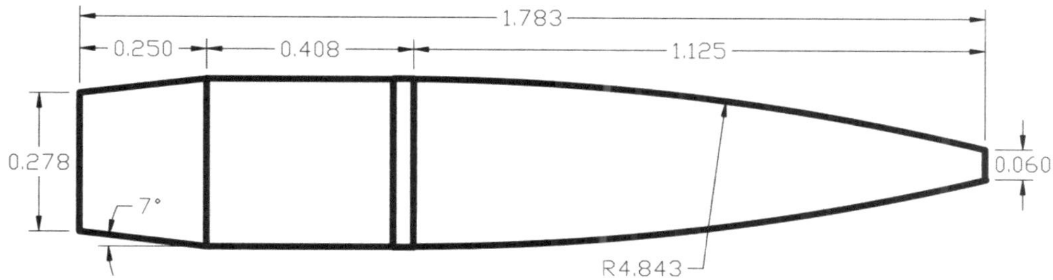

Lot #:	Unknown	Ogive Radius:	14.34 calibers
Sectional Density:	0.315 lb/in^2	Rt/R:	0.96

Ballistic Performance

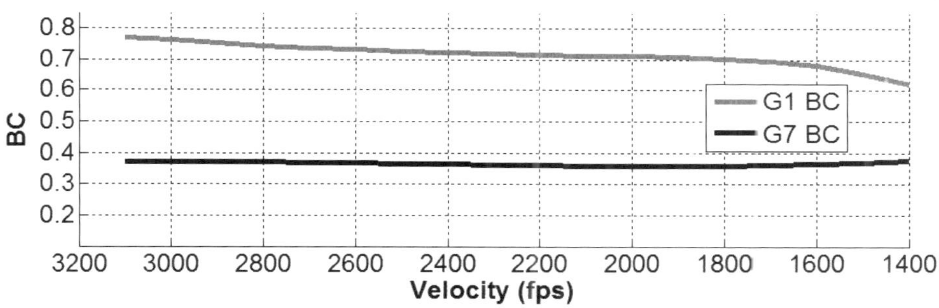

	Above 3000	3000-2500	2500-2000	2000-1500	Below 1500
G1 BC	0.789	0.745	0.720	0.684	0.656
G7 BC	0.373	0.369	0.362	0.363	0.369

Averages from 3000 fps to 1500 fps:
G1 BC: 0.714, variation of G1 BC: 0.106
G7 BC: 0.366, variation of G7 BC: 0.015
G7 form factor: 0.860

Rifling Twist Rate and Stability

Barrel Twist	Stability Factor Best case	Stability Factor Nominal	Stability Factor Worst case
1:9"	2.08	1.82	1.57
1:10"	1.68	1.47	1.27
1:11"	1.39	1.22	1.05
1:12"	1.17	1.02	0.88

Minimum twist required for stability is 1:11 1"
Recommended twist for optimal performance is 1:9.9" or faster

Cutting Edge 338 caliber 265 grain MTH

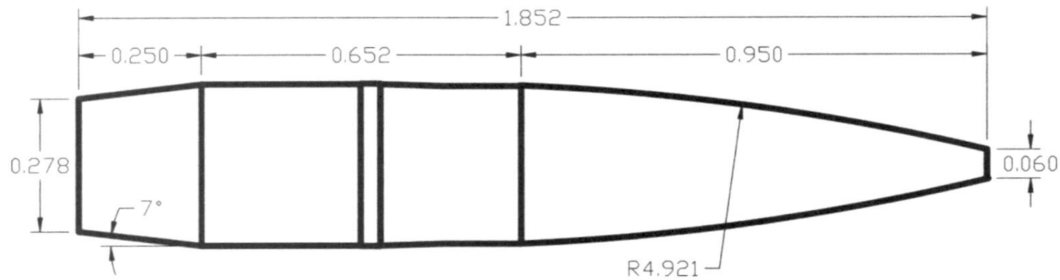

Lot #:	Unknown	Ogive Radius:	14.56 calibers
Sectional Density:	0.331 lb/in^2	Rt/R:	0.70

Ballistic Performance

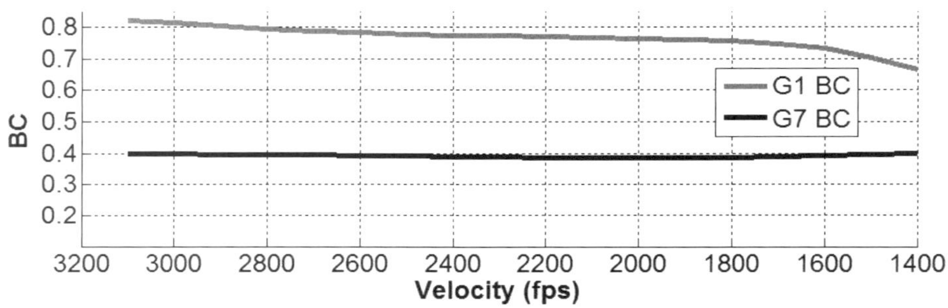

	Above 3000	3000-2500	2500-2000	2000-1500	Below 1500
G1 BC	0.840	0.796	0.772	0.735	0.705
G7 BC	0.398	0.395	0.388	0.390	0.397

Averages from 3000 fps to 1500 fps:
G1 BC: 0.765, variation of G1 BC: 0.109
G7 BC: 0.393, variation of G7 BC: 0.013
G7 form factor: 0.844

Rifling Twist Rate and Stability

Barrel Twist	Stability Factor Best case	Nominal	Worst case
1:9"	1.96	1.71	1.48
1:10"	1.58	1.38	1.20
1:11"	1.31	1.14	0.99
1:12"	1.10	0.96	0.83

Minimum twist required for stability is 1:10.7"
Recommended twist for optimal performance is 1:9.6" or faster

Cutting Edge 338 caliber 275 grain MTH

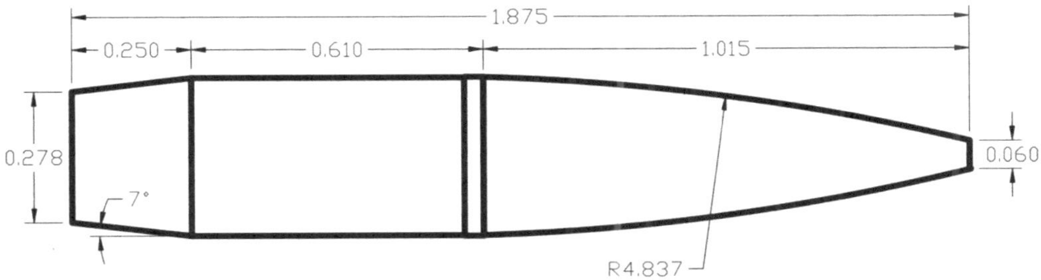

Lot #:	Unknown	Ogive Radius:	14.31 calibers
Sectional Density:	0.344 lb/in^2	Rt/R:	0.79

Ballistic Performance

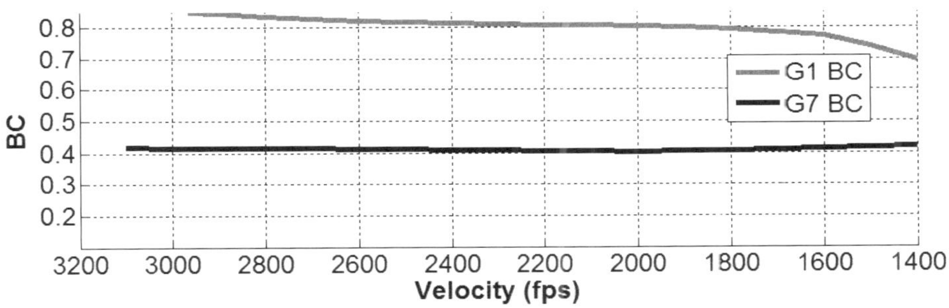

	Above 3000	3000-2500	2500-2000	2000-1500	Below 1500
G1 BC	0.881	0.836	0.811	0.772	0.740
G7 BC	0.417	0.414	0.408	0.410	0.417

Averages from 3000 fps to 1500 fps:
G1 BC: 0.804, variation of G1 BC: 0.113
G7 BC: 0.412, variation of G7 BC: 0.014
G7 form factor: 0.834

Rifling Twist Rate and Stability

Barrel Twist	Stability Factor Best case	Stability Factor Nominal	Stability Factor Worst case
1:9"	1.96	1.71	1.48
1:10"	1.59	1.38	1.20
1:11"	1.31	1.14	0.99
1:12"	1.10	0.96	0.83

Minimum twist required for stability is 1:10.7"
Recommended twist for optimal performance is 1:9.6" or faster

GS Custom 338 caliber 200 grain SP

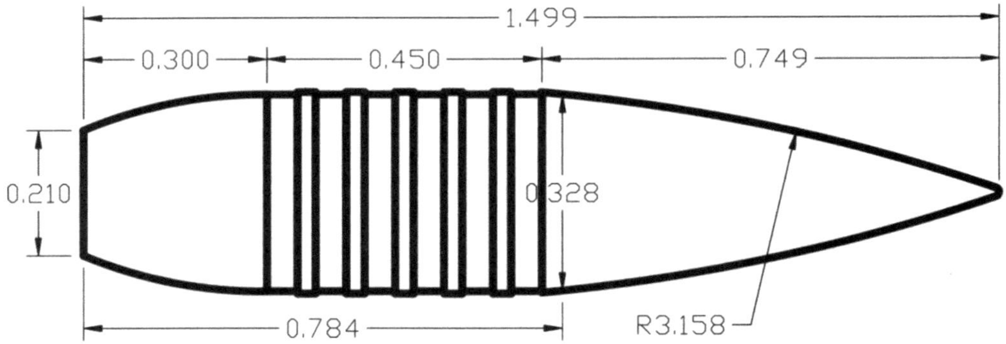

Lot #:	Unknown	Ogive Radius:	3.16 calibers
Sectional Density:	0.250 lb/in^2	Rt/R:	0.57

Ballistic Performance

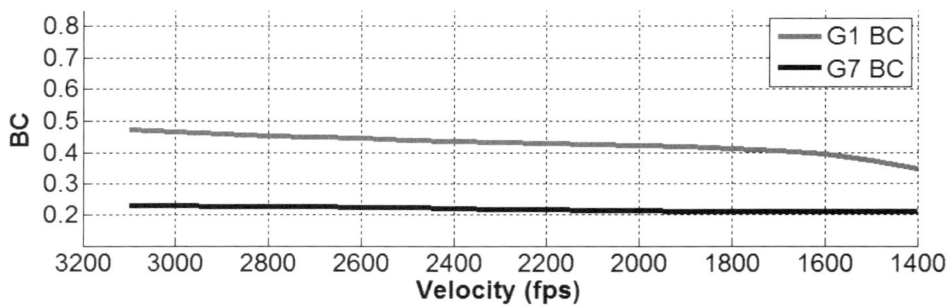

	Above 3000	3000-2500	2500-2000	2000-1500	Below 1500
G1 BC	0.485	0.454	0.432	0.399	0.374
G7 BC	0.230	0.225	0.217	0.211	0.210

Averages from 3000 fps to 1500 fps:
G1 BC: 0.426, variation of G1 BC: 0.093
G7 BC: 0.218, variation of G7 BC: 0.018
G7 form factor: 1.147

Rifling Twist Rate and Stability

Barrel Twist	Stability Factor Best case	Nominal	Worst case
1:9"	2.74	2.39	2.07
1:10"	2.22	1.94	1.67
1:11"	1.83	1.60	1.38
1:12"	1.54	1.34	1.16

Minimum twist required for stability is 1:12.7"
Recommended twist for optimal performance is 1:11.4" or faster

GS Custom 338 caliber 232 grain SP

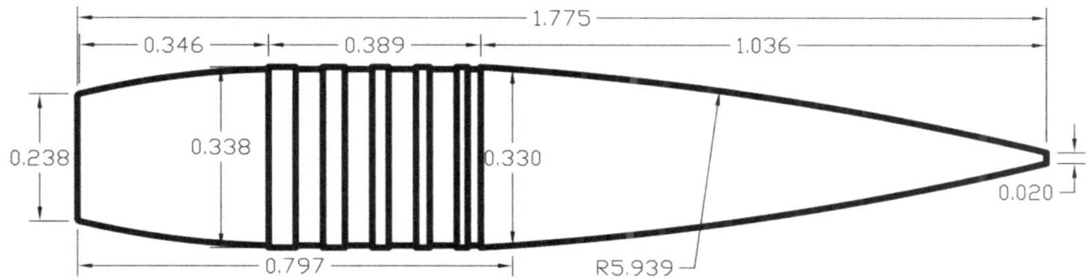

Lot #:	Unknown	Ogive Radius:	17.57 calibers
Sectional Density:	0.290 lb/in^2	Rt/R:	0.60

Ballistic Performance

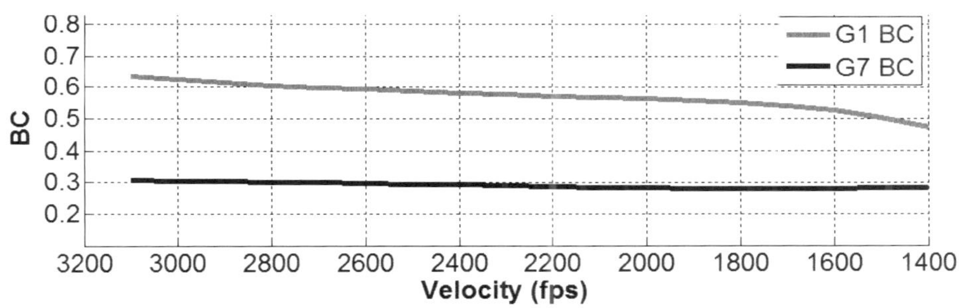

	Above 3000	3000-2500	2500-2000	2000-1500	Below 1500
G1 BC	0.652	0.606	0.575	0.533	0.504
G7 BC	0.309	0.301	0.289	0.283	0.283

Averages from 3000 fps to 1500 fps:
G1 BC: 0.570, variation of G1 BC: 0.122
G7 BC: 0.292, variation of G7 BC: 0.023
G7 form factor: 0.995

Rifling Twist Rate and Stability

Barrel Twist	Stability Factor Best case	Nominal	Worst case
1:9"	1.95	1.71	1.47
1:10"	1.58	1.38	1.19
1:11"	1.31	1.14	0.99
1:12"	1.10	0.96	0.83

Minimum twist required for stability is 1:10.7"
Recommended twist for optimal performance is 1:9.6" or faster

GS Custom 338 caliber 267 grain SP

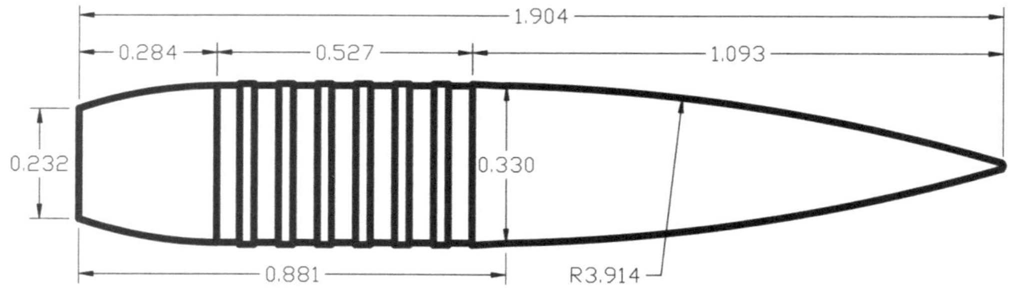

Lot #:	Unknown	Ogive Radius:	11.56 calibers
Sectional Density:	0.334 lb/in²	Rt/R:	0.84

Ballistic Performance

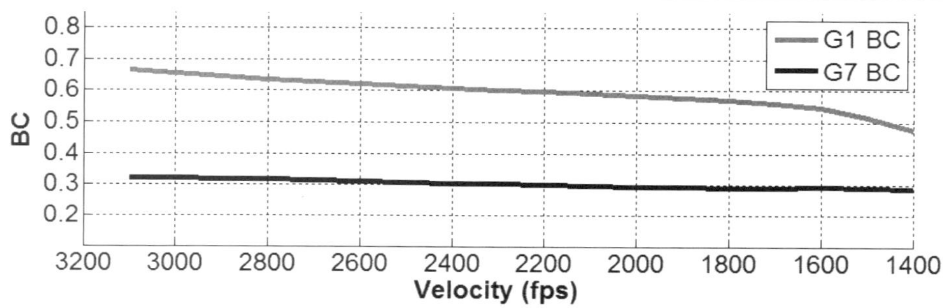

	Above 3000	3000-2500	2500-2000	2000-1500	Below 1500
G1 BC	0.684	0.635	0.600	0.552	0.517
G7 BC	0.324	0.315	0.302	0.292	0.290

Averages from 3000 fps to 1500 fps:
G1 BC: 0.593, variation of G1 BC: 0.138
G7 BC: 0.304, variation of G7 BC: 0.030
G7 form factor: 1.102

Rifling Twist Rate and Stability

Barrel Twist	Stability Factor Best case	Nominal	Worst case
1:8"	2.30	2.01	1.74
1:9"	1.82	1.59	1.37
1:10"	1.47	1.29	1.11
1:11"	1.22	1.06	0.92

Minimum twist required for stability is 1:10.3"
Recommended twist for optimal performance is 1:9.3" or faster

Hornady 338 caliber 285 grain Amax

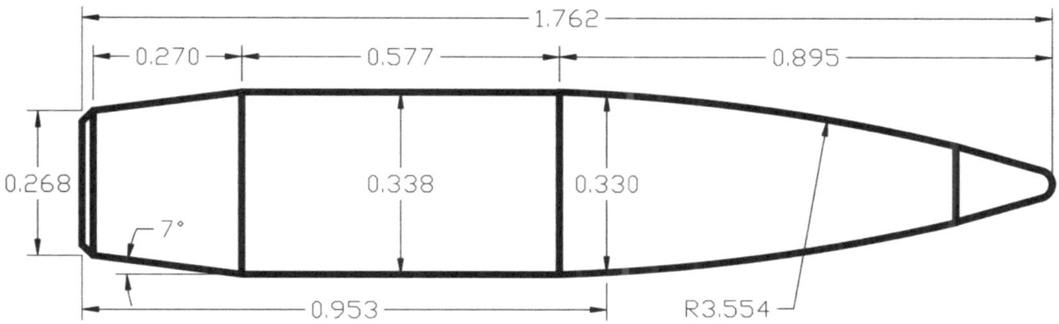

Lot #:	214032	Ogive Radius:	10.52 calibers
Sectional Density:	0.356 lb/in^2	Rt/R:	0.75

Ballistic Performance

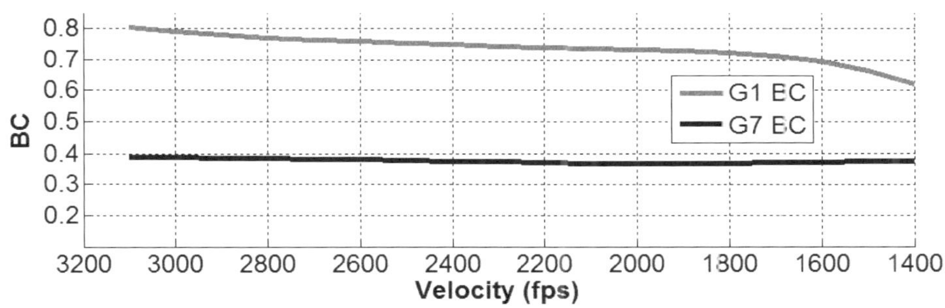

	Above 3000	3000-2500	2500-2000	2000-1500	Below 1500
G1 BC	0.821	0.773	0.744	0.699	0.665
G7 BC	0.389	0.383	0.374	0.371	0.374

Averages from 3000 fps to 1500 fps:

G1 BC: 0.736, variation of G1 BC: 0.128

G7 BC: 0.377, variation of G7 BC: 0.019

G7 form factor: 0.945

Rifling Twist Rate and Stability

Barrel Twist	Stability Factor Best case	Nominal	Worst case
1:9"	2.43	2.13	1.84
1:10"	1.97	1.72	1.49
1:11"	1.63	1.42	1.23
1:12"	1.37	1.20	1.03

Minimum twist required for stability is 1:12.0"

Recommended twist for optimal performance is 1:10.7" or faster

Hornady 338 caliber 250 grain BTHP Match

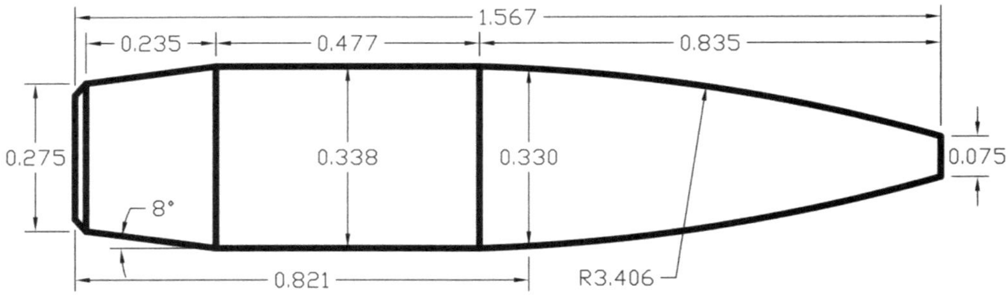

Lot #:	**2070301**	**Ogive Radius:**	**10.08 calibers**
Sectional Density:	**0.313 lb/in²**	**Rt/R:**	**0.82**

Ballistic Performance

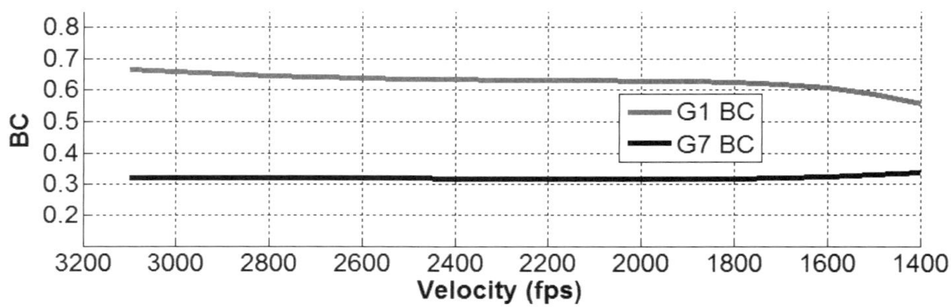

	Above 3000	3000-2500	2500-2000	2000-1500	Below 1500
G1 BC	0.676	0.647	0.633	0.609	0.587
G7 BC	0.320	0.321	0.318	0.324	0.331

Averages from 3000 fps to 1500 fps:
G1 BC: 0.628, variation of G1 BC: 0.071
G7 BC: 0.322, variation of G7 BC: 0.014
G7 form factor: 0.970

Rifling Twist Rate and Stability

Barrel Twist	Stability Factor Best case	Nominal	Worst case
1:9"	3.01	2.63	2.27
1:10"	2.44	2.13	1.84
1:11"	2.01	1.76	1.52
1:12"	1.69	1.48	1.28

Minimum twist required for stability is 1:13.3"
Recommended twist for optimal performance is 1:11.9" or faster

Hornady 338 caliber 285 grain BTHP Match

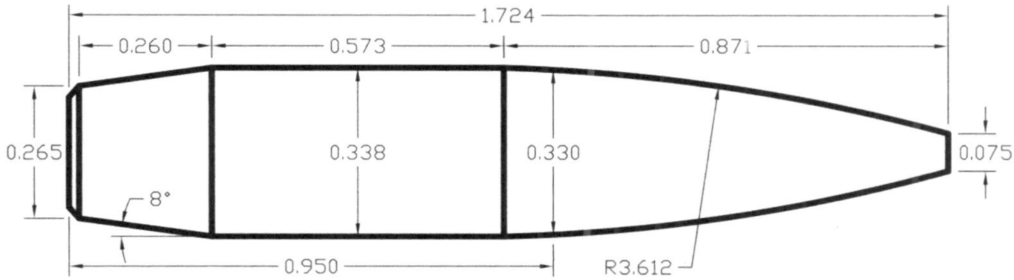

Lot #:	2100445	Ogive Radius:	10.69 calibers
Sectional Density:	0.356 lb/in^2	Rt/R:	0.82

Ballistic Performance

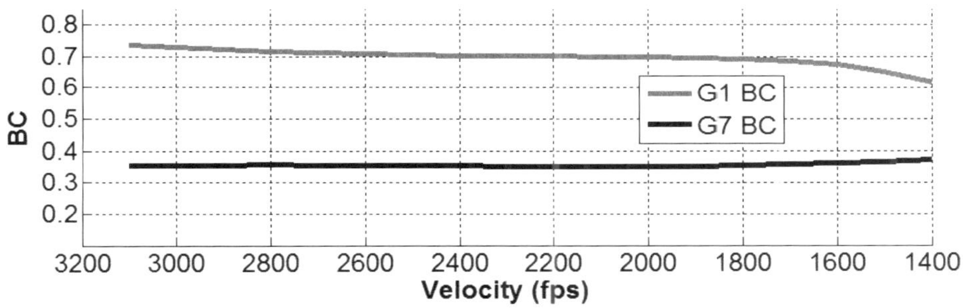

	Above 3000	3000-2500	2500-2000	2000-1500	Below 1500
G1 BC	0.748	0.717	0.703	0.675	0.651
G7 BC	0.355	0.356	0.353	0.359	0.367

Averages from 3000 fps to 1500 fps:
G1 BC: 0.696, variation of G1 BC: 0.077
G7 BC: 0.357, variation of G7 BC: 0.016
G7 form factor: 0.997

Rifling Twist Rate and Stability

Barrel Twist	Stability Factor Best case	Nominal	Worst case
1:9"	2.59	2.27	1.96
1:10"	2.10	1.84	1.59
1:11"	1.74	1.52	1.31
1:12"	1.46	1.27	1.10

Minimum twist required for stability is 1:12.4"
Recommended twist for optimal performance is 1:11.1" or faster

Hornady 338 caliber 200 grain FXT

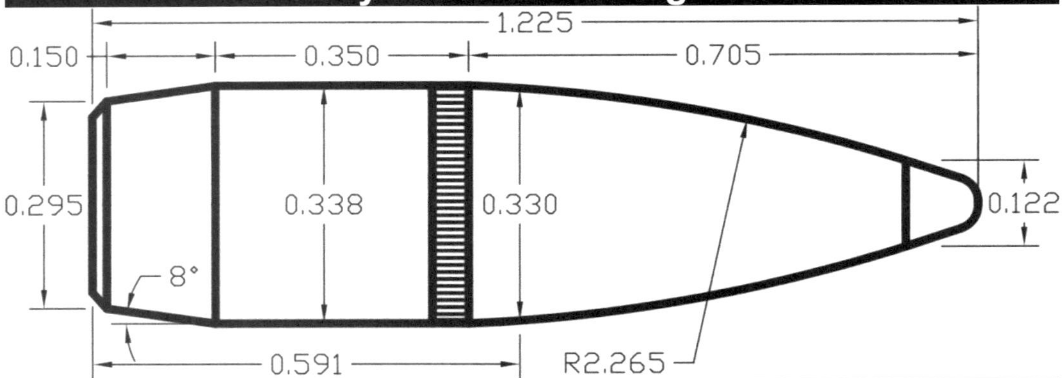

Lot #:	Unknown	Ogive Radius:	6.70 calibers
Sectional Density:	0.250 lb/in^2	Rt/R:	0.77

Ballistic Performance

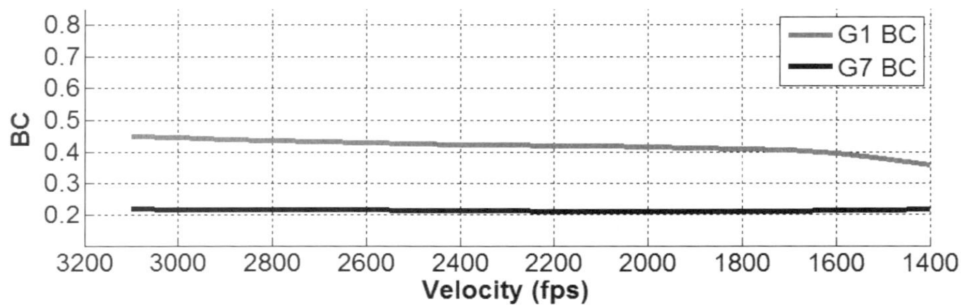

	Above 3000	3000-2500	2500-2000	2000-1500	Below 1500
G1 BC	0.460	0.436	0.421	0.398	0.380
G7 BC	0.218	0.216	0.212	0.211	0.214

Averages from 3000 fps to 1500 fps:
G1 BC: 0.417, variation of G1 BC: 0.066
G7 BC: 0.214, variation of G7 BC: 0.009
G7 form factor: 1.170

Rifling Twist Rate and Stability

Barrel Twist	Stability Factor Best case	Stability Factor Nominal	Stability Factor Worst case
1:9"	4.90	4.28	3.70
1:10"	3.97	3.47	3.00
1:11"	3.28	2.86	2.48
1:12"	2.75	2.41	2.08

Minimum twist required for stability is 1:17.0"
Recommended twist for optimal performance is 1:15.2" or faster

Hornady 338 caliber 185 grain GMX

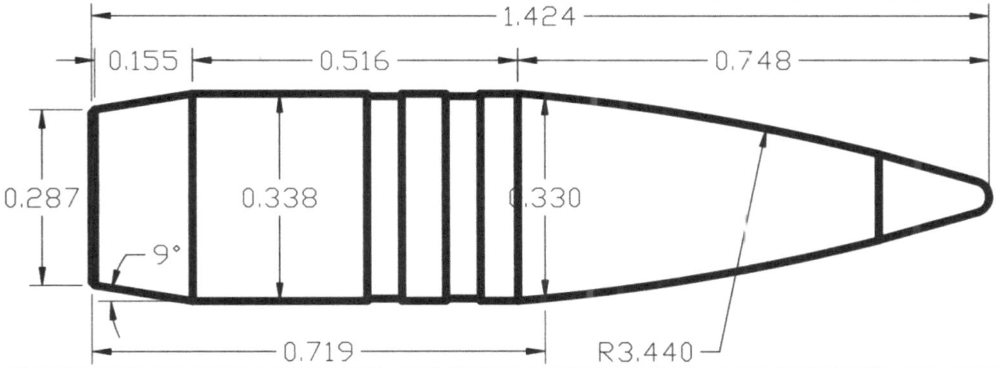

Lot #:	2121341	Ogive Radius:	3.44 calibers
Sectional Density:	0.231 lb/in^2	Rt/R:	0.50

Ballistic Performance

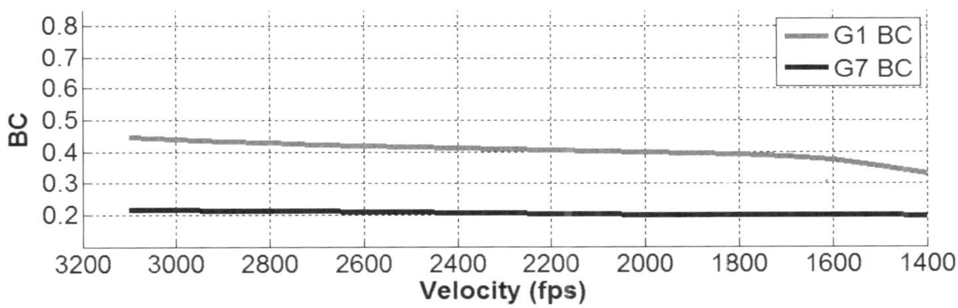

	Above 3000	3000-2500	2500-2000	2000-1500	Below 1500
G1 BC	0.458	0.428	0.408	0.378	0.356
G7 BC	0.217	0.212	0.205	0.201	0.200

Averages from 3000 fps to 1500 fps:
G1 BC: 0.403, variation of G1 BC: 0.084
G7 BC: 0.206, variation of G7 BC: 0.015
G7 form factor: 1.122

Rifling Twist Rate and Stability

Barrel Twist	Stability Factor Best case	Nominal	Worst case
1:9"	2.94	2.57	2.22
1:10"	2.38	2.08	1.80
1:11"	1.97	1.72	1.49
1:12"	1.65	1.44	1.25

Minimum twist required for stability is 1:13.2"
Recommended twist for optimal performance is 1:11.8" or faster

Hornady 338 caliber 200 grain SST

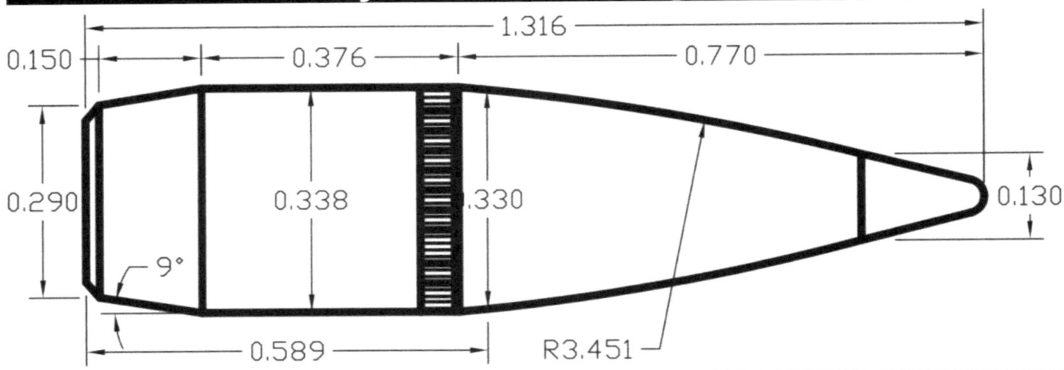

Lot #:	**Unknown**	**Ogive Radius:**	**10.21 calibers**
Sectional Density:	**0.250 lb/in^2**	**Rt/R:**	**0.48**

Ballistic Performance

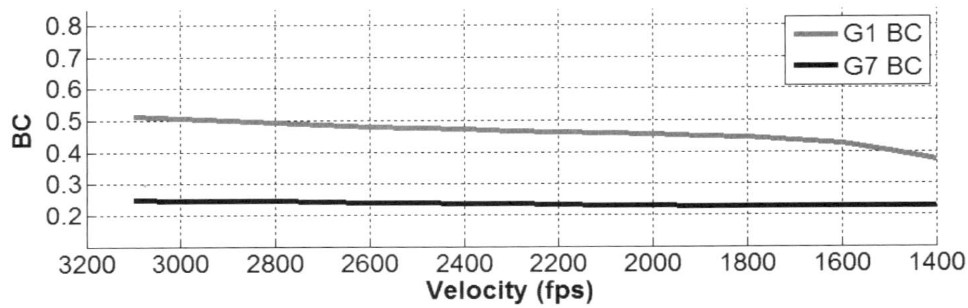

	Above 3000	3000-2500	2500-2000	2000-1500	Below 1500
G1 BC	0.529	0.492	0.466	0.430	0.404
G7 BC	0.250	0.244	0.234	0.228	0.227

Averages from 3000 fps to 1500 fps:
G1 BC: 0.461, variation of G1 BC: 0.104
G7 BC: 0.236, variation of G7 BC: 0.021
G7 form factor: 1.062

Rifling Twist Rate and Stability

	Stability Factor		
Barrel Twist	**Best case**	**Nominal**	**Worst case**
1:9"	3.54	3.10	2.68
1:10"	2.87	2.51	2.17
1:11"	2.37	2.07	1.79
1:12"	1.99	1.74	1.51

Minimum twist required for stability is 1:15.3"
Recommended twist for optimal performance is 1:13.7" or faster

Hornady 338 caliber 225 grain SST

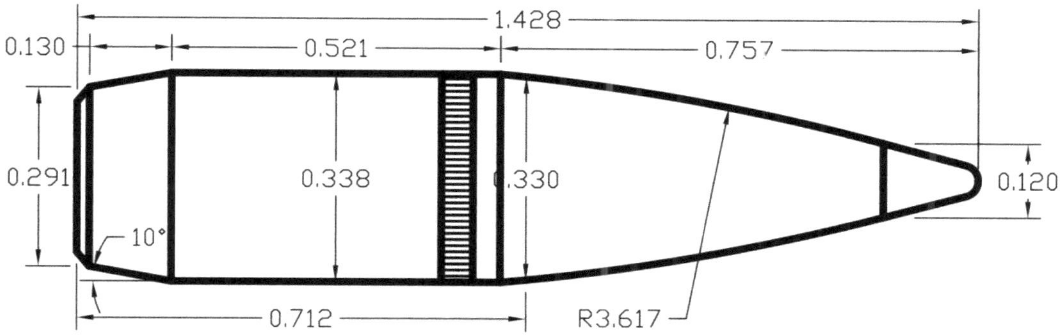

Lot #:	2070151	Ogive Radius:	10.7 calibers
Sectional Density:	0.281 lb/in^2	Rt/R:	0.55

Ballistic Performance

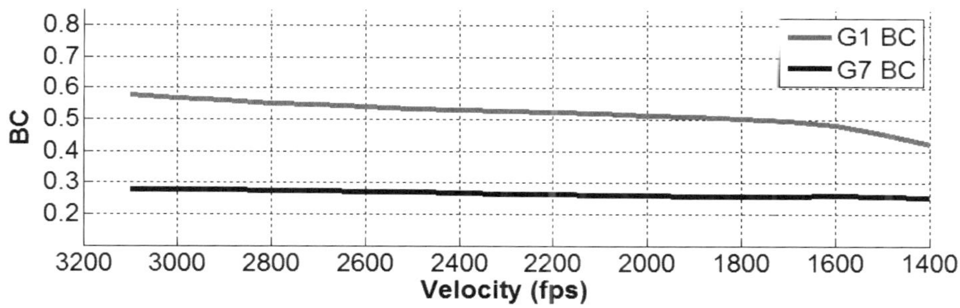

	Above 3000	3000-2500	2500-2000	2000-1500	Below 1500
G1 BC	0.591	0.552	0.525	0.487	0.458
G7 BC	0.280	0.274	0.264	0.258	0.257

Averages from 3000 fps to 1500 fps:
G1 BC: 0.519, variation of G1 BC: 0.110
G7 BC: 0.266, variation of G7 BC: 0.021
G7 form factor: 1.060

Rifling Twist Rate and Stability

Barrel Twist	Stability Factor Best case	Nominal	Worst case
1:9"	3.54	3.10	2.68
1:10"	2.87	2.51	2.17
1:11"	2.37	2.07	1.79
1:12"	1.99	1.74	1.51

Minimum twist required for stability is 1:14.5"
Recommended twist for optimal performance is 1:12.9" or faster

Lapua 338 caliber 250 grain FMJBT

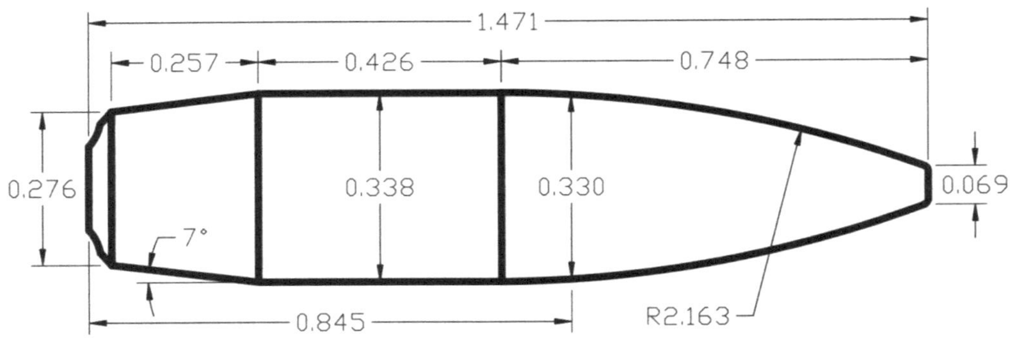

Lot #:	P00612501/3134512	Ogive Radius:	6.40 calibers
Sectional Density:	0.313 lb/in^2	Rt/R:	0.99

Ballistic Performance

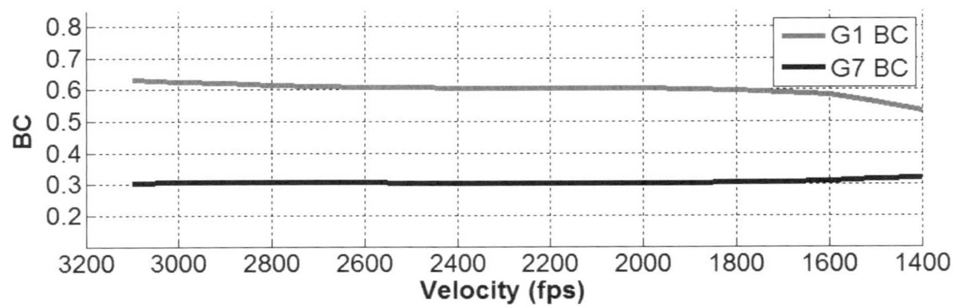

	Above 3000	3000-2500	2500-2000	2000-1500	Below 1500
G1 BC	0.642	0.617	0.606	0.583	0.563
G7 BC	0.304	0.306	0.304	0.310	0.317

Averages from 3000 fps to 1500 fps:
G1 BC: 0.600, variation of G1 BC: 0.064
G7 BC: 0.308, variation of G7 BC: 0.014
G7 form factor: 1.015

Rifling Twist Rate and Stability

Barrel Twist	Stability Factor Best case	Nominal	Worst case
1:9"	3.61	3.16	2.73
1:10"	2.93	2.56	2.21
1:11"	2.42	2.11	1.83
1:12"	2.03	1.78	1.54

Minimum twist required for stability is 1:14.6"
Recommended twist for optimal performance is 1:13.1" or faster

Lapua 338 caliber 250 grain Scenar

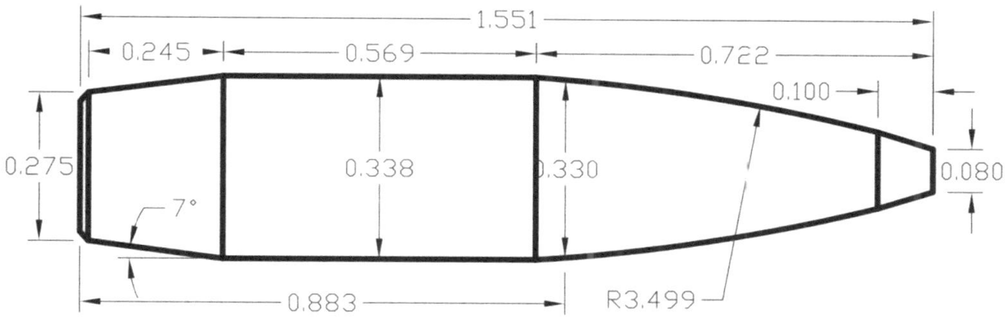

Lot #:	P00437502	Ogive Radius:	10.35 calibers
Sectional Density:	0.313 lb/in²	Rt/R:	0.65

Ballistic Performance

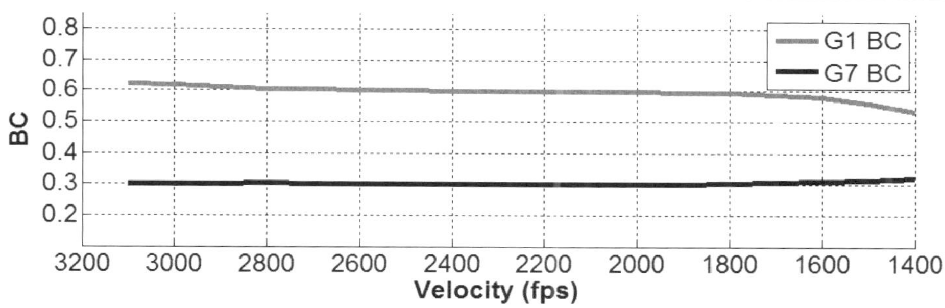

	Above 3000	3000-2500	2500-2000	2000-1500	Below 1500
G1 BC	0.633	0.609	0.599	0.579	0.561
G7 BC	0.300	0.302	0.301	0.308	0.316

Averages from 3000 fps to 1500 fps:
G1 BC: 0.594, variation of G1 BC: 0.057
G7 BC: 0.305, variation of G7 BC: 0.016
G7 form factor: 1.025

Rifling Twist Rate and Stability

Barrel Twist	Stability Factor Best case	Nominal	Worst case
1:9"	3.10	2.71	2.34
1:10"	2.51	2.19	1.90
1:11"	2.07	1.81	1.57
1:12"	1.74	1.52	1.32

Minimum twist required for stability is 1:13.5"
Recommended twist for optimal performance is 1:12.1" or faster

Lapua 338 caliber 300 grain Scenar

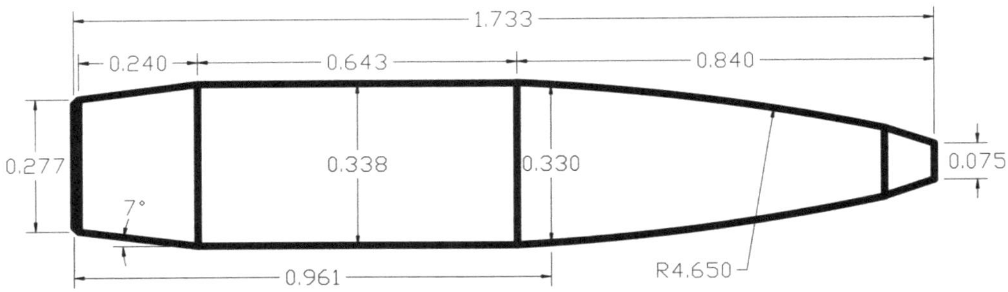

Lot #:	P00468901	Ogive Radius:	13.76 calibers
Sectional Density:	0.375 lb/in^2	Rt/R:	0.71

Ballistic Performance

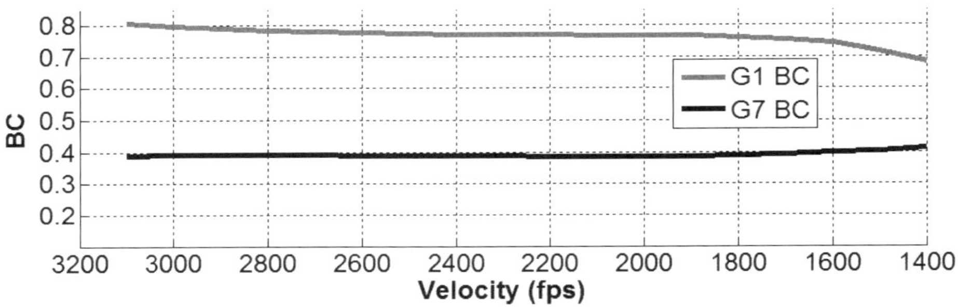

	Above 3000	3000-2500	2500-2000	2000-1500	Below 1500
G1 BC	0.821	0.786	0.771	0.742	0.717
G7 BC	0.389	0.390	0.387	0.395	0.404

Averages from 3000 fps to 1500 fps:
G1 BC: 0.764, variation of G1 BC: 0.082
G7 BC: 0.392, variation of G7 BC: 0.018
G7 form factor: 0.956

Rifling Twist Rate and Stability

Barrel Twist	Stability Factor Best case	Nominal	Worst case
1:9"	2.69	2.35	2.03
1:10"	2.18	1.90	1.65
1:11"	1.80	1.57	1.36
1:12"	1.51	1.32	1.14

Minimum twist required for stability is 1:12.6"
Recommended twist for optimal performance is 1:11.3" or faster

Lehigh 338 caliber 230 grain Match Solid

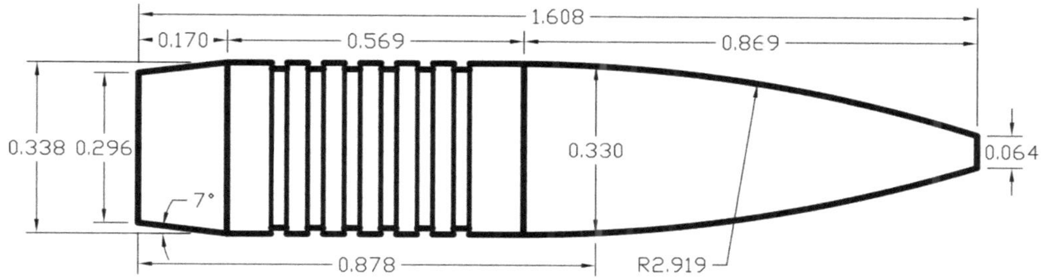

Lot #:	Unknown	Ogive Radius:	8.64 calibers
Sectional Density:	0.288 lb/in^2	Rt/R:	0.97

Ballistic Performance

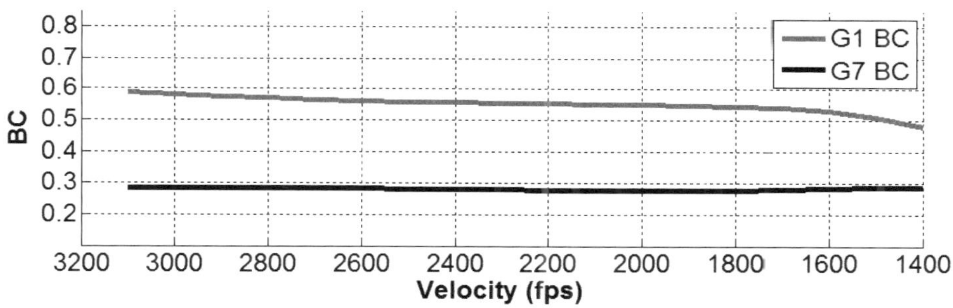

	Above 3000	3000-2500	2500-2000	2000-1500	Below 1500
G1 BC	0.601	0.571	0.555	0.530	0.510
G7 BC	0.285	0.283	0.279	0.282	0.287

Averages from 3000 fps to 1500 fps:
G1 BC: 0.551, variation of G1 BC: 0.073
G7 BC: 0.282, variation of G7 BC: 0.010
G7 form factor: 1.018

Rifling Twist Rate and Stability

Barrel Twist	Stability Factor Best case	Nominal	Worst case
1:9"	2.57	2.24	1.94
1:10"	2.08	1.82	1.57
1:11"	1.72	1.50	1.30
1:12"	1.44	1.26	1.09

Minimum twist required for stability is 1:12.3"
Recommended twist for optimal performance is 1:11.0" or faster

Lehigh 338 caliber 245 grain Match Solid

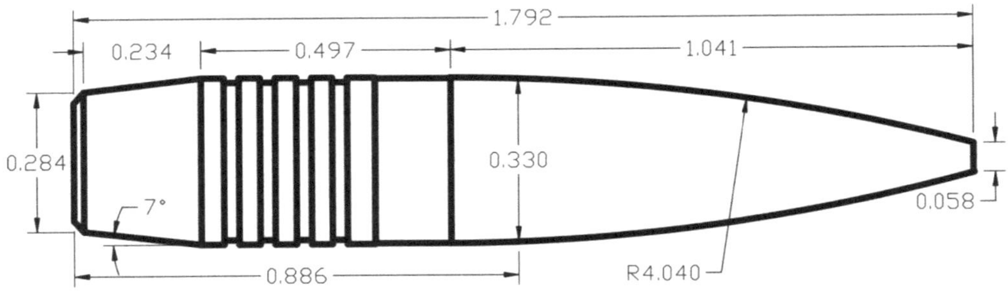

Lot #:	Unknown	Ogive Radius:	11.95 calibers
Sectional Density:	0.306 lb/in^2	Rt/R:	0.98

Ballistic Performance

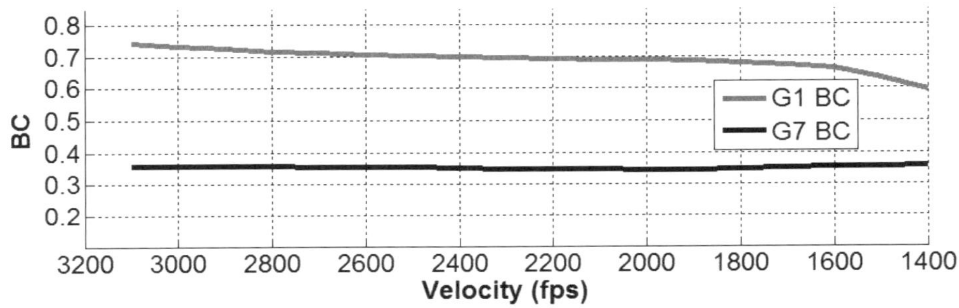

	Above 3000	3000-2500	2500-2000	2000-1500	Below 1500
G1 BC	0.760	0.719	0.696	0.661	0.633
G7 BC	0.360	0.356	0.350	0.351	0.356

Averages from 3000 fps to 1500 fps:
G1 BC: 0.690, variation of G1 BC: 0.101
G7 BC: 0.354, variation of G7 BC: 0.013
G7 form factor: 0.866

Rifling Twist Rate and Stability

Barrel Twist	Stability Factor Best case	Nominal	Worst case
1:9"	1.99	1.74	1.50
1:10"	1.61	1.41	1.22
1:11"	1.33	1.16	1.01
1:12"	1.12	0.98	0.85

Minimum twist required for stability is 1:10.8"
Recommended twist for optimal performance is 1:9.7" or faster

Nosler 338 caliber 180 grain Accubond

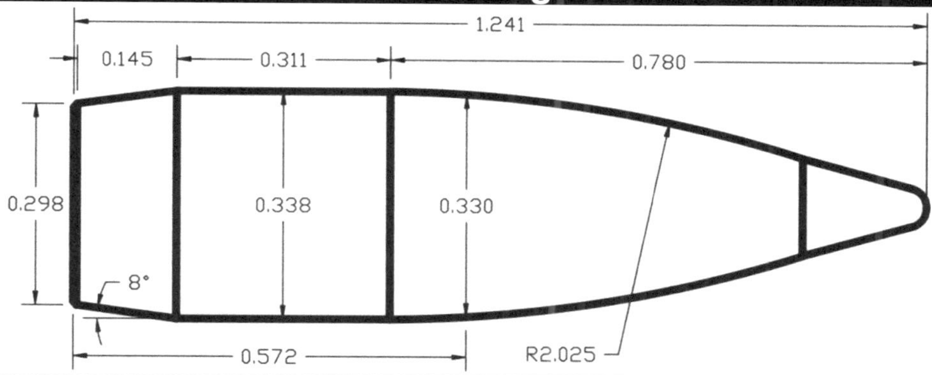

Lot #:	Unknown	Ogive Radius:	5.99 calibers
Sectional Density:	0.225 lb/in^2	Rt/R:	0.95

Ballistic Performance

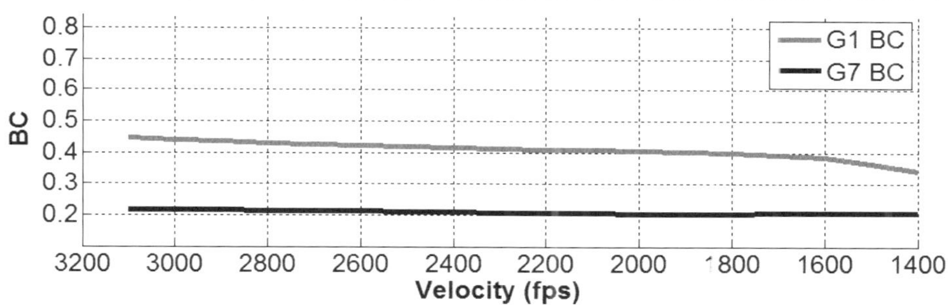

	Above 3000	3000-2500	2500-2000	2000-1500	Below 1500
G1 BC	0.457	0.430	0.412	0.386	0.367
G7 BC	0.217	0.213	0.207	0.205	0.206

Averages from 3000 fps to 1500 fps:
G1 BC: 0.408, variation of G1 BC: 0.074
G7 BC: 0.209, variation of G7 BC: 0.012
G7 form factor: 1.077

Rifling Twist Rate and Stability

Barrel Twist	Stability Factor Best case	Nominal	Worst case
1:9"	4.25	3.71	3.21
1:10"	3.44	3.00	2.60
1:11"	2.84	2.48	2.15
1:12"	2.39	2.09	1.80

Minimum twist required for stability is 1:15.8"
Recommended twist for optimal performance is 1:14.2" or faster

Nosler 338 caliber 200 grain Accubond

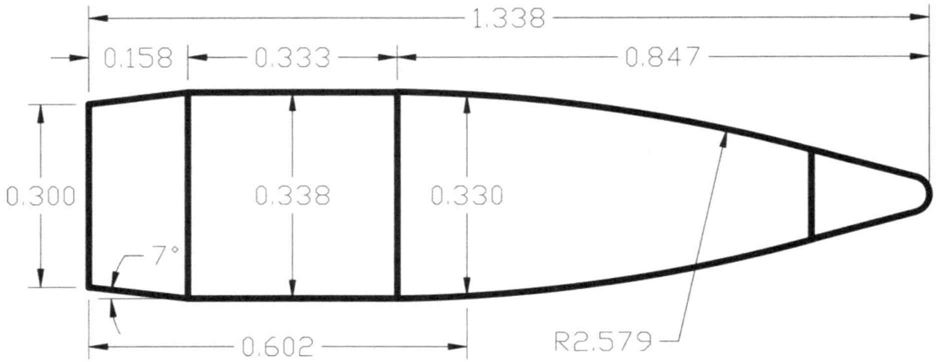

Lot #:	EM06F10	Ogive Radius:	7.63 calibers
Sectional Density:	0.250 lb/in^2	Rt/R:	0.98

Ballistic Performance

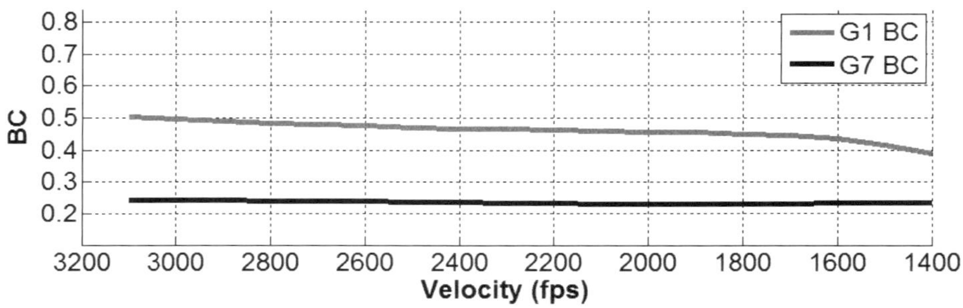

	Above 3000	3000-2500	2500-2000	2000-1500	Below 1500
G1 BC	0.516	0.484	0.465	0.437	0.416
G7 BC	0.244	0.240	0.234	0.232	0.234

Averages from 3000 fps to 1500 fps:
G1 BC: 0.461, variation of G1 BC: 0.082
G7 BC: 0.236, variation of G7 BC: 0.013
G7 form factor: 1.060

Rifling Twist Rate and Stability

Barrel Twist	Stability Factor Best case	Nominal	Worst case
1:12"	2.14	1.87	1.62
1:13"	1.82	1.59	1.38
1:14"	1.57	1.37	1.19
1:15"	1.37	1.20	1.03

Minimum twist required for stability is 1:15.0"
Recommended twist for optimal performance is 1:13.4" or faster

Nosler 338 caliber 225 grain Accubond

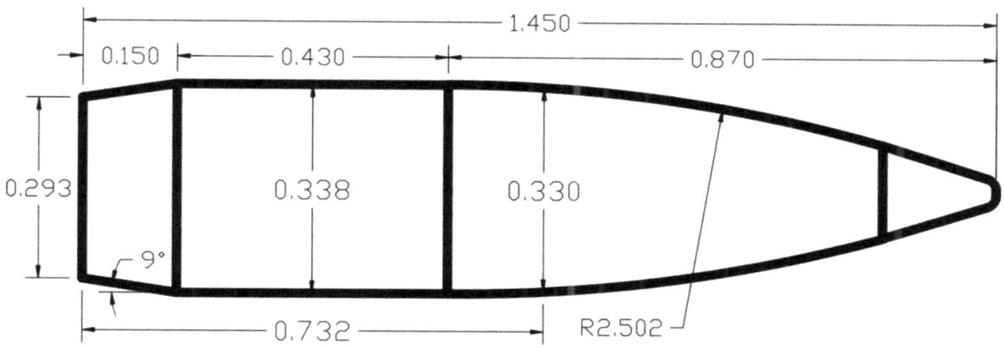

Lot #:	GH15F10	Ogive Radius:	7.40 calibers
Sectional Density:	0.281 lb/in²	Rt/R:	0.80

Ballistic Performance

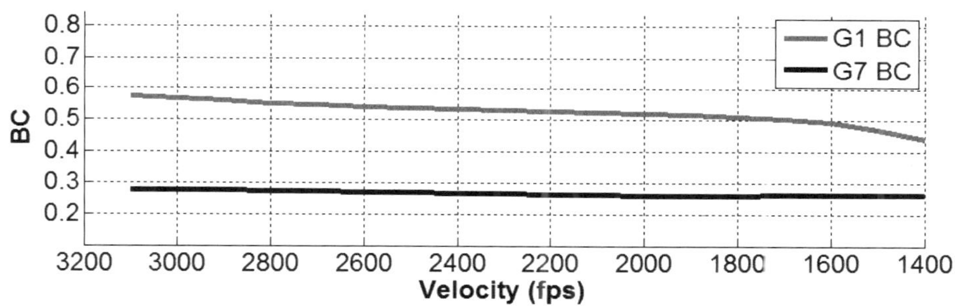

	Above 3000	3000-2500	2500-2000	2000-1500	Below 1500
G1 BC	0.589	0.552	0.529	0.495	0.470
G7 BC	0.279	0.274	0.266	0.263	0.264

Averages from 3000 fps to 1500 fps:
G1 BC: 0.524, variation of G1 BC: 0.098
G7 BC: 0.268, variation of G7 BC: 0.016
G7 form factor: 1.049

Rifling Twist Rate and Stability

Barrel Twist	Stability Factor Best case	Nominal	Worst case
1:9"	3.39	2.96	2.56
1:10"	2.75	2.40	2.07
1:11"	2.27	1.98	1.71
1:12"	1.91	1.67	1.44

Minimum twist required for stability is 1:14.1"
Recommended twist for optimal performance is 1:12.6" or faster

Nosler 338 caliber 250 grain Accubond

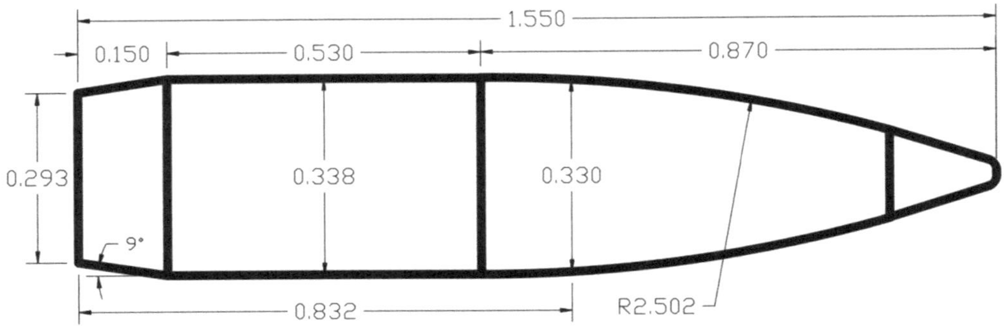

Lot #:	BI10F10 2	Ogive Radius:	7.40 calibers
Sectional Density:	0.313 lb/in^2	Rt/R:	0.80

Ballistic Performance

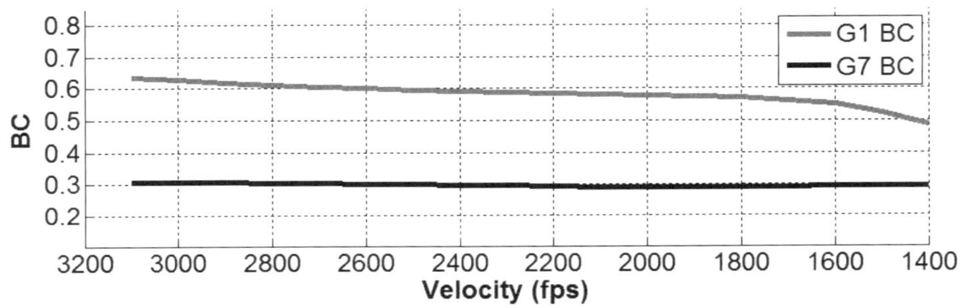

	Above 3000	3000-2500	2500-2000	2000-1500	Below 1500
G1 BC	0.652	0.612	0.588	0.551	0.524
G7 BC	0.309	0.304	0.296	0.293	0.294

Averages from 3000 fps to 1500 fps:
G1 BC: 0.582, variation of G1 BC: 0.105
G7 BC: 0.298, variation of G7 BC: 0.016
G7 form factor: 1.049

Rifling Twist Rate and Stability

Barrel Twist	Stability Factor Best case	Nominal	Worst case
1:9"	3.10	2.71	2.34
1:10"	2.51	2.20	1.90
1:11"	2.08	1.82	1.57
1:12"	1.75	1.53	1.32

Minimum twist required for stability is 1:13.5"
Recommended twist for optimal performance is 1:12.1" or faster

Nosler 338 caliber 300 grain Custom Competition

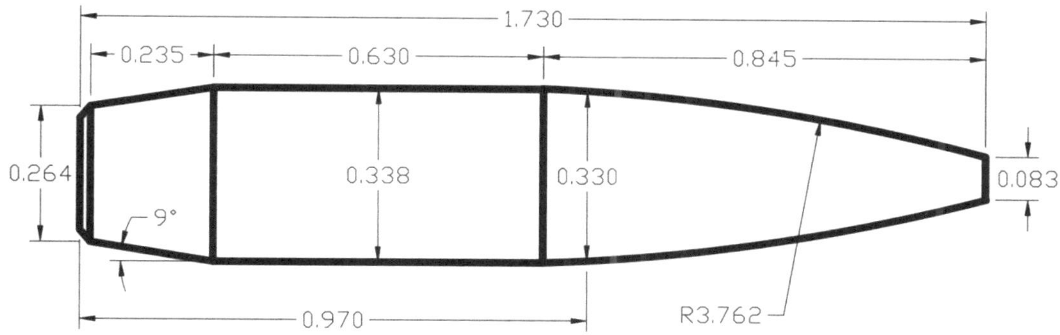

Lot #:	**EM20F20**	**Ogive Radius:**	**11.13 calibers**
Sectional Density:	**0.375 lb/in²**	**Rt/R:**	**0.76**

Ballistic Performance

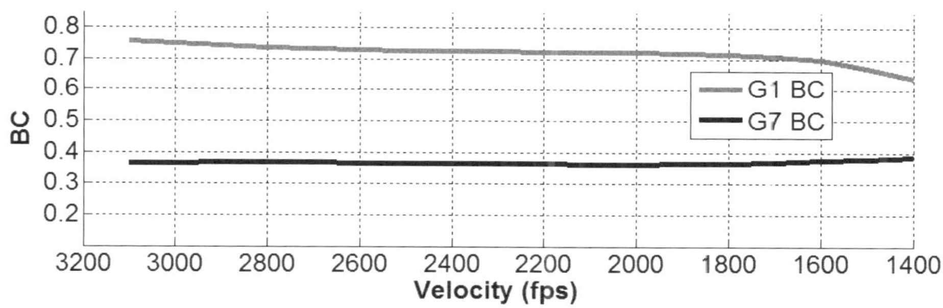

	Above 3000	3000-2500	2500-2000	2000-1500	Below 1500
G1 BC	0.768	0.739	0.725	0.698	0.674
G7 BC	0.364	0.366	0.365	0.371	0.380

Averages from 3000 fps to 1500 fps:
G1 BC: 0.718, variation of G1 BC: 0.075
G7 BC: 0.369, variation of G7 BC: 0.017
G7 form factor: 1.017

Rifling Twist Rate and Stability

Barrel Twist	Stability Factor: Best case	Stability Factor: Nominal	Stability Factor: Worst case
1:10"	2.19	1.91	1.65
1:11"	1.81	1.58	1.37
1:12"	1.52	1.33	1.15
1:13"	1.30	1.13	0.98

Minimum twist required for stability is 1:12.6"
Recommended twist for optimal performance is 1:11.3" or faster

Nosler 338 caliber 200 grain E-Tip

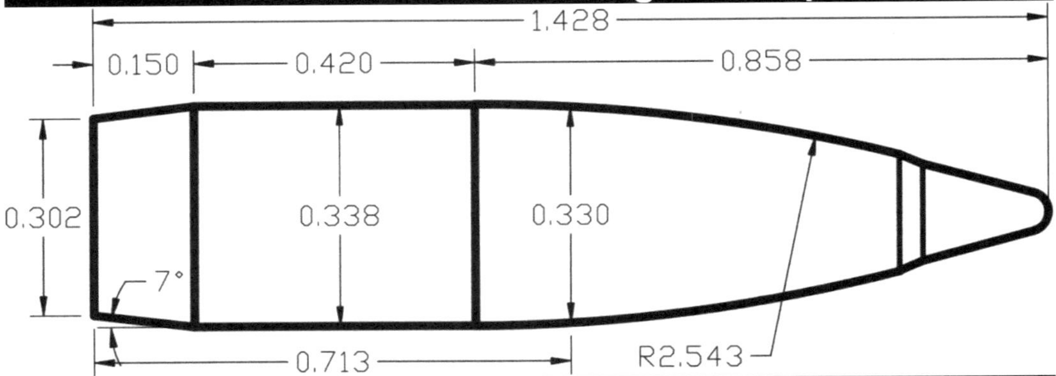

Lot #:	CG31F14	**Ogive Radius:**	**7.52 calibers**
Sectional Density:	**0.250 lb/in²**	**Rt/R:**	**1.00**

Ballistic Performance

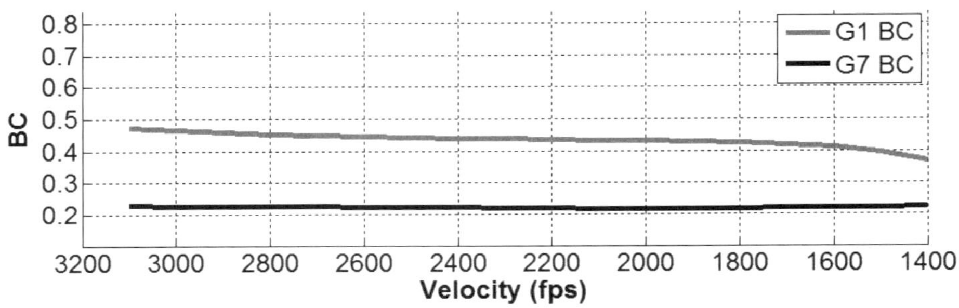

	Above 3000	3000-2500	2500-2000	2000-1500	Below 1500
G1 BC	0.483	0.454	0.438	0.413	0.394
G7 BC	0.229	0.225	0.220	0.219	0.222

Averages from 3000 fps to 1500 fps:

G1 BC: 0.434, variation of G1 BC: 0.072

G7 BC: 0.222, variation of G7 BC: 0.011

G7 form factor: 1.125

Rifling Twist Rate and Stability

	Stability Factor		
Barrel Twist	**Best case**	**Nominal**	**Worst case**
1:9"	3.15	2.75	2.38
1:10"	2.55	2.23	1.93
1:11"	2.11	1.84	1.59
1:12"	1.77	1.55	1.34

Minimum twist required for stability is 1:13.6"

Recommended twist for optimal performance is 1:12.2" or faster

Rocky Mountain 338 caliber 325 grain RBT

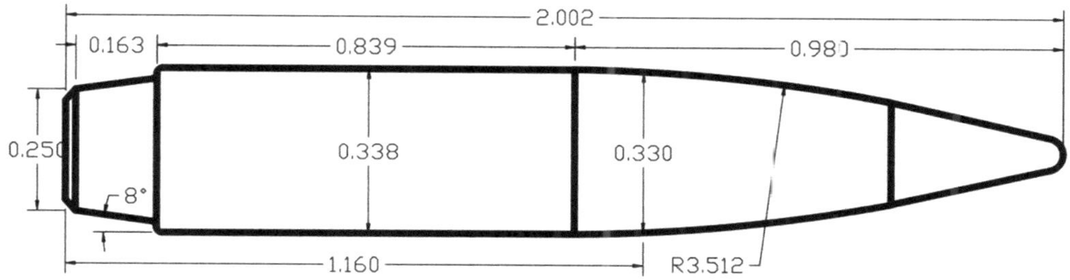

Lot #:	Unknown	Ogive Radius:	10.39 calibers
Sectional Density:	0.406 lb/in^2	Rt/R:	0.91

Ballistic Performance

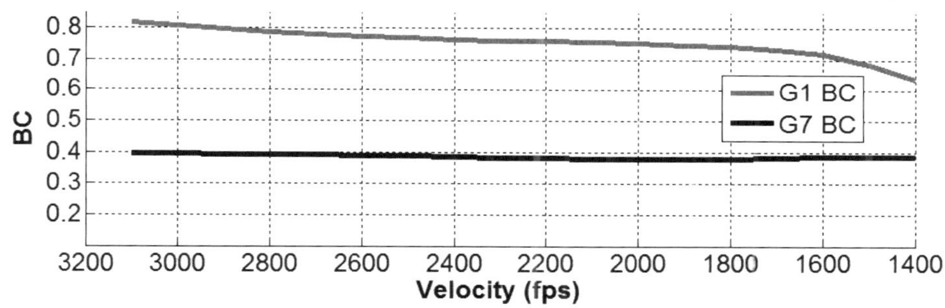

	Above 3000	3000-2500	2500-2000	2000-1500	Below 1500
G1 BC	0.835	0.788	0.761	0.719	0.686
G7 BC	0.396	0.391	0.383	0.382	0.386

Averages from 3000 fps to 1500 fps:
G1 BC: 0.754, variation of G1 BC: 0.121
G7 BC: 0.386, variation of G7 BC: 0.017
G7 form factor: 1.052

Rifling Twist Rate and Stability

Barrel Twist	Stability Factor Best case	Nominal	Worst case
1:9"	1.91	1.67	1.44
1:10"	1.54	1.35	1.17
1:11"	1.28	1.12	0.96
1:12"	1.07	0.94	0.81

Minimum twist required for stability is 1:10.6"
Recommended twist for optimal performance is 1:9.5" or faster

RUAG 338 caliber 300 grain Swiss P Target

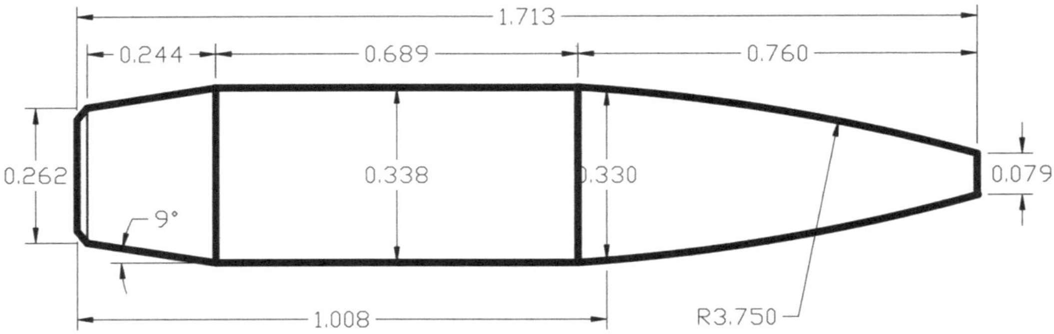

Lot #:	Unknown	Ogive Radius:	11.10 calibers
Sectional Density:	0.375 lb/in^2	Rt/R:	0.61

Ballistic Performance

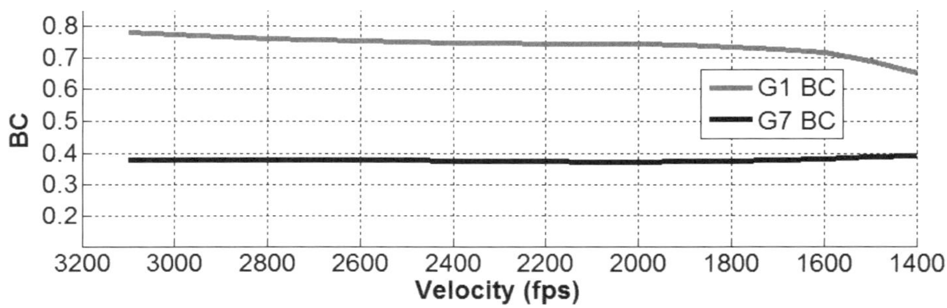

	Above 3000	3000-2500	2500-2000	2000-1500	Below 1500
G1 BC	0.793	0.761	0.746	0.715	0.688
G7 BC	0.376	0.378	0.375	0.380	0.387

Averages from 3000 fps to 1500 fps:
G1 BC: 0.738, variation of G1 BC: 0.085
G7 BC: 0.379, variation of G7 BC: 0.015
G7 form factor: 0.990

Rifling Twist Rate and Stability

Barrel Twist	Stability Factor Best case	Nominal	Worst case
1:10"	2.25	1.97	1.70
1:11"	1.86	1.63	1.41
1:12"	1.57	1.37	1.18
1:13"	1.33	1.17	1.01

Minimum twist required for stability is 1:12.8"
Recommended twist for optimal performance is 1:11.5" or faster

Sierra 338 caliber 250 grain SP GameKing

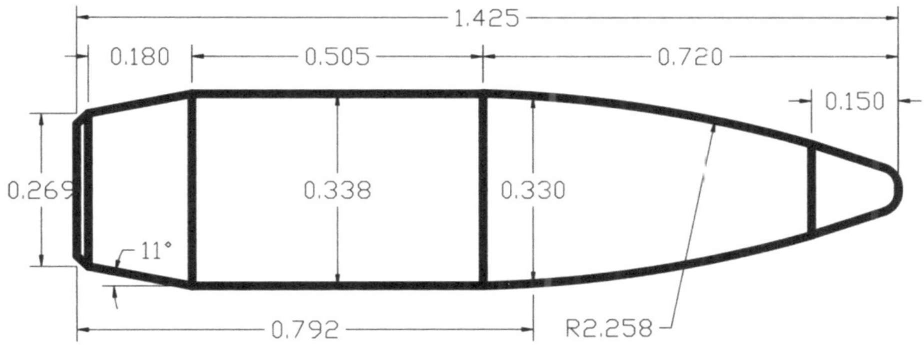

Lot #:	0034574375858	Ogive Radius:	6.68 calibers
Sectional Density:	0.313 lb/in^2	Rt/R:	0.90

Ballistic Performance

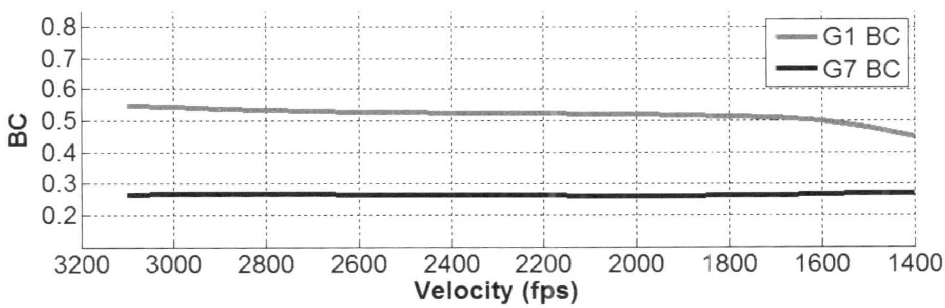

	Above 3000	3000-2500	2500-2000	2000-1500	Below 1500
G1 BC	0.557	0.535	0.524	0.500	0.480
G7 BC	0.264	0.265	0.263	0.266	0.270

Averages from 3000 fps to 1500 fps:

G1 BC: 0.518, variation of G1 BC: 0.063

G7 BC: 0.266, variation of G7 BC: 0.009

G7 form factor: 1.177

Rifling Twist Rate and Stability

Barrel Twist	Stability Factor Best case	Nominal	Worst case
1:9"	3.96	3.46	2.99
1:10"	3.21	2.80	2.42
1:11"	2.65	2.32	2.00
1:12"	2.23	1.95	1.68

Minimum twist required for stability is 1:15.3"

Recommended twist for optimal performance is 1:13.7" or faster

Sierra 338 caliber 250 grain MatchKing

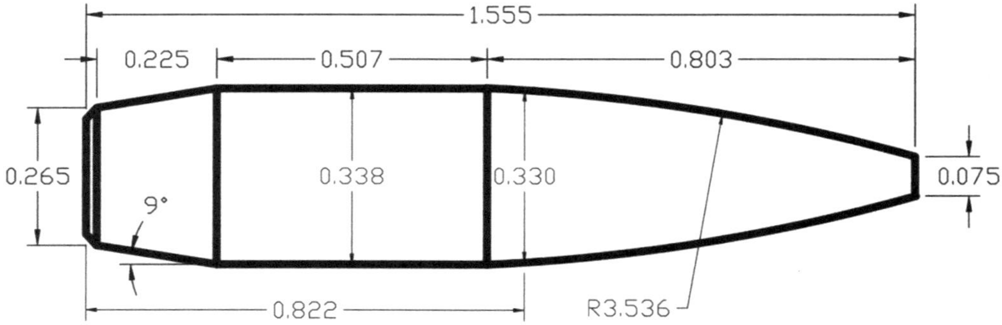

Lot #:	0033461351313	Ogive Radius:	10.46 calibers
Sectional Density:	0.313 lb/in^2	Rt/R:	0.75

Ballistic Performance

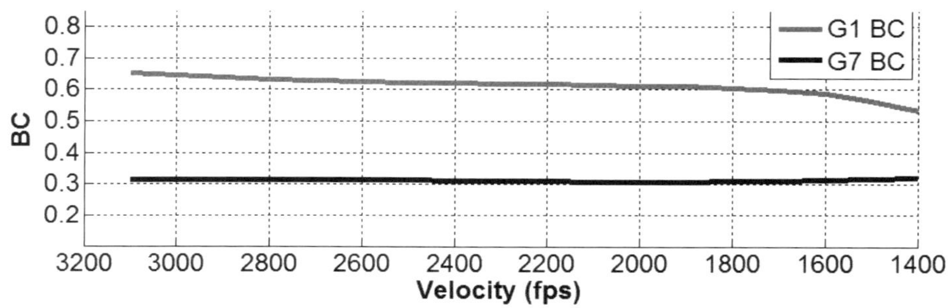

	Above 3000	3000-2500	2500-2000	2000-1500	Below 1500
G1 BC	0.665	0.635	0.618	0.589	0.565
G7 BC	0.315	0.315	0.311	0.313	0.318

Averages from 3000 fps to 1500 fps:
G1 BC: 0.612, variation of G1 BC: 0.081
G7 BC: 0.314, variation of G7 BC: 0.010
G7 form factor: 0.996

Rifling Twist Rate and Stability

Barrel Twist	Stability Factor Best case	Nominal	Worst case
1:9"	3.07	2.69	2.32
1:10"	2.49	2.18	1.88
1:11"	2.06	1.80	1.56
1:12"	1.73	1.51	1.31

Minimum twist required for stability is 1:13.5"
Recommended twist for optimal performance is 1:12.0" or faster

Sierra 338 caliber 300 grain MatchKing

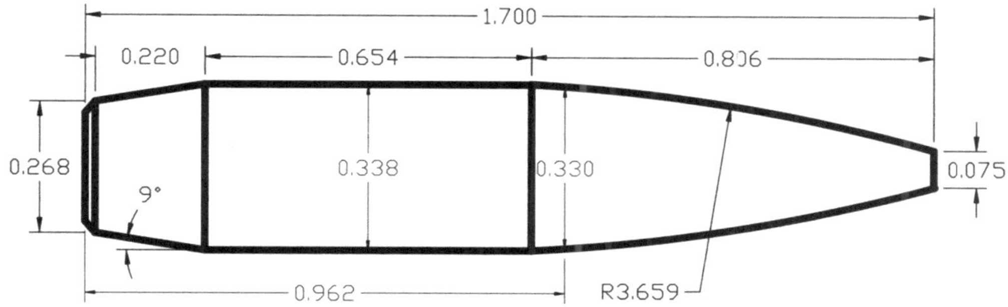

Lot #:	0042168123162	Ogive Radius:	10.82 calibers
Sectional Density:	0.375 lb/in^2	Rt/R:	0.72

Ballistic Performance

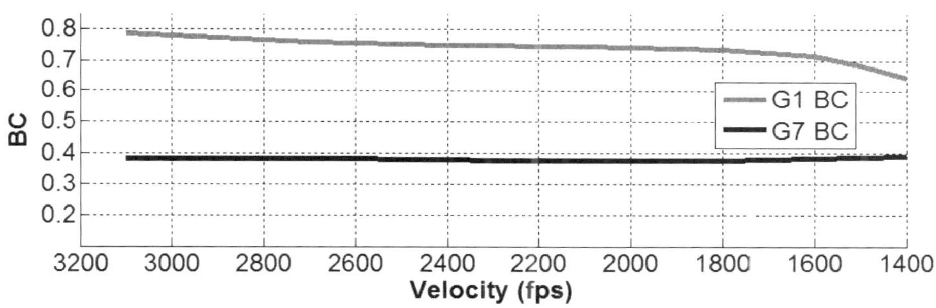

	Above 3000	3000-2500	2500-2000	2000-1500	Below 1500
G1 BC	0.803	0.768	0.749	0.716	0.687
G7 BC	0.380	0.381	0.377	0.380	0.386

Averages from 3000 fps to 1500 fps:
G1 BC: 0.742, variation of G1 BC: 0.094
G7 BC: 0.380, variation of G7 BC: 0.013
G7 form factor: 0.986

Rifling Twist Rate and Stability

Barrel Twist	Stability Factor Best case	Nominal	Worst case
1:9"	2.84	2.49	2.15
1:10"	2.30	2.01	1.74
1:11"	1.90	1.66	1.44
1:12"	1.60	1.40	1.21

Minimum twist required for stability is 1:13.0"
Recommended twist for optimal performance is 1:11.6" or faster

.375 Caliber

.375 caliber bullets can range from 250 to 415 grains, although most bullets in this caliber are made of solid copper or brass and are somewhat lighter. The image below relates bullet weight to sectional density for .375 caliber bullets.

Range of Weight and Sectional Density for 0.375 Caliber

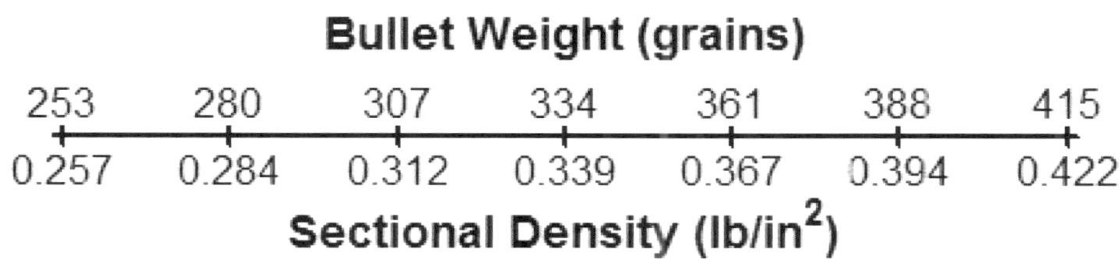

Sectional Density (SD) is a measure of how much mass is behind the bullets frontal area. Higher SD means a greater ability for the bullet to penetrate targets on impact, and also retain velocity in flight.

Spectrum of G7 form factors

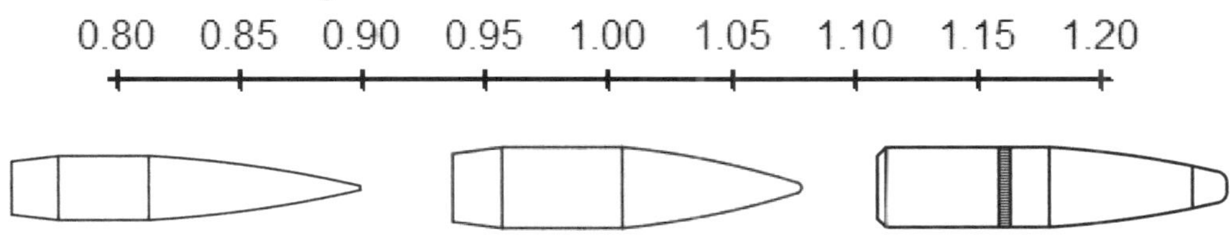

G7 Form factors range from 0.80 for the lowest drag shapes, up to 1.20+ for blunter shapes.

G7 Ballistic Coefficients are simply Sectional Density (SD) divided by G7 form factor.

You can determine a bullets' suitability for long range performance by seeing where it lies on the SD and form factor lines. Ideally, you want a bullet with a high SD and a low G7 form factor.

As an example, a 350 grain .375 caliber bullet has an SD of 0.356. If that bullet had a G7 form factor of 0.90, it's G7 BC would be:

G7 BC = SD/FF = 0.356/0.90 = 0.396.

Chey Tac 375 caliber 350 grain Solid

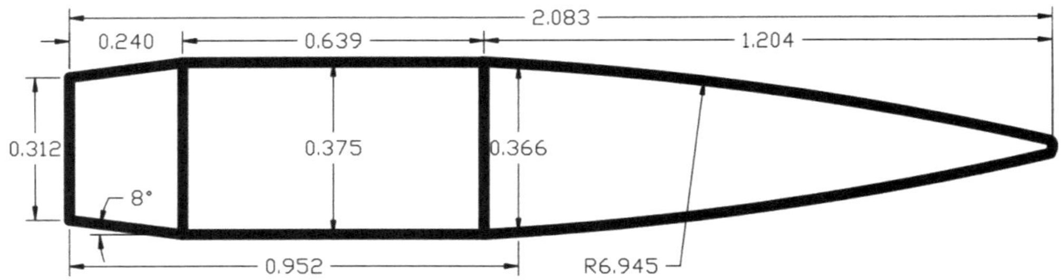

Lot #:	Unknown	Ogive Radius:	18.52 calibers
Sectional Density:	0.356 lb/in^2	Rt/R:	0.70

Ballistic Performance

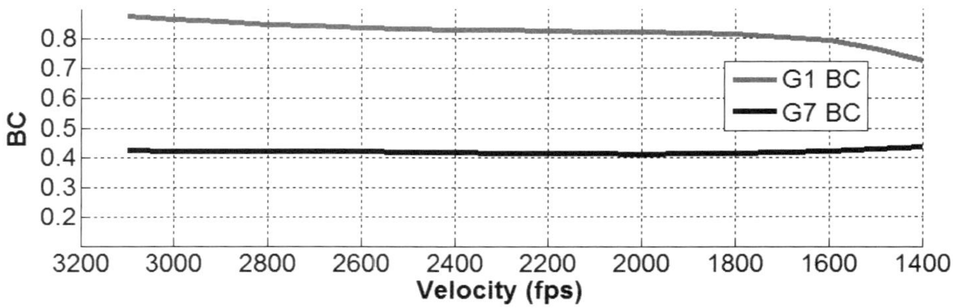

	Above 3000	3000-2500	2500-2000	2000-1500	Below 1500
G1 BC	0.894	0.851	0.828	0.793	0.764
G7 BC	0.423	0.422	0.416	0.422	0.430

Averages from 3000 fps to 1500 fps:
G1 BC: 0.822, variation of G1 BC: 0.103
G7 BC: 0.422, variation of G7 BC: 0.018
G7 form factor: 0.843

Rifling Twist Rate and Stability

Barrel Twist	Stability Factor Best case	Nominal	Worst case
1:10"	1.81	1.58	1.37
1:11"	1.50	1.31	1.13
1:12"	1.26	1.10	0.95
1:13"	1.07	0.94	0.81

Minimum twist required for stability is 1:11.8"
Recommended twist for optimal performance is 1:10.6" or faster

Cutting Edge .375 caliber 325 grain Lazer

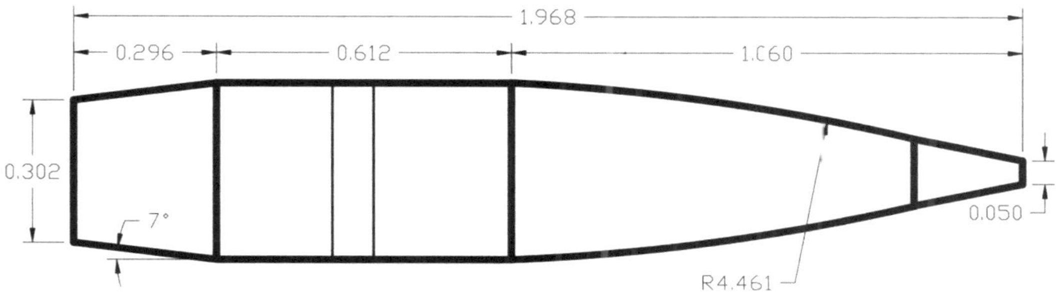

Lot #:	03132014	Ogive Radius:	11.9 calibers
Sectional Density:	0.330 lb/in^2	Rt/R:	0.79

Ballistic Performance

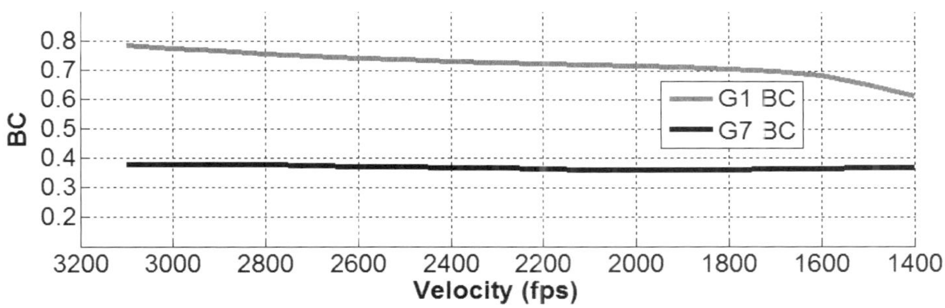

	Above 3000	3000-2500	2500-2000	2000-1500	Below 1500
G1 BC	0.805	0.758	0.729	0.687	0.654
G7 BC	0.381	0.376	0.367	0.365	0.368

Averages from 3000 fps to 1500 fps:
G1 BC: 0.722, variation of G1 BC: 0 123
G7 BC: 0.370, variation of G7 BC: 0.018
G7 form factor: 0.892

Rifling Twist Rate and Stability

Barrel Twist	Stability Factor Best case	Stability Factor Nominal	Stability Factor Worst case
1:9"	2.45	2.14	1.85
1:10"	1.99	1.74	1.50
1:11"	1.64	1.43	1.24
1:12"	1.38	1.21	1.04

Minimum twist required for stability is 1:12.0"
Recommended twist for optimal performance is 1:10.8" or faster

Cutting Edge .375 caliber 350 grain Lazer

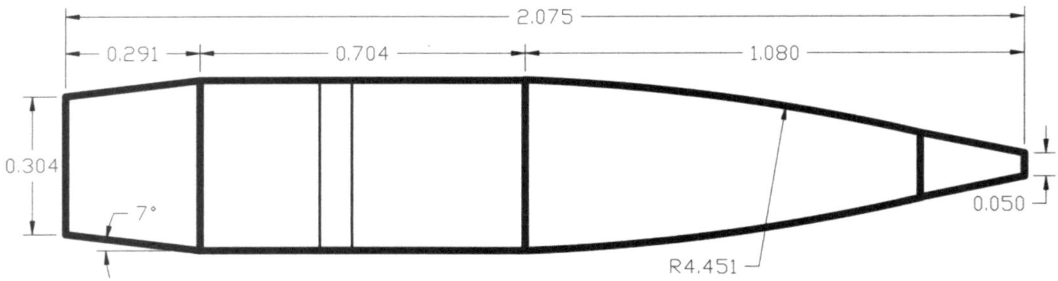

Lot #:	**01202015**	**Ogive Radius:**	**11.87 calibers**
Sectional Density:	**0.356 lb/in^2**	**Rt/R:**	**0.83**

Ballistic Performance

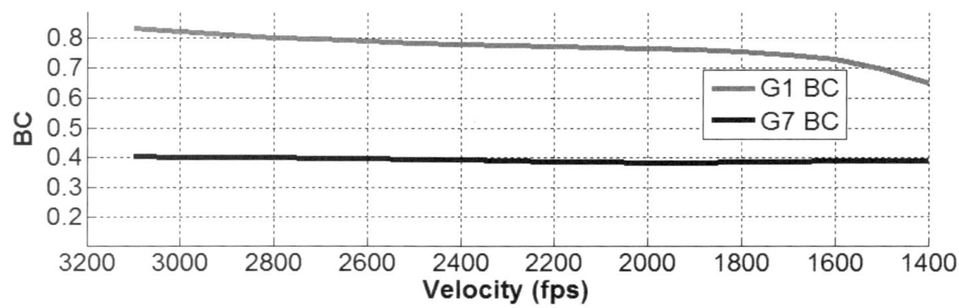

	Above 3000	3000-2500	2500-2000	2000-1500	Below 1500
G1 BC	0.853	0.803	0.774	0.730	0.696
G7 BC	0.404	0.398	0.389	0.387	0.391

Averages from 3000 fps to 1500 fps:
G1 BC: 0.767, variation of G1 BC: 0.128
G7 BC: 0.393, variation of G7 BC: 0.019
G7 form factor: 0.905

Rifling Twist Rate and Stability

Barrel Twist	Stability Factor Best case	Nominal	Worst case
1:9"	2.26	1.98	1.71
1:10"	1.83	1.60	1.38
1:11"	1.51	1.32	1.14
1:12"	1.27	1.11	0.96

Minimum twist required for stability is 1:11.5"
Recommended twist for optimal performance is 1:10.3" or faster

Cutting Edge 375 caliber 352 grain MTAC

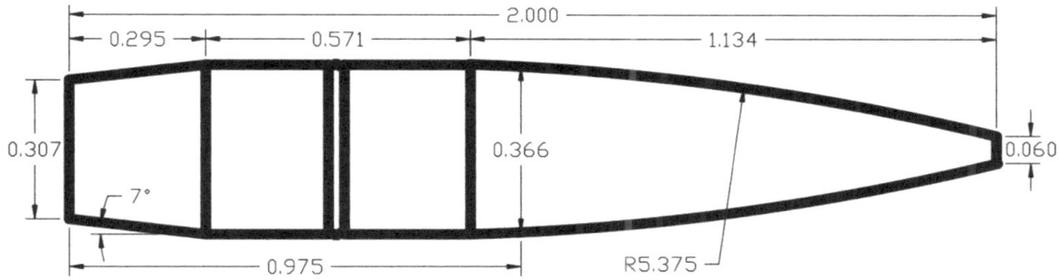

Lot #:	**Unknown**		**Ogive Radius:**	**14.33 calibers**
Sectional Density:	**0.358 lb/in^2**		**Rt/R:**	**0.77**

Ballistic Performance

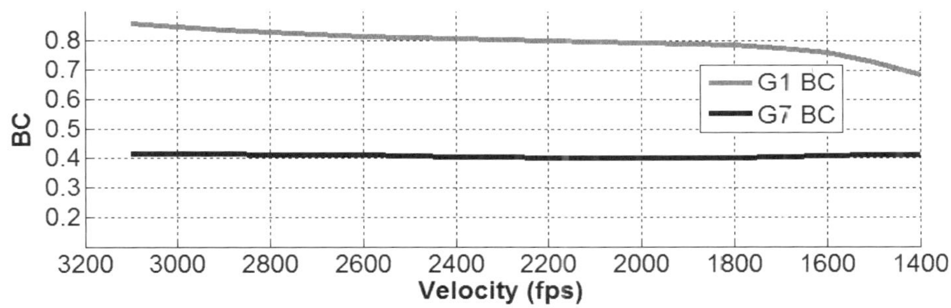

	Above 3000	3000-2500	2500-2000	2000-1500	Below 1500
G1 BC	0.877	0.830	0.803	0.762	0.730
G7 BC	0.415	0.411	0.404	0.405	0.411

Averages from 3000 fps to 1500 fps:
G1 BC: 0.796, variation of G1 BC: 0.118
G7 BC: 0.408, variation of G7 BC: 0.015
G7 form factor: 0.876

Rifling Twist Rate and Stability

	Stability Factor		
Barrel Twist	**Best case**	**Nominal**	**Worst case**
1:10"	2.05	1.79	1.55
1:11"	1.70	1.48	1.28
1:12"	1.43	1.25	1.08
1:13"	1.21	1.06	0.92

Minimum twist required for stability is 1:12.2"
Recommended twist for optimal performance is 1:10.9" or faster

Cutting Edge .375 caliber 300 grain MTH

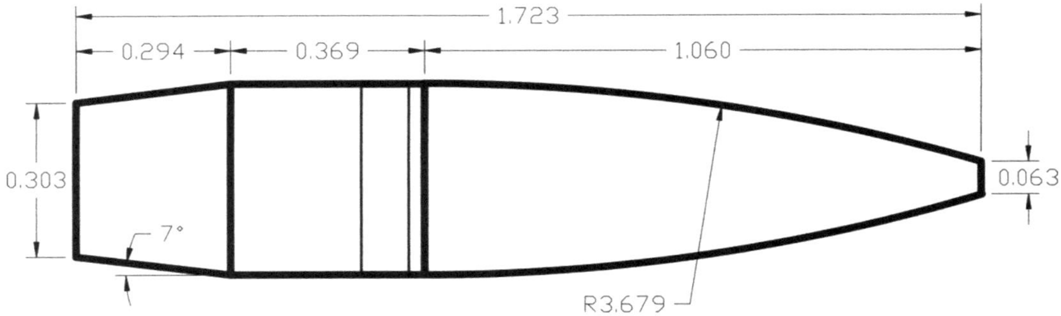

Lot #:	unknown	Ogive Radius:	9.81 calibers
Sectional Density:	0.305 lb/in^2	Rt/R:	1.00

Ballistic Performance

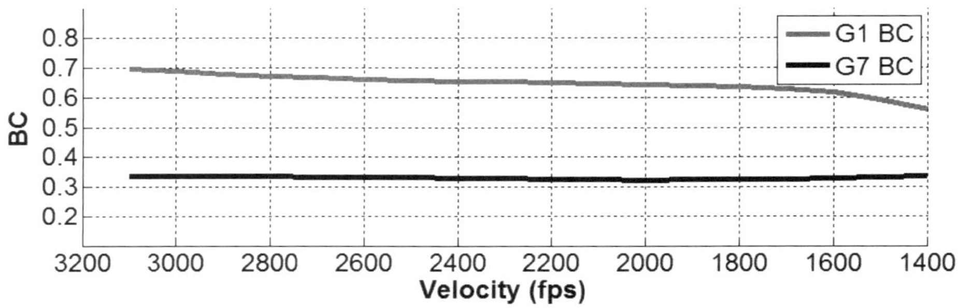

	Above 3000	3000-2500	2500-2000	2000-1500	Below 1500
G1 BC	0.711	0.673	0.651	0.618	0.591
G7 BC	0.337	0.334	0.327	0.328	0.333

Averages from 3000 fps to 1500 fps:
G1 BC: 0.645, variation of G1 BC: 0.096
G7 BC: 0.331, variation of G7 BC: 0.013
G7 form factor: 0.921

Rifling Twist Rate and Stability

Barrel Twist	Stability Factor Best case	Nominal	Worst case
1:11"	2.23	1.95	1.69
1:12"	1.88	1.64	1.42
1:13"	1.60	1.40	1.21
1:14"	1.38	1.21	1.04

Minimum twist required for stability is 1:14.0"
Recommended twist for optimal performance is 1:12.5" or faster

Cutting Edge .375 caliber 320 grain MTH

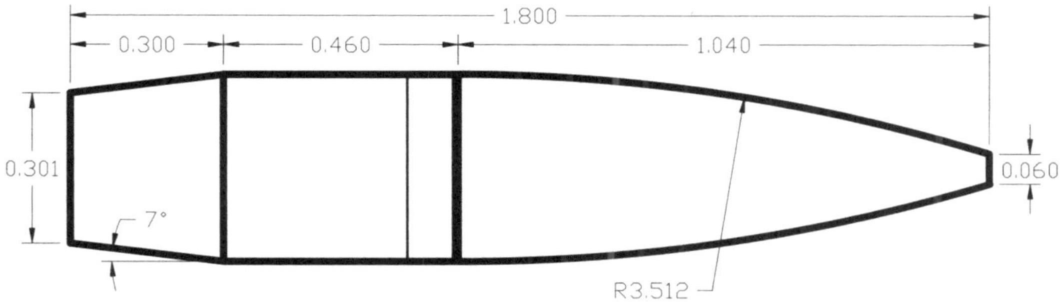

Lot #:	05152014	Ogive Radius:	9.37 calibers
Sectional Density:	0.325 lb/in^2	Rt/R:	1.00

Ballistic Performance

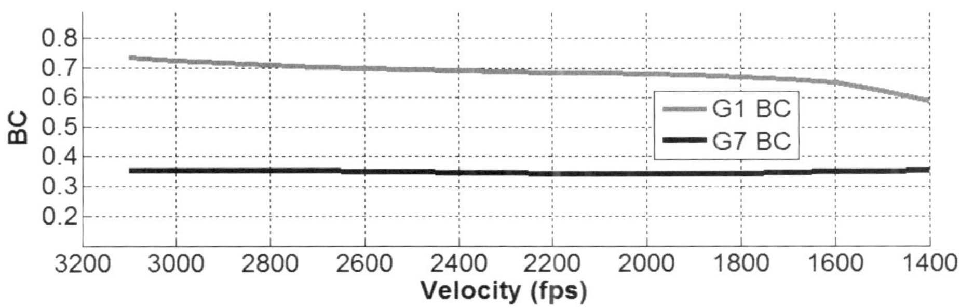

	Above 3000	3000-2500	2500-2000	2000-1500	Below 1500
G1 BC	0.750	0.711	0.689	0.654	0.626
G7 BC	0.355	0.353	0.346	0.347	0.352

Averages from 3000 fps to 1500 fps:
G1 BC: 0.682, variation of G1 BC: 0.101
G7 BC: 0.350, variation of G7 BC: 0.013
G7 form factor: 0.929

Rifling Twist Rate and Stability

Barrel Twist	Stability Factor Best case	Nominal	Worst case
1:11"	2.10	1.83	1.59
1:12"	1.76	1.54	1.33
1:13"	1.50	1.31	1.14
1:14"	1.30	1.13	0.98

Minimum twist required for stability is 1:13.6"
Recommended twist for optimal performance is 1:12.2" or faster

Cutting Edge .375 caliber 330 grain MTH

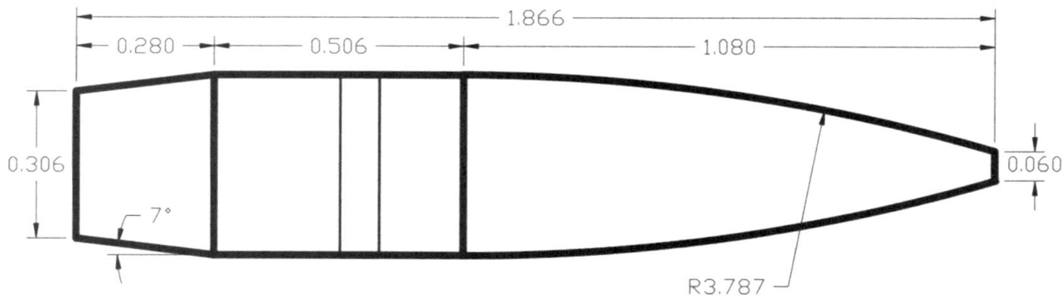

Lot #:	08142014	Ogive Radius:	10.10 calibers
Sectional Density:	0.335 lb/in^2	Rt/R:	1.00

Ballistic Performance

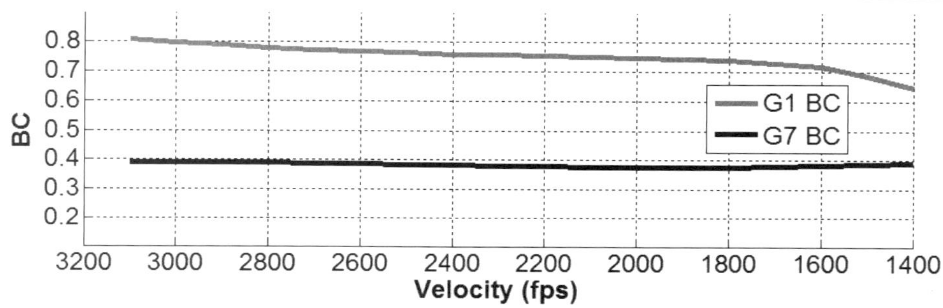

	Above 3000	3000-2500	2500-2000	2000-1500	Below 1500
G1 BC	0.825	0.781	0.756	0.717	0.686
G7 BC	0.391	0.387	0.380	0.381	0.386

Averages from 3000 fps to 1500 fps:
G1 BC: 0.749, variation of G1 BC: 0.113
G7 BC: 0.384, variation of G7 BC: 0.015
G7 form factor: 0.873

Rifling Twist Rate and Stability

Barrel Twist	Stability Factor Best case	Nominal	Worst case
1:10"	2.36	2.06	1.78
1:11"	1.95	1.70	1.47
1:12"	1.64	1.43	1.24
1:13"	1.39	1.22	1.05

Minimum twist required for stability is 1:13.1"
Recommended twist for optimal performance is 1:11.7" or faster

Cutting Edge 375 caliber 350 grain MTH

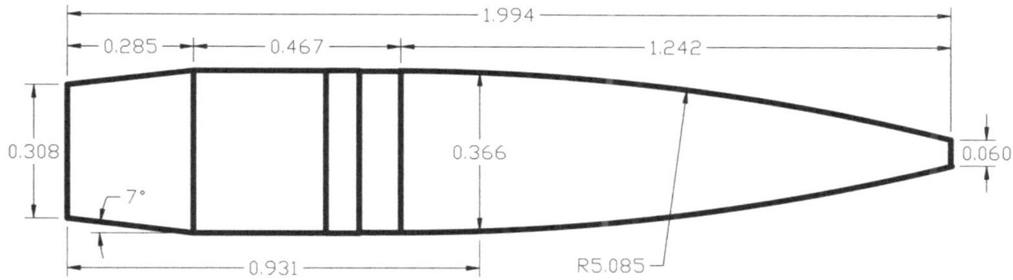

Lot #:	Unknown	Ogive Radius:	13.56 calibers
Sectional Density:	0.356 lb/in^2	Rt/R:	0.99

Ballistic Performance

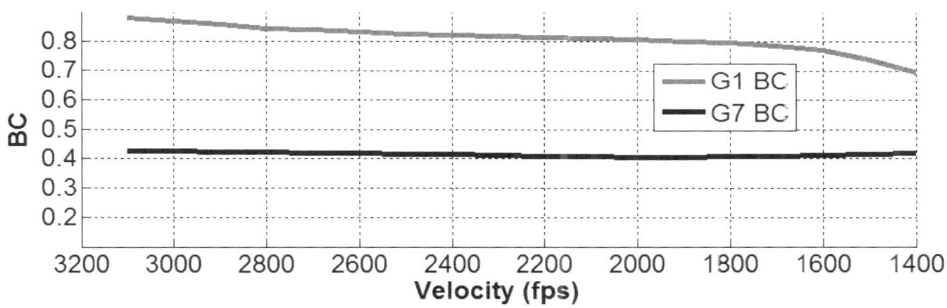

	Above 3000	3000-2500	2500-2000	2000-1500	Below 1500
G1 BC	0.899	0.847	0.817	0.773	0.739
G7 BC	0.426	0.420	0.411	0.411	0.416

Averages from 3000 fps to 1500 fps:
G1 BC: 0.810, variation of G1 BC: 0.129
G7 BC: 0.415, variation of G7 BC: 0.019
G7 form factor: 0.856

Rifling Twist Rate and Stability

Barrel Twist	Stability Factor Best case	Nominal	Worst case
1:10"	2.06	1.80	1.56
1:11"	1.70	1.49	1.29
1:12"	1.43	1.25	1.08
1:13"	1.22	1.06	0.92

Minimum twist required for stability is 1:12.2"
Recommended twist for optimal performance is 1:11.0" or faster

GS Custom 375 caliber 291 grain SP

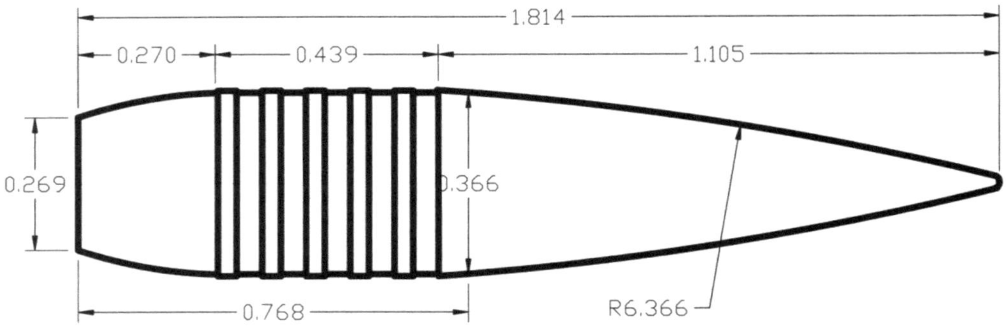

Lot #:	Unknown	Ogive Radius:	16.98 calibers
Sectional Density:	0.296 lb/in²	Rt/R:	0.55

Ballistic Performance

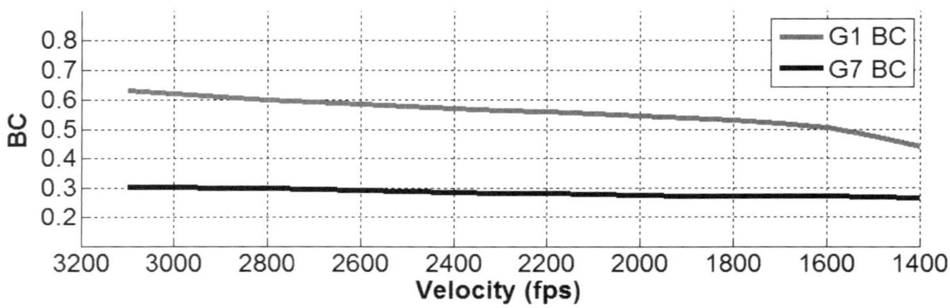

	Above 3000	3000-2500	2500-2000	2000-1500	Below 1500
G1 BC	0.650	0.598	0.561	0.512	0.478
G7 BC	0.307	0.297	0.282	0.271	0.268

Averages from 3000 fps to 1500 fps:
G1 BC: 0.555, variation of G1 BC: 0.143
G7 BC: 0.284, variation of G7 BC: 0.035
G7 form factor: 1.044

Rifling Twist Rate and Stability

Barrel Twist	Best case	Stability Factor Nominal	Worst case
1:10"	2.26	1.97	1.71
1:11"	1.87	1.63	1.41
1:12"	1.57	1.37	1.18
1:13"	1.34	1.17	1.01

Minimum twist required for stability is 1:12.8"
Recommended twist for optimal performance is 1:11.5" or faster

GS Custom 375 caliber 330 grain SP

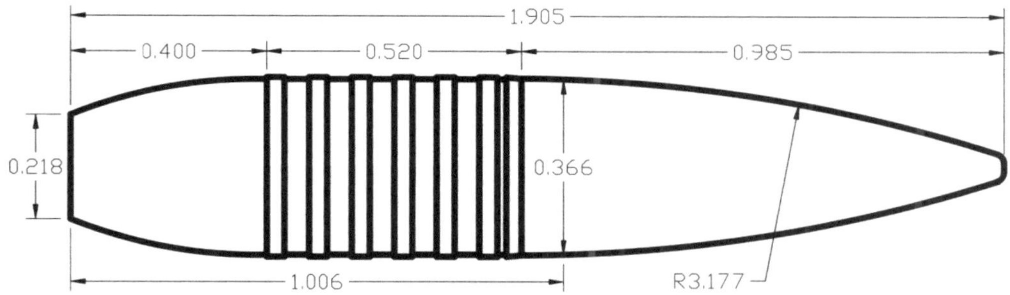

Lot #:	**Unknown**	**Ogive Radius:**	**8.47 calibers**
Sectional Density:	**0.335 lb/in^2**	**Rt/R:**	**0.98**

Ballistic Performance

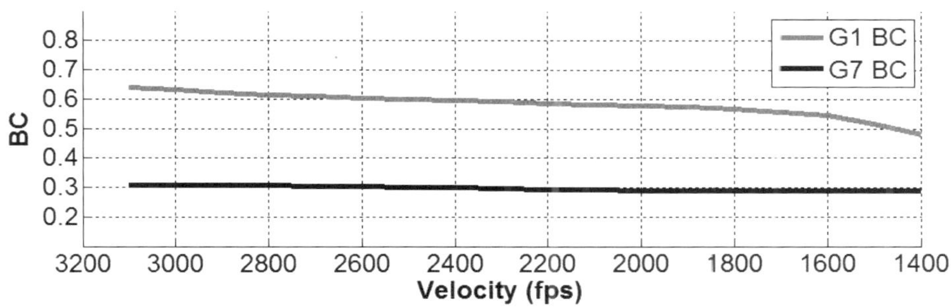

	Above 3000	3000-2500	2500-2000	2000-1500	Below 1500
G1 BC	0.652	0.614	0.588	0.547	0.517
G7 BC	0.309	0.305	0.296	0.290	0.290

Averages from 3000 fps to 1500 fps:
G1 BC: 0.581, variation of G1 BC: 0.113
G7 BC: 0.297, variation of G7 BC: 0.018
G7 form factor: 1.128

Rifling Twist Rate and Stability

Barrel Twist	Stability Factor Best case	Nominal	Worst case
1:10"	2.22	1.94	1.68
1:11"	1.83	1.60	1.39
1:12"	1.54	1.35	1.16
1:13"	1.31	1.15	0.99

Minimum twist required for stability is 1:12.7"
Recommended twist for optimal performance is 1:11.4" or faster

GS Custom 375 caliber 355 grain SP

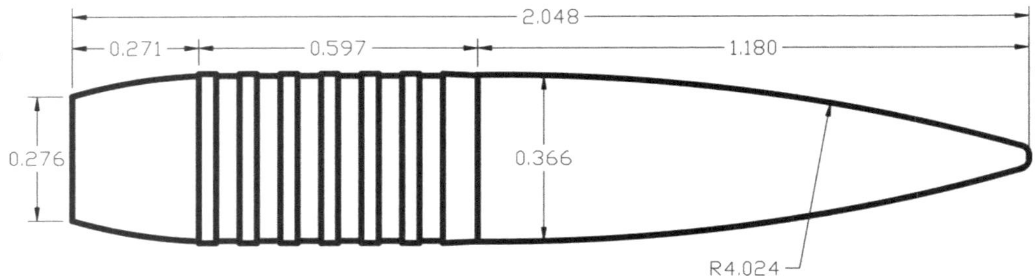

Lot #:	Unknown	Ogive Radius:	4.02 calibers
Sectional Density:	0.361 lb/in^2	Rt/R:	1.00

Ballistic Performance

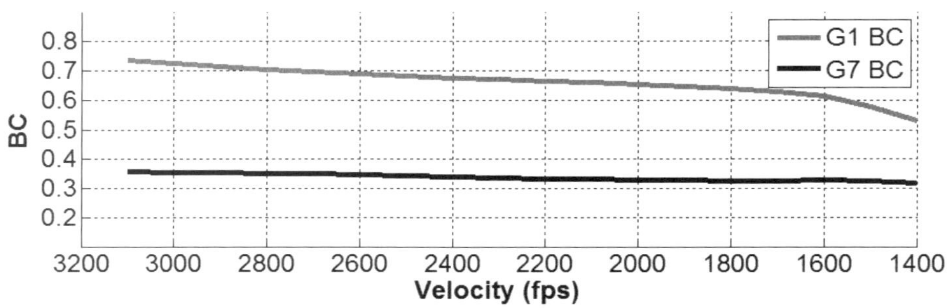

	Above 3000	3000-2500	2500-2000	2000-1500	Below 1500
G1 BC	0.757	0.705	0.668	0.616	0.578
G7 BC	0.358	0.349	0.336	0.327	0.324

Averages from 3000 fps to 1500 fps:
G1 BC: 0.660, variation of G1 BC: 0.148
G7 BC: 0.338, variation of G7 BC: 0.031
G7 form factor: 1.069

Rifling Twist Rate and Stability

Barrel Twist	Stability Factor Best case	Nominal	Worst case
1:10"	1.93	1.69	1.46
1:11"	1.60	1.39	1.21
1:12"	1.34	1.17	1.01
1:13"	1.14	1.00	0.86

Minimum twist required for stability is 1:11.9"
Recommended twist for optimal performance is 1:10.6" or faster

Lehigh 375 caliber 330 grain Match Solid

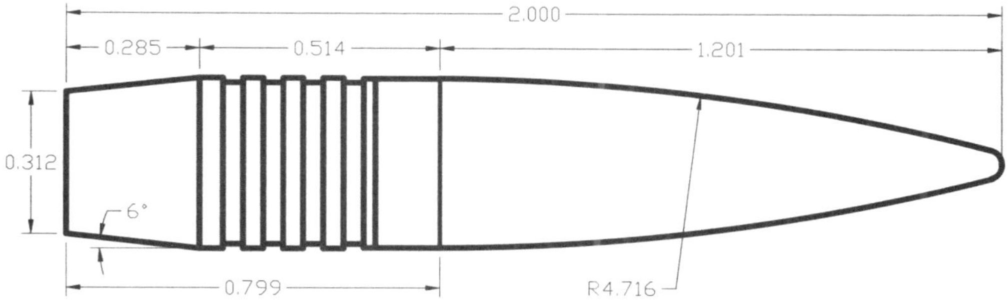

Lot #:	**Unknown**	**Ogive Radius:**	**12.58 calibers**
Sectional Density:	**0.335 lb/in^2**	**Rt/R:**	**1.00**

Ballistic Performance

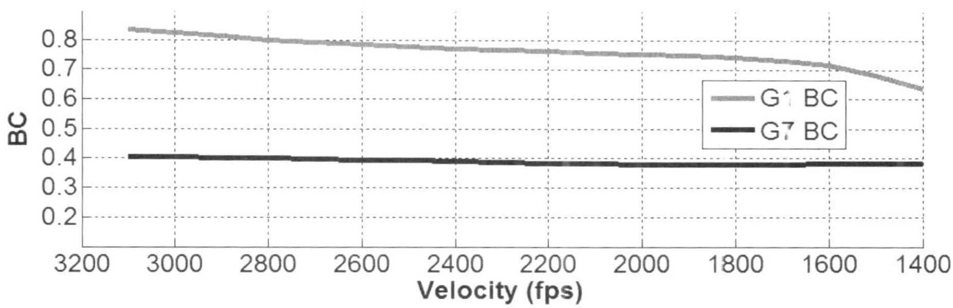

	Above 3000	3000-2500	2500-2000	2000-1500	Below 1500
G1 BC	0.860	0.803	0.767	0.719	0.682
G7 BC	0.407	0.398	0.386	0.381	0.384

Averages from 3000 fps to 1500 fps:
G1 BC: 0.761, variation of G1 BC: 0.143
G7 BC: 0.390, variation of G7 BC: 0.025
G7 form factor: 0.861

Rifling Twist Rate and Stability

	Stability Factor		
Barrel Twist	**Best case**	**Nominal**	**Worst case**
1:10"	1.92	1.68	1.45
1:11"	1.59	1.39	1.20
1:12"	1.34	1.17	1.01
1:13"	1.14	0.99	0.86

Minimum twist required for stability is 1:11.8"
Recommended twist for optimal performance is 1:10.6" or faster

Nosler .375 caliber 300 grain Accubond

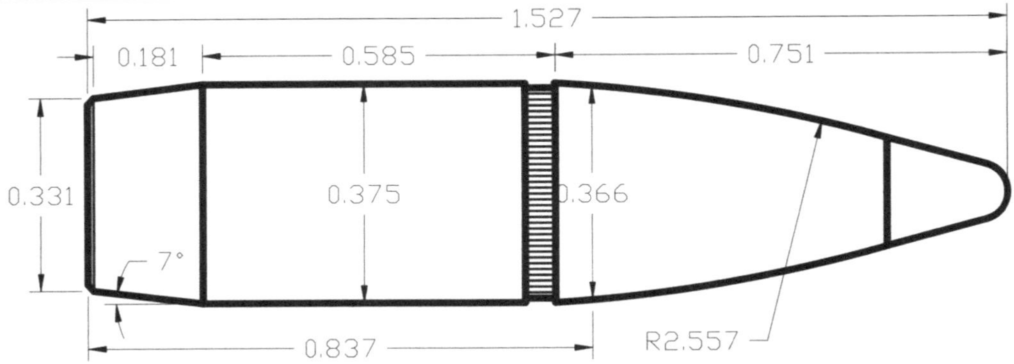

Lot #:	BM17F10-1	Ogive Radius:	6.82 calibers
Sectional Density:	0.305 lb/in^2	Rt/R:	0.77

Ballistic Performance

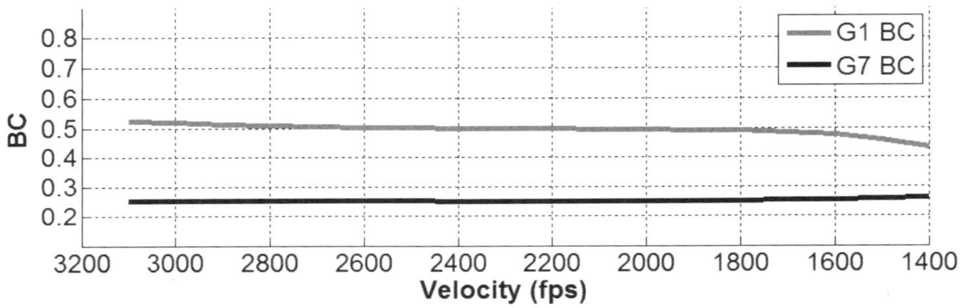

	Above 3000	3000-2500	2500-2000	2000-1500	Below 1500
G1 BC	0.534	0.510	0.498	0.476	0.457
G7 BC	0.253	0.253	0.250	0.253	0.257

Averages from 3000 fps to 1500 fps:
G1 BC: 0.493, variation of G1 BC: 0.062
G7 BC: 0.253, variation of G7 BC: 0.009
G7 form factor: 1.205

Rifling Twist Rate and Stability

Barrel Twist	Stability Factor Best case	Nominal	Worst case
1:13"	2.27	1.98	1.72
1:14"	1.96	1.71	1.48
1:15"	1.71	1.49	1.29
1:16"	1.50	1.31	1.13

Minimum twist required for stability is 1:16.7"
Recommended twist for optimal performance is 1:14.9" or faster

Sierra 375 caliber 300 grain GameKing

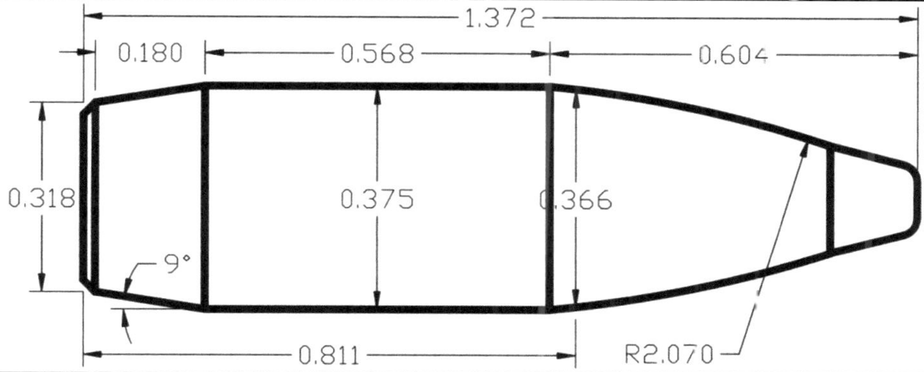

Lot #:	006076278KL60	Ogive Radius:	5.52 calibers
Sectional Density:	0.305 lb/in^2	Rt/R:	0.56

Ballistic Performance

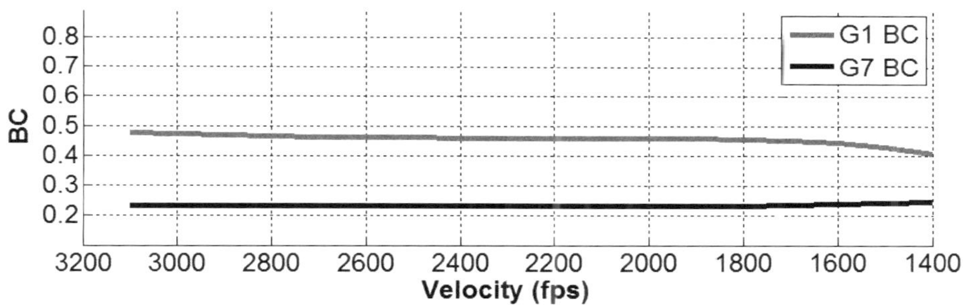

	Above 3000	3000-2500	2500-2000	2000-1500	Below 1500
G1 BC	0.483	0.467	0.460	0.444	0.429
G7 BC	0.229	0.231	0.231	0.236	0.241

Averages from 3000 fps to 1500 fps:
G1 BC: 0.455, variation of G1 BC: 0.044
G7 BC: 0.234, variation of G7 BC: 0.011
G7 form factor: 1.305

Rifling Twist Rate and Stability

Barrel Twist	Stability Factor Best case	Nominal	Worst case
1:10"	5.22	4.56	3.94
1:11"	4.31	3.77	3.26
1:12"	3.62	3.17	2.74
1:13"	3.09	2.70	2.33

Minimum twist required for stability is 1:19.5"
Recommended twist for optimal performance is 1:17.4" or faster

Sierra 375 caliber 350 grain MatchKing

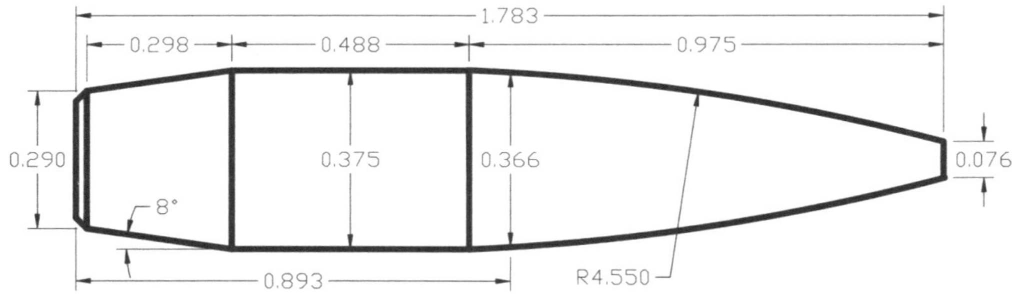

Lot #:	70737079857	Ogive Radius:	12.13 calibers
Sectional Density:	0.356 lb/in^2	Rt/R:	0.72

Ballistic Performance

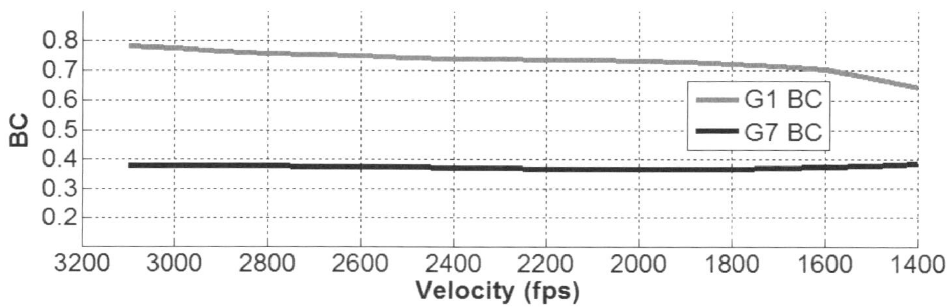

	Above 3000	3000-2500	2500-2000	2000-1500	Below 1500
G1 BC	0.798	0.760	0.738	0.704	0.676
G7 BC	0.378	0.377	0.371	0.374	0.380

Averages from 3000 fps to 1500 fps:
G1 BC: 0.732, variation of G1 BC: 0.099
G7 BC: 0.375, variation of G7 BC: 0.013
G7 form factor: 0.947

Rifling Twist Rate and Stability

Barrel Twist	Stability Factor Best case	Nominal	Worst case
1:10"	2.86	2.49	2.16
1:11"	2.36	2.06	1.78
1:12"	1.98	1.73	1.50
1:13"	1.69	1.48	1.28

Minimum twist required for stability is 1:14.4"
Recommended twist for optimal performance is 1:12.9" or faster

.408 Caliber

.408 caliber bullets typically range from 300 to 535 grains, although most bullets in this caliber are made of solid copper or brass and are somewhat lighter. The image below relates bullet weight to sectional density for .408 caliber bullets.

Range of Weight and Sectional Density for 0.408 Caliber

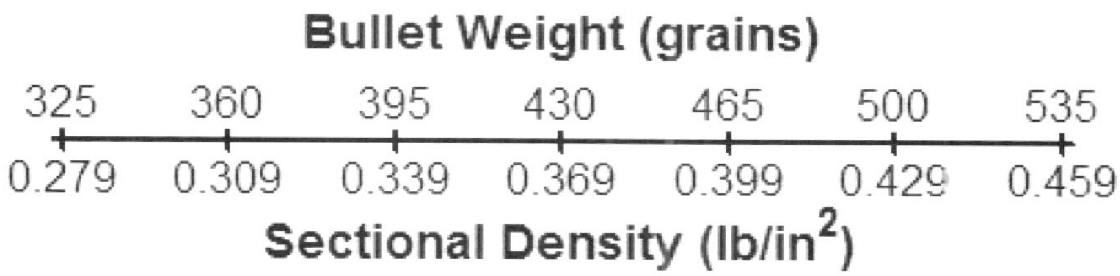

Sectional Density (SD) is a measure of how much mass is behind the bullets frontal area. Higher SD means a greater ability for the bullet to penetrate targets on impact, and also retain velocity in flight.

Spectrum of G7 form factors

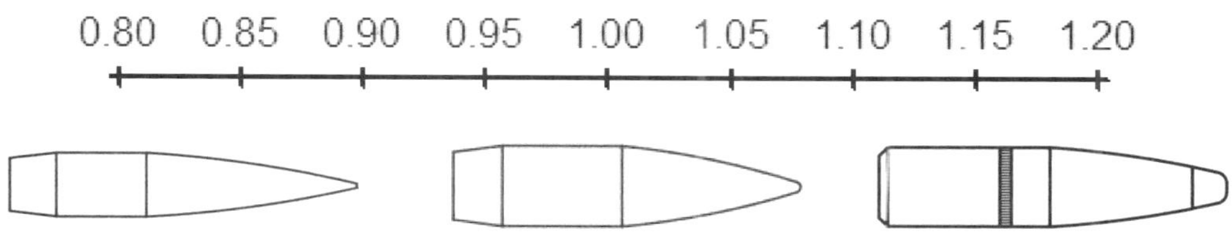

G7 Form factors range from 0.80 for the lowest drag shapes, up to 1.20+ for blunter shapes.

G7 Ballistic Coefficients are simply Sectional Density (SD) divided by G7 form factor.

You can determine a bullets' suitability for long range performance by seeing where it lies on the SD and form factor lines. Ideally, you want a bullet with a high SD and a low G7 form factor.

As an example, a 419 grain .408 caliber bullet has an SD of 0.360. If that bullet had a G7 form factor of 0.83, it's G7 BC would be:

G7 BC = SD/FF = 0.360/0.83 = 0.434.

Chey Tac 408 caliber 305 gr Solid

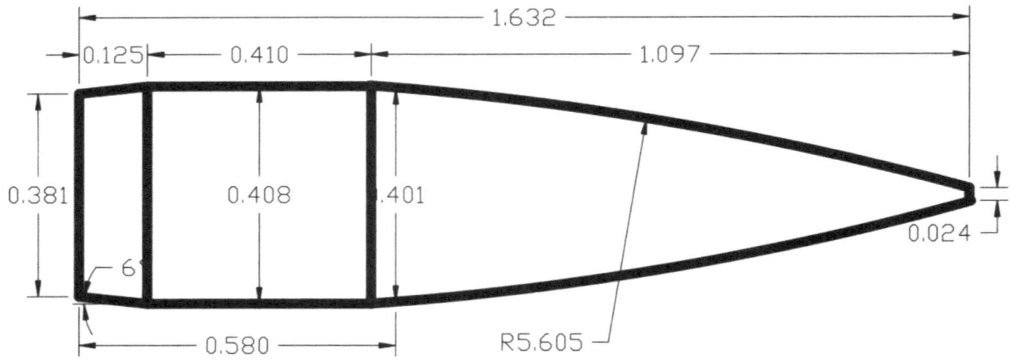

Lot #:	Unknown	Ogive Radius:	13.74 calibers
Sectional Density:	0.262 lb/in^2	Rt/R:	0.58

Ballistic Performance

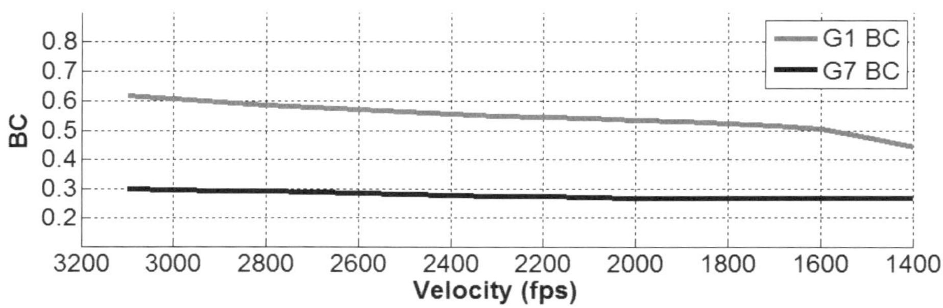

	Above 3000	3000-2500	2500-2000	2000-1500	Below 1500
G1 BC	0.636	0.584	0.549	0.507	0.478
G7 BC	0.301	0.289	0.276	0.269	0.268

Averages from 3000 fps to 1500 fps:
G1 BC: 0.545, variation of G1 BC: 0.128
G7 BC: 0.279, variation of G7 BC: 0.028
G7 form factor: 0.940

Rifling Twist Rate and Stability

Barrel Twist	Stability Factor Best case	Nominal	Worst case
1:11"	3.12	2.73	2.36
1:12"	2.62	2.29	1.98
1:13"	2.23	1.95	1.69
1:14"	1.93	1.68	1.46

Minimum twist required for stability is 1:16.6"
Recommended twist for optimal performance is 1:14.8" or faster

Chey Tac 408 caliber 419 gr Solid

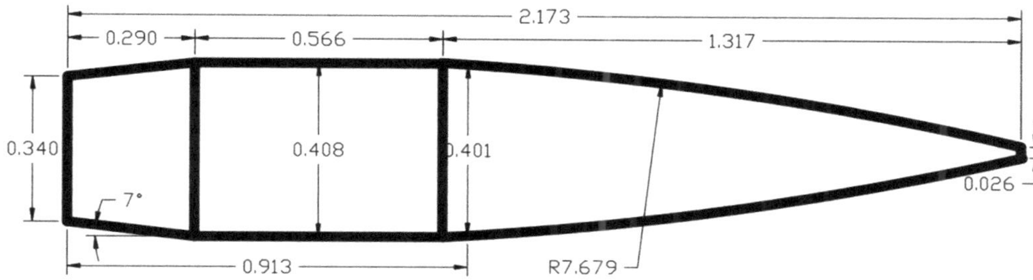

	Lot #:	Unknown	Ogive Radius:	18.82 calibers
	Sectional Density:	0.360 lb/in^2	Rt/R:	0.60

Ballistic Performance

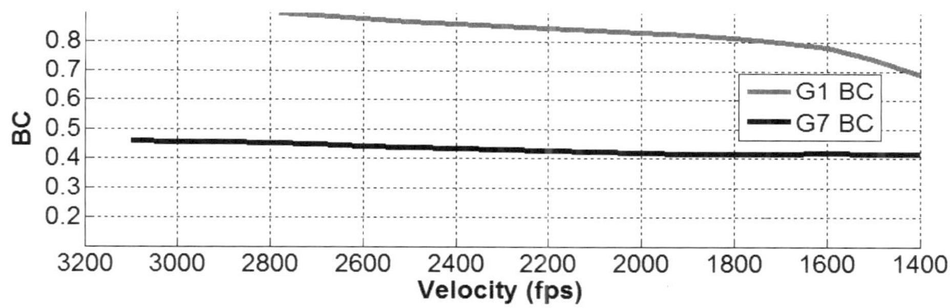

	Above 3000	3000-2500	2500-2000	2000-1500	Below 1500
G1 BC	0.977	0.902	0.852	0.787	0.742
G7 BC	0.462	0.447	0.428	0.418	0.417

Averages from 3000 fps to 1500 fps:
G1 BC: 0.845, variation of G1 BC: 0.192
G7 BC: 0.432, variation of G7 BC: 0.040
G7 form factor: 0.833

Rifling Twist Rate and Stability

Barrel Twist	Stability Factor Best case	Nominal	Worst case
1:11"	1.86	1.63	1.41
1:12"	1.57	1.37	1.18
1:13"	1.33	1.17	1.01
1:14"	1.15	1.01	0.87

Minimum twist required for stability is 1:12.8"
Recommended twist for optimal performance is 1:11.5" or faster

Cutting Edge 408 caliber 390 grain MTH

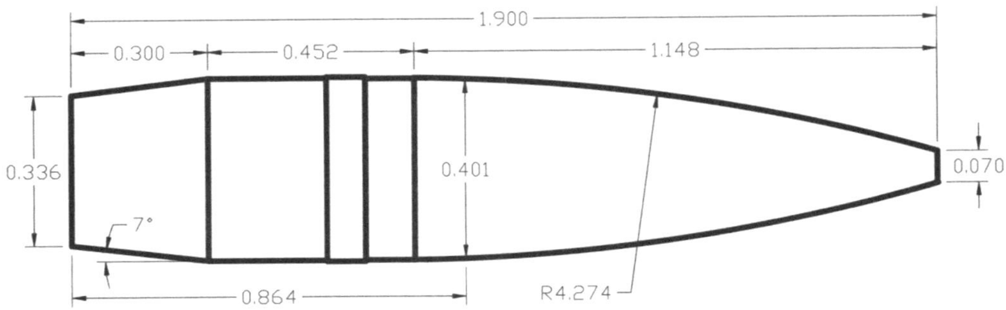

Lot #:	Unknown	Ogive Radius:	10.48 calibers
Sectional Density:	0.335 lb/in²	Rt/R:	0.93

Ballistic Performance

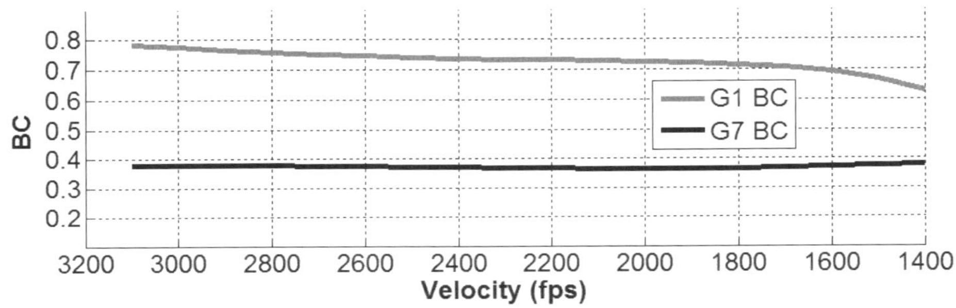

	Above 3000	3000-2500	2500-2000	2000-1500	Below 1500
G1 BC	0.801	0.758	0.733	0.696	0.666
G7 BC	0.380	0.376	0.368	0.370	0.375

Averages from 3000 fps to 1500 fps:
G1 BC: 0.727, variation of G1 BC: 0.109
G7 BC: 0.373, variation of G7 BC: 0.015
G7 form factor: 0.898

Rifling Twist Rate and Stability

Barrel Twist	Stability Factor Best case	Stability Factor Nominal	Stability Factor Worst case
1:11"	2.57	2.24	1.94
1:12"	2.16	1.89	1.63
1:13"	1.84	1.61	1.39
1:14"	1.59	1.38	1.20

Minimum twist required for stability is 1:15.0"
Recommended twist for optimal performance is 1:13.5" or faster

Cutting Edge 408 caliber 415 grain MTH

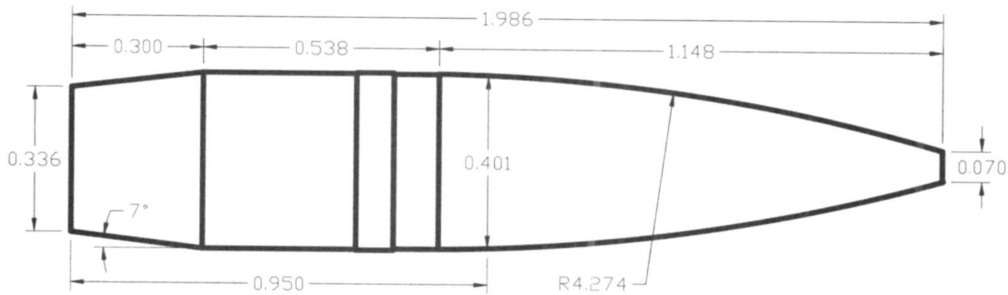

Lot #:	Unknown	Ogive Radius:	10.48 calibers
Sectional Density:	0.356 lb/in^2	Rt/R:	0.93

Ballistic Performance

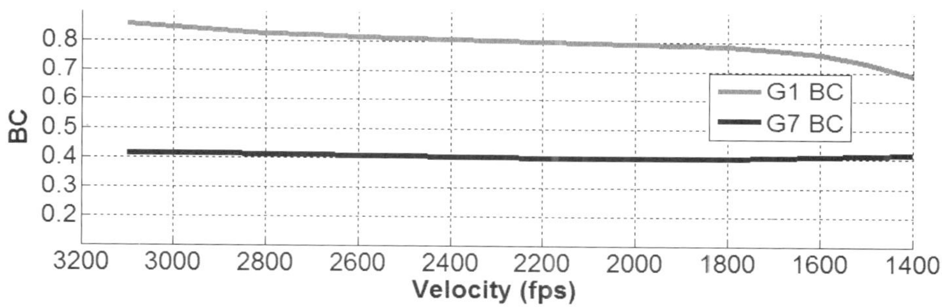

	Above 3000	3000-2500	2500-2000	2000-1500	Below 1500
G1 BC	0.876	0.829	0.803	0.763	0.731
G7 BC	0.415	0.411	0.404	0.406	0.412

Averages from 3000 fps to 1500 fps:
G1 BC: 0.796, variation of G1 BC: 0.116
G7 BC: 0.408, variation of G7 BC: 0.015
G7 form factor: 0.872

Rifling Twist Rate and Stability

Barrel Twist	Stability Factor Best case	Nominal	Worst case
1:11"	2.40	2.10	1.81
1:12"	2.02	1.76	1.52
1:13"	1.72	1.50	1.30
1:14"	1.48	1.30	1.12

Minimum twist required for stability is 1:14.5"
Recommended twist for optimal performance is 1:13.0" or faster

GS Custom 408 caliber 385 grain SP

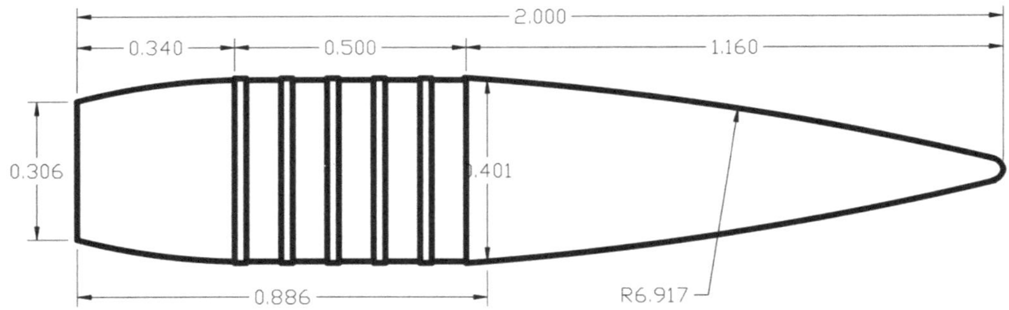

Lot #:	Unknown	Ogive Radius:	16.95 calibers
Sectional Density:	0.330 lb/in^2	Rt/R:	00.54

Ballistic Performance

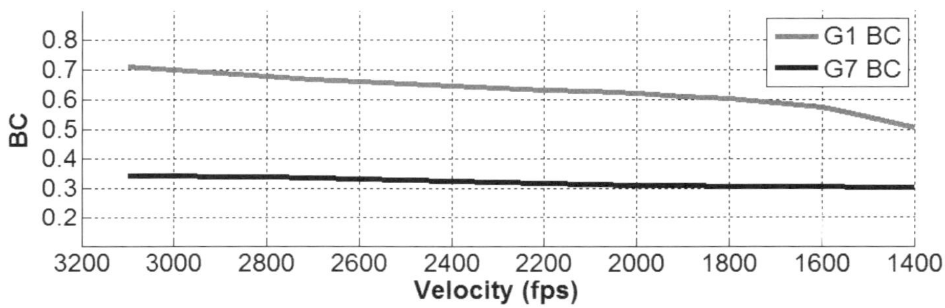

	Above 3000	3000-2500	2500-2000	2000-1500	Below 1500
G1 BC	0.733	0.677	0.636	0.581	0.543
G7 BC	0.347	0.336	0.320	0.308	0.305

Averages from 3000 fps to 1500 fps:
G1 BC: 0.629, variation of G1 BC: 0.157
G7 BC: 0.322, variation of G7 BC: 0.037
G7 form factor: 1.029

Rifling Twist Rate and Stability

Barrel Twist	Stability Factor Best case	Nominal	Worst case
1:11"	2.18	1.91	1.65
1:12"	1.83	1.60	1.39
1:13"	1.56	1.37	1.18
1:14"	1.35	1.18	1.02

Minimum twist required for stability is 1:13.9"
Recommended twist for optimal performance is 1:12.4" or faster

HSM .408 caliber 400 grain T50 Solid

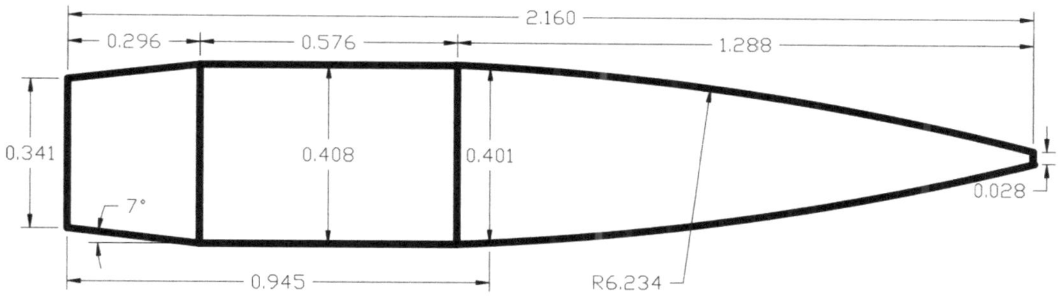

Lot #:	unknown	Ogive Radius:	15.23
Sectional Density:	0.343 lb/in²	Rt/R:	0.72

Ballistic Performance

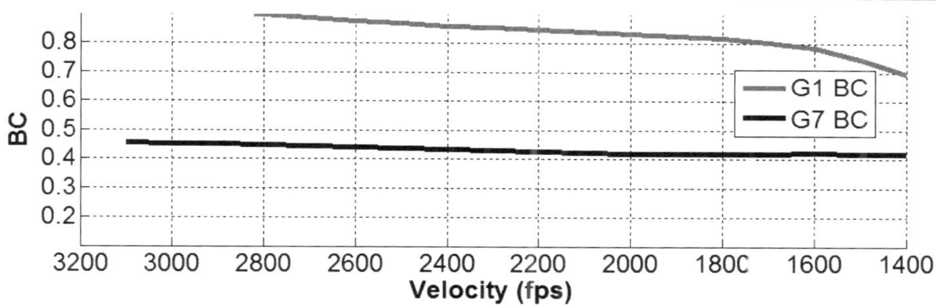

	Above 3000	3000-2500	2500-2000	2000-1500	Below 1500
G1 BC	0.968	0.897	0.851	0.791	0.748
G7 BC	0.458	0.445	0.428	0.420	0.420

Averages from 3000 fps to 1500 fps:
G1 BC: 0.844, variation of G1 BC: 0.178
G7 BC: 0.432, variation of G7 BC: 0.034
G7 form factor: 0.795

Rifling Twist Rate and Stability

Barrel Twist	Stability Factor Best case	Nominal	Worst case
1:10"	2.19	1.91	1.65
1:11"	1.81	1.58	1.37
1:12"	1.52	1.33	1.15
1:13"	1.30	1.13	0.98

Minimum twist required for stability is 1:12.6"
Recommended twist for optimal performance is 1:11.3" or faster

Lehigh 408 caliber 400 grain Match Solid

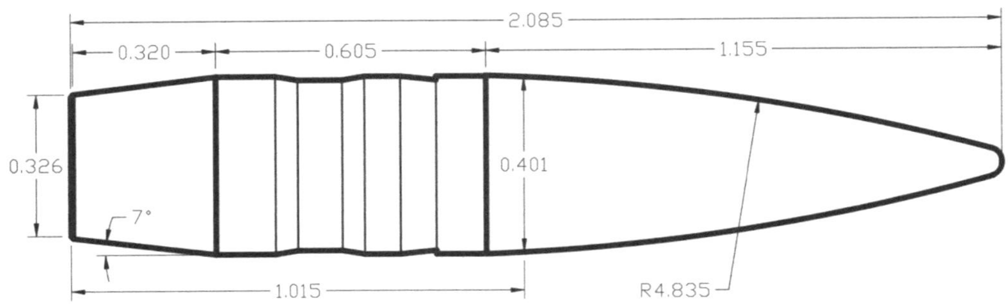

Lot #:	Unknown	Ogive Radius:	11.85 calibers
Sectional Density:	0.343 lb/in^2	Rt/R:	0.78

Ballistic Performance

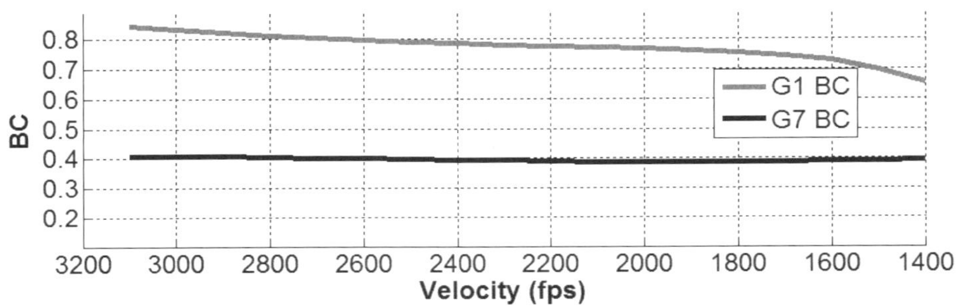

	Above 3000	3000-2500	2500-2000	2000-1500	Below 1500
G1 BC	0.865	0.812	0.779	0.732	0.696
G7 BC	0.410	0.403	0.392	0.388	0.391

Averages from 3000 fps to 1500 fps:
G1 BC: 0.772, variation of G1 BC: 0.138
G7 BC: 0.396, variation of G7 BC: 0.022
G7 form factor: 0.868

Rifling Twist Rate and Stability

Barrel Twist	Best case	Stability Factor Nominal	Worst case
1:11"	2.01	1.75	1.52
1:12"	1.69	1.47	1.27
1:13"	1.44	1.26	1.09
1:14"	1.24	1.08	0.94

Minimum twist required for stability is 1:13.3"
Recommended twist for optimal performance is 1:11.9" or faster

.416 Caliber

.416 caliber bullets typically range from 400 to 450 grains, which is on the light end of the scale compared to other calibers. All of the bullets that were tested in this caliber are made of solid copper or brass. The image below relates bullet weight to sectional density for .416 caliber bullets.

Range of Weight and Sectional Density for 0.416 Caliber

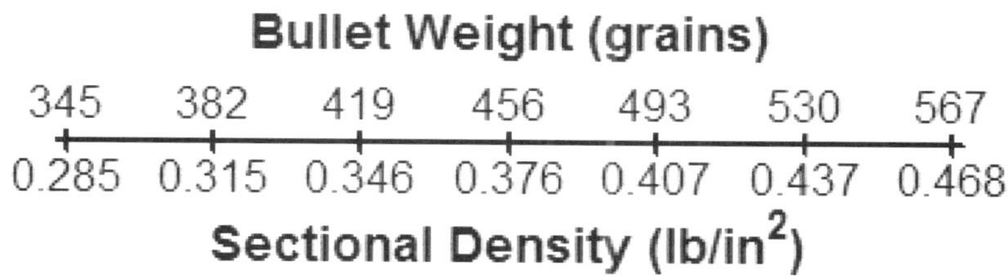

Sectional Density (SD) is a measure of how much mass is behind the bullets frontal area. Higher SD means a greater ability for the bullet to penetrate targets on impact, and also retain velocity in flight.

Spectrum of G7 form factors

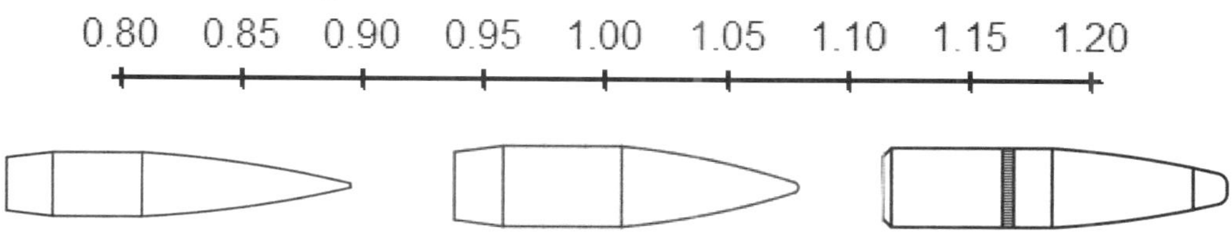

G7 Form factors range from 0.80 for the lowest drag shapes, up to 1.20+ for blunter shapes.

G7 Ballistic Coefficients are simply Sectional Density (SD) divided by G7 form factor.

You can determine a bullets' suitability for long range performance by seeing where it lies on the SD and form factor lines. Ideally, you want a bullet with a high SD and a low G7 form factor.

As an example, suppose a 398 grain .416 caliber bullet has an SD of 0.329. If that bullet had a G7 form factor of 0.92, it's G7 BC would be:

G7 BC = SD/FF = 0.329/0.920 = 0.357.

Barrett .416 caliber 398 grain Brass Solid

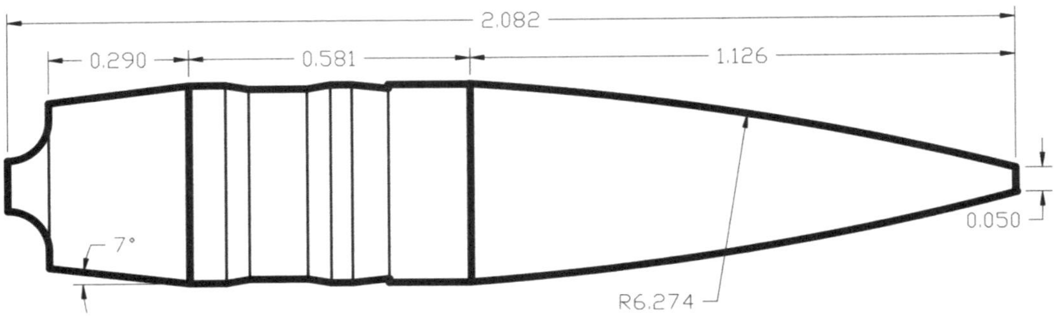

Lot #:	unknown	Ogive Radius:	15.08 calibers
Sectional Density:	0.329 lb/in^2	Rt/R:	0.57

Ballistic Performance

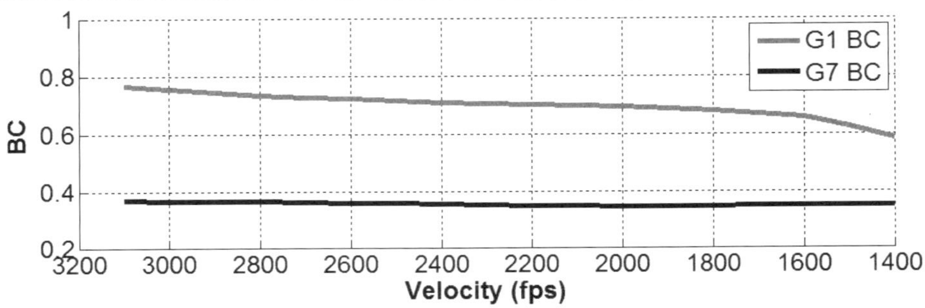

	Above 3000	3000-2500	2500-2000	2000-1500	Below 1500
G1 BC	0.786	0.736	0.704	0.659	0.626
G7 BC	0.372	0.365	0.354	0.350	0.352

Averages from 3000 fps to 1500 fps:
G1 BC: 0.697, variation of G1 BC: 0.131
G7 BC: 0.357, variation of G7 BC: 0.022
G7 form factor: 0.920

Rifling Twist Rate and Stability

Barrel Twist	Stability Factor Best case	Nominal	Worst case
1:11"	2.08	1.82	1.57
1:12"	1.75	1.53	1.32
1:13"	1.49	1.30	1.13
1:14"	1.29	1.12	0.97

Minimum twist required for stability is 1:13.5"
Recommended twist for optimal performance is 1:12.1" or faster

Cutting Edge .416 caliber 422 grain MTAC

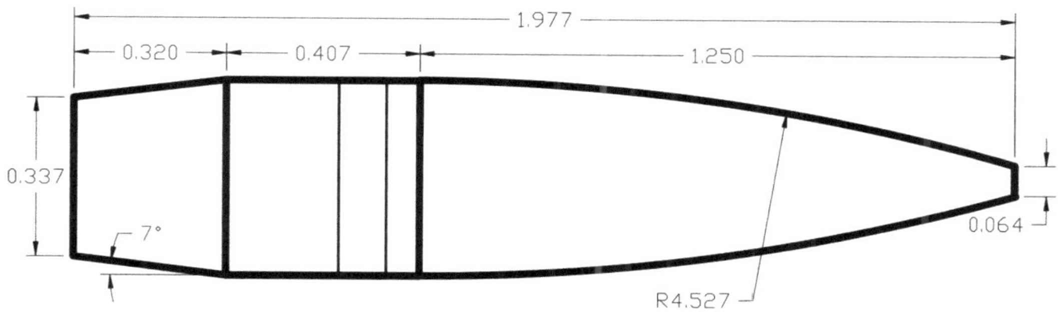

Lot #:	12042014	Ogive Radius:	10.88 calibers
Sectional Density:	0.348 lb/in^2	Rt/R:	1.00

Ballistic Performance

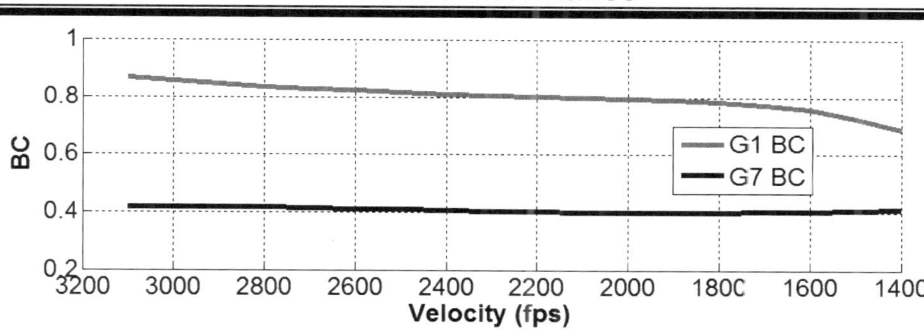

	Above 3000	3000-2500	2500-2000	2000-1500	Below 1500
G1 BC	0.888	0.837	0.806	0.762	0.727
G7 BC	0.421	0.415	0.405	0.405	0.409

Averages from 3000 fps to 1500 fps:
G1 BC: 0.799, variation of G1 BC: 0.130
G7 BC: 0.410, variation of G7 BC: 0.020
G7 form factor: 0.850

Rifling Twist Rate and Stability

Barrel Twist	Stability Factor Best case	Nominal	Worst case
1:12"	2.16	1.89	1.63
1:13"	1.84	1.61	1.39
1:14"	1.59	1.39	1.20
1:15"	1.38	1.21	1.04

Minimum twist required for stability is 1:15.0"
Recommended twist for optimal performance is 1:13.5" or faster

Cutting Edge .416 caliber 446 grain MTAC

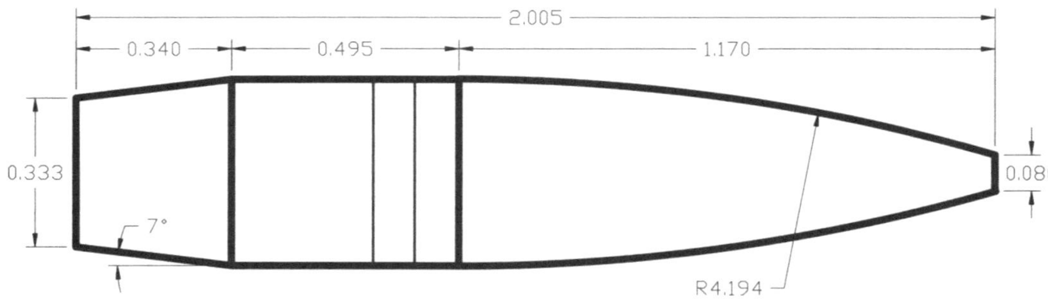

Lot #:	unknown	Ogive Radius:	10.08 calibers
Sectional Density:	0.368 lb/in^2	Rt/R:	0.99

Ballistic Performance

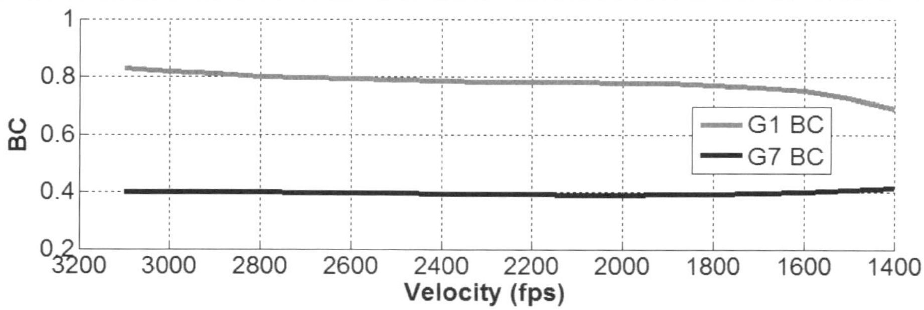

	Above 3000	3000-2500	2500-2000	2000-1500	Below 1500
G1 BC	0.842	0.803	0.783	0.752	0.725
G7 BC	0.399	0.398	0.394	0.400	0.408

Averages from 3000 fps to 1500 fps:
G1 BC: 0.777, variation of G1 BC: 0.093
G7 BC: 0.399, variation of G7 BC: 0.018
G7 form factor: 0.923

Rifling Twist Rate and Stability

Barrel Twist	Stability Factor Best case	Nominal	Worst case
1:12"	2.19	1.91	1.65
1:13"	1.87	1.63	1.41
1:14"	1.61	1.41	1.22
1:15"	1.40	1.22	1.06

Minimum twist required for stability is 1:15.1"
Recommended twist for optimal performance is 1:13.6" or faster

Cutting Edge .416 caliber 420 grain MTH

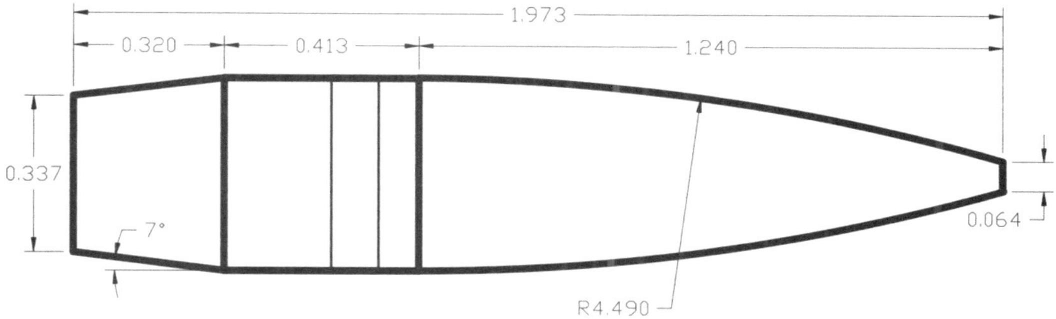

Lot #:	03132015	Ogive Radius:	10.79 calibers
Sectional Density:	0.347 lb/in^2	Rt/R:	1.00

Ballistic Performance

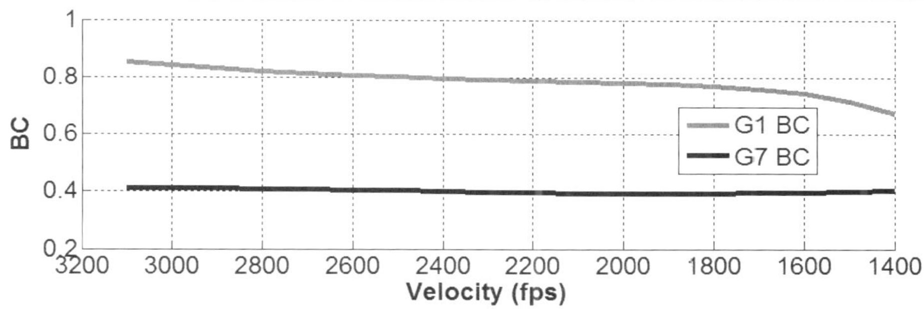

	Above 3000	3000-2500	2500-2000	2000-1500	Below 1500
G1 BC	0.871	0.821	0.791	0.748	0.714
G7 BC	0.413	0.407	0.398	0.397	0.402

Averages from 3000 fps to 1500 fps:
G1 BC: 0.785, variation of G1 BC: 0.127
G7 BC: 0.402, variation of G7 BC: 0.019
G7 form factor: 0.862

Rifling Twist Rate and Stability

Barrel Twist	Stability Factor Best case	Nominal	Worst case
1:12"	2.16	1.89	1.63
1:13"	1.84	1.61	1.39
1:14"	1.59	1.39	1.20
1:15"	1.38	1.21	1.04

Minimum twist required for stability is 1:15.1"
Recommended twist for optimal performance is 1:13.5" or faster

Cutting Edge .416 caliber 444 grain MTH

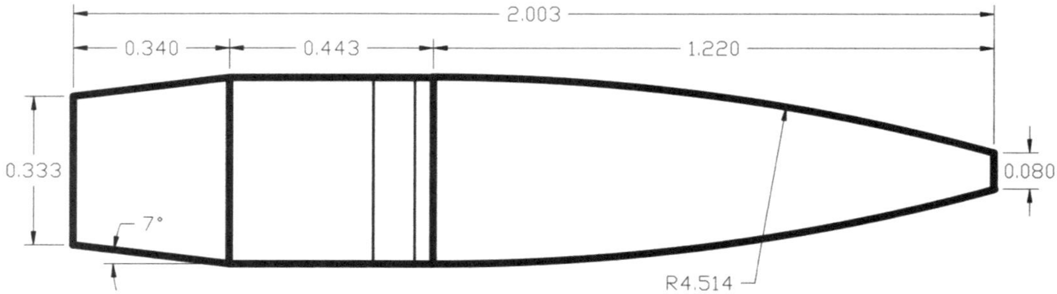

Lot #:	03132015	Ogive Radius:	10.85 calibers
Sectional Density:	0.367 lb/in^2	Rt/R:	1.00

Ballistic Performance

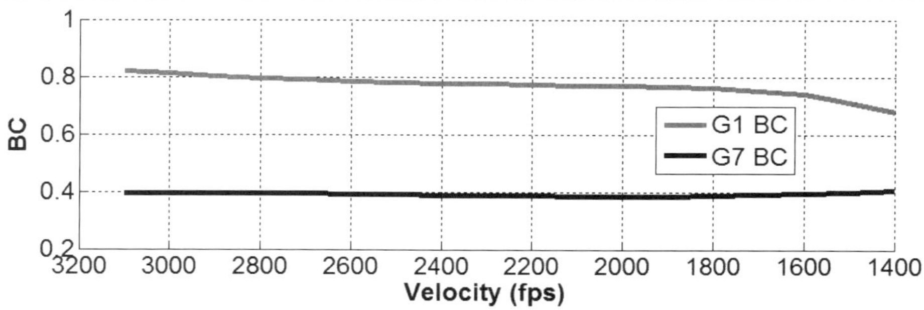

	Above 3000	3000-2500	2500-2000	2000-1500	Below 1500
G1 BC	0.838	0.797	0.776	0.742	0.714
G7 BC	0.397	0.395	0.390	0.394	0.402

Averages from 3000 fps to 1500 fps:
G1 BC: 0.770, variation of G1 BC: 0.099
G7 BC: 0.395, variation of G7 BC: 0.016
G7 form factor: 0.928

Rifling Twist Rate and Stability

Barrel Twist	Stability Factor Best case	Nominal	Worst case
1:12"	2.19	1.91	1.65
1:13"	1.86	1.63	1.41
1:14"	1.61	1.40	1.21
1:15"	1.40	1.22	1.06

Minimum twist required for stability is 1:15.1"
Recommended twist for optimal performance is 1:13.5" or faster

Lehigh Defense .416 caliber 416 grain Solid Match

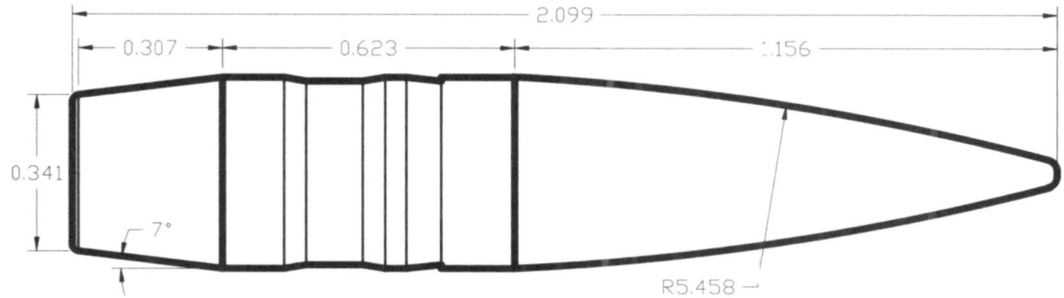

Lot #:	**unknown**	**Ogive Radius:**	**13.12 calibers**
Sectional Density:	**0.343 lb/in^2**	**Rt/R:**	**0.69**

Ballistic Performance

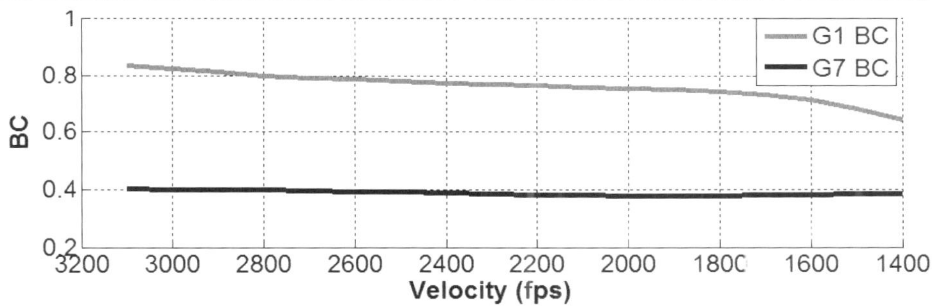

	Above 3000	3000-2500	2500-2000	2000-1500	Below 1500
G1 BC	0.856	0.802	0.768	0.720	0.685
G7 BC	0.405	0.398	0.386	0.382	0.385

Averages from 3000 fps to 1500 fps:
G1 BC: 0.761, variation of G1 BC: 0.139
G7 BC: 0.390, variation of G7 BC: 0.023
G7 form factor: 0.881

Rifling Twist Rate and Stability

Barrel Twist	**Stability Factor** Best case	Nominal	Worst case
1:11"	2.16	1.89	1.64
1:12"	1.82	1.59	1.37
1:13"	1.55	1.35	1.17
1:14"	1.34	1.17	1.01

Minimum twist required for stability is 1:13.8"
Recommended twist for optimal performance is 1:12.4" or faster

.510 Caliber

.510 caliber bullets typically range from 650 to 800+ grains, which, proportionally, is on the light end of the scale compared to smaller calibers. The image below relates bullet weight to sectional density for .510 caliber bullets.

Range of Weight and Sectional Density for 0.51 Caliber

Bullet Weight (grains)

636	704	772	840	908	976	1044
0.349	0.387	0.424	0.461	0.499	0.536	0.574

Sectional Density (lb/in^2)

Sectional Density (SD) is a measure of how much mass is behind the bullets frontal area. Higher SD means a greater ability for the bullet to penetrate targets on impact, and also retain velocity in flight.

Spectrum of G7 form factors

0.80 0.85 0.90 0.95 1.00 1.05 1.10 1.15 1.20

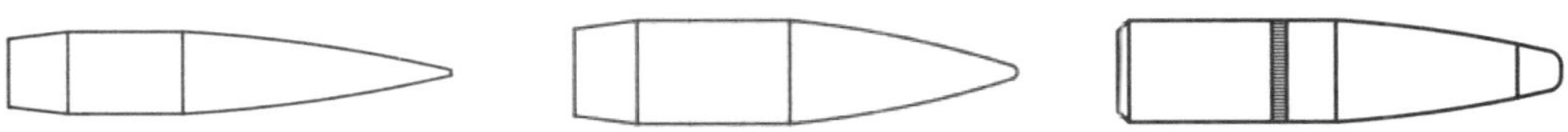

G7 Form factors range from 0.80 for the lowest drag shapes, up to 1.20+ for blunter shapes.

G7 Ballistic Coefficients are simply Sectional Density (SD) divided by G7 form factor.

You can determine a bullets' suitability for long range performance by seeing where it lies on the SD and form factor lines. Ideally, you want a bullet with a high SD and a low G7 form factor.

As an example, suppose a 750 grain .510 caliber bullet has an SD of 0.412. If that bullet had a G7 form factor of 0.811, it's G7 BC would be:

G7 BC = SD/FF = 0.412/0.811 = 0.508.

Barnes .510 caliber 750 grain LR Solid Bore Rider

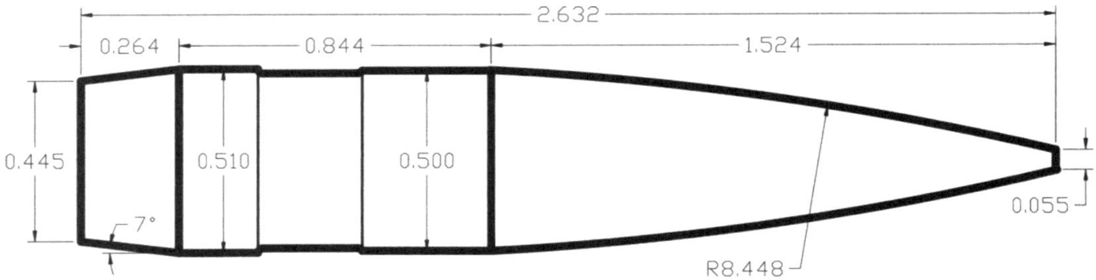

Lot #:	6368781-1	Ogive Radius:	16.57 calibers
Sectional Density:	0.412 lb/in^2	Rt/R:	0.62

Ballistic Performance

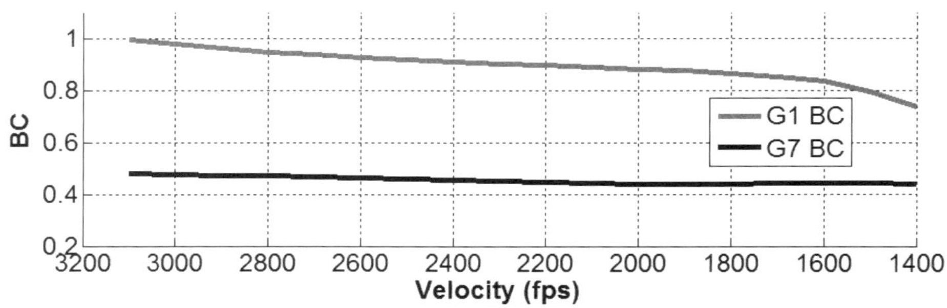

	Above 3000	3000-2500	2500-2000	2000-1500	Below 1500
G1 BC	1.022	0.948	0.901	0.839	0.795
G7 BC	0.484	0.470	0.453	0.445	0.446

Averages from 3000 fps to 1500 fps:
G1 BC: 0.894, variation of G1 BC: 0.184
G7 BC: 0.458, variation of G7 BC: 0.034
G7 form factor: 0.901

Rifling Twist Rate and Stability

Barrel Twist	Stability Factor Best case	Nominal	Worst case
1:13"	2.10	1.83	1.58
1:14"	1.81	1.58	1.36
1:15"	1.57	1.37	1.19
1:16"	1.38	1.21	1.04

Minimum twist required for stability is 1:16.1"
Recommended twist for optimal performance is 1:14.4" or faster

Barnes .510 caliber 800 grain LR Solid

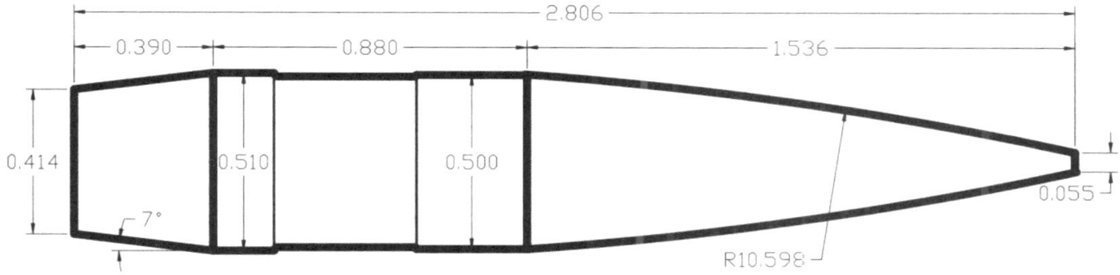

Lot #:	**6459446-1**	**Ogive Radius:**	**20.78 calibers**
Sectional Density:	**0.439 lb/in²**	**Rt/R:**	**0.47**

Ballistic Performance

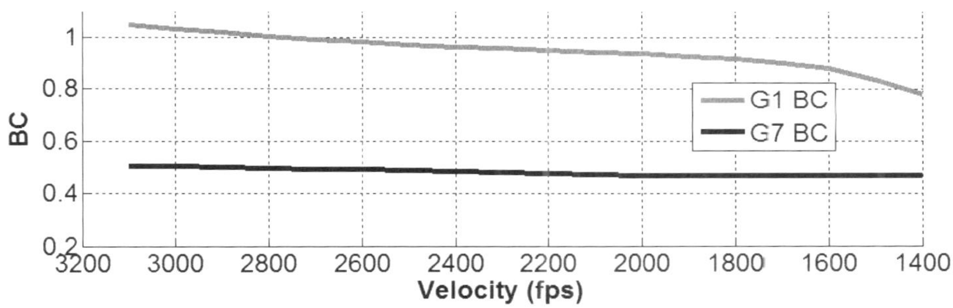

	Above 3000	3000-2500	2500-2000	2000-1500	Below 1500
G1 BC	1.077	1.004	0.955	0.888	0.839
G7 BC	0.510	0.498	0.480	0.471	0.472

Averages from 3000 fps to 1500 fps:
G1 BC: 0.946, variation of G1 BC: 0.195
G7 BC: 0.484, variation of G7 BC: 0.035
G7 form factor: 0.908

Rifling Twist Rate and Stability

	Stability Factor		
Barrel Twist	**Best case**	**Nominal**	**Worst case**
1:12"	2.17	1.90	1.64
1:13"	1.85	1.62	1.40
1:14"	1.60	1.40	1.21
1:15"	1.39	1.22	1.05

Minimum twist required for stability is 1:15.1"
Recommended twist for optimal performance is 1:13.5" or faster

Barnes .510 caliber 750 grain Tac-LR

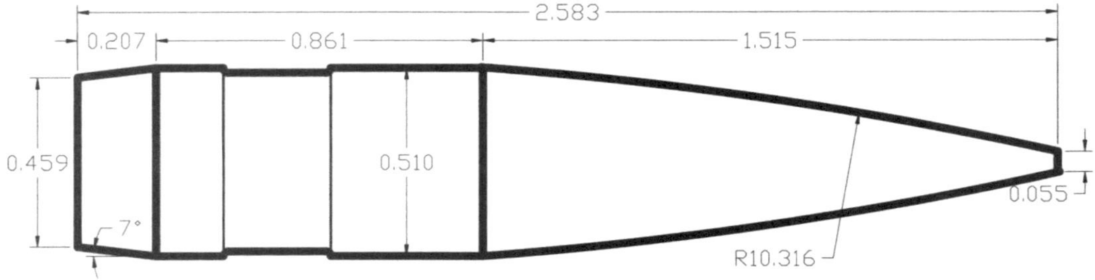

Lot #:	6464861-1	Ogive Radius:	22.23 calibers
Sectional Density:	0.412 lb/in^2	Rt/R:	0.49

Ballistic Performance

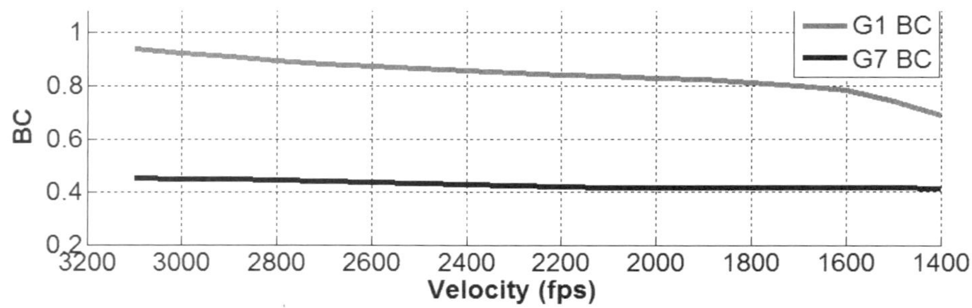

	Above 3000	3000-2500	2500-2000	2000-1500	Below 1500
G1 BC	0.967	0.894	0.847	0.788	0.745
G7 BC	0.458	0.443	0.426	0.418	0.419

Averages from 3000 fps to 1500 fps:
G1 BC: 0.841, variation of G1 BC: 0.179
G7 BC: 0.431, variation of G7 BC: 0.035
G7 form factor: 0.958

Rifling Twist Rate and Stability

Barrel Twist	Stability Factor Best case	Nominal	Worst case
1:13"	2.21	1.93	1.67
1:14"	1.91	1.67	1.44
1:15"	1.66	1.45	1.26
1:16"	1.46	1.28	1.10

Minimum twist required for stability is 1:16.5"
Recommended twist for optimal performance is 1:14.8" or faster

Barnes .510 caliber 647 grain Tac-X, TSX

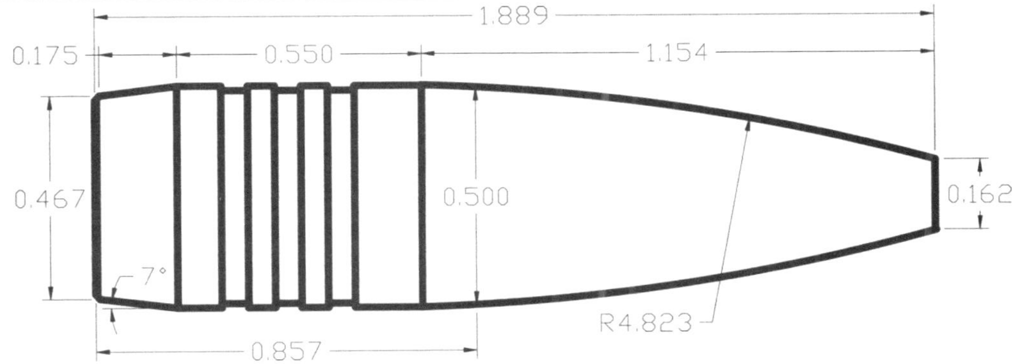

Lot #:	6397453-1	Ogive Radius:	9.46 calibers
Sectional Density:	0.355 lb/in^2	Rt/R:	0.81

Ballistic Performance

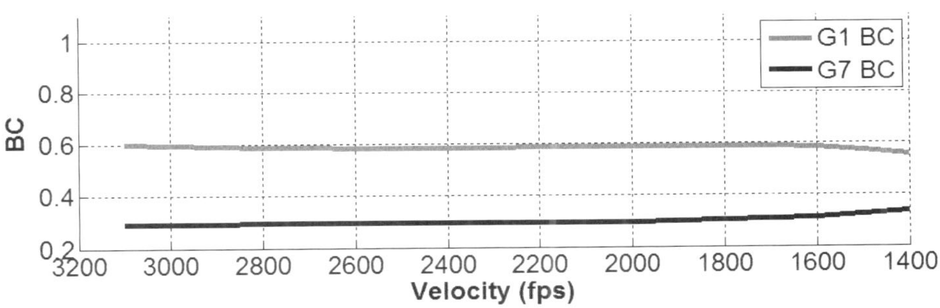

	Above 3000	3000-2500	2500-2000	2000-1500	Below 1500
G1 BC	0.610	0.591	0.587	0.581	0.571
G7 BC	0.289	0.293	0.295	0.310	0.323

Averages from 3000 fps to 1500 fps:
G1 BC: 0.586, variation of G1 BC: 0.025
G7 BC: 0.301, variation of G7 BC: 0.031
G7 form factor: 1.181

Rifling Twist Rate and Stability

Barrel Twist	Stability Factor Best case	Nominal	Worst case
1:20"	2.00	1.75	1.51
1:21"	1.81	1.58	1.37
1:22"	1.65	1.44	1.25
1:23"	1.51	1.32	1.14

Minimum twist required for stability is 1:24.1"
Recommended twist for optimal performance is 1:21.6" or faster

Cutting Edge .510 caliber 720 grain MTAC_X03

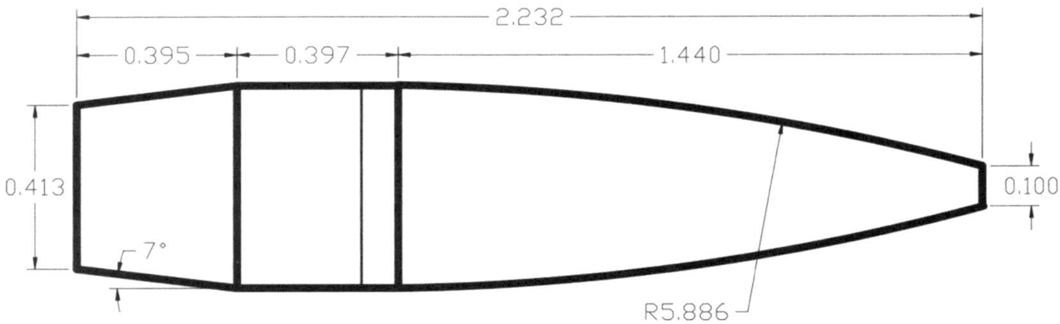

Lot #:	08252014	Ogive Radius:	11.54 calibers
Sectional Density:	0.396 lb/in^2	Rt/R:	0.88

Ballistic Performance

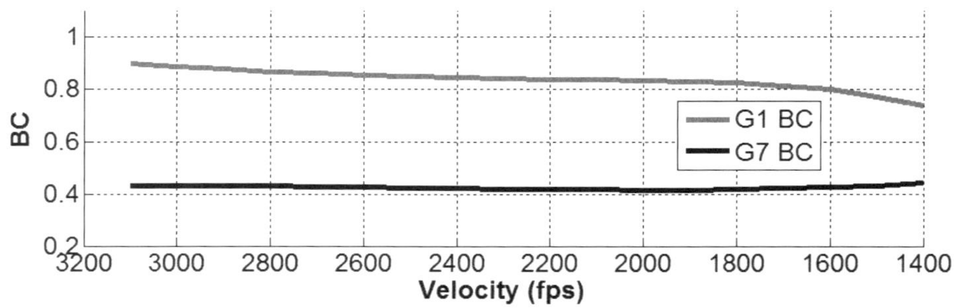

	Above 3000	3000-2500	2500-2000	2000-1500	Below 1500
G1 BC	0.915	0.868	0.842	0.804	0.774
G7 BC	0.434	0.430	0.423	0.427	0.436

Averages from 3000 fps to 1500 fps:
G1 BC: 0.836, variation of G1 BC: 0.113
G7 BC: 0.429, variation of G7 BC: 0.017
G7 form factor: 0.922

Rifling Twist Rate and Stability

Barrel Twist	Stability Factor Best case	Nominal	Worst case
1:16"	2.15	1.88	1.62
1:17"	1.90	1.66	1.44
1:18"	1.70	1.48	1.28
1:19"	1.52	1.33	1.15

Minimum twist required for stability is 1:20.0"
Recommended twist for optimal performance is 1:17.9" or faster

Cutting Edge .510 caliber 762 grain MTAC_X02

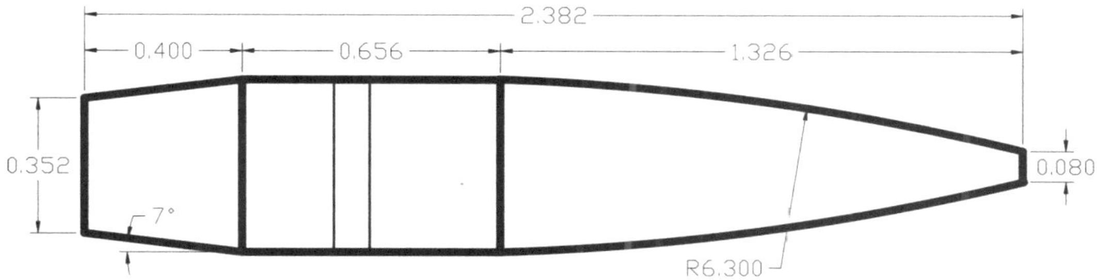

Lot #:	08252014	Ogive Radius:	12.35 calibers
Sectional Density:	0.419 lb/in^2	Rt/R:	0.67

Ballistic Performance

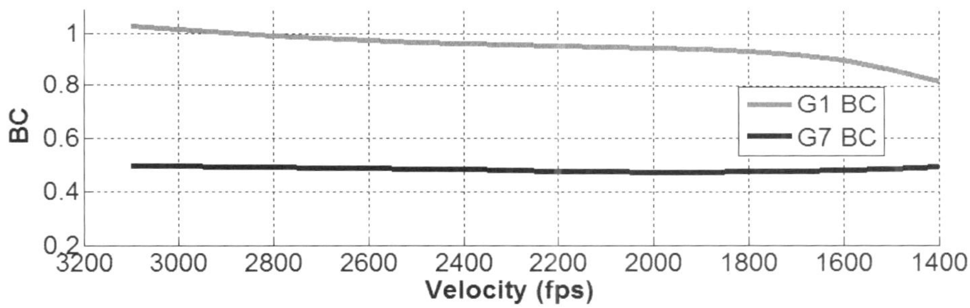

	Above 3000	3000-2500	2500-2000	2000-1500	Below 1500
G1 BC	1.053	0.993	0.958	0.907	0.867
G7 BC	0.499	0.493	0.482	0.482	0.488

Averages from 3000 fps to 1500 fps:
G1 BC: 0.950, variation of G1 BC: 0.150
G7 BC: 0.487, variation of G7 BC: 0.022
G7 form factor: 0.859

Rifling Twist Rate and Stability

	Stability Factor		
Barrel Twist	**Best case**	**Nominal**	**Worst case**
1:15"	2.14	1.87	1.62
1:16"	1.88	1.64	1.42
1:17"	1.67	1.46	1.26
1:18"	1.49	1.30	1.12

Minimum twist required for stability is 1:18 7"
Recommended twist for optimal performance is 1:16.7" or faster

Cutting Edge .510 caliber 802 grain MTAC_X01

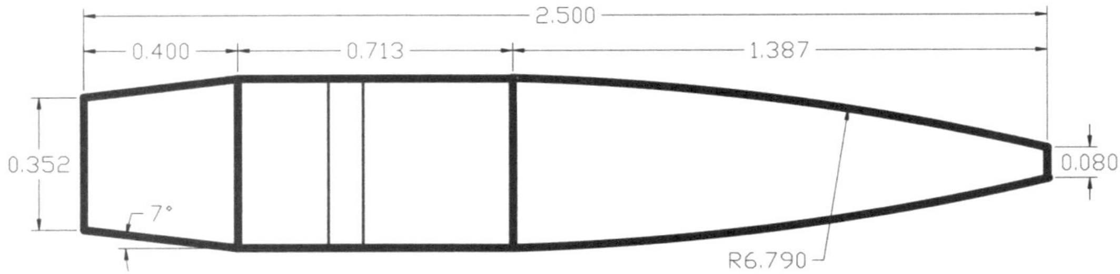

Lot #:	01212015	Ogive Radius:	13.31 calibers
Sectional Density:	0.441 lb/in^2	Rt/R:	0.68

Ballistic Performance

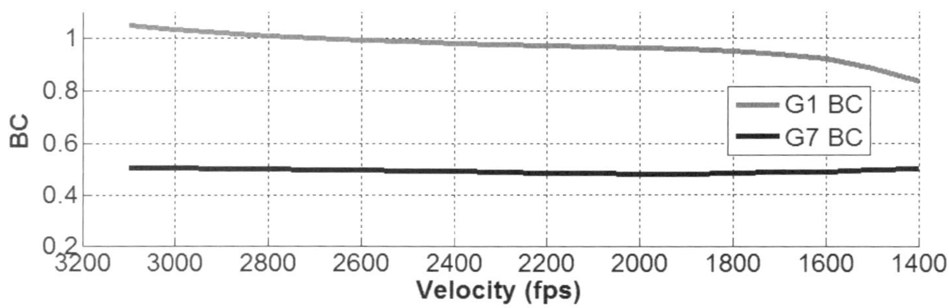

	Above 3000	3000-2500	2500-2000	2000-1500	Below 1500
G1 BC	1.071	1.010	0.975	0.925	0.885
G7 BC	0.507	0.501	0.490	0.491	0.498

Averages from 3000 fps to 1500 fps:
G1 BC: 0.967, variation of G1 BC: 0.149
G7 BC: 0.496, variation of G7 BC: 0.021
G7 form factor: 0.888

Rifling Twist Rate and Stability

Barrel Twist	Stability Factor Best case	Nominal	Worst case
1:15"	1.96	1.71	1.48
1:16"	1.72	1.50	1.30
1:17"	1.52	1.33	1.15
1:18"	1.36	1.19	1.03

Minimum twist required for stability is 1:17.9"
Recommended twist for optimal performance is 1:16.0" or faster

Hornady .510 caliber 750 grain Amax

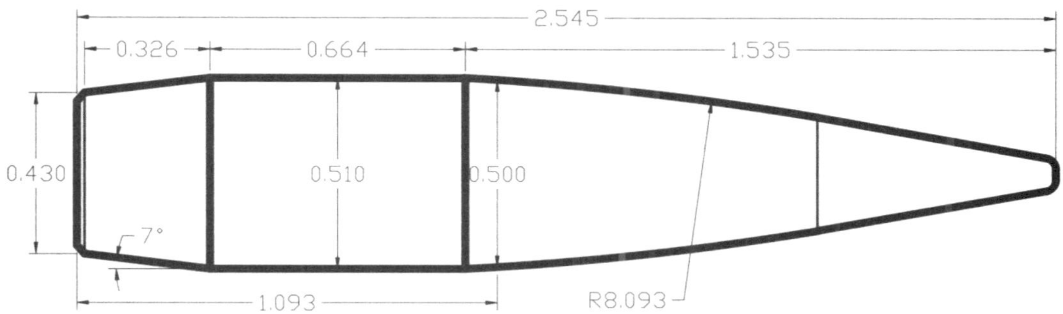

Lot #:	21307134	Ogive Radius:	15.89 calibers
Sectional Density:	0.412 lb/in^2	Rt/R:	0.69

Ballistic Performance

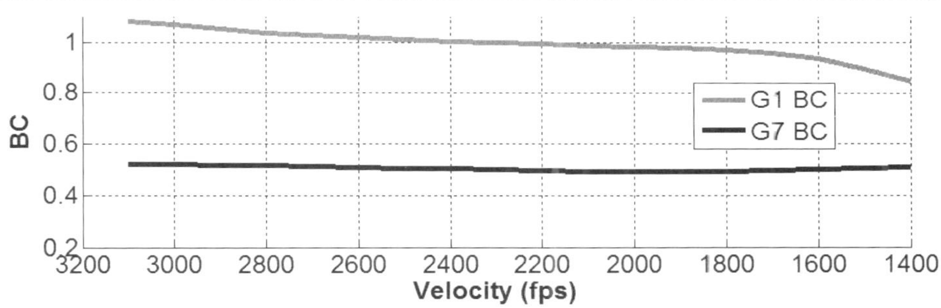

	Above 3000	3000-2500	2500-2000	2000-1500	Below 1500
G1 BC	1.108	1.040	0.998	0.942	0.900
G7 BC	0.525	0.516	0.502	0.501	0.506

Averages from 3000 fps to 1500 fps:
G1 BC: 0.991, variation of G1 BC: 0.168
G7 BC: 0.508, variation of G7 BC: 0.027
G7 form factor: 0.811

Rifling Twist Rate and Stability

Barrel Twist	Stability Factor Best case	Nominal	Worst case
1:14"	1.99	1.74	1.51
1:15"	1.74	1.52	1.31
1:16"	1.53	1.33	1.15
1:17"	1.35	1.18	1.02

Minimum twist required for stability is 1:16.9"
Recommended twist for optimal performance is 1:15.1" or faster

Lapua .510 caliber 800 grain Bullex-N

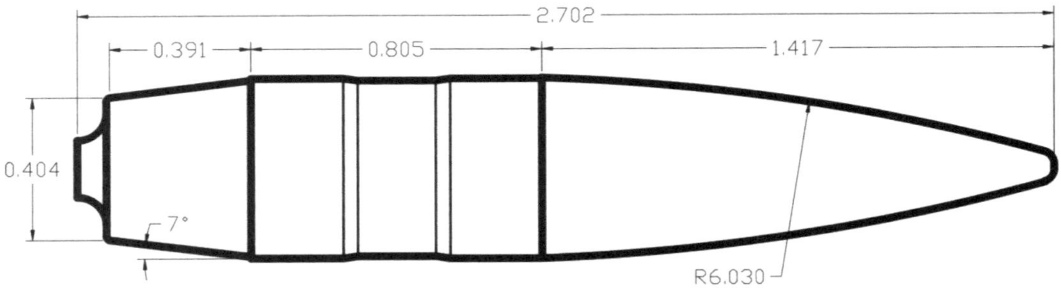

Lot #:	unknown	Ogive Radius:	11.82 calibers
Sectional Density:	0.439 lb/in²	Rt/R:	0.78

Ballistic Performance

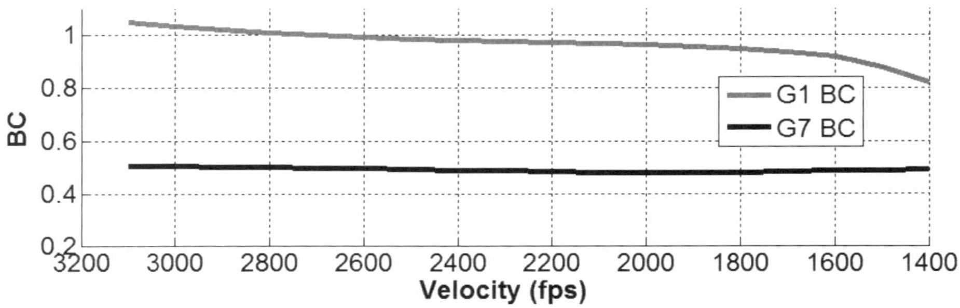

	Above 3000	3000-2500	2500-2000	2000-1500	Below 1500
G1 BC	1.071	1.010	0.973	0.919	0.877
G7 BC	0.507	0.501	0.489	0.488	0.493

Averages from 3000 fps to 1500 fps:
G1 BC: 0.964, variation of G1 BC: 0.157
G7 BC: 0.494, variation of G7 BC: 0.023
G7 form factor: 0.889

Rifling Twist Rate and Stability

Barrel Twist	Stability Factor Best case	Nominal	Worst case
1:13"	2.07	1.81	1.56
1:14"	1.78	1.56	1.35
1:15"	1.55	1.36	1.17
1:16"	1.37	1.19	1.03

Minimum twist required for stability is 1:16.0"
Recommended twist for optimal performance is 1:14.3" or faster

Lehigh Defense .510 caliber 750 grain Controlled Chaos

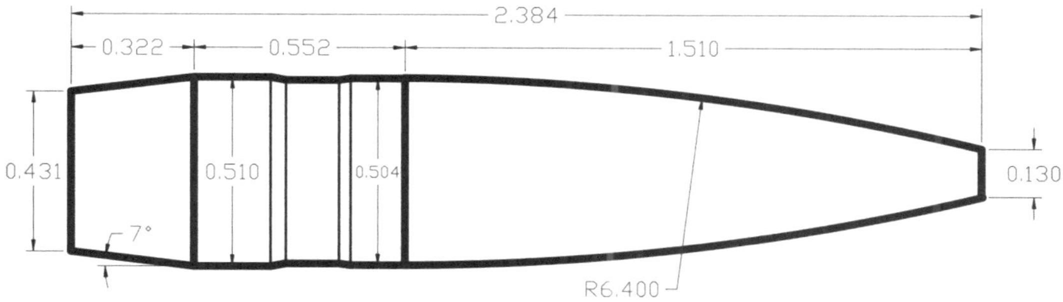

Lot #:	unknown	Ogive Radius:	12.55 calibers
Sectional Density:	0.412 lb/in^2	Rt/R:	0.95

Ballistic Performance

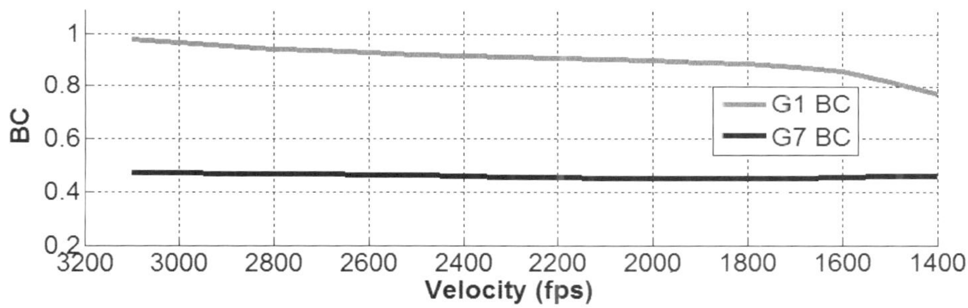

	Above 3000	3000-2500	2500-2000	2000-1500	Below 1500
G1 BC	1.004	0.946	0.913	0.862	0.822
G7 BC	0.475	0.469	0.459	0.457	0.462

Averages from 3000 fps to 1500 fps:
G1 BC: 0.904, variation of G1 BC: 0.148
G7 BC: 0.463, variation of G7 BC: 0.021
G7 form factor: 0.889

Rifling Twist Rate and Stability

Barrel Twist	Stability Factor Best case	Nominal	Worst case
1:19"	1.91	1.66	1.44
1:20"	1.72	1.50	1.30
1:21"	1.56	1.36	1.18
1:22"	1.42	1.24	1.07

Minimum twist required for stability is 1:22.4"
Recommended twist for optimal performance is 1:20.0" or faster

Lehigh Defense .510 caliber 650 grain Solid Match

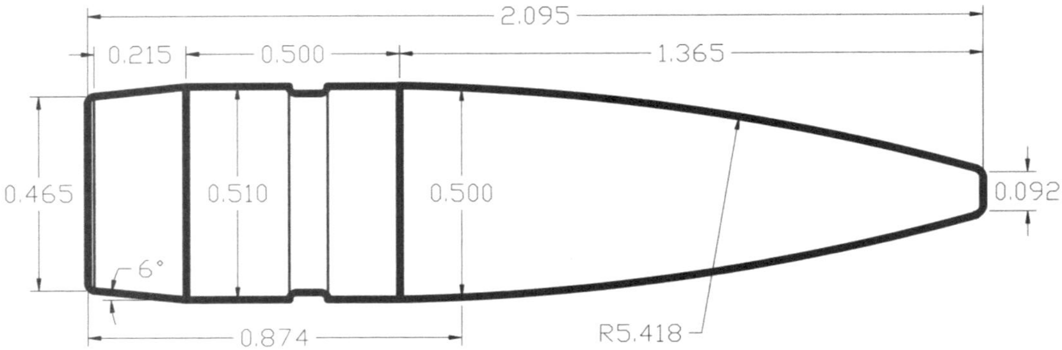

Lot #:	14209	Ogive Radius:	10.62 calibers
Sectional Density:	0.357 lb/in^2	Rt/R:	0.86

Ballistic Performance

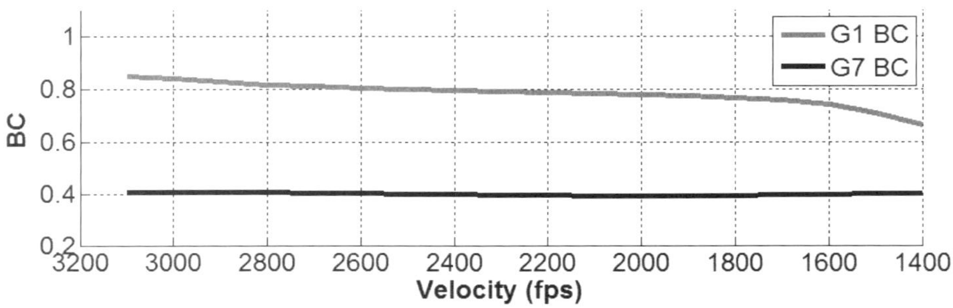

	Above 3000	3000-2500	2500-2000	2000-1500	Below 1500
G1 BC	0.870	0.820	0.791	0.747	0.712
G7 BC	0.412	0.407	0.398	0.396	0.401

Averages from 3000 fps to 1500 fps:
G1 BC: 0.784, variation of G1 BC: 0.128
G7 BC: 0.402, variation of G7 BC: 0.018
G7 form factor: 0.889

Rifling Twist Rate and Stability

Barrel Twist	Stability Factor Best case	Nominal	Worst case
1:17"	2.06	1.80	1.56
1:18"	1.84	1.61	1.39
1:19"	1.65	1.44	1.25
1:20"	1.49	1.30	1.13

Minimum twist required for stability is 1:20.8"
Recommended twist for optimal performance is 1:18.6" or faster

Lehigh Defense .510 caliber 750 grain Solid Match

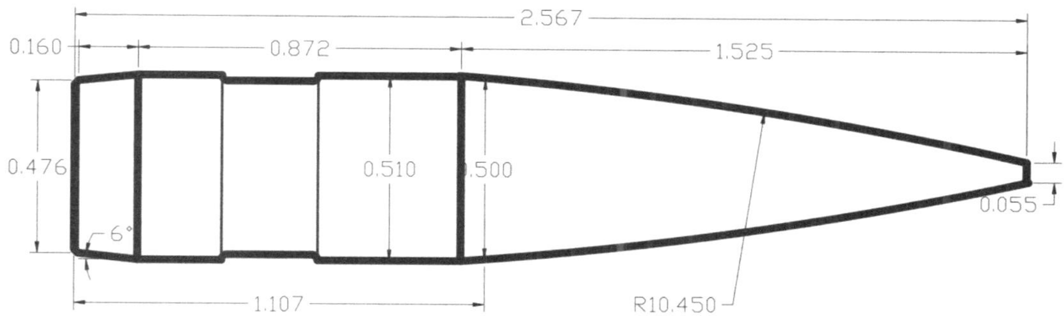

Lot #:	unknown	Ogive Radius:	20.49 calibers
Sectional Density:	0.412 lb/in^2	Rt/R:	0.39

Ballistic Performance

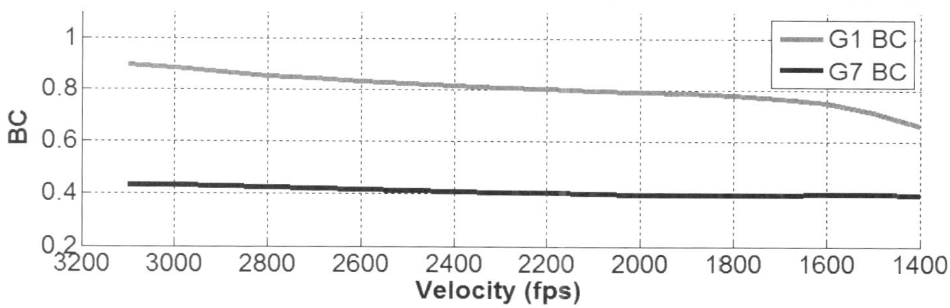

	Above 3000	3000-2500	2500-2000	2000-1500	Below 1500
G1 BC	0.927	0.855	0.809	0.753	0.714
G7 BC	0.438	0.424	0.407	0.400	0.401

Averages from 3000 fps to 1500 fps:
G1 BC: 0.804, variation of G1 BC: 0.171
G7 BC: 0.412, variation of G7 BC: 0.034
G7 form factor: 1.002

Rifling Twist Rate and Stability

Barrel Twist	Stability Factor Best case	Nominal	Worst case
1:13"	2.25	1.97	1.70
1:14"	1.94	1.70	1.47
1:15"	1.69	1.48	1.28
1:16"	1.49	1.30	1.12

Minimum twist required for stability is 1:16.7"
Recommended twist for optimal performance is 1:14.9" or faster

Lehigh Defense .510 caliber 808 grain Solid Match

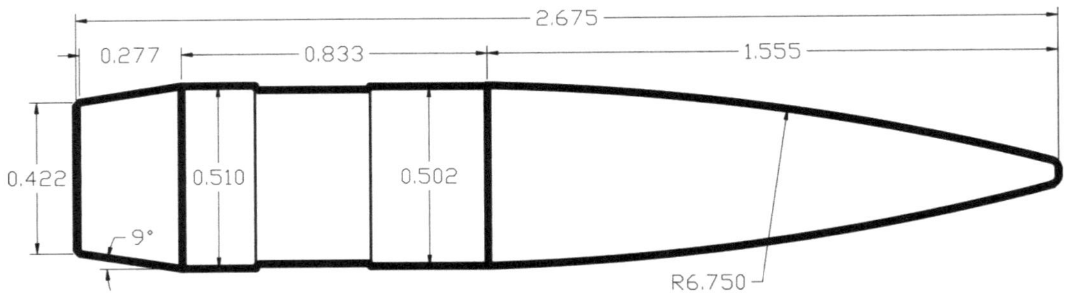

Lot #:	unknown	Ogive Radius:	13.24 calibers
Sectional Density:	0.444 lb/in^2	Rt/R:	0.81

Ballistic Performance

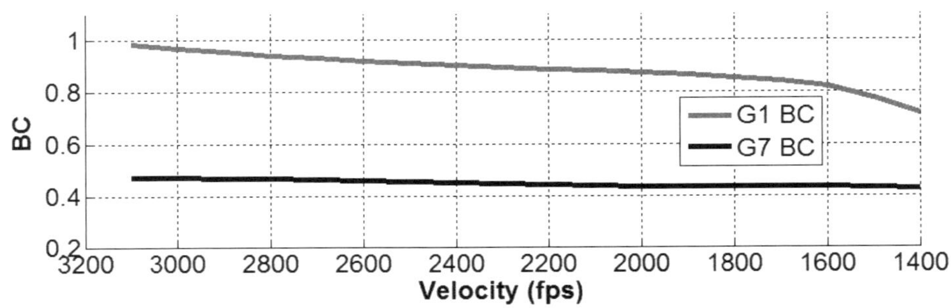

	Above 3000	3000-2500	2500-2000	2000-1500	Below 1500
G1 BC	1.011	0.940	0.892	0.825	0.777
G7 BC	0.479	0.466	0.448	0.437	0.436

Averages from 3000 fps to 1500 fps:
G1 BC: 0.882, variation of G1 BC: 0.193
G7 BC: 0.452, variation of G7 BC: 0.038
G7 form factor: 0.984

Rifling Twist Rate and Stability

Barrel Twist	Stability Factor Best case	Nominal	Worst case
1:13"	2.15	1.88	1.63
1:14"	1.86	1.62	1.40
1:15"	1.62	1.41	1.22
1:16"	1.42	1.24	1.07

Minimum twist required for stability is 1:16.3"
Recommended twist for optimal performance is 1:14.6" or faster

.224 Caliber bullets

Bullet	SD	i_7	G7 BC	G1 BC
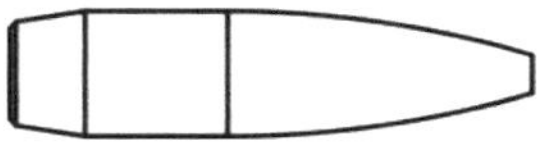 *Barnes 69 Match Burner*	0.197	1.172	0.168	0.326
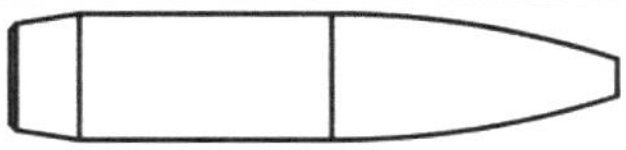 *Barnes 85 Match Burner*	0.242	1.188	0.204	0.397
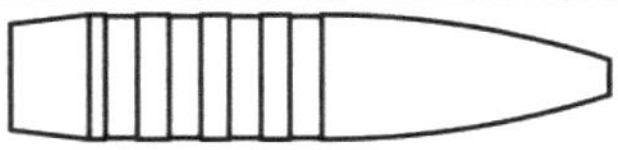 *Barnes 70 TacX MIL/LE*	0.199	1.295	0.154	0.300
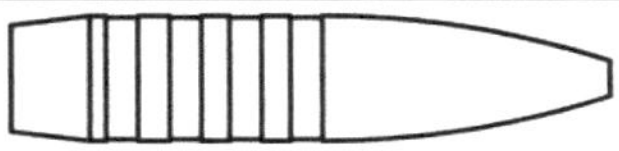 *Barnes 70 TSXBT*	0.199	1.260	0.158	0.308
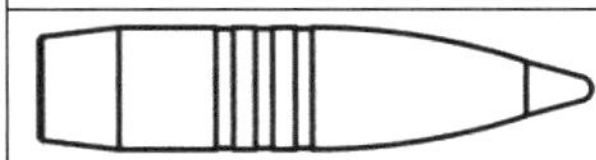 *Barnes 62 TTSXBT*	0.177	1.243	0.142	0.277
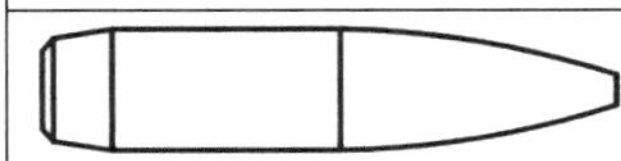 *Berger 77 Tactical OTM*	0.219	1.142	0.192	0.374
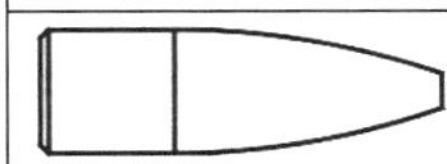 *Berger 52 Target FB*	0.148	1.231	0.120	0.234
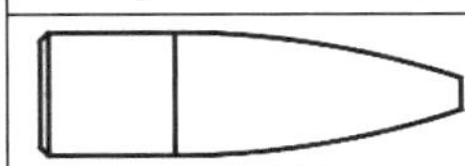 *Berger 55 Target FB*	0.157	1.205	0.130	0.253
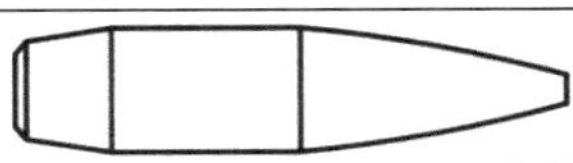 *Berger 70 Target VLD*	0.199	1.042	0.191	0.373
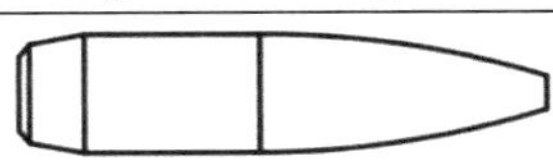 *Berger 73 Target Boat Tail*	0.208	1.168	0.178	0.347
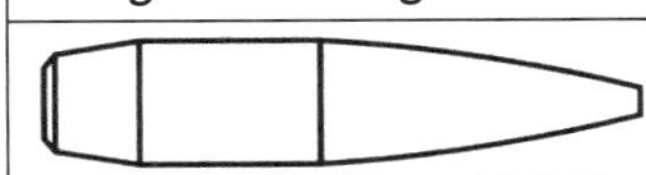 *Berger 75 Target VLD*	0.214	0.989	0.216	0.421
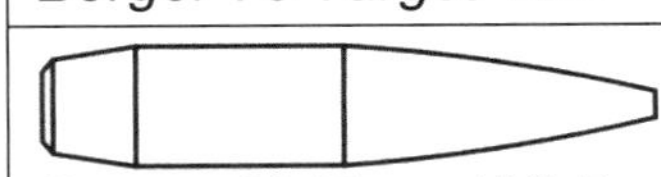 *Berger 80 Target VLD*	0.228	0.977	0.233	0.455

(.224 Caliber bullets continued)				
Bullet	**SD**	**i_7**	**G7 BC**	**G1 BC**
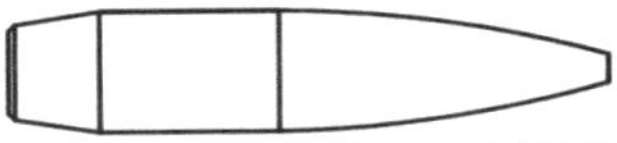 *Berger 80.5 Target FULLBORE*	0.229	1.015	0.277	0.443
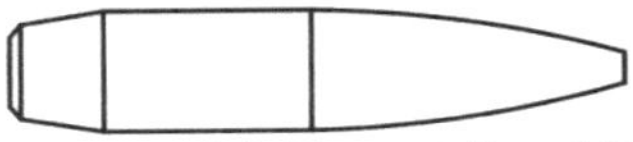 *Berger 82 Target Boat Tail*	0.233	1.038	0.225	0.438
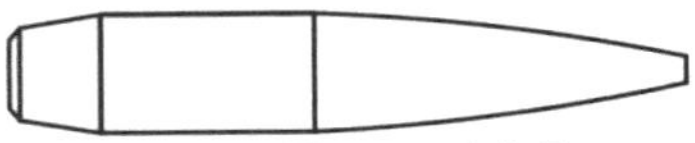 *Berger 90 Target VLD*	0.256	0.937	0.273	0.533
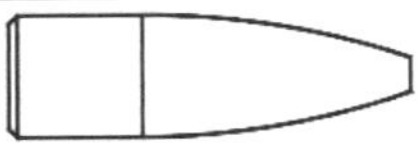 *Berger 52 Varmint FB*	0.148	1.437	0.103	0.201
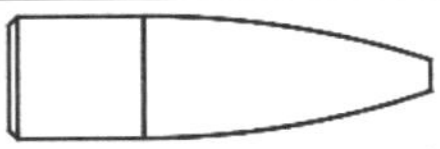 *Berger 55 Varmint FB*	0.157	1.463	0.107	0.209
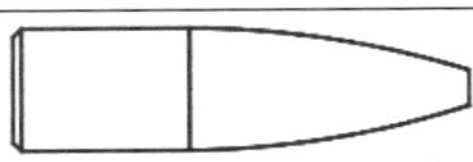 *Berger 60 Varmint FB*	0.171	1.229	0.139	0.271
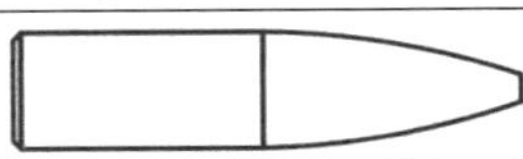 *DRT 79 Frangible*	0.225	1.182	0.190	0.371
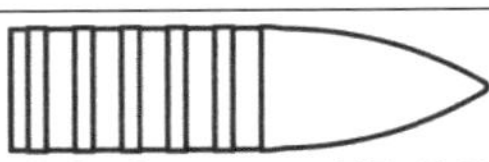 *GS Custom 57 SP*	0.162	1.639	0.099	0.193
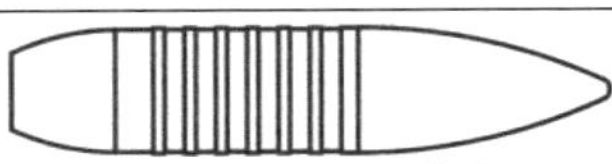 *GS Custom 69 SP*	0.197	1.373	0.143	0.279
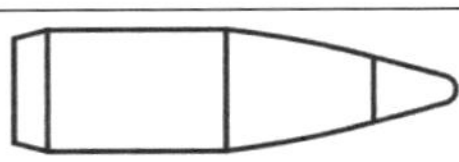 *Hornady 52 Amax*	0.148	1.240	0.120	0.234
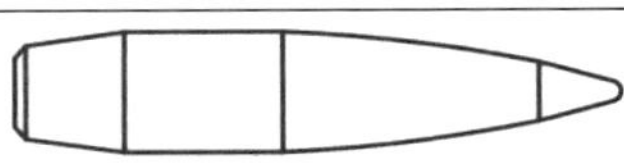 *Hornady 75 Amax*	0.214	0.997	0.214	0.418
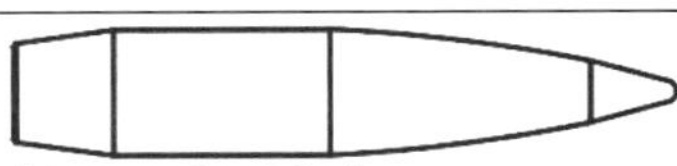 *Hornady 80 Amax*	0.228	0.987	0.231	0.450

(.224 Caliber bullets continued)				
Bullet	**SD**	**i_7**	**G7 BC**	**G1 BC**
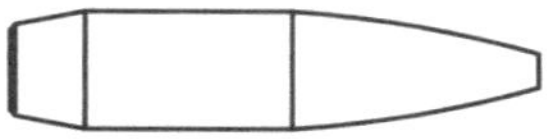 Hornady 68 BTHP	0.194	1.107	0.175	0.341
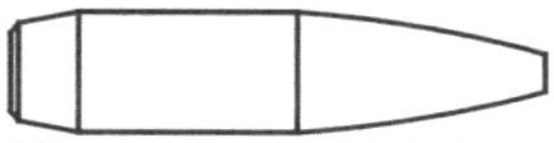 Hornady 75 BTHP	0.214	1.170	0.183	0.356
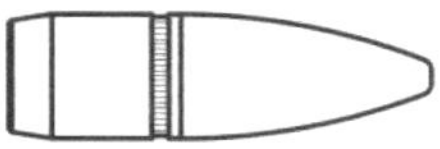 Hornady 55 FMJBT	0.157	1.264	0.124	0.241
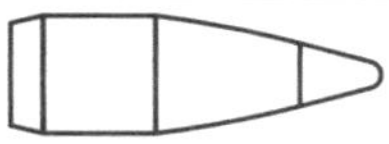 Hornady 40 Vmax	0.114	1.308	0.087	0.170
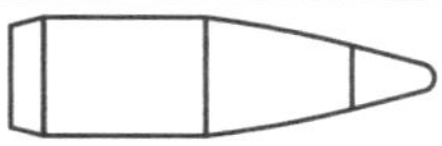 Hornady 50 Vmax	0.142	1.293	0.110	0.216
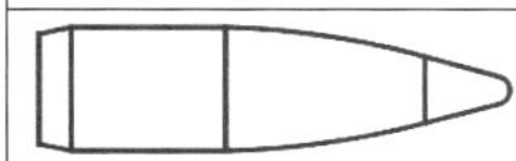 Hornady 53 Vmax	0.151	1.161	0.130	0.254
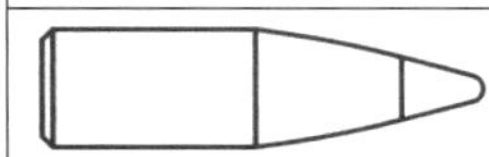 Hornady 55 Vmax	0.157	1.284	0.122	0.238
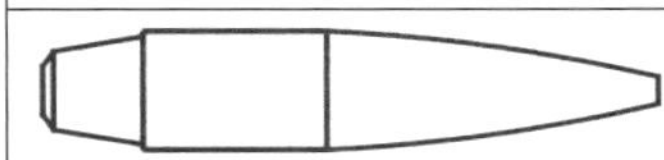 JLK 80 VLD	0.228	0.989	0.230	0.449
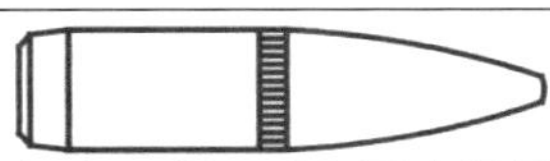 Lake City 62 FMJBT	0.177	1.148	0.154	0.300
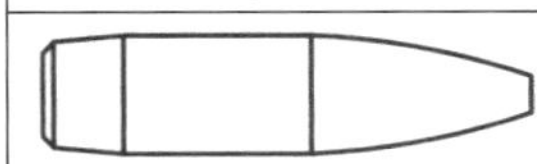 Lapua 69 Scenar	0.197	1.164	0.169	0.329
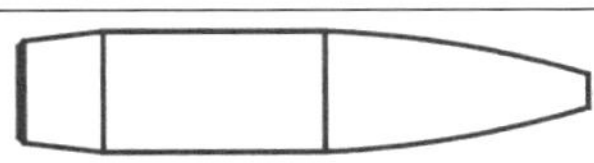 Lapua 77 Scenar	0.219	1.063	0.206	0.401
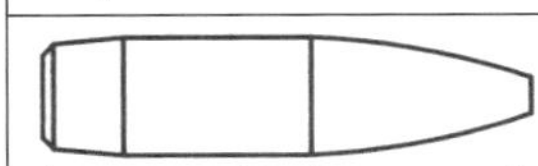 Lapua 69 Scenar L	0.197	1.098	0.179	0.348

(.224 Caliber bullets continued)				
Bullet	**SD**	**i_7**	**G7 BC**	**G1 BC**
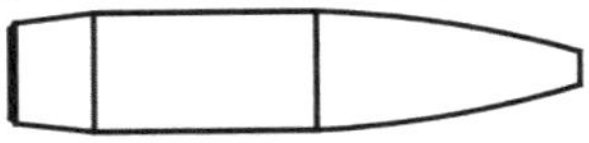 Lapua 77 Scenar L	0.219	1.082	0.203	0.394
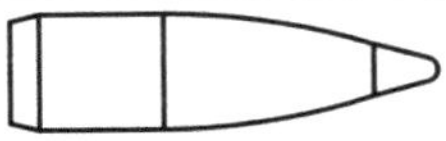 Nosler 55 Ballistic Tip	0.157	1.205	0.130	0.254
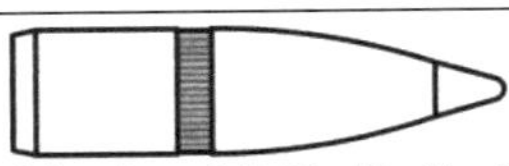 Nosler 60 Ballistic Tip	0.171	1.288	0.133	0.259
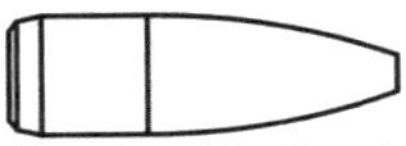 Nosler 52 Custom Competition	0.148	1.311	0.113	0.220
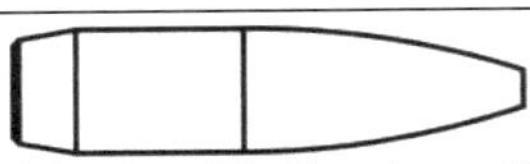 Nosler 69 Custom Competition	0.197	1.239	0.159	0.309
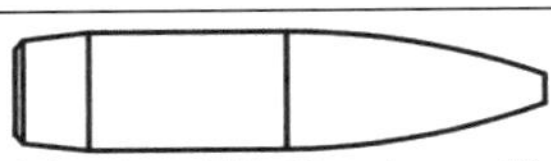 Nosler 77 Custom Competition	0.219	1.138	0.193	0.375
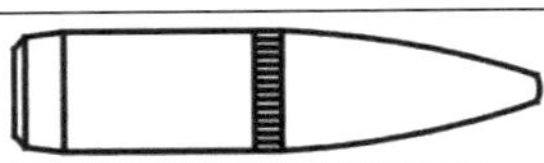 PMC 62 FMJBT Green Tip Ammo	0.177	1.096	0.161	0.314
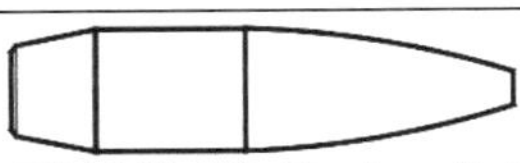 RUAG 69 Swiss P Target	0.197	1.191	0.165	0.322
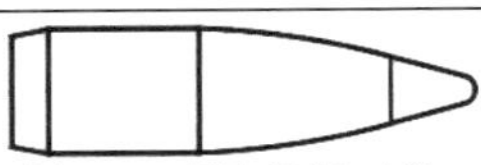 Sierra 55 BlitzKing	0.157	1.214	0.129	0.252
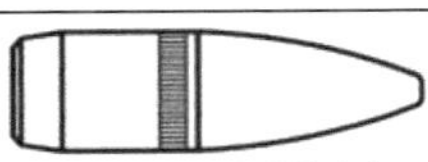 Sierra 55 FMJ	0.157	1.194	0.131	0.255
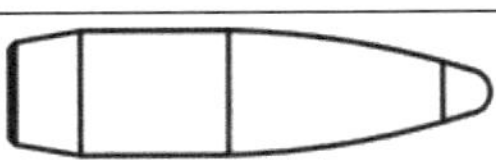 Sierra 65 GameKing	0.185	1.217	0.152	0.296
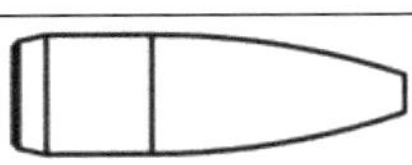 Sierra 52 MatchKing	0.148	1.328	0.112	0.217

(.224 Caliber bullets continued)				
Bullet	**SD**	**i_7**	**G7 BC**	**G1 BC**
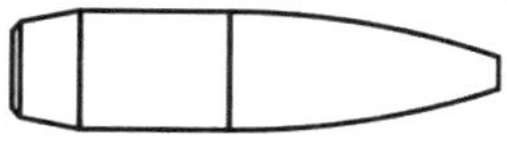 *Sierra 69 MatchKing*	0.197	1.161	0.169	0.330
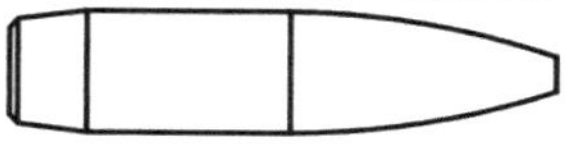 *Sierra 77 MatchKing*	0.219	1.135	0.193	0.376
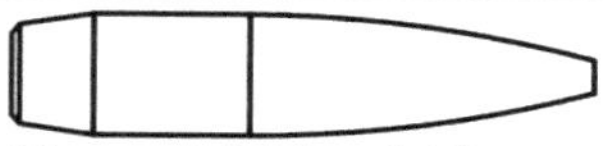 *Sierra 80 MatchKing*	0.228	1.030	0.221	0.431
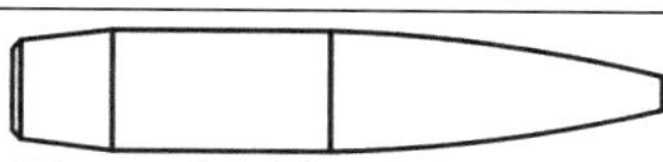 *Sierra 90 MatchKing*	0.256	0.999	0.257	0.499
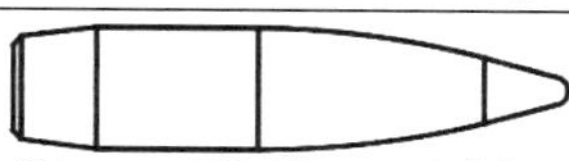 *Sierra 69 Tipped MatchKing*	0.197	1.082	0.182	0.354
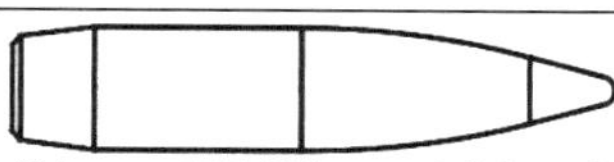 *Sierra 77 Tipped MatchKing*	0.219	1.085	0.202	0.394

.243 Caliber (6mm) bullets				
Bullet	**SD**	**i_7**	**G7 BC**	**G1 BC**
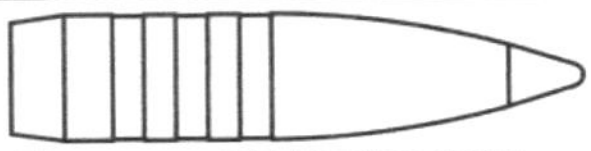 Barnes 80 TTSXBT	0.194	1.214	0.159	0.311
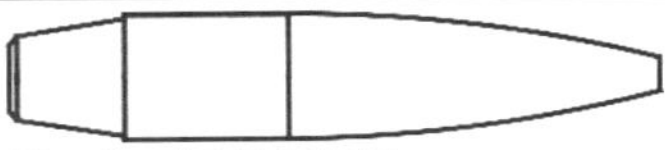 Barts 105 RBT	0.254	0.988	0.257	0.501
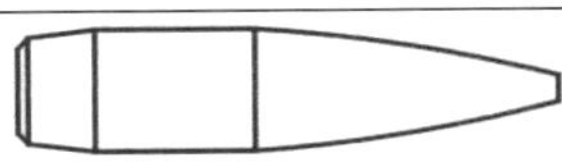 Berger 87 Hunting VLD	0.211	0.961	0.219	0.427
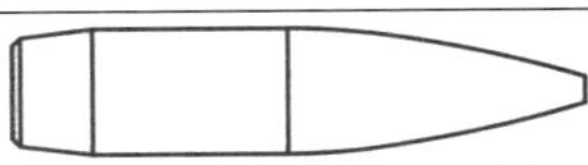 Berger 95 Classic Hunter	0.230	1.033	0.223	0.434
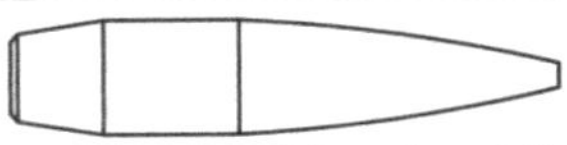 Berger 95 Hunting VLD	0.230	0.962	0.239	0.465
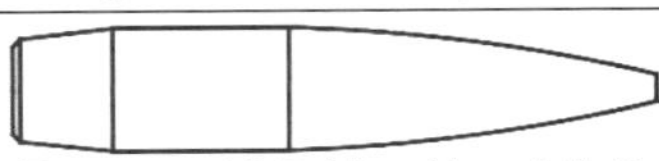 Berger 105 Hunting VLD	0.254	0.915	0.278	0.541
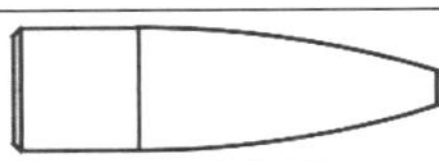 Berger 62 Target EUWIN	0.150	1.293	0.116	0.226
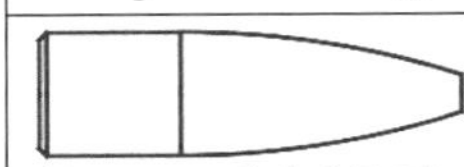 Berger 64.25 Target Column	0.155	1.318	0.118	0.229
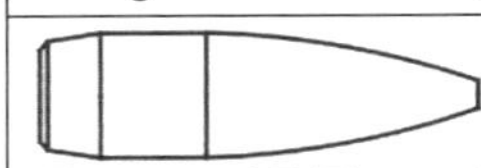 Berger 65 Target BT	0.246	1.248	0.126	0.245
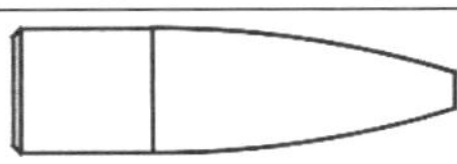 Berger 65 Target WEB BR	0.246	1.248	0.126	0.245
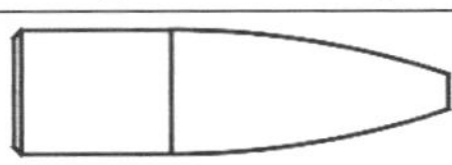 Berger 68 Target FB	0.165	1.256	0.131	0.255
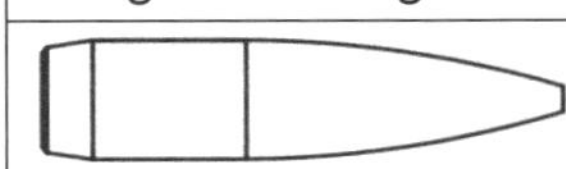 Berger 90 Target BT	0.218	1.037	0.210	0.409

(6mm bullets continued)				
Bullet	**SD**	**i_7**	**G7 BC**	**G1 BC**
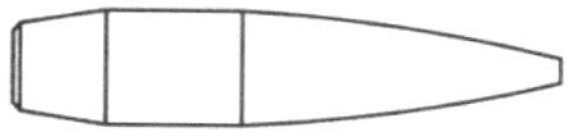 *Berger 95 Target VLD*	0.230	0.958	0.240	0.467
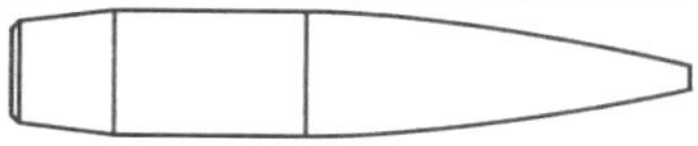 *Berger 105 Target Hybrid*	0.254	0.925	0.275	0.535
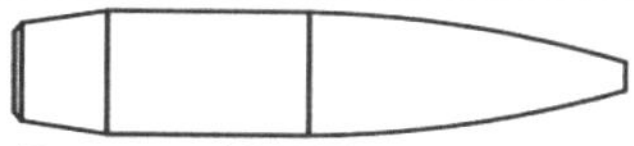 *Berger 105 Target Boat Tail*	0.254	1.004	0.253	0.493
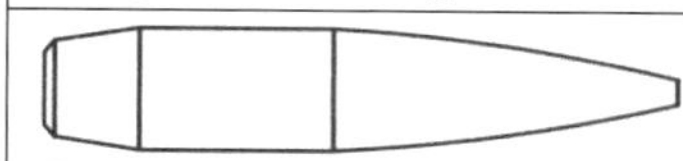 *Berger 105 Target VLD*	0.254	0.961	0.265	0.515
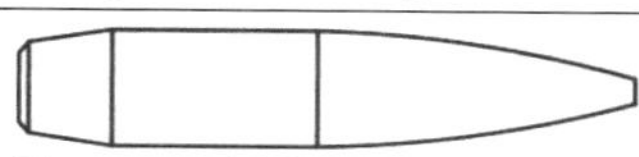 *Berger 108 Target Boat Tail*	0.261	0.997	0.268	0.521
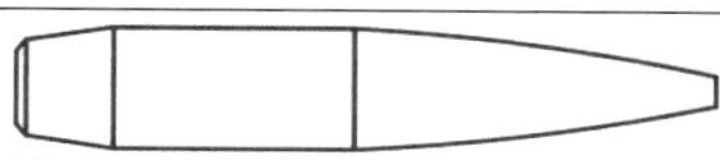 *Berger 115 Target VLD*	0.278	0.963	0.289	0.563
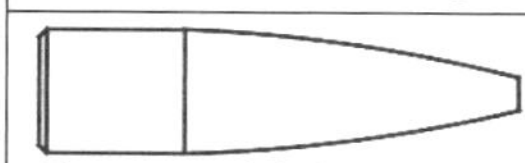 *Berger 69 Varmint High BC FB*	0.167	1.128	0.148	0.288
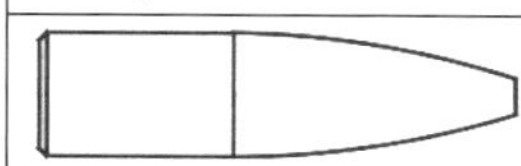 *Berger 80 Varmint FB*	0.194	1.257	0.154	0.300
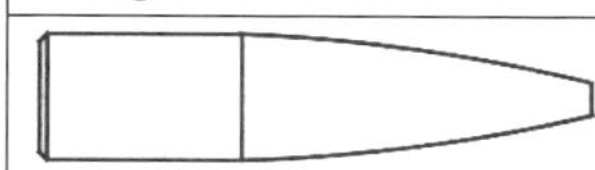 *Berger 88 Varmint HBC FB*	0.213	1.092	0.195	0.380
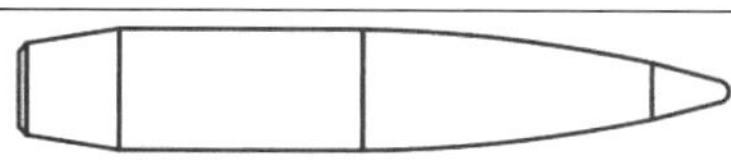 *DTAC 117 BT*	0.283	1.000	0.283	0.553
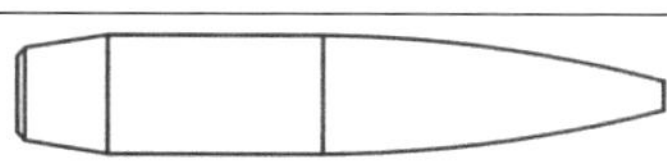 *DTAC 115 BTHP*	0.278	1.007	0.276	0.538
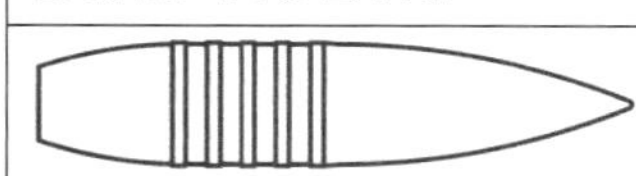 *GS Custom 81 SP*	0.196	1.195	0.164	0.320

(6mm bullets continued)				
Bullet	**SD**	**i_7**	**G7 BC**	**G1 BC**
 *GS Custom 109 SP*	0.264	1.054	0.251	0.490
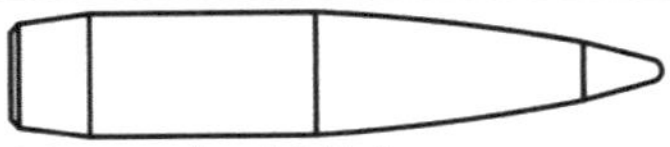 *Hornady 105 Amax*	0.254	1.010	0.252	0.491
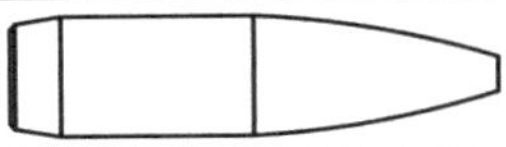 *Hornady 87 BTHP*	0.211	1.181	0.178	0.347
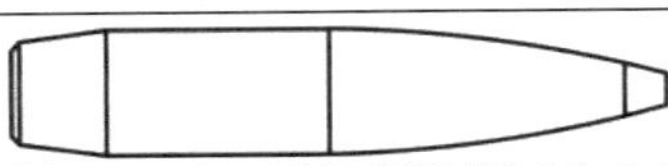 *Hornady 105 BTHP Match*	0.254	1.032	0.246	0.479
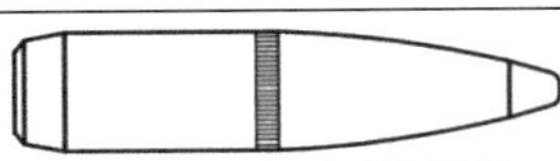 *Hornady 100 BTSP*	0.242	1.225	0.198	0.385
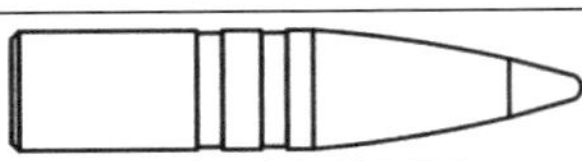 *Hornady 80 GMX*	0.194	1.347	0.144	0.280
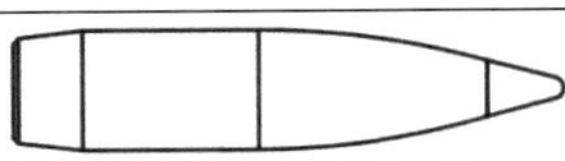 *Hornady 85 Interbond*	0.206	1.081	0.190	0.371
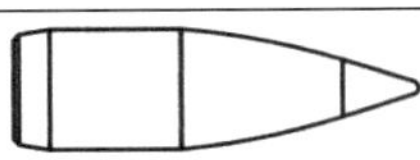 *Hornady 58 Vmax*	0.140	1.177	0.119	0.233
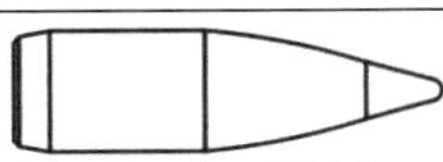 *Hornady 65 Vmax*	0.157	1.178	0.134	0.261
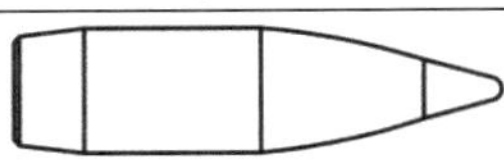 *Hornady 75 Vmax*	0.182	1.115	0.163	0.317
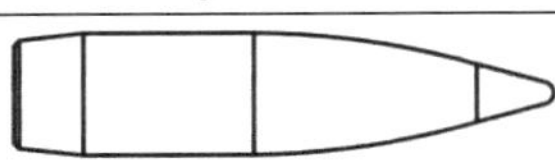 *Hornady 87 Vmax*	0.211	1.076	0.196	0.381
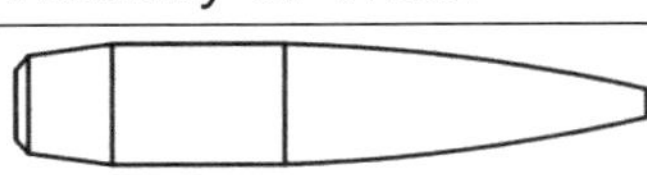 *JLK 105 VLD*	0.254	0.950	0.267	0.521

(6mm bullets continued)				
Bullet	**SD**	**i_7**	**G7 BC**	**G1 BC**
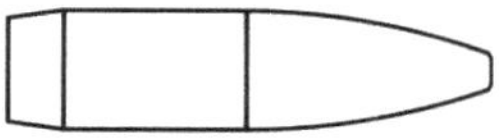 *Lapua 90 FMJBT*	0.218	1.180	0.185	0.360
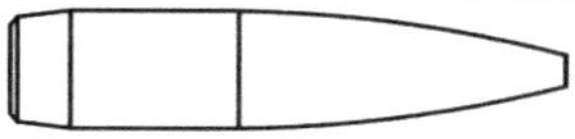 *Lapua 90 Scenar*	0.218	0.999	0.218	0.424
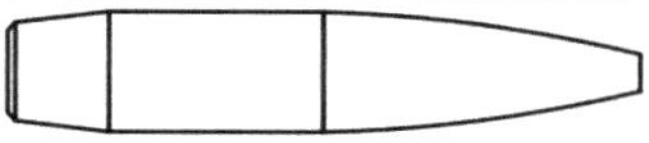 *Lapua 105 Scenar*	0.254	1.025	0.248	0.483
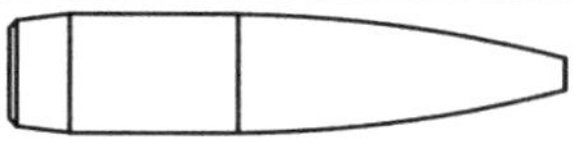 *Lapua 90 Scenar L*	0.218	1.031	0.211	0.411
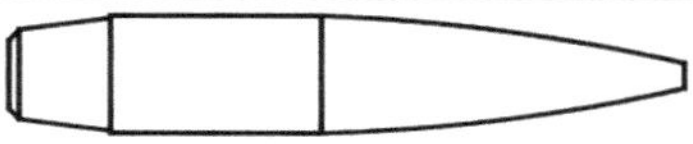 *Matrix 107 VLD*	0.259	0.982	0.264	0.513
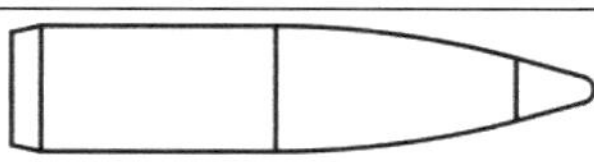 *Nosler 90 Accubond*	0.218	1.198	0.182	0.355
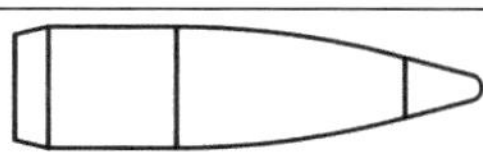 *Nosler 70 Ballistic Tip*	0.169	1.184	0.143	0.280
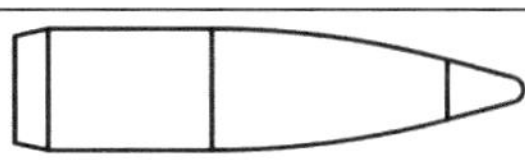 *Nosler 80 Ballistic Tip*	0.194	1.172	0.165	0.323
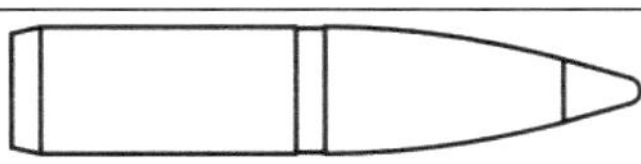 *Nosler 90 E-Tip*	0.218	1.328	0.164	0.320
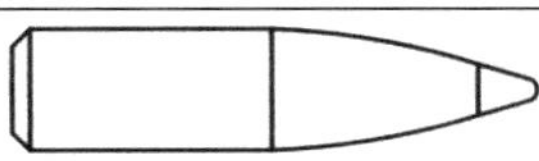 *Nosler 95 Partition*	0.230	1.303	0.177	0.343
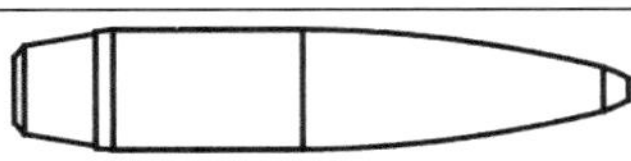 *Precision Ballistics 103 VLD*	0.249	1.067	0.234	0.456
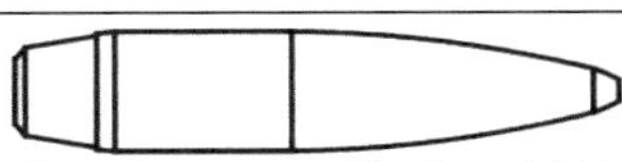 *Precision Ballistics 105 VLD*	0.254	1.058	0.240	0.469

(6mm bullets continued)				
Bullet	**SD**	**i_7**	**G7 BC**	**G1 BC**
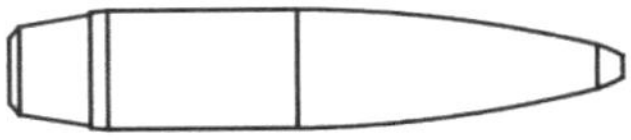 *Precision Ballistics 108 VLD*	0.261	1.037	0.252	0.492
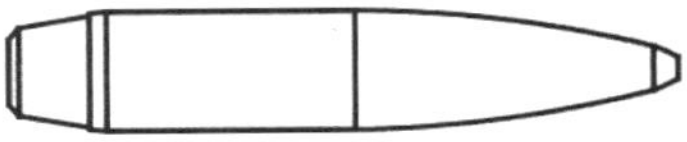 *Precision Ballistics 113 VLD*	0.274	1.070	0.256	0.499
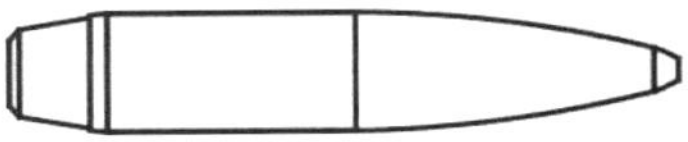 *Precision Ballistics 115 VLD*	0.278	1.096	0.254	0.495
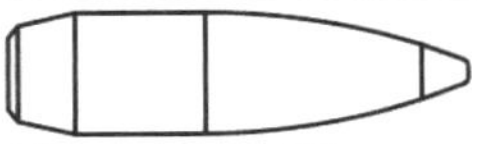 *Sierra 80 Blitz*	0.194	1.256	0.154	0.300
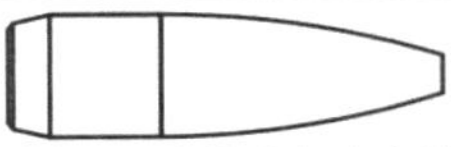 *Sierra 70 MatchKing*	0.169	1.286	0.132	0.257
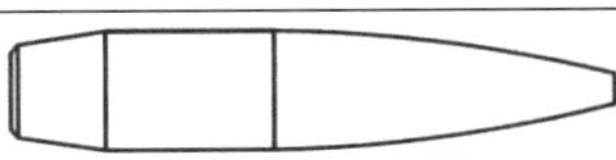 *Sierra 95 MatchKing*	0.230	0.993	0.231	0.451
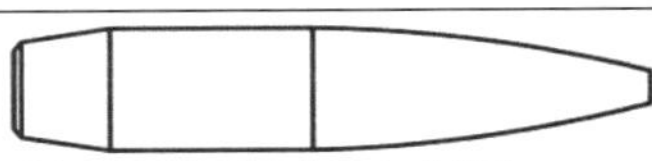 *Sierra 107 MatchKing*	0.259	0.994	0.260	0.507
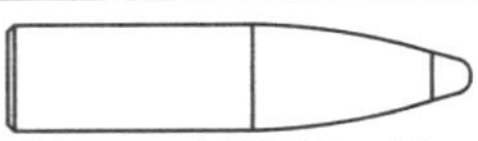 *Sierra 100 ProHunter*	0.242	1.323	0.183	0.356
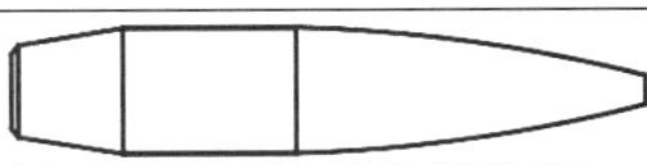 *Vapor Trail 103 BTHP*	0.249	0.958	0.260	0.507

25 Caliber (.257") bullets

Bullet	SD	i_7	G7 BC	G1 BC
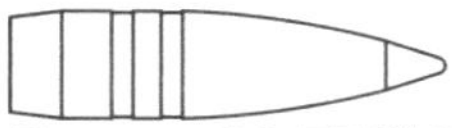 Barnes 80 TTSXBT	0.173	1.184	0.146	0.285
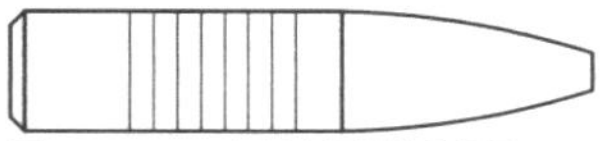 Barnes 115 TSX FB	0.249	1.481	0.168	0.327
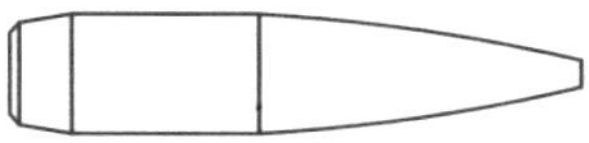 Berger 115 VLD	0.249	1.006	0.247	0.482
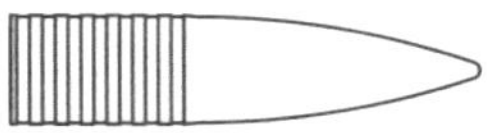 GS Custom 90 SP	0.195	1.240	0.157	0.307
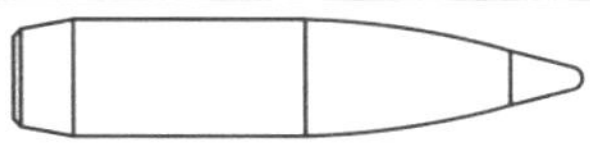 Hornady 110 Interbond	0.238	1.169	0.204	0.398
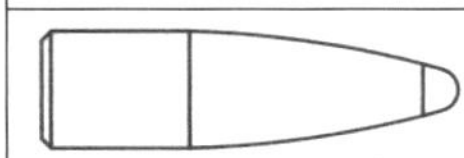 Hornady 87 Spitzer	0.188	1.298	0.145	0.282
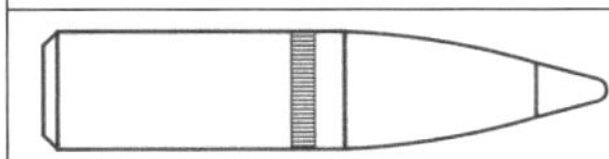 Hornady 117 SST	0.253	1.298	0.195	0.380
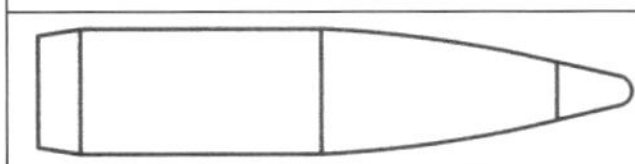 Nosler 110 Accubond	0.238	1.166	0.204	0.399
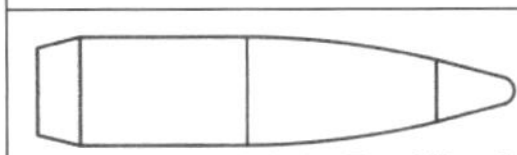 Nosler 100 Ballistic Tip	0.216	1.227	0.176	0.345
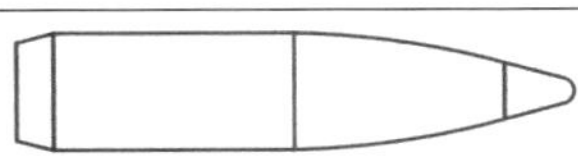 Nosler 115 Ballistic Tip	0.249	1.219	0.204	0.399
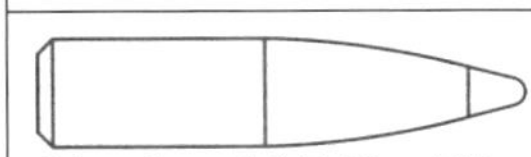 Nosler 115 Partition	0.249	1.257	0.198	0.386
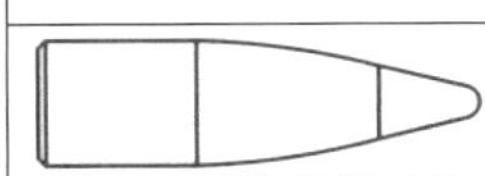 Sierra 70 BlitzKing	0.151	1.296	0.117	0.228

(257 caliber bullets continued)				
Bullet	**SD**	**i_7**	**G7 BC**	**G1 BC**
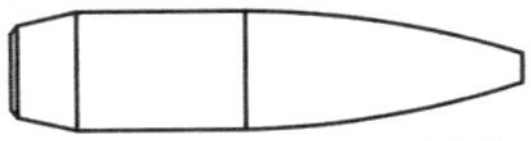 *Sierra 100 MatchKing*	0.216	1.173	0.184	0.360
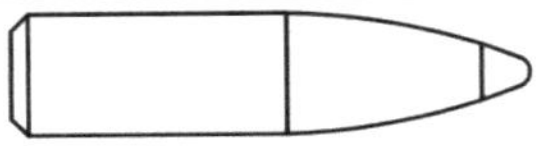 *Sierra 117 ProHunter*	0.253	1.410	0.180	0.349
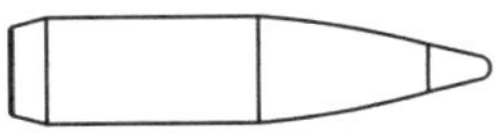 *Swift 100 Scirocco II*	0.216	1.249	0.174	0.340

.264 Caliber (6.5mm) bullets				
Bullet	**SD**	**i_7**	**G7 BC**	**G1 BC**
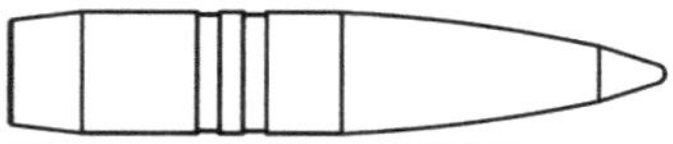 *Barnes 127 LRXBT*	0.260	1.141	0.228	0.446
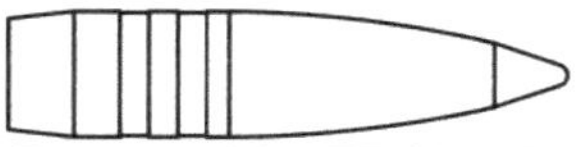 *Barnes 100 TTSXBT*	0.205	1.172	0.175	0.341
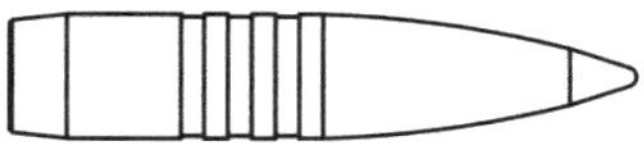 *Barnes 120 TTSXBT*	0.246	1.212	0.203	0.396
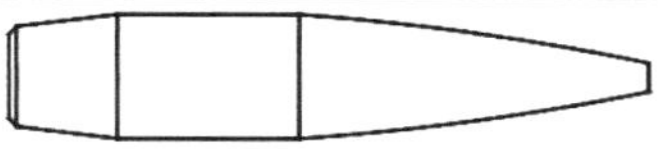 *Berger 130 Hunting VLD*	0.266	0.926	0.288	0.561
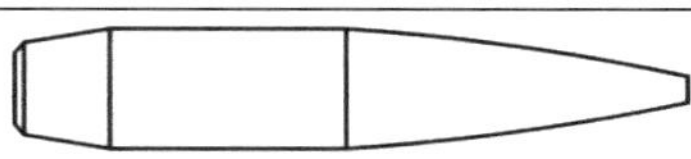 *Berger 140 Hunting VLD*	0.287	0.935	0.307	0.598
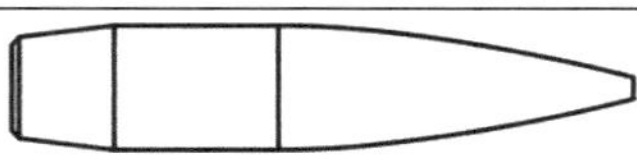 *Berger 130 Tactical AR Hybrid*	0.267	0.929	0.287	0.559
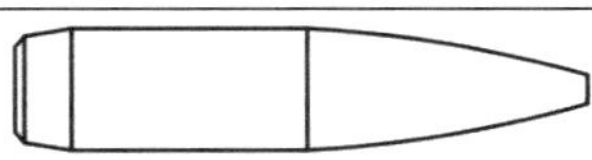 *Berger 120 Target BT*	0.246	1.025	0.240	0.468
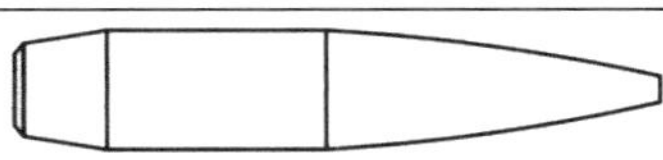 *Berger 130 Target VLD*	0.267	0.927	0.288	0.560
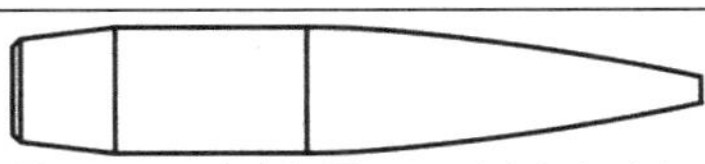 *Berger 140 Target Hybrid*	0.287	0.923	0.311	0.606
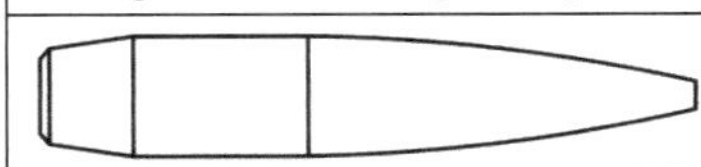 *Berger 140 Target LR BT*	0.287	0.944	0.304	0.592
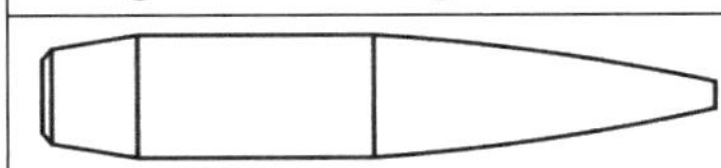 *Berger 140 Target VLD*	0.287	0.945	0.304	0.591
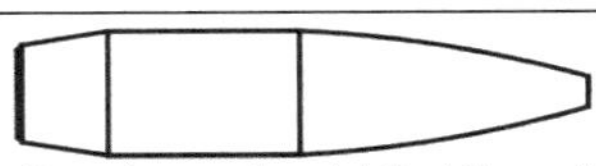 *Cauterucio 119 10-ogive*	0.244	1.012	0.241	0.470

(6.5mm bullets continued)

Bullet	SD	i_7	G7 BC	G1 BC
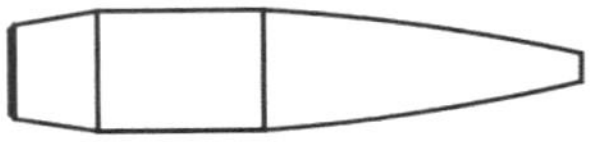 Cauterucio 119 15-ogive	0.244	0.916	0.266	0.519
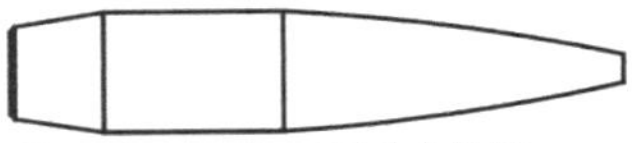 Cauterucio 131 VLD	0.269	0.907	0.296	0.577
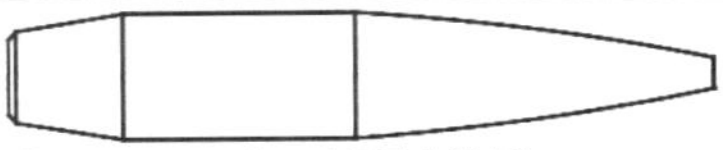 Cauterucio 155 VLD	0.318	0.946	0.336	0.654
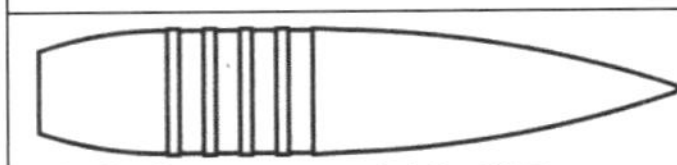 GS Custom 112 SP	0.230	1.219	0.189	0.368
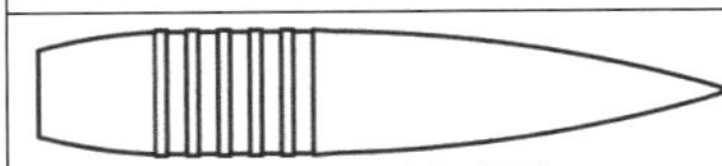 GS Custom 119 SP	0.244	1.145	0.213	0.417
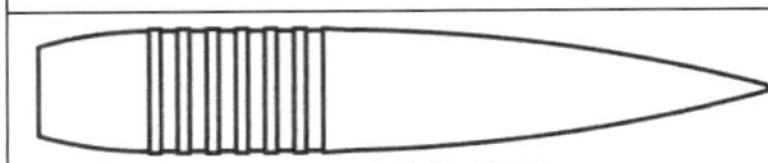 GS Custom 128 SP	0.262	1.056	0.249	0.487
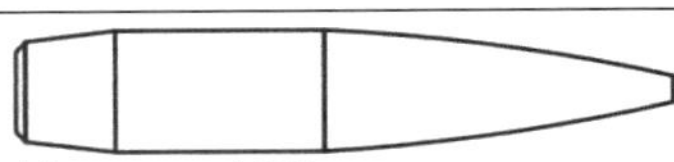 Hoover 136	0.279	0.888	0.314	0.612
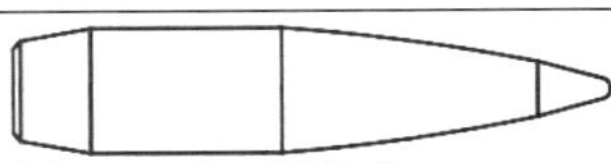 Hornady 123 Amax	0.252	1.055	0.239	0.467
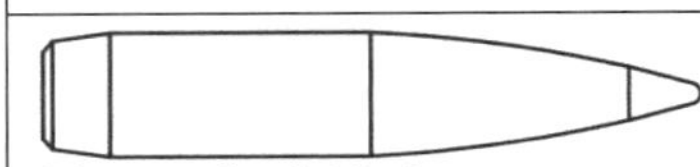 Hornady 140 Amax	0.287	0.996	0.288	0.562
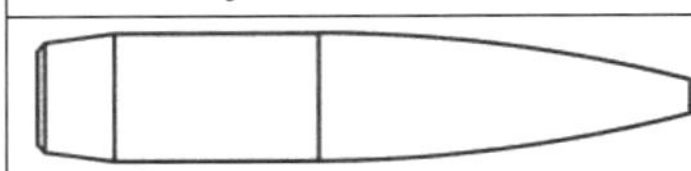 Hornady 140 BTHP Match	0.287	1.008	0.285	0.554
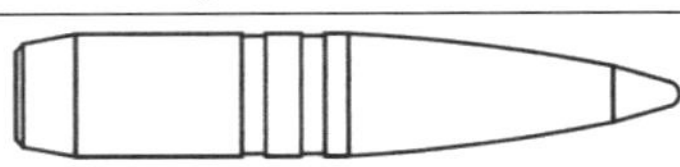 Hornady 120 GMX	0.246	1.089	0.226	0.442
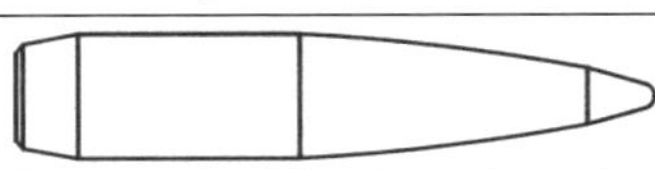 Hornady 129 Interbond	0.264	0.999	0.265	0.518

				(6.5mm bullets continued)
Bullet	**SD**	**i7**	**G7 BC**	**G1 BC**
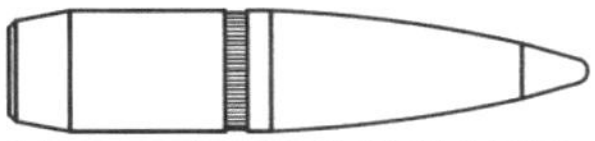 *Hornady 123 SST BT*	0.252	1.069	0.236	0.462
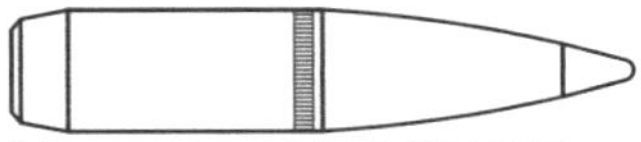 *Hornady 129 SST BT*	0.264	1.072	0.247	0.482
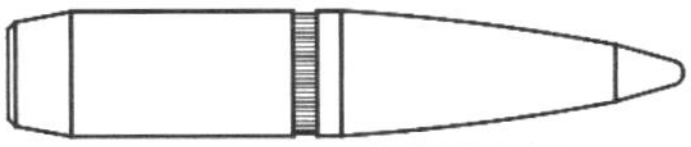 *Hornady 140 SST BT*	0.287	1.085	0.265	0.517
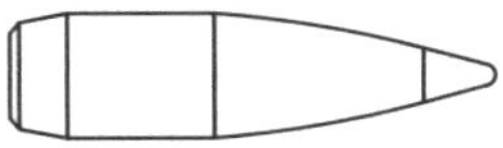 *Hornady 95 Vmax*	0.195	1.069	0.182	0.356
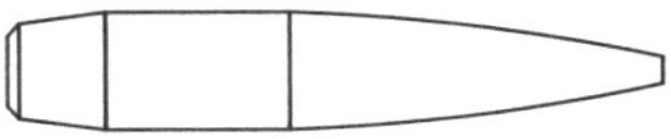 *JLK 140 VLD*	0.287	0.895	0.321	0.625
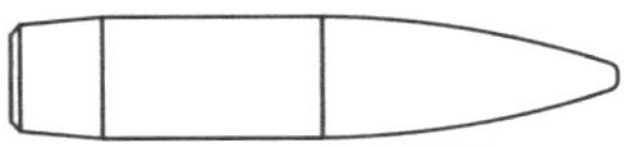 *Lapua 144 FMJBT*	0.295	1.023	0.289	0.562
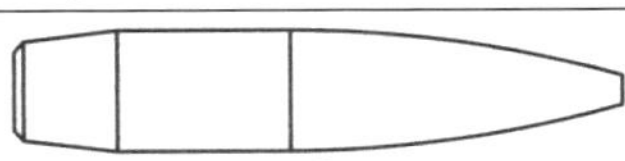 *Lapua 108 Scenar*	0.221	0.985	0.225	0.438
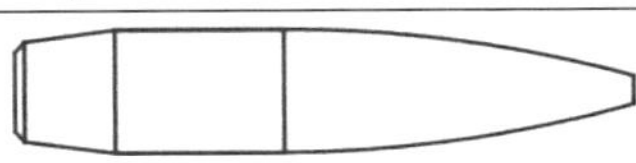 *Lapua 123 Scenar*	0.252	0.950	0.265	0.517
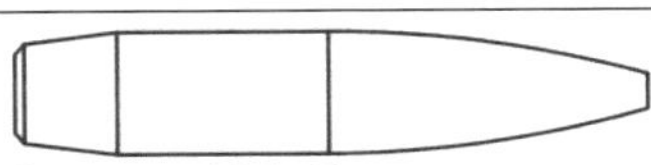 *Lapua 139 Scenar*	0.285	0.983	0.290	0.564
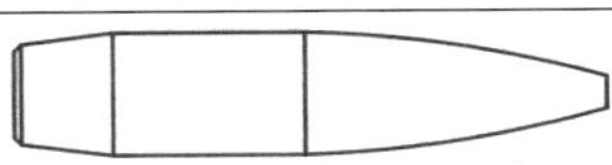 *Lapua 120 Scenar L*	0.246	1.018	0.242	0.470
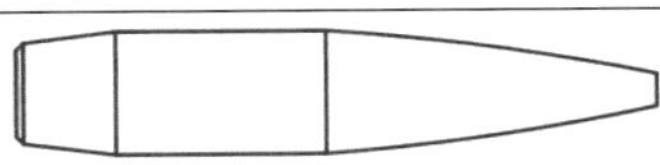 *Lapua 136 Scenar L*	0.279	0.969	0.288	0.560
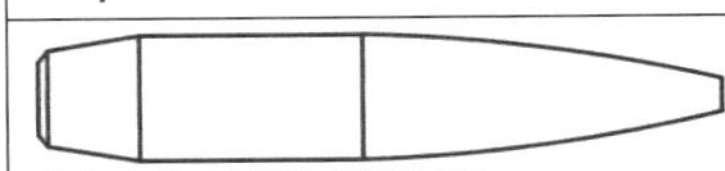 *Matrix 150 BTHP*	0.308	0.992	0.310	0.604

(6.5mm bullets continued)				
Bullet	**SD**	**i_7**	**G7 BC**	**G1 BC**
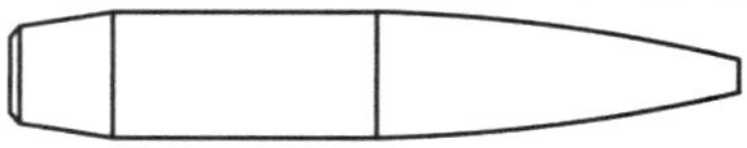 *Matrix 160 BTHP*	0.328	0.989	0.332	0.646
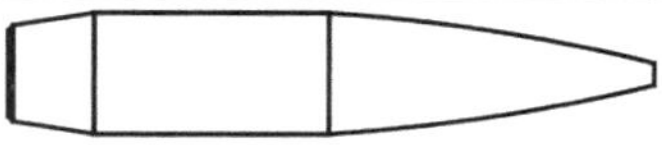 *Norma 130 Diamond*	0.267	0.905	0.294	0.575
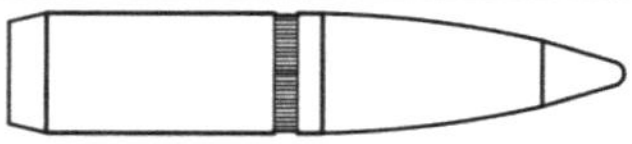 *Nosler 130 Accubond*	0.267	1.185	0.225	0.440
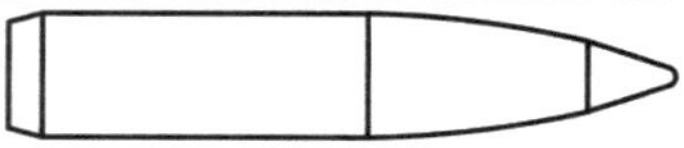 *Nosler 140 Accubond*	0.287	1.214	0.237	0.463
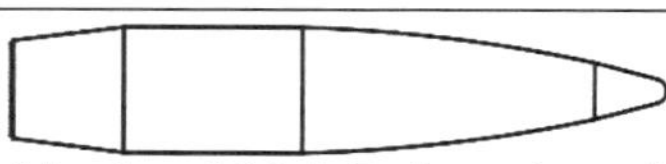 *Nosler 129 LR Accubond*	0.264	0.935	0.283	0.552
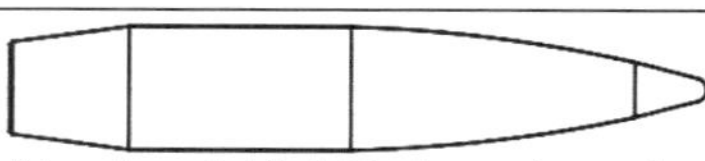 *Nosler 142 LR Accubond*	0.291	0.993	0.293	0.572
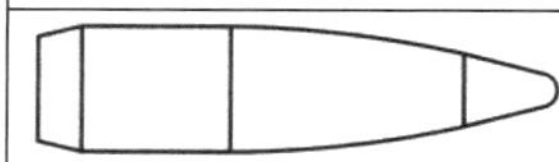 *Nosler 100 Ballistic Tip*	0.205	1.193	0.172	0.336
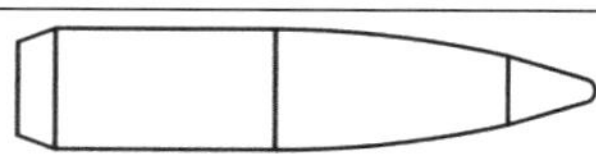 *Nosler 120 Ballistic Tip*	0.246	1.149	0.214	0.419
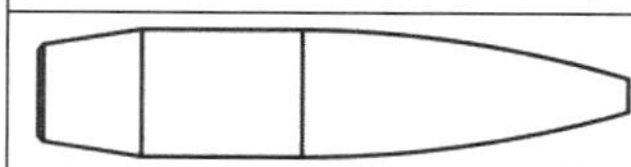 *Nosler 123 Custom Competition*	0.252	1.048	0.241	0.468
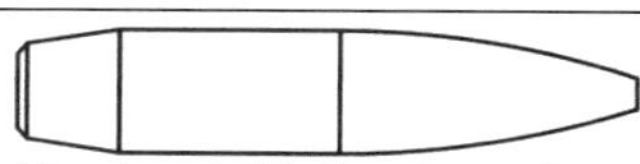 *Nosler 140 Custom Competition*	0.287	1.021	0.281	0.547
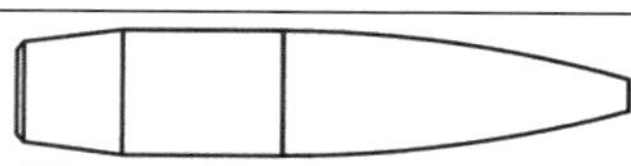 *Sierra 107 MatchKing*	0.219	0.954	0.230	0.448
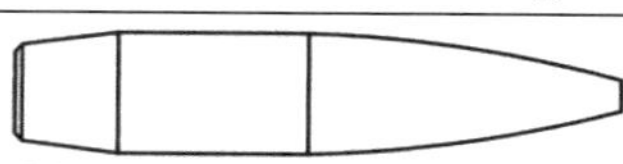 *Sierra 123 MatchKing*	0.252	0.970	0.260	0.506

(6.5mm bullets continued)				
Bullet	**SD**	**i_7**	**G7 BC**	**G1 BC**
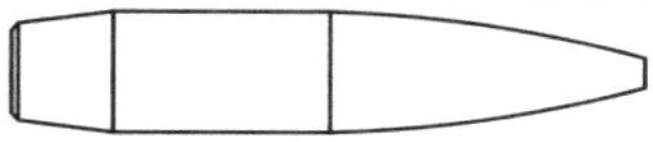 *Sierra 142 MatchKing*	0.291	0.968	0.301	0.585
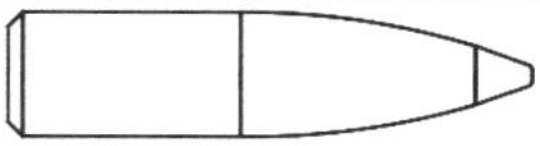 *Sierra 120 ProHunter*	0.246	1.207	0.204	0.397
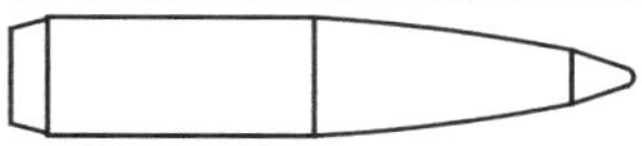 *Swift 130 Scirocco*	0.267	1.061	0.251	0.491

270 Caliber (.277") bullets				
Bullet	**SD**	**i_7**	**G7 BC**	**G1 BC**
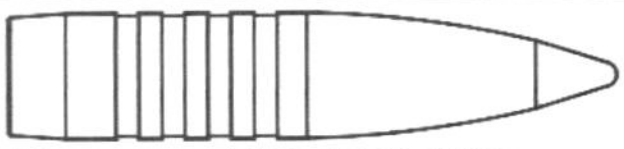 *Barnes 130 TSXBT*	0.242	1.248	0.194	0.378
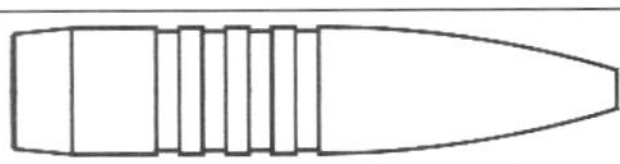 *Barnes 140 TSXBT*	0.261	1.272	0.205	0.399
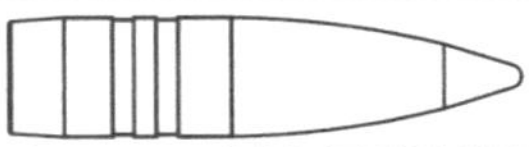 *Barnes 110 TTSXBT*	0.205	1.189	0.172	0.336
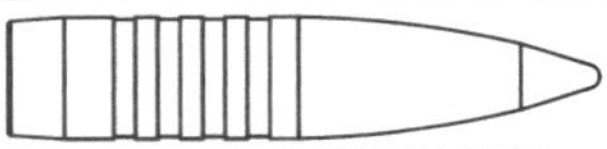 *Barnes 130 TTSXBT*	0.242	1.266	0.191	0.373
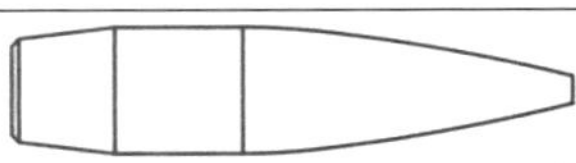 *Berger 130 Classic Hunter*	0.242	0.961	0.252	0.491
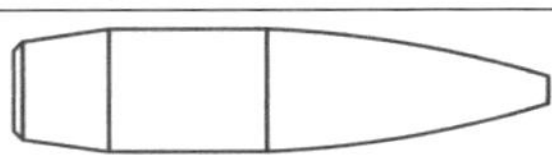 *Berger 130 Hunting VLD*	0.242	1.025	0.236	0.460
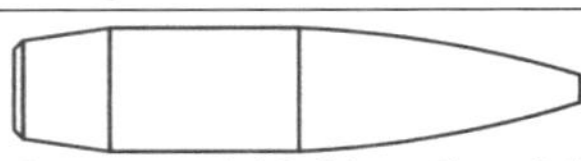 *Berger 140 Hunting VLD*	0.261	1.011	0.258	0.502
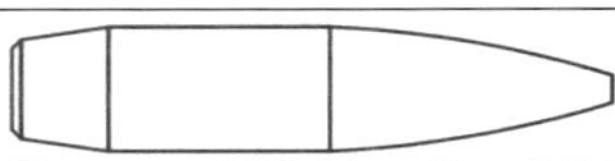 *Berger 150 Hunting VLD*	0.279	1.054	0.265	0.516
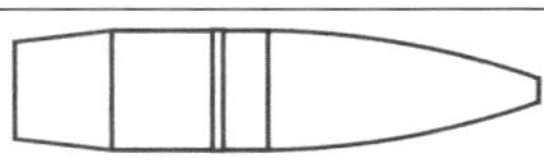 *Cutting Edge 120 HPBT*	0.223	1.070	0.209	0.407
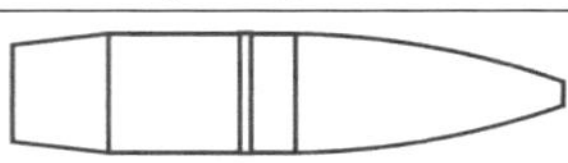 *Cutting Edge 130 HPBT*	0.242	1.065	0.227	0.443
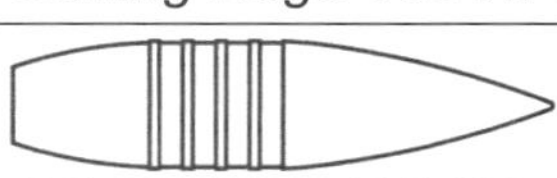 *GS Custom 105 SP*	0.196	1.258	0.156	0.304
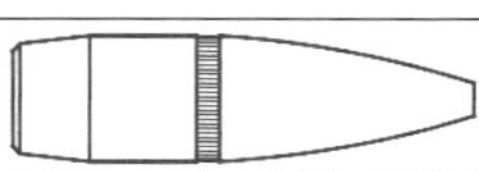 *Hornady 110 BTHP*	0.205	1.132	0.181	0.352

(.270 caliber bullets continued)				
Bullet	**SD**	**i_7**	**G7 BC**	**G1 BC**
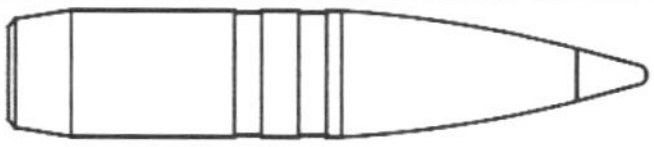 *Hornady 130 GMX*	0.242	1.142	0.212	0.414
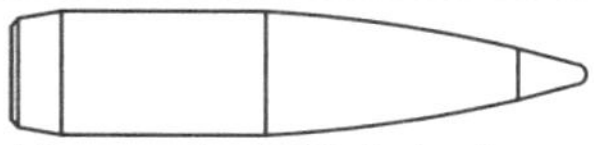 *Hornady 130 Interbond*	0.242	1.059	0.229	0.447
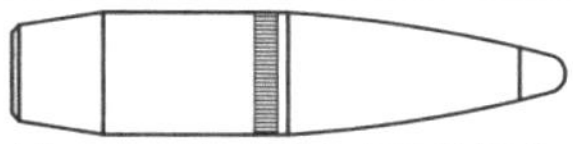 *Hornady 140 BTSP Interlock*	0.261	1.145	0.228	0.443
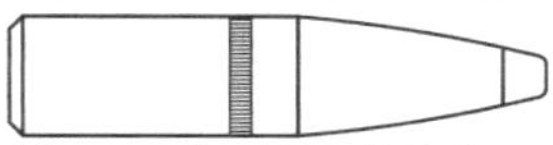 *Hornady 150 SP Interlock*	0.279	1.348	0.207	0.403
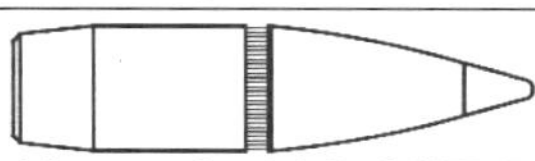 *Hornady 120 SST BT*	0.223	1.176	0.190	0.370
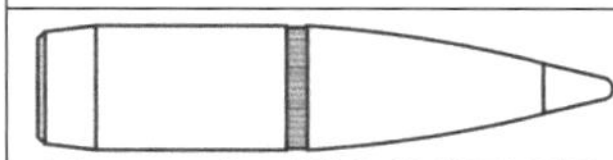 *Hornady 130 SST BT*	0.242	1.084	0.223	0.436
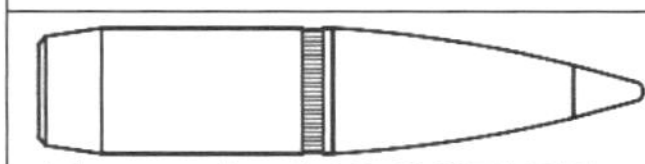 *Hornady 140 SST BT*	0.261	1.086	0.240	0.469
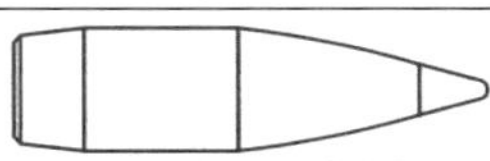 *Hornady 110 Vmax*	0.205	1.138	0.180	0.351
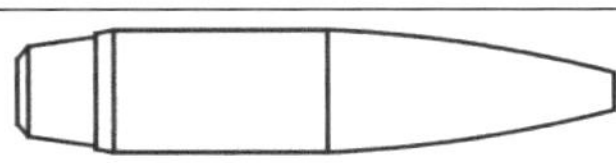 *Matrix 150 RBT*	0.279	1.211	0.231	0.449
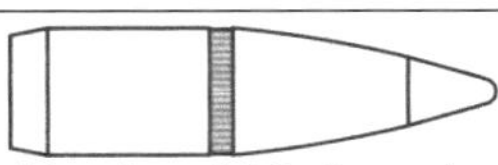 *Nosler 100 Accubond*	0.186	1.255	0.149	0.290
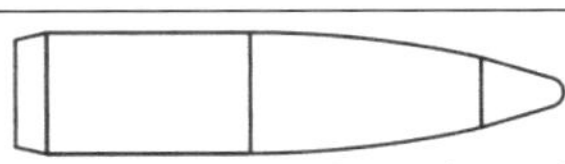 *Nosler 130 Accubond*	0.242	1.224	0.198	0.387
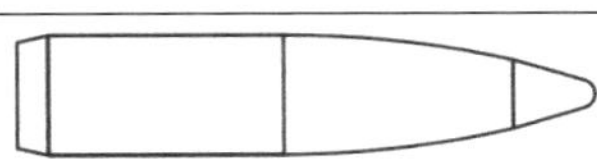 *Nosler 140 Accubond*	0.261	1.200	0.217	0.424

(.270 caliber bullets continued)				
Bullet	**SD**	**i_7**	**G7 BC**	**G1 BC**
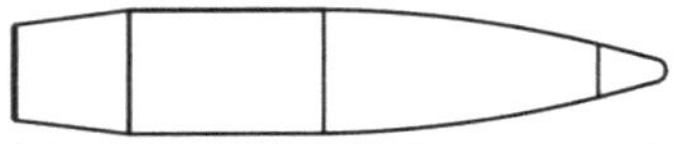 *Nosler 150 LR Accubond*	0.279	0.959	0.291	0.568
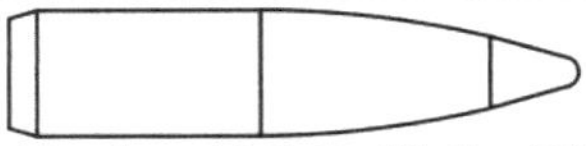 *Nosler 130 Ballistic Silver Tip*	0.242	1.215	0.199	0.390
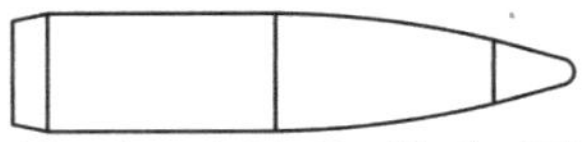 *Nosler 140 Ballistic Tip*	0.261	1.148	0.227	0.444
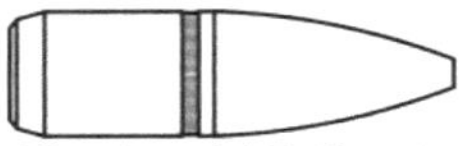 *Nosler 115 Custom Competition*	0.214	1.347	0.159	0.311
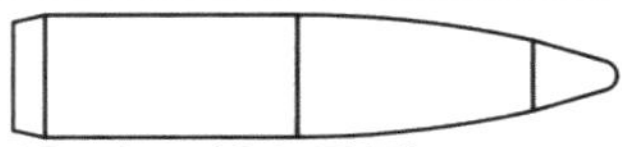 *Nosler 130 E-Tip*	0.242	1.258	0.192	0.376
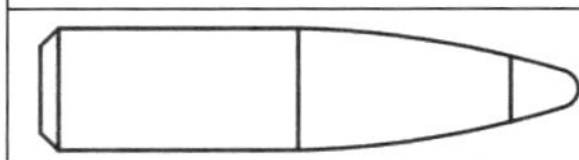 *Nosler 140 Partition*	0.261	1.257	0.208	0.404
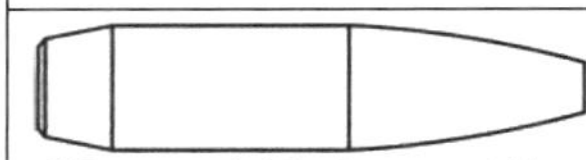 *Sierra 140 GameKing*	0.261	1.394	0.187	0.364
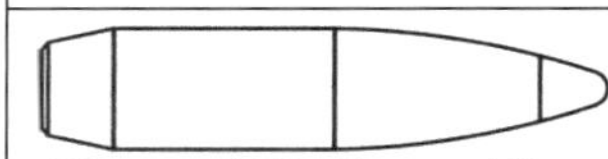 *Sierra 150 GameKing*	0.279	1.180	0.237	0.461
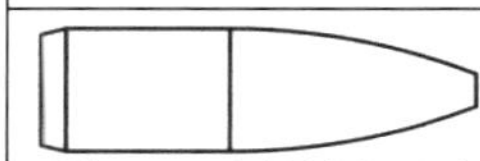 *Sierra 115 MatchKing*	0.214	1.325	0.162	0.315
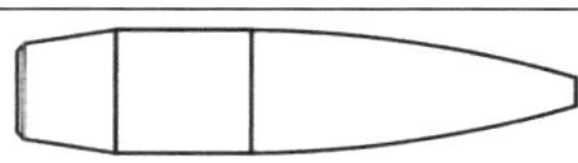 *Sierra 135 MatchKing*	0.251	0.992	0.253	0.494
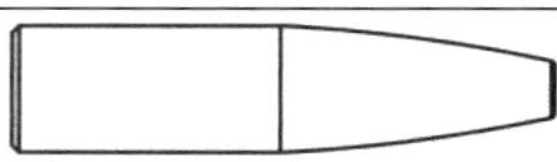 *Swift 140 A-Frame*	0.261	1.531	0.172	0.332
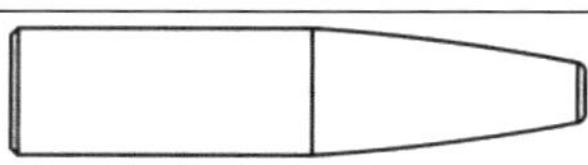 *Swift 150 A-Frame*	0.279	1.516	0.186	0.360

.284 Caliber (7mm) bullets				
Bullet	**SD**	**i_7**	**G7 BC**	**G1 BC**
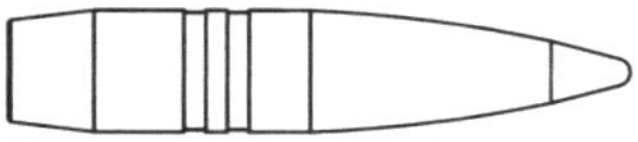 *Barnes 145 LRXBT*	0.257	1.078	0.238	0.465
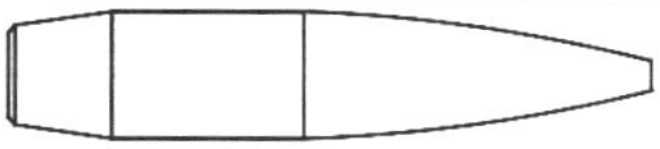 *Barnes 171 Match Burner*	0.303	0.977	0.310	0.604
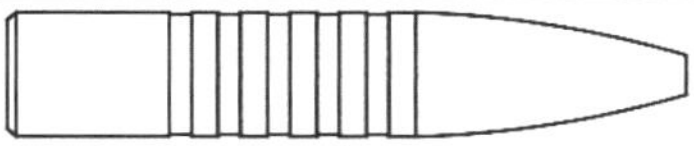 *Barnes 175 TSX FB*	0.310	1.497	0.207	0.403
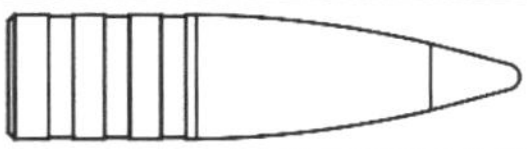 *Barnes 110 TTSX FB*	0.195	1.257	0.155	0.303
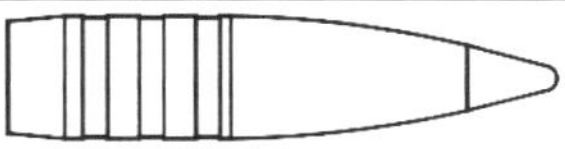 *Barnes 120 TTSXBT*	0.213	1.138	0.187	0.364
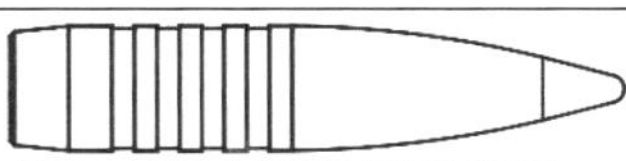 *Barnes 140 TTSXBT*	0.248	1.188	0.209	0.407
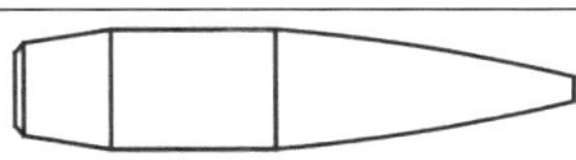 *Berger 140 Hunting VLD*	0.248	0.970	0.256	0.498
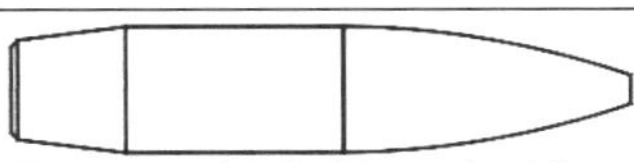 *Berger 168 Classic Hunter*	0.298	1.026	0.290	0.565
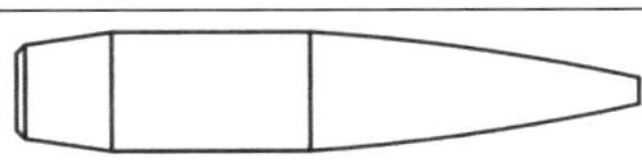 *Berger 168 Hunting VLD*	0.298	0.941	0.316	0.616
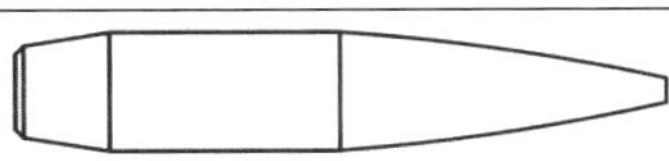 *Berger 180 Hunting VLD*	0.319	0.925	0.345	0.671
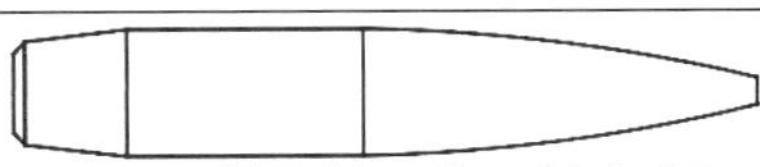 *Berger 195 Hunting Hybrid*	0.345	0.893	0.387	0.754
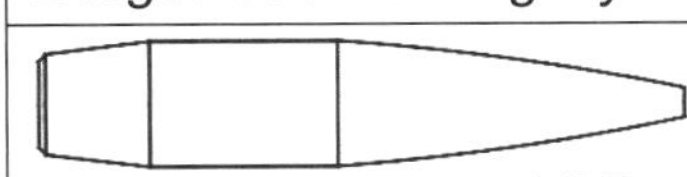 *Berger 168 Target VLD*	0.298	0.926	0.322	0.626

(.284 caliber bullets continued)				
Bullet	**SD**	i_7	**G7 BC**	**G1 BC**
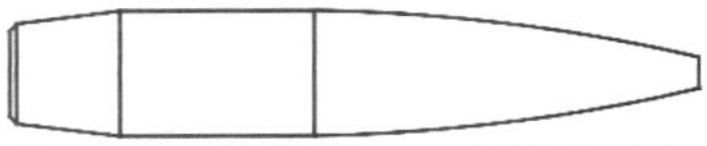 *Berger 180 Target Hybrid*	0.319	0.915	0.349	0.679
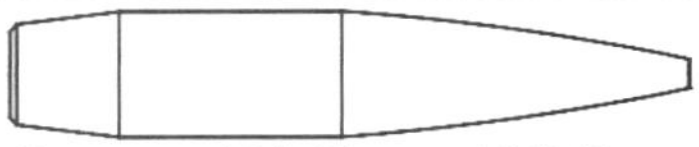 *Berger 180 Target VLD*	0.319	0.912	0.350	0.681
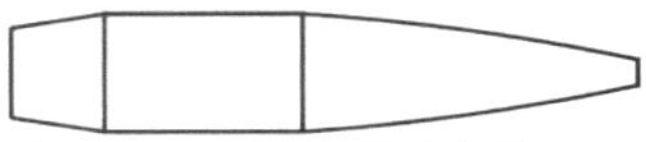 *Cauterucio 177 VLD*	0.314	0.897	0.349	0.681
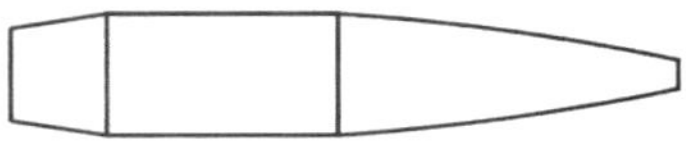 *Cauterucio 189 VLD*	0.335	0.930	0.360	0.701
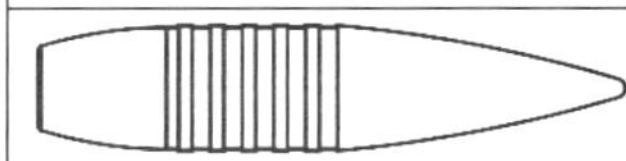 *GS Custom 125 SP*	0.221	1.188	0.187	0.364
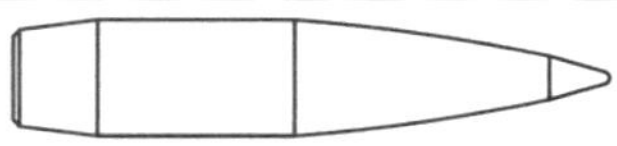 *Hornady 162 Amax*	0.287	0.936	0.307	0.598
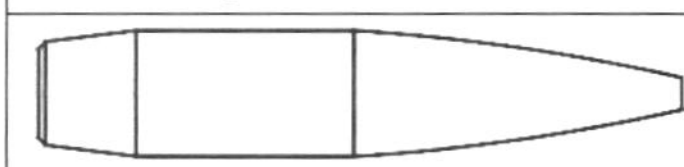 *Hornady 162 BTHP Match*	0.287	0.966	0.297	0.578
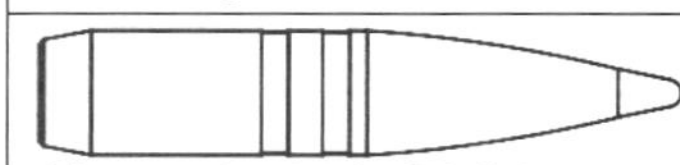 *Hornady 139 GMX*	0.246	1.102	0.224	0.437
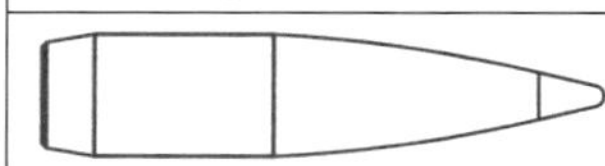 *Hornady 139 Interbond*	0.246	0.993	0.248	0.485
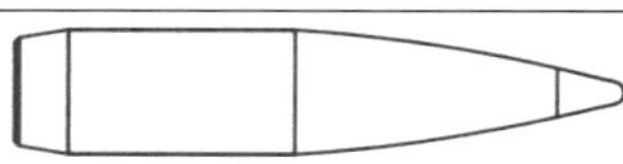 *Hornady 154 Interbond*	0.273	1.001	0.273	0.533
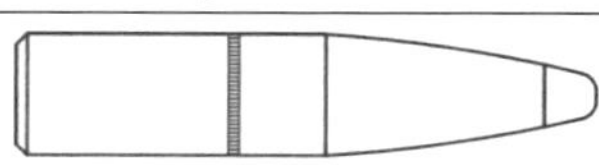 *Hornady 175 Interlock*	0.310	1.314	0.236	0.459
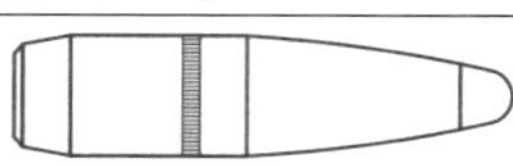 *Hornady 139 BTSP Interlock*	0.246	1.207	0.204	0.397

(.284 caliber bullets continued)				
Bullet	**SD**	**i_7**	**G7 BC**	**G1 BC**
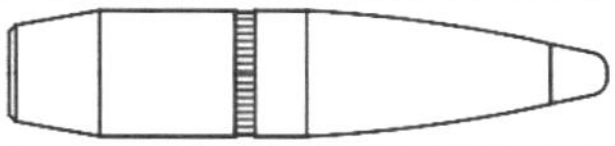 *Hornady 162 BTSP Interlock*	0.287	1.165	0.246	0.480
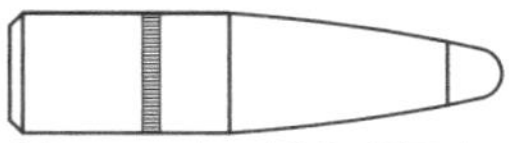 *Hornady 139 SP Interlock*	0.246	1.261	0.195	0.380
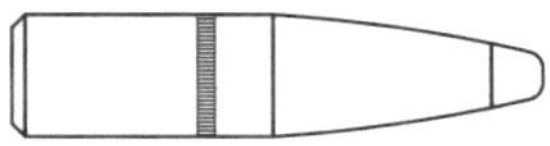 *Hornady 154 SP Interlock*	0.273	1.300	0.210	0.409
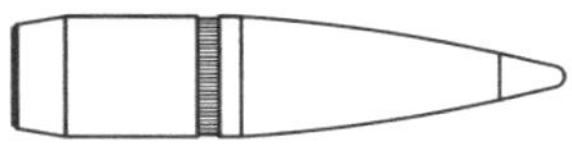 *Hornady 139 SST BT*	0.246	1.068	0.231	0.451
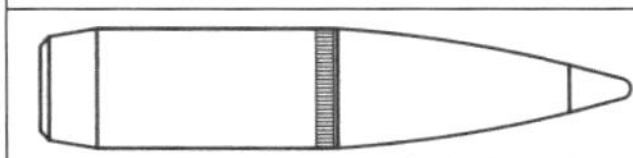 *Hornady 154 SST BT*	0.273	1.086	0.251	0.491
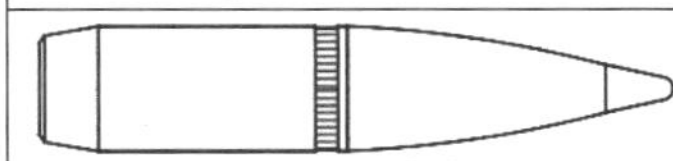 *Hornady 162 SST BT*	0.287	1.048	0.274	0.535
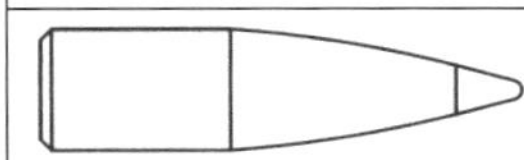 *Hornady 120 Vmax*	0.213	1.155	0.184	0.359
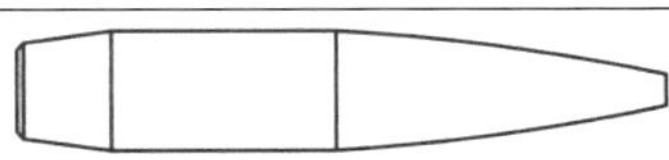 *JLK 180 BTHP*	0.319	0.966	0.330	0.643
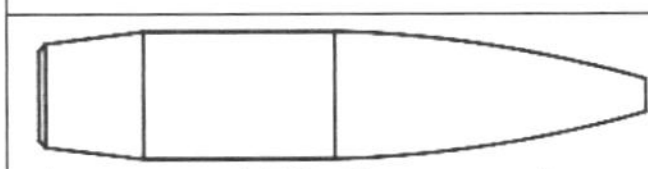 *Lapua 150 Scenar L*	0.266	0.911	0.268	0.522
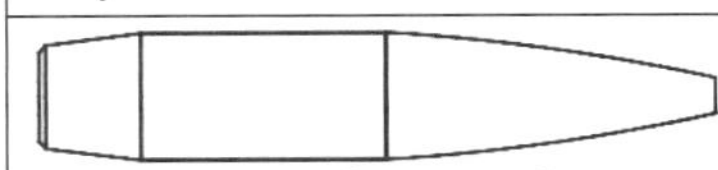 *Lapua 180 Scenar L*	0.319	0.999	0.319	0.621
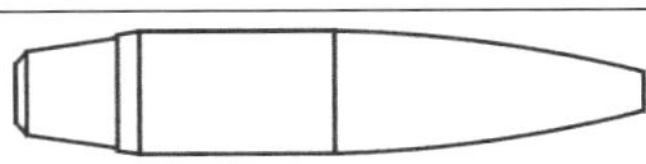 *Matrix 175 RBT*	0.310	1.096	0.283	0.551
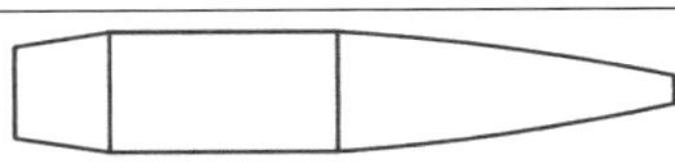 *Matrix 168 VLD*	0.298	0.956	0.311	0.606

(.284 caliber bullets continued)				
Bullet	**SD**	**i_7**	**G7 BC**	**G1 BC**
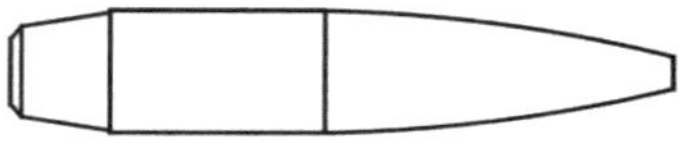 *Matrix 190 VLD*	0.337	0.978	0.344	0.670
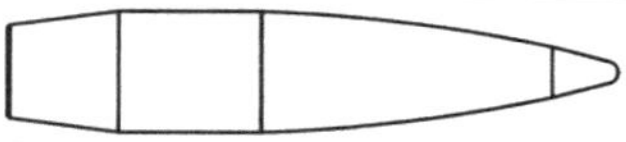 *Nosler 150 LR Accubond*	0.266	0.902	0.295	0.575
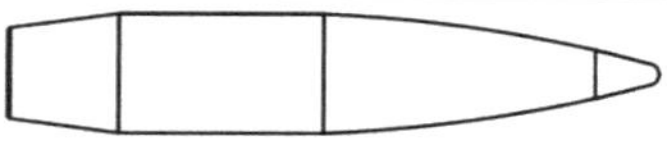 *Nosler 168 LR Accubond*	0.298	0.896	0.332	0.648
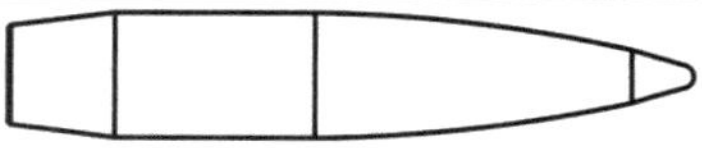 *Nosler 175 LR Accubond*	0.310	0.940	0.330	0.643
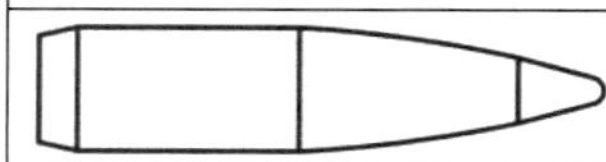 *Nosler 140 Ballistic Silver Tip*	0.248	1.163	0.213	0.417
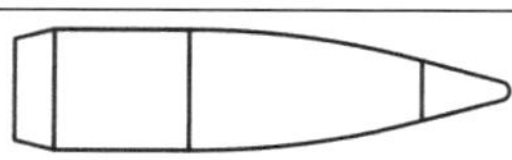 *Nosler 120 Ballistic Tip*	0.213	1.137	0.187	0.365
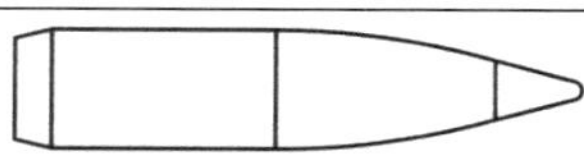 *Nosler 150 Ballistic Tip*	0.266	1.131	0.235	0.458
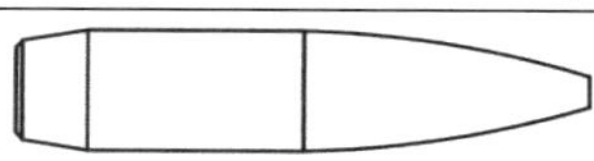 *Nosler 168 Custom Competition*	0.298	1.106	0.269	0.524
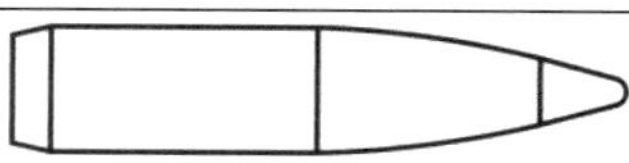 *Nosler 140 E-Tip*	0.248	1.200	0.207	0.405
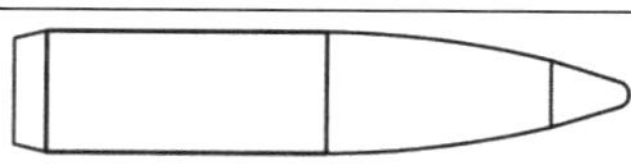 *Nosler 150 E-Tip*	0.266	1.204	0.221	0.432
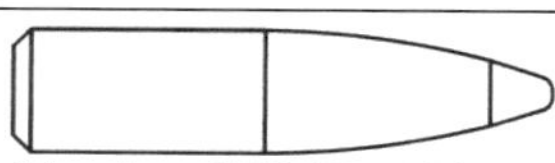 *Nosler 150 Partition*	0.266	1.256	0.211	0.411
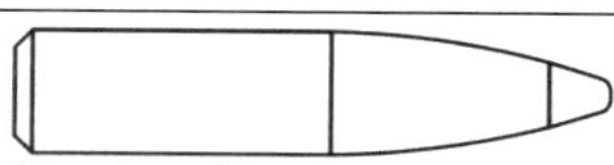 *Nosler 175 Partition*	0.310	1.265	0.245	0.477

(.284 caliber bullets continued)				
Bullet	**SD**	**i_7**	**G7 BC**	**G1 BC**
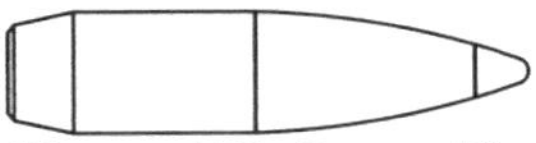 *Sierra 150 GameKing*	0.266	1.209	0.220	0.428
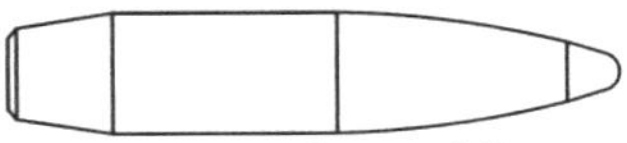 *Sierra 175 GameKing*	0.310	1.058	0.293	0.570
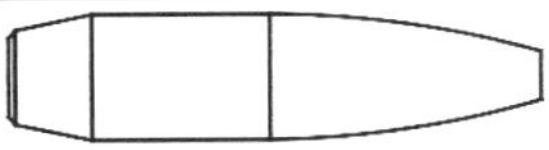 *Sierra 140 HPBT GameKing*	0.248	1.356	0.183	0.356
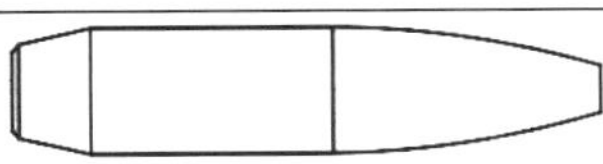 *Sierra 160 HPBT GameKing*	0.283	1.350	0.210	0.409
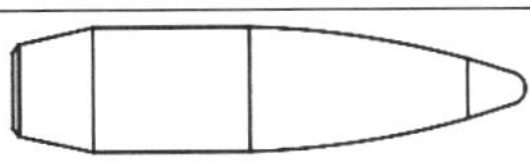 *Sierra 140 SBT GameKing*	0.248	1.198	0.207	0.404
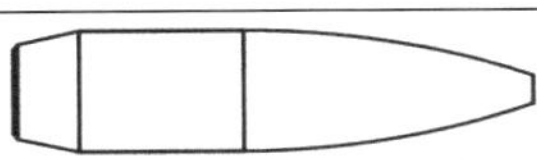 *Sierra 130 MatchKing*	0.230	1.133	0.203	0.397
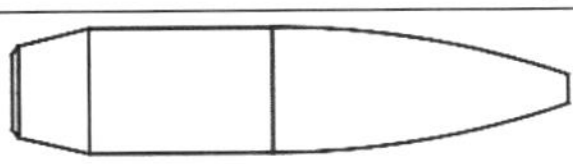 *Sierra 150 MatchKing*	0.266	1.144	0.232	0.454
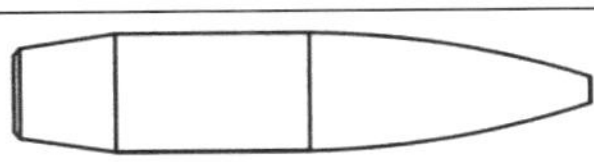 *Sierra 168 MatchKing*	0.298	1.030	0.289	0.563
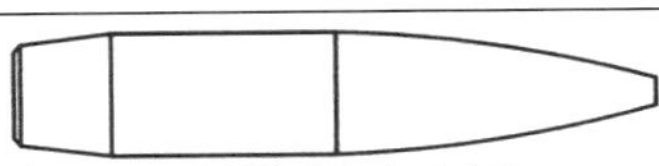 *Sierra 175 MatchKing*	0.310	0.948	0.327	0.637
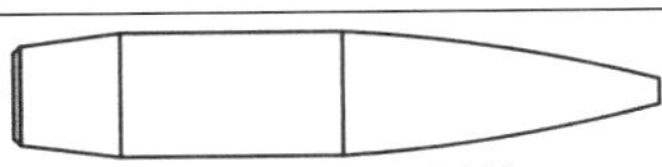 *Sierra 180 MatchKing*	0.319	0.946	0.337	0.657
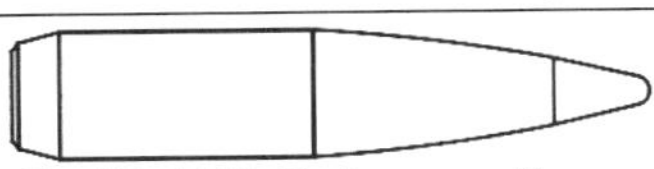 *Swift 150 Scirocco II*	0.266	1.100	0.242	0.474
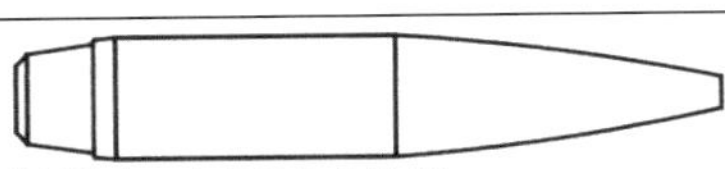 *Wildcat 200 ULD*	0.354	0.985	0.360	0.700

.308 Caliber (7.62mm) bullets

Bullet	SD	i_7	G7 BC	G1 BC
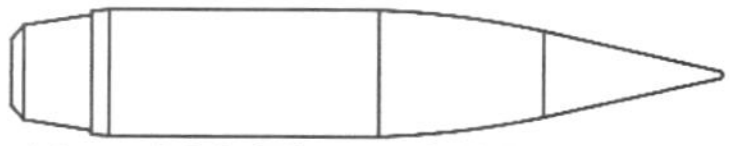 Alco 220 Tipped RBBT	0.331	1.029	0.322	0.630
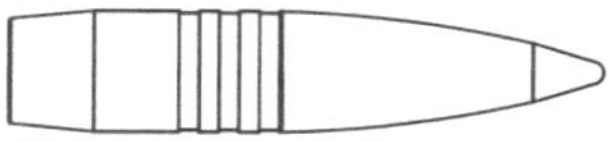 Barnes 175 LRXBT	0.264	1.068	0.247	0.481
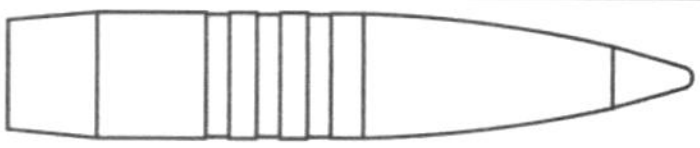 Barnes 200 LRXBT	0.301	1.070	0.281	0.549
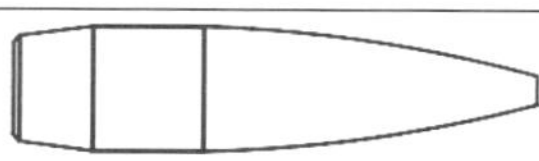 Barnes 155 Match Burner	0.233	0.977	0.239	0.465
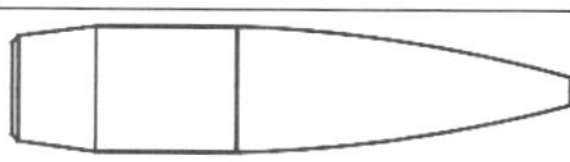 Barnes 175 Match Burner	0.264	0.989	0.266	0.519
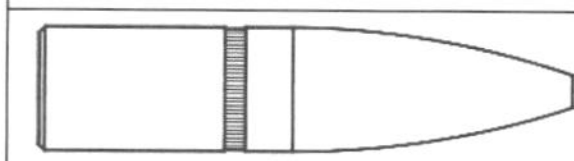 Barnes 150 Tac RRLP FB	0.226	1.255	0.180	0.351
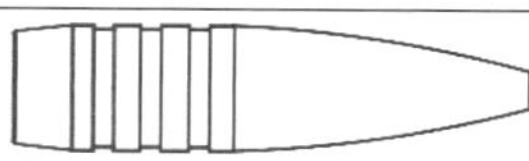 Barnes 150 Tac-X BT	0.226	1.170	0.193	0.376
 Barnes 110 Tac-X FB M/LE	0.166	1.200	0.138	0.270
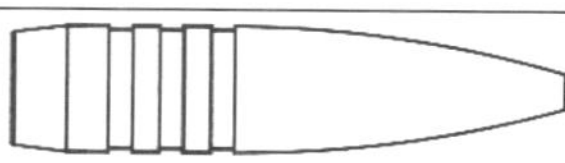 Barnes 168 Tac-X M/LE	0.253	1.182	0.214	0.417
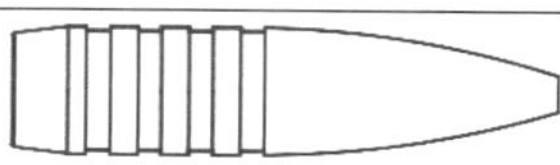 Barnes 165 TSX BT	0.249	1.255	0.198	0.385
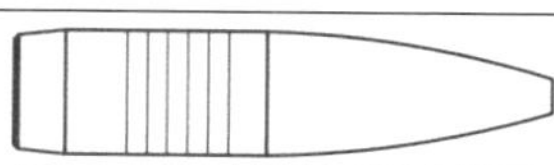 Barnes 168 TSX BT	0.253	1.264	0.200	0.390
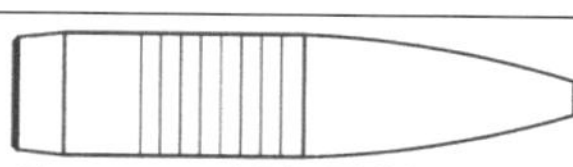 Barnes 180 TSX BT	0.271	1.197	0.227	0.441

(.308 caliber (7.62mm) bullets continued)				
Bullet	**SD**	**i_7**	**G7 BC**	**G1 BC**
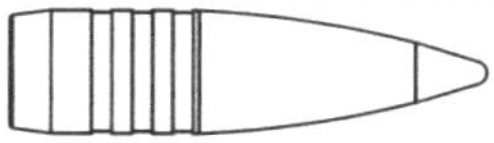 *Barnes 130 TTSXBT*	0.196	1.128	0.174	0.339
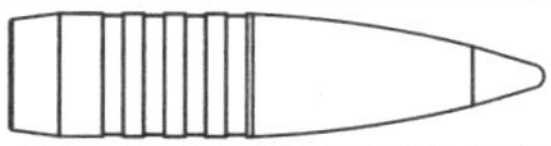 *Barnes 150 TTSXBT*	0.226	1.142	0.198	0.386
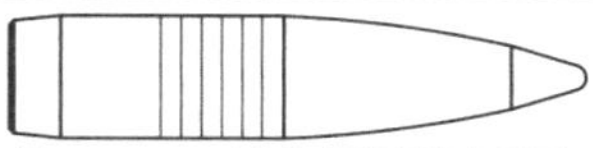 *Barnes 168 TTSX BT*	0.253	1.086	0.233	0.454
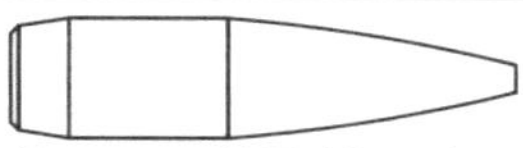 *Berger 155 Hunting VLD*	0.233	0.981	0.238	0.464
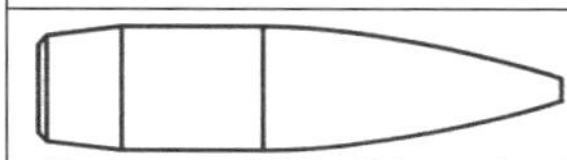 *Berger 168 Classic Hunter*	0.253	1.010	0.251	0.488
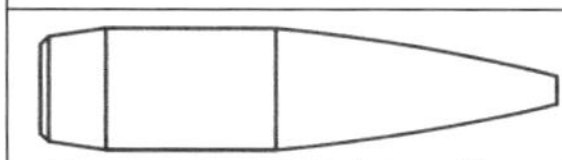 *Berger 168 Hunting VLD*	0.253	0.992	0.255	0.497
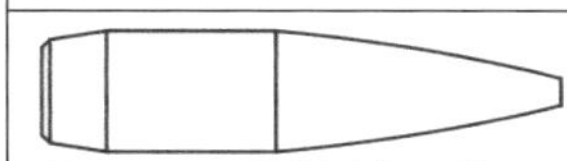 *Berger 175 Hunting VLD*	0.264	0.995	0.265	0.516
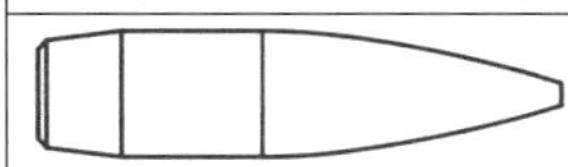 *Berger 185 Classic Hunter*	0.279	1.021	0.273	0.532
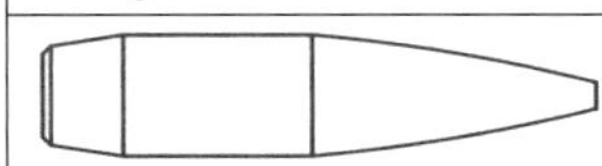 *Berger 185 Hunting VLD*	0.279	0.976	0.285	0.556
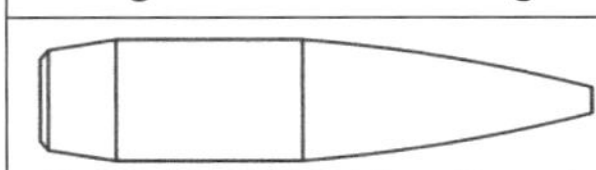 *Berger 190 Hunting VLD*	0.286	0.987	0.290	0.565
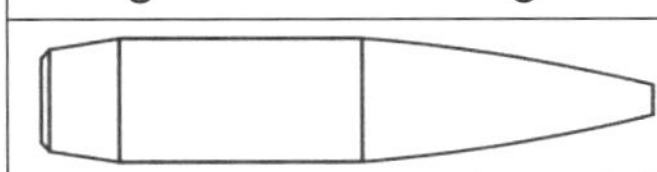 *Berger 210 Hunting VLD*	0.316	0.989	0.320	0.623
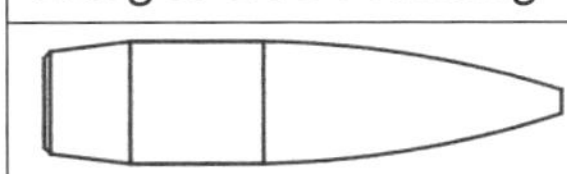 *Berger 175 Tactical OTM*	0.264	1.002	0.263	0.512

(.308 caliber (7.62mm) bullets continued)				
Bullet	**SD**	**i_7**	**G7 BC**	**G1 BC**
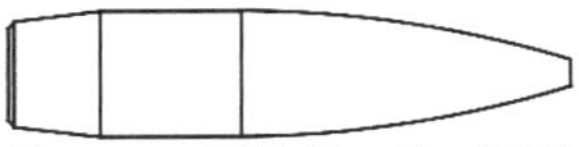 *Berger 185 Tactical OTM Juggernaut*	0.279	0.986	0.283	0.551
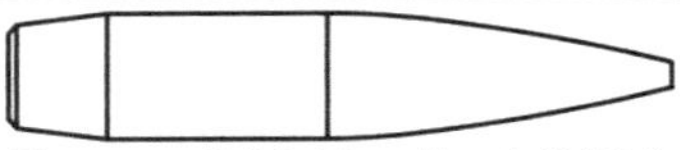 *Berger 230 Tactical OTM Hybrid*	0.346	0.953	0.364	0.708
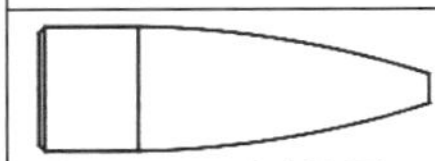 *Berger 115 Target FB*	0.173	1.168	0.148	0.289
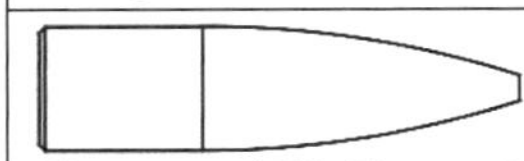 *Berger 150 Target FB*	0.226	1.124	0.201	0.392
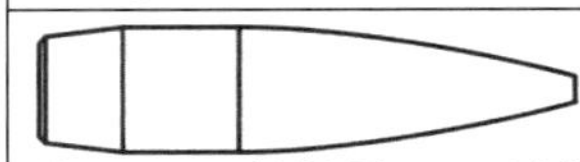 *Berger 155 Target Hybrid*	0.233	0.953	0.245	0.477
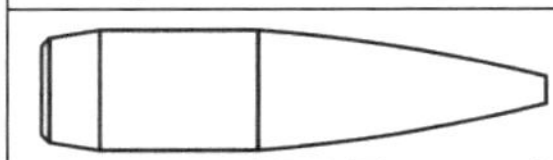 *Berger 155 Target VLD*	0.233	0.981	0.238	0.464
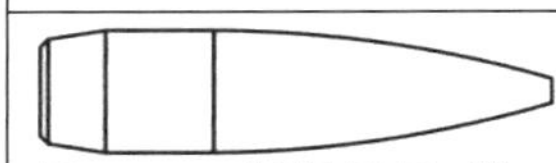 *Berger 155.5 Fullbore*	0.234	0.968	0.241	0.470
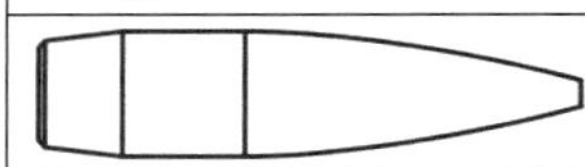 *Berger 168 Target Hybrid*	0.253	0.960	0.263	0.513
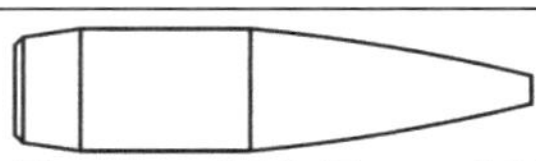 *Berger 168 Target VLD*	0.253	0.973	0.260	0.507
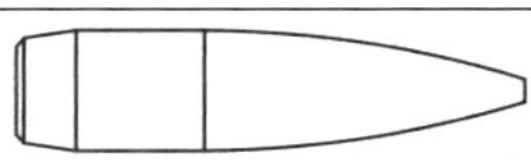 *Berger 175 Target Long Range BT*	0.264	1.005	0.262	0.511
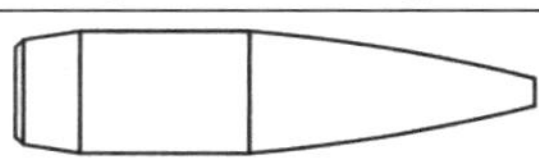 *Berger 175 Target VLD*	0.264	1.006	0.262	0.511
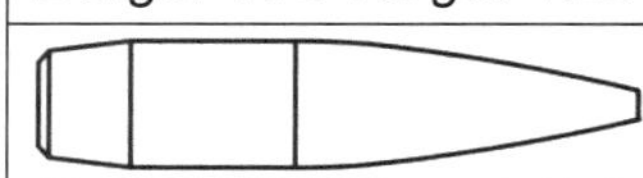 *Berger 185 Target Hybrid*	0.279	0.945	0.295	0.575

(.308 caliber (7.62mm) bullets continued)				
Bullet	**SD**	**i_7**	**G7 BC**	**G1 BC**
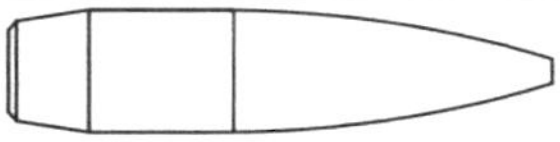 *Berger 185 Target Long Range BT*	0.279	0.981	0.284	0.554
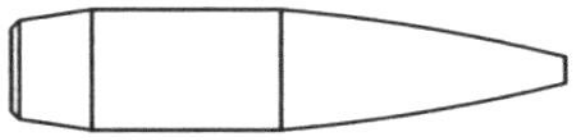 *Berger 185 Target VLD*	0.279	1.002	0.278	0.542
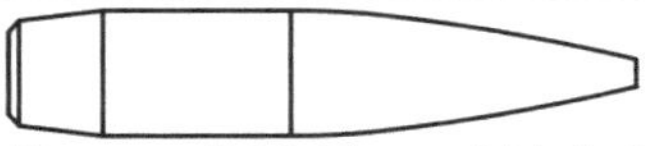 *Berger 200 Target Hybrid*	0.301	0.955	0.316	0.615
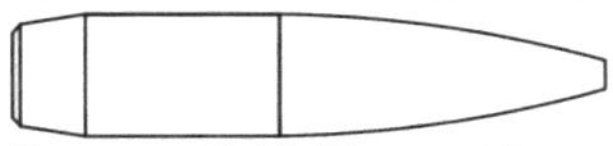 *Berger 210 Target Long Range BT*	0.316	0.988	0.320	0.624
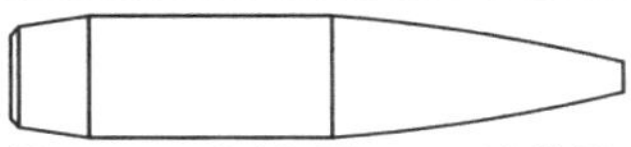 *Berger 210 Target VLD*	0.316	0.995	0.318	0.619
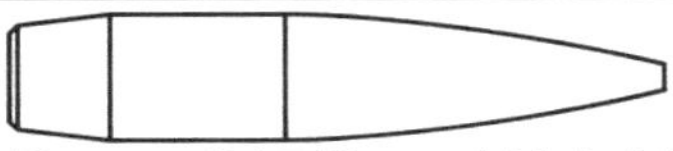 *Berger 215 Target Hybrid*	0.324	0.915	0.354	0.689
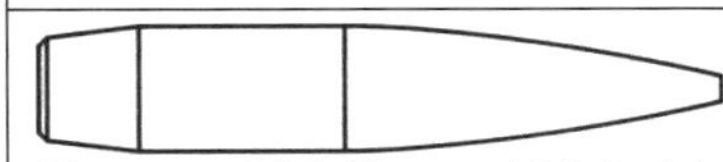 *Berger 230 Target Hybrid*	0.346	0.942	0.368	0.717
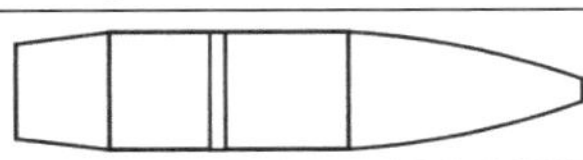 *Cutting Edge 180 HPBT*	0.271	1.108	0.245	0.477
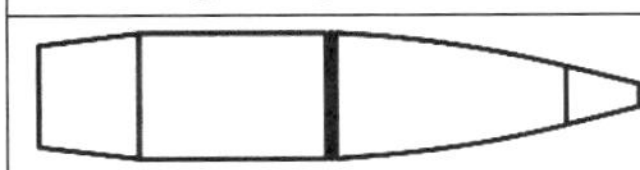 *Cutting Edge 165 Lazer*	0.249	1.005	0.247	0.482
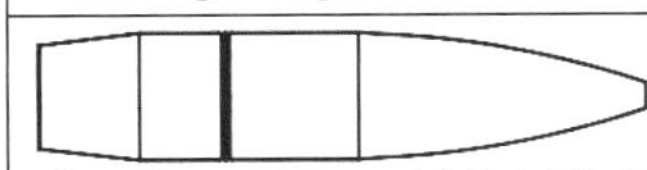 *Cutting Edge 180 LRHPBT-MAX*	0.271	1.041	0.260	0.507
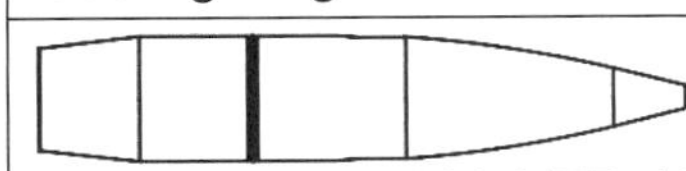 *Cutting Edge 180 LZR 180 MAX*	0.271	0.973	0.279	0.543
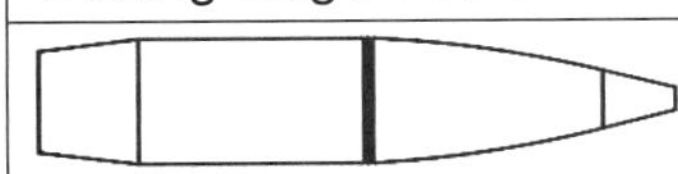 *Cutting Edge 180 Lazer 180 STD*	0.271	1.008	0.269	0.524

(.308 caliber (7.62mm) bullets continued)				
Bullet	**SD**	**i_7**	**G7 BC**	**G1 BC**
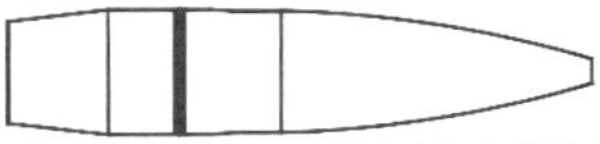 Cutting Edge 165 MTH	0.249	0.972	0.256	0.498
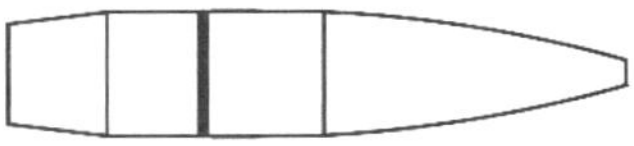 Cutting Edge 180 MTH C21	0.271	0.959	0.283	0.551
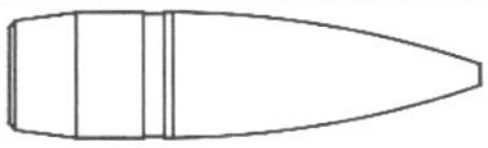 DAG 147 FMJ BT	0.221	1.020	0.217	0.423
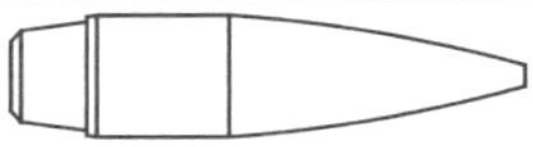 Dietlein 155.5 ULD	0.234	1.004	0.233	0.455
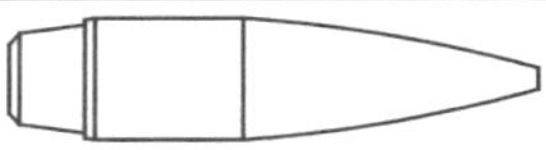 Dietlein 168 ULD	0.253	1.001	0.253	0.493
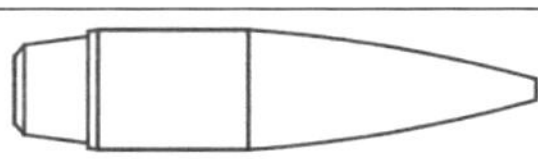 Dietlein 173 ULD	0.261	1.070	0.243	0.475
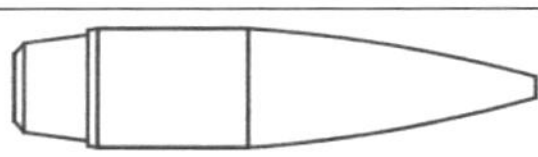 Dietlein 175 ULD	0.264	1.069	0.246	0.481
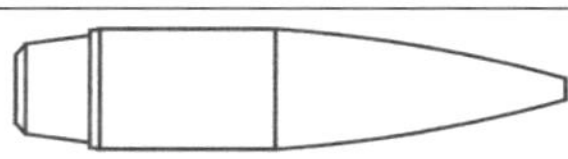 Dietlein 185 ULD	0.279	1.023	0.272	0.531
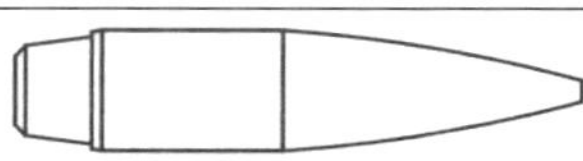 Dietlein 190 ULD	0.286	1.027	0.279	0.544
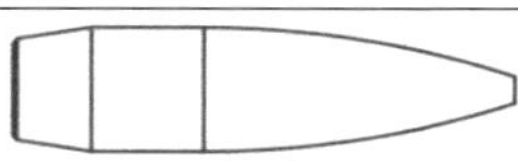 DRT 175 Frangible (non-lead)	0.264	1.113	0.237	0.462
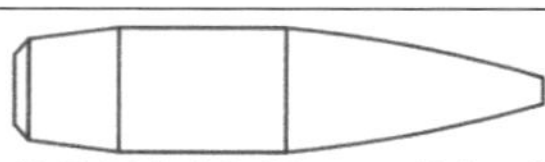 DRT 200 Frangible (non-lead)	0.301	1.020	0.295	0.575
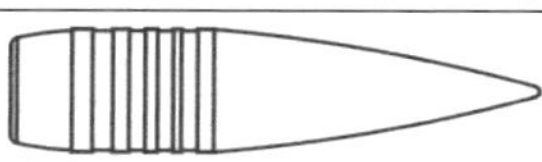 GS Custom 137 SP	0.206	1.023	0.202	0.395

(.308 caliber (7.62mm) bullets continued)				
Bullet	**SD**	**i_7**	**G7 BC**	**G1 BC**
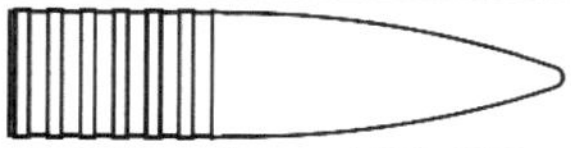 GS Custom 158 SP	0.238	1.335	0.178	0.348
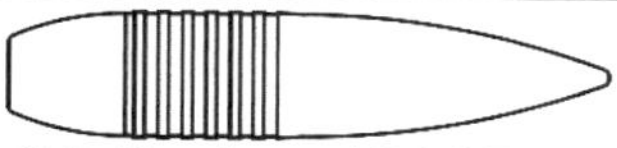 GS Custom 167 SP	0.252	1.226	0.205	0.400
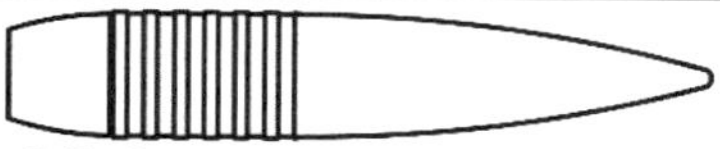 GS Custom 197 SP	0.297	1.094	0.271	0.530
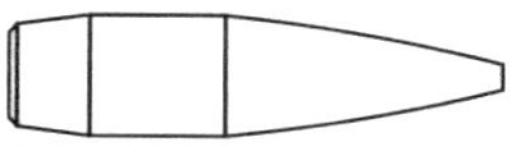 155 HBC (Australia)	0.233	0.989	0.236	0.460
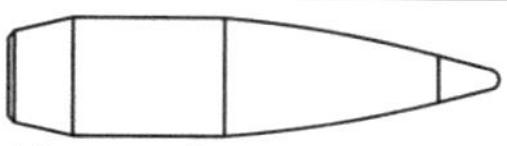 Hornady 155 Amax	0.233	1.099	0.213	0.417
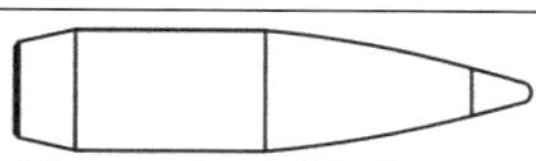 Hornady 168 Amax	0.253	1.100	0.230	0.451
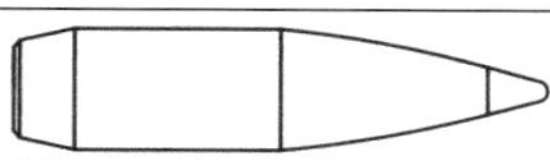 Hornady 178 Amax	0.268	1.117	0.240	0.470
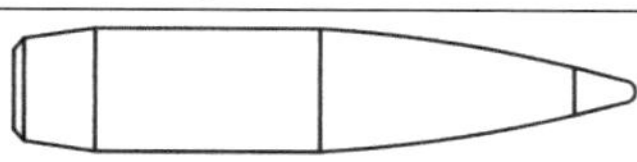 Hornady 208 Amax	0.313	0.966	0.324	0.633
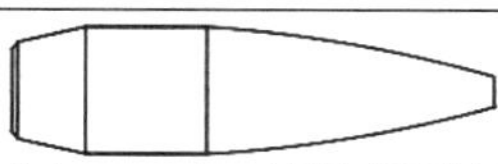 Hornady 155 BTHP Match	0.233	1.133	0.206	0.403
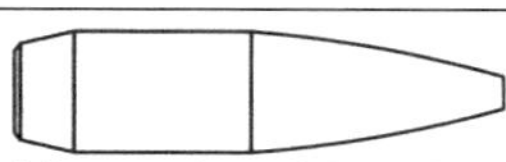 Hornady 168 BTHP Match	0.253	1.143	0.221	0.432
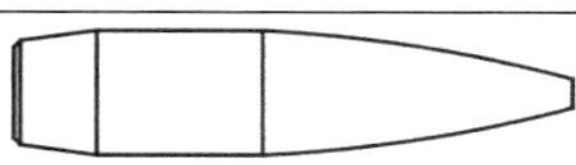 Hornady 178 BTHP Match	0.268	1.043	0.257	0.501
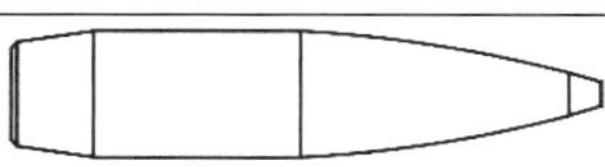 Hornady 195 BTHP Match	0.294	0.997	0.295	0.574

(.308 caliber (7.62mm) bullets continued)				
Bullet	**SD**	**i_7**	**G7 BC**	**G1 BC**
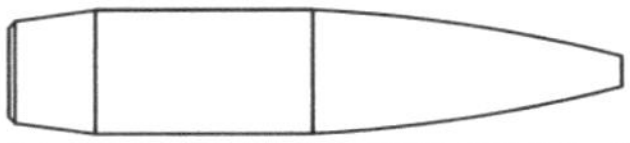 *Hornady 208 BTHP Match*	0.313	0.982	0.319	0.621
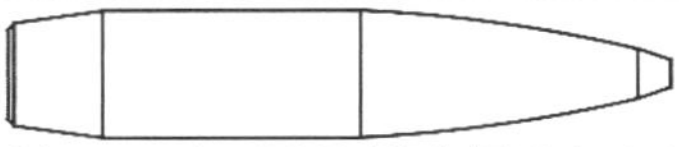 *Hornady 225 BTHP Match*	0.339	0.988	0.343	0.668
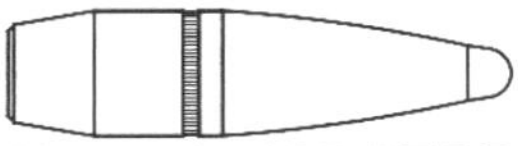 *Hornady 165 BTSP Interlock*	0.249	1.224	0.203	0.395
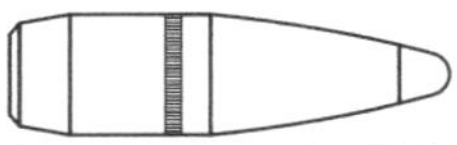 *Hornady 150 BTSP*	0.226	1.262	0.179	0.349
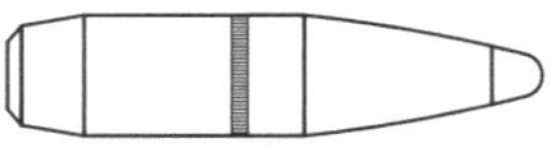 *Hornady 190 BTSP*	0.286	1.205	0.237	0.463
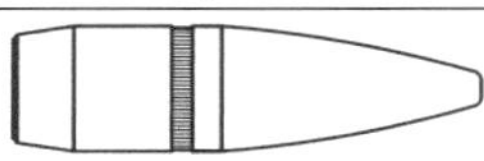 *Hornady 150 FMJBT*	0.226	1.100	0.205	0.400
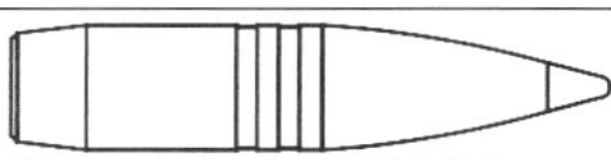 *Hornady 165 GMX*	0.316	1.111	0.224	0.437
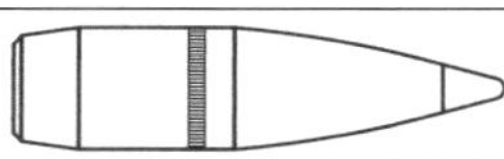 *Hornady 150 SST BT*	0.226	1.088	0.208	0.405
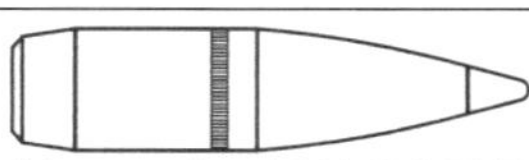 *Hornady 165 SST BT*	0.249	1.110	0.224	0.437
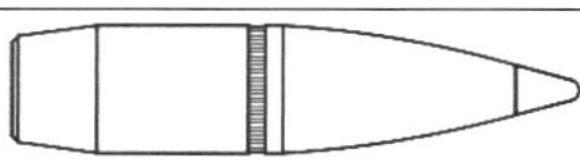 *Hornady 180 SST BT*	0.271	1.071	0.253	0.494
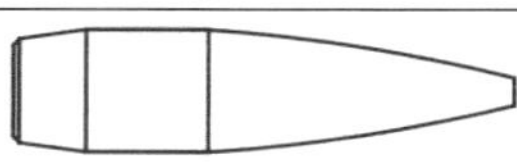 *JLK 155 VLD*	0.233	0.981	0.238	0.464
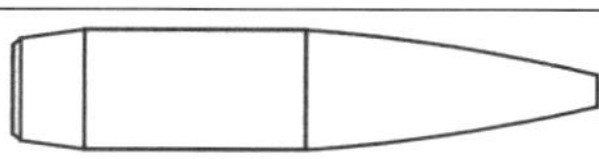 *JLK 210 VLD*	0.316	1.025	0.309	0.601

(.308 caliber (7.62mm) bullets continued)				
Bullet	**SD**	**i_7**	**G7 BC**	**G1 BC**
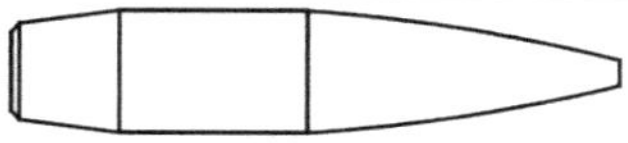 *JLK 210 VLD Long BT*	0.316	0.921	0.343	0.669
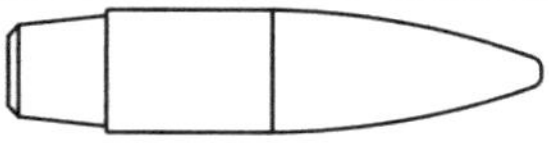 *Lapua 185 FMJBT (D46)*	0.279	1.067	0.261	0.508
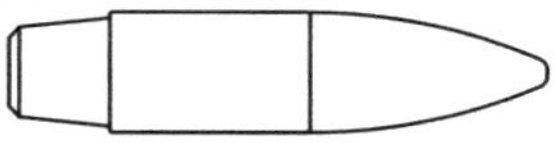 *Lapua 200 FMJBT (subsonic)*	0.301	1.210	0.249	0.485
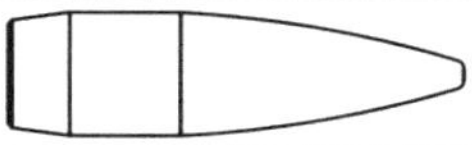 *Lapua 150 FMJBT*	0.226	1.052	0.215	0.419
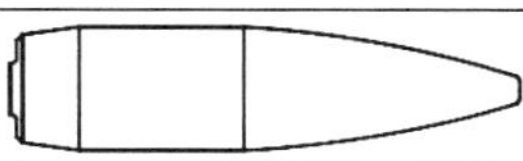 *Lapua 170 FMJBT*	0.256	1.076	0.238	0.464
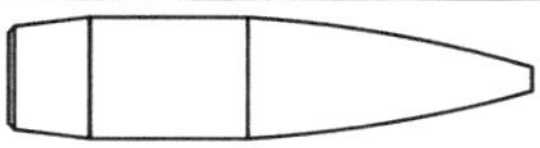 *Lapua 155 Scenar*	0.233	0.988	0.236	0.461
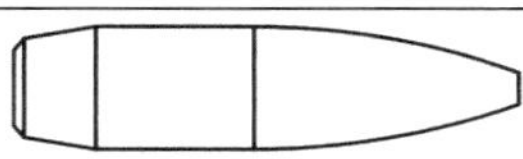 *Lapua 167 Scenar*	0.252	1.163	0.216	0.421
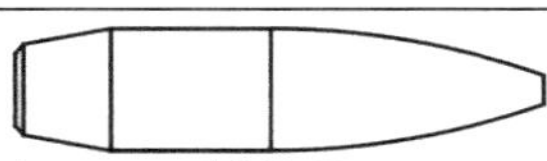 *Lapua 185 Scenar*	0.279	1.130	0.246	0.480
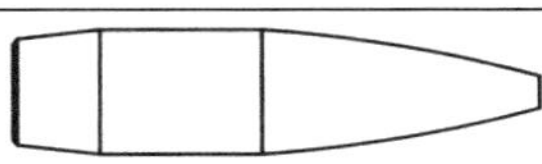 *Lapua 155 Scenar L*	0.233	1.056	0.221	0.431
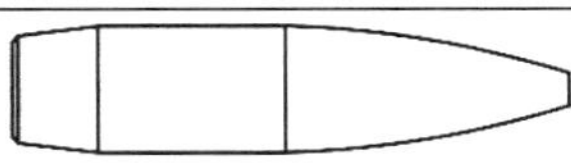 *Lapua 175 Scenar L*	0.264	1.067	0.247	0.481
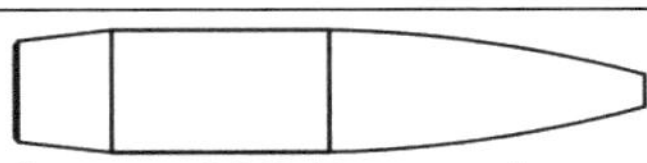 *Lapua 220 Scenar L*	0.331	1.033	0.321	0.624
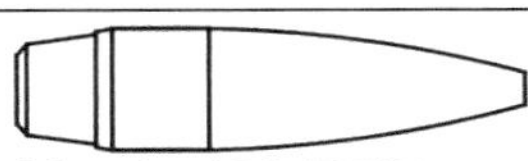 *Matrix 168 RBT*	0.253	1.056	0.240	0.466

(.308 caliber (7.62mm) bullets continued)				
Bullet	**SD**	**i_7**	**G7 BC**	**G1 BC**
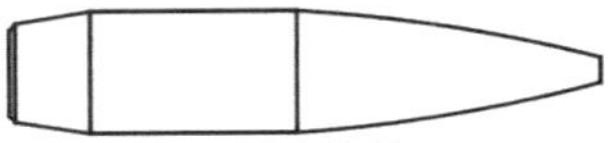 *Matrix 210 VLD*	0.316	0.971	0.326	0.635
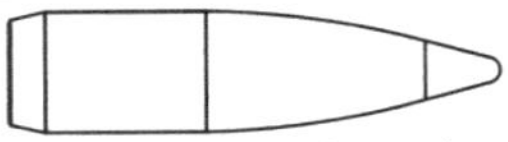 *Nosler 150 Accubond*	0.226	1.146	0.197	0.386
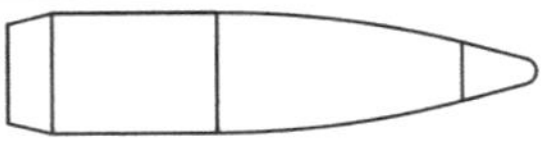 *Nosler 165 Accubond*	0.249	1.091	0.228	0.446
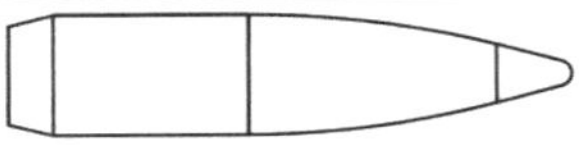 *Nosler 180 Accubond*	0.271	1.101	0.247	0.482
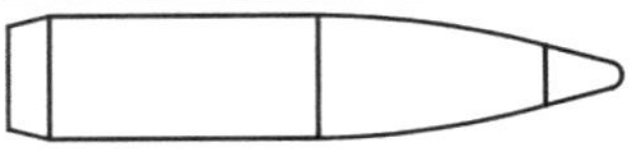 *Nosler 200 Accubond*	0.301	1.123	0.268	0.525
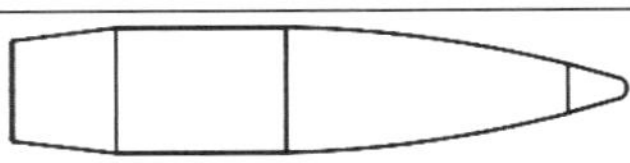 *Nosler 190 Long Range Accubond*	0.286	0.981	0.292	0.569
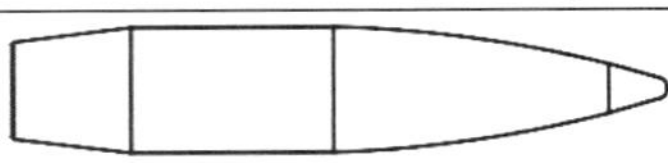 *Nosler 210 Long Range Accubond*	0.316	0.954	0.332	0.647
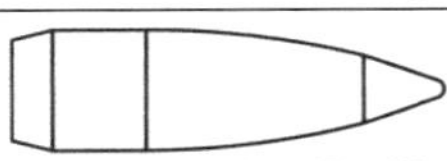 *Nosler 125 Ballistic Tip*	0.188	1.127	0.167	0.327
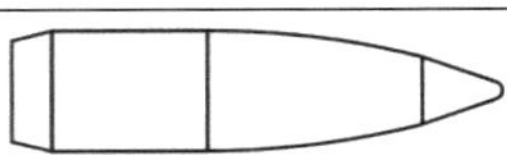 *Nosler 150 Ballistic Tip*	0.226	1.114	0.203	0.397
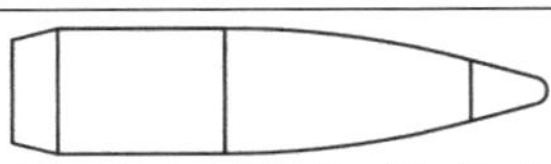 *Nosler 165 Ballistic Tip*	0.249	1.094	0.227	0.445
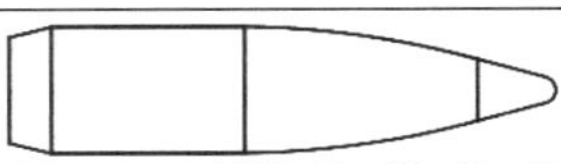 *Nosler 168 Ballistic Tip*	0.253	1.159	0.219	0.427
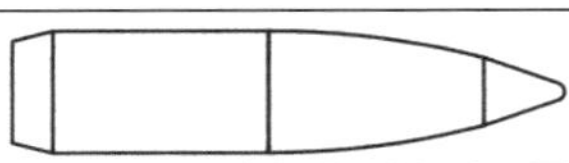 *Nosler 180 Ballistic Tip*	0.271	1.125	0.241	0.472

(.308 caliber (7.62mm) bullets continued)				
Bullet	**SD**	**i_7**	**G7 BC**	**G1 BC**
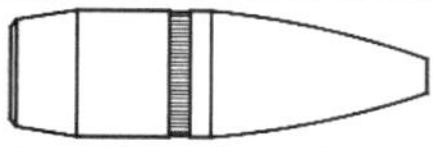 Nosler 140 Custom Competition	0.211	1.271	0.166	0.323
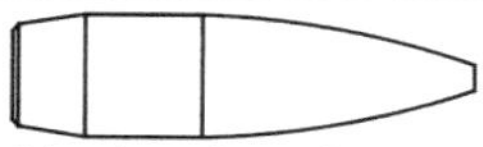 Nosler 155 Custom Competition	0.233	1.098	0.213	0.414
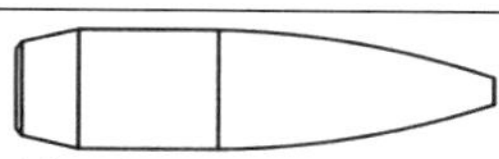 Nosler 168 Custom Competition	0.253	1.159	0.218	0.427
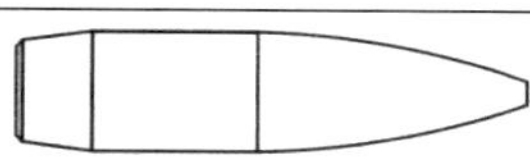 Nosler 175 Custom Competition	0.264	1.059	0.249	0.485
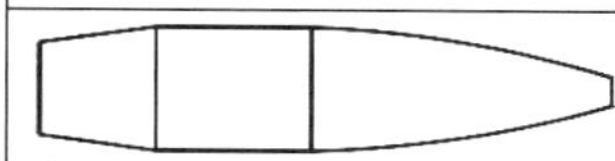 Nosler 190 Custom Competition	0.286	1.030	0.278	0.541
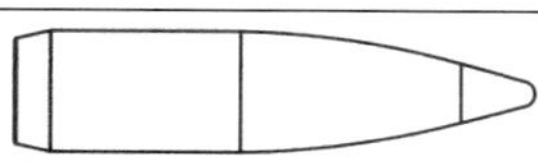 Nosler 150 E-Tip	0.226	1.193	0.189	0.370
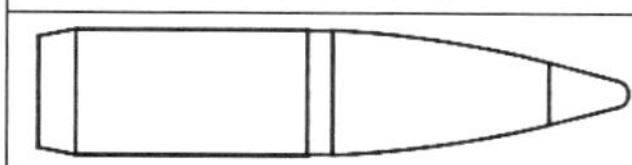 Nosler 168 E-Tip	0.253	1.159	0.218	0.427
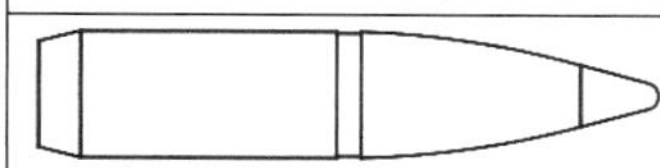 Nosler 180 E-Tip	0.271	1.208	0.225	0.439
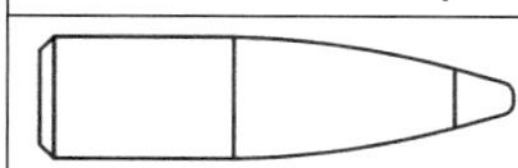 Nosler 165 Partition	0.249	1.218	0.204	0.397
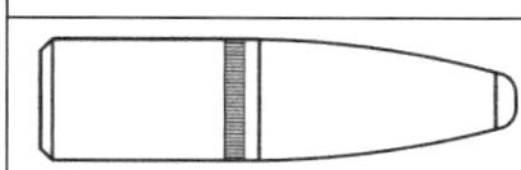 Nosler 180 Partition	0.271	1.475	0.184	0.358
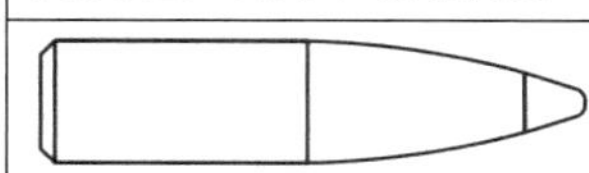 Nosler 200 Partition	0.301	1.210	0.249	0.485
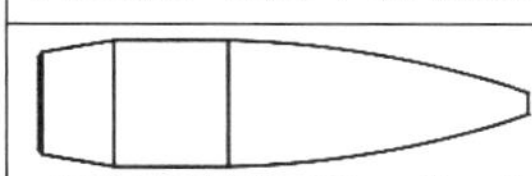 155 PMP (South Africa)	0.233	1.041	0.224	0.438

(.308 caliber (7.62mm) bullets continued)				
Bullet	**SD**	**i_7**	**G7 BC**	**G1 BC**
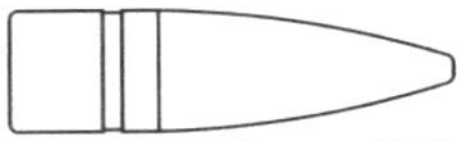 *Prvi Partizan 150 FMJ*	0.226	1.146	0.197	0.384
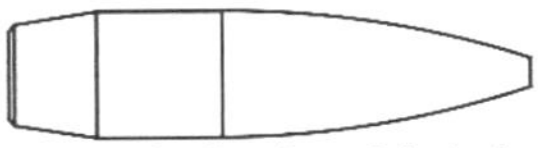 *RUAG Swiss Match 175 HPBT*	0.264	1.071	0.246	0.479
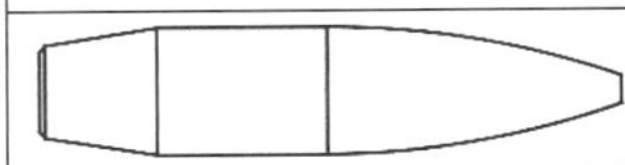 *RUAG Swiss Match 200 HPBT*	0.301	1.061	0.284	0.553
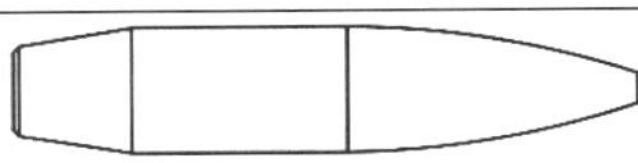 *RUAG Swiss Match 220 HPBT*	0.331	1.062	0.312	0.608
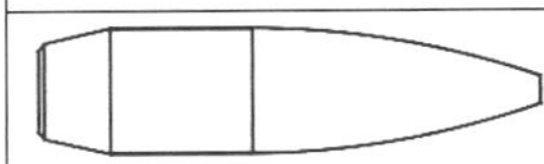 *RUAG Swiss P Target 168 HPBT*	0.253	1.150	0.220	0.430
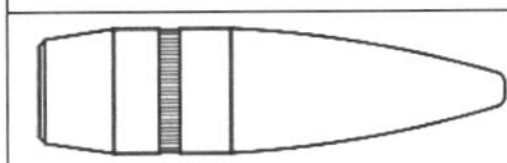 *Sierra 150 FMJ BT*	0.226	1.091	0.207	0.403
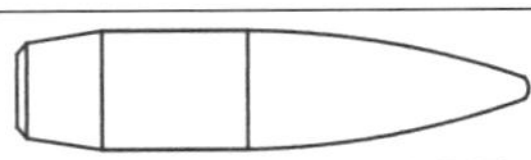 *Sierra 173 FMJ BT*	0.261	1.039	0.251	0.489
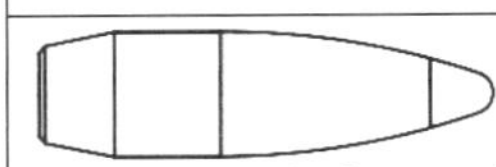 *Sierra 150 GameKing*	0.226	1.189	0.190	0.371
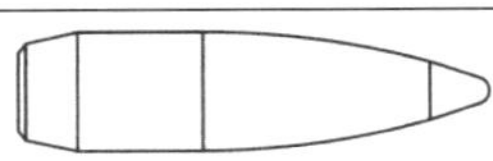 *Sierra 165 GameKing*	0.249	1.169	0.213	0.415
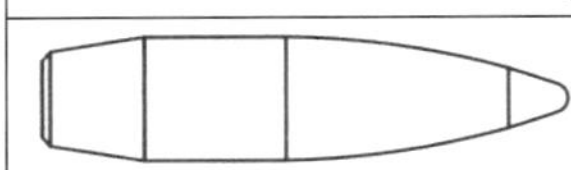 *Sierra 180 GameKing*	0.271	1.120	0.242	0.472
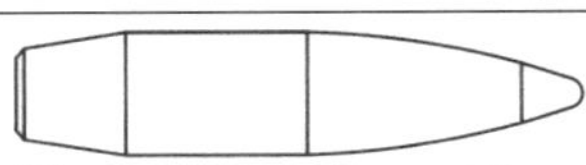 *Sierra 200 GameKing*	0.301	1.041	0.289	0.564
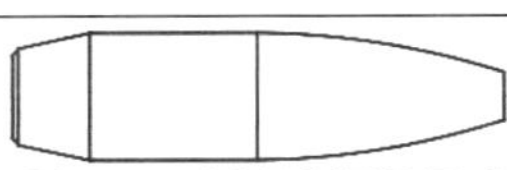 *Sierra 165 HPBT GameKing*	0.249	1.397	0.178	0.347

(.308 caliber (7.62mm) bullets continued)

Bullet	SD	i_7	G7 BC	G1 BC
 *Sierra 125 MatchKing*	0.188	1.063	0.177	0.346
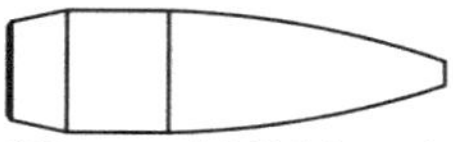 *Sierra 135 MatchKing*	0.203	1.116	0.182	0.356
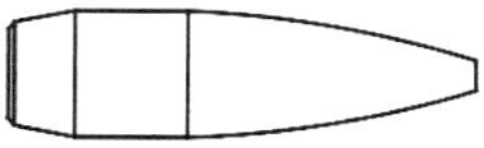 *Sierra 150 MatchKing*	0.226	1.128	0.200	0.391
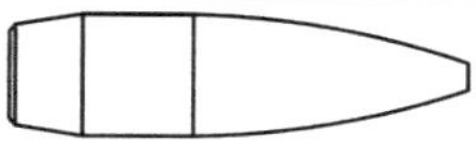 *Sierra 155 MatchKing Palma (2155)*	0.233	1.092	0.214	0.417
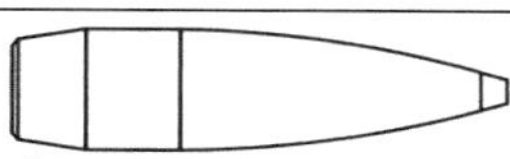 *Sierra 155 MatchKing Palma (2156)*	0.233	0.985	0.237	0.462
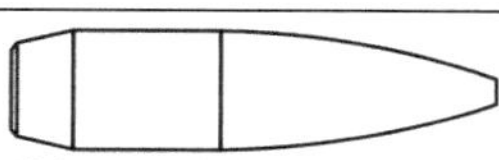 *Sierra 168 MatchKing*	0.253	1.160	0.218	0.426
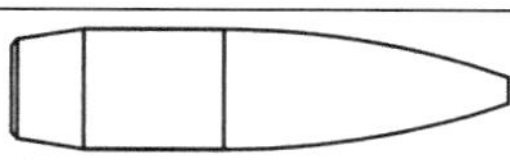 *Sierra 175 MatchKing*	0.264	1.085	0.243	0.473
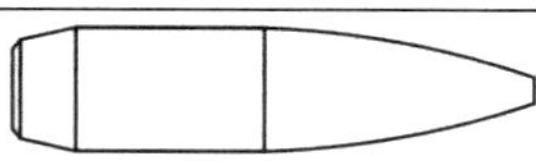 *Sierra 180 MatchKing*	0.271	1.099	0.247	0.482
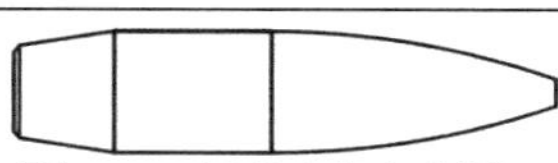 *Sierra 190 MatchKing*	0.286	1.070	0.267	0.521
No image available *Sierra 200 MatchKing*	0.301	1.058	0.285	0.555
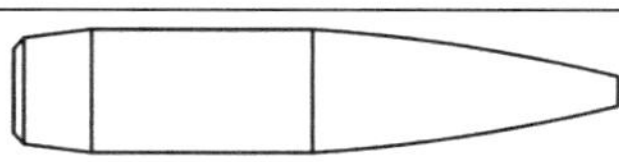 *Sierra 210 MatchKing*	0.316	1.000	0.316	0.616
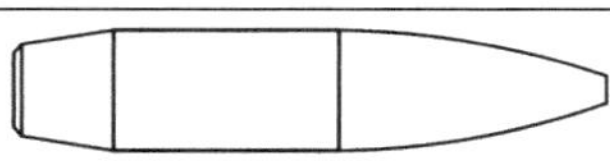 *Sierra 220 MatchKing*	0.331	1.068	0.310	0.604

(.308 caliber (7.62mm) bullets continued)				
Bullet	**SD**	i_7	**G7 BC**	**G1 BC**
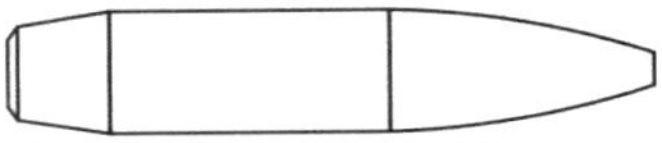 *Sierra 240 MatchKing*	0.361	1.093	0.331	0.644
 *Sierra 125 Tipped MatchKing*	0.188	1.070	0.176	0.344
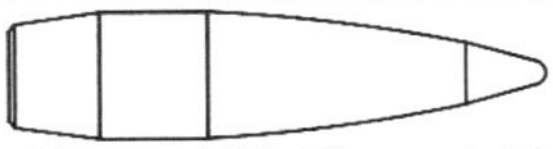 *Sierra 155 Tipped MatchKing*	0.233	0.983	0.238	0.464
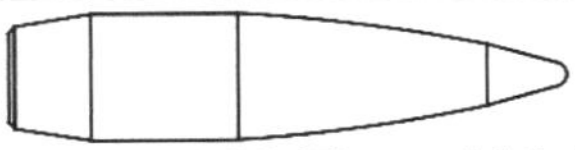 *Sierra 168 Tipped MatchKing*	0.253	0.984	0.257	0.503
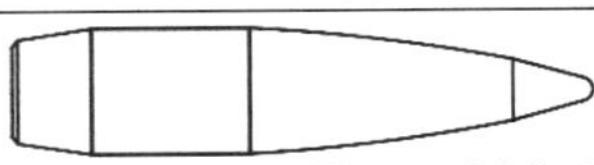 *Sierra 175 Tipped MatchKing*	0.264	0.991	0.266	0.520
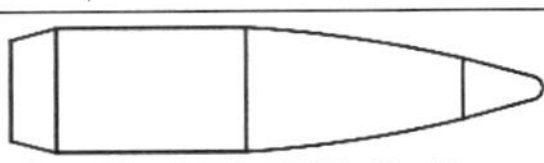 *Swift 150 BT Spitzer*	0.226	1.132	0.200	0.392
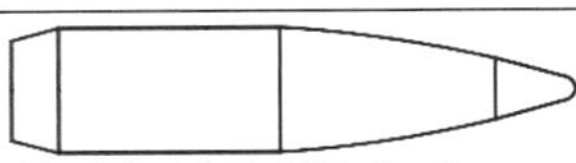 *Swift 165 BT Spitzer*	0.249	1.126	0.221	0.433
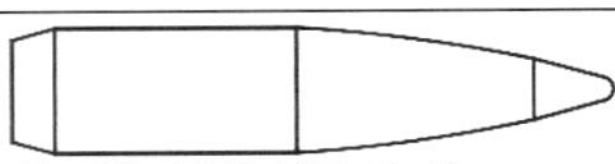 *Swift 180 BT Spitzer*	0.271	1.103	0.246	0.483
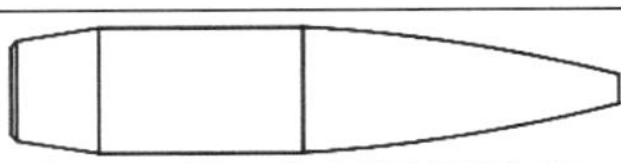 *Vapor Trail 207 BTHP*	0.312	1.009	0.309	0.602

.338 Caliber bullets				
Bullet	**SD**	**i_7**	**G7 BC**	**G1 BC**
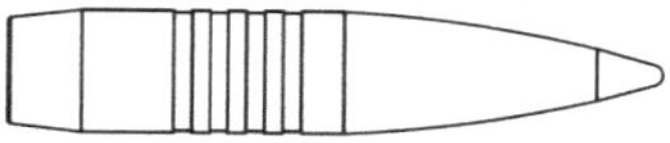 *Barnes 265 LRX BT*	0.331	1.069	0.310	0.605
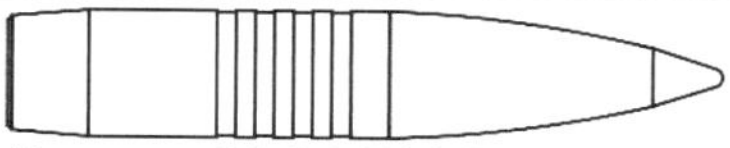 *Barnes 280 LRX BT*	0.350	1.086	0.322	0.629
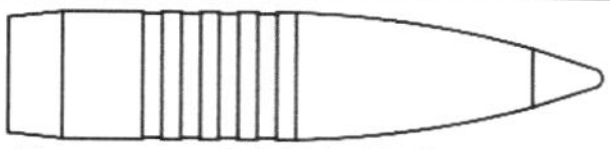 *Barnes 225 TAC-TX BT*	0.281	1.134	0.248	0.484
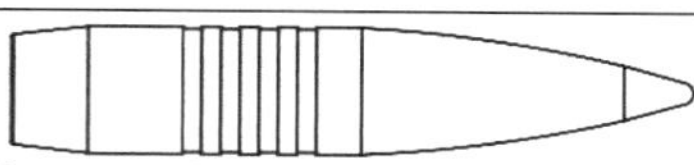 *Barnes 265 TAC-TX BT*	0.331	1.108	0.299	0.583
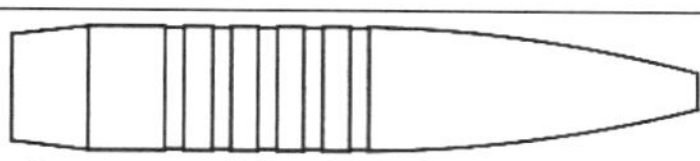 *Barnes 285 TAC-X BT*	0.356	1.168	0.306	0.594
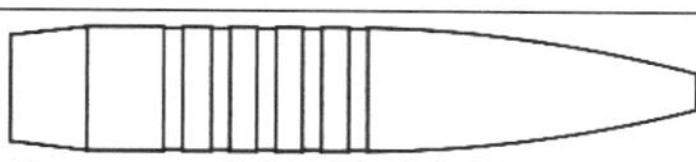 *Barnes 285 TSX BT*	0.356	1.180	0.302	0.588
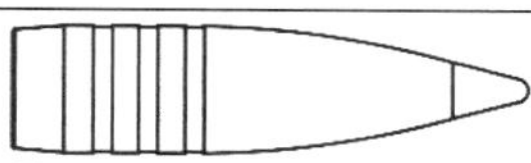 *Barnes 185 TTSX BT*	0.231	1.128	0.205	0.400
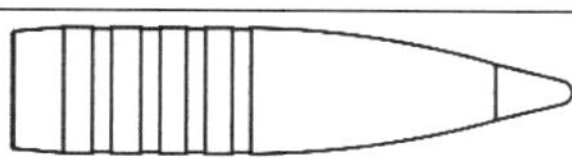 *Barnes 210 TTSX BT*	0.263	1.132	0.232	0.453
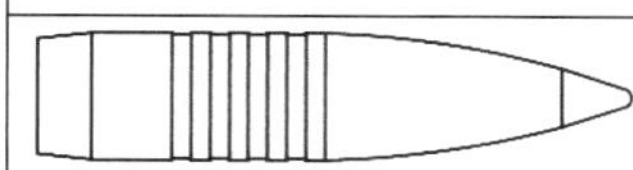 *Barnes 225 TTSX BT*	0.281	1.110	0.253	0.494
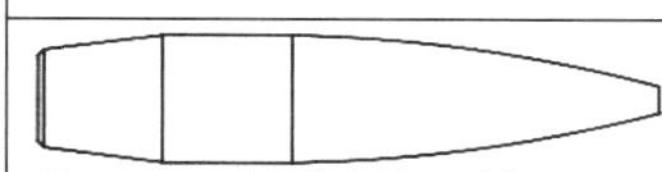 *Berger 250 Elite Hunter*	0.313	0.891	0.351	0.684
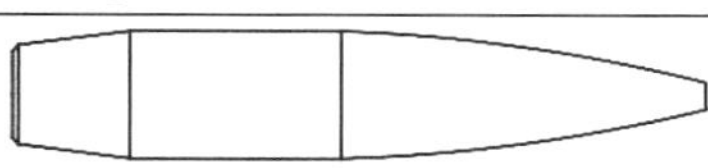 *Berger 300 Elite Hunter*	0.375	0.900	0.417	0.812
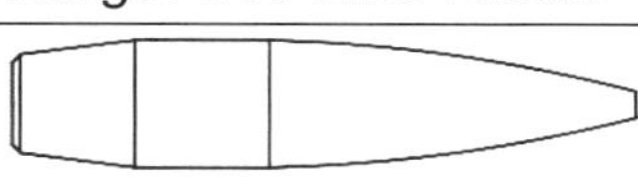 *Berger 250 Tactical OTM*	0.313	0.895	0.349	0.680

(.338 caliber bullets continued)				
Bullet	**SD**	**i_7**	**G7 BC**	**G1 BC**
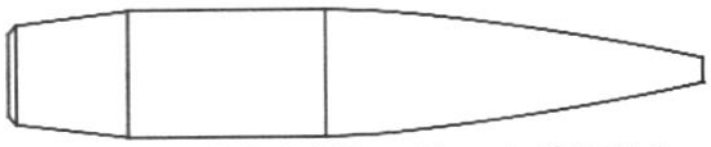 *Berger 300 Tactical OTM*	0.375	0.900	0.417	0.812
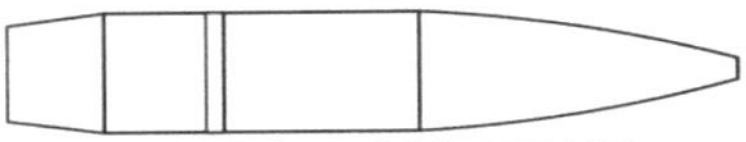 *Cutting Edge 300 HPBT*	0.375	0.927	0.405	0.789
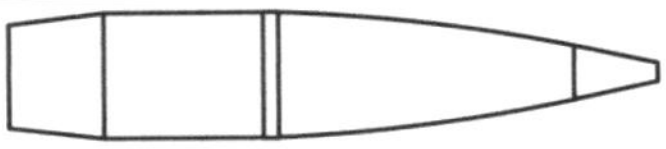 *Cutting Edge 225 Lazer*	0.281	0.902	0.312	0.609
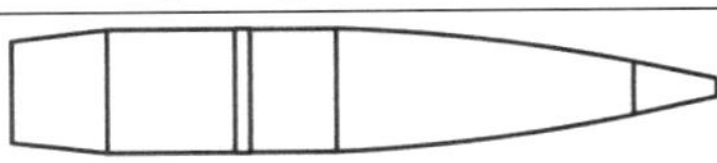 *Cutting Edge 250 Lazer*	0.313	0.877	0.357	0.696
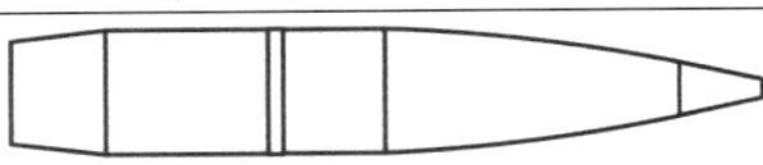 *Cutting Edge 275 Lazer*	0.344	0.860	0.400	0.780
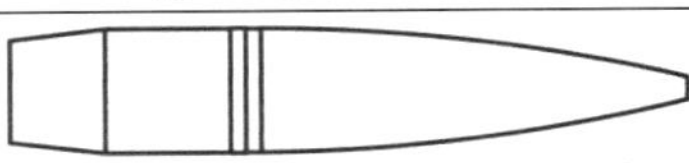 *Cutting Edge 254 MTAC*	0.318	0.862	0.368	0.718
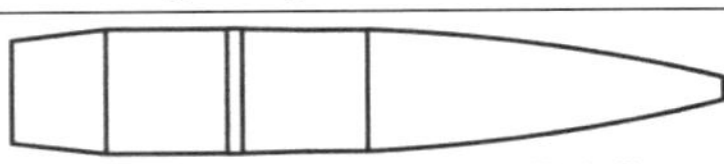 *Cutting Edge 275 MTAC*	0.344	0.829	0.415	0.809
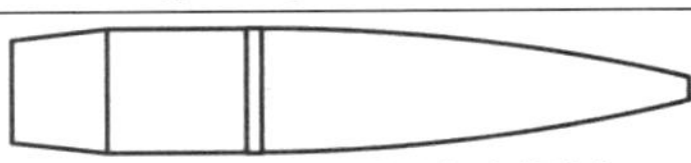 *Cutting Edge 252 MTH*	0.315	0.860	0.366	0.714
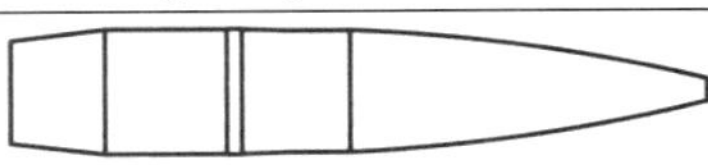 *Cutting Edge 265 MTH*	0.331	0.844	0.393	0.765
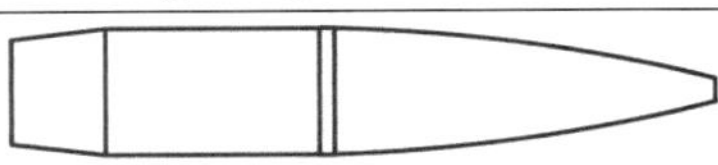 *Cutting Edge 275 MTH*	0.344	0.834	0.412	0.804
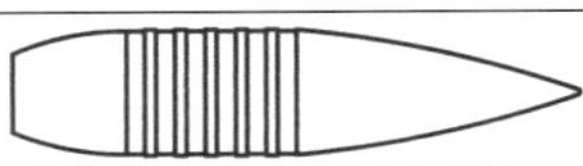 *GS Custom 200 SP*	0.250	1.147	0.218	0.426
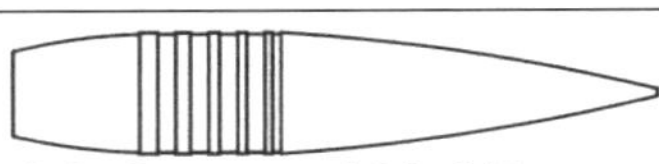 *GS Custom 232 SP*	0.290	0.995	0.292	0.570

(.338 caliber bullets continued)				
Bullet	**SD**	**i_7**	**G7 BC**	**G1 BC**
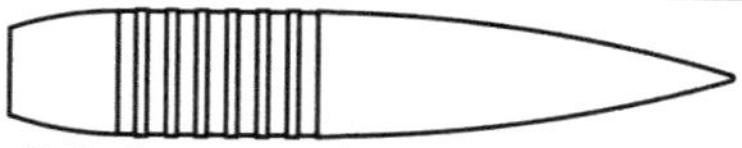 *GS Custom 267 SP*	0.334	1.102	0.304	0.593
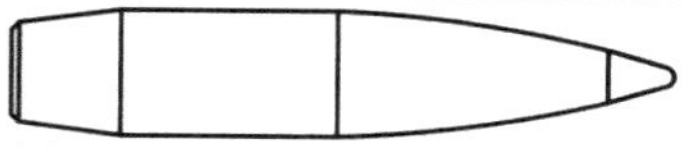 *Hornady 285 Amax*	0.356	0.945	0.377	0.736
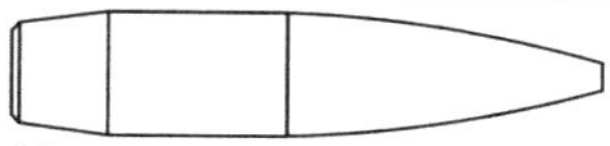 *Hornady 250 BTHP Match*	0.313	0.970	0.322	0.628
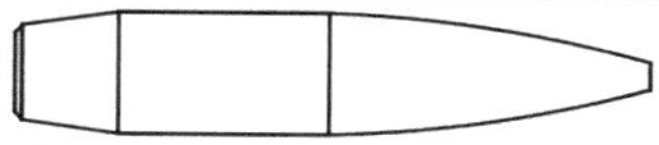 *Hornady 285 BTHP Match*	0.356	0.997	0.357	0.696
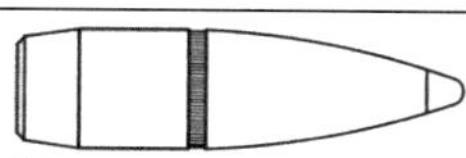 *Hornady 200 FXT*	0.250	1.170	0.214	0.417
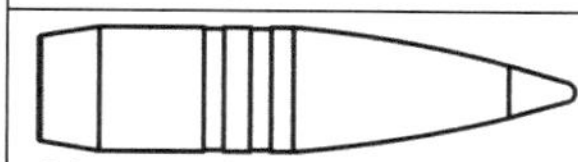 *Hornady 185 GMX*	0.231	1.122	0.206	0.403
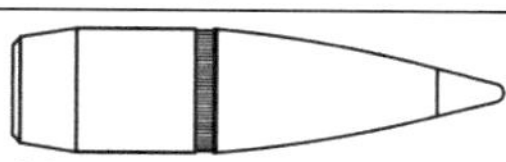 *Hornady 200 SST*	0.250	1.062	0.236	0.461
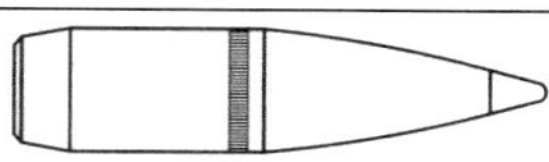 *Hornady 225 SST*	0.281	1.060	0.266	0.519
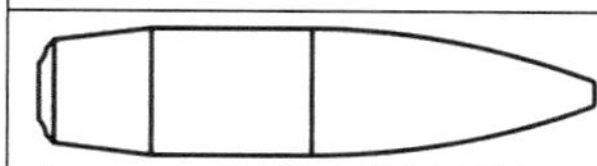 *Lapua 250 FMJBT*	0.313	1.015	0.308	0.600
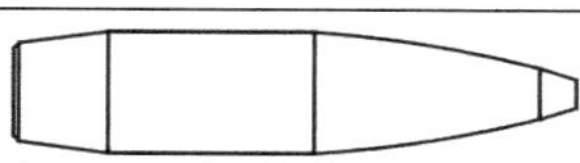 *Lapua 250 Scenar*	0.313	1.025	0.305	0.594
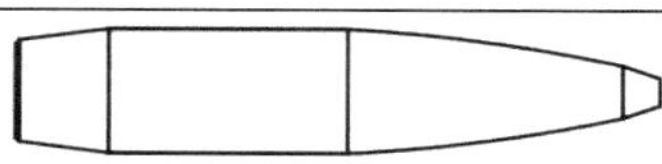 *Lapua 300 Scenar*	0.375	0.956	0.392	0.764
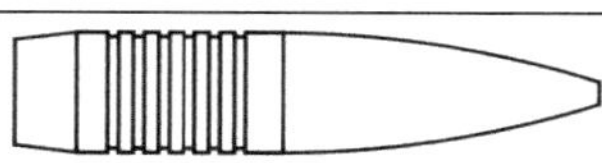 *Lehigh 230 Match Solid*	0.288	1.018	0.282	0.551

(.338 caliber bullets continued)				
Bullet	**SD**	**i_7**	**G7 BC**	**G1 BC**
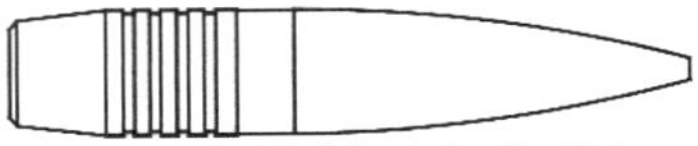 *Lehigh 245 Match Solid*	0.306	0.866	0.354	0.690
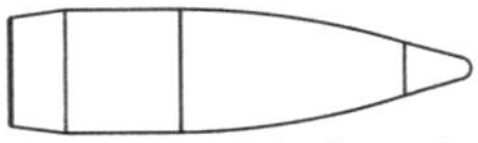 *Nosler 180 Accubond*	0.225	1.077	0.209	0.408
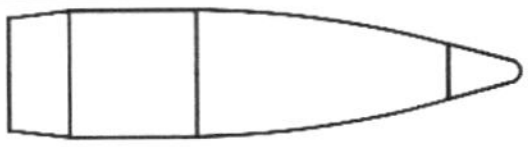 *Nosler 200 Accubond*	0.250	1.060	0.236	0.461
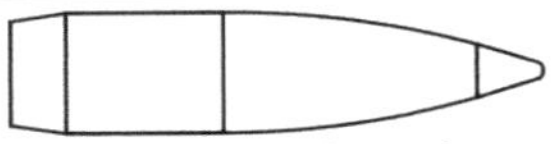 *Nosler 225 Accubond*	0.281	1.049	0.268	0.524
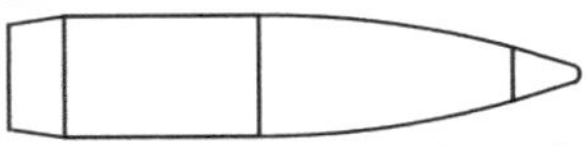 *Nosler 250 Accubond*	0.313	1.049	0.298	0.582
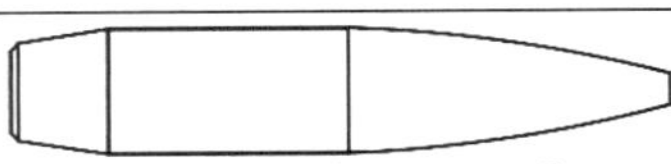 *Nosler 300 Custom Competition*	0.375	1.017	0.369	0.718
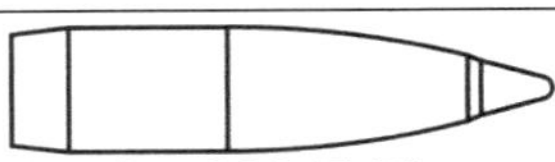 *Nosler 200 E-Tip*	0.250	1.125	0.222	0.434
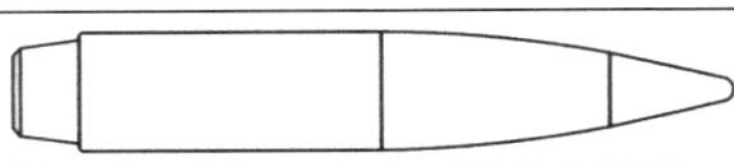 *Rocky Mountain 325 RBT*	0.406	1.052	0.386	0.754
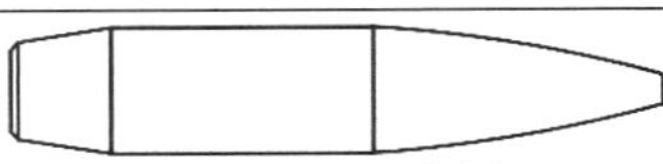 RUAG Swiss P 300	0.375	0.990	0.379	0.738
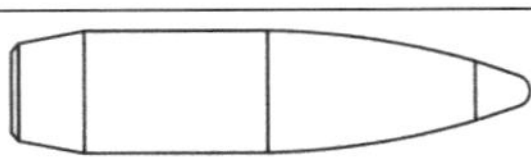 *Sierra 250 Game King*	0.313	1.177	0.266	0.518
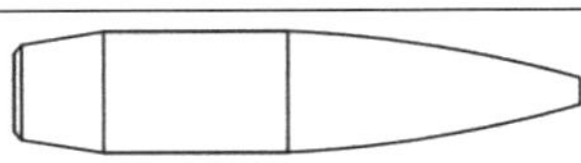 *Sierra 250 MatchKing*	0.313	0.996	0.314	0.612
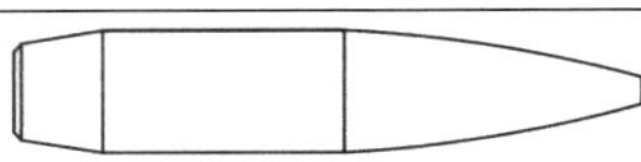 *Sierra 300 MatchKing*	0.375	0.986	0.380	0.742

.375 Caliber bullets

Bullet	SD	i_7	G7 BC	G1 BC
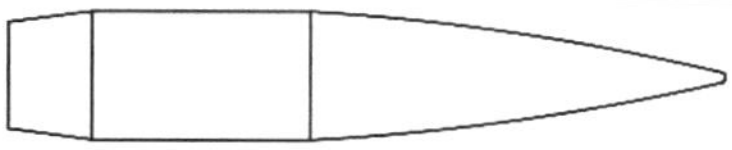 *Chey Tac 350 Solid*	0.356	0.843	0.422	0.822
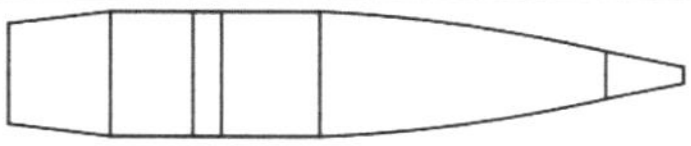 *Cutting Edge 325 Lazer*	0.330	0.892	0.370	0.722
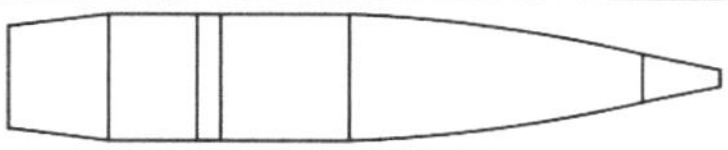 *Cutting Edge 350 Lazer*	0.356	0.905	0.393	0.767
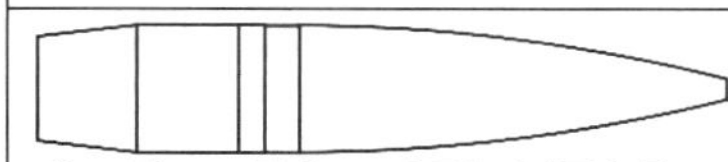 *Cutting Edge 352 MTAC*	0.358	0.876	0.408	0.796
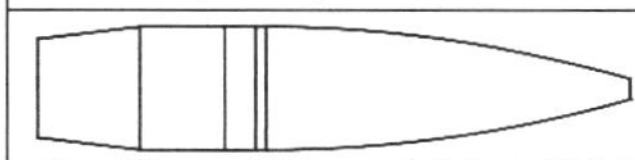 *Cutting Edge 300 MTH*	0.305	0.921	0.331	0.645
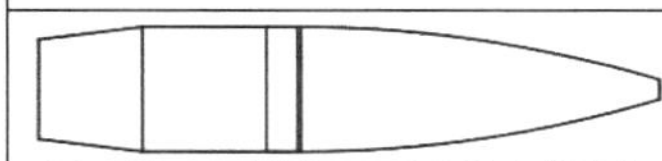 *Cutting Edge 320 MTH*	0.325	0.929	0.350	0.682
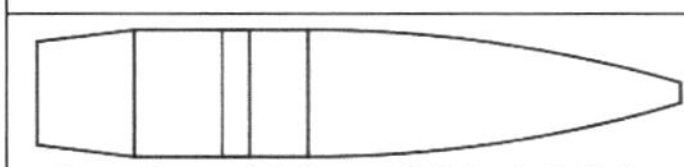 *Cutting Edge 330 MTH*	0.355	0.873	0.384	0.749
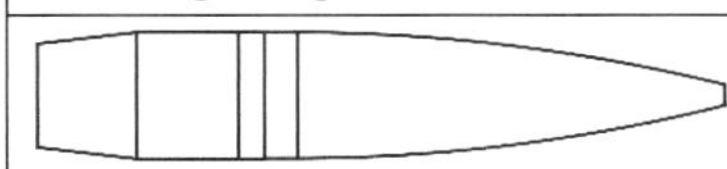 *Cutting Edge 350 MTH*	0.356	0.856	0.415	0.810
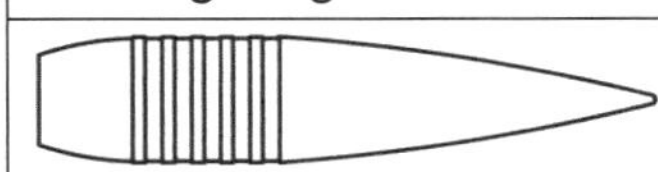 *GS Custom 291 SP*	0.296	1.044	0.284	0.555
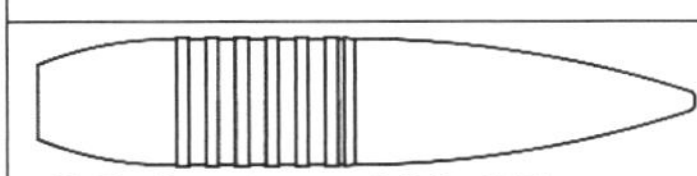 *GS Custom 330 SP*	0.335	1.128	0.297	0.581
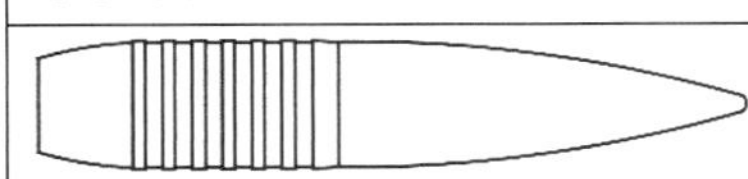 *GS Custom 355 SP*	0.361	1.069	0.338	0.660
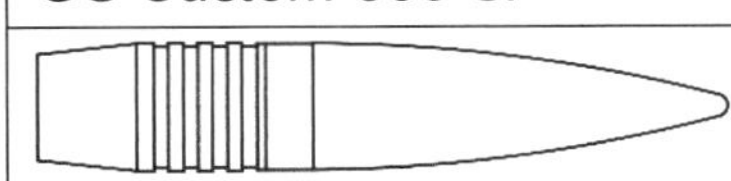 *Lehigh 330 Match Solid*	0.335	0.861	0.390	0.761

(.375 caliber bullets continued)				
Bullet	**SD**	**i_7**	**G7 BC**	**G1 BC**
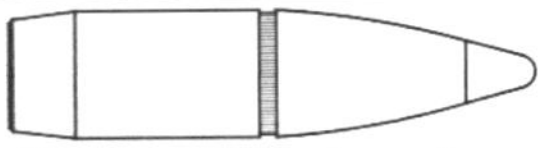 *Nosler 300 Accubond*	0.305	1.205	0.253	0.493
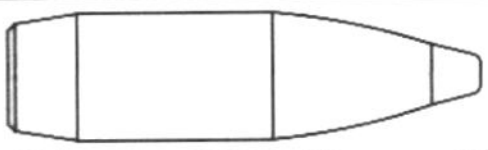 *Sierra 300 GameKing*	0.305	1.305	0.234	0.455
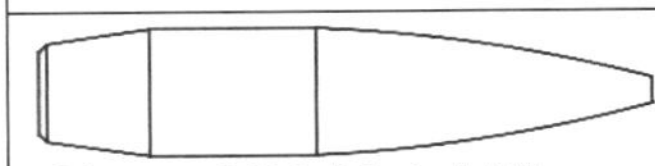 *Sierra 350 MatchKing*	0.356	0.947	0.375	0.732

.408 Caliber bullets

Bullet	SD	i_7	G7 BC	G1 BC
 *Chey Tac 305 Solid*	0.262	0.940	0.279	0.545
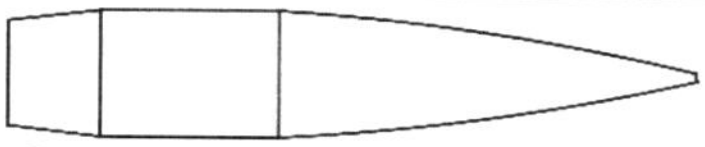 *Chey Tac 419 Solid*	0.360	0.833	0.432	0.845
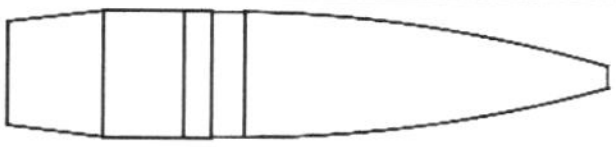 *Cutting Edge 390 MTH*	0.335	0.898	0.373	0.727
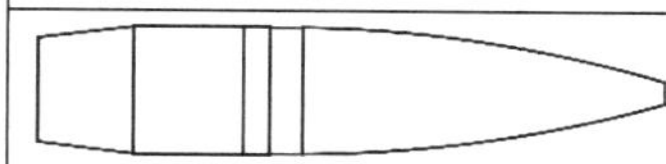 *Cutting Edge 415 MTH*	0.356	0.872	0.408	0.796
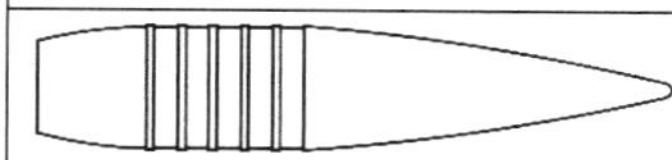 *GS Custom 385 gr SP*	0.330	1.029	0.322	0.629
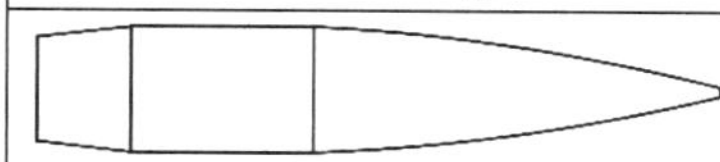 *HSM 400 T50 Solid*	0.343	0.795	0.432	0.844
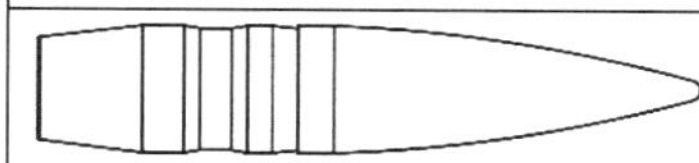 *Lehigh 400 Match Solid*	0.343	0.868	0.396	0.772

.416 Caliber bullets

Bullet	SD	i_7	G7 BC	G1 BC
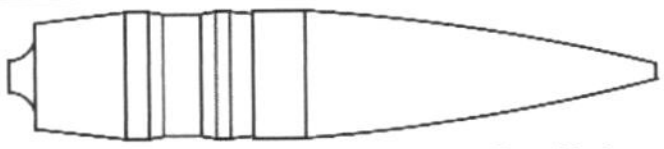 *Barrett 398 Brass Solid*	0.329	0.920	0.357	0.697
 *Cutting Edge 422 MTAC*	0.348	0.850	0.410	0.799
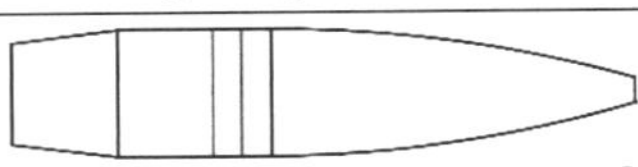 *Cutting Edge 446 MTAC*	0.368	0.923	0.399	0.777
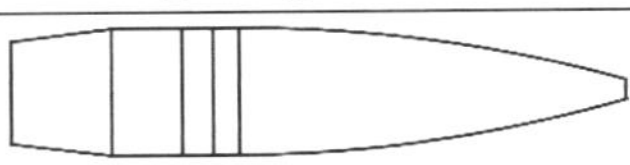 *Cutting Edge 420 MTH*	0.347	0.862	0.402	0.785
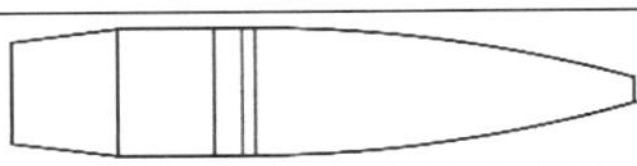 *Cutting Edge 444 MTH*	0.367	0.928	0.395	0.770
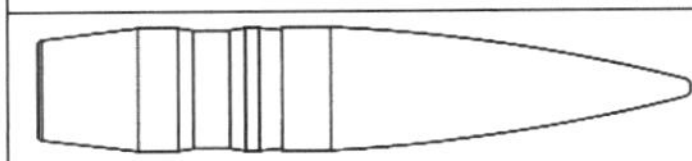 *Lehigh 416 Solid Match*	0.343	0.881	0.390	0.761

.510 Caliber bullets				
Bullet	**SD**	**i_7**	**G7 BC**	**G1 BC**
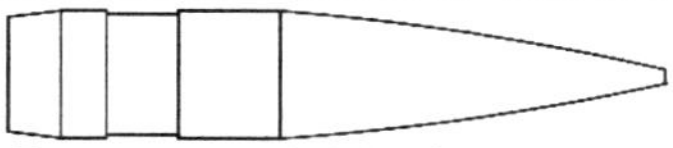 *Barnes 750 LR Solid Bore*	0.412	0.901	0.458	0.894
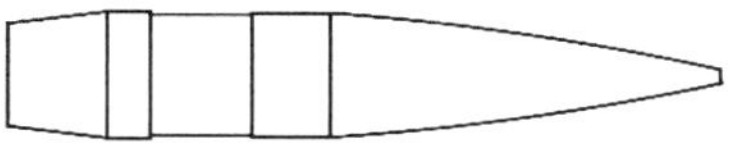 *Barnes 800 LR Solid*	0.439	0.908	0.484	0.946
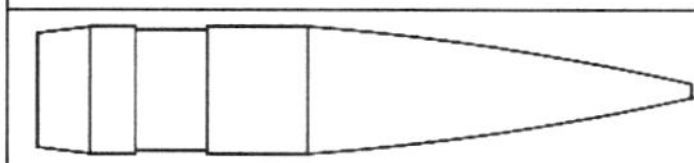 *Barnes 750 TAC-LR*	0.412	0.958	0.431	0.841
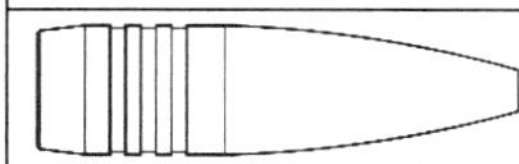 *Barnes 647 TAC-X/TSX*	0.355	1.181	0.301	0.586
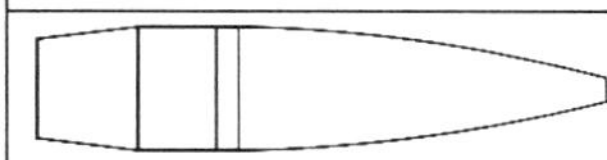 *Cutting Edge 720 MTAC_X03*	0.396	0.922	0.429	0.836
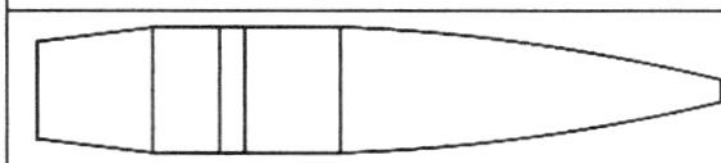 *Cutting Edge 762 MTAC_X02*	0.419	0.859	0.487	0.950
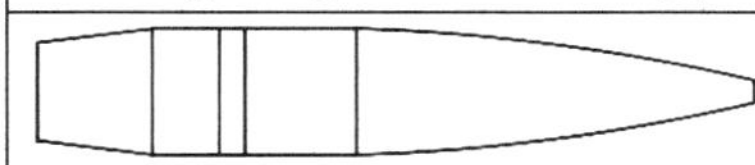 *Cutting Edge 802 MTAC_X01*	0.441	0.888	0.496	0.967
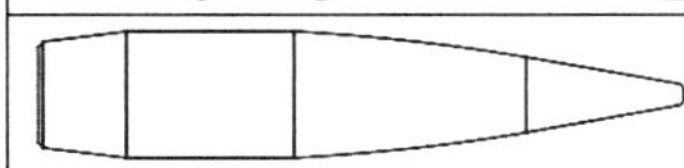 *Hornady 750 Amax*	0.412	0.811	0.508	0.991
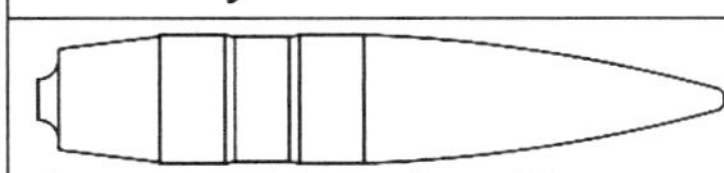 *Lapua 800 Bullex-N*	0.439	0.889	0.494	0.964
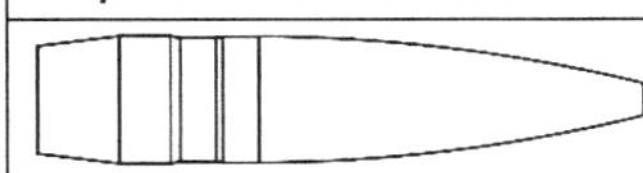 *Leigh 750 Controlled Chaos*	0.412	0.889	0.463	0.904
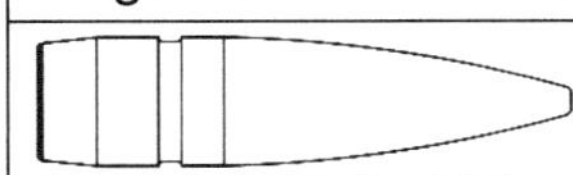 *Lehigh 650 Solid Match*	0.357	0.889	0.402	0.784
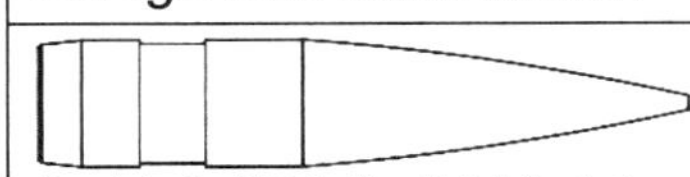 *Lehigh 750 Solid Match*	0.412	1.002	0.412	0.804

(.510 caliber bullets continued)				
Bullet	**SD**	**i_7**	**G7 BC**	**G1 BC**
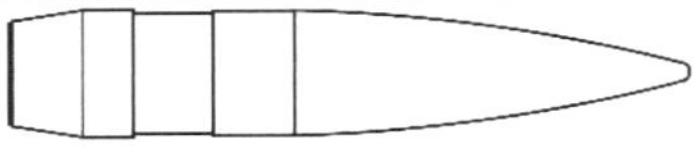 *Lehigh 808 solid Match*	0.444	0.984	0.452	0.882

Notes

Notes

Notes

Notes

Notes

Notes

Notes

Notes

Notes

Notes

Notes

Notes

Notes